T0328213

SUB-SEASONAL TO SEASONAL PREDICTION

SUB-SEASONAL TO SEASONAL PREDICTION

The Gap Between Weather and Climate Forecasting

Edited by

ANDREW W. ROBERTSON

Senior Research Scientist, International Research Institute for Climate and Society (IRI), Earth Institute, Columbia University, NY, United States

FRÉDÉRIC VITART

Senior Research Scientist, European Centre for Medium-Range Weather Forecasts (ECMWF), United Kingdom

Elsevier
Radarweg 29, PO Box 211, 1000 AE Amsterdam, Netherlands
The Boulevard, Langford Lane, Kidlington, Oxford OX5 1GB, United Kingdom
50 Hampshire Street, 5th Floor, Cambridge, MA 02139, United States

Notices
Knowledge and best practice in this field are constantly changing. As new research and experience broaden our understanding, changes in research methods, professional practices, or medical treatment may become necessary.

Practitioners and researchers must always rely on their own experience and knowledge in evaluating and using any information, methods, compounds, or experiments described herein. In using such information or methods they should be mindful of their own safety and the safety of others, including parties for whom they have a professional responsibility.

To the fullest extent of the law, neither the Publisher nor the authors, contributors, or editors, assume any liability for any injury and/or damage to persons or property as a matter of products liability, negligence or otherwise, or from any use or operation of any methods, products, instructions, or ideas contained in the material herein.

Library of Congress Cataloging-in-Publication Data
A catalog record for this book is available from the Library of Congress

British Library Cataloguing-in-Publication Data
A catalogue record for this book is available from the British Library

ISBN: 978-0-12-811714-9

For information on all Elsevier publications visit our website at https://www.elsevier.com/books-and-journals

Publisher: Candice Janco
Acquisition Editor: Laura Kelleher
Editorial Project Manager: Emily Thomson
Production Project Manager: Paul Prasad Chandramohan
Cover Designer: Christian Bilbow

Typeset by SPi Global, India

Contents

6. Extratropical Sub-seasonal to Seasonal Oscillations and Multiple Regimes: The Dynamical Systems View

MICHAEL GHIL, ANDREAS GROTH, DMITRI KONDRASHOV, ANDREW W. ROBERTSON

7. Tropical-Extratropical Interactions and Teleconnections

HAI LIN, JORGEN FREDERIKSEN, DAVID STRAUS, CRISTIANA STAN

8. Land Surface Processes Relevant to Sub-seasonal to Seasonal (S2S) Prediction

PAUL A. DIRMEYER, PIERRE GENTINE, MICHAEL B. EK, GIANPAOLO BALSAMO

9. Midlatitude Mesoscale Ocean-Atmosphere Interaction and Its Relevance to S2S Prediction

R. SARAVANAN, P. CHANG

10. The Role of Sea Ice in Sub-seasonal Predictability

MATTHIEU CHEVALLIER, FRANÇOIS MASSONNET, HELGE GOESSLING, VIRGINIE GUÉMAS, THOMAS JUNG

11. Sub-seasonal Predictability and the Stratosphere

AMY BUTLER, ANDREW CHARLTON-PEREZ, DANIELA I.V. DOMEISEN, CHAIM GARFINKEL, EDWIN P. GERBER, PETER HITCHCOCK, ALEXEY YU. KARPECHKO, AMANDA C. MAYCOCK, MICHAEL SIGMOND, ISLA SIMPSON, SEOK-WOO SON

PART III

S2S MODELING AND FORECASTING

12. Forecast System Design, Configuration, and Complexity

YUHEI TAKAYA

13. Ensemble Generation: The TIGGE and S2S Ensembles

ROBERTO BUIZZA

14. GCMs With Full Representation of Cloud Microphysics and Their MJO Simulations

IN-SIK KANG, MIN-SEOP AHN, HIROAKI MIURA, ANEESH SUBRAMANIAN

15. Forecast Recalibration and Multimodel Combination

STEFAN SIEGERT, DAVID B. STEPHENSON

16. Forecast Verification for S2S Timescales

CAIO A.S. COELHO, BARBARA BROWN, LAURIE WILSON, MARION MITTERMAIER, BARBARA CASATI

PART IV

S2S APPLICATIONS

17. Sub-seasonal to Seasonal Prediction of Weather Extremes

FRÉDÉRIC VITART, CHRISTOPHER CUNNINGHAM, MICHAEL DEFLORIO, EMANUEL DUTRA, LAURA FERRANTI, BRIAN GOLDING, DEBRA HUDSON, CHARLES JONES, CHRISTOPHE LAVAYSSE, JOANNE ROBBINS, MICHAEL K. TIPPETT

Contributors

S. Abhilash Department of Atmospheric Science, Cochin University of Science and Technology, Kochi, India

Min-Seop Ahn School of Earth Environment Sciences, Seoul National University, Seoul, Republic of Korea

Walter E. Baethgen International Research Institute of Climate and Society (IRI), Columbia University, Palisades, NY, United States

Gianpaolo Balsamo European Centre for Medium-Range Weather Forecasts, Reading, United Kingdom

Juan Bazo Red Cross Red Crescent Climate Centre, The Hague, Netherlands

Barbara Brown Research Applications Laboratory, National Center for Atmospheric Research, Boulder, CO, United States

Gilbert Brunet Meteorological Research Division, Environment and Climate Change Canada, Dorval, QC, Canada

Roberto Buizza Scuola Superiore Sant'Anna, Pisa, Italy; ECMWF, Reading, United Kingdom

Amy Butler Cooperative Institute for Research in Environmental Sciences (CIRES)/National Oceanic and Atmospheric Administration (NOAA), Boulder, CO, United States

Barbara Casati Environment and Climate Change Canada, Montreal, QC, Canada

P. Chang Department of Atmospheric Sciences and Department of Oceanography, Texas A&M University, College Station, TX, United States

Andrew Charlton-Perez Department of Meteorology, University of Reading, Reading, United Kingdom

Rajib Chattopadhyay Indian Institute of Tropical Meteorology, Pune, India

Matthieu Chevallier CNRM, Université de Toulouse, Météo France, CNRS, Toulouse, France; Division of Marine Forecasting and Oceanography, Météo France, Toulouse, France

Caio A.S. Coelho Centro de Previsão de Tempo e Estudos Climáticos, Instituto Nacional de Pesquisas Espaciais, Cachoeira Paulista, SP, Brazil

Erin Coughlan de Perez Red Cross Red Crescent Climate Centre, International Research Institute for Climate and Society, Columbia University, Palisades, NY, United States; Institute for Environmental Studies, VU University Amsterdam, Netherlands

Christopher Cunningham National Centre for Monitoring and Early Warning of Natural Disasters (CEMADEN), São José dos Campos, Brazil

Rutger Dankers Met Office, Exeter, United Kingdom

Michael DeFlorio NASA Jet Propulsion Laboratory/California Institute of Technology, Pasadena, CA, United States

Matthew DeGennaro Florida International University, Miami, United States

Mathieu Destrooper German Red Cross, Berlin, Germany

Paul A. Dirmeyer Center for Ocean-Land-Atmosphere Studies, George Mason University, Fairfax, VA, United States

Giovanni Dolif National Centre for Monitoring and Early Warning of Natural Disasters (CEMADEN), São José dos Campos, Brazil

Daniela I.V. Domeisen Institute for Atmospheric and Climate Science, ETH Zürich, Zürich, Switzerland

Robyn Duell Australian Bureau of Meteorology, Melbourne, VIC, Australia

Emanuel Dutra Instituto Dom Luiz, IDL, Faculty of Sciences, University of Lisbon, Lisbon, Portugal

Michael B. Ek National Center for Atmospheric Research, Boulder, CO, United States

Laura Ferranti European Centre for Medium-Range Weather Forecasts (ECMWF), Reading, United Kingdom

Jorgen Frederiksen CSIRO Oceans and Atmosphere, Aspendale, Victoria, Australia

Chaim Garfinkel Freddy and Nadine Hermann Institute of Earth Science, Hebrew University of Jerusalem, Jerusalem, Israel

Pierre Gentine Columbia University, New York, NY, United States

Edwin P. Gerber Courant Institute of Mathematical Sciences, New York University, New York, NY, United States

Michael Ghil Department of Atmospheric and Oceanic Sciences, University of California at Los Angeles, Los Angeles, CA, United States; Geosciences Department and Laboratoire de Météorologie Dynamique (CNRS and IPSL), Ecole Normale Supérieure and PSL Research University, Paris, France

Helge Goessling Alfred Wegener Institute for Polar and Marine Research, Bremerhaven, Germany

Brian Golding Met Office, Exeter, United Kingdom

Andreas Groth Department of Atmospheric and Oceanic Sciences, University of California at Los Angeles, Los Angeles, CA, United States

Virginie Guémas CNRM, Université de Toulouse, Météo France, CNRS, Toulouse, France; Barcelona Supercomputing Centre, Barcelona, Spain

Peter Hitchcock École Polytechniqe, Saclay, Paris, France

Debra Hudson Bureau of Meteorology, Melbourne, Australia

Charles Jones University of California, Santa Barbara, (UCSB), Santa Barbara, CA, United States

Susmitha Joseph Indian Institute of Tropical Meteorology, Pune, India

Thomas Jung Alfred Wegener Institute for Polar and Marine Research, Bremerhaven, Germany

In-Sik Kang Indian Ocean Operational Oceanographic Research Center, SOED/ Second Institute of Oceanography, Hangzhou, China; Center of Excellence of Climate Change Research, King Abdulaziz University, Jeddah, Saudi Arabia

Alexey Yu. Karpechko Finnish Meteorological Institute, Helsinki, Finland

Dmitri Kondrashov Department of Atmospheric and Oceanic Sciences, University of California at Los Angeles, Los Angeles, CA, United States

Phani M. Krishna Indian Institute of Tropical Meteorology, Pune, India

Christophe Lavaysse University of Grenoble Alpes, CNRS, IRD, G-INP, IGE, Grenoble, France

Hai Lin Environment and Climate Change Canada, Dorval, Quebec, Canada

Rachel Lowe Department of Infectious Disease Epidemiology, Center for the Mathematical Modelling of Infectious Diseases, London School of Hygiene and Tropical Medicine, London, United Kingdom; Climate and Health Programme, Barcelona Institute for Global Health (ISGLOBAL), Barcelona, Spain

Victor Marchezini National Centre for Monitoring and Early Warning of Natural Disasters (CEMADEN), São José dos Campos, Brazil

Nadège Martiny Center de Recherches de Climatologie, Biogéosciences, UMR 6282 CNRS, Université Bourgogne Franche-Comté, Besançon, France

François Massonnet Georges Lemaître Centre for Earth and Climate Research, Earth and Life Institute, Université catholique de Louvain, Louvain-la-Neuve, Belgium; Barcelona Supercomputing Centre, Barcelona, Spain

Amanda C. Maycock School of Earth and Environment, University of Leeds, Leeds, United Kingdom

John Methven Department of Meteorology, University of Reading, Reading, United Kingdom

Brian Mills Meteorological Research Division, Environment and Climate Change Canada, Waterloo, ON, Canada

Marion Mittermaier UK Met Office, Exeter, United Kingdom

Hiroaki Miura Department of Earth and Planetary Science, University of Tokyo, Tokyo, Japan

Vincent Moron Aix-Marseille Univ, CNRS, IRD, INRA, Coll France, CEREGE, Aix en Provence, France

Tetsuo Nakazawa Typhoon Research Department, Meteorological Research Institute, Japan Meteorological Agency, Tokyo, Japan

Hannah Nissan International Research Institute for Climate and Society (IRI), Columbia University, Palisades, NY, United States

D.R. Pattanaik India Meteorological Department, New Delhi, India

Joanne Robbins Met Office, Exeter, United Kingdom

Andrew W. Robertson International Research Institute for Climate and Society (IRI), Columbia University, Palisades, NY, United States

Pascal Roucou Center de Recherches de Climatologie, Biogéosciences, UMR 6282 CNRS, Université Bourgogne Franche-Comté, Besançon, France

A.K. Sahai Indian Institute of Tropical Meteorology, Pune, India

R. Saravanan Department of Atmospheric Sciences and Department of Oceanography, Texas A&M University, College Station, TX, United States

Juan Pablo Sarmiento Florida International University, Miami, United States

Stefan Siegert Department of Mathematics, University of Exeter, Devon, United Kingdom

Michael Sigmond Canadian Centre for Climate Modelling and Analysis, Environment and Climate Change Canada, Victoria, BC, Canada

Amber Silver University of Albany, New York, United States

Isla Simpson Climate and Global Dynamics Laboratory, National Center for Atmospheric Research, Boulder, CO, United States

Roop Singh Red Cross Red Crescent Climate Centre, The Hague, Netherlands

Seok-Woo Son School of Earth and Environmental Sciences, Seoul National University, Seoul, Republic of Korea

Cristiana Stan George Mason University, Fairfax, VA, United States

David B. Stephenson Department of Mathematics, University of Exeter, Devon, United Kingdom

David Straus George Mason University, Fairfax, VA, United States

Aneesh Subramanian AOPP, Department of Physics, University of Oxford, Oxford, United Kingdom; Scripps Institution of Oceanography, UCSD, San Diego, CA, United States

Yuhei Takaya Meteorological Research Institute, Japan Meteorological Agency, Ibaraki, Japan

Rafael Terra Department of Fluid Mechanics and Environmental Engineering, School of Engineering, Universidad de la República, Montevideo, Uruguay

Madeleine C. Thomson International Research Institute for Climate and Society (IRI), Columbia University, Palisades, NY, United States

Michael K. Tippett Columbia University, New York, NY, United States

Adrian M. Tompkins Abdus Salam International Center for Theoretical Physics (ICTP), Trieste, Italy

Zoltan Toth NOAA, Boulder, CO, United States

Rachel Trajber Florida International University, Miami, United States

Frédéric Vitart European Centre for Medium-Range Weather Forecasts (ECMWF), Reading, United Kingdom

Lei Wang Institute of Atmospheric Sciences, Fudan University, Shangai, People's Republic of China

Andrew Watkins Australian Bureau of Meteorology, Melbourne, VIC, Australia

Laurie Wilson Environment and Climate Change Canada, Montreal, QC, Canada

Steven J. Woolnough National Centre for Atmospheric Science, University of Reading, Reading, United Kingdom

Preface

The World Meteorological Organization (WMO) launched the Sub-seasonal to Seasonal Prediction Project (S2S) in November 2013 as a collaboration between the World Weather Research Programme (WWRP) and the World Climate Research Programme (WCRP). The primary goals of this project were improving forecast skill and understanding the dynamics and climate drivers on the S2S timescale (i.e., from 2 weeks to an entire season). The S2S Implementation Plan was developed by a group that included representatives from WWRP/THORPEX (The Observing System Research and Predictability Experiment), WCRP, the Commission for Basic Systems (CBS), and the Commission for Climatology (CCl); it gives a high priority to establishing collaboration and coordination between operational centers and the research community. The S2S project has a special emphasis on high-impact weather events, with a "Ready-Set-Go" approach, featuring coordination among operational centers and promoting the uptake of S2S products and information by the application communities. This principle links strongly to the seamless prediction approach in the WMO research community.

The grand challenge of accelerating progress in Earth-system observation, analysis, and prediction capabilities was postulated by Shapiro et al. (2010). In this context, seamless prediction was introduced for S2S prediction to span the boundary between weather and climate (Brunet et al., 2010). These authors extended the concept of seamlessness beyond the realm of atmospheric predictions to include the consideration of biophysical and socioeconomic factors pertinent to successful decision-making.

At the World Weather Open Science Conference (WWOSC) in 2014, seamless prediction was featured prominently to cover timescales from minutes to months, considering all compartments of the Earth-system, including hydrology and air quality, and linking to users, applications, and social sciences. Seamlessness is a useful concept to address the need for information for users, stakeholders, and decision-makers that is smooth and consistent across the artificial barriers that exist across timescales and space scales, observing systems, modeling approaches, disciplines, and communities. The outcome of the WWOSC led to the development of the WWRP Implementation plan, in which the S2S project is a very clear example of such seamless collaboration between WWRP and WCRP.

As the WCRP embarks on a new strategic era of more integrative science to respond to the international climate agenda toward a society resilient to climate variability and change, the S2S project represents a very concrete example of how diverse research communities can tackle issues of common interest and deliver very tangible and measurable outcomes in a short time frame.

In the context of WMO, seamless prediction considers not only all compartments of the Earth-system, but also all disciplines of the weather-climate-water-environment value cycle (monitoring and observation, models, forecasting, end-user products, dissemination and communication, perception and interpretation, decision-making, and feedback to research requirements) to deliver tailored weather, climate, water, and environmental information covering minutes to centuries and spanning the local to global scale.

Phase I of the S2S Project (ending in 2018) has accomplished several major achievements and important milestones, but there is also a clear recognition that much of the research, product development, and uptake by the application communities are still in their infancy. Much remains to be done to fully realize the vision of S2S timescale prediction, both in terms of improving the skill of the forecasts, as well as creating forecast products to help inform user decisions in the range of 2 weeks to a season. These issues will be the focus of Phase II of the S2S project.

This special book aims at summarizing some key progress achieved during Phase I of the S2S Project. Part I introduces S2S in the broader context of limits of weather forecast horizon and the predictability of weather within climate and planetary wave dynamics. Part II provides a systematic review of potential sources of predictability, including the Madden-Julian Oscillation (MJO), various teleconnections, land, ocean, sea ice, and stratosphere. Part III covers some important implementation aspects of S2S prediction systems, ensemble generation, model development issues, recalibration, and multimodel combination and verification. Finally, Part IV illustrates the societal benefits of such research to improve the prediction of extremes and to enhance preparedness for humanitarian actions, communication and dissemination of products, seamless prediction of monsoons, sectoral decisions with probabilistic approaches, and climate impacts on health.

Progress made to date in Phase I of the S2S Project is paving the way for a pilot transition of some of those research outcomes into operations in a seamless fashion, which will be the focus of Phase II of the project (which is now approved by the respective governing bodies of WWRP and WCRP and Executive Council of WMO). Phase II will enhance the S2S database as a backbone to S2S research and follow the entire value chain to also strengthen user applications and decision-making. Fundamental research to improve prediction skill in the S2S time range will be pursued on MJO prediction and teleconnections; land, ocean, and stratosphere initialization and configuration; ensemble generation; and the role of atmospheric composition, capitalizing further on the very successful collaboration between the weather and climate research communities.

WMO is extremely grateful for the tremendous contribution of the S2S community, in particular the S2S Steering Committee, its co-chairs Frédéric Vitart from the European Centre for Medium-range Weather Forecasting (ECMWF) and Andrew Robertson from the International Research Institute (IRI), Columbia University and the International Coordination Office hosted by the Korea Meteorological Administration (KMA), in implementing this project. It also acknowledges the financial and in-kind support provided by numerous countries and institutions during Phase I. Noteworthy is also the crucial support of the S2S database provided by ECMWF and the China Meteorological Administration (CMA).

Phase II of the S2S project is certainly something to look forward to, and the work and support of the international community are very much appreciated and encouraged.

Co-authors:
Paolo Ruti (Chief World Weather Research Programme, WMO),
Michel Rixen (Senior Scientific Officer World Climate Research Programme, WMO)
Estelle de Coning (Scientific Officer, World Weather Research Programme, WMO)

Acknowledgements

The editors would like to express their deep appreciation to all the chapter authors for their enthusiasm and willingness to contribute their knowledge and experience across the diverse aspects of the S2S science, modeling and societal applications, without which this book would not have been possible. In addition, the comments on the original book proposal from several anonymous reviewers helped substantially improve the original concept. We also wish to thank the publishers and editors at Elsevier for proposing this volume, and for their guidance, technical assistance and patience in bringing this project to fruition over an almost two year period, in particular Emily Thomson and Louisa Hutchins. Finally, the editors would like to thank the IRI and ECMWF respectively for their support; AWR acknowledges the support of a fellowship from Columbia University's Center for Climate and Life.

PART I

SETTING THE SCENE

CHAPTER

1

Introduction: Why Sub-seasonal to Seasonal Prediction (S2S)?

Frédéric Vitart, Andrew W. Robertson†*

*European Centre for Medium-Range Weather Forecasts (ECMWF), Reading, United Kingdom
†International Research Institute for Climate and Society (IRI), Columbia University, Palisades, NY, United States

OUTLINE

A rapid evolution is taking place in weather and climate prediction (Shapiro et al., 2010; Bauer et al., 2015). Historically, there has been a clear separation between weather and climate prediction, despite the fact that both use similar numerical tools. *Weather prediction* refers to the prediction of daily weather patterns from a few days up to about 2 weeks in advance, whereas *climate forecasting* refers to the prediction of climate fluctuations averaged over a season and beyond.[1] This time-scale separation between weather and climate prediction has been accompanied by a divide in the weather and climate research communities, for

[1]The terms *forecasting* and *prediction* are used synonymously here; in some cases, *forecasting* is preferred in the context of forecast use and verification, while *prediction* is more general.

Sub-seasonal to Seasonal Prediction
https://doi.org/10.1016/B978-0-12-811714-9.00001-2

the historical reasons described later in this chapter. However, a convergence is taking place, spurred by the growing realization that weather and climate take place on a continuum of time and space scales. Coherent phenomena on a range of scales along this continuum lead to predictability on scales from subdaily, to weeks, months, years, decades, and beyond (Hoskins, 2012). The sub-seasonal to seasonal time range (abbreviated as *S2S*), the focus of this book, sits where the weather and climate scales meet and corresponds to predictions beyond 2 weeks, but less than a season. It is also a key time range for *seamless* weather/climate prediction, in which a single model is used to make forecasts all the way from weather scales to seasonal or longer climate scales or, in a more limited interpretation, that the underlying predictability is seamless across timescales, even if pragmatism dictates the use of different models for forecasting at different lead times (Brunet et al., 2010). We note that the term *S2S* has recently been used more broadly to include seasonal forecasts up to 12 months ahead (NAS, 2016).

Objective weather and climate forecasting can be divided into two branches: empirical (or statistical) forecasting and numerical weather/climate prediction with dynamical models. Empirical forecasting has been practiced in one form or other for more than a hundred years, if not thousands of years (e.g., Taub, 2003); it consists of making forecasts based on past experience or, in the modern era, by using observational data of current and past states of the weather or climate to fit (or train) a statistical model. Empirical methods can be simple (e.g., persistence, where the current weather/climate is predicted to persist for a certain period of time) or more sophisticated (e.g., regression models or discriminant analysis). For example, the analog method of weather forecasting involves examining today's state of the atmosphere and finding days in the past with similar weather patterns (analogs). The forecaster then would predict the weather based on these past analog days. Correspondingly, an analog seasonal climate forecast might be based on past years with similar phases of the El Niño—Southern Oscillation (ENSO).

Numerical weather and climate prediction, on the other hand, uses mathematical (dynamical) models of the atmosphere or Earth system to predict the weather or the climate. The mathematical models used for short-range weather prediction or long-range climate prediction are based on the same physical principles and set of equations, called the *primitive equations,* with the primary distinction being that climate models need to include additional components of the climate system, such as the ocean, depending on the forecast lead time being targeted. These equations are used to evolve the density, pressure, and potential temperature scalar fields and the air velocity (wind) vector field of the atmosphere either on a latitude-longitude grid or in spectral space, through time. The effects of subgrid-scale processes, including convection, radiation, and interactions with the underlying surface, are not treated explicitly but instead are parameterized in terms of the resolved-scale variables. Although empirical methods can be used for sub-seasonal to seasonal prediction, this book focuses largely on dynamical numerical prediction models.

We begin this introductory chapter with a brief history of dynamical weather and climate prediction, together with the World Meteorological Organization (WMO) programs that were created to coordinate these activities and that gave birth to S2S. We then introduce sub-seasonal to seasonal forecasting research and practice, from the discovery of S2S predictability sources, improvements in numerical weather prediction (NWP), development of seamless forecasting, and the demand from applications. The chapter concludes with a summary of the structure of this book.

1 HISTORY OF NUMERICAL WEATHER AND CLIMATE FORECASTING

Numerical weather prediction (NWP) had its roots early in the 20th century, when a better understanding of atmospheric physics led to the establishment of the primitive equations of the atmosphere (Abbe, 1901; Bjerknes, 1904). The first numerical weather forecast was attempted in 1922 by the English scientist Lewis Fry Richardson, in a report called "Weather Prediction by Numerical Process"; he performed his study while working as an ambulance driver in World War I. In this publication, he described how small terms in the prognostic fluid dynamical equations governing atmospheric flow could be neglected, and a finite differencing scheme in time and space could be devised, to find numerical prediction solutions. He constructed a 6-hour forecast of pressure at two points in central Europe by hand. It took him about 6 weeks to produce this 6-hour forecast; unfortunately, it falsely predicted a surge in sea level pressure when in reality, the pressure remained about the same. The number of calculations required to perform weather forecasts is so huge that it was only with the advent of digital computers and the development of numerical methods that weather forecasting in real time became possible.

The first successful computer weather forecast was produced in 1950 with the ENIAC (Electronic Numerical Integrator and Computer) digital computer, taking almost 24 hours to make the 24-hour forecast (Charney et al., 1950). From there, Carl-Gustav Rossby produced the first operational weather forecast (i.e., routine predictions for practical use) based on the barotropic equation in September 1954. Numerical weather forecasting began shortly afterward on a regular basis in the United States, and at around the same time in other countries. For many years, weather forecasts were issued only from a single integration of the atmospheric model from the best estimate of the atmospheric initial condition. Following Edward Lorenz's groundbreaking 1963 paper "Deterministic Nonperiodic Flow," published in *Journal of the Atmospheric Sciences*, which showed how small changes in the initial conditions could lead to very different forecasts due to the nonlinearity of the primitive equations, ensemble forecasts started to be produced operationally in the 1990s. Instead of making a single forecast of the most likely weather pattern, a set (or *ensemble*) of forecasts were produced, giving an indication of the range of possible future states of the atmosphere, and thus the uncertainty of the forecast, stemming from imperfect knowledge of the initial conditions and shortcomings in model formulation. Today, ensemble weather forecasts are initialized using large numbers of perturbed initial conditions, and the model output is often presented in the form of probabilities.

Getting the best possible estimate of the atmospheric initial conditions is central to NWP, and advances in forecast skill over the past 10 years have come in roughly equal parts from improving these estimates and developing models (Bauer et al., 2015). Various methods are used to gather observational data for forecast initialization (radiosondes, weather satellites, and commercial aircraft and ship reports). These observations are generally irregularly spaced and contain errors, so they need to be processed to perform quality control and obtain values at locations that are usable by the model's mathematical algorithms; this process is called *data assimilation and objective analysis*. Then the numerical model can predict how the weather will evolve from its initial state as an initial value problem.

Numerical weather forecasting has improved significantly since the 1950s thanks to improved scientific knowledge, huge improvement in computing capacity, and the advent of satellite data. Computing power has increased by about an order of magnitude every 5 years since the 1980s. Data assimilation algorithms employ the forecast model and use the order of 10^7 observations per day to derive initial conditions that are physically consistent (Bauer et al., 2015). Improvements in forecast skill have been objectively and quantitatively assessed against verifying observations. In the range of 3–10 days ahead, skill has increased by about 1 day per decade, so that today's 10-day forecast is as accurate as the 7-day forecast in the early 1980s, as shown in Fig. 1 of Bauer et al. (2015). The predictive skill in the Northern and Southern hemispheres is almost equal today thanks to the effective use of satellite data providing global coverage.

The first seasonal forecasts issued by a government office were empirical, and they were probably those issued by the Indian Meteorological Department (IMD) in the 1880s; they used Himalayan snow cover as a statistical predictor for the summer monsoon. The work of Henry Blanford and Sir Gilbert Walker, both early directors of the IMD in colonial times, were motivated by the devastating droughts and famines in India in the late 19th century, which, it has been argued (Davis, 2000), gave birth to the modern field of tropical meteorology.

The first dynamical climate model was developed in 1956 by Norman Phillips, who developed a mathematical model that realistically depicted monthly and seasonal tropospheric circulation patterns. Following Phillips's work, several groups began working to create general circulation models (GCMs) based on the atmospheric primitive equations on the sphere. This development closely paralleled that of NWP models, but with lower horizontal resolution to enable longer simulations, and with parameterizations strictly conserving mass and energy necessary for studies of seasonal to interannual climate variability and climate change. The first GCM that combined both oceanic and atmospheric processes was developed in the late 1960s at the Geophysical Fluid Dynamics Laboratory at the National Oceanic and Atmospheric Administration (NOAA).

Climate predictability comes from the relatively slow evolution (i.e., taking months and even longer) of the atmospheric lower boundary conditions such as sea surface temperature (SST), sea ice, soil moisture, and snow cover. For instance, SST anomalies associated with El Niño or its opposite, La Niña, can be predicted a few months in advance (Barnston et al., 2012), leading to predictability in the impact of these anomalies on the atmosphere, such as a reduction of tropical storm activity in the Atlantic, or rainfall over many parts of the tropics, associated with El Niño (Gray, 1984; Ropelewski and Halpert, 1987). However, this impact of SST on daily weather is not deterministic, and the resulting predictability in seasonal averaged weather was called "predictability of the second kind" by Lorenz (1975).

This fundamental distinction between weather and seasonal climate prediction led to the introduction of probabilistic concepts—including ensemble prediction—into the latter well before the former. Seasonal climate forecasts are (or should be) issued in terms of changes or shifts in climatological probability distribution of weather parameters such as temperature and precipitation; climate forecasting on seasonal timescales and longer is not about predicting the exact weather several months or years in advance, but rather about predicting future changes in its probability distribution over large averaging time periods (ranging from seasons to multiple decades). The importance of ENSO-related, tropical SST anomalies as boundary forcing on the atmosphere led to the development of two-tier seasonal forecasting systems, consisting of separate components for (1) predicting the evolution of tropical Pacific SST, and (2) simulating the atmospheric response using ensembles of atmospheric GCMs in multimodel combination. This two-tier approach popularized the paradigm of seasonal forecasting as a boundary value problem, and it was used in real-time seasonal climate forecasting at several centers, including at the International Research Institute for Climate and Society (IRI) between 1998 and 2016 (Mason et al., 1999). Fundamentally, however, all dynamical prediction is an initial value problem for the evolution of the phenomena that have predictability on the relevant timescale (Hoskins, 2012).

As mentioned already, climate and NWP models are both based on the same set of numerical representations of the primitive equations. However, climate models need to include additional components of the Earth system in order to represent sources of climate predictability on longer timescales. These include the ocean, land surface, and cryosphere, as well as atmospheric chemistry (including aerosols, ozone, and greenhouse gases) and a more detailed representation of the stratosphere. The coupling of GCMs of the atmosphere and ocean, typically developed separately by research groups of atmospheric scientists and oceanographers, remains a big challenge for climate modelers because small imbalances in the surface fluxes between the models can lead to large drifts in climate when the models are coupled together.

The time evolution of the other components of the climate system is usually assumed to be too small to have a significant impact on weather forecasts a few days in advance. This is why weather forecasts historically are based on only an atmospheric global or regional circulation model, in which sea-ice and SST fields are simply persisted from the initial conditions, and with other components of the Earth system set at their climatological values (e.g., aerosols). However, this additional complexity of climate models has been offset by greater intricacy in the formulation of initial conditions in weather forecasting. Atmospheric observation and data assimilation are traditionally associated with the weather forecasting community rather than the climate forecasting community because of the key importance of good initialization for weather forecasting, while seasonal climate forecasts largely rely on predictability of the second kind, associated with the evolution of the SST boundary conditions. Another key difference between weather and seasonal forecasting is the resolution of the atmospheric model. Because the integrations are much shorter in weather forecasting than in seasonal forecasting, weather forecasts are usually produced with much finer horizontal and vertical resolution than climate models. The typical resolution of seasonal climate forecasts, such as in the North American Multimodel Ensemble (NMME; Kirtman et al., 2014), and simulations for the Climate Model Intercomparison Project Phase 5 (CMIP5; Taylor et al., 2012) is around 100 km, whereas global weather forecasts are now produced routinely at a resolution of up to 8 km, and short-range forecasts with regional models have a resolution of a few hundred meters.

Seasonal forecasting today is carried out routinely by 12 WMO-designated Global Producing Centers (GPCs), as well as by a consortium of research and operational centers in North America—the NMME—and other nongovernmental centers, including the APEC Climate Centre (APCC) and IRI; typically, the forecasts are issued toward the middle of every month. All these centers now use coupled ocean-atmosphere (one-tier) models, in which the initial conditions of the ocean, land, and (in some cases) sea ice are prescribed; seasonal climate prediction is largely an initial value problem in these models, as opposed to a boundary value problem in the two-tier approach. The key boundary conditions in these coupled ocean-atmosphere-land-ice models are prescribed greenhouse gas concentrations; these variations are especially important for making hindcasts for past years, which are needed for assessing forecast skill.

Climate prediction is also carried out on longer timescales to make decadal predictions, and especially to make projections of anthropogenic climate change. The IPCC has coordinated successive climate change assessments based on increasingly sophisticated Earth system models, which are GCMs to which further components of the Earth system relevant to longer timescales have been added (e.g., ice sheets).

2 SUB-SEASONAL TO SEASONAL FORECASTING

As mentioned previously, sub-seasonal to seasonal prediction (forecasts from about 2 weeks to a season ahead) addresses the gap between medium-range weather forecasting and seasonal forecasting. According to the WMO definitions (http://www.wmo.int/pages/prog/www/DPS/GDPS-Supplement5-AppI-4.html), the S2S scale corresponds to extended-range weather forecasting (10–30 days), and the first part of long-range forecasting (30 days up to 2 years). These ranges are approximate, and a committee formed by the National Academy of Sciences in the United States recently defined the S2S range as between 2 weeks and 12 months (NAS, 2016). As discussed previously, there are good historical reasons for the split between weather and seasonal climate forecasting: S2S was considered a difficult time range for weather forecasting, being both too long for much memory of the atmospheric initial conditions and too short for SST anomalies to be felt sufficiently strongly, making it difficult to beat persistence and leading to the notion of a gap between the two ranges.

A pioneering sub-seasonal forecast attempt was made by Miyakoda et al. (1983). This paper showed how the pronounced blocking event of 1977, which generated exceptional snowy conditions over Florida, was successfully reproduced in 1-month forecasts produced by a GCM (Fig. 6 in Miyakoda et al., 1983). In addition, Miyakoda et al. (1986) found some marginal skill in eight January 1-month integrations using a 10-day running mean filter applied to the prognoses. The use of 10-day low-pass filtering is significant because it implicitly recognizes the importance of time aggregation, which introduces a climate forecasting element. The report of successful forecasts beyond day 10 triggered a great deal of interest at that time, and many of the world's operational prediction centers experimented with extended-range forecasts (from 10 to 30 days ahead) (Tracton et al., 1989; Owen and Palmer, 1987; Molteni et al., 1986; Déqué and Royer, 1992).

The European Centre for Medium-Range Weather Forecasts (ECMWF) used its operational forecast model to produce a pair of 31-day forecasts starting at 2 consecutive days for every month from April 1985 to January 1989 (Palmer et al., 1990). These experiments generally showed some moderate skill after 10 days (Miyakoda et al., 1986; Déqué and Royer, 1992; Brankovic et al., 1988), particularly when comparing the forecast to climatology. However, a particularly tough test for extended-range forecasting is to beat the skill of persistence forecasts. At ECMWF, the extended-range experiments described in Molteni et al. (1986) failed to produce forecasts after 10 days that were significantly better than persisting the medium-range operational forecasts. As a consequence, this experiment did not lead to an operational extended-range forecasting system at ECMWF.

Anderson and Van den Dool (1994) added another pessimistic note to this problem, demonstrating that some apparent high-quality forecasts in the extended range that triggered the initial enthusiasm for monthly forecasting could have occurred by chance. Using the extended-range model from the National Centers for Environmental Prediction (NCEP) Dynamical Extended-Range Forecasting (DERF) (Tracton et al., 1989), they found that after 12 days, the model did not produce better forecasts than a no-skill control. These disappointing results reinforced for many years the idea that the sub-seasonal to seasonal timescale was a "predictability desert." However, interest in the S2S time range revived in the last decade thanks to four factors.

2.1 The Discovery of Sources of Sub-seasonal to Seasonal Predictability Associated With Atmosphere, Ocean, and Land Processes

Although they are not yet fully understood, the most important sources to date are the following:

- The Madden-Julian Oscillation (MJO): As the dominant mode of intraseasonal variability of organized convective activity, the MJO has a considerable impact not only in the tropics, but also in the middle and high latitudes. In addition, it is considered a major source of global predictability on the sub-seasonal timescale (e.g., Waliser, 2011).
- Soil moisture: Memory in soil moisture can last several weeks and influence the atmosphere through changes in evaporation and the surface energy budget, affecting sub-seasonal forecasts of air temperature and precipitation over certain regions during certain seasons (e.g., Koster et al., 2010b).
- Snow cover: The radiative and thermal properties of widespread snow cover anomalies have the potential to modulate local and remote climate over monthly to seasonal timescales (e.g., Sobolowski et al., 2010; Lin and Wu, 2011).
- Stratosphere-troposphere interaction: Signals of changes in the polar vortex and the Northern Annular Mode/Arctic Oscillation (NAM/AO) are often seen to propagate downward from the stratosphere, with the anomalous tropospheric flow lasting up to about 2 months (Baldwin et al., 2003).
- Ocean conditions: Anomalies in SST lead to changes in air-sea heat flux and convection that affect atmospheric circulation. Forecasts of tropical intraseasonal variability are found to improve when a coupled model is used (e.g., Woolnough et al., 2007; Fu et al., 2007).

2.2 Improvements in Numerical Weather Forecasting

The skill of medium-range forecasting has improved continuously over the past two decades, due to model improvements and better data and forecast initialization. These improvements have not been limited to the first 2 weeks. In particular, dynamical models have shown remarkable improvements in MJO forecast skill scores in recent years (Fig. 1). About 10 years ago, the forecast skill of the MJO by dynamical models was considerably less than that of empirical models (e.g., Chen and Alpert, 1990; Jones et al., 2000a; Hendon et al., 2000),

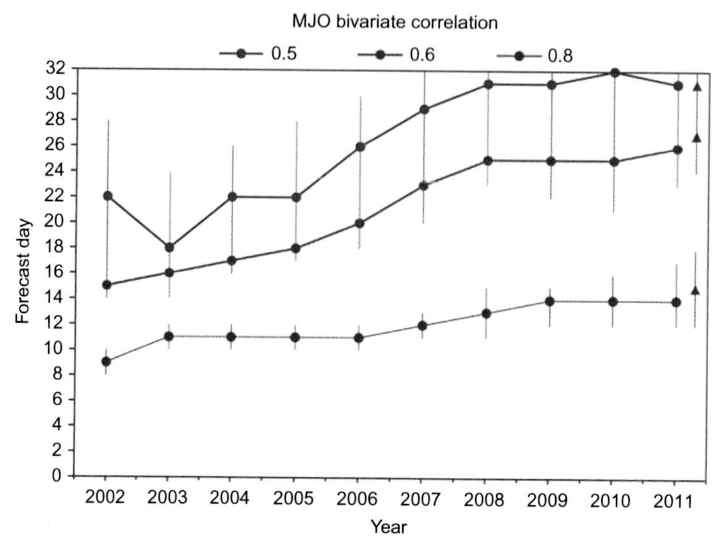

FIG. 1 Evolution of the MJO skill scores (bivariate correlations applied to Wheeler-Hendon index) since 2002. The MJO skill scores have been computed on the ensemble mean of the ECMWF reforecasts produced during a complete year. The blue, red, and brown lines indicate the day when the MJO bivariate correlation reaches 0.5, 0.6 and 0.8, respectively. The triangles show the skill scores obtained when rerunning the 2011 reforecasts with the version of the IFS that was implemented operationally in June 2012 (cycle 38R1). The vertical bars represent the 95% confidence interval (CI) computed using a 10,000-bootstrap resampling procedure (Vitart, 2014).

with skill only up to days 7–10. Recently, skillful MJO forecasts have been reported well beyond 10 days (e.g., Kang and Kim, 2010; Rashid et al., 2011; Vitart and Molteni, 2010; Wang et al., 2014; Vitart, 2014). This progress can be attributed to model improvement (e.g., Bechtold et al., 2008a) and better initial conditions, as well as the availability of historical reforecasts to calibrate the forecasts. Vitart (2014) also reported a significant improvement in 2-m temperature weekly mean prediction in the extratropics for weeks 3 and 4. Newman et al. (2003) found some strong predictability of week 2 and week 3 averages in some regions of the Northern Hemisphere using a statistical linear inverse model (LIM). These improvements in numerical prediction have provided an important stimulus for operational centers to revisit the sub-seasonal to seasonal prediction problem.

2.3 Development of Seamless Prediction

As alluded to already, the unifying concept of weather-climate predictability across multiple timescales has become increasingly prevalent over the last decade, as witnessed by several recent publications (Hurrell et al., 2009; Brunet et al., 2010; Shapiro et al., 2010; Hoskins, 2012). Fig. 1 in Hurrell et al. (2009) illustrates this concept, in which slower larger-scale climate phenomena provide the background for smaller and faster scales, while the integrated effects of the latter can exert important feedback about the former. This concept is epitomized at the S2S scale, which bridges between planetary scale phenomena (including ENSO and the MJO) and local daily weather conditions. Weather and climate have always been applied sciences, and indeed the quest for better early warning of high-impact weather events has contributed to the revival of interest in S2S. There is a long history of studies of so-called low-frequency variability (T > 10 days) of midlatitude weather, beginning with the work of Rossby and his contemporaries on index cycles describing sub-seasonal vacillations between blocked and zonal flows in the Northern Hemisphere midlatitudes. This early work led to studies on multiple equilibria and weather regimes (Charney and DeVore, 1979; Charney and Straus, 1980; Reinhold and Pierrehumbert, 1982) and the application of dynamical systems theory to S2S timescales (Ghil and Robertson, 2002).

In terms of forecasting, atmospheric models are run at the highest-possible resolution to better simulate the representation of weather fronts. Climate forecasting, on the other hand, is based on more complete Earth system models to better represent the evolution of atmospheric boundary conditions with less emphasis on high resolution because the simulation of day-to-day weather variability was not assumed to be fundamental. This difference between climate and weather forecasting is starting to disappear for the reasons summarized in this chapter and pursued in more depth in the other chapters in this book.

How can the theoretical 2-week limit of Lorenz be broken? Two aspects of the seamless prediction paradigm come into play. The first is that the Lorenz limit was derived in the context of midlatitude atmospheric dynamics of baroclinic waves, which have life cycles of about a week. The key to predictability on longer timescales is the existence of predictable phenomena on those timescales, such as the MJO. The second aspect is that averaging on the relevant timescale is critical; while the details of the weather on a specific day will not be predictable beyond 1–2 weeks, weekly or longer aggregates of weather statistics may be predictable in many cases, in the probabilistic sense of climate forecasts. What should the averaging period be for S2S forecasts? Zhu et al. (2014) have suggested that the averaging period should increase in tandem with the lead time, with a 1-week averaging corresponding to a 1-week lead time, and so on, as shown in Fig. 2.

Because weather forecast models are become increasingly skillful and are able to produce skillful forecasts up to 9 days in advance, the weather community has shown increasing interest in using more complex models that include other components of the Earth system to push the limit of predictive skill out a bit more. For example, most operational weather

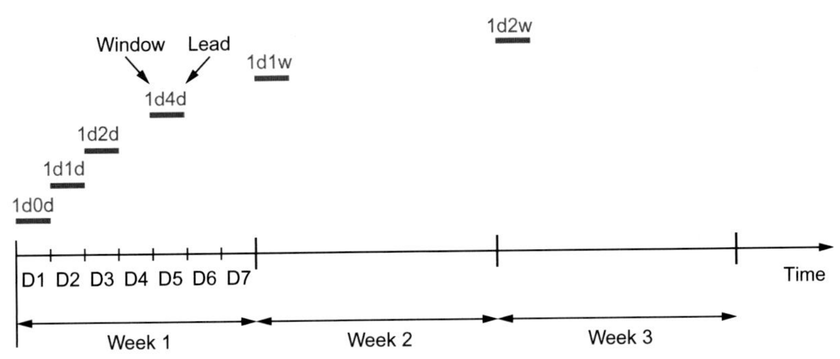

FIG. 2 Schematic of time window and lead time definitions. The horizontal axis represents the forecast time from the initial condition. The expression "1d1d" refers to an averaging window of 1 day at a lead time of 1 day, "2d2d" represents an averaging window of 2 days at a lead time of 2 days, and so on. Note that 1d1d is what is usually called "day 2" in some papers, and "1w1w" is what is usually called "week 2" (Zhu et al., 2014).

forecasting systems still use persisted SSTs (this means that the atmospheric model sees the same SST patterns as in the initial conditions during the full length of the model integration) because it was assumed that SST variations are too slow and small to affect weather forecasting in a significant way. However, ocean-atmosphere interaction has been found more recently to have a significant impact on some atmospheric phenomena, like the MJO (e.g., Woolnough et al., 2007) and tropical cyclone intensity (Bender and Ginis, 2000). As a consequence, some operational weather forecasting systems now include an ocean and a sea ice model, such as at ECMWF (Janssen et al., 2013), which previously had been used only in climate models.

Conversely, there is a growing interest in the climate community to better represent mesoscale weather events in long-range simulation. This is for two main reasons:

- A good representation of weather within climate (synoptic scale events) can feed back into a better representation of the large-scale climate.
- Being able to represent synoptic scale events may allow the direct prediction of impact of climate change on the statistics of the weather. For instance, this would help answer question such as: will global warming impact the number and severity of winter storms over Europe?

There is also a strong interest in testing climate models in weather configurations to help identify systematic errors in models and improve their ability to predict weather events (e.g., a WMO project called Transpose-AMIP, in which climate models are used experimentally for weather prediction). There have also been efforts to run weather forecast models in climate mode to test the evolution of systematic errors associated with slowly varying boundary conditions (e.g., Hazeleger et al., 2010).

From the physical perspective, there are no overriding reasons why weather and climate models should be different, and some operational centers like the United Kingdom's Met Office (UKMO) already use the same atmospheric model for weather and climate forecasting (the unified model, see http://www.metoffice.gov.uk/research/modelling-systems/unified-model). This evolution toward seamless prediction benefits sub-seasonal prediction, in which atmospheric predictability comes from both the initial conditions and boundary conditions. On the other hand, for practical reasons, it may be more efficient to run lower-resolution models with more ensemble members for longer lead times, and the optimal initialization strategy may also depend on the lead time and the phenomena that are the sources of predictability for that lead time.

2.4 Demand From Users for S2S Forecasts

As we have seen, the developments of both weather and climate science and forecasting have strongly use-inspired histories. Thus, it can come as no surprise that societal factors play an important role in answering the question "Why S2S?" The research program of the WMO, a specialized agency of the United Nations (UN) whose mandate covers weather, climate, and water resources with the 191 UN member-states and territories, is organized around two components: the World Climate Research Programme (WCRP), charged with determining the predictability of climate and the effect of humans on climate; and the World Weather Research Programme (WWRP), charged with advancing society's ability to cope with high-impact weather through research focused on improving the accuracy, lead time, and utilization of weather forecasts. Both WWRP and WCRP have been strong proponents for the development of S2S forecasts for societal benefit, and the WWRP/WCRP S2S Prediction joint research project is the result of those combined forces of improving early warning of climate extremes and understanding human influence on their generation.

While many end-users benefit from applying weather and climate forecasts in their decision-making, many studies suggest that such information is underutilized across a wide range of economic sectors (e.g., Morss et al., 2008b; Rayner et al., 2005; O'Connor et al., 2005; Pielke Jr. and Carbone, 2002; Hansen, 2002). This may be explained partly by the presence of gaps in forecasting capabilities, such as at the sub-seasonal scale of prediction, as well as the large gap between the physical science and end-user domains, which includes the complex task of bringing forecast information into the sphere of multifaceted decision-making.

The sub-seasonal to seasonal scale is especially relevant, as it has the potential to bridge between applications at daily weather timescales and much longer seasonal through decadal climate timescales, where in both cases considerably more societal and economic research has been conducted (e.g., decision and economic valuation studies and climate change impact and adaptation studies). It is therefore an ideal scale to improve forecasts and to evaluate the development, use, and value of predictive information in decision-making. Extending downward from the seasonal scale, a seasonal forecast might inform a crop-planting choice, while sub-monthly forecasts could help inform tactical farming decisions such as when to irrigate a crop or apply fertilizer or pesticides. This would make the cropping calendar a function of the sub-seasonal to seasonal forecast, and thus dynamic in time. In situations where seasonal forecasts are already in use, sub-seasonal ones could be used as updates, such as estimating end-of-season crop yields. Sub-seasonal forecasts may play an especially important role where initial conditions and intraseasonal oscillation yield strong sub-seasonal predictability, while seasonal predictability is weak, such as in the case of the Indian summer monsoon. For example, extending upward from the application of NWP, which is often routine, there is the potential opportunity to extend flood forecasting with rainfall-runoff hydraulic models beyond days to weeks.

In the context of humanitarian aid and disaster preparedness, the Red Cross Red Crescent Climate Centre/IRI have proposed a Ready-Set-Go concept for using forecasts from weather to seasonal (Goddard et al., 2014). In this formulation, seasonal forecasts are used to begin the monitoring of mid- and short-range forecasts, update contingency plans, train volunteers, and enable early warning systems (Ready); sub-monthly forecasts are used to alert volunteers and warn communities (Set); and weather forecasts are used to activate volunteers, distribute instructions to communities, and evacuate areas if needed (Go). This paradigm could be useful in other sectors as well, as a means to frame the contribution of sub-seasonal forecasts to climate service development within the Global Framework Climate Services (GFCS; Vaughan and Dessai, 2014), which provides a worldwide mechanism for coordinated actions to enhance the quality, quantity, and application of climate services; these services aim to equip decision-makers in climate-sensitive sectors with higher-quality information to help them make climate-smart decisions, helping societies better adapt to climate change.

In principle, advanced notification, on the order of 2 to several weeks, of tropical storms, severe heat or cold waves, the onset or uncharacteristic behavior of the monsoonal rains, and other potentially high-impact events, could yield substantial benefits through reductions in mortality and morbidity and economic efficiencies across a broad range of sectors. Realization of the potential value of such information, however, is a function of several variables, including the sensitivity of an individual, group, enterprise, or organization (or something it values) to particular weather events; the extent and qualities of its exposure to the hazard; its capacity to act to mitigate or manage the impacts such that losses are avoided and benefits are enhanced; and the ability of predictive information to influence its decisions to take action. Unlocking value, therefore, involves much more than creating a new or more accurate prediction, product, or service.

A type of S2S climate forecast has already been popular in applied settings for some time, in the form of *weather-within-climate* seasonal forecasts. For example, the overall frequency of rainy days over the growing season is a key variable for rainfed crops because evenly distributed rainfall is much more beneficial to plants than a few intense downpours with long dry spells in between (Hansen, 2002). It has been shown that in the tropics, seasonal predictability of daily rainfall frequency is often higher at local scale than the seasonal rainfall total (Moron et al., 2007), potentially increasing the usefulness of forecasts by increasing both their salience and credibility (Meinke et al., 2006). Seasonal forecasts tailored to agricultural use thus have started to target the number of rainy days occurring over a particular 3-month season rather than the usual 3-month rainfall average. These weather-within-climate forecasts have seasonal lead time, but the target variable is sub-seasonal. This topic is discussed further in Chapter 3.

3 RECENT NATIONAL AND INTERNATIONAL EFFORTS ON SUB-SEASONAL TO SEASONAL PREDICTION

For the reasons mentioned up to now in this chapter, there recently has been a growing interest in sub-seasonal to seasonal prediction. Ten years ago, only two operational centers, the Japan Meteorological Agency (JMA) and the ECMWF, were producing forecasts at the sub-seasonal time range. Today, at least ten operational centers and most of the WMO GPCs are issuing sub-seasonal to seasonal forecasts routinely.

In 2013, the WWRP and WCRP launched a 5-year joint research initiative, the Sub-seasonal to Seasonal Prediction Project (S2S), with the goal of improving forecast skill and understanding of the sub-seasonal to seasonal timescale, as well as promoting its uptake by operational centers and exploitation by the application communities (Vitart et al., 2012b). A major outcome of this project has been the establishment of a database of near-real-time forecasts (3 weeks behind real time) and reforecasts from 11 operational centers across the world (Vitart et al., 2017): the Australian Bureau of Meteorology (BoM), China Meteorological Administration (CMA), ECMWF, Environment and Climate Change Canada (ECCC), the Institute of Atmospheric Sciences and Climate (ISAC) with the Italian National Research Council, Hydro-meteorological Centre of Russia (HMCR), JMA, Korea Meterological Administration (KMA), Météo-France/Centre National de Recherche Meteorologique (CNRM), National Centers for Environmental Prediction (NCEP) and the UKMO. This database provides an important tool to advance our understanding of the S2S time range and help evaluate the benefit of multimodel sub-seasonal prediction. Sub-seasonal to seasonal prediction is also central to several current U.S. initiatives, such as the NOAA/MAPP initiative funded by NOAA. There are also efforts in the United States to enhance collaboration between agencies such as the U.S. navy, NOAA, the National Aeronautical and Space Administration (NASA) and the National Science Foundation (NSF) for the development and implementation of an improved Earth system prediction capability (ESPC) on timescales ranging from a few days to weeks, months, seasons, and beyond.

4 STRUCTURE OF THIS BOOK

This book has four parts. Part I, "Setting the Scene," addresses the question of the reasons to use sub-seasonal prediction, which has been briefly discussed in this first chapter. It provides background on NWP and an introduction to ensemble prediction methods. From that basis, it introduces the continuum spatial-scale dependence of the forecast time horizon, with larger spatiotemporal scales predictable for longer into the S2S range and beyond, and discusses the concept of climate predictability of weather statistics based on aggregation in time and space (Chapters 2 and 3). Part I concludes with a theoretical consideration of the potentially predictable modes on S2S scales from the point of view of atmospheric dynamics (Chapter 4).

Part II, the largest part of the book, discusses many of the sources of sub-seasonal predictability identified so far: the MJO (Chapter 5); extratropical waves, oscillations and regimes (Chapter 6); tropical-extratropical teleconnections (Chapter 7); land surface processes (Chapter 8); midlatitude ocean-atmosphere interaction (Chapter 9); sea ice (Chapter 10); and the stratosphere (Chapter 11). Chapter 6 shows an example of how the theoretical framework of dynamical systems provides practical tools for low-order empirical modeling and prediction of S2S variability.

Part III of the book is devoted to several S2S modeling and forecasting issues: the design of forecasting systems used for sub-seasonal prediction (Chapter 12); the generation of ensemble forecasts and data assimilation (Chapter 13); the importance of high-resolution modeling (Chapter 14); the development and testing of S2S forecast products through forecast calibration and multimodel combination (Chapter 15); and verification methods (Chapter 16). A detailed overview of the medium-range and sub-seasonal systems currently used at operational forecasting centers around the world, including their initialization and generation methods, is provided in Chapter 13.

Part IV of the book is dedicated to the use of sub-seasonal forecasts in applications, beginning with the potential to provide early warning of extreme weather events (Chapter 17); seamless prediction of monsoon onset and active/break phases (Chapter 20). This is followed by a chapter on the seamless framework for the early-action use of sub-seasonal forecasts (Ready-Set-Go concept) developed in the humanitarian aid community (Chapter 18); communication and dissemination of forecasts and engaging user communities (Chapter 19); lessons learned from 25 years informing sectoral decisions with probabilistic climate forecasts in the agricultural and energy sectors in Uruguay (Chapter 21); and predicting climate impacts on health at S2S timescales (Chapter 22).

The book concludes with a brief epilogue on prospects for the future of S2S in Chapter 23.

While each chapter is largely self-contained, the references have been consolidated at the end of the book since many are cited in multiple chapters.

CHAPTER

2

Weather Forecasting: What Sets the Forecast Skill Horizon?

*Zoltan Toth**, *Roberto Buizza*[†,‡]

[*]NOAA, Boulder, CO, United States [†]Scuola Superiore Sant'Anna, Pisa, Italy [‡]ECMWF, Reading, United Kingdom

OUTLINE

1 INTRODUCTION

Weather and climate are two aspects of a single reality, the time-evolving atmosphere, as it interacts with the surrounding geospheres. Simplistically, *weather* can be defined as the instantaneous manifestation of this reality, while *climate* refers to weather conditions or their

Sub-seasonal to Seasonal Prediction
https://doi.org/10.1016/B978-0-12-811714-9.00002-4

statistics over extended (typically seasonal or longer) time periods. As discussed later in this chapter, the conditions of the atmosphere and its surrounding spheres can be predicted scientifically. In general, more specifics of the expected weather can be foreseen at short lead times, while fewer details of the instantaneous weather are predictable at longer ranges. In particular, specificity about the nature, timing, and position of weather events becomes increasingly elusive as the lead time of forecasts increases.

With advances in the science and technology of prediction, the quality of weather forecasts also has improved, extending the time range for which specific weather forecasts can be made. For example, over the Northern Hemisphere extra tropics, 10-day forecasts of synoptic-scale features are as skillful today as 7-day forecasts were 30 years ago (see Fig. 1).[1] *Sub-seasonal to seasonal (S2S) forecasting* refers to the time range beyond which prediction of weather with finer granularity is lost (today, around 15 days lead time), but lower, sub-seasonal time-frequency and larger spatial-scale variations are still predictable (up to a season or so). After a discussion in Section 2 on the scientific basis for and the evolution of methodologies used in weather forecasting, Section 3 will review, in a historical context, how improved forecast techniques have extended the practical limit of weather forecasting. Forecast techniques used in low-skill

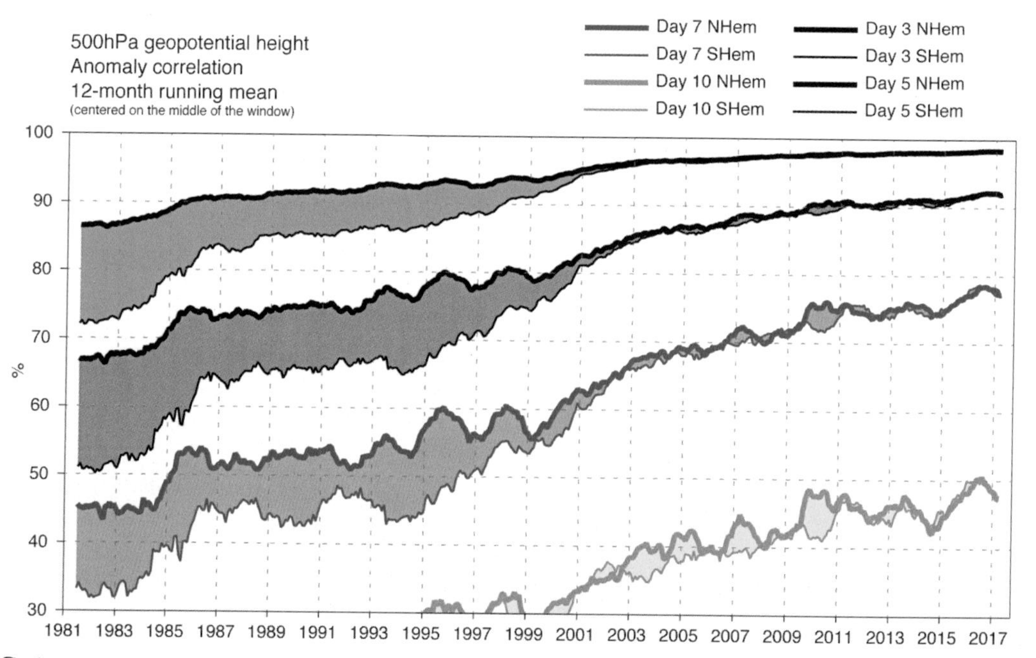

FIG. 1 Monthly averaged forecast skill measured by anomaly correlation coefficient for the 500-hPa geopotential height high-resolution operational forecasts issued by the ECMWF. The pair of *blue, red, green, and yellow lines* show the skill of the 3-, 5-, 7-, and 10-day forecasts over the Northern Hemisphere *(thick lines)* and Southern Hemisphere *(thin lines)*; the shading between the pairs of lines indicates the difference between the skill over the two hemispheres.

[1]Predictability is further explored throughout this chapter, with more formal discussions given in Sections 2 and 6.

environments will also be discussed in Section 4, with a special emphasis on ensemble techniques used so ubiquitously today covered in Section 5. Section 6 reviews how lessons learned from past improvements in weather forecasting may inform S2S efforts to expand the practical limits of predictability and to better exploit forecast skill in the extended range. In particular, we distinguish between *predictability* as conventionally defined related to the spatial and temporal phase of individual weather events (*traceable* predictability), versus predictability of the frequency of such events conditioned on larger-scale (and hence traceable for longer time periods) regimes (*climatic* predictability).

2 THE BASICS OF NUMERICAL WEATHER PREDICTION

The weather that we experience every day depends on atmospheric processes. The atmosphere, of course, is not isolated from, but rather influenced by, its surroundings. Solar insolation, varying primarily on an annual basis, is one of the primary factors driving the general circulation of the atmosphere. Many other slowly varying external factors such as ocean and land surface processes impart an additional level of predictability through their coupling to the atmosphere, which is particularly noticeable on the S2S timescales. Unless noted otherwise, by *predictability,* we refer to current or future scientifically based capacities to skillfully predict the evolution of the atmosphere or its surrounding spheres. Before delving into forecasting the state of coupled systems, however, first we turn our focus to the atmosphere itself. As we will see, some general lessons learned about the predictability of the atmosphere carryover to the more complex coupled ocean, ice, land, and atmosphere systems.

2.1 The Atmosphere as a Dynamical System

Atmospheric processes have been the subject of intense scientific studies. A well-established and critical characteristic of the atmosphere is that its time evolution follows specific rules, and therefore it behaves like a dynamical system. On macroscales, the evolution of the atmosphere is also deterministic, governed by specific physical laws. Importantly, if we know the state of the atmosphere at one point in time, with the use of these natural laws we can predict its state at future times as well. The deterministic nature of the atmosphere (e.g., Richardson, 1922) thus provides the basis for its prediction, which since the 1970s has been done mostly by computers. *Numerical weather prediction (NWP),* as the process is referred to today, will be discussed further later in this chapter.

2.2 Predictability

The behavior of periodic or quasi-periodic deterministic systems such as the solar system can be well predicted for long periods of time relative to the system's characteristic timescale (e.g., a solar year). Under some circumstances, the behavior of periodic or quasi-periodic deterministic systems, however, becomes irregular. This is due to the emergence of instabilities. Forces that previously could balance each other well in a stable fashion become imbalanced, giving rise to a new, dynamically evolving behavior. Interestingly, the atmosphere appears to behave like that: at high levels of viscosity, its laboratory and numerical models follow regular, periodic, and hence highly predictable behavior that turns aperiodic and much less predictable when viscosity is lowered below a certain level (see, e.g., Ghil et al., 2010 and references therein).

Deterministic dynamical systems with at least one unstable relationship or instability are called chaotic systems. A pendulum suspended via an elastic band (e.g., bungee-jumping) or spring (e.g., Lynch, 2002) is a simple example of a system with aperiodic motions where the centrifugal and gravitational forces temporarily overtake the force of suspension until the spring or band is sufficiently stretched so with its reduced elasticity it can counteract the other two forces. Elasticity in this example is a nonlinear function of the length of a spring or band. Temporally unstable developments in finite size chaotic systems are kept in check by such nonlinear interactions.

If both the governing laws and state of a deterministic system at an instant (such as the atmosphere)[2] are exactly known, the future states, even if the system is chaotic, can be predicted perfectly in perpetuity. The governing laws of real-life systems, of course, are not exactly known. And although the error variance in analysis fields can be estimated, the actual state of natural systems is not known either, due to observational uncertainties. In practice, we can forecast only with imperfect numerical models, from imperfect initial conditions. Errors from the initial condition will amplify in such forecasts due to the instabilities in the atmospheric system, and get convoluted with errors from the use of imperfect models.

Because the true states of natural systems are never known, the actual error patterns in the analysis fields are not known either. The evolution of hypothetical forecast errors, nevertheless, can be explored by studying the evolution of various perturbations to the state of a system. Linear perturbation models and their inverse or adjoint versions are derived by the linearization of the governing equations of a dynamical system (e.g., Errico, 1997). Such models can explore the evolution of infinitesimally small initial perturbations. Linear perturbation studies reveal the innate nature of instabilities of chaotic systems. Without the nonlinear interactions present in the full systems, linear perturbations associated with system instabilities have an exponential, unlimited growth (e.g., Lorenz, 1963; see Fig. 2) that is controlled by a single speed or growth parameter S:

$$v(t) = e^{St} \tag{1}$$

where $v(t)$ is perturbation or error variance at time t. As mentioned earlier, chaotic systems have at least one such instability. Interestingly, in complex dynamical systems, almost any error pattern has a finite projection in all directions in the multidimensional phase space of the system. This includes unstable directions as well. No matter how small, almost any error pattern will project onto instabilities, eventually leading to a complete loss of predictability in chaotic systems.

The exponential growth of linear perturbations clearly indicates that chaotic systems like the atmosphere have only finite predictability. The nature and time after which predictability is lost in realistic systems, however, can be assessed only through the consideration of nonlinear interactions. While nonlinear perturbations are less amenable to theoretical

[2] An estimate of the instantaneous state of the atmosphere, used as an initial condition for NWP forecasting, is called an *analysis*.

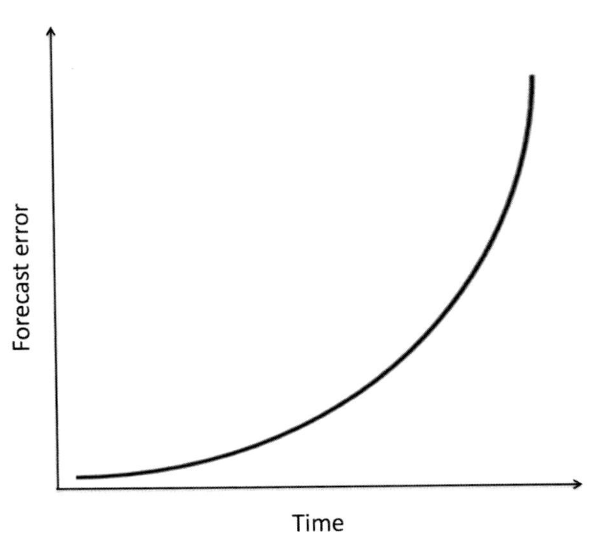

FIG. 2 The exponential growth of linearly evolving perturbation or error variance associated with instabilities in chaotic systems (with arbitrary units).

inquiries, they can be studied by comparing numerical forecasts made with the same model but with slightly different initial conditions (e.g., Yuan et al., 2018). Estimates of error variance between past forecasts and reality, the latter represented by the corresponding verifying analysis fields, also can be used to assess the statistics of error behavior.

Such studies reveal that nonlinear interactions limit the growth of otherwise exponentially amplifying perturbations and errors. Interestingly, in chaotic dynamical systems, the complex, nonlinearly saturating behavior of the initially exponentially growing error variance (v) in time (t) can be well characterized by the following simple and general relationship (Lorenz, 1982):

$$v(t) = \frac{R}{1 + e^{-St}} \tag{2}$$

Notably, the logistic error growth shown in Eq. (2) and displayed in Fig. 3 requires only one additional parameter beyond the speed parameter S that reflects the intensity of the underlying instabilities in linear error growth in Eq. (1). This new parameter is R, the total range or the level of variance at which error growth saturates at long lead times, as all predictability is lost. Note that R, as one of its basic global characteristics, reflects the overall size of the attractor.[3]

2.3 Scale-Dependent Behavior

Atmospheric motions can be documented using different basis functions or phase space coordinate systems. The state of the variables can be described, for example, at selected points in space (i.e., grid depiction), or in special combinations of the gridded variables, such as

[3]The term *attractor* here refers to the collection of all time trajectories that the system can ever visit naturally.

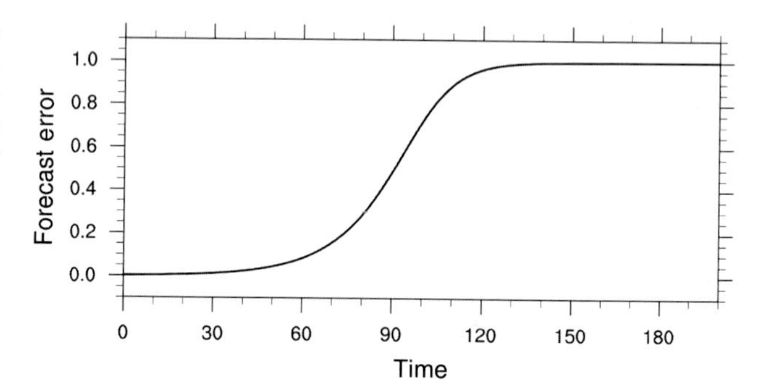

FIG. 3 Logistic growth of nonlinearly evolving normalized perturbation or error variance ($R = 1$, with arbitrary time units) associated with instabilities in chaotic systems. Note the similarity in the initial phase of error growth to exponential growth, as displayed in Fig. 2.

empirical orthogonal functions (EOFs; e.g., Lorenz, 1956), or a coordinate system based on a Fourier decomposition of waves according to their geographical scale (e.g., Orszag, 1969). Using the scale decomposition approach, for example, Rossby (1939) showed that the propagation of atmospheric and ocean features, since called *Rossby waves,* are dynamically dependent on their scale: smaller waves, in general, travel faster.

It turns out that the intensity of instabilities, and hence the speed or rate of error growth (S), also depend on the scale of the motions. In fact, both S and the range (R) parameters in Eq. (2) are a strong function of the spatiotemporal scale of features with which they are associated. We first note that weather is manifested in spatiotemporally coherent features. It follows that the spatial and temporal scales of atmospheric phenomena are connected. Smaller-scale (on the order of hundreds of meters) cumulus clouds, for example, develop as perturbations upon the prevailing environment much faster (in tens of minutes) than large-scale (hundreds of kilometers) extratropical cyclones that evolve over several days. For smaller systems, this results in much faster nonlinear perturbation growth (i.e., larger S) that, simply due to their size, saturates at a low energy level (smaller R).

The same instabilities that force the development of features in the flow are also responsible for the emergence of errors that result in modified, more or less intense, or lack of features. Therefore, faster perturbation growth corresponds with faster error growth, as well as a faster loss in predictability for smaller-scale features. A simple measure of forecast or predictive skill is the correlation between forecast and verifying analysis anomalies taken from the time mean or climatological conditions. An anomaly correlation (AC) or pattern correlation (see, e.g., AMS, 2000) of 1 corresponds to forecasts that perfectly capture the analyzed variability of weather, while zero indicates no skill at all relative to climatological information. As seen in Fig. 4, while features with total wave number 10 (approximately 3000 km horizontal scale, such as a large-scale extratropical cyclone) can be predicted with a useful level of skill (i.e., above 0.6 AC) out to 6 days ahead, smaller-scale features (total wave number 60, about 500 km, e.g., convective clouds organized into mesoscale convective systems) can be skillfully predicted only 1 day in advance. The predictability of individual clouds (not shown in Fig. 4) is even shorter.

AC measures skill in terms of how well forecast and verifying analysis patterns are aligned; therefore, it is is insensitive to the amplitude of errors. Analysis and forecast error

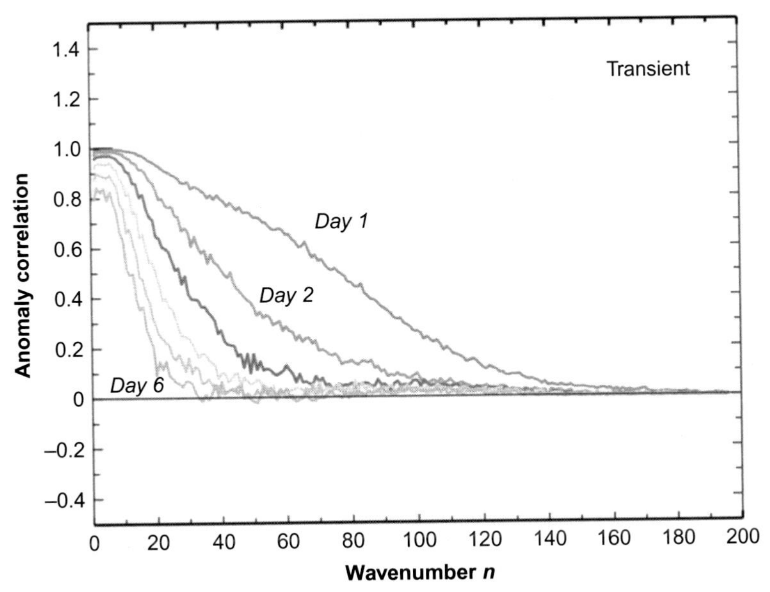

FIG. 4 The AC between Canadian Meteorological Center/Meteorological Service of Canada global 500-hPa forecast anomalies, valid for days 1 (right) through 6 (left), and the verifying analysis anomalies, as a function of total wave number (from the largest scales on the globe to 200 km at $n = 200$), for January 2002. *Adapted from Fig. 6 of Boer, G.J., 2003. Predictability as a function of scale. Atmos. Ocean, 41 (3), 203–215.*

variance,[4] on the other hand, measures the amplitude of errors at each grid point. As the forecast features become uncorrelated, forecast error variance saturates at an R value twice the level of climatological variance for the scales in question (e.g., Peña and Toth, 2014). As seen in Fig. 4, predictability is lost faster on smaller scales. Fig. 5 offers a quantitative assessment of the predictability of the remaining NWP forecasts with progressively longer lead time as a function of total wave number. As noted earlier, forecasts for smaller-scale features (characterized with larger wave numbers) have a smaller climatological variance, and therefore a lower saturation value R (see the heavy solid line on the right side of Fig. 5), corresponding at each scale to the sum of the climatological variance for the model representing nature, and another model used to produce forecasts. The dashed and lighter solid curves in Fig. 5 stand for the error variance in the initial (analysis) and 1–14 day forecast fields, respectively; for synoptic and smaller scales (wave number > 8),[5] the difference between these and the heavy solid curve can be interpreted as the information or predictability remaining in the NWP fields.

[4] After Peña and Toth (2014), analysis and forecast error is interpreted here as the difference between reality interpolated to the NWP grid and the analysis or forecast fields.

[5] Note that due to the reference data used by Privé and Errico (2015) to define anomalies, the thick line in Fig. 5 does not represent an accurate estimate of climatological random error variance for the largest scales.

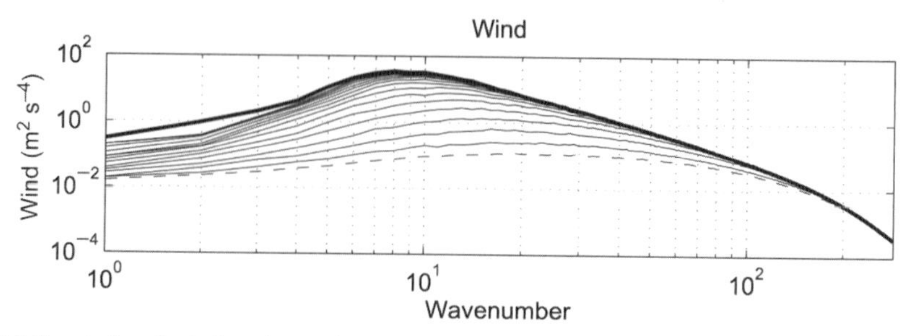

FIG. 5 A 356-hPa rotational wind analysis (dashed line), 1–14-day lead time (*solid line*, from low to high), and random forecast (*R*, *heavy solid line*) error variance as a function of the total wave number for a July–August observing system simulation experiment (OSSE, adapted from Fig. 1 of Privé and Errico, 2015), expressed as anomalies from the respective 2-month means (nature run, analysis, or forecast).

As we move from smaller to larger scales (from right to left in Fig. 5), the analysis error variance curve departs from the random error curve around total wave number 200 about 150 km), indicating that in this OSSE experiment, the analysis retains information about simulated reality down to those scales. Forecasts after 2 days, however, already lose information and predictability on scales finer than wave number 100 (about 300 km), and 4 or 7 days into the forecast, predictability is lost on scales finer than 750 or 1500 km (wave numbers 40 and 20), respectively. As we reach a lead time of 14 days (the shorter end of the S2S time range), only a relatively small amount of information remains on planetary scales greater than 5000 km. Discernable differences between the top thin lines indicate that in this OSSE, some predictability still may be left on the largest scales out to 14 days lead time, even though nature was simulated without full coupling with slowly changing oceanic or ice conditions. If we apply the generic logistic relationship for the description of error growth on progressively larger scales, *S* monotonically decreases, while *R* increases up to, and attains its maximum value at, around about 4000 km (wave number 8).

In summary, while determinism makes predictions possible, chaos, in the presence of initial errors and model imperfections, places limits on the extent of predictability. In particular, with the current level of analysis errors, predictive information on individual events beyond 14 days is restricted to slowly evolving, planetary-scale waves. The position of finer-scale details is predictable only on shorter timescales. The implications of these limitations on S2S forecasting will be further discussed in Section 6.

2.4 Coupled Systems

So far, we have discussed predictability and forecasting in the atmosphere with prescribed boundary conditions, uncoupled from its surrounding spheres. In reality, the atmosphere is in two-way interactions with the ocean, land, and cryosphere. Such two-way coupling with these other systems make the atmospheric circulation more complex. However, the higher level of complexity, due to the rather long "memory" of some high-energy oceanic, ice, and land surface processes, may actually extend the predictability of weather or its statistics in the coupled system. In fact, many of the processes in the coupled system act on a slow timescale compared to those in the atmosphere. A significant portion of the energy of the coupled system, for example, is locked into annual or decadal scale variations in deep ocean circulation, deep soil moisture and temperature, or sea and land ice fluctuations. When the atmosphere is coupled to these systems, its energy spectrum also shifts to slower scales, giving rise to potential predictability in the atmospheric portion of the full coupled system.[6]

In particular, the coupled ocean-atmosphere system enables the emergence, or enhances the influence of, some slower-evolving instabilities. These in turn led to the formation of longer-lived phenomena such as the El Niño-Southern Oscillation (ENSO). As is the case with the uncoupled atmosphere, the dynamics of the coupled system both supports (due to determinism) and limits (due to chaotic behavior) the predictability of the entire coupled ocean-land-ice-atmosphere system (and, within it, the atmospheric portion of the system). Many of these phenomena, arising from coupling with slower-evolving processes in surrounding spheres and associated with slower-growing instabilities and errors, will be studied in more detail in upcoming chapters.

3 THE EVOLUTION OF NWP TECHNIQUES

Information about future weather originates from observations of its current state. Observations are made either in situ or remotely, from space-, air-, or ground-based platforms. Observing systems typically provide sparse (in space and time) and incomplete (in terms of variables) coverage, as well as inaccurate data about the natural system. To mitigate this situation, data assimilation (DA) must statistically spread information about the state of the system from observed to unobserved sites, times, and variables. Temporal extrapolation of information from past observations is a critical part of the DA process because the first guess for each analysis is provided by a short-range NWP forecast from the previous analysis. In a continual DA-forecast cycle, the DA step statistically combines information from the latest observations with the first guess forecast.

Analyses contain both random (originating from the observations, first guess forecast, and statistical procedures) and growing (originating from the first guess forecast) errors. Growing errors, by definition, amplify in NWP forecasts of chaotic systems, while random errors typically decay. As Toth and Kalnay (1993, 1997) noted, DA-forecast cycles also act as a retaining filter or amplifier of the growing type of error. A major emphasis in NWP development is the reduction of errors in the analysis, especially the growing errors that by definition, influence most the quality of forecasts. Atmospheric analyses are discussed more in Chapter 13.

NWP models are used to approximate the temporal evolution of nature in spatiotemporally discretized fashion on a finite grid. At the core of NWP models are prognostic equations that describe the relationship between the state of the system at two consecutive time steps. Forecasts are produced by the temporal integration of such equations. Natural processes are truncated in space, time, and in the range of represented physical processes. The effect on the resolved scales of some physical processes happening on finer scales unresolved by the

[6]Note that the predictability of the atmosphere can be realistically assessed only in a coupled ocean-land-ice-atmosphere system. If the boundary conditions of the atmosphere are prescribed, the time range of skillful forecasts for the atmosphere will be distinctly different (e.g., Goswami and Shukla, 1991).

models are statistically parameterized (i.e., physical parameterization of convection), while others are ignored (e.g., electrical processes). NWP models are thus imperfect.

Beyond the chaotically amplifying initial errors mentioned previously, forecasts are also affected by errors related to the use of such imperfect models. A smaller part of random (e.g., truncation-related) errors introduced at each time step projects onto growing directions and will amplify as initial errors do. Other model-related errors are due to the model's attractor being displaced from that of nature and manifest as a systematic drift of forecasts started with observed initial conditions, evolving until the model reaches a state close to its attractor.

How have today's NWP systems developed? Next, we offer a brief overview from 1950 onward. For more information, the readers are referred to Lynch (2006, 2008), Deutsche Meteorologische Gesellshaft e.V. (2000), and the papers listed in this chapter.

3.1 Computational Infrastructure

The concept of and scientific underpinning for NWP were created in the 1920s (Richardson, 1922). Due to the large amount of calculations involved, its practical implementation, however, had to wait for the invention of electronic computers in the 1950s. Computational power has continuously increased since, allowing data processing, DA, and forecast applications to become higher in spatial and temporal resolution and ever more sophisticated. Until the 1990s, this increase manifested in ever-faster processing speed, while the number of processors for parallel computing has increased dramatically since.

3.2 Observing Systems

Both surface and upper-air observations (Fig. 6) are indispensable for the initialization of three-dimensional (3D) NWP models. Surface in situ observing networks have been developed since the late 1800s, while the upper-air network was developed during World War II and in the 1950s and 1960s, partly in support of NWP forecasting. A key development that made NWP possible was the decision to share observations. Thanks to the work of the World Meteorological Organization (WMO, https://www.wmo.int/pages/index_en.html), observation standards were defined, and agreements were put in place to collect and exchange observational data in a timely manner. This led to a further expansion of the global network of observations that allowed NWP centers to generate more accurate estimates of the initial state of the atmosphere, which is required to generate operational weather predictions.

In particular, it is worth discussing two key projects. The first is the World Weather Watch (WWW, http://www.wmo.int/pages/prog/www/index_en.html), established in 1963. Since then, the WWW has been one of the WMO's core programs, combining observing systems, telecommunication facilities, and data-processing and forecasting centers. Today, the WWW, operated by WMO member-states, makes meteorological and related environmental information freely available for the provision of efficient services in all countries. The second project, started in 1967, was the Global Atmospheric Research Program (GARP). GARP, which ran until 1982, helped advance research in the field of weather prediction, including the organization and coordination of several important field experiments, such as the GARP Atlantic Tropical Experiment in 1974 and the Alpine Experiment (ALPEX) in 1982. One key

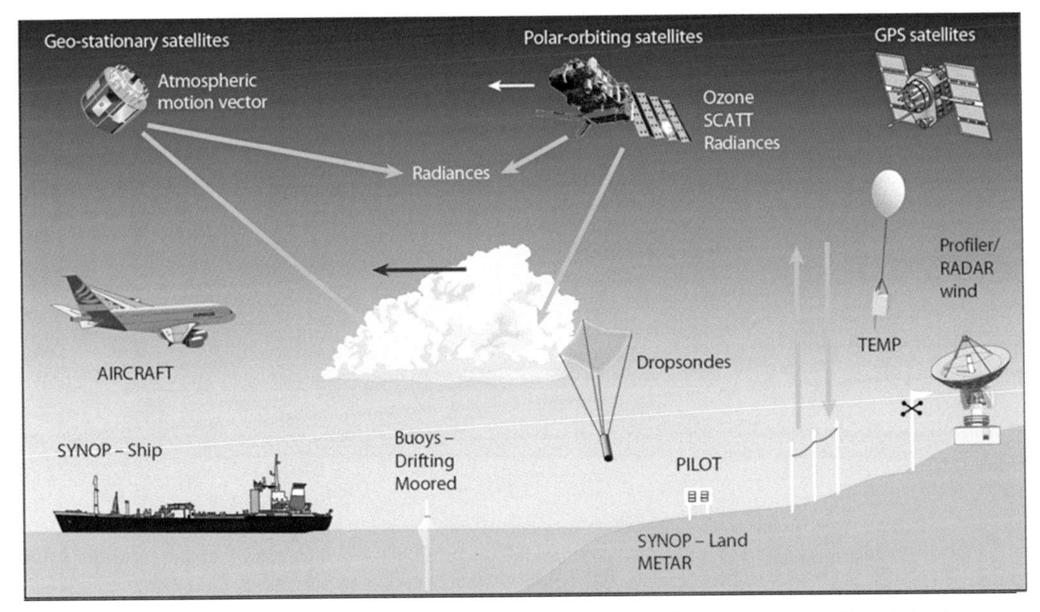

FIG. 6 Schematic of the most common atmospheric in situ and remote observing systems. *From https://www.ecmwf.int/en/research/data-assimilation/observations.*

component of GARP was the First GARP Global Experiment (FGGE), which took place in 1979. FGGE contributed to the extension of the observations of the WWW. These field experiments contributed to the availability of a larger number of, and more accurate, observations, allowing scientists to better test and validate their models.

No credible cost-benefit analysis exists as to the optimal configuration of existing or possibly planned observing systems, so observing system-related developments have been somewhat ad hoc. Although there may be other competitive solutions, governments of developed nations made strategic investments in the design and implementation of satellite platforms. An important milestone of the 1960s was the launch of TIROS-1 (TIROS stands for "Television Infrared Observation Satellite"; see https://science.nasa.gov/missions/tiros), the first weather satellite launched by the National Aeronautics and Space Administration (NASA). TIROS paved the way for the Nimbus program, whose technology and findings are the heritage of most of the Earth-observing satellites that NASA and the National Oceanic and Atmospheric Administration (NOAA) have launched since then. As seen in Fig. 7, the volume of observations in general (including surface, upper-air, and satellite-based measurements) has increased dramatically since the start of NWP in the 1950s. Of particular note is the superexponential growth of satellite-based, mostly radiance observations since the 1980s. Today, about 95% of observations come from instruments onboard satellites. Many of these observations, just as with other remotely sensed data such as lidar and radar observations, contain nonnegligible systematic errors and are indirect (i.e., not of the NWP model variables themselves, but only related to them). Their assimilation, therefore, poses special challenges. The relative role of the observing versus the DA-modeling systems in NWP will be revisited in Section 3.5.

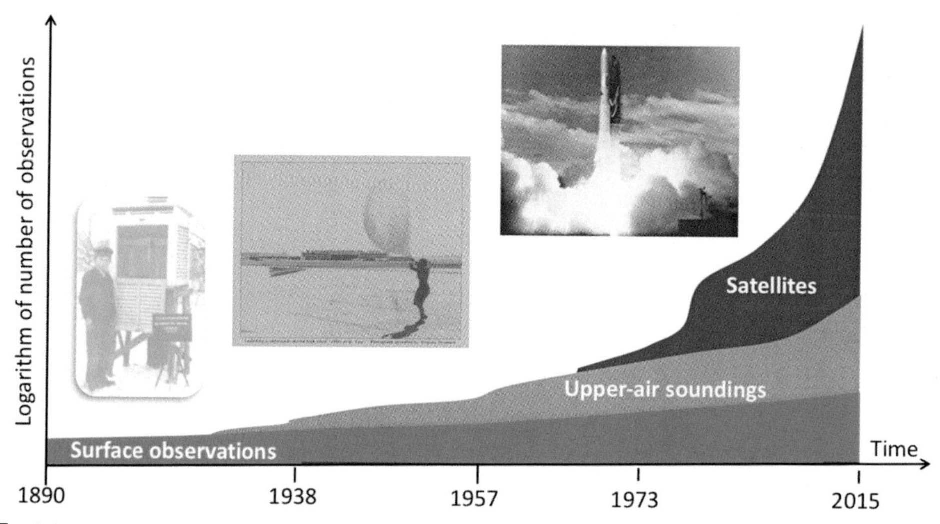

FIG. 7 Schematic illustration of the increasing number of daily surface *(green)*, upper-air *(orange)*, and satellite *(burgundy)* observations available for use in ECMWF reanalysis projects (shown on a logarithmic scale), as a function of time. For reference, in 2015, 44 M of a total of 600 M observations per day were assimilated. *Adapted from Dee, D., 2009. Representation of climate signals in reanalysis. In: Presentation at the Fifth International Symposium on Data Assimilation, Melbourne, Australia, 5–9 October 2009. See at: https://www.dropbox.com/s/ifge2r5wimiyc3h/Dee_2009_Melbourne.pdf?dl=0.*

3.3 Data Assimilation

Perhaps the most critical element in DA is the spatiotemporal propagation of observational information from observations to unobserved analyzed variables. Initial attempts at DA discarded information in all previously taken observations and used persistence or climatology as background fields to fill the large gaps in observational coverage (Bergthorsson and Doos, 1955). Only in the 1960s were NWP forecasts introduced as first guesses in the context of DA-forecast cycles that are still used today. This change led to major forecast improvements, as well as to the birth of modern data assimilation.

For the spatial propagation of information, at each observation site, Gaussian correlation models were used separately in schemes called "optimal interpolation" (Gandin, 1963). Relatively simple relationships served to ensure dynamical balance between the model variables (e.g., normal mode initialization; Errico and Rasch, 1988). A big step forward was the introduction of 3D variational DA (3DVAR). These methods solve a global minimization problem using covariance information across space and variables, derived from a climatology of differences between similar forecast states (e.g., Parrish and Derber, 1992).

It was not until the introduction of four-dimensional (4D) variational DA (4DVAR) that observational information was propagated across variables, space, and time (within an assimilation window) in a flow-dependent manner (e.g., Courtier et al., 1994). This revolutionized the main technique of DA that has not been surpassed since. A plethora of mostly sequential, ensemble-based DA methods have also been developed, but their performance lags that of 4DVAR (see, e.g., Bonavita et al., 2017 and the references therein). Although the basic 4DVAR methods have not been surpassed, various improvements, including ensemble-based methods to propagate covariance information across assimilation windows, have been developed over the past decade.

A key attribute to the variational approach is that it allows the assimilation of often remotely taken observations of nonmodel variables such as radiation or radar reflectivity. In such cases, observational models (called *observation operators*) connecting the observed and analyzed variables become part of the DA scheme. In this era of proliferating satellite-based radiance and other remotely sensed indirect observations, this feature has invaluable benefits.

3.4 Modeling

Beyond DA enhancements, model development is another key area for improving NWP forecast performance. In 1950, history was made at the University of Pennsylvania by the creation of the first numerical weather prediction using a simplified set of equations on the Electronic Numerical Integrator and Computer (ENIAC). In this effort, a barotropic vorticity model developed at the Massachusetts Institute of Technology (MIT; Charney et al., 1950) was used. Follow-on research in the United States led in 1955 to the start of routine numerical weather prediction under the Joint Numerical Weather Prediction Unit (JNWPU), a joint project between the U.S. Air Force, Navy, and Weather Bureau.

Across the Atlantic, Carl-Gustaf Rossby's group at the Swedish Meteorological and Hydrological Institute used a similar model to produce the first European operational forecast in 1954 (Persson, 2005; Harper et al., 2007). As computers advanced, more complex models capable of simulating a wider range of physical processes became established. Phillips (1956) developed the first primitive-equation numerical model capable of realistically depicting the main features of the troposphere. As a prelude to S2S, the first coupled ocean-atmosphere general circulation model was developed at the NOAA's Geophysical Fluid Dynamics Laboratory (GFDL, https://www.gfdl.noaa.gov/climate-modeling/) in 1960. As scientists developed better models and computers continued to increase in power, more nations started producing operational weather forecasts. In 1966, Germany and the United States began producing operational forecasts based on primitive-equation models. The United Kingdom and Australia followed suit in 1972 and 1977, respectively.

In 1967, a working group of the European Commission suggested that stronger collaboration in natural sciences, and more specifically in weather forecasting, could advance the science and technology in this area. Eventually, this led to the establishment in 1975 of the European Centre for Medium-Range Weather Forecasts (ECMWF, https://www.ecmwf.int/). ECMWF joined the ranks of the already established national meteorological services of the United States, Germany, United Kingdom, Australia, and many other countries, with the aim of tackling the medium-range problem, to assess whether it would be possible to issue skillful forecasts 5–10 days ahead.

Because a simple decrease in grid spacing (i.e., increase in spatial resolution) directly reduces truncation errors, it is not surprising that it leads to the reduction of both analysis and forecast error variance. Changes in horizontal resolution, however, require at least a cubic increase in computer power because a proportional reduction in the length of time steps is also required. Fig. 8 shows, for example, that a single forecast with a 64-km resolution model implemented at ECMWF in 1991 used a sustained computer power of about 0.001 teraflops (TF). This model configuration had about 3 million grid points. In contrast, the 2017 version of ECMWF's high-resolution global model has about 300 million grid points, spaced 9 km apart, and uses a sustained computer power of about 300 TF.

Numerics such as the rather technical choices for time-step and interpolation schemes, horizontal and vertical grids or other representations, and model variables is aimed at minimizing noise and maximizing the conservation of selected quantities in NWP forecasts. Continued experimentation so far yielded no clear or generally favorable solutions in this area.

Beyond reducing spatial/temporal truncation-related errors, the introduction of new or enhanced physical processes or related parameterization packages is another source of significant gains in the quality of NWP performance. The first numerical models in the 1950s and 1960s had no or very simple physics, while today's models represent an increasingly growing set of interacting physical processes (see Fig. 9).

Coupling an atmospheric model with models of surrounding spheres is another critical path for improved forecast performance. Two-way coupled Earth system modeling can be especially important for the S2S range, as slowly varying processes carry relatively large energy in the ocean, cryosphere, and land surface. Fig. 10 shows the composition of a state-of-the-art Earth system model used for weather and climate studies.

3.5 Improvements in Forecast Performance

Forecast improvements like those shown in Fig. 1 earlier in this chapter are due to the combined effects from all the main components of NWP highlighted here: (1) more and higher-

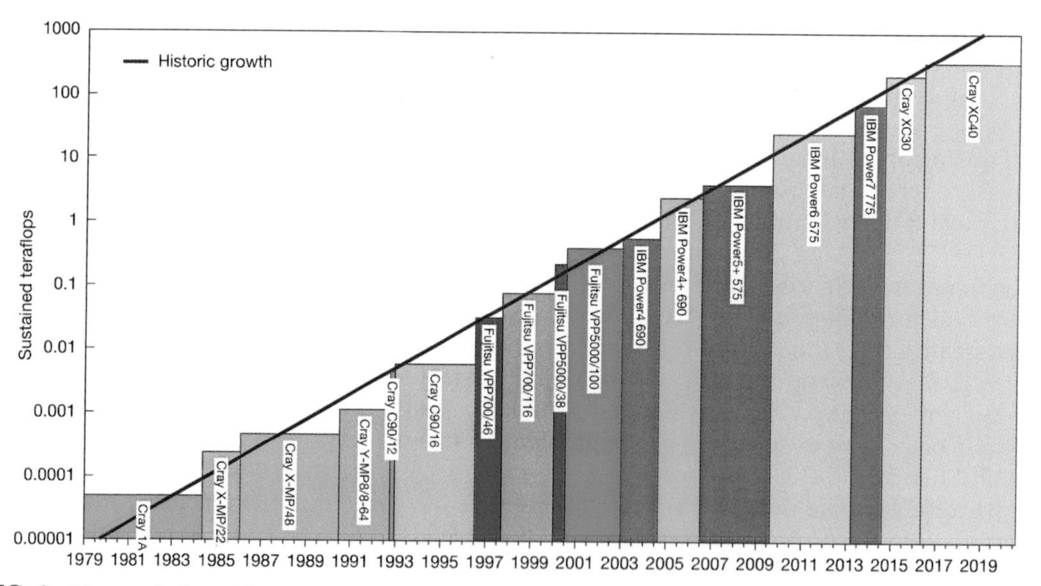

FIG. 8 Time evolution of the computer power installed at ECMWF, expressed in terms of sustained TF (i.e., 10^{12} floating point operations per second) on a logarithmic scale from 1979 to the present.

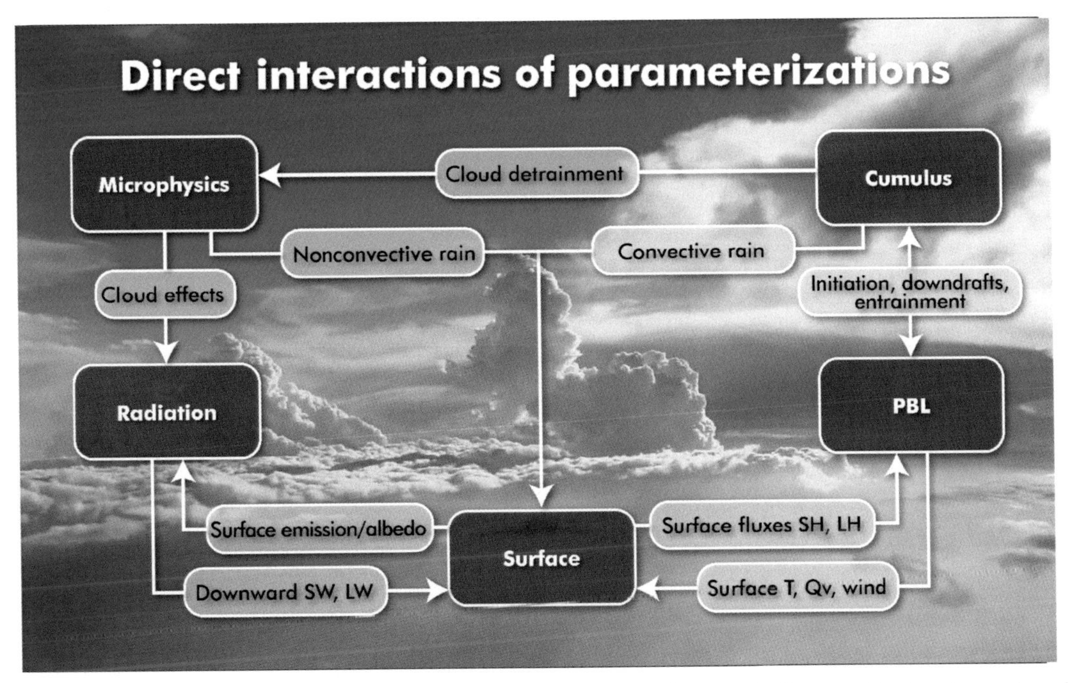

FIG. 9 Interactions *(gray)* between five physical parameterization schemes *(black)* of the Weather Research and Forecasting (WRF) model. Courtesy of the Developmental Testbed Center (DTC).

quality observations, (2) improved DA methods, (3) more realistic numerical models, and (4) more computer power that facilitates many of the advances seen in the other areas. We must note that either by design or necessity, NWP centers often introduce changes in more than one area at the same time, which makes attribution more difficult.

Fig. 11 may shed some light on the relative roles of changes in the observing versus DA-modeling systems on forecast skill. The 15-year period of 1984–98 saw a dramatic increase in the skill of Northern Hemisphere operational forecasts. This period also witnessed a dramatic rise in the availability of satellite observations (cf. Fig. 7). Because the development and maintenance of satellite-based observing systems are arguably the most expensive components of the NWP observing—data assimilation—modeling infrastructure, a naturally arising question is whether the approximately 17-percentage-point gain in operational Northern Hemisphere anomaly correlation scores (shown by the black dashed line in Fig. 11) is primarily due to the costly investments in satellite observing systems, or whether it reflects more the combined effect of less expensive developments in DA and modeling science and computational technology.

Changes in the quality of reforecasts (solid black line in Fig. 11), beyond any natural variability in atmospheric predictability, are only due to observing system changes because the reanalyses and reforecasts were made with a frozen, 1995 National Centers for Environmental Prediction (NCEP) DA-modeling system (Kistler et al., 2001). Remarkably, the Northern Hemisphere reforecasts exhibit no (or only minimal, if any) trend over the 1984–98 period, indicating that DA-modeling improvements dominate the gain in skill, probably at a small

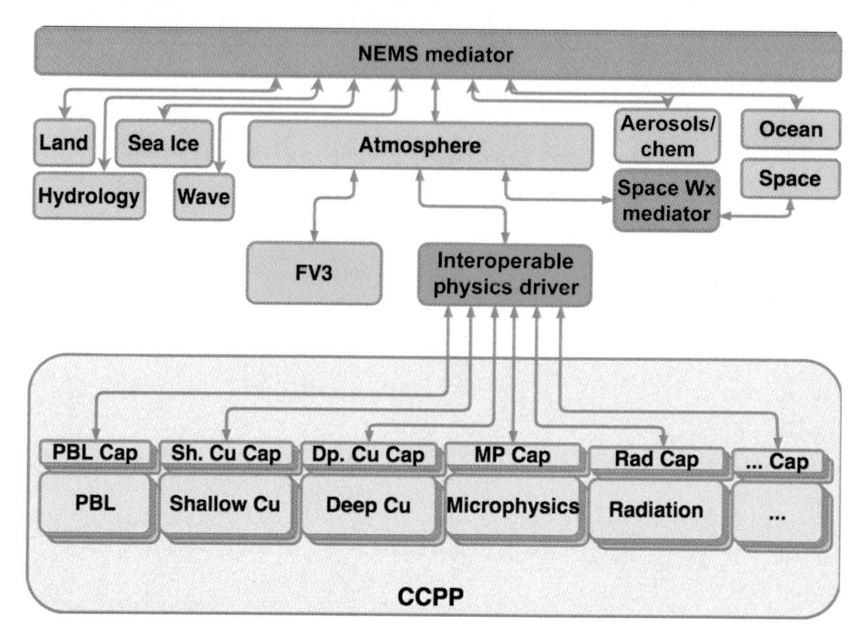

FIG. 10 Schematic diagram of the main models *(yellow)*, their overall mediators *(coral)*, and coupling connections *(lines with arrows)* for the Earth system model of the U.S. Next-Generation Global Prediction System (NGGPS). The dynamics (FV3), physics driver, and physics components of the atmospheric model are shown in light and dark green and blue, respectively. *Courtesy of the Global Model Test Bed of the DTC.*

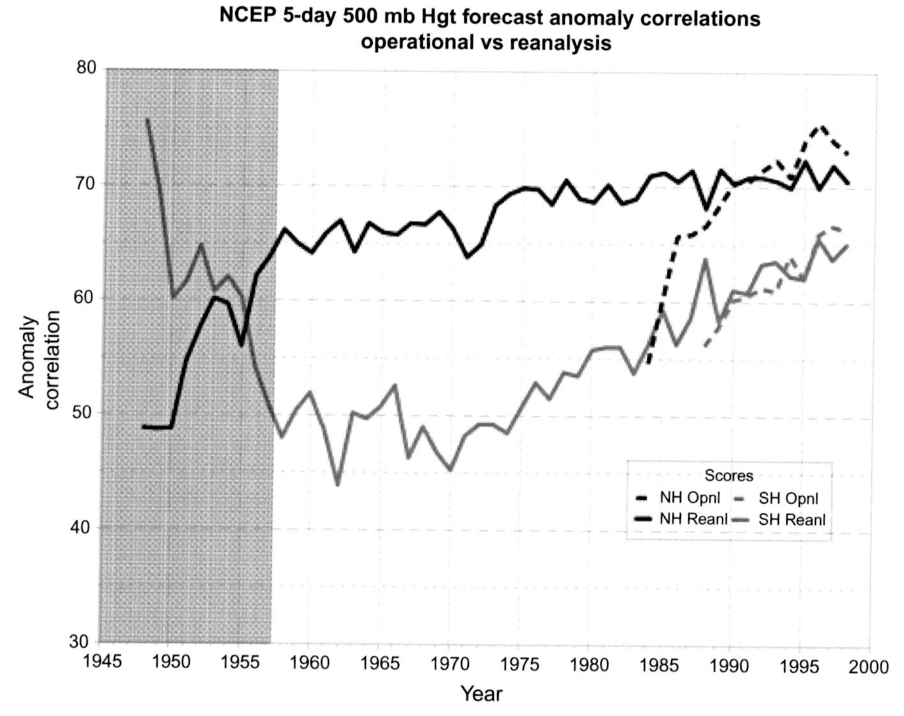

FIG. 11 Annually averaged 5-day anomaly correlation for forecasts from the NCEP-NCAR 50-year reanalysis *(solid lines)* and from NCEP operations—*dashed*, available only after 1984 for the Northern (NH, *black*), and from 1988 for the Southern Hemisphere (SH, *gray*) extratropics. Note that SH scores are artificially high before the 1960s *(shaded)* as verifying analysis fields have too few observations to deviate from the numerical forecasts used as background fields. *Adapted from Kistler, R., Kalnay, E., Collins, W., Saha, S., White, G., Woollen, J., Chelliah, M., Ebisuzaki, W., Kanamitsu, M., Kousky, V., van den Dool, H., Jenne, R., Fiorino, M., 2001. The NCEP-NCAR 50-year reanalysis: monthly means CD-ROM and documentation. Bull. Am. Meteorol. Soc. 82, 247–268.*

fraction of the cost of concurrent observing system developments. On the other hand, over the in situ, data-sparse Southern Hemisphere, improvements in DA and modeling techniques in the operational forecasts (gray dashed line) appear to bring no added gain beyond what the expansion of the observing network offered in the reforecasts (gray solid line).

Fig. 12 offers further insight into the attribution of skill gains to various changes in DA and modeling techniques. Fig. 12 shows the time evolution of the skill of operational, high-resolution forecasts relative to the skill in a reanalysis-reforecast system that was implemented and "frozen" in 2003 (and never upgraded afterward), thus reflecting 2003 knowledge and technology. Both synoptic-scale (500 hPa geopotential height, 850 hPa temperature, and mean sea-level pressure/MSL), and local weather parameters (2-m temperature, cloud cover, and the 10-m wind speed) are represented in the graph.

The vertical lines in Fig. 12 indicate dates when NWP upgrades were introduced; the red lines show when resolution also changed. It is important to point out that an increase in model resolution is associated with an increase in the resolution used in data assimilation as well, typically allowing more observations to be used. Some highlights from the graph include:

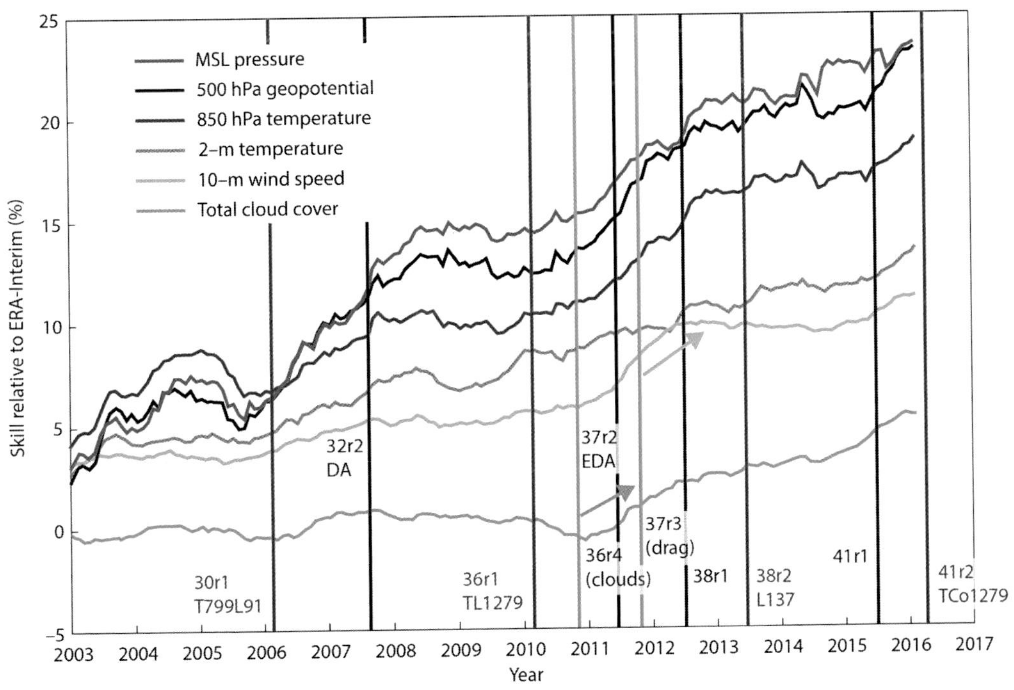

FIG. 12 Time evolution of monthly average skill for operational 5-day high-resolution ECMWF Northern Hemisphere extratropics forecasts from 2003 to 2017, relative to reforecasts made with the frozen ERA-Interim DA-modeling system (Dee et al., 2011). *Blue, black, red, orange, green, and cyan lines* stand for MSL, 500 hPa geopotential height, 850 hPa temperature, 2-m temperature, 10-m wind speed, and total cloud cover. Relative skill is defined as the percentage change of skill in the operational forecasts compared to ERA-Interim reforecasts, both verified against ERA-Interim analyses. Vertical *(red)* lines indicate dates when model (resolution) upgrades were introduced.

- Changes including resolution upgrades (marked 30r1 in 2006, 36r1 in 2010, and 38r2 in 2013) usually bring the largest improvements, but because other model changes are also included, one cannot attribute the improvements only to resolution.
- Upgrades in the data assimilation (e.g., versions 32r2 in 2007 and 37r2 in 2010) also bring substantial improvement.
- Some model changes without an increase in resolution (e.g., 36r4 in 2010 and 37r3 in 2011) can also bring substantial improvement, especially if measured in the variables closely linked with the specific model upgrades (e.g., the upgraded cloud scheme in version 36r4 substantially improved the skill in the prediction of total cloud cover).
- Every change is carefully selected for implementation counts and can lead to improvements.

3.6 Weather Versus Climate Prediction

At the beginning, NWP forecasts were made out to only 24 hours of lead time. With improved techniques, operational weather forecasts were extended out to 3, and then 5–10 and 15 days into the future. These forecasts exploited information primarily about the initial condition of the atmosphere on one hand, and improvements in the modeling of relatively fast weather processes on the other. Ocean, land, and ice processes, if included, were often simplified in NWP models. Meanwhile, the climate community's modeling efforts focused on the initialization and modeling of the slower ocean, land, and ice components of the coupled system. The first successful real-time seasonal predictions of the 1997–98 El Niño event at NCEP (Barnston et al., 1999) gave further impetus for advancing coupled modeling in both the weather and climate communities. With time, it became evident that improved initialization and modeling of the fast and slow components of the coupled system are not exclusive. In fact, both scientific and practical arguments indicated potential gains from a comprehensive and seamless weather-climate forecasting effort, benefiting both weather and climate prediction (Toth et al., 2007). The biggest beneficiary of a converged weather-climate approach, however, has been the intermediary S2S forecasting, where success, without a coupled model initialized well *both* for the fast atmospheric and slow ocean/land/ice components, had been elusive.

The integration of weather and climate forecasting, first started in research, soon reached operations. The full coupling of the earlier atmosphere only models with ocean, sea ice, and land surface models, as well as the significant extension of the NWP forecasts into the weekly range, effectively blurred the line between operational NWP and what originally started as separate monthly or seasonal climate modeling efforts. Today's ECMWF sub-seasonal system, for example, is a seamless extension of its medium-range ensemble, complete with a coupled land, ocean, and atmosphere model out to 46 days twice a week (Vitart et al., 2014a). While the trend is clear, the full integration of forecast effort still has not reached the seasonal time range. For various reasons, including the prevalence of significant biases in fully coupled models, both research and operations in seasonal prediction use somewhat different models and procedures from those applied in weather and S2S forecasting.

As for operational seasonal prediction, it started at NWP and climate forecasting centers in the middle to late 1990s. Since its first implementation, ECMWF has upgraded its seasonal ensemble prediction system four times (Molteni et al., 2011). The latest version, system-5 (SEAS5), was implemented into operation in November 2017. This system, typical of state-of-the-art efforts, uses a coupled ocean, land, sea ice, and atmosphere model, and it produces forecasts once a month up to 7 months ahead (once a quarter, forecasts are extended up to 13 months). In addition to their weather prediction suites, several national meteorological centers are running seasonal ensembles out to a year. See Chapter 13 for an overview of the global ensembles that are operational today.

4 ENHANCEMENT OF PREDICTABLE SIGNALS

As discussed earlier, predictability is quickly lost on the fine scale, and in a few days even on the moderately large scales. How can we extract a relatively small signal (i.e., the predictable component in a forecast), or alternatively, filter out the large amount of uninformative noise from extended-range weather, S2S, and even longer-range forecasts? Bias-corrected output from an NWP forecast looks just as realistic at a 20-day lead time as at shorter (say, 3-day) lead time. Putting the unlabeled forecast maps side by side, we would not be able to tell which was which. However, we know that the skill of a 20-day forecast is very low at best, whereas the skill of a 3-day forecast is usually high enough for many users to take specific actions. This indeed poses a real challenge for both lay and professional users of weather forecasts, as the possible extraction of any useful information from longer-range forecasts is nontrivial. The presence of spatiotemporally coherent features in the atmosphere, however, offers different ways for the removal of higher-frequency forecast variability that, at a given lead time, may have lost predictability.

4.1 Spatiotemporal Aggregation

Because neither the exact timing nor the location of fine-scale weather features is predictable beyond very short lead times, as an alternative to using the actual prediction valid at a point in time and space, it is a common practice to consider forecast values collected from or aggregated in a spatiotemporal neighborhood of the point of interest (refer to Figs. 4 and 5 for spatial examples, or Roads (1986) for temporal averaging). This is an inexpensive way of generating a range of forecast values, so long as the surrounding terrain is reasonably uniform. One can compute, for example, the spatial and/or temporal mean of forecasts (or their anomalies from long-term means) as a way of removing the unpredictable noise from the forecasts. Alternatively, neighboring forecast values can be used for the inexpensive generation of multiple, a range of, or probabilistic forecasts based on a single NWP prediction (Atger, 2001).

The spatial size or temporal domain of the averaging can be chosen so that a large part of the variance associated with unpredictable scales is filtered out, while most of the predictable signal is retained. Also, 5-, 7- (weekly), 10-, or 30-day (monthly), and district, state, or continental means are various ways that researchers and operational forecasters have attempted to extract and present to users the signal from low-skill forecasts.

It is well understood that due to aliasing effects, box-shaped filters or aggregation areas (with weights of zero outside the zone of interest) used in simple spatial or temporal averaging will introduce some noise into the forecast (see, e.g., Fig. 15-2 in Smith, 2013). Some of the aliasing and noise can be mitigated by using filters with different shapes (e.g., with Gaussian-shaped weights), and by repositioning the filter over each element of interest (i.e., shifting filters or running averages over grid points or time instances).

By combining information from adjacent areas or times, spatiotemporal averaging introduces some errors in a forecast for the selected point of interest. Also, filtering parameters must be chosen to correspond to the actual, lead-time-dependent level of forecast skill. To avoid underfiltering or overfiltering, one must carefully choose filtering parameters to match the actual level of skill.

4.2 Ensemble Averaging

Ensemble forecasting offers a dynamical solution to the abovementioned problems with spatiotemporal averaging. The concept of ensemble forecasting, as discussed further in Section 5 later in this chapter, is rather simple. The analysis of the state of the atmosphere is intentionally degraded by introducing multiple realizations of perturbations that are all within the bounds of the estimated uncertainty in the initial state. In addition to the control forecast from the best analysis, NWP forecasts are made from all perturbed states. The ensemble mean of such forecasts, albeit at very significant computational cost [depending on the number of ensemble members, with O(10) being more costly], offers a flow-dependent filter that, so long as the perturbations are consistent with error statistics, reflects the actual level of forecast skill. An ensemble also offers alternative forecast scenarios valid for each point in time and space, without neighborhood aggregation.

4.3 Removal of Systematic Errors

Numerical models represent, in an approximate way, the dynamics and the physical processes of the real atmosphere. Natural processes are truncated spatially to the model grid space, temporally by the model time steps, and also by the parameterization of some (and ignorance of other) physical processes. Forecast errors thus accumulate due not only to chaotic amplification, but also differences between the dynamics of reality and our models. Some of these differences result in small random errors imparted at each time step of the model into the forecast state, after which it undergoes chaotic amplification just as errors in the initial condition do. Other model-related errors, however, are systematic in nature and manifest as a drift in model forecasts initialized from analyzed conditions. The systematic drift from observed to model-preferred conditions also hinders the use of weather forecasts. A host of statistical methods have been proposed to mitigate this problem. Depending on the level of sophistication and the available of required pairs of forecast-verifying analysis sample data, systematic errors can be estimated over the entire climatological attractor (i.e., by comparing the time means or climatologies of unconditional analysis versus different lead time forecasts), or over different regimes (i.e., via comparing conditional climate means).

Due to the various model truncations mentioned here, many user-relevant atmospheric variables are not part of NWP models. This poses yet another challenge for the use of NWP forecasts. Separate statistical or physical relationships need to be developed between the model prognostic variables and the desired user variables (often on a finer spatial scale than the model's grid) to mitigate the situation. For further discussion of the postprocessing of NWP forecasts, there is a plethora of related literature (see, e.g., Li et al., 2017), and also see Chapter 15 of this volume.

5 ENSEMBLE TECHNIQUES: BRIEF INTRODUCTION

In the past 25 years, NWP has seen a shift from a deterministic approach, based on single numerical integrations, to a probabilistic one, with ensembles of numerical integrations used to estimate the probability distribution function of the future state of the atmospheric variables. This section offers a brief overview of the ensemble approach, which is so ubiquitous now in NWP practice. See Chapter 13 for more details on operational and other near-real-time ensemble forecast systems.

5.1 Background

From the early days of NWP, it was clear that there are some cases when forecast errors remain small even for longer forecast ranges, and others when even shorter range forecasts can have large errors. This operational experience was supported by scientific studies pointing out that due to the chaotic nature of the atmosphere, in the presence of strong instabilities, even small initial errors can grow rapidly and affect forecast quality at shorter ranges.

Yet until the 1990s, the prevailing thinking expressed succinctly by Bengtsson (1991) was that "[w]eather prediction is a well-defined deterministic problem. Starting from a given initial state, any future state can be obtained by integrating the classical Navier-Stokes equations forward in time. Therefore, a weather forecast can, in principle, be calculated in the same way as the motion of the planets or the trajectory of a missile."

However, Bengtsson himself, along with a number of pioneers from the 1950s on (see, e.g., Lewis, 2005) recognized that errors from both the observations and the imperfect models could "contaminate" a forecast. In the 1970s and 1980s, various groups investigated whether case-dependent variations in the quality of forecasts could be estimated in advance (say, when a forecast is issued) to see whether future weather was easier or more difficult to predict than average. In other words, scientists searched for an objective method to associate each forecast with a level of confidence. At the time, various approaches were tested at the major NWP centers. By the early 1990s, a number of groups converged on the idea of using ensemble forecasts. Indeed, 1992 saw the implementation of the first operational ensemble prediction systems at ECMWF and NCEP, following the early work of Lorenz (1975) and others. This development started a new era in NWP prediction, providing multiple scenarios alternative to a single traditional forecast started from the best estimate of the initial state. The implementations at NCEP and ECMWF were to be followed by many others, first at the CMC of the Canadian Meteorological Service in 1995, and then elsewhere.

5.2 Methodology

The main concept behind the ensemble approach is rather simple: generate a set of *N* perturbed forecasts, each designed to simulate the effect of possible uncertainties associated with the unperturbed (or control) forecast. Then use the *N* perturbed forecasts to estimate the range of possible outcomes, most probable set of values, and/or the probability that a future parameter (say, temperature at a point in time and space) will be higher or lower than a certain value (Fig. 13).

In the 1980s, different techniques were developed and tested for the generation of reliable and accurate ensembles. One of the approaches tested at the predecessor of the NCEP used lagged forecasts: forecasts initialized recently (say, every 6 hours in the past 2 days) were considered as a lagged ensemble (Ebisuzaki and Kalnay, 1991). The results showed reasonable quality for the medium forecast range, beyond about a week. However, the inclusion of less skillful, older forecasts degraded ensemble quality for the shorter time ranges. Scientists at ECMWF (Hollingsworth, 1980) and NCAR (Errico and Baumhefner, 1987; Tribbia and Baumhefner, 2003) generated ensembles starting all at the same time, but with initial

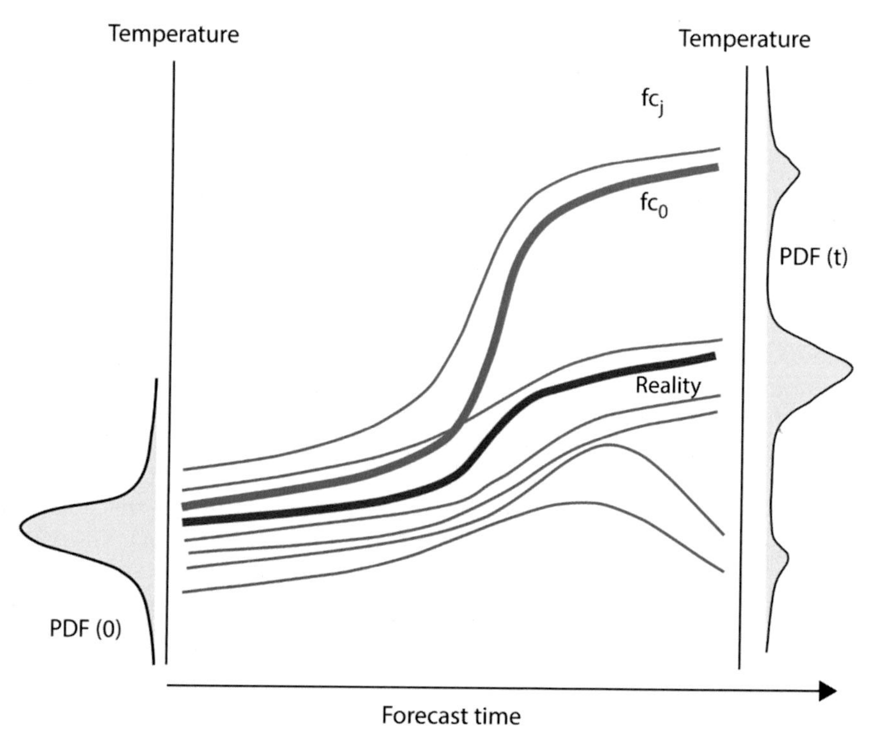

FIG. 13 Schematic of the ensemble approach to the prediction of the time evolution of a selected weather parameter *(red curve)*. Until 1992, NWP forecasts were issued using a single model integration *(bold blue line)* starting from initial conditions with the smallest possible error. Due to initial uncertainties and model approximations, forecasts of a chaotic system diverge from reality. The ensemble approach introduces a set of perturbed forecasts *(thin blue lines)* to estimate a multitude of possible initial and forecast states that in turn can be used to estimate the associated probability distribution functions (PDF, *black curves with yellow shading* at initial (left) and a selected forecast time (right).

conditions perturbed in a random way. This method did not deliver good results either, because the random perturbations did not lead to sufficient forecast diversity: ensemble members remained too similar to provide valuable information on possible future scenarios.

The beginning of the 1990s saw the development and testing of more promising methods both at ECMWF and at NCEP. There are different ways that ensembles can simulate the initial condition and model-related uncertainties. In the first version of the ECMWF global ensemble (Molteni et al., 1996), initial uncertainties were simulated using singular vectors (SVs), which are the perturbations with the fastest growth over a finite time interval (Buizza and Palmer, 1995). SVs remained the only type of initial perturbations used in the ECMWF ensemble until 2008, when perturbations from multiple data assimilation cycles, known as *Ensembles of Data Assimilations (EDAs)*, were also incorporated in addition to SVs (Buizza et al., 2008). Today, SVs remain an essential component of the ECMWF ensemble. They provide dynamically relevant information about initial uncertainties that are linked with forecast errors.

In the first version of NCEP's global ensemble, bred vectors (BVs) were used to simulate initial uncertainties instead of SVs (Toth and Kalnay, 1993). Perturbations in the BV cycle aim to emulate errors in the analysis-forecast cycle (see also the discussion in Section 5). The BV method is based on the notion that due to perturbation dynamics, growing errors have a tendency to accumulate in analysis fields generated by data assimilation (Toth and Kalnay, 1997). Assuming that errors introduced in the assimilation step project in both growing and decaying perturbation directions, one observes that the growing errors amplify, while decaying errors diminish in the pursuant forecast step. Consecutive applications of the analysis-forecast cycle amount to a natural selection (or breeding) of fast-growing errors. The result, confirmed by OSSE studies (e.g., Errico and Prive, 2014), is that growing perturbations dominate analysis error variance compared to a prior expectation of random (neutral) errors.

The ensemble introduced operationally at the Meteorological Service of Canada (MSC) in 1995 adopted a Monte Carlo approach, designed to include as many sources of error as possible. They simulated initial uncertainties due to both observational errors and data assimilation assumptions, as well as, for the first time, model uncertainties (Houtekamer et al., 1996). For the estimation and representation of uncertainties in initial conditions, MSC, similar to Evensen (1994), designed a new data assimilation scheme, a forerunner of a large number of ensemble-based data assimilation plans. Following the Canadian example, a stochastic model perturbation scheme designed to simulate model uncertainties was introduced in the ECMWF ensemble (Buizza et al., 1999).

After the early implementations at NCEP, ECMWF, and MSC, most other operational centers have also introduced ensemble techniques on the global and regional scales, often including schemes to simulate model uncertainties. See Chapter 13 (Ensemble generation: the TIGGE and S2S ensembles) in this volume for a description of the main characteristics of global ensemble forecast systems operational in 2017. These implementations amounted to a paradigm shift in operational NWP from a deterministic approach, based on a single forecast, to a probabilistic one, in which ensembles are used to estimate the probability density function of initial and forecast states.

5.3 Use of Ensembles

Today, it is generally accepted that forecasts must include estimates of uncertainty or confidence that on any day allow forecasters to assess how predictable the future is. Short- and medium-range forecasts, monthly and seasonal forecasts, and even decadal forecasts and climate projections are based on ensembles, providing not only the most likely scenario, but also the uncertainty associated with it. Ensembles of short-range forecasts or data assimilation cycles are also used to estimate the uncertainty in the initial state (analysis).

Probabilistic forecasts, where the probability of the occurrence of a predefined event is predicted, are among the most common ensemble-based products. As ensembles are built around and use the same techniques as other NWP forecasts, it is not surprising that their performance follows similar patterns. Analogous to Fig. 1 (which shows the time evolution of the skill of the ECMWF single, high-resolution forecasts of the 500 hPa geopotential height from 1979 to the present), Fig. 14 shows the time evolution of ensemble-based probabilistic forecasts of the 500 hPa geopotential height over the Northern Hemisphere, from 1995 to the present. Both graphs reveal a steady improvement of forecast accuracy over the years, for example, indicating that a 7-day forecast is as good today as a 5-day forecast was 20 years ago.

Because forecast skill decreases and predictable signals are gradually lost with increasing lead time, a probabilistic approach is paramount. Probabilistic forecasts can be generated statistically based on an ensemble, or even a single unperturbed forecast (e.g., Glahn and Lowry, 1972). Many users, however, are affected by a multitude of weather parameters spread across lead time, space, and variables. Given the sheer number of (and sometimes unforeseen type of)

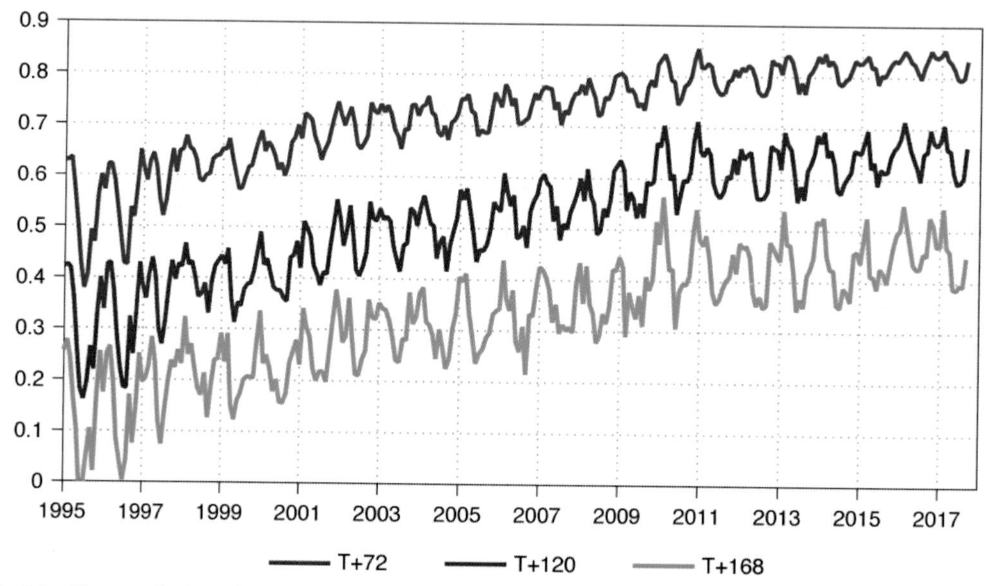

FIG. 14 Time evolution of monthly-average skill using the continuous rank probability skill score (CRPSS) of 3- *(blue)*, 5- *(red)*, and 7-day *(green)* ECMWF probabilistic predictions of the 500-hPa geopotential height over the Northern Hemisphere, measured in comparison with climatologically based probabilistic information.

user applications, the pregeneration of single forecast- or ensemble-based joint or aggregate probabilistic products using statistical calibration methods is simply not plausible. A properly formulated and statistically calibrated ensemble, however, supports innovative applications, including the use of a range of plausible spatiotemporal scenarios, given the initial condition and its uncertainty.

Each realistic ensemble scenario can be fed through user applications such as a model of energy demand from a customer base as a function of various weather and other parameters. Assuming equiprobable sampling by the ensemble members, a rational decision can be made as to optimal, weather forecast–dependent staging for the production and distribution of energy.

Admittedly, many users may lack the sophistication required for the construction of such application models. Nevertheless, an approach like this may offer a quantitative and comprehensive way of capturing case dependent predictability so critical to real-life weather-dependent activities. Ensemble forecasts can also help to prepare for threatening high-impact events in geospheres coupled with the atmosphere that are triggered by weather, where possibly highly nonlinear relationships may limit the use of other, statistically oriented postprocessing methods.

6 EXPANDING THE FORECAST SKILL HORIZON

Recall from the earlier discussion that the term *weather* refers to the time sequence of atmospheric events. As we have seen in Figs. 4 and 5, our ability to pinpoint smaller-scale features in time and space rapidly diminishes as lead time increases. Consequently, at longer lead times, only larger spatial-scale features are traceable in space and time. This confirms Shukla's (1981) early surmise that "the evolution of long waves remains sufficiently predictable at least up to one month." He also suggested that improvements in model resolution and physical parameterizations could extend the predictability of time and space averages even beyond 1 month.

Buizza and Leutbecher (2015) explain that "'forecast skill horizons' beyond 2 weeks are now achievable thanks to major advances in numerical weather prediction. More specifically, they are made possible by the synergies of better and more realistic models, which include more accurate simulation of relevant physical processes (e.g., the coupling of the atmosphere to dynamical ocean and ocean wave models), improved data-assimilation methods that allow a more accurate estimation of the initial conditions, and advances in ensemble techniques." This explanation is consistent with earlier discussions from Shukla (1998) and Hoskins (2013). Shukla (1998) referred to "predictability in the midst of chaos" to explain how skillful long-range predictions of phenomena like El Niño were possible despite fast error-growth rates from small to large scales. Hoskins (2013) wrote about "discriminating between the music and the noise," and introduced the concept of a predictability chain, whereby, for example, "a large anomaly in the winter stratospheric vortex gives some predictive power for the troposphere in the following months."

The reader is referred to Buizza and Leutbecher (2015) for a discussion on the sensitivity of the skill of ensemble-based forecast fields to spatial and temporal filtering, as covered in

Section 4. By applying the same metric to ECMWF ensemble forecasts with increasingly coarser spatial and temporal scales, they showed that forecasts of instantaneous, grid-point fields are skillful up to 16–23 days, while forecasts of large-scale, time-averaged fields have skill up to 23–32 days (because they used ensembles with a maximum forecast length of 32 days, they could not comment on whether the forecasts were skillful for even longer times). These ensemble-based results are consistent with skill estimates for single forecasts, as reviewed in Figs. 4 and 5.

The scale dependency of forecast skill is illustrated in Fig. 15, which shows the forecast time up to which skillful forecasts can be generated (y-axis, logarithmic scale) as a function of the scale of the different phenomena (x-axis, in kilometers). The graph includes not only the ensemble-based estimates from Buizza and Leutbecher (2015), but also other estimates for surface variables such as total precipitation, and for large-scale/low-frequency patterns such as the North Atlantic Oscillation (NAO) and the Madden Julian Oscillation (MJO) and El Niño. The vertical line at 36 km indicates the horizontal resolution of the ECMWF ensemble forecasts in 2013–14, used by Buizza and Leutbecher (2015) to divide the resolved and the unresolved scales. The red lines relative to the instantaneous and finer-scale surface variables are closer to the x-axis, indicating that surface variables are less predictable. By contrast, the blue lines related to teleconnection patterns (e.g., NAO and MJO) and to average Pacific region sea-surface temperature anomalies (SSTA) affected by El Niño are further away from the x-axis and closer to the top-right part of the diagram, illustrating that these large-scale patterns can be skillfully predicted months ahead.

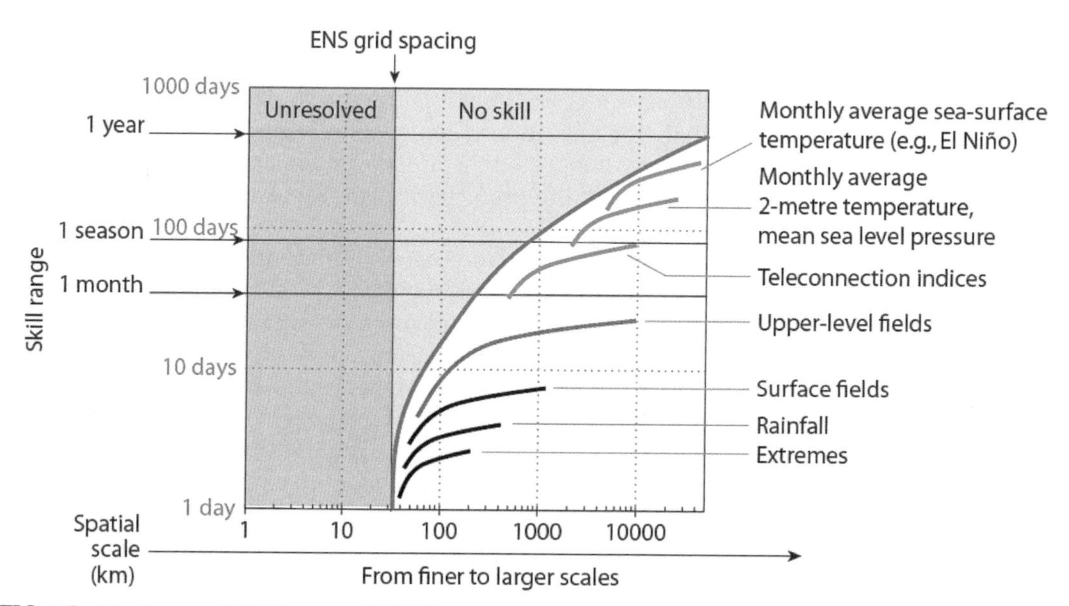

FIG. 15 The Forecast Skill Diagram, which illustrates up to which forecast time ensemble-based, probabilistic forecasts are skillful, is shown (from Buizza et al., 2015). The vertical line labeled "ENS grid spacing" denotes the grid spacing of the ECMWF ensemble used in the Buizza and Leutbecher (2015) study (which was about 36 km), which was used to generate some of the curves in this diagram.

Two further features have been added to the diagram: a blue-line area, drawn schematically to include all the individual lines, and a pink "no-skill: region. The blue line shows the forecast skill horizon for the ECMWF ensemble at the time of the Buizza and Leutbecher (2015) and Buizza et al. (2015) studies. It is still much less than 10 days for very detailed forecasts. The fact that the blue line curves to the right indicates very clearly that the forecast skill horizon, as discussed in earlier sections (cf. the discussion of Boer, 2003), is scale- and variable-dependent. Forecast skill is also a function of geographical location and the season of the year (see Buizza and Leutbecher, 2015). Forecast performance results presented in Fig. 15 reflect the state of the art in science and technology as of the mid-2010s and are a function of errors in the analysis field (i.e., size of initial errors), and errors introduced due to approximations in numerical modeling. Should data assimilation and modeling techniques continue improving, which will be partly due to future S2S research, forecast performance statistics, or practical predictability, will likewise keep improving.

Fig. 15 demonstrates that a portion of large-scale and low-frequency variability can be successfully predicted with today's NWP systems. It is also evident (see the heavy solid line showing an estimate of error variance in random forecasts for rotational wind in Fig. 5) that the energy spectrum of errors, and thus the potential forecast signal, peak around wave number 8 and rapidly drop over the more predictable planetary scales. The predictable forecast signal associated with the planetary scale temperature or other variables of common interest is even lower, while the socioeconomic impact of such slow changes may be only marginal. The introduction of proper coupling of the atmosphere with surrounding spheres in the coming years is expected to raise predictable variance for larger-scale and longer-term motions, but only to a limited extent. It is well understood that high-impact weather events are often triggered by finer scale and much more rapidly evolving features with no traceability beyond a few days whatsoever. Such atmospheric features are often the cause of sudden or catastrophic events in other spheres as well, such as inland flooding, mudslides, snowdrifts, surges, high ocean waves, and the breaking off of ice shelves. In summary, predictability of the more critical small- and moderate-scale motions is quickly lost, while the low level of the remaining predictability in the planetary scales over the extended range is associated with low variability.

Lorenz (1975) and follow-on investigators (e.g., Chu, 1999) distinguish between first and second kinds of predictability, the first influenced by the initial condition of a system itself, while the second by its boundary conditions. Recognizing that in the context of a coupled system, the conditions at the boundary between two subsystems—just as any other variable of a coupled system—are also initial value dependent, we offer a somewhat different perspective on various types of predictability. Following the discussion given so far here, we define the first kind, or *traceable*, predictability as the ability to continuously follow the propagation, emergence, and demise of features in a forecast from the initial time on (see Fig. 16). Fig. 15 assesses this type of predictability in practice, with today's NWP systems. For small-scale features, traceability, as we saw it, is lost at early lead times. Due to the nonlinear interactions between various scales of motion, the statistics (i.e., time frequency) of finer-scale phenomena may be different under distinct larger-scale conditions.[7]

[7]Unlike traceable predictability, this type of behavior is not assessed in Fig. 15.

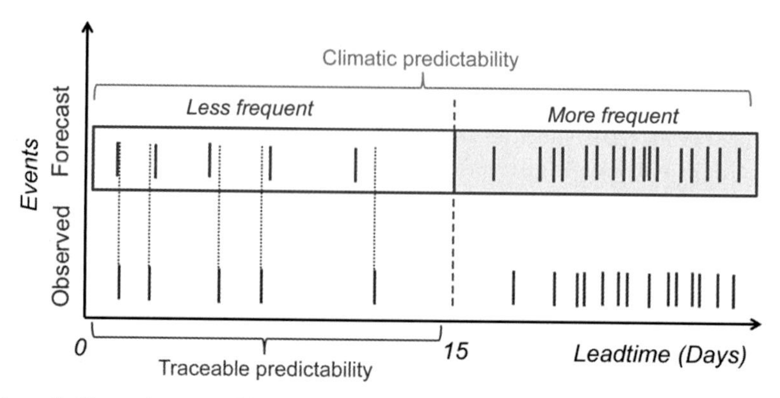

FIG. 16 A schematic illustrating traceable and climatic predictability. On the shorter, weather timescales, individual observed events *(red bars)* are predicted *(blue bars)* with acceptable timing errors (traceable predictability). Low-frequency changes in the frequency or other statistical characteristics of the observed events associated with large-scale regimes, however, can be predicted for longer periods (climatic predictability).

High-impact weather event statistics, for example, are modulated by slowly varying and somewhat predictable phenomena. In other words, high-frequency weather modulations are conditioned on slower varying changes. We refer to the predictability of variations in such frequency and other statistics of finer-scale features, conditioned on the first, traceable kind of predictability of larger-scale motions, as *second kind*, or *climatic predictability*.

The second type of predictability or prediction, therefore, is concerned about climatic frequencies of finer-scale weather phenomena conditioned on the presence of still somewhat predictable large-scale phenomena, as compared to the full climatological frequency of such events. The frequency of tornadoes (see, e.g., the 4–8-day outlooks issued by NCEP's Storm Prediction Center, http://www.spc.noaa.gov/products/outlook/) or hurricanes (see, e.g., seasonal outlooks issued by NCEP's Climate Prediction Center, http://www.cpc.ncep.noaa.gov/products/outlooks/hurricane.shtml) may be below or above overall climatological frequencies, depending on large-scale weather or seasonal regimes that remain somewhat predictable even after the traceability of smaller-scale individual features at any specific place or time is lost. See Chapter 17 of this book for examples of, and more discussion about, the predictability of certain statistics of some of these events. An ensemble of forecasts, for example, may reveal some skillful information on the phase of some still predictable, larger-scale features (i.e., traceable predictability), while beyond their own traceable predictability, the frequency of finer-scale phenomena in ensemble members may reflect frequency statistics consistent with the predicted larger scales (i.e., climatic predictability). This second type of predictability may play a significant role, especially in extended-range S2S predictions.

7 CONCLUDING REMARKS: LESSONS FOR S2S FORECASTING

This chapter reviewed the basis for and limits of weather predictability. We saw how systematic efforts aimed at exploiting the first kind or traceable predictability led the NWP community to the extension of the weather forecast horizon. Numerical modeling and data assimilation technique development has focused on capturing short timescale behavior primarily in the atmosphere and near its boundaries that are most critical to weather forecasting. Building on these successes, inroads also have been made in predictions beyond the limit of traceable weather forecasting. As in the past decade, NWP adopted an Earth system modeling approach that fully couples atmospheric motions with slower processes, and clear evidence has emerged about the possibility of predicting conditional statistics or the climatology of weather on extended ranges. Aiming for a more thorough exploitation of this second, climatic type of weather predictability, and building on the experience and successes of weather forecasting, S2S must embark on its own systematic path to realistically describe and exploit in both numerical modeling and initial state estimation slowly varying processes in adjoining spheres that in the future can impart significant additional skill in predicting large-scale atmospheric regimes. Statistics of associated high-impact weather can then be derived by ensemble or other methods for more widespread and quantitative socioeconomic applications.

Acknowledgments

The authors acknowledge helpful discussions with Drs. Nikki Prive of NASA, Ligia Bernardet of CIRES at NOAA/GSD, Malaquias Pena of the University of Connecticut, Thomas Auligne of the Joint Center for Satellite Data Assimilation, and Lars Isaksen of ECMWF. Comments by Drs. Shan Sun and Benjamin Green of CIRES at NOAA/GSD, and by the editors, Drs. Andrew Robertson and Frederic Vitart, on earlier versions of the text led to significant improvements in both the presentation and content. Fig. 3 was kindly provided by Dr. Jie Feng, University of Oklahoma.

CHAPTER

3

Weather Within Climate: Sub-seasonal Predictability of Tropical Daily Rainfall Characteristics

*Vincent Moron**, *Andrew W. Robertson*[†], *Lei Wang*[‡]

*Aix-Marseille Univ, CNRS, IRD, INRA, Coll France, CEREGE, Aix en Provence, France
[†]International Research Institute for Climate and Society (IRI), Columbia University, Palisades, NY, United States [‡]Institute of Atmospheric Sciences, Fudan University, Shangai, People's Republic of China

OUTLINE

1 INTRODUCTION

Sub-seasonal characteristics of tropical rainfall such as rainfall occurrence and monsoon onset date are important for rain-fed agriculture, where long dry spells can ruin a crop and where the onset of the rainy season is a common planting time (Sivakumar, 1988).

Sub-seasonal to Seasonal Prediction
https://doi.org/10.1016/B978-0-12-811714-9.00003-6

Sometimes referred to as "weather within climate," it is the daily statistics of weather that ultimately lead to the societal impacts of climate anomalies, such as floods and agricultural droughts. While the timing of individual wet and dry spells cannot be predicted beyond the weather scale, skillful sub-seasonal to seasonal (S2S) forecasts of their frequency of occurrence may be feasible and could be of great societal value.

The seasonal rainfall amount ($\overline{R}$ hereafter, where the overbar denotes time-averaging) is the simplest and most general characteristic of a rainy season because it sums all rainy events during a season. It is generally assumed that $\overline{R}$ is the most predictable precipitation quantity on seasonal time scales at regional scales ($L \sim 100-1000$ km), and seasonal forecasts of $\overline{R}$ (usually on 3-month periods) are currently issued by many forecasting centers on a regular basis (Goddard et al., 2001; Gong et al., 2003; Barnston et al., 2010; Kirtman et al., 2014; Tompkins et al., 2017). Seasonal rainfall predictability is primarily associated with sea surface temperature (SST) anomalies and coupled ocean-atmosphere modes of variation (primarily El Niño Southern Oscillation [ENSO]), but anomalies in soil moisture (The GLACE Team et al., 2004; Douville and Chauvin, 2000), snow cover and sea ice (Cohen and Entekhabi, 1999), and stratosphere-troposphere interactions (Thompson et al., 2002; Cohen et al., 2010) also contribute. The persistent and large-scale atmospheric responses to these anomalous forcings lead to near-homogeneous anomalies of $\overline{R}$ at regional scales. This link between predictability and spatial scale of $\overline{R}$ anomalies comes from the systematic repetition of a near-constant forcing and response across the season, enabling the emergence of a predictable "signal" above the unpredictable "noise." In this context, noise is a statistical quantity and may be seen as the impacts of all atmospheric weather motions that are canceled out through the temporal summation across a season. This signal-to-noise ratio concept is analogous to the familiar ensemble approach used in general circulation model (GCM) simulations forced by boundary conditions (Rowell, 1998). The mean across ensemble members (stations or grid points, respectively) isolates the signal through repetition of the same dynamical response to the forcing(s) in each ensemble member (stations or grid points), independent of its different initial conditions (e.g., various locations), while the spread among the members (stations or grid points) measures the noise; that is, the fraction of the response not determined by the forcing (Shukla, 1998). Thus, metrics of potential predictability based on spatial coherence of observed anomalies, such as the correlogram, decorrelation distance, number of degrees of freedom, etc., are based on a similar concept to potential predictability revealed by GCM ensembles. Note that we do not attempt to answer the downscaling question of how the signal may be locally modified within a near-homogeneous region.

A simple yet instructive decomposition of $\overline{R}$ involves the product of the number of wet days (N_R) (e.g., ≥ 1 mm), and the mean rainfall intensity ($\overline{I} = \frac{\overline{R}}{N_R}$) on wet days. $\overline{I}$ reflects both the instantaneous rain rate (Le Barbé et al., 2002), which is high for convective rainfall, and the duration of rainy events (Ricciardulli and Sardeshmukh, 2002; Smith et al., 2005; Dai et al., 2009), which is related to the spatial scale and the movement of the rain-bearing systems; these range from localized thunderstorms to larger organized systems such as tropical cyclones and mesoscale convective complexes lasting from several hours to a day or more at a fixed location. Ricciardulli and Sardeshmukh (2002) have shown that tropical wet events estimated from satellite images last in mean 4.9 hours over the continents and 6.2 hours over the oceans. Individual thunderstorms can produce very high local rainfall rates (Dai et al.,

2009; Trenberth et al., 2017), which make the interannual variations of $\bar{I}$ noisy even after averaging over a season (i.e., Moron et al., 2007). By contrast, the rainfall occurrence field tends to reflect spatiotemporal hierarchical organization of convection (Orlanski, 1975). Previous analyses of rain gauges (or 0.25 degrees grid points) in several tropical regions, including India and tropical Africa (Moron et al., 2006, 2007, 2009b, 2017), found that the spatially coherent interannual covariations of $\overline{R}$ are mostly conveyed by N_R, while $\bar{I}$ is a far noisier characteristic; the spatial autocorrelation of $\bar{I}$ decays more quickly than for either $\overline{R}$ or N_R due to its dependency on the wettest few days of each season. A similar finding has been obtained with finer space-time data that allow the number and intensity of subdaily wet events to be quantified, finding that interannual-to-decadal variations of precipitation in Sahelian-Sudanian and Guinean Africa are mostly related to changes in the number of wet events, primarily due here to mesoscale convective systems (MCSs) (Le Barbé et al., 2002; Lebel and Ali, 2009). This suggests that the predictability of interannual variations of seasonal amounts at regional-scale over these regions of Africa is likely to stem mostly from changes in the frequency of the MCSs rather than from changes in their intensity.

The basic decomposition of $\overline{R}$ in the previous paragraph could be refined by considering onset and end dates of the rainy season, enabling differentiation, for example, between a dry spell associated with a delayed onset from those occurring within the monsoon period itself (Moron et al., 2015a). $\overline{R}$, then, may be more fully decomposed in terms of the daily rainfall statistics (e.g., rainfall frequency and mean intensity) during each temporal phase of the monsoon season separately. The sources of predictability for these daily rainfall characteristics involves modulation of the various rain-bearing systems (from the individual thunderstorms to MCSs and tropical depressions) by the slow phenomena mentioned here. For example, regional-scale monsoon onset over the Maritime Continent in September-November is almost systematically delayed (advanced) during warm (cold) ENSO events (Haylock and McBride, 2001; Moron et al., 2009a), while interannual variations in rainfall during the core of monsoon season, around December to February, are less spatially coherent and less potentially predictable (Moron et al., 2010). Another shorter source of predictability is provided by convectively coupled equatorial waves (CCEW; Wheeler and Kiladis, 1999; Lubis and Jacobi, 2015), including the Madden Julian Oscillation (MJO; Waliser et al., 2003; Zhang, 2005).

For sub-seasonal forecasts, the time aggregation period is generally only 1 or 2 weeks (Zhu et al., 2014), so a stronger signal or reduced noise will be required in order to obtain a comparable signal-to-noise ratio to the seasonal case. There is increasing evidence that the popular 2-week "weeks 3 + 4" (i.e., 15–28 forecast days after the starting dates) sub-monthly range represents such an opportunity, at least in some cases. Certain constellations of the MJO and ENSO may give rise to windows of forecast opportunity where the signal is sufficiently enhanced (Li and Robertson, 2015). Summing over 2 weeks reduces the weather noise for certain stages of the monsoon's seasonal evolution and emphasizes any intraseasonal mode of variation that is not canceled out over 2 consecutive weeks. Even if the impact of a particular MJO phase over a given region may typically last only less than a week, its impact can still be appreciable over a 2-week period because the opposite phase will not be included in the same 2 weeks.

This chapter presents an analysis of tropical rainfall weather-within-climate predictability based on estimates of spatial coherence calculated from gridded observed rainfall datasets. Running 15-day time windows are used to identify the sub-seasonal modulation. With these estimates of potential predictability in hand, we harness them to interpret the patterns of anomaly correlation skill seen in European Center for Medium-Range Weather Forecasts (ECMWF) week 3 + 4 hindcasts.

2 DATA AND METHODS

2.1 Daily Rainfall and OLR

Two primary rainfall datasets are investigated. The first is the Indian Meteorological Department (IMD) high-resolution (0.25 degrees × 0.25 degrees) gridded daily rainfall data for the Apr.-Nov. (extended summer monsoon season) 1901–2014 (Pai et al., 2014; Moron et al., 2017). These gridded data were prepared by spatially interpolating daily rainfall from 6955 Indian stations (with varying data availability periods) using distance-weighted interpolation (Shepard, 1968). The interpolated values were computed as the weighted sum of the station data within a search radius of 1.5 degrees. The scheme was locally modified by including directional effects and barriers (Shepard, 1968). The second is the daily Global Precipitation Climatology Project (GPCP) 1.3 (beta version) dataset (1 degree × 1 degree) from Oct. 1996 to Sep. 2016. Daily precipitation values are estimated from multisatellite observations and are calibrated versus rain gauges at a monthly time scale (Huffman et al., 2001). Some computations are also made using rainfall estimates from the pentad CPC Merged Analysis of Precipitation (CMAP) (2.5 degrees × 2.5 degrees) from Jan. 1979 to Dec. 2015 (Xie and Arkin, 1996). The pentad data are just copied to the daily time scales. Lastly, we use the interpolated daily National Oceanic and Atmospheric Administration Outgoing Longwave Radiation (NOAA OLR) dataset (2.5 degrees × 2.5 degrees) from Jun. 1974 to Dec. 2016 (Liebmann and Smith, 1996).

2.2 S2S Forecasts

Reforecasts of total precipitation are evaluated for the ECMWF Variable-Resolution Ensemble Prediction System monthly forecast system (VarEPS-monthly; Vitart et al., 2008), prepared for the WWRP/WCRP S2S project database (Vitart et al., 2017). The atmospheric component of the ECMWF model (version CY41R1) has 91 vertical levels and a horizontal resolution of TCo639 (16 km) up to day 10 and TCo319 (32 km) after day 10. More model details are given in Vitart et al. (2017). Semi-weekly reforecasts of the ECMWF model over the 20-year (Jun. 1995–May 2014) reforecast period are analyzed, corresponding to real-time forecast start dates every Monday and Thursday from Jul. 2015 to Jun. 2016. The ECMWF reforecasts consist of 1 control and 10 perturbed forecasts on each start date, and the ensemble mean skill is evaluated. The GPCP version 2.1 (Huffman et al., 2009) daily precipitation estimates on a 1 × 1 grid are used for forecast validation. The daily data are averaged from day 15 to day 28 to generate week 3 + 4 time series for both forecasts and observations. The ECMWF total precipitation reforecast is interpolated from 1.5-degrees into 1-degree resolution so it can be compared with the GPCP dataset.

2.3 Method to Estimate the Spatial Coherence

The spatial coherence is empirically estimated using the spatial autocorrelation of the interannual ranks of running 15-day amounts (and frequency of wet days receiving ≥1 mm) in a 500-km radius. Considering the ranks instead of the amounts themselves reduces the skewness. Other radii, from 150 to 1000 km, leads to similar spatial and temporal modulations of the spatial coherence. In the same way, the main results are not very sensitive to the data because OLR or CMAP datasets lead to similar results even if their lower resolution (i.e., 2.5 degrees vs 1 degree grid) tends to increase the average spatial autocorrelation. Lastly, the spatial autocorrelation in a 500-km radius introduces a weak negative bias about the spatial coherence, where the spatial autocorrelation pattern is anisotropic as around the Pacific and Atlantic Inter Tropical Convergence Zone (ITCZ) (not shown), but computing the area of spatial autocorrelation above a given threshold as $1/e$ (Ricciardulli and Sardeshmukh, 2002) is far more time-consuming. In summary, our main results about the spatial and temporal modulations of the spatial coherence appear to be rather insensitive to data, horizontal resolution, and empirical estimation.

3 RESULTS

3.1 Daily Rainfall Characteristics of the Indian Summer Monsoon

Fig. 1 gives an example of the diversity of sub-seasonal scenarios for two anomalously dry (1986 and 2002) and two wet (1983 and 1988) monsoon seasons, at two 0.25 degrees grid points near Mumbai (18 degrees 56′N, 72 degrees 50′E) and New Delhi (28 degrees 37′N, 77 degrees 14′E). At Mumbai, all four onsets are quite close to the long-term mean (Jun. 13), while the end is anomalously late in the wet years and early in the dry ones. At New Delhi, it is not possible to define onset and end dates due to erratic rain in 1986 (i.e., only three significant wet spells separated by long dry spells; Fig. 1F), while the 2002 season is shorter and appears shifted later than usual, mostly due to a very long dry spell from late June to late July (Fig. 1H). Both dry seasons at New Delhi illustrate the challenge of defining the true onset of the rainy season as soon as fairly long dry spells occur after (or between) the first wet spells, such as around late June in 2002 in New Delhi (Fig. 1H). This uncertainty in onset definition makes prediction challenging. Note also that both dry years share a very long break in July at New Delhi and Mumbai despite the distance of 1100 km between them. The anomalously wet season in 1983 (Fig. 1B) is related to a longer season than usual in New Delhi, but also to more very wet days. Such behavior is also observed in 1988 (Fig. 1D), except for the end, which occurs close to the climatological long-term mean. Unlike at Mumbai, the wet days at New Delhi are clearly more intense during anomalously wet seasons.

Fig. 1 illustrates that a good (i.e., anomalously wet) monsoon on the subcontinental scale (Sontakke et al., 2008), such as in 1983 or 1988, can have very different sub-seasonal evolutions on the local scale (Moron et al., 2017), and emphasizes the importance of predicting these daily statistics. This example also illustrates the difficulty of calculating daily rainfall statistics strictly between the monsoon onset and end dates because these dates themselves are subject

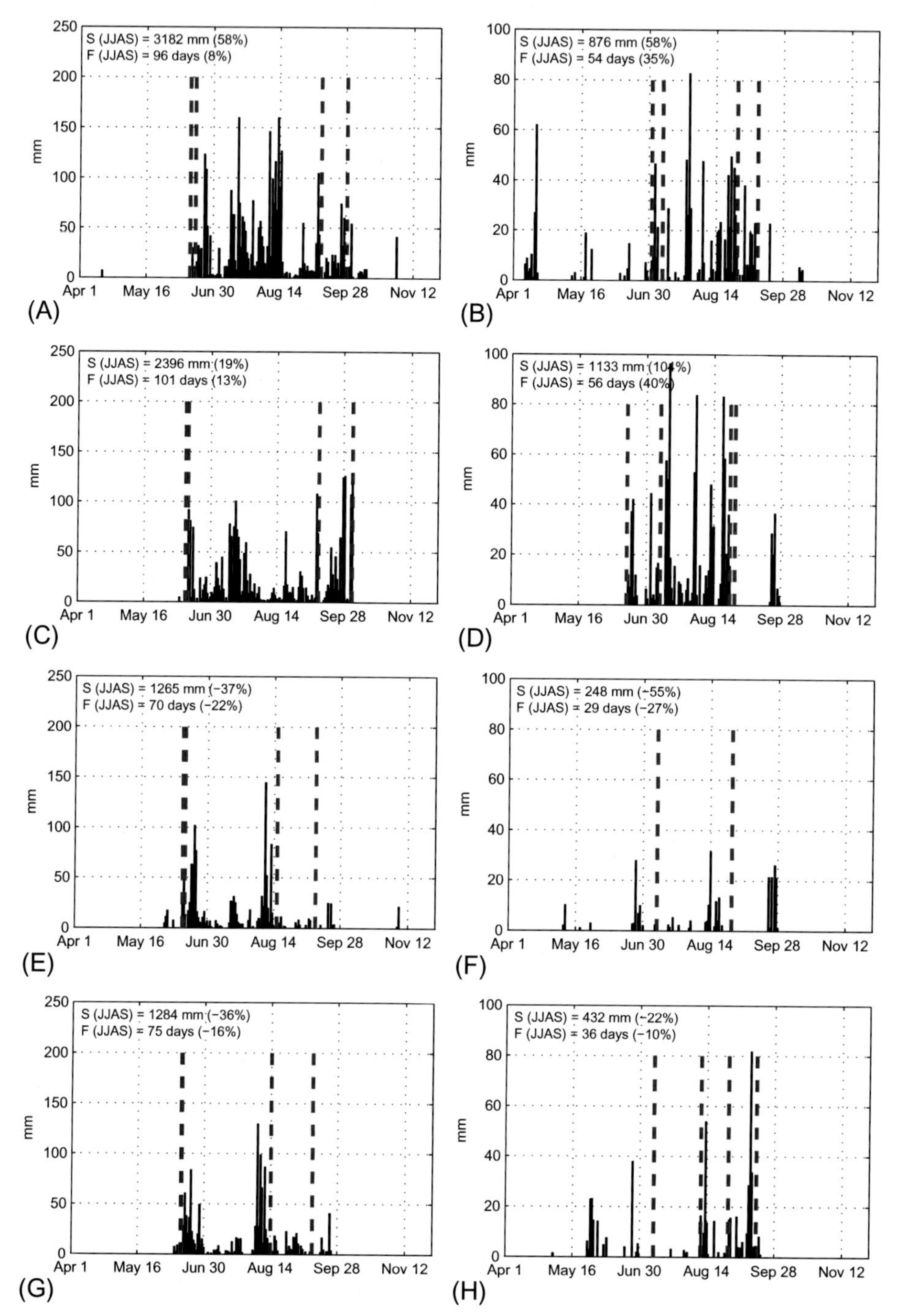

FIG. 1 Daily mean rainfall (*bars*) in millimeters for two 0.25 degrees grid points from the IMD dataset corresponding to the locations of Mumbai (left column) and New Delhi (right column) for two "good" (anomalously wet) (i.e., 1983 and 1988) JJAS monsoons and "bad" (anomalously dry) JJAS monsoons (i.e., 1986 and 2002) at the Indian scale (Sontakke et al., 2008). The *vertical dashed red lines* denote the climatological onset and end dates of the season (defined as the first and last wet day of a 5-day wet spell receiving at least the climatological amount received during 5-day wet spells (from April 1 to November 30) without any 10-day dry spell receiving less than 5 mm in the following 30 days [and previous for end days]) (Moron et al., 2017). The onset is computed from Apr. 1 and the end is computed retrospectively from Nov. 30. The *vertical blue dashed lines* denote the onset and end dates for the given seasons and locations. (A) Mumbai 1983, (B) New Delhi 1983, (C) Mumbai 1988, (D) New Delhi 1988, (E) Mumbai 1986, (F) New Delhi 1986, (G) Mumbai 2002, (H) New Delhi 2002.

to great uncertainty. A flexible alternative is to consider the sub-seasonal modulation of spatial coherence within a sliding window across the season. The window should be short enough to properly distinguish between the various stages of the monsoon, such as onset, core, and end, but long enough to filter out the shortest time scales associated with synoptic systems. It should also be short enough to capture different phases of the MJO and related phenomena, such as the northward-propagating intraseasonal oscillation (ISO) (Krishnamurthy and Shukla, 2000, 2008; Moron et al., 2012).

3.2 Sub-seasonal Modulation of Spatial Coherence Across India

The sub-seasonal evolution of spatial coherence over India is shown in Fig. 2A for $\overline{R}$, N_R over Monsoonal India (defined by black contour in panels B–E; Moron et al., 2017). Both curves have a similar evolution, with a small, short-lived peak around the mean onset date and a larger and longer one around the mean withdrawal ones (Fig. 2A). The spatial coherence of both rainfall amounts and frequency is at minimum during the core of the rainy season, when the climatological mean amounts reach their largest annual values. Such behavior has been observed using other metrics of spatial coherence, including the number of degrees of freedom (Moron et al., 2017). The spatial coherence is larger for frequency than for amounts, especially during the core of the season (thin vs thick red curves; Fig. 2A).

The four lower subpanels of Fig. 2 show the spatial patterns of the leading empirical orthogonal function (EOF) of the interannual ranks of amounts at all India grid points for four 15-day subperiods around the onset, during the core and then around the withdrawal (vertical red lines in Fig. 2A). The leading EOF pattern is dominated by the interannual variability over the core monsoonal zone, including the Western Ghats, the northern part of the Peninsula, most of the Indo-Gangetic plain, and the desert areas in the northwest. However, positive loadings are generally considerably higher around the onset (Fig. 2B) and withdrawal (Fig. 2E) than during the core of the monsoon in July (Fig. 2C) and August (Fig. 2D). The explained variance drops significantly during July-August. The interannual variations of the 15-day amount and frequency around the onset date will partly convey the anomalous advance or the delay of onset, consistent with the sharp peak in spatial coherence seen near the beginning of June in Fig. 2A. However, the more gradual increase from late August to early October may be accounted for by changes in withdrawal date, which takes more time to achieve across India than onset (Moron and Robertson, 2014) and/or the buildup of spatial coherence during the monsoon season itself. Overall, this sub-seasonal modulation suggests that a significant portion of precipitation in July-August is due to intense rainfall (Stephenson et al., 1999) embedded in small to mesoscale features, which have a smaller spatial scale than the large-scale atmospheric circulation patterns responsible for the monsoon onset and withdrawal (Moron et al., 2017).

3.3 Sub-seasonal Modulation of Spatial Coherence Over the Whole Tropical Zone

Do these findings for India generalize to the whole tropical zone (30 degrees N to 30 degrees S)? Fig. 3 shows the results of a similar analysis of the global tropics (including oceans) using

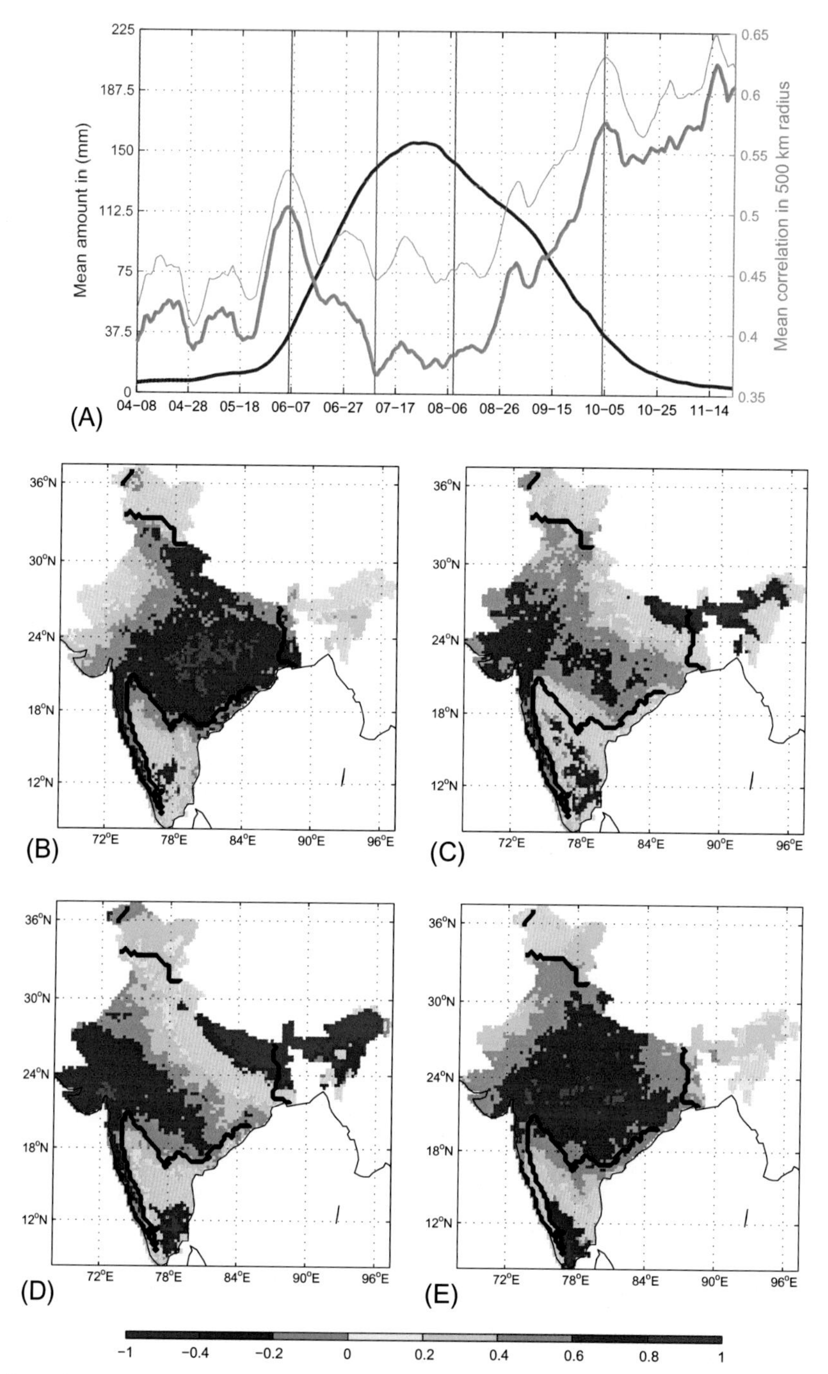

FIG. 2 See legend on opposite page.

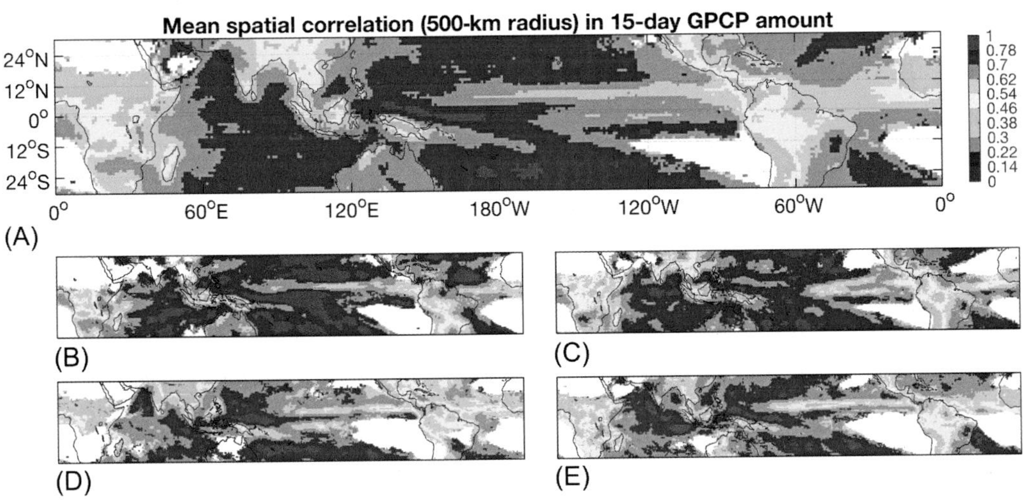

FIG. 3 Mean spatial autocorrelations in the 500-km radius for GPCP. The spatial autocorrelations are computed using ranks of amounts of rainfall between the central grid points and all grid points in a 500-km radius (including latitudes north and south of the latitudinal limits of the map). The running 15-day windows where the climatological mean amount ≪10 mm are not considered in the time average. *Blank areas* never reach the threshold of a climatological mean ≥10 mm in any of the running 15-day windows. All computations are done over Oct. 1996 to Sep. 2016. For the seasonal maps (four lower panels), all running 15-day belonging to a given season are considered (i.e., for December-February, the seventy-six 15-day periods from Dec. 1-Dec. 15 to Feb. 14-Feb. 28 are considered). (A) Year, (B) DJF, (C) MAM, (D) JJA, (E) SON.

daily rainfall amounts from GPCP. The analysis is applied over the whole calendar year using 365 running 15-day windows (top panel) and those belonging to the four usual meteorological seasons (lower panels).

The spatial coherence of 15-day GPCP rainfall anomalies is much lower over land than over the oceans in general. Area averages are given in Table 1, including for coastal areas (defined as sea ≤500 km to land) and open ocean (≫500 km from land). Some of the smallest values of spatial coherence are associated with high mean daily intensity (not shown), especially over the continents (South and Southeast Asia, Western Amazonia, etc.), although over the oceans, the equatorial eastern Indian Ocean and West Pacific warm pool ocean regions exhibit both

FIG. 2, CONT'D (A) Mean seasonal variations of spatially averaged amounts (*blue line*, left ordinate) and spatial autocorrelations of interannual ranks of mean amount (*bold red*, right ordinate) and frequency of wet days receiving ≥1 mm (*thin red*, right ordinate) in a 500-km radius in running 15-day windows (centered on the dates shown on the *abscissa*) averaged over the summer monsoonal regime (Gadgil, 2003) defined in Moron et al. (2017) from a clustering of the mean annual cycle across India and *underlined* by a black contour in panels (B)–(E). The grid points and 15-day windows receiving ≪ 10 mm in mean (over the 1901–2014 period) are excluded from the spatial average of autocorrelations. The *vertical red line* in panel (A) shows the center of the four 15-day windows used to extract the leading EOF of amount ranks in panels (B)–(E) computed over all India grid points. The EOFs are shown as loadings; that is, the correlations between the amount ranks and the leading principal component and the explained variance is noted in the panel (B) to (E) titles. Correlations that are not significant at the two-sided 95% level according to a random-phase test are indicated by *gray areas*. (A) Mean amount and spatial autocorrelation in 500-km radius [amount, frequency]. (B) EOF 1, May 30–June 13, $V = 27\%$. (C) EOF 1, Jul. 2–16, $V = 19\%$. (D) EOF 1, Aug. 1–15, $V = 23\%$. (E) EOF 1, Sep. 28–Oct. 12, $V = 29\%$.

TABLE 1

	Land	Coast	Open Sea	OLR	CMAP
Year	0.53	0.64	0.68	0.61	0.65
Dec.-Feb.	0.59	0.66	0.71	0.66	0.69
Mar.-May	0.55	0.64	0.69	0.67	0.68
Jun.-Aug.	0.49	0.62	0.66	0.70	0.67
Sep.-Oct.	0.53	0.64	0.67	0.64	0.64

Notes: *Columns 2–4: Spatial averages of the mean autocorrelation in a 500-km radius for land, coast (≤500 km from land) and open sea (≫500 km from land). The areas where mean rainfall over sliding 15-day windows is always ≪10 mm during the periods shown in Column 1 are not considered in the spatial averages; columns 5–6: pattern correlation between GPCP maps (linearly interpolated over CMAP and OLR grids) and the mean autocorrelation in a 500-km radius for the whole tropical zone. The areas where mean rainfall over sliding 15-day window is always ≪10 mm during the periods shown in column 1 are not considered in the pattern correlations.*

high mean intensity and strong spatial coherence (Fig. 3). Relative minima in spatial coherence coincide with the ITCZ over the northern equatorial central and eastern Pacific and Atlantic oceans (Fig. 3), where the pattern of the spatial autocorrelation tends to be primarily zonal rather than isotropic. Seasonal values of spatial coherence peak in austral summer and are at minimum during boreal summer, especially across land in the Northern Hemisphere. The seasonal modulation may be partly related to the amplitude of ENSO events peaking near the end of the calendar year. Several pockets of large spatial coherence coincide with regions of high seasonal rainfall predictability, including the Maritime Continent south of the equator around the austral summer monsoon onset in SON (September-November) (Haylock and McBride, 2001; Moron et al., 2009b), and near the boreal summer monsoon onset over the South China Sea and Philippines in MAM (March-May) (Moron et al., 2009a). Low values of spatial coherence may be partly due to cancelation between positive and negative correlations within a 500-km radius, such as those between northeast India and the Indo-Gangetic plain during the core of the rainy season (Fig. 2C and D), or they may reflect increased ascent in a deep convection center and increased subsidence nearby. Table 1 also includes the pattern correlations between GPCP spatial autocorrelations and those computed using CMAP and OLR datasets, demonstrating that the results are quite robust to the choice of rainfall or deep convection dataset. The lower spatial coherence across land versus ocean may be, at least partly, explained by the interactions promoting small-scale deep convection and involving the diurnal cycle, land-sea, and mountain-valley breezes, as well as gravity wave effects (Yang and Slingo, 2001; Slingo et al., 2003), which are far stronger over the landmasses and the coastal areas than over the open seas.

The sub-seasonal evolution of spatial coherence across the tropics is depicted in Fig. 4, and constructed by identifying the timing of the minimum and maximum of local spatial coherence conditioned on the local seasonal cycle of the 15-day mean rainfall amount. The spatial coherence and mean rainfall are low-pass-filtered using a recursive digital filter with a cut-off at 1/90 cycle/day. Most of the tropical zone is associated with a unimodal regime and two distinct rainy seasons are rare (not shown). Fig. 4 shows the timings of minimum and maximum spatial coherence during the year, considering only periods when the local climatological mean amount reaches at least 10 mm per 15 days, for all tropical grid points (panels A and D), as well

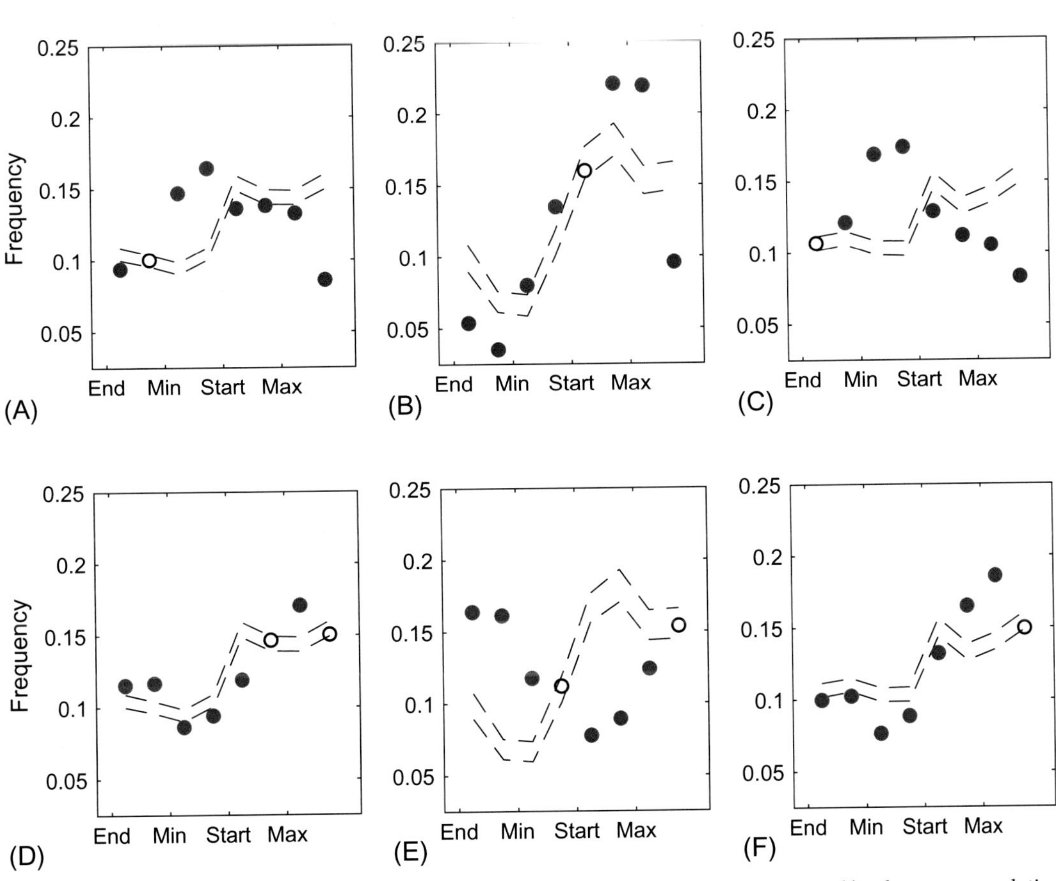

FIG. 4 Frequency of (A–C) minimum and (D–F) maximum spatial coherence (as estimated by the mean correlation between the interannual ranks of amounts on sliding 15-day windows at a central grid point and those at the surrounding ones in a 500-km radius) versus the eight phases of the mean amount of rainfall. The search is limited to the time when the mean amount ≥10 mm per 15-day windows, but the eight phases are computed on the whole year from the low-pass filtered (cut-off = 1/90 cycle/day) annual cycle of mean amount. The *abscissa* indicates the approximative end, minimum, start, and maximum of the local rainy season. In cases of two (or indistinct) rainy seasons, phases 2–3 correspond to the lowest annual rainfall and phases 6–7 correspond to the highest ones. The observed frequencies are indicated by *circles*, while the *dashed lines* are 95% confidence interval (CI; *dashed lines*) computed from 500 random resamplings of the time series of spatial coherence. *Red and blue circles* indicate the significant positive and negative anomalies at the two-sided 95% levels, respectively. The *first column* is for the whole tropical zone. The *second and third columns* are, respectively, for land and ocean. (A) Min. sp. coh. (all), (B) Min. sp. coh. (land), (C) Min. sp. coh. (sea), (D) Max. sp. coh. (all), (E) Max. sp. coh. (land), (F) Max. sp. coh. (sea).

as for land (panels B and E) and ocean (panels C and F) grid points separately. The minimum spatial coherence tends to occur around the time of the highest rainfall, especially over land. Maximum spatial coherence tends to occur around the onset and end of the wet season, while it is less common around the peak of the local seasonal cycle of rainfall, especially over land (Fig. 4). The sub-seasonal modulation is less clear for the ocean.

Fig. 5 shows the mean seasonal cycle of low-pass-filtered mean rainfall amount and spatial coherence for 12 regions chosen across the tropics, constructed from GPCP data. India (Fig. 5B) shows a similar behavior as those shown on Fig. 2A despite a different resolution and period covered. It is also similar to the Sahel (Fig. 5A), with the minimum in spatial coherence coinciding with the highest mean rainfall amount around early August, with the spatial coherence peaking on either side, near the start and end of the rainy season. Similar behavior is seen for most of the regions that include only land points (Fig. 5A–D, F, J, K), as well as over the North Atlantic ITCZ (Fig. 5L). Equatorial East Africa displays a different behavior during the "long rains" in MAM, with spatial coherence peaking near the start in March and then decreasing to mid-June (Camberlin et al., 2009; Moron et al., 2013, 2015a); spatial coherence is then broadly positively correlated with the seasonal evolution of rainfall during the short rains in OND (October-December) (Fig. 5E). Similar behavior is also seen for South Africa (Fig. 5K) and Amazonia (Fig. 5J), but with a weak amplitude. The remaining oceanic boxes over Western, Central, and Eastern equatorial Pacific (Fig. 5G–I) do not show a clear seasonal modulation of the (usually strong) spatial coherence.

Figs. 4 and 5 show that the spatial coherence in rainfall amount is usually low around the core of the wet season and tends to peak near its end or during its transition to dry season. They show also that the modulation is different between land and oceanic grid points, with the clearest season modulation found over land, where the monsoons are themselves strongest (note that completely dry periods (i.e., any 15-day window receiving $\ll 10$ mm in mean) are excluded from the calculations). These figures suggest that the anomalous advance and delay of the *end* of wet seasons (and secondarily their onsets) exhibit a larger scale than the core of the season, when small-scale intense, wet events may decrease the spatial coherence, at least over several continental areas such as those shown in Fig. 5A–D and F. This behavior does not appear across most of the oceans, and several continents do not exhibit the common modulation revealed by the continental regions quoted previously. As before, the explanations may involve either the variable amplitude of the impact of CCEWs (including MJO) or the variable influence of diurnal cycle, small-scale land-sea and mountain breezes and related gravity waves on local-scale rainfall. A larger (weaker) spatial coherence is expected when the first (second) processes dominate the overall variance of daily rainfall.

3.4 Skill and Spatial Coherence of S2S Reforecasts

This section presents maps of S2S model precipitation spatial coherence (Fig. 6A) and forecast skill (Figs. 6B and 7) estimated from ECMWF reforecasts, at a forecast lead time of 15–28 days (labeled "week 3 + 4"). Skill maps are constructed of the temporal correlation of anomalies (CORA), using the ensemble means of fortnight week 3 + 4 averages of the ECMWF reforecasts. The anomalies are calculated by removing the observational long-term mean of each season, as well as the models' week 3 + 4 reforecast climatology, to exclude the mean bias and model drift. ECMWF reforecasts from all the semiweekly start dates that fall in each season are lumped together as a time series in order to compare with their observed counterparts.

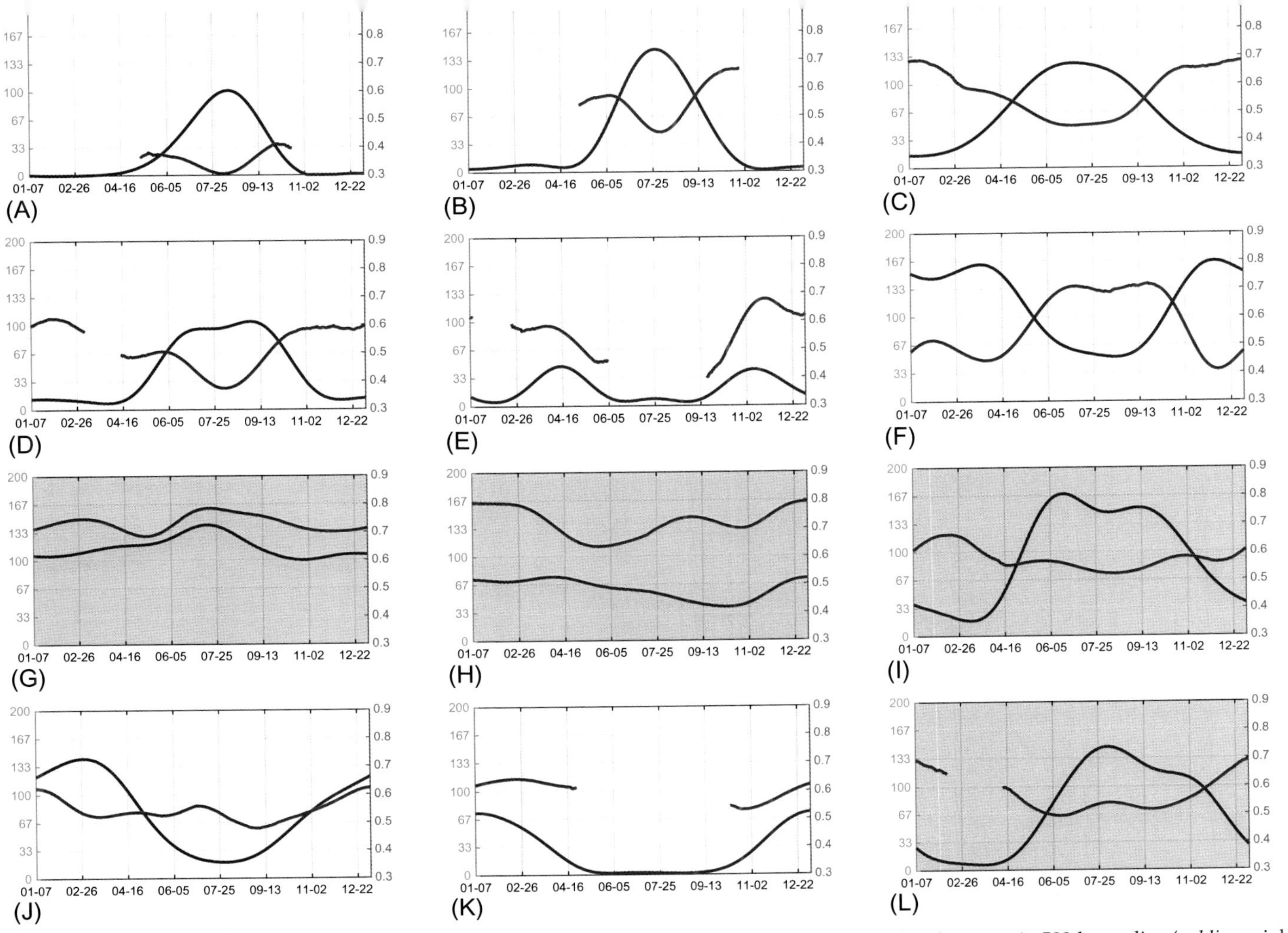

FIG. 5 Mean amount (in millimeters, *blue lines*, left ordinate) and spatial correlation of interannual ranks of amount in 500-km radius (*red lines*, right ordinate) spatially averaged over 12 tropical regions defined below. The amounts and correlations are computed using GPCP data on running 15-day windows (the center of each 15-day window is indicated on the *abscissa*) and the time series are low-pass-filtered (cut-off = 1/90 cycle/day). Data where the mean amount ≪10 mm in running 15-day windows are not used in the spatial averages. The *gray panels* use only oceanic grid points, while the *blank panels* use only landmass grid points. (A) Sahel: 12–16 degrees N, 15 degrees W-10 degrees E; (B) India: 17–30 degrees N, 72–87 degrees E; (C) SE Asia: 12–30 degrees N, 100–122 degrees E; (D) Central Am.: 12–25 degrees N, 240–270 degrees E; (E) Equat. E. Afr.: 5 degrees S-5 degrees N, 35–50 degrees E; (F) S. Indonesia: 10 degrees S-0, 90–120 degrees E; (G) ITCZ W. Pacific: 0–10 degrees N, 135–160 degrees E; (H) C. Pacific: 5 degrees S-5 degrees N, 160–210 degrees E; (I) ITCZ E. Pacific: 5–13 degrees N, 250–280 degrees E; (J) Amazonia: 15 degrees S-0, 290–320 degrees E; (K) S. Africa: 28–15 degrees S, 20–40 degrees E; (L) ITCZ Atlantic: 5–13 degrees N, 320–345 degrees E.

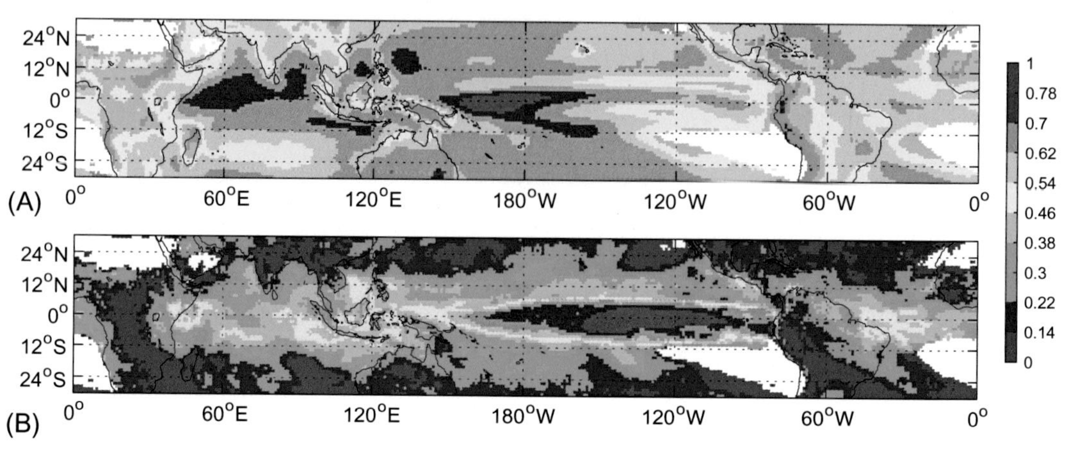

FIG. 6 (A) Mean spatial autocorrelations in 500-km radius for ECWMF week 3 + 4 reforecast precipitation anomalies. The spatial autocorrelations are computed using ranks of interannual anomalies of rainfall between the central grid points and all grid points in a 500-km radius (including latitudes north and south of the latitudinal limits of the map); (B) ECMWF week 3 + 4 anomaly correlation coefficient (CORA) for total precipitation for all year. Any CORA value >0.3 is statistically significant at the 99% CI by a one-tailed *t*-test (as only positive ACC is considered skillful). The areas receiving ≪ 10 mm in Fig. 3A are marked as blank areas.

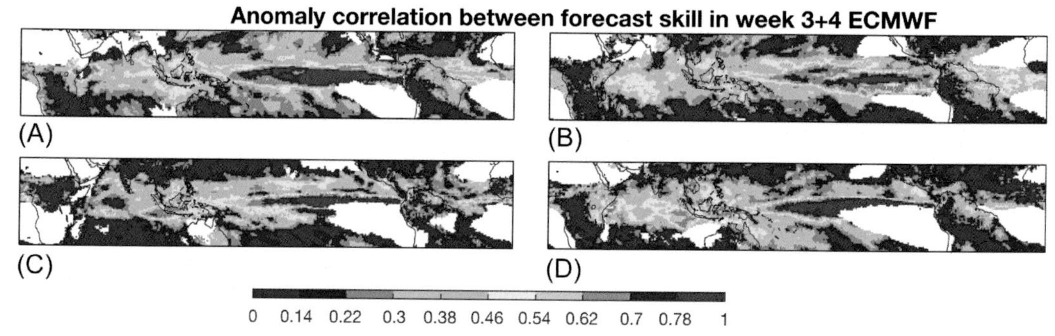

FIG. 7 ECMWF week 3 + 4 anomaly correlation coefficient (CORA) for total precipitation for (A) DJF, (B) MAM, (C) JJA, and (D) SON. Any CORA value >0.3 is statistically significant at the 99% CI by a one-tailed *t*-test (as only positive ACC is considered skillful). The areas receiving ≪ 10 mm in Fig. 3B–E are marked as blank areas.

The spatial coherence of week 3 + 4 precipitation anomalies from the ECMWF control run (Fig. 6A) (computed from the 20 reforecast years corresponding to each Monday real-time forecast start date) is very similar to the observed estimate (Fig. 3A). The pattern correlation between both maps is 0.73, reproducing the contrast between land and sea. The spatial averages of ECMWF spatial coherence equal, respectively, 0.49, 0.57, and 0.62 for land, coastal, and open ocean (close to the GPCP estimates in Table 1).

The geographical distribution of ECMWF week 3 + 4 forecast skill broadly resembles that of the spatial coherence, although the pattern correlation between them is low (0.27). The similarity in patterns is mostly related to the much higher skill over ocean than over land, especially over the tropical Pacific due to the high local persistence there of ENSO-related SST anomalies (Li and Robertson, 2015). The geographical extent of this highly predictable belt region changes with the season (Fig. 7), being relatively narrower in JJA (June-August) (red regions in the Pacific sector of Fig. 7). Good skill can also be found over the western Indian Ocean, which extends westward to Africa, in boreal winter and fall (Fig. 7A and D). The western tropical Atlantic and the adjacent continents such as eastern Amazonia also show good skill in boreal winter and spring (Fig. 7A and B). In contrast, the eastern tropical Indian Ocean and Maritime Continent are more predictable in austral winter and spring (Fig. 7C and D).

These seasonal changes in skill are generally consistent with those in the spatial coherence as shown in Fig. 3, with higher spatial coherence being associated with higher forecast skill. This is especially clear over the Maritime Continent, where low skill over the islands tends to coincide with lower values of spatial coherence there compared to the surrounding ocean. However, this correspondence between forecast skill and spatial coherence of rainfall anomalies is less clear elsewhere. For example, skill is often high over the central and eastern Pacific ITCZ and around the Atlantic ITCZ despite relatively low spatial coherence. The lack of skill over most continents for week 3 + 4 forecasts indicates that much room remains for progress in dynamical models in capturing sub-seasonal variability in the tropical rainfall. However, the comparison with spatial coherence suggests that some of these regions are often intrinsically less predictable.

4 DISCUSSION AND CONCLUDING REMARKS

Most previous predictability studies of tropical rainfall have focused primarily on interannual variations of seasonal rainfall amount, typically associated with the atmospheric response to slow boundary forcings, primarily ENSO and other coupled ocean-atmosphere modes of variation. The temporal summation across the season smooths the characteristics of individual rainfall events, thus emphasizing the impact of systematic (i.e., near-constant in time) "slow" forcings. The progressive temporal summation across the season also increases the spatial coherence of rainfall anomalies due to the near-constant modulation of either intensity, size, or frequency of the instantaneous wet events across a region. We focused here on quantifying the spatial coherence of sub-seasonal (2-week average) rainfall anomalies, allowing different sub-seasonal stages during a wet season to be distinguished. The spatial coherence is estimated with the mean spatial autocorrelation in a 500-km radius around each grid points. These observational sub-seasonal estimates of predictable spatial scale are then compared with estimates of forecast skill from ECMWF week 3+4 reforecasts from the S2S database.

The analysis of 0.25 degrees daily Indian rainfall illustrates the fact that two "good" monsoons (1983 and 1988) and two "bad" monsoons (1986 and 2002) at the Indian scale exhibit very different sub-seasonal evolutions at the local scale (Fig. 1). The spatial coherence of biweekly rainfall anomalies peaks near onset and withdrawal of the summer monsoon and reach its minimum during the core of the rainy season, in JA (Fig. 2A). This illustrates an example where the seasonal (JJAS) amount—which is dominated by the largest rainfall in JA—may blur smaller predictable signals in June or September. The leading EOF of ranks of

TABLE 2

	All	Land	Coast	Open Sea
Year	0.30	0.20	0.31	0.34
Dec. Feb.	0.33	0.23	0.32	0.37
Mar.-May	0.31	0.21	0.31	0.36
Jun.-Aug.	0.27	0.19	0.30	0.29
Sep.-Oct.	0.29	0.21	0.32	0.32

Notes: *Columns 2–5: Spatial averages of the mean skill for weeks 3 + 4 for all tropics, land, coast (<500 km from land) and open sea (>500 km from land). The areas where mean rainfall over sliding 15-day windows is always ≪ 10 mm during the periods shown in column 1 are not used in the spatial averages.*

amounts in JA (Fig. 2C and D) shows an out-of-phase pattern between most of the peninsula and western India on one hand and the Himalayan foothills and northeastern parts of India on the other hand, which may be viewed as the fingerprint of the main intraseasonal mode of variation (i.e., ISO; Krishnamurthy and Shukla, 2000, 2008; Moron et al., 2012), but the largest loadings ≥ 0.6 also cover a smaller area in JA than around the onset (Fig. 2B) and withdrawal (Fig. 2E) stages of the boreal monsoon. This sub-seasonal modulation with the smallest spatial coherence recorded around the core of the wet season appears to be rather general across the tropical continents (Fig. 4B and E), even if there are some exceptions (such as Equatorial Eastern Africa, Southern Africa, and Amazonia; Fig. 5E, J, and K). Moreover, the spatial scales are systematically larger across the ocean than over the landmasses (Figs. 3, 4, 6A and Tables 1 and 2). This land-sea contrast is also seen in S2S forecast skill (Fig. 6B and Table 2).

Regarding the general spatial difference between land and ocean, we can make a first hypothesis: area of subdaily wet events defined as contiguous wet grid points or those recording deep convection above a given threshold are systematically smaller over continents than over oceans (Ricciardulli and Sardeshmukh, 2002; Smith et al., 2005; Dai et al., 2009; Trenberth et al., 2017). It is also well established that the impact of the diurnal cycle and related processes, including land-sea and mountain-valley breezes and gravity waves (Yang and Slingo, 2001; Slingo et al., 2003), is stronger across continents and leads to shorter/smaller wet events across continents than over oceans. An ocean imposes, by definition, an homogeneous large-scale boundary forcing on the atmosphere, and the diurnal cycle is reduced due to thermal inertia. An open question is the interaction between the strong diurnal cycle and the modulation provided by various CCEWs (Wheeler and Kiladis, 1999; Lubis and Jacobi, 2015). For example, the amplitude of CCEWs decreases rather radically from the Gulf of Guinea to the Central Africa in boreal spring (Kamsu-Tamo et al., 2014). In that context, a strong impact of larger wet patterns, as tropical depressions or tropical-temperate-trough (TTT) systems (as those related to the South India Convergence Zone in southeast Africa and to the South Atlantic Convergence Zone in eastern Brazil), or where the CCEWs strongly affect the intensity and occurrence of local-scale rainfall may lead to a larger spatial scale of 15-day rainfall anomaly. A discrepancy is then possible between spatial coherence and predictability since a larger daily or subdaily wet pattern is not necessarily predictable at the S2S timescale. For example, a rather large spatial coherence over subtropical southern Africa (Figs. 3A, 6A), especially in

DJF (December-February) (Fig. 3B) does not match with an high predictability (Fig. 7A) during this season. In that case, the larger spatial coherence may be primarily an effect of TTTs generating a large wet pattern at the daily time scale (Macron et al., 2014) but it may be not really predictable by ECMWF with a lead time of 2 weeks. On the contrary, the combination of a large spatial coherence and high S2S predictability in ECMWF (Figs. 3 and 6), as the one over equatorial East Africa in SON (Figs. 3E, 5E, and 6) may primarily reflect a significant impact of MJO synchronizing efficiently the local-scale rainfall and being predictable at the S2S timescale (Pohl and Camberlin, 2006; Berhane and Zaitchik, 2014). The combination of relatively large spatial coherence with high S2S predictability is also observed over northeastern South America, and smaller continental pockets such as northern Australia or southern China.

Regarding the temporal modulation of spatial coherence (and S2S forecasts), several hypotheses may be emphasized. First, the seasonal modulation of the large-scale source of predictability could be a trivial source. We have shown here that the spatial coherence peaks and has high predictability in DJF, in phase with the usual peaks of ENSO events (Rasmusson and Carpenter, 1982; Ropelewski and Halpert, 1987, 1996). But the minimum spatial coherence is not observed in MAM at the usual time of shift between warm and cold ENSO events, but rather in JJA, when the ITCZ reaches its farthest location from the equator and when it is located over several landmasses in Central America, and between Western Sahel-Sudan and Southeast Asia. The exact role of the latitudinal distance of the main heating source from the equator and of the area of these heating sources on the scale and predictability of atmospheric anomalies remains to be established. A second source may be the role of extreme wet events that are known as significantly decreasing the spatial coherence and predictability of seasonal amounts in India (Stephenson et al., 1999; Moron et al., 2017). A third source of subseasonal modulation could be related to the multiscale interaction involving the slow forcings, the basic atmospheric state, and the diurnal cycle across the seasonal cycle. For example, it has been shown that warm ENSO promotes regional-scale subsidence over the Maritime Continent, but the interaction between this signal and local-scale rainfall varies over time. During warm ENSO events, the regional-scale, anomalous, low-level easterlies tend to counteract the usual monsoon flow and thus promote the occurrence of a quiet weather type (Moron et al., 2015b) characterized by anomalously weak, low-level winds during the core of the wet season (i.e., DJF). These weak winds emphasize the role of the diurnal cycle that lead to localized positive rainfall anomalies restricted to small parts of the islands despite the regional-scale subsidence anomalies (Qian et al., 2010). In that case, the spatial fragmentation of the rainfall anomaly field occurs during the core of the wet season, when the westerly basic flow is strong enough to be almost canceled by the anomalous easterlies associated with the ENSO forcing. Such varying interaction may be possible over other archipelagos as the Caribbean basin.

Finally, the fact that the spatial coherence usually peaks either near onset or withdrawal needs further analysis. First, the anomalous delay or advance of regional-scale onset or withdrawal may be the primary signal of the peak in spatial coherence around these times. We can also hypothesize a role played by land-atmosphere interaction: For most of the tropical continents except for some equatorial areas with very long, or even constant, moist conditions, the soils are fully dried up during 6–10 months. We can then hypothesize that the onset of the season over the continents is mostly forced by large-scale atmospheric circulation because soils cannot provide any local moisture that can trigger deep convection. In other words, a tropical landmass at the end of the dry season would wait for the favorable conditions associated with a combination of any source of slow and S2S predictability and large-scale phenomena conveying enough moisture to get the first rains. The intense warming of the dry surfaces lowers the static stability of the lower atmosphere. It is possible that the combination of local dry and moist instability conveyed by large-scale atmospheric circulation is very efficient close to the start of the season. The first rains in India are indeed very intense (Moron et al., 2017). The first wet events either related to local-scale thunderstorms or MCS would have two effects: (i) moistening the soils and starting the local recycling of water, which is indeed a positive feedback (Meehl, 1997; Douville et al., 2001, 2007); and (ii) increasing the heterogeneity of surface temperatures and humidity at the regional scale. As time goes by during the rainy season, the second effect will progressively vanish due to the different locations (or tracks) of wet events (from thunderstorms to MCSs) due to atmospheric circulation (including S2S phenomena). It could provide a partial explanation of the increase of spatial coherence toward the end of the season, at least over India (Douville et al., 2001, 2007), and which seems to be rather general across most of the tropical continents. This effect may be superimposed on the increasing power of ENSO events, at least for the northern tropics.

CHAPTER

4

Identifying Wave Processes Associated With Predictability Across Time Scales: An Empirical Normal Mode Approach

Gilbert Brunet, John Methven*†

*Meteorological Research Division, Environment and Climate Change Canada, Dorval, QC, Canada †Department of Meteorology, University of Reading, Reading, United Kingdom

OUTLINE

Sub-seasonal to Seasonal Prediction
https://doi.org/10.1016/B978-0-12-811714-9.00004-8

1 INTRODUCTION

It can be demonstrated that predictive skill ranges from weeks out to seasons when the forecast metric involves the statistical treatment of weather variables. Typically, the range of regional forecast skill increases with the scale of the region considered, as well as the length of the time window used for verification (see Chapter 2). Comparisons can be made in terms of averages over forecast lead times, over ensemble members, or, in a probabilistic sense, using ensemble forecasts. These extended-range forecasts go beyond the limit of predictability for point forecasts (the value of an atmospheric variable at a particular time and location). Such forecasts depend on the nature of the variable being forecast and the phenomena dominating its fluctuations. For example, geopotential on a pressure surface is a smooth field dominated by synoptic-scale (or larger) weather systems and is predictable out to 7–15 days, depending upon the flow configuration. Finer-scale fields, like vorticity, have shorter predictability limits. For example, Frame et al. (2015) showed that the predictive skill for the strike probability of cyclonic vorticity centers, within a given radius of locations in the Euro-Atlantic sector, increases with feature intensity and scale. Precipitation typically has a much shorter limit to its predictability, owing to its dependence on vertical motion and convective-scale features in the flow.

There are many plausible sources for sub-seasonal to seasonal (S2S) predictability, including slowly varying boundary conditions for tropospheric weather systems: coupling with the ocean (see Chapters 5 and 9), land surface processes (see Chapter 8), the cryosphere (see Chapter 10), and the stratosphere (see Chapter 11). However, sub-seasonal regimes are characterized by large-scale patterns of variability that exhibit internal variability over long time scales, and they can be oscillatory in nature or characterized by long-range teleconnections. Consider three contrasting examples:

1. The weather in the tropics and the extratropics is influenced by the phase of the Madden-Julian Oscillation (MJO). The MJO is the dominant tropical mode of sub-seasonal variability, which propagates eastward along the equator but has strong remote influences (see Chapters 5 and 7).
2. Unprecedented extreme summer precipitation events in western Europe have occurred in the last decade, and they have been associated with quasi-stationary Rossby wave patterns on the midlatitude jet stream (Blackburn et al., 2008), prompting new theories of resonant excitation of Rossby modes (Petoukhov et al., 2013; Coumou et al., 2014).
3. The Russian heatwave that occurred in 2010 was associated with a persistent midlatitude blocking regime (Dole et al., 2011).

In these examples, a distinct dynamical phenomenon is involved, and the properties of that phenomenon influence the weather in a predictable manner. Here, we focus on isolating oscillatory phenomena or slowly propagating modes responsible for enhanced predictability in the S2S range. In addition to decoupling the phenomena from the data, it is important to gain insight into their intrinsic dynamical properties and interactions in order to anticipate how they will contribute to predictability. Also, in the context of a changing climate, if we understand how modes of variability relate to the climate state, we will be able to anticipate how variability and predictability may vary with climate change.

Energy spectra in the wave-number space (calculated by transform of spatial distributions and averaging over many realizations) show a smooth continuum from the planetary to the kilometer scale, indicating that there is no spectral gap distinguishing large-scale phenomena from smaller-scale ones. Similarly, frequency spectra exhibit a smooth continuum from seasons to hours. However, statistical techniques that utilize both spatial and temporal information, such as the popular empirical orthogonal function (EOF) technique, show that covariance in time is typically dominated by large-scale patterns of variability. The EOF technique for a given variable maximizes the variance, which is explained by a series truncation of fixed length.

A remarkable property of an EOF mathematical construction is that EOF patterns are orthogonal to one another (i.e., the integral over the domain of two different EOF patterns multiplied together equals zero) and their corresponding time series are orthogonal as well, making them a complete basis that can be used to project variability from a discrete data set. Hence, the first EOF is the spatial pattern that explains most temporal variance; the first and second EOFs form the two-dimensional (2D) orthogonal basis that explains the most variance, and so on for higher terms in the series. The disadvantage of using a purely statistical approach is that the spatial structures obtained (the EOFs) and their corresponding principal component time series do not have distinct physical properties; therefore, it is difficult to anticipate their behavior beyond the limits of the time series examined.

A relatively unexploited approach is to combine conservation laws derived from consideration of atmospheric dynamics with the orthogonality approach to identify distinct patterns of variability that can be linked to characteristic physical properties (e.g., an intrinsic frequency or a phase speed). This approach can be called *empirical normal mode (ENM) analysis.* Just as an analysis of the shape and physical construction of a bell can be used to anticipate the frequency at which it will ring when struck, the spatial structure of an ENM can be used to predict its intrinsic frequency or phase speed from conservation properties. If it can be shown in general that a small number of such modes dominate S2S variability, as was demonstrated by Brunet (1994) for the 315 K isentropic surface, a huge reduction in the dimensional size of the system to be solved will result. Hence, the ENM basis could be the natural one to use to study waves and weather regimes in low-order dynamical systems, as discussed in Chapter 6.

Other disturbances may perturb the modes over a wide range of frequencies or stochastically, which causes the observed frequency spectrum to resemble a continuum. Therefore, knowledge of their intrinsic frequencies provides potentially useful information. The well-known fluctuation-dissipation theorem (FDT) describes how the time-average response of a dynamic system to random perturbations will resemble the structure of the slowest (longest-time-scale), unforced mode of variability. For example, Ring and Plumb (2008) studied the response of the Southern Annular Mode to forcing by drawing on the FDT. They used the principal oscillation pattern (POP) analysis developed by Hasselman (1988) and Penland (1989), which relies on the calculation of lag covariances to obtain the temporal behavior of spatial patterns. An advantage of the ENM technique is that lag covariances are not required because the frequency information stems from the dynamical properties.

A popular approach used in both weather and climate sciences is *composite sampling* to study long geophysical time series. With this method, statistical properties (e.g., average and standard deviation) of similar segments of a time series are examined, and a given segment is included in the composite if a characteristic event occurs within it. There is a vast body of literature describing such approaches; these studies use atmospheric reanalysis time series to link weather-climate events (e.g., Molteni et al., 1988; Robertson and Metz, 1989; Cotton et al., 1989; Vautard, 1990; Ferranti et al., 1989; Lin and Brunet, 2009). For example, Asaadi et al. (2016a, 2017) used this approach to identify fundamental dynamical and physical processes related to hurricane genesis. They showed that the coexistence of an African easterly wave nonlinear critical layer and a region of a weak meridional potential vorticity (PV) gradient over several days might be a major factor determining whether tropical disturbances develop into hurricanes. This finding answered the long-standing question of why only a small fraction of African easterly waves contribute to hurricane genesis. The study showed a way how S2S variability can modulate the hurricane season.

In general, the utilization of models of varying complexity is also needed for understanding and identifying sources of predictability. For example, with regard to sub-seasonal variability, this approach made it possible to demonstrate for the first time a two-way linkage between the Madden Julian Oscillation (MJO) and the Arctic Oscillation (AO) in a simplified general circulation model (GCM), numerical weather prediction (NWP) systems, and observations. These studies have pointed the way toward improving S2S predictive skill and provided examples of how global teleconnections are influencing regional weather in more complex modeling systems (Lin et al., 2007, 2009, 2010a; Lin and Brunet, 2011); also see Chapter 7.

A key message arising from this work is that even though the atmosphere is chaotic and stirring by large-scale waves and eddies generate fine-scale structures in air masses and conserved properties such as PV, the large-scale dynamics may be closer to linear behavior than expected. Characteristic spatial patterns that vary slowly but are not periodic in nature are often treated as oscillations. Teleconnections typically fall in this category. There are many situations in which interactions between wave phenomena can be identified and explained mechanistically using linear dynamics. The purpose of this chapter is to provide a theoretical and statistical framework for studying the sub-seasonal predictability in observational and model data using these concepts in an integrated manner.

Section 2 introduces the components of the framework, including the notional partition between a background state and perturbations to it, the concept of wave activity conservation, and the implications of conservation for modes of variability. Section 3 outlines the ENM technique and presents some implications for the behavior of perturbations that can be deduced from the approach. Several applications of the approach to global atmospheric data are presented to illustrate the technique and its potential. Conclusions are presented in Section 4.

2 PARTITIONING ATMOSPHERIC BEHAVIOR USING ITS CONSERVATION PROPERTIES

There are two major aims of the approach to understanding atmospheric variability presented here:

- Development of a theoretical framework that is capable of isolating a slowly varying component of the atmosphere that is influenced by slow processes, such as radiative forcing, and described by well-known equations. This phenomenon may be considered to be linked with climate.

- Development of a technique to isolate coherent, dynamical modes of variability from observed global data and models. These modes have intrinsic properties that can be deduced from theory.

In spite of the fact that there is a continuum of complex behavior (including nonlinear interactions) across scales, our goal is to take the underpinning theory as far as possible in terms of isolating processes from observations and representing them in forecasts. The robustness of the approach will be tested with global reanalysis data. Possible applications that result from the deductions about forcing of variability, dynamical processes and wave resonance also will be discussed.

Areas where this approach could be useful include the following:

- Anticipating how variability might change with changing climate by increasing our understanding of the properties of dynamical modes of variability and their dependence on the background state
- Identifying the physical links between large-scale modes of variability and high-impact weather that typically occurs on smaller scales
- Using these links to forecast the likelihood of high-impact weather, even though representing high-impact weather itself in models may be very challenging
- Diagnostics to identify model errors in the representation of dynamics

Examples include the risk of extreme precipitation that is conditional on the phase of the MJO and persistent midlatitude weather extremes associated with a particular phase of quasi-stationary Rossby waves, such as the extremely wet summers that occurred in western Europe in 2007 and 2012 (Blackburn et al., 2008; de Leeuw et al., 2016), or the Russian heatwave in 2010 (Dole et al., 2011).

2.1 Partitioning Variability: Background State and Wave Activity

Typically, atmospheric variability is identified through statistical analysis that does not explicitly use the properties of atmospheric dynamics. A starting point is to identify an anomaly: it could be some form of readily recognized coherent structure in the atmosphere (such as a tropical cyclone), but more often it is obtained by subtraction of some form of mean state to define perturbation fields at every point in the model:

$$q' = q - q_0 \quad (1)$$

These perturbations are then analyzed statistically. In such an approach, the definition of perturbations and their properties clearly depends on the mean state, q_0, that is chosen.

The three most popular approaches for defining mean state are as follows:

1. *Global average*. In this case, the system can be described by the global integral of the evolution equations, but all the dynamics are contained entirely within the perturbations that dominate the atmospheric response to forcing (e.g., the global temperature response to greenhouse gas forcing or volcanic eruption).
2. *Eulerian time average (at fixed locations)*. The mean state can readily be calculated from data, but it is not a complete solution of the governing equations. Forcing from eddy fluxes

must be added. All the time dependence rests within the perturbations. One example of this approach is forecasting a regional seasonal temperature anomaly in contrast to numerical weather prediction (NWP) of the total temperature field.

3. *Eulerian zonal average (average around latitude circles)*. Although it is often used in dynamic meteorology, as in the wave-mean flow interaction problem, this approach has a disadvantage: The mean can vary as quickly as the perturbations because it may be changed by adiabatic eddy fluxes.

Research in the 1960s, 1970s, and 1980s showed that the evolution of the mean meridional circulation (a zonal mean, usually time-filtered) depends crucially on the variables chosen to describe the data (Andrews et al., 1987). In particular, there is a very marked difference in the mean state deduced by averaging pressure-level data (the easiest coordinate to use in assimilating observational profile data), compared with averaging along isentropic levels (surfaces of constant potential temperature). The key reason is that potential temperature is materially conserved following adiabatic motion, so isentropic-coordinate averages partition diabatic behavior from adiabatic behavior. In contrast, vertical motion across pressure surfaces can occur through both adiabatic and diabatic processes. A major example is the Ferrell cell, which appears in the midlatitudes as a thermally indirect circulation in the pressure-coordinate average, but is absent in the isentropic-coordinate average (Townsend and Johnson, 1985). The origin of this structure is predominantly adiabatic motion along sloping isentropic surfaces within midlatitude weather systems.

An extension of the isentropic-coordinate approach involves using coordinates where two of the variables are properties that are approximately materially conserved. The most common example is to use PV in conjunction with potential temperature (e.g., Nakamura, 1995). If motions are adiabatic and frictionless, then the PV and potential temperature cannot be modified along trajectories. If surfaces of constant potential temperature and PV intersect, those intersections must be transported around by the fluid as material lines. For example, the midlatitude tropopause is often described as a particular PV surface (2 PVU), and the position of the tropopause on each isentropic surface that intersects it is then stirred by the fluid motion along that surface (Hoskins et al., 1985). However, for conservative motion, mass cannot be transported across isentropic surfaces or PV surfaces, nor can it be transported across the intersections between the surfaces. This is a strong constraint on fluid behavior.

In a conceptual sense, we can describe the entire atmosphere in conserved variable coordinates (i.e., a 2D plane with potential temperature and PV as the axes), where motion of atmospheric mass is possible only through the action of diabatic or frictional processes. This framework is described as a *modified Lagrangian mean (MLM)* (McIntyre, 1980). A true Lagrangian mean would require calculation of the trajectories of all air parcels and some form of time-averaging along trajectories and over trajectories from similar initial conditions. In contrast, the MLM state is obtained by using the approximately conserved variables as markers of air masses, and therefore tracers of fluid motion. This has a distinct advantage. The time-dependent winds stir tracers so that fine-scale structures are developed through chaotic advection, and there is a cascade to ever-finer-scale structures in the absence of nonconservative processes. However, real tracers including chemical constituents, as well as PV and potential temperature, are subject to nonconservative processes that act to dissipate the smallest features (halting the scale cascade) and also act to maintain large-scale contrasts (so that the entire atmosphere does not become well-mixed). The MLM approach takes advantage of this property.

Another crucial aspect of introducing a partition between a background state and perturbations about that state is that it is necessary to predict the evolution of both components using the equations of motion and thermodynamics. First, we will consider the background state (the evolution of the perturbations will be considered in Section 2.2). An important way to predict background state evolution is to use the integral conservation properties of the full state. The definition of the MLM state uses two material conservation properties. From this, it is possible to deduce that mass cannot cross surfaces of constant potential temperature (θ) or constant Ertel PV (q) if the flow is conservative (adiabatic and frictionless). Therefore, the mass enclosed in a volume bounded above and below by two neighboring isentropic surfaces ($\Delta\theta$) and laterally by a PV contour (with value Q) must be conserved:

$$M(Q,\theta) = \frac{1}{\Delta\theta}\iiint r\,dA\,d\theta \tag{2}$$

where r is the density in isentropic coordinates and A is the area enclosed.

Kelvin's circulation theorem also states that the circulation within any closed material contour on an isentropic surface is invariant if the flow is adiabatic and frictionless. A set of circulation integrals can be defined by the integral of the tangential absolute velocity (in an inertial frame) around all closed PV contours (i.e., varying Q):

$$C(Q,\theta) = \oint_{q=Q} \underline{u}.d\underline{l} = \langle u_t \rangle L_Q \tag{3}$$

where the average tangential speed of the flow around the circuit, $\langle u_t \rangle$, must depend inversely upon the length of the circuit, L_Q. Although the mass enclosed by the circuit is invariant, the length of its boundary depends upon the degree of contortion of the PV contour by the flow. Therefore, the average speed and local velocity must depend on the shape of PV contours. To specify the background state and perturbations to it completely, it is necessary to make an assumption about the shape of the PV contours that define the background state.

One option is to define the MLM background state as zonally symmetric and to require it to obey the same equations of motion as the full flow. This can be achieved by an adiabatic rearrangement, in which the mass and circulation enclosed by every PV contour in isentropic layers are the same as in the full state. The geometry of the contours is made to be concentric around latitude circles (McIntyre, 1980; Methven and Berrisford, 2015). Using Stokes's theorem, the circulation also can be expressed as a volume integral of PV:

$$C(Q,\theta) = \frac{1}{\Delta\theta}\iiint rq\,dA\,d\theta \tag{4}$$

because the absolute vorticity on an isentropic surface is given by rq.

Consider a thought experiment where there is a distorted polar vortex in an isentropic layer characterized by uniform high PV (value Q) inside the vortex and zero PV outside. For illustration purposes, perturbations in isentropic density along the layer are assumed to be small compared with the mean density, R. In this case, the circulation within the wavy contour is approximately RQA, where A is the area enclosed. Therefore, the background state zonal flow around the vortex edge is

$$u_0 = \frac{RQA - C_p}{L_0} \tag{5}$$

where L_0 is the length of the latitude circle encompassing a vortex of area A that is concentric about the pole in the background state. It is usual to define an equivalent latitude for the contour such that $L_0 = 2\pi a(^{\pi}/_2 - \phi_e)$ and $A = 2\pi a^2(1 - \sin\phi_e)$, where a denotes the Earth's radius. The area integral of the vertical component of the planetary vorticity ($2\Omega \sin\phi$) within the background state contour is $C_p = \Omega.\, 2\pi a^2(1 - (\sin\phi_e)^2)$. It is subtracted to obtain the zonal flow in the rotating frame of the Earth. The zonal average of the flow around latitude ϕ_e is then given by

$$[u] = \frac{RQ(A-B) - C_p}{L_0} \tag{6}$$

where B is the area within the latitude circle that is external to the disturbed vortex where the PV is zero. Two properties are immediately apparent. If there is any disturbance, we expect $[u] < u_0$ because $0 < B < A$, and we also can anticipate that the Eulerian zonal average $[u]$ will fluctuate with the disturbance amplitude, as characterized by the area occupied by low PV ridges (B).

The flow and density can be obtained by inversion of the PV distribution (Methven and Berrisford, 2015). In this way, we can define the distribution of dynamical atmospheric variables and their evolution. Because the PV and θ distributions of the MLM state are zonally symmetric, the zonal flow obtained from PV inversion must be symmetric as well. Furthermore, because the zonal flow is parallel to the PV contours along isentropic surfaces, there can be no change associated with advection. Without additional approximations, the solution to the primitive equations on the sphere in this situation is a state that is in hydrostatic and gradient wind balance. Because the zonal integral of the full state conserves zonal angular momentum, so must the zonally symmetric background state due to translational invariance in the zonal direction. The perturbations also must obey a pseudoangular momentum conservation law, as described in Section 2.2.

An alternative would be to define the background as a strictly steady state. In this case, the time invariance implies energy conservation of the background, as well as global energy conservation of the full state, and the perturbations obey a pseudoenergy conservation law. However, there are several disadvantages to this approach. The background state is not exactly in balance if the flow is zonally asymmetric, so the PV distribution cannot, in general, be inverted to obtain the flow and density. Furthermore, except in a special situation where the flow happens to be parallel to PV contours everywhere, the background could not be steady without the continuous action of forcing introduced into the evolution equations.

Therefore, a compromise must be made. Either we identify the background state with zonal symmetry, in which case even stationary waves and zonal variations in climate must be regarded as part of the perturbation field, or we identify the background with a steady state (time symmetry), in which case the evolution of the background is not considered (by definition), and the maintenance of the background also would require a forcing term in the equations. Most people would define the term *climate* using some notional component that is slowly varying and inherently large-scale, but this approach loses the advantages of precise symmetry in space or time.

In the analysis of perturbations that follows in this chapter, we will use a zonally symmetric background state, but also assume that it evolves much more slowly than the perturbations. As explained previously, this is approximately true for the MLM state because it can evolve only through nonconservative processes, and these modify global circulation only slowly. Fig. 1 illustrates the evolution of the MLM state for the Northern Hemisphere in June 2007, obtained using the equivalent latitude iteration by PV inversion (ELIPVI) method of Methven and Berrisford (2015).

It is immediately apparent that $u_0 > [u]$, and also that $[u]$ varies more rapidly, as predicted previously. The solid black contour on both plots marks the tropopause (PV = 2 PVU;

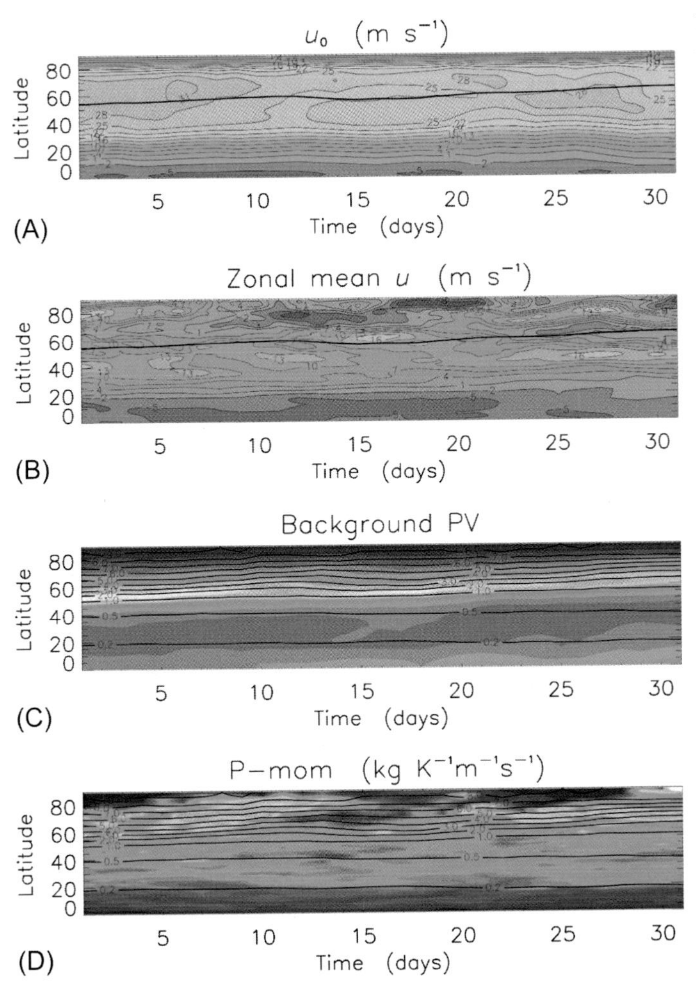

FIG. 1 Evolution of the atmosphere on the 320K isentropic surface calculated from daily ERA-Interim data for June 2007. (A) MLM background state zonal flow. (B) Eulerian zonal mean flow. (C) PV (PVU) in contours; color shading indicates the meridional PV gradient (the largest values are pink). (D) Wave activity (pseudomomentum density) in color shading overlain on background state PV contours (CI = 5000 up to 153,000 kg K^{-1} m^{-1} s^{-1} shown in lightest *pink*).

PVU $= 10^{-6}$ kg^{-1} s^{-1} K m^2) on the 320K isentropic surface. It is obvious that it migrates poleward over the month, aside from a brief period in the middle. Fig. 1C shows the MLM state PV distribution on the 320K surface. All the midlatitude PV contours are migrating poleward, which can occur only through nonconservative (diabatic or frictional) processes. The meridional gradient of PV (color shading) also migrates with the tropopause on this surface. This is the seasonal march of the tropopause.

The mechanism is "vortex erosion" (Legras and Dritschel, 1993), where continuous filamentation of PV by the breaking of Rossby waves on the polar vortex edge transports mass within PV filaments away from the edge region into the surf zone, where the PV is mixed (McIntyre and Palmer, 1984). The net result is less mass within the vortex, and thus a poleward displacement. The high PV in the lower stratosphere (polar regions) is maintained by radiative cooling (Haynes, 2005), but this is weakest at the summer solstice; therefore, the high PV is not maintained. This occurs until late August, when the cooling begins to strengthen again and the high PV reservoir builds. The tropopause progresses slowly equatorward during the autumn. The time scale of delay behind the cycle in solar insolation, therefore, is related to the radiative equilibrium time scale of the troposphere (30 days, James, 1994). Fig. 1D shows a measure of wave activity on the 320K surface, which will be explained in the next section. It can be seen that the marked variations in the zonal mean are related to variations in wave activity, as argued earlier in this chapter.

2.2 Wave Activity Conservation Laws

Once we have isolated the background state, the partition is useful only if we also can anticipate the properties for evolution of the perturbations from it. The aim is to find a definition for wave activity that satisfies a conservation law of the following form:

$$\frac{\partial A}{\partial t} + \nabla . \underline{F} = D \tag{7}$$

which stems from the conservation laws obeyed by the full system combined with those satisfied by the background state due to its symmetry. In Eq. (7), A is a wave activity density, F is the wave activity flux, and D stands for the effects of nonconservative processes only. McIntyre and Shepherd (1987) set out a systematic approach to find the conservation laws for perturbations by combining globally conserved properties (such as angular momentum or energy) with properties, called *Casimirs*, which depend only on materially conserved properties. Two key examples are the mass and circulation enclosed by PV contours within isentropic layers, which can be described as functions of the PV and potential temperature (θ) coordinates.

The method proceeds by defining the pseudo(angular)momentum density as

$$P = -r(Z+S) + r_0(Z_0 + S_0) \tag{8}$$

where Z is the specific zonal angular momentum, $S(q,\theta)$ is the Casimir density (as yet unspecified), and the subscript zero refers to the same quantities in the background state. A central aspect is that there is a whole continuum of conserved properties, but the approach is to identify the property where the first-order contribution (in wave amplitude) is zero by construction, therefore ensuring that the resulting wave activity is second order (or higher).

Haynes (1988) showed the full nonlinear result for the primitive equations on the sphere which, in the limit of small wave slope, reduces to the more familiar form of wave activity density (see, e.g., Vallis, 2006, Section 7, for a simple derivation of the first term):

$$P = \frac{1}{2} r_0 Q_y \eta^2 - r' u' \cos\phi + \left(\frac{1}{2} r_0{}^2 q_0 \eta_b{}^2 - r_0 u' \eta_b \right)_b \cos\phi \frac{\partial \theta_{0b}}{\partial y} \quad (9)$$

where $Q_y = r_0 \cos\phi \partial q_0 / \partial y$ is the appropriate mass-weighted meridional PV gradient on the sphere, $y = a\phi$ is the meridional coordinate, and $\eta = -q' / (\partial q_0 / \partial y)$ is defined as the meridional displacement of a PV contour relative to its latitude in the background state. Because the interior PV gradient of the background state is positive, the first term in Eq. (9) is positive definite; therefore, it is a useful measure of the amplitude of Rossby wave activity. Also, southward (negative) displacements give rise to positive PV anomalies through the advection of PV. The second pseudomomentum term, $-r'u' \cos\phi$, is often described as the *gravity wave term* because it is absent under quasi-geostrophic balanced dynamics and does not involve meridional PV fluxes. However, it can be an important player in some large-scale motions. For example, it is the dominant term in equatorial Kelvin wave activity.

The last term is proportional to the meridional gradient of potential temperature along the lower boundary. Because $\frac{1}{2} r_0 \eta_b{}^2$ is positive definite, the boundary term takes the sign of $r_0 q_0 \partial \theta_{0b} / \partial y$, which is typically negative (in both hemispheres) and opposes the interior term (although the terms involving u' are not sign definite). Note that η_b represents the meridional displacement of θ contours along the lower boundary in the full state relative to their position in the background state.

Fig. 2 shows the PV anomaly pattern (color shading) for a particular snapshot of the atmosphere on an isentropic surface (320K) that intersects the tropopause. The Ertel PV anomalies are weighted by the background-state isentropic density because this quantity, $r_0(q - q_0)$, reduces to the quasi-geostrophic PV anomaly field under the approximations of QG theory (see Section 12.4 of Hoskins and James, 2014). Because the density is much smaller at high latitudes in the stratosphere, this downweights the high-latitude negative anomalies; consequently, the positive anomalies are much more prominent. The positive anomalies are in tropopause troughs where the air has been displaced far equatorward and the tropopause is lower than its surroundings. Note that although each trough is distorted differently by advection, it is clear that there are seven centers of action around the midlatitudes. However, they are not equally spaced, with the strong anomalies over the eastern United States and Western Europe being separated the most, and the anomalies over Alaska and the west coast of North America being closest together.

Overall, the pattern projects most strongly onto zonal wave number 6 and the correspondence is highlighted by overplotting the wave number 6 Fourier component of the PV field. This serves as an indication that despite the nonlinearities introduced by advection and Rossby wave–breaking, the dynamics of the large-scale pattern may be interpreted in terms of wave propagation and interaction. This hypothesis will be tested using the ENM approach, which is derived by combining statistical analysis of data with wave-activity-conservation properties.

A similar, but less frequently used, conservation law exists for pseudoenergy. The large-amplitude derivation (Haynes, 1988) begins with a definition similar to Eq. (8), but using the

FIG. 2 Snapshot of the atmosphere (00UT June 23, 2007) on the 320K isentropic surface. The contour lines show PV, Fourier-filtered to zonal wave number 6 (interval 0.5 PVU), and the shading shows the PV anomaly, $r_0(q - q_0)$, from the unfiltered data overlain (negative anomalies in pink; the most positive anomalies in *red*; interval $10^{-5}\,s^{-1}$). This pattern recurred throughout June and July 2007, and the trough (positive PV anomaly) over Western Europe gave rise to extreme monthly rainfall (Blackburn et al., 2008).

energy density in place of the zonal angular momentum. The pseudoenergy conservation law is then obtained using the time symmetry (i.e., steadiness) of the background state. Methven (2013) derived the small-amplitude expression for pseudoenergy, including perturbations near the lower boundary:

$$H = \frac{1}{2} r_0 \left(u'^2 + v'^2 \right) + \frac{1}{2} \frac{h_0}{g p_0 \theta} p'^2 - \frac{u_0}{\cos\phi} P \tag{10}$$

where the first term is the kinetic energy of the perturbations, the second is the available potential energy, and the third is called the *Doppler term* (for reasons which will become apparent), and is proportional to P (including the boundary terms). Note that perturbations are defined relative to the background state by using Eq. (1), where position is identified in isentropic coordinates (λ, ϕ, θ). For example, background pressure can differ from the full state at the same position in θ-coordinates.

Even at large amplitudes, the conservation laws imply certain properties for the perturbations. Consider a coherent disturbance that is neither growing nor decaying, but rather translating primarily along latitude circles. If the background can be defined as both zonally symmetric and steady, then the disturbance must have both a conserved pseudomomentum and pseudoenergy. The ratio of pseudoenergy to pseudomomentum gives the translation speed of the reference frame from which the disturbance appears steady to the observer. In other words, it defines the phase speed of the disturbance (Held, 1985; Zadra, 2000). This is the one of the central properties that we will use to characterize modes of variability in data.

2.3 The Implications of Wave-Activity Conservation for Modes of Variability

The small-amplitude limit of wave activity in Eq. (9) is precisely quadratic in disturbance amplitude, and this property has been exploited to make deductions about the disturbances in general and the properties of normal modes of atmospheric dynamics in particular (Held, 1985):

1. Because the wave activity is globally conserved, its rate of change is zero. Therefore, if the background state is also steady, disturbance amplitude can grow everywhere only if the global wave activity is identically zero. From Eq. (9), this leads to the celebrated Charney-Stern necessary condition for shear instability: The PV gradient must change sign somewhere within the domain. This argument includes the possibility that the negative pseudomomentum could be associated with the boundary wave activity in Eq. (9), as first described for baroclinic instability by Bretherton (1966).
2. Consequently, growing normal modes must have zero pseudomomentum, while neutral modes could have nonzero values.
3. Because normal modes evolve independently, each conserves pseudomomentum on its own. Therefore, if the global pseudomomentum of a superposition of modes is to be conserved, then the normal modes must be orthogonal with respect to pseudomomentum.
4. If the background state is also steady, pseudoenergy will be conserved and conclusions 2 and 3 also pertain to pseudoenergy. Also, because the disturbance energy is positive definite, from Eq. (10), we can obtain the Fjortoft necessary condition for shear instability: the zonal flow and meridional PV gradient must be positively correlated on average across the domain.
5. As argued in Section 2.2, the phase speed of neutral modes is given by the ratio of pseudoenergy to pseudomomentum. Considering the small-amplitude quadratic forms, we see that there are two distinct influences on phase speed:

$$c_p = -\frac{\langle H \rangle}{\langle P \rangle} = \left\langle \frac{u_0}{\cos\phi} P \right\rangle / \langle P \rangle - \langle E \rangle / \langle P \rangle \quad (11)$$

Here, the angle brackets denote an integral over the entire domain. The Doppler term in pseudoenergy gives the rate of advection of the disturbance by the zonal flow, even in the presence of shear. Because the wave frequency can be defined as $\omega = c_p k$, the Doppler term represents the shift in frequency associated with zonal advection by the background flow. Its sign depends only on the sign of the background zonal flow weighted by locations where the wave activity is largest. The second term describes propagation relative to the zonal flow and is proportional to the disturbance energy E. Because the energy is positive definite, the direction of propagation depends on the sign of the mode's pseudomomentum.

Note that it is not immediately obvious how to predict the phase speed of growing normal modes because they must have zero pseudomomentum and pseudoenergy. However, a solution to this problem was obtained by Heifetz et al. (2004), who used wave activity orthogonality to recast the growing and decaying normal modes obtained from complex conjugate normal mode solutions in terms of the linear superposition of a pair of counterpropagating Rossby waves (CRWs). By construction, CRWs are orthogonal with respect to pseudomomentum, and therefore, one CRW has positive and the other CRV negative pseudomomentum such that their sum is zero when only the growing normal mode is present. Therefore, they describe disturbances that propagate in opposite directions (relative to the flow where wave activity is large). However, they are not orthogonal with respect to energy that can grow or decay as a CRW pair evolves.

They are used to give a mechanistic explanation for baroclinic or barotropic instability (a generalization of the Eady model for any unstable parallel zonal flow). The growth rate of normal modes can be expressed in terms of the energy of interaction between the CRWs. Perhaps most important, an expression for the phase speed of a growing normal mode is obtained as the average of the intrinsic phase speeds of the two CRWs:

$$c_{NM} = -\frac{1}{2}\left(\frac{\langle H_1 \rangle}{\langle P_1 \rangle} + \frac{\langle H_2 \rangle}{\langle P_2 \rangle}\right)$$

where $\langle P_1 \rangle = -\langle P_2 \rangle$ in the CRW construction.

3 THE ENM APPROACH TO OBSERVED DATA AND MODELS AND ITS RELEVANCE TO S2S DYNAMICS AND PREDICTABILITY

One way of diagnosing and characterizing the atmospheric S2S variability is to use a *phase space approach*, which has been shown to be very valuable in mathematics, physics, and atmospheric dynamics. The phase space of a geophysical fluid is a space where the state of the flow at a given time corresponds to one unique point. Usually, the phase space for a geophysical fluid can be represented by a steady basic state and superposed wave disturbances that are each represented individually as an oscillation in a 2D-phase plane (with characteristic amplitude and phase). In a nonlinear flow, this decomposition is nonunique (as discussed in Section 2.1), and the phase space trajectory can be complicated.

Here, we will focus on the evolution of waves and the insights that come from the small-amplitude limit for disturbances where linear wave theory applies. This is motivated by the prevalence of wave propagation on larger scales (e.g., Fig. 3), even though stirring by the large scales results in a continuous cascade of PV to smaller scales and the wavelike patterns associated with teleconnections. Even with such a drastic assumption, the proposed ENM diagnostic framework is very insightful and can be applied often to various type of flows with success, including nonlinear flow. Evidence from examples will be discussed next.

In general, waves transfer energy and momentum in flows through dynamical processes that obey the conservation laws introduced in Section 2. In Section 3.1, we will show that conservation laws constrain fundamentally the space-time characteristics of waves and are central to normal mode theory. For a given dissipative and stochastically forced flow, the conservation laws can be used to augment the physical relevance of the statistical principal component analysis (PCA) and its associated EOFs. In the context of waves relative to a steady basic state, if PCA is performed using wave-activity conservation laws, we demonstrate that the EOFs are the normal modes obtained from the linear wave theory.

The latter result permits the development of a statistical and empirical diagnostic framework with a built-in linear wave theory interpretation. It has been named ENM analysis by

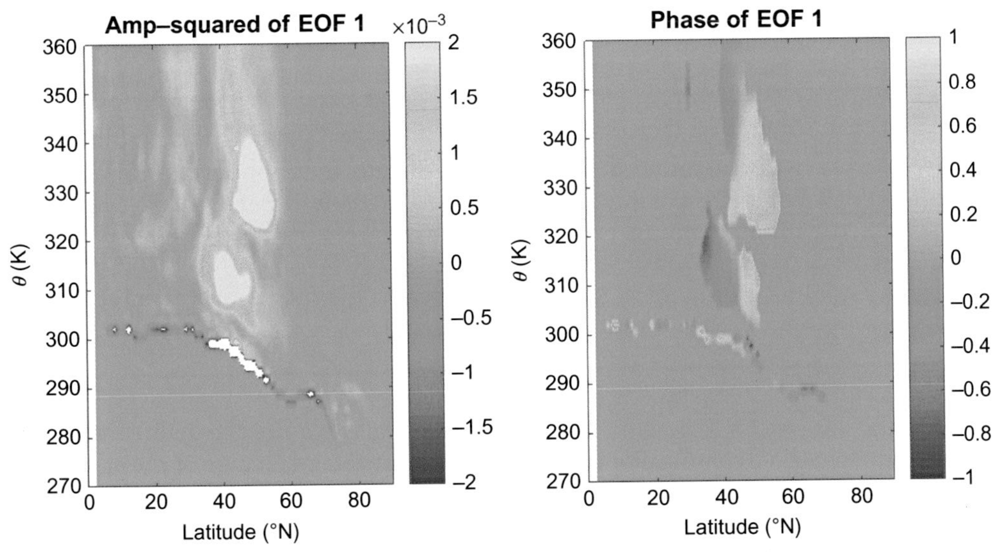

FIG. 3 The structure of the leading ENM pair for zonal wave number 6 during June and July 2007. (A) The amplitude squared of the ENM is its pseudomomentum structure, which is positive in the interior and negative in the boundary domain, which spans the space between the wavy and background state θ contours on the lower boundary (for each θ value). ENM is normalized such that the integral of amplitude squared over the latitude-θ section is unity (units rad^{-1} K^{-1}). (B) The phase of the meridional air parcel displacement associated with the ENM pair (radians/π).

Brunet (1994). In circumstances when the phase speeds obtained from ENM properties are consistent with the phase speed of waves observed by tracking (the ENM phase speed condition; see Eq. 19), we also will show in Section 3.1 that some aspects of ENM analysis are still conceptually and quantitatively relevant to stochastically forced and damped nonlinear flows. In Section 3.2, we will discuss the potential of ENM analysis for diverse applications, and in Section 3.3, we will focus on the ENM analysis of S2S variability.

3.1 ENMs: Bridging Principal Component, Normal Modes, and Conservation Laws

In Section 2.2, we discussed two well-known conservation laws for wave activity in geophysical fluid dynamics (Haynes, 1988). These wave activities are the total pseudoenergy for a steady basic state and the total pseudomomentum for a zonal symmetric basic state. When the basic-state wind is uniform, pseudoenergy and pseudomomentum reduce to total energy and total enstrophy, respectively, but shear in the basic state flow changes the dynamics fundamentally.

In the past, the quest for optimal bases has given rise to many independent rediscoveries of what is now known as *EOF space-time biorthogonal expansions.* A historical review of this topic is given by Sirovich and Everson (1992). The physical interpretation in terms of normal mode bases was explored by North (1984) for an atmospheric linear dynamical system with normal modes satisfying a self-adjoint equation. This work was extended by Brunet (1994) to normal modes of the primitive atmospheric equation for sheared flows (in general not a self-adjoint problem) using the conserved pseudomomentum and pseudoenergy wave activities. In this more general case, the ENMs are not statistical eigenfunctions of a covariance matrix, but rather a solution of a generalized symmetric eigenvalue problem.

The statements about dynamical systems, normal modes, phase speed, and conservation laws discussed here can be found in Brunet (1994), Brunet and Vautard (1996), and Charron and Brunet (1999). For the rest of this discussion, we will assume that a zonal basic state exists, ensuring that pseudomomentum and pseudoenergy are conserved for an unforced and inviscid flow.

Consider a nonlinear dynamical system expressed in the following general form:

$$\frac{\partial X}{\partial t} = G(X) \tag{12}$$

where X is the state vector (e.g., the distribution of variables required to define the flow state), and $G(X)$ represents a dynamical operator that gives the rate of change of the state given the current state. If we linearize around a time-independent zonal basic state X_0 solution of Eq. (12), we obtain

$$\frac{\partial X'}{\partial t} = iG_0X' = iH_AAX' \tag{13}$$

where H_A and nonsingular A are time-independent Hermitian operators if and only if $W_A = \langle X', AX' \rangle$ is a conserved quantity (Charron and Brunet, 1999). This was demonstrated explicitly for the shallow-water model on the sphere for Rossby and gravity waves in Brunet and Vautard (1996). For the rest of this discussion, we will assume that W_A is the total pseudoenergy.

The bracketed term $\langle f, g \rangle = \int f^{\dagger} g dv$ represents an integral that can be over a line, area, or volume. The specific example is given by Eq. (10) for the dynamics described by the primitive equations on the sphere, where the state vector would need to be described by $X' = (u', v', p', r', \eta)$, including the perturbations in the interior and appearing in the lower-boundary terms, and the bracket represents the volume integral over the atmosphere.

Considering the normal mode expansion, in which each Z_n is a monochromatic wave solution of Eq. (13), then

$$X' = \sum_n a_n(t) Z_n = \sum_n a_{n,0} e^{i\omega_n t} Z_n$$

$$\omega_n Z_n = H_A A Z_n \text{ where } \quad \langle Z_n, AZ_m \rangle = \alpha_n \delta_{n,m} \text{ and } \quad \overline{a_n a_m^*} = \delta_{nm} \tag{14a}$$

in which α_n is the total pseudoenergy of the normal mode Z_n, a_n. ω_n is the natural frequency of the normal mode, and the overbar is the time average. This is a generalized eigenvalue problem, and because A and H_A are Hermitian operators and A is nonsingular, we can show for a space of finite dimension that the set of $\{Z_n\}$ forms a complete orthogonal basis under the pseudoenergy metric $M_A = \langle f, Ag \rangle$ (Bai et al., 2000). Hence, the normal mode expansion provides a full solution to the initial value problem associated with Eq. (13). The normal modes span a phase plane in phase space. They represent propagating waves except for stationary waves (i.e., $\omega_n = 0$).

If we linearize around a zonally symmetric basic-state X_0 solution of Eq. (12), then pseudomomentum $W_J = \langle X', JX' \rangle$ is conserved and normal mode solutions satisfy:

$$\omega_n Y_n = H_J J Y_n \text{ where } \langle Y_n, JY_m \rangle = \beta_n \delta_{n,m} \text{ and } \overline{a_n a_m^*} = \delta_{nm} \tag{14b}$$

where β_n is the total pseudomomentum of the normal mode Y_n. Note that the normal modes $\{Z_n\}$ and $\{Y_n\}$ are identical and form a unique basis if the eigenvalue pairs (α_n, β_n) are nondegenerate. We will assume for the rest of this discussion that this is the case. It follows from the framework of complete ensemble of commutating operators (CECO), where a set of many operators have the same eigenvectors if they all commute and their eigenvalues are determined uniquely for each eigenvector (Cohen-Tannoudji et al., 1973). For the primitive equations on the sphere with pseudoenergy and pseudomomentum conservation laws, we can write:

$W_A = cW_J$, and hence, for each normal mode, we have

$$c_n = \frac{\langle Z'_n, AZ'_n \rangle}{\langle Z'_n, JZ'_n \rangle} \tag{15}$$

where c is the mean phase speed of the flow (Held, 1985; Zadra, 2000), as discussed in Section 2.2 with respect to Eq. (11). This relationship also holds true for modified wave activities (by adding a divergent term that is conserved by construction) for individual isentropic surfaces that do not intersect with the surface (Zadra, 2000).

In the geophysical context, it is often more representative of observed flows to have a damped and stochastic forced model (for more on this, see Chapter 6). For the sake of simplicity, we will assume in the following that we have already decomposed Eq. (13) for each zonal wave number k.

Then the forced-dissipative version of Eq. (13) for each zonal wave number (ignore the subscript) can be expressed in the following form:

$$\frac{\partial X'}{\partial t} = iH_A AX' - \gamma X' + \varepsilon \tag{16}$$

where γ is a Raleigh-damping coefficient, and ε is a random forcing that is uncorrelated in time (e.g., Wiener process) and should be square-summable in space.

If we expand Eq. (16) in terms of normal modes $\{Z_n\}$, the complete time series is

$$X'(x,t) = \sum a_n(t) Z_n(x)$$

and the Fourier transform of the time domain in the equation yields the following coefficients:

$$\tilde{a}_n = \frac{\tilde{\varepsilon}_n}{i(\omega - \omega_n) + \gamma} \text{ where } \alpha_n \tilde{\varepsilon}_n = \langle Z_n, A\tilde{\varepsilon} \rangle \tag{17}$$

where the Fourier transformed variables, $\tilde{g}(\omega) = (2\pi)^{-1} \frac{1}{T} \int_{-T/2}^{T/2} g(t) e^{-i\omega t} dt$ and α_n, are defined by Eq. (14a).

When $\gamma = 0$, a solution for Eq. (17) exists only if the Fredholm alternative (Riesz and Sz-Nagy, 1953) is satisfied; hence, $\tilde{\varepsilon}_n|_{\omega=\omega_n} = 0$ for all n.

If the pair (α_n, β_n) as defined in Eqs. (14a), (14b) is nondegenerate, then in the limit $T \rightarrow \infty$ and using Eqs. (14a), (17), we can show that if A and B are nonsingular and we have a time series X' solution of the stochastically forced and damped dynamical system (Eq. 16), then for each zonal wave number k:

$$CJX_n = \beta_n X_n \text{ and } CAX_n = \alpha_n X_n \text{ with } \omega_n = k\frac{\alpha_n}{\beta_n} \tag{18}$$

where $\{X_n\}$ are the normal modes of Eq. (13) and the covariance matrix elements are defined by $C(x, x') = \overline{X'(x)X'(x')^*}$. A normal mode X_n obtained using a covariance matrix approach as in Eq. (18) is named an ENM.

The phase speed relationship (Eq. 15) implies that by knowing the X' time series, we have completely solved the initial value problem of Eq. (13), including its nonhomogeneous damped version with stochastic forcing (Eq. 16). This is possible because the unforced and nondissipative evolution equation (Eq. 13) is a completely integrable Hamiltonian system, which for a given truncation N has $2N$ constants of motion. Associated with the complete ENM basis $\{X_n\}$, we also have a complete orthonormal basis (i.e., principal components) in time $\{a_n\}$, where $\overline{a_n a_m^*} = \delta_{n,m}$. It means that each individual wave/ENM of a different type (e.g., gravity and Rossby waves) obtained through an ENM analysis will span biorthogonal subspaces in space and time with a clear dynamical interpretation.

ENM analysis permits, in practice, the ability to diagnose a specific wave spatial structure and its time evolution, which is not contaminated by other waves present in the flow. In particular, it will efficiently partition fast and slow modes without using any time-filtering technique. In Brunet and Vautard (1996), this has been shown to be very advantageous relative to a standard EOF analysis for simulated linear and nonlinear upper-tropospheric barotropic flows. It should be noted that other statistical techniques are available for studying and predicting atmospheric oscillations in the presence of damping and stochastic forcing. Two examples are the principal oscillation pattern (POP) method (Penland, 1989) and the constructed analog (CA) method (Van den Dool, 1994). They have been successfully used in long-range forecasting (e.g., Van den Dool and Barnston, 1995).

These two statistical methods rely fundamentally on temporal lag techniques for a given variable *time series* in which a linear *regression* equation is used to predict future values based on both the current variable values and the lagged (past period) values. It can be readily shown that, in general, these two are mathematically equivalent. The ENM analysis is fundamentally different because it is not based on time-lag correlation, but only on the existence of conserved wave activities. This makes ENM analysis very robust when dealing with noise. For example, the ENMs derived from Eq. (19) are not affected at all by a random reordering of the time series. This is not the case for POP and CA analyses because the covariance matrix depends only on time-averaging operations and hence is invariant under a reordering of the time series.

Of course, when performing an ENM analysis in practice, we need to assess the validity of the underlying normal mode assumptions for a given time series, such as the choice of basic state, conservation laws, and small-amplitude wave activities. The small-amplitude approximation can be relaxed by using finite-amplitude wave activities, but as discussed in Brunet (1994), the interpretation of the ENM analysis results is definitely more problematic. With the exception of the research done by Brunet (1994), all ENM analyses to date have been performed using small-amplitude wave activities.

In general, we can objectively falsify the effectiveness of ENMs at representing the dynamic of atmospheric flows (whether simulated or observed). It can be done in the context of CECO theory and using temporal lag techniques, as in CA and POP analyses. One important aspect of such an evaluation methodology are the ENM phase speed conditions:

$$\overline{\Omega_n} = -i\overline{\frac{da}{dt}a^*} = k\frac{\langle X_n, AX_m\rangle}{\langle X_n, JX_m\rangle} = \omega_n \tag{19}$$

where $\overline{\Omega_n}$ are the observed mean natural frequencies. As discussed previously, these relations are necessary to demonstrate that an atmospheric flow is an integrable dynamical system. In practice, the principal components are generally not monochromatic, but the ENM phase speed conditions require that for each ENM, the observed mean natural frequency $\overline{\Omega_n}$ (derived from the principal component time series) is equal to its intrinsic natural frequency ω_n. These phase speed conditions have been verified within statistical estimation errors (e.g., dependence on the length of time series) for a wide variety of geophysical flows spanning mesoscale to planetary scales and have provided significant insights for many problems.

3.2 ENM in Applications Relevant to Predictability Across Time Scales

First, we will illustrate ENM analysis in application with a relative simple and explicit example taken from (Brunet and Vautard, 1996): the shallow-water model on the sphere in spherical coordinate. In many aspects, this model is relevant to the S2S prediction problem. It is a global barotropic model that supports Rossby, Rossby-gravity, Kelvin, and gravity waves typical of the midlatitude and tropical-upper-troposphere regions. Then for the ENM analysis of the evolution equation (Eq. 12), we have the following perturbation state vector for a given zonal wave number s:

$$X' = X - X_0 = \begin{pmatrix} u' \\ v' \\ \sigma' \\ P' \end{pmatrix} \text{ with basic state } X_0 = \begin{pmatrix} u_0 \\ 0 \\ \sigma_0 \\ P_0 \end{pmatrix} \tag{20}$$

where u', v', σ' and P' are the nondimensional zonal wind, meridional wind, height, and PV perturbations, respectively. The pseudoenergy A and pseudomomentum J operators are explicitly

$$A = \frac{1}{2}\begin{pmatrix} \sigma_0 & 0 & u_0 & 0 \\ 0 & \sigma_0 & 0 & 0 \\ u_0 & 0 & 1/F_R & 0 \\ 0 & 0 & 0 & -\dfrac{u_0\sigma_0^2}{\dfrac{dP_0}{d\phi}} \end{pmatrix} \text{ and } J = \frac{\cos(\varphi)}{2}\begin{pmatrix} 0 & 0 & 1 & 0 \\ 0 & 0 & 0 & 0 \\ 1 & 0 & 0 & 0 \\ 0 & 0 & 0 & -\dfrac{\sigma_0^2}{\dfrac{dP_0}{d\phi}} \end{pmatrix} \tag{21}$$

where F_R is the Froude number and ϕ the latitude. Hence, from the considerations of the previous section, and with the sole knowledge of these operators, we can perform an ENM analysis for a given shallow-water model time series X by solving the generalized eigenvalue problem (Eq. 18).

Note that the uniqueness and completeness of the ENM basis depend only on the rank of these two matrices (and of the covariance matrix). From their determinant we can show that this tantamount to have bounds a and j for which, for any latitude,

$$0<\frac{\sigma_0^2}{\frac{dP_0}{d\varphi}}<j \text{ and } 0<-\frac{u_0\sigma_0^3}{\frac{dP_0}{d\varphi}}\left(\frac{\sigma_0}{F_R}-u_0^2\right)<a \tag{22}$$

If they are not satisfied, the first and second conditions are related to the Charney-Stern and Fjortoft necessary conditions for shear instability, respectively, as discussed in Section 2.3. But the second condition also states that the wind should not match the local value of the gravity wave phase speed. The second stability criterion was first derived in a somewhat different manner by Ripa (1983), and it guarantees that there is no unstable normal mode. In the presence of unstable normal modes, the ENM analysis is still valid, but it needs a different approach, along the line explained for CRWs in Section 2.3. Of course, the unstable ENMs also can be obtained directly from the kernel of the pseudomomentum generalized eigenvalue problem (Eq. 18) (Martinez et al., 2010a).

Using the spectral shallow model on the sphere, Brunet and Vautard (1996), the ENM phase speed conditions (Eq. 19) were tested for linear and nonlinear regimes and various metrics (e.g., pseudomomentum versus the square of height) for typical Northern Hemisphere winter jets. The EOF diagnostics based on different metrics has clearly shown the advantage of using wave activities.

The time and zonal mean basic state of the time series was also the best choice. It minimizes the perturbation variance, and hence it is the closest to the small-amplitude limit. Of note, the phase speed conditions were satisfied for sheared, low-frequency Rossby waves in the linear regime only, for a relatively high spectral resolution of T100 compared to typical climate model resolutions in the 1990s (T32-64). The modes of variability of the shallow-water model with a realistic radius of deformation were well tuned only for sufficiently high spectral resolution. This example highlights the potential of ENM analysis as a suitable tool for assessing the numerical accuracy of dynamical processes in climate and prediction models. Zadra et al. (2002b) extended the method to multi-layer data from a primitive equation model and applied it to the Canadian Global Environmental Multiscale (GEM) NWP model dynamical core.

An important finding is that ENM phase speed conditions were verified even for nonlinear simulations with wave-breaking events (Brunet and Vautard, 1996). This indicates that over a sufficiently long period of time, the cumulative effect of nonlinear terms can be considered negligible when verifying the phase speed conditions, even if the ENMs are interacting nonlinearly. For example, although Rossby wave–breaking and the complexity of the flow results in the stretching and folding of individual troughs in different ways in Fig. 2, the propagation of the underlying wave pattern of PV anomalies is quantitatively predicted by Eq. (19). Therefore, ENMs are in some respects dynamically relevant to nonlinear problems. The framework for studying nonlinear interaction of normal modes in sheared flows and their relation to wave activities has been established in Vanneste and Vial (1994). This may provide an avenue for studying the nonlinear interactions of ENMs with applications to the study of the predictability and dynamical processes of the S2S phase space and could pave the way for the empirical application of different chaos theory techniques to identify routes to chaos (e.g., KAM theory and period doubling).

In the literature so far, ENM phase speed conditions have been verified when analyzing global atmospheric North Hemisphere atmospheric variability on the 315K isentropic surface (Brunet, 1994), shallow-water-model Rossby waves (Brunet and Vautard, 1996), multi-layer NCEP winter reanalyses (Zadra et al., 2002a), simulated gravity waves in Charron and Brunet (1999), and for hurricane vortex, Rossby wave dynamics (Chen et al., 2003; Martinez et al., 2010a,b, 2011). In these studies, the ENM phase speed conditions were demonstrated within reasonable margins for a large portion of the studied wave activity variances.

Most recently, Methven et al. (2018) have extended the technique to the stratified problem, including the lower boundary, which introduces considerable complexity in using reanalysis data; but it is important due to the negative term introduced in the pseudomomentum and the corresponding term in pseudoenergy. In other words, the propagation of the potential temperature wave along the lower boundary modifies the phase speed of baroclinic waves, as is most familiar in the Eady and Charney models of baroclinic instability (Heifetz et al., 2004). Fig. 3 illustrates the structure of the largest amplitude ENM at zonal wave number 6 obtained from ERA-I data for June–July 2007. Note that in the calculation, the perturbations are defined relative to the MLM background state calculated by Methven and Berrisford (2015). The mode is propagating, and thus described by a pair of ENM structures in quadrature.

It has the largest amplitude in the upper troposphere and the lower stratosphere (straddling the tropopause) and a distinct negative pseudomomentum contribution associated with the boundary terms (here seen sloping along the lower boundary of the background state). Therefore, it is a baroclinic wave, although it has much stronger interior wave activity than the boundary term, and so it does not have zero pseudomomentum, as would be expected for a baroclinic growing normal mode. Its phase changes across the midlatitude jet in the troposphere, which is a signature of the different wave-breaking directions on either side of the jet. Evidence for this can be seen in Fig. 2 in the Fourier-filtered PV field. This particular structure meets the ENM phase speed conditions within a few meters per second (Methven et al., 2018). These studies clearly show the relevance of combining PCA, wave activities (including their associated Elliasen-Palm flux), and normal mode theory to diagnose atmospheric dynamics.

The advantage of the biorthogonal ENMs is evident: It permits a systematic approach for examining the statistics of data and exploring whether a modal perspective is dynamically relevant or not. The ENM phase speed conditions provide important nontrivial information on the dynamic processes (e.g., gravity versus Rossby waves) found in each of the subspaces spanned by the ENMs, as in hurricanes, where there is no separation of time scales between gravity and vortex Rossby waves due to a finite Rossby number (Chen et al., 2003). Similarly, the technique can be used to distinguish among types of wave modes when they are not separated, in terms of spatial scale. In this regard, it holds promise for addressing tropical dynamics in particular.

Falsification of the ENM phase speed conditions can happen for the following reasons: (1) nonlinear effects cannot be neglected; (2) the damping is not Raleigh or stochastic forcing is not Wiener; (3) the wave variability is not well simulated in the model; and (4) some dynamical and physical processes are not represented properly in the wave activities. The latter can be sensitive to the basic state choice, as demonstrated in Brunet and Vautard (1996) with a shallow-water model. In some situations (e.g., baroclinic development), the important contributions of the boundary terms have been neglected (see Zadra et al., 2002a), as pointed out by Methven (2013).

The ENM approach can be used to study the response of a conservative dynamical system to arbitrary forcing. Nowadays, this is a key issue in climate science. According to Cooper and Haynes (2011), a study about the FDT, the response of a conservative dynamical system to steady forcing applied for $t > 0$, and in the limit, $t \to \infty$ would be equal in the ENM framework to $\delta x = -T^{-1}\delta f$, where T is simply the diagonal matrix composed of the intrinsic natural frequency of the ENMs. The advantage relative to the POP approach is that the response can be studied individually for each biorthogonal ENM with a direct physical interpretation. Of course, one of the focuses of climate studies should be on low-frequency dynamical processes like S2S variability.

3.3 ENM Application to the Atmospheric S2S Variability

In Brunet (1994) and Zadra et al. (2002a), large-scale atmospheric variability was shown to be spanned by ENMs, with intrinsic natural oscillation periods ranging from days to months. The ENM phase speed conditions were verified for almost all the wave activity except for some specific ENMs. For example, Brunet (1994) showed that the ENMs associated with Atlantic blocking did not satisfy the ENM phase speed conditions, possibly due to the nonlinear transient feedback and omission of boundary terms in the wave activities.

In Brunet (1994), the S2S North Hemisphere variability on the 315K isentropic surface for 24 winters was characterized quantitatively and empirically within an ENM framework. Fig. 4 shows the distribution of the total observed wave activity per day as a function of the ENM intrinsic natural oscillation period ($\omega_n = c_n k$) for various truncation thresholds. Eight discretelike ENMs were identified, which show a finite contribution to the total wave activity (over 1% individually), with distinct intrinsic periods from 14 to 200 days spanning the S2S time range. They represent around 20%–30% of the total wave activity and are closely linked to large-scale patterns like the AO and Atlantic blocking.

The rest of the wave activity (70%–80%) is represented by a continuous spectrum of ENMs with a peak around 3–5 days associated with transients, storms, and baroclinic waves (diagnosed as very distinct propagating ENM pairs, as shown in Fig. 3). The discretelike ENM pairs were shown to be relatively predictable, with a large e-folding time of 3–5 days. Here, the e-folding time is defined for each ENM pairs as the time average of its amplitude tendency (growth or decay) in the phase plane (Brunet, 1994), which is a good measure of predictability. The continuous spectrum has e-folding times of less than 3 days, which is consistent with predictability theory for baroclinic wave activity (Leith, 1978). The eight discretelike ENM pairs partly control the evolution of the continuous spectrum and the distribution of high-impact weather because they are large-scale features dominating the advection of PV. They are good candidates to span the phase space of a low-order model of S2S variability (see Chapter 6).

For example, Fig. 5 shows the phase-space-probability density of the zonal wave number 2 ENM pair, with an intrinsic period of 35 days for the winters from 1963 to 1987. An asymmetric bimodal signature with one localized small peak and one wide peak can be clearly

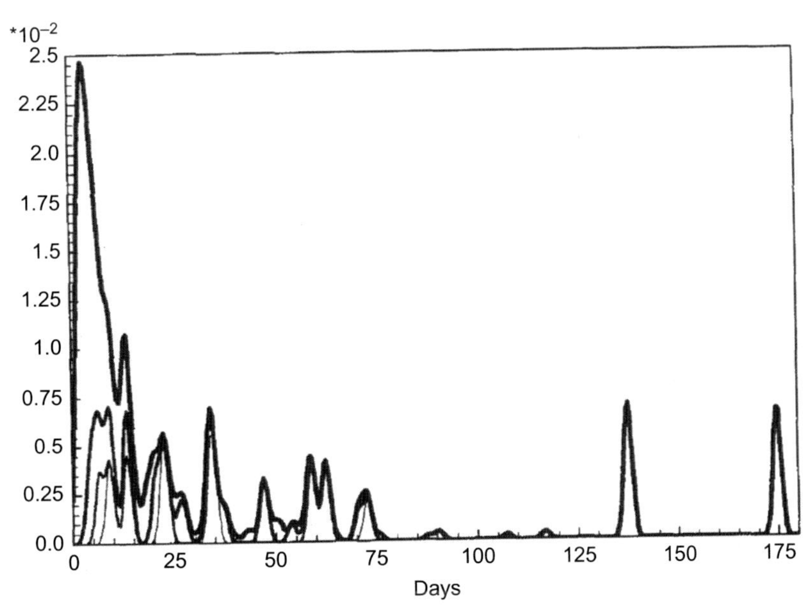

FIG. 4 Distribution of the observed wave-activity spectral density as a function of the ENM oscillation period for the NH 315K isentropic surface. The solid curves, from the thicker to the thinner, correspond to the total wave activity in percentage with individual ENM contribution higher than 0.1%, 0.4%, 0.7% and 1%, respectively. *From Brunet, G., 1994. Empirical normal mode analysis of atmospheric data. J. Atmos. Sci., 51, 932–952.*

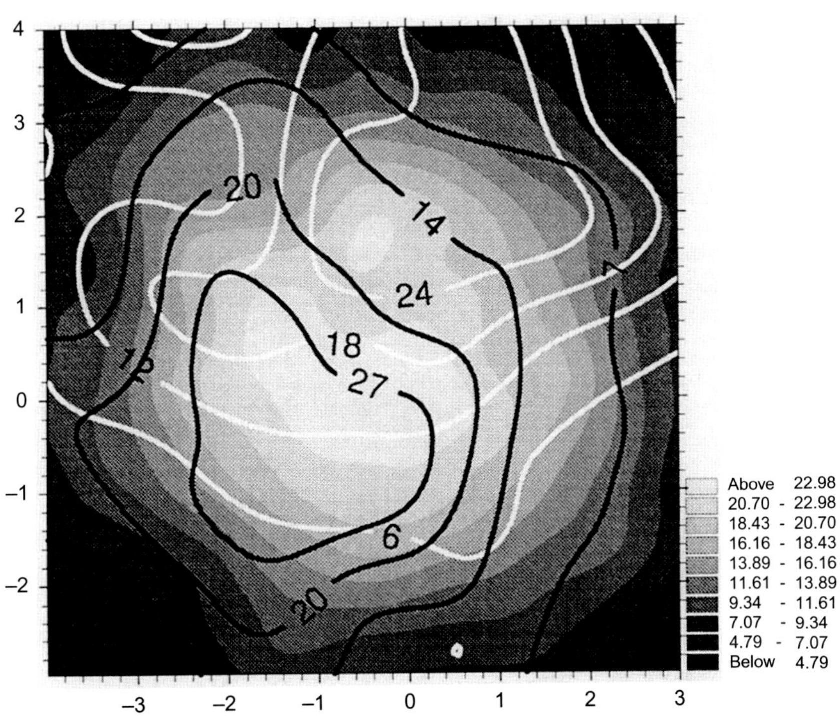

FIG. 5 Phase-space-probability density of the zonal wave number 2 ENM, with an intrinsic period of 35 days. The black and white contour lines correspond to the ZO and BL weather regime events, respectively. The probability density is shaded (units are 10^{-3}). *From Brunet, G., 1994. Empirical normal mode analysis of atmospheric data. J. Atmos. Sci., 51, 932–952.*

observed. A composite of wave activity maps shows that the small peak is associated with Atlantic blocking. This is confirmed in the same image by the probability densities of the zonal (ZO) and Atlantic blocking (BL) weather regime events obtained by Vautard (1990), where the ZO event density has a pattern similar to the main structure of the ENM bimodality.

It is noteworthy that the ZO and BL densities in Fig. 5 show a strong amplification of the ENM pair wave activity for the BL relative to ZO regimes, with an approximate π phase change in the amplitude that can be readily observed by noting that the BL maximum density is farther from the phase-plane origin than the ZO maximum density. This is observed for two other discretelike ENM pairs, and is typical of a resonant process. The presence of weakly unstable hemispherical normal modes in the slow variability is predicted by the wave-mean flow interaction theory of Charney and DeVore (1979); also see Chapter 6. They proposed orographically induced linear and nonlinear resonance mechanisms to explain phase locking and multiple equilibria of weather regimes.

It is recognized that the North Atlantic Oscillation (NAO) temporal variability spans many time scales and is subject to a wide range of atmospheric and oceanic forcing. For example, Molteni et al. (2015) has shown extratropical teleconnections with the Indo-Pacific region that have common atmospheric responses associated with different forcing spanning weeks to interdecadal time scales, which is consistent with resonant behavior. More in-depth studies of discretelike ENM phase space, weather regimes, and resonance mechanisms are needed to make progress on these S2S issues. One step toward this objective was achieved by the three-dimensional (3D) ENM analyses of the observed and simulated global atmosphere by Zadra et al. (2002a,b). The focus of Zadra et al. (2002a) was not S2S variability per se, but the upper-troposphere and lower-stratosphere observed variability along the tropopause. It has been shown that a large part of the wave activity around the tropopause was spanned by ENMs with intrinsic oscillation periods of less than 14 days, and this can be explained using the theory of quasi-modes (Rivest and Farrell, 1992).

Quasi-modes are defined as superpositions of singular modes that are sharply peaked in the phase speed domain but have large-scale (delocalized) structures, as opposed to singular modes, which represent sheared disturbances advected by the flow. They are often reminiscent of monochromatic discrete modes (neutral or unstable) that have been displaced into the continuous spectrum by the modification of the wind basic state (e.g., addition of wind shear) or dynamical processes like f-plane versus β-plane (Zadra, 2000). Quasi-modes are often weakly damped by critical layer stirring (Briggs et al., 1970; Schecter et al., 2000, 2002; Schecter and Montgomery, 2006; Martinez et al., 2010a) and maintain their energy for relatively long periods. They are easily excited by external forcing due to their large spatial structure, which is typical of discrete modes. They are not easily computed or identified by numerical or analytical methods (e.g., identification of relevant Landau poles), but they have been readily identified by ENM analyses in hurricane simulation (Martinez et al., 2010a) and atmospheric reanalysis (Zadra et al., 2002a) diagnostics. The latter study was able to identify leading modes with dipolar pressure patterns along the summer hemisphere tropopause that have well-defined phase speeds and decay rates of a few days, which can be explained by the theory of quasi-modes (e.g., wave number 5 with phase speed of 12 m/s and 3-day decay rate).

It is quite possible that the eight discretelike ENMs spanning the S2S variability in the 2D barotropic study of Brunet (1994) are quasi-modes. Further studies are required to confirm this hypothesis, and these will probably need to be 3D ENM analyses because Zadra et al. (2002a,b) have identified a relatively larger number of discretelike ENMs with intrinsic periods of over 14 days spanning the S2S variability (e.g., more than a half-dozen for the zonal wave number 1 alone).

The ENM also could be used in the atmosphere-ocean S2S context where, in general, pseudoenergy is expected to be conserved, but not pseudomomentum. For example, the latter is not conserved when an oceanic basin has irregular boundaries because the zonal symmetry is broken. The ocean ENM diagnostic could be used to look at the MJO atmosphere-ocean coupled problem, with the ocean limited to the mixing layer. This would be the first step toward looking at ENMs of the ocean-atmosphere with sub-seasonal to multidecadal time scales. It is noteworthy that wave-activity study of the ocean is almost nonexistent except for the work of P. Ripa (Shepherd, 2003).

4 CONCLUSION

The first important objective of this chapter was to tackle the problem of diagnosing S2S atmospheric variability by splitting the diabatic and adiabatic flow components using fundamental principles from geophysical fluid dynamics. This was shown to be possible by taking advantage of the MLM theory based on the conservation of PV and potential temperature. The MLM partitioning of S2S variability in terms of slow diabatic processes, such as radiative forcing, and large adiabatic dynamical processes leads to the second important goal of our proposed methodology, which is to be able to extract dynamical modes of S2S variability from observed global data and model simulations with coherent space-time characteristics using fundamental properties that can be deduced from theory.

We demonstrate in this chapter that ENM analysis, with its built-in characteristics based on conservation laws, PCAs, and normal mode theory, provides an appropriate theoretical framework. ENM analysis is able to frame S2S scientific studies suitably and bring new perspectives (e.g., partitioning the S2S variability in fast and slow modes based on the ENM intrinsic phase speed). For example, the use of the ENM technique to date has revealed that a small number of structures dominate the observed variability at lower phase speeds than baroclinic waves. However, there are many unanswered questions regarding the nature of the modes. Are the ENM structures consistent with the structure of normal modes? Can we confirm or rule out their quasi-modal interpretation? Are they robust in a very long time series? The background state has a strong seasonal variation, but the calculation assumes a steady background, so should the modes be obtained for each season separately? What is the role of boundary wave activity?

Once the discretelike ENM phase space spanning the S2S variability has been characterized properly, we will be well positioned to address the S2S predictability problem. To address the predictability challenges, we need to understand the following points better:

- The extent to which a low number of distinct modes describe variability on S2S time scales and the robustness of those structures from year to year.
- How these discretelike modes interact with one another, with faster disturbances, and with the background state through nonlinear interactions. The role of physical mechanisms,

such as wave resonance, multiple equilibria, attractor sets, and stable and unstable limit cycles (see Chapter 6 for more on this topic), is also important.

- The role of the slow modes in predictability on S2S time scales.
- How the ENM approach can be used to examine tropical-extratropical interactions and the teleconnections that result in longer-term predictability in the extratropics.
- The degree to which ENMs present a reduction of atmospheric dynamics in terms of the average response of the system to stochastic and other forcings and use the phase space of ENMs to understand changes in S2S variability with climate change, including weather-regime responses to climate change (structure and occurrence), by performing intercomparisons of global and regional climate models as recommended by Palmer (1999).

Using statistical and theoretical research programs together will improve our knowledge of the S2S forecast problem and point the way toward exploiting new sources of predictability. For example, Brunet (1994) and the subsequent applications of ENM analysis to the observed and simulated global atmosphere (Zadra et al., 2002a,b) provided the guidance needed to look at the problem of the MJO and the NAO two-way interaction problem through teleconnections (Lin et al., 2009) and its impact on NWP skill (Lin et al., 2010a; Lin and Wu, 2011). As in the previous example, to make future advances in the S2S forecast problem, we also will need to use a hierarchy of GCMs of increasing complexity to gain the necessary dynamical and physical insights (e.g., Derome et al., 2005; Lin et al., 2007).

We believe that the S2S forecast problem is at the forefront of the weather and climate predictability continuum, where S2S variability can be represented by a finite number of relatively large-scale discretelike modes. These discretelike modes evolve in a complex manner through nonlinear interactions with themselves and transient eddies and weak dissipative processes. The sources of predictability are a mixture of fast adiabatic and slow diabatic processes that can be differentiated and diagnosed properly with a phase space approach based on ENM and MLM theories. Although the approach described here is not unique, the key to better prediction of S2S variability and weather regimes in a changing climate lies in improved understanding of the fundamental nature of S2S phase space structure and associated predictability arising from dynamical processes.

Acknowledgments

We thank Martin Charron, Yongsheng Chen, Yosvany Martinez, and Ayrton Zadra for their important contributions to ENM analyses. G. Brunet would like to express thanks to Hai Lin for his continuous leadership, discussions, and support throughout the years on the S2S diagnostic and prediction problem. Also, many thanks for the work of Tom Frame in cosupervision of the master's dissertation projects of Lina Boljka and Carlo Cafaro, which have advanced the ENM analysis, including the MLM background-state and lower-boundary-wave activity, and led to Fig. 3, and Paul Berrisford, for his contributions to the development of the background-state calculation. We thank Fréderic Vitart and Andrew Robertson for their helpful comments.

PART II

SOURCES OF S2S PREDICTABILITY

CHAPTER

5

The Madden-Julian Oscillation

Steven J. Woolnough

National Centre for Atmospheric Science, University of Reading, Reading, United Kingdom

OUTLINE

1 INTRODUCTION

The Madden-Julian oscillation (MJO), sometimes known as the intraseasonal oscillation (ISO), is the leading mode of intraseasonal variability in the tropical climate system. It is a planetary scale, eastward-moving disturbance with a broad spectral peak around 40–60 days, which modulates the tropical deep convection, and hence the tropical precipitation. The heating and divergent circulation associated with the anomalous deep convection provides a tropical source of Rossby waves, which can propagate into the extratropics and thus influence midlatitude weather systems. The global scale of the impacts of the MJO makes it a leading source of potential predictability on sub-seasonal timescales.

Sub-seasonal to Seasonal Prediction
https://doi.org/10.1016/B978-0-12-811714-9.00005-X

The Intra-Seasonal Oscillation (ISO) was first identified by Madden and Julian (1971, 1972) through spectral analysis of upper-air wind observations. They identified a broad spectral peak at around 40–50 days in the zonal wind at Canton Island (3 degrees S, 17 degrees W) with coherent, out-of-phase oscillations between upper- and lower-tropospheric winds. Analysis of additional tropical stations allowed Madden and Julian to identify the planetary wave number 1–2 scale and eastward propagation of the disturbance, which they described as "an eastward movement of large-scale circulation cells oriented in the equatorial (zonal) plane." From the divergent circulation, they hypothesized variations in convection associated with the circulation anomalies and Fig. 16 of Madden and Julian (1972) provides a remarkably accurate description of the zonal structure of the MJO and its associated convective anomalies. Early analysis of satellite remotely sensed clouds (Gruber, 1974; Zangvil, 1975) supported their description of the convective anomalies associated with the MJO.

During boreal summer, the ISO in the Indian and West Pacific oceans has both strong eastward- and northward-propagating components (e.g., Lau and Chen, 1986; Lee et al., 2013). This northward propagation is closely associated with the active/break cycles of the Asian summer monsoon. These northward-propagating events are often, but not always, associated with eastward-moving MJO events; Wang and Rui (1990b) estimated that about 50% of the northward-moving events are associated with MJO events. The distinct nature of the intraseasonal variability during boreal summer has led to this variability being considered a separate mode of variability, usually known as the Boreal Summer Intraseasonal Oscillation (BSISO). While there are clearly strong links between the MJO and BSISO, this chapter will focus on the eastward-propagating MJO, noting its distinct seasonal characteristics where appropriate.

This chapter will summarize the observed characteristics of the MJO, its impacts on the weather in the tropics and extratropics, theories about its maintenance and propagation, its representation in weather and climate models, and the current capability of operational sub-seasonal prediction systems for MJO prediction. The concluding section will highlight some current priorities for MJO research relevant to sub-seasonal prediction.

Since its discovery in the 1970s, there has been an extensive body of literature on all aspects of the MJO, and it is beyond the scope of this chapter to cover the breadth and depth of that literature, but over this period, there have been a number of reviews, including Madden and Julian (1994) and Zhang (2005); two major field campaigns, TOGA-COARE (Webster and Lukas, 1992) and CINDY/DYNAMO (e.g., Gottschalck et al., 2013); and a comprehensive book on the subject called *Intraseasonal Variability in the Atmosphere-Ocean Climate System* by Lau and Waliser (2005), and published in a second edition Lau and Waliser (2012).

2 THE REAL-TIME MULTIVARIATE MJO INDEX

Characterizing the nature of the MJO requires a method of identifying the MJO in observational (or model) data. Wave-number frequency decomposition of the equatorially averaged outgoing longwave radiation (OLR) (or related variables; Fig. 1) shows the characteristic organization of tropical convection by the theoretical equatorial wave modes of Matsuno (1966), as well as a distinct maximum in wave numbers 1–5 and periods 40–60 days

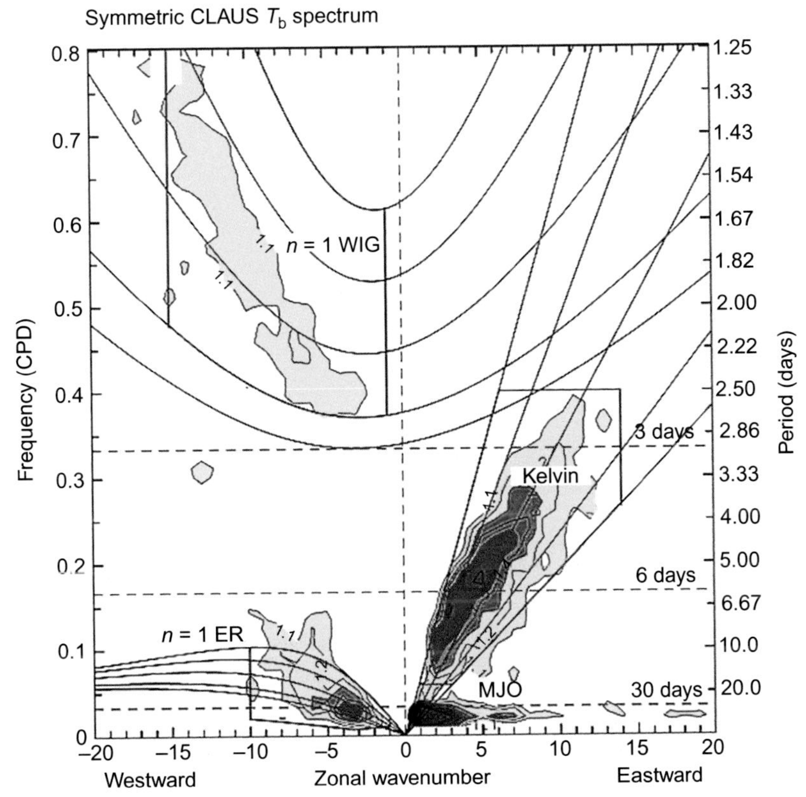

FIG. 1 Wave-number frequency power spectrum of the symmetric component of the Cloud Archive User Service (CLAUS) brightness temperature (T_b) for Jul. 1983 to Jun. 2005, summed from 15 degrees N to 15 degrees S, plotted as the ratio between the T_b power and the power in a smoothed, *red-noise background*. Contours are every 0.1 beginning at 1.1, where the signal is significant at greater than the 95% level. Dispersion *curves* for the $n = 1$ equatorial Rossby (ER) wave, Kelvin wave, and $n = 1$ inertia gravity (IG) wave are plotted for equivalent depths of 8, 12, 25, 50, and 90 m. *Reproduced from Kiladis, G., Wheeler, M., Haertel, P., Straub, H., Roundy, P., 2009. Convectively coupled equatorial waves. Rev. Geophys. 47, RG2003.*

associated with the MJO. The distinct spectral peak in frequency of the MJO led to a number of approaches based on band-pass-filtered data, with a range of filter widths from 20–100 days (e.g., Slingo et al., 1996) to 30–60 days (e.g., Knutson and Weickmann, 1987). The simplest approach to characterize the nature of the intraseasonal variability is to regress fields against the band-passed-filtered variable, usually the OLR at a base point (e.g., Hendon and Salby, 1994), to reveal both the temporal and spatial structures of the associated variability, and this approach is still often used to characterize the nature of the ISO in climate models (e.g., Jiang et al., 2015) and then compare it to observations. Other techniques have used empirical orthogonal function (EOF) analysis (e.g., Wilks, 2011), typically of OLR, to identify the leading spatial modes of variability. When performing this analysis on the intraseasonal-filtered OLR, the two leading principal components form a pair that describes the eastward-propagating MJO. The principal component time series can then be used to construct regression maps for other variables as a function of the phase of the MJO (e.g., Matthews, 2000).

While these techniques have yielded an enormous amount of insight into the temporal and spatial characteristics of the MJO, their reliance on filtering in time makes them unsuitable for identifying the MJO in real time. The desire for an MJO index that could be used for real-time monitoring and forecasting applications led to the development by Wheeler and Hendon (2004) of the real-time multivariate MJO (RMM) index.

The RMM index is based on the two leading EOFs of combined fields of daily anomalies of near-equatorial (15 degrees S-15 degrees N) averaged OLR and zonal winds at 850 and 200 hPa. First, the raw daily fields are converted to anomalies from the seasonal cycle by the removal of a smooth annual cycle computed from the mean and the first three harmonics of the annual cycle. The influence of interannual variability is removed by linear regression against a sea surface temperature (SST) time series related to the El Niño Southern Oscillation (ENSO), and then the removal of the mean of the previous 120 days.

The dominance of the MJO in the tropical sub-seasonal variability and the coherent relationship between the baroclinic zonal wind structure and the convection as identified by Madden and Julian (1971, 1972) means that the anomaly structure associated with the MJO emerges as the leading pair of EOFs (Fig. 2) without the need for temporal filtering.

The planetary scale, structure of the anomalies, and out-of-phase relationship between the upper- and lower-level winds in near-quadrature are in clear agreement with the structure identified in Madden and Julian (1972). EOF 2 describes a dipole of active and suppressed convection over the Indian Ocean and West Pacific, while EOF 1 describes active and suppressed phases of the MJO over the Maritime Continent region.

The RMM indices are obtained by projecting daily observed anomalies, computed as in the construction of the original EOFs,[1] onto these EOFs to give two indices, RMM1 and RMM2, which describe the phase and amplitude of the MJO. The indices are typically shown in an RMM phase diagram, as in Fig. 3 for the season Oct. to Mar. 2003–04. In this example, an MJO event emerges in phase 5 in the middle of October and propagates through to phase 8 before decaying at the beginning of November. At the beginning of December, a new MJO event develops in the Western Indian Ocean (phase 2) and propagates around the globe, strengthening in phases 7–8. This event continues to propagate back through the Indian Ocean and subsequently decays as the active convection reaches the Maritime Continent (phase 5). Around the second week of March, a new MJO event emerges, with its active phase in the Western Indian Ocean (phase 3) and propagates through the Western Pacific. Fig. 4 shows the equatorial (15 degrees N-15 degrees S) OLR, and 850- and 200-hPa zonal wind anomalies from the climatological seasonal cycle for the same 2003–04 winter season. The eastward-propagating OLR and baroclinic zonal wind structure associated with these MJO events are clearly visible.

A number of studies have noted that the RMM index is dominated by the projection onto the wind fields (e.g., Kiladis et al., 2014; Kerns and Chen, 2016) and that the convective signal contributes only a small component to the RMM index. Wheeler and Hendon (2004) noted that the benefit of a multivariate index is that there is stronger coherence and better spectral

[1] In operational forecasting practice, following Gottschalck et al. (2010), the step removing the ENSO signal by linear regression onto the ENSO SST anomalies is omitted.

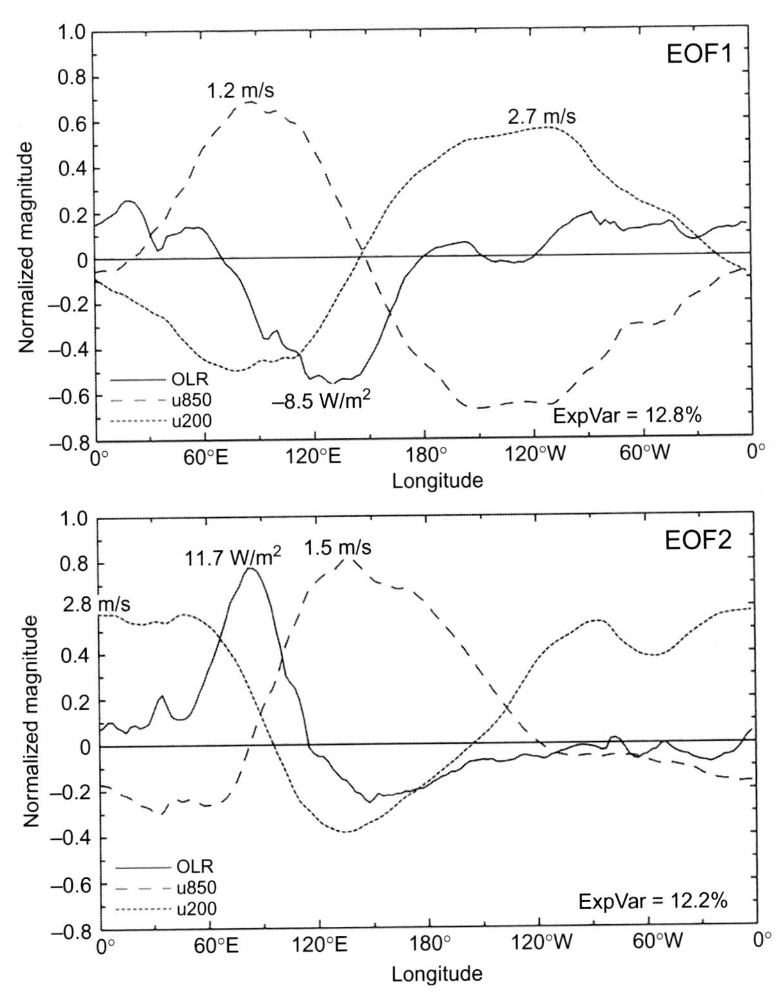

FIG. 2 Spatial structures of EOFs 1 and 2 of the combined analysis of OLR and 850- and 200-hPa zonal winds (u850, u200). As each field is normalized by its global (all longitudes) variance before the EOF analysis, its magnitude may be plotted on the same relative axis. Multiplying each normalized magnitude by its global variance gives the field anomaly that occurs for a 1-standard-deviation perturbation of the PC, as given for the absolute maxima of each field. The variance explained by the respective EOFs is 12.8% and 12.2%. *Reproduced from Wheeler, M., Hendon, H., 2004. An all-season real-time multivariate MJO index: development of an index for monitoring and prediction. Mon. Weather Rev. 132, 1917–1932.*

characteristics compared to single-variable indices. To better capture the convective signal of the MJO, Kiladis et al. (2014) developed an OLR-based index (OLR MJO Index; OMI), which could be applied in real time. While the general characteristics of the MJO are similar between the two indices, differences in individual MJO events can be quite large (even missing in one or the other). This OMI may be more useful for the RMM for applications in which the precipitation or diabatic heating are of primary importance, such as, in the role of the MJO as a Rossby wave source.

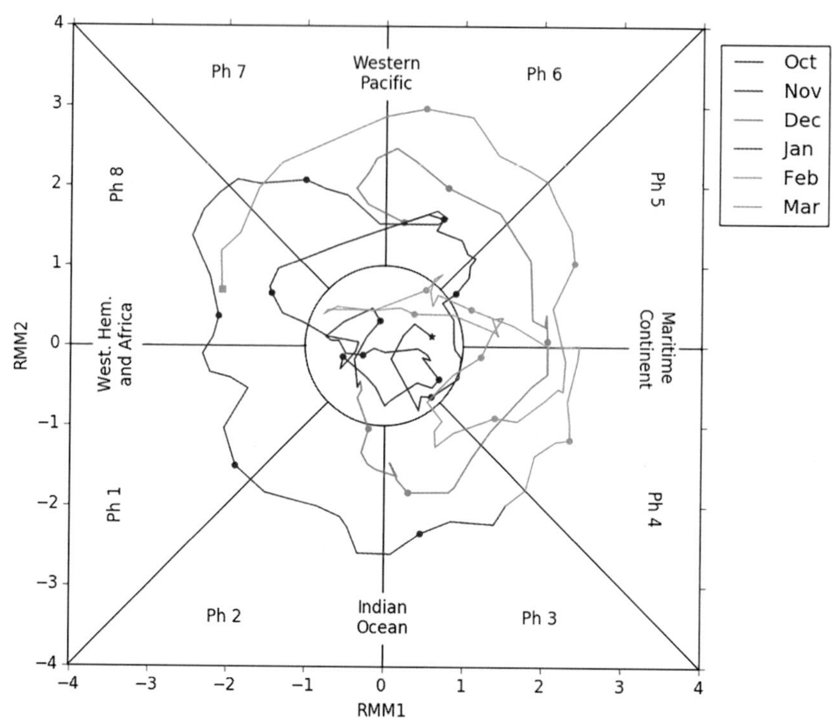

FIG. 3 Example of RMM phase diagram for Oct. 1, 2003 (the *blue star*) to Mar. 31, 2004 (the *cyan square*). The *colored circles* are plotted every 7 days from Oct. 1. Days inside the *unit circle* are considered to have weak MJO activity. An MJO event propagates counterclockwise around the diagram. The diagram is divided into eight phases, and the regions of active convection associated with the MJO are labelled.

3 OBSERVED MJO STRUCTURE

A number of studies have characterized the structure and propagation characteristics of the MJO, using a variety of statistical techniques, including linear regression and composites (e.g., Rui and Wang, 1990; Hendon and Salby, 1994; Matthews, 2000; Sperber, 2003; Kiladis et al., 2005; Waliser et al., 2009; Wang et al., 2018). Waliser et al. (2009) described an activity of the CLIVAR Madden-Julian Oscillation Working Group to develop a set of diagnostics of the MJO for comparison to model simulations, and the associated website[2] has figures with an extensive set of diagnostics. While such diagnostics present a statistical or mean view of the MJO, each individual event varies in its initiation location, amplitude, phase speed, and duration.

Matthews (2008) analyzed MJO events between 1974 and 2005 and characterized them as either primary or secondary MJOs. Secondary MJOs are those that have already passed through each phase of the MJO and return to their starting phase, while primary MJOs are those that have no preexisting MJO activity. Matthews (2008) found that about 40% of

[2]See http://climate.snu.ac.kr/mjo/_diagnostics/index.htm.

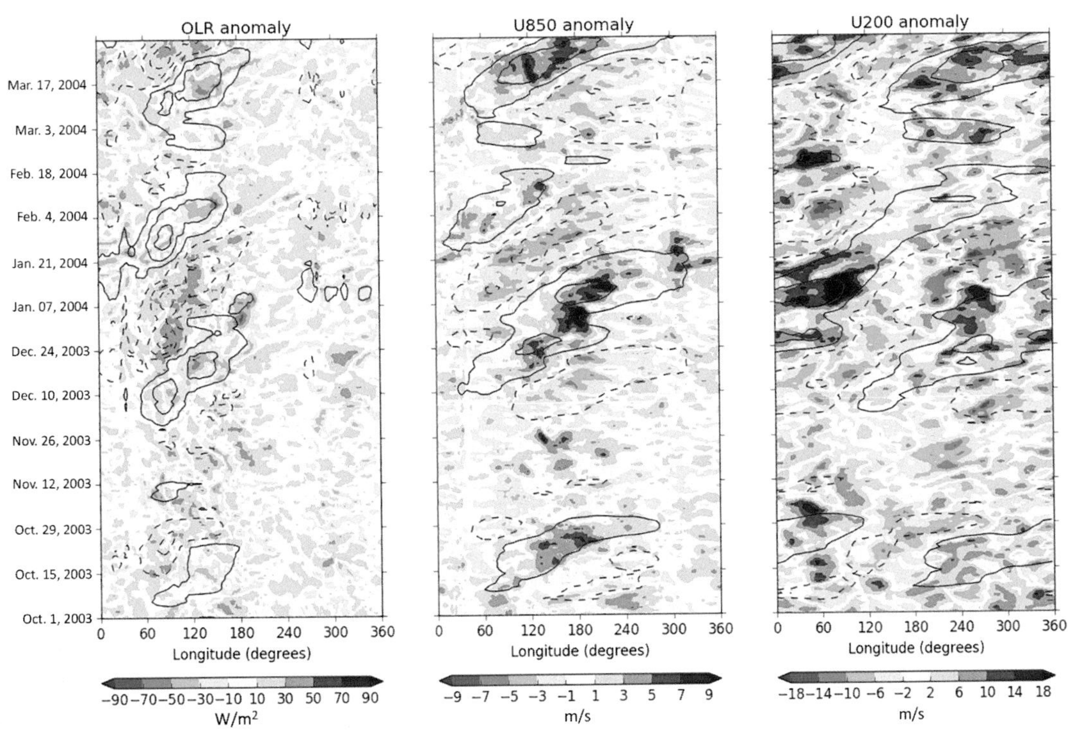

FIG. 4 Longitude-time plots of the equatorially averaged (15 degrees N-15 degrees S) OLR (Lee and NOAA CDR Program, 2011), and 850- and 200-hPa ERA-Interim (Dee et al., 2011) zonal winds for Oct. 1, 2003 to Mar. 31, 2004. The *colors* represent the total anomaly from the climatological seasonal cycle (mean + 3 harmonics) of Oct. 1996 to Sep. 2015. The *contours* represent the projection of the RMM indices onto these fields. For the zonal winds, the contours are as for the full anomaly with the negative contours *dashed*, while for the OLR, the contours are every 10 W/m^2 beginning at 5 W/m^2 with the negative contours (enhanced convection) *solid*.

primary MJO events initiate in the Indian Ocean. Analysis of the starting phase of events with RMM >1 for at least 15 days beginning in Oct.-Mar. of 1980–81 through 2015–16 has around 43% beginning in phases 1–3. This fraction is relatively insensitive to the minimum duration of an event (between 10–20 days) or whether the RMM amplitude can drop below 1 for a short 1- to 2-day period within the event, although the total number of events does vary. Kiladis et al. (2014) compared the initiation of MJO events using the RMM and the OMI indices and found that using an OLR-based index increases the fraction of events that begin in the Indian Ocean. While the Indian Ocean is marginally preferred over other locations, MJO events can and do initiate in any phase. Matthews (2008) went on to identify precursor signals for the initiation of primary MJO events and found that for primary MJO events that initiate in the Indian Ocean, some (but not all), are preceded by a suppressed convective anomaly that develops in situ, peaking about 15 days prior to the initiation of the MJO event. Later, Straub (2013) investigated the initiation of the MJO using both OLR-only and OLR- and circulation-based indices. They found that contrary to Matthews (2008), there is no suppressed convective signal over the Indian Ocean, but rather a weak, suppressed signal over the Maritime

Continent; and that although there is no convective signal, much of the circulation structure is similar to that for successive events, and particularly for initiation in the Western Indian Ocean, there are lower-level easterly and upper-level westerly wind anomalies in the Indian Ocean. Straub (2013) raised a note of caution in that there are significant differences in the dates and locations of MJO initiation depending on the choice of index, which makes diagnosing the precursors of MJO initiation difficult. While Matthews (2008) found some evidence of equatorward propagation of extratropical disturbances prior to successive MJO events in the Indian Ocean, he found no evidence of such a link prior to primary events. However, a number of studies have shown evidence of links between the extratropical circulation, including Lin et al. (2009), who showed a link between the NAO and subsequent MJO phases over the Atlantic and Africa; and Vitart and Jung (2010), who showed improved MJO prediction skill in the European Centre for Medium-Range Weather Forecasts (ECMWF) model, when the Northern Hemisphere extratropics is relaxed back to analysis.

Figs. 5 and 6 show the composite structure of each phase of the MJO based on the RMM indices of Wheeler and Hendon (2004). The composites are produced from daily anomalies, constructed as for the calculation of the RMM indices (i.e., removing the mean seasonal cycle and the average of the previous 120 days) for the extended winter (November-April) of the seasons 1996–97 to 2013–14. Each phase composite is constructed by averaging over the days in that phase with RMM amplitude >1.

While each MJO event varies, the "canonical" MJO develops as a positive convective anomaly in the Indian Ocean with suppressed convection over the Western Pacific and the South Pacific Convergence Zone (SPCZ) (e.g., phases 1–2 in Figs. 5 and 6). This convective anomaly and its circulation anomalies intensify and propagate eastward at about 5 m/s. Over the complex topography of the Maritime Continent, where the MJO interacts with the steep orography and the strong diurnal cycle of land-sea breezes and convection (e.g., Wu and Hsu, 2009; Peatman et al., 2014), the convective signal weakens, particularly over land (e.g., phases 4–5). In the Maritime Continent region, the convective signal of the MJO over land often leads the oceanic convection by a phase (Peatman et al., 2014). About 20–30 days after phase 1, the active convection is in the West Pacific and SPCZ and the convection is suppressed in the Indian Ocean (e.g., phases 5–6), although not all MJO events propagate from the Indian Ocean into the West Pacific.[3] Once over the cooler waters of the central and eastern Pacific, the convective anomalies of the MJO weaken, but the circulation anomalies persist, particularly at upper levels, and continue to propagate eastward. Away from the convectively active region over the Indian Ocean and West Pacific, the propagation speed increases to around 10 m/s.

The OLR anomalies in the composites show the envelope of active convection associated with the MJO in a composite sense, but within that envelope of activity is embedded a range of smaller-scale convective systems (e.g., Nakazawa, 1998; Chen et al., 1996), including equatorial Kelvin waves propagating eastward at phase speeds of 15–20 m/s, faster than the MJO convective envelope, and westward-moving waves with lifetimes of 1–2 days.

[3]Of the 52 events with RMM amplitude >1 for at least 15 days which initiate in phases 1–3 in Oct.-Mar. of 1980–81 to 2015–16, only 90% reach phase 4 and 63% reach phase 6 with amplitude >1, these percentages are relatively insensitive to the minimum duration of the event or allowing the MJO to decay into the unit circle for 1–2 days. Kim et al. (2014a) found that the active phase of convection in the Indian Ocean is more likely to propagate through the Maritime Continent if it is accompanied by strongly suppressed convection in the West Pacific.

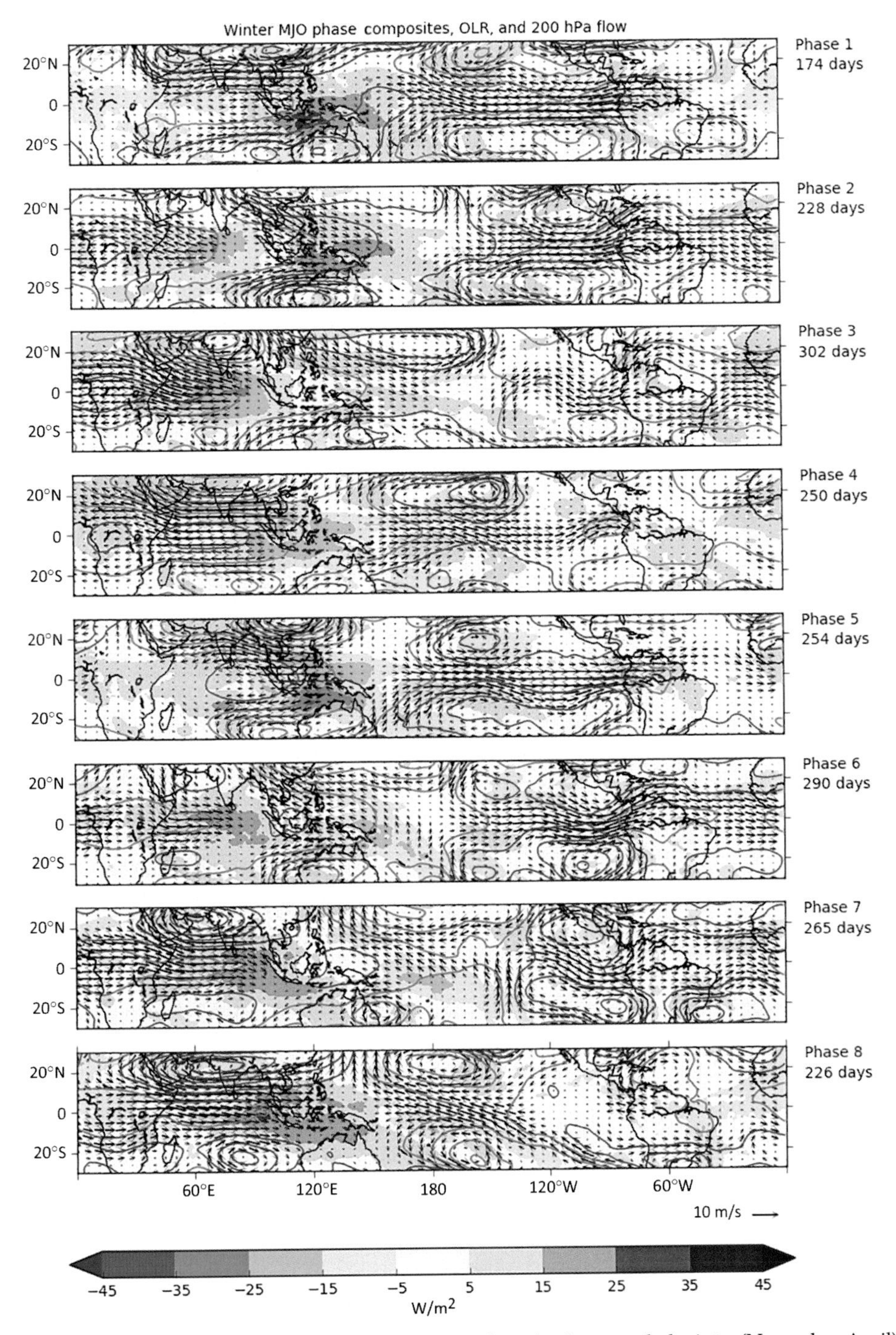

FIG. 5 Composite anomaly structure of the MJO by RMM phase for the extended winter (November-April) for the seasons 1996–97 to 2013–14. The *colors* show the OLR (Lee and NOAA CDR Program, 2011). The *contours* show the 200-hPa ERA-Interim (Dee et al., 2011; European Centre for Medium-Range Weather Forecasts, 2014) geopotential, where the contours begin at $\pm 50\ m^2/s^2$ and at $100\ m^2/s^2$ intervals, with positive contours *green* and negative contours *magenta*; vectors show the 200-hPa ERA-Interim winds.

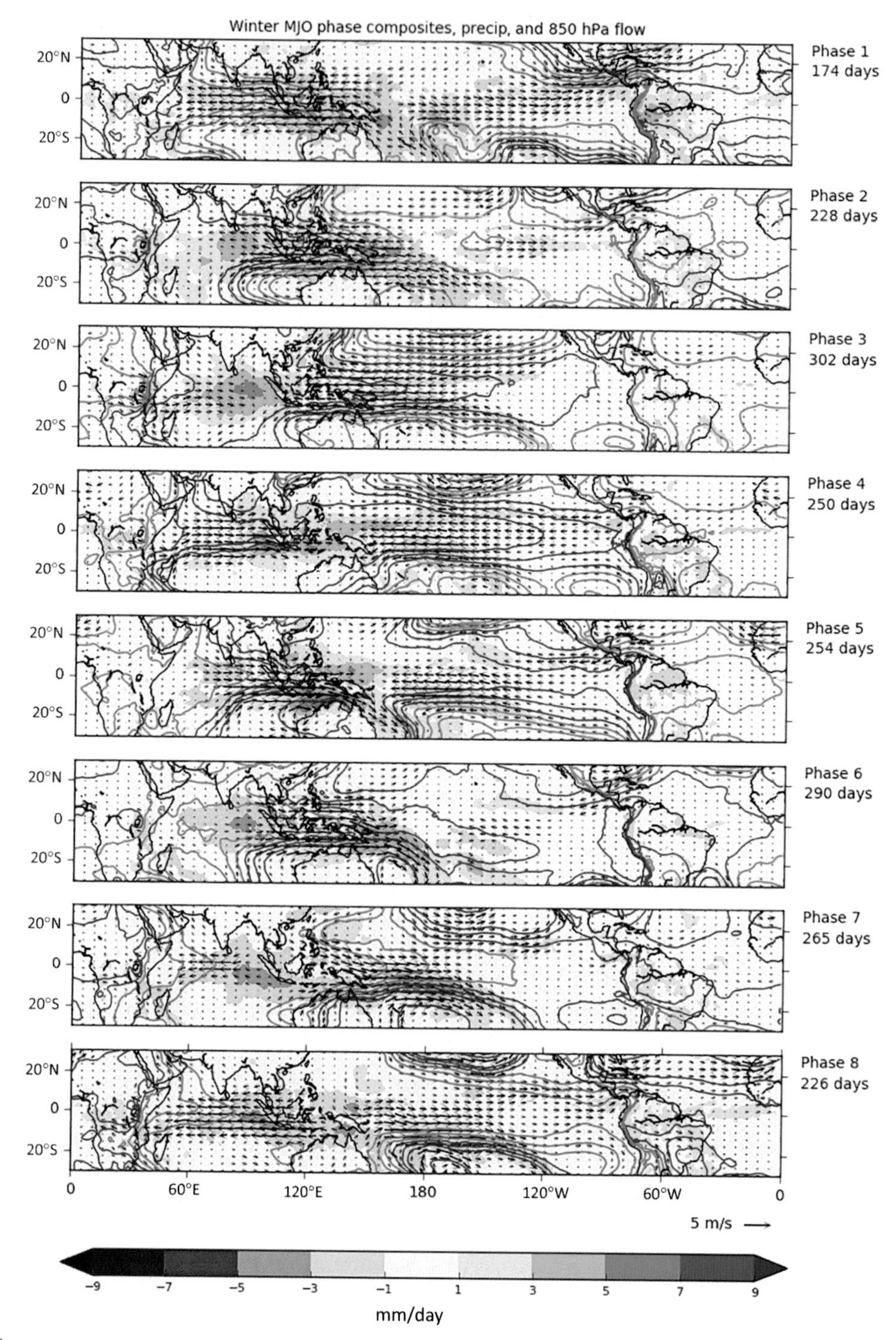

FIG. 6 Composite anomaly structure of the MJO by RMM phase for the extended winter (November-April) for the seasons 1996–97 to 2013–14. The *colors* show the GPCP precipitation (Huffman et al., 2001). The *contours* show the 850-hPa ERA-Interim (Dee et al., 2011) geopotential, where the contours begin at ±10, 30, 50 m^2/s^2 and then are at 100 m^2/s^2 intervals, with positive contours *green* and negative contours *magenta*; vectors show the 850 hPa ERA-Interim winds.

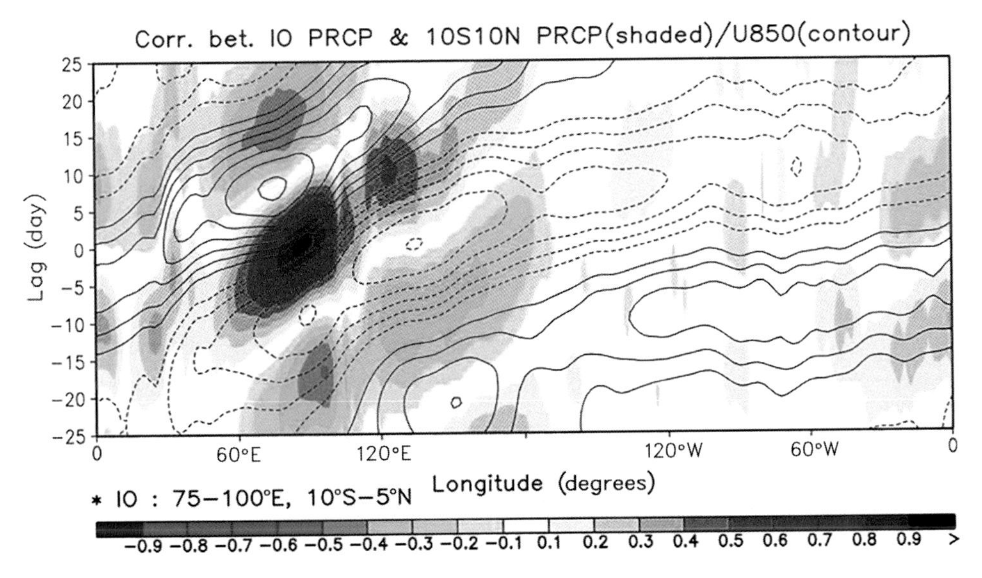

FIG. 7 November to April lag-longitude diagram of 10 degrees N-10 degrees S averaged intraseasonal (20–100-day) precipitation anomalies (*colors*) and zonal wind anomalies (*contours*) against intraseasonal precipitation averaged over the Indian Ocean (10 degrees S-5 degrees N, 75–100 degrees E). *Contours and colors* are plotted every 0.1. The zero line is not shown. *Reproduced from Waliser, D., Hendon, H., Kim, D., Maloney, E., Wheeler, M., Weickmann, K., Zhang, C., Donner, L., Gottschalck, J., Higgins, W., Kang, I.S., Legler, D., Moncrieff, M., Schubert, S., Stern, W., Vitart, F., Wang, B., Wang, W., Woolnough, S., 2009. MJO simulation diagnostics. J. Climate 22, 3006–3030.*

These composites do not show the characteristics of the propagation of the MJO, but the average time to pass through each phase is around 6 days, corresponding to a period of about 48 days. Fig. 7 shows the eastward propagation of the MJO through a lag correlation of 20- to 100-day band-passed precipitation and 850-hPa zonal wind against precipitation at a base point in the Indian Ocean, the slow propagation of the convective signal through the Indian Ocean and West Pacific, the weakening of the convective signal over the Maritime Continent, and the weak convective anomalies and faster propagation speed over the Western Hemisphere are clearly apparent. Seo and Kumar (2008) show that the stronger MJO events tend to propagate more slowly and have longer periods than weaker MJO events.

Associated with the enhanced convection, there is a tongue of low, 850-hPa geopotential height (and surface pressure) that extends to the east of the enhanced convection, reminiscent of the Gill (1980) response to tropical heating, and the low-level wind anomalies are easterly to the east of the active convection, and westerly to the west of the active convection. Although not obviously apparent in the geopotential or wind anomalies, the stream function shows the characteristic low-level Rossby gyres to the west of the convection (e.g., Kiladis et al., 2005). While the low-level structure is broadly reminiscent of the Gill (1980) response, Wang and Chen (2017) note that the zonal extent of the easterlies compared to the westerlies is smaller than for the Gill (1980) response, and their strength compared to the westerlies is increased. The surface low-pressure anomaly is associated with boundary-layer convergence and moisture convergence to the east of the convection. At upper levels, the sign of the zonal flow is reversed, with westerly flow to the east of the active convection and easterly flow to the west.

The geopotential field has the classical quadrupole response to tropical heating, with a pair of upper-level anticyclones to the west of the convection and cyclones to the east of the convection. The vertical structure of the MJO shows a westward tilt in the height of humidity, vertical velocity anomalies (e.g., Sperber, 2003; Kiladis et al., 2005), and diabatic heating (e.g., Jiang et al., 2011), with an increase in low-level humidity to the east of the convection and gradual deepening of the convection from shallow clouds, through cumulus congestus to deep convective and stratiform clouds (e.g., Kikuchi and Takayabu, 2004), as the MJO-associated convection transitions from its suppressed phase to an active phase.

For comparison, Fig. 8 shows the precipitation and low-level circulation composites for the boreal summer (June-September) season. Over the Indian Ocean and West Pacific, there is clear evidence of both the northward and eastward propagation characteristic of the BSISO, with a northwest-southeast-tilted precipitation band.

There is considerable interannual variability in MJO activity. Fig. 9 shows the 91-day running mean variance in the RMM indices (or, equivalently, the 91-day mean-squared RMM amplitude). The larger amplitude in the boreal winter is clearly evident, as is the year-to-year variability. Hendon et al. (1999) found that this interannual variability in MJO activity can be largely explained by the number of MJO events in a season rather than by the amplitude of events. Hendon et al. (1999) and Slingo et al. (1999) both examined the relationship between MJO activity and interannual SST variability and found no significant linear correlation between MJO activity and ENSO, and a more recent study (Son et al., 2017) confirmed this overall relationship between ENSO and MJO activity. A number of studies (e.g., Feng et al., 2015; Pang et al., 2016) have examined the relationship between the MJO and central or eastern Pacific El Niño events and found that the boreal winter MJO became stronger than normal during central Pacific El Niño events and weaker than normal during eastern Pacific El Niño events. Hendon et al. (1999) also noted this weakening of the MJO for the strong eastern Pacific El Niño events of 1982–83 and 1997–98. However, the small sample size, nonlinear nature of these relationships, and event-to-event variability of El Niño makes it difficult to be certain about these relationships. While ENSO overall does not affect MJO activity, it does have an impact on the characteristics of the MJO events. In El Niño events, MJO activity is enhanced over the central Pacific, with an eastward shift in the extent of the convective signal of the MJO by about 20 degrees (e.g., Kessler, 2001), which is consistent with the SST warming and eastward shift of the mean convection; however, once again, these changes in MJO propagation characteristics may depend on the nature of the El Niño event (e.g., Feng et al., 2015; Pang et al., 2016).

Although there is no clear relationship between MJO activity and interannual SST variability, Yoo and Son (2016) and Son et al. (2017) have identified a correlation between MJO activity and the Quasi-Biennial Oscillation (QBO; e.g., Baldwin et al., 2001) in boreal winter (DJF), with linear correlations between measures of MJO amplitude (including RMM and an OLR-based measure) and the equatorial zonal mean wind at 50 hPa of about −0.55, with increased MJO activity during easterly phases of the QBO and reduced activity during the westerly phases. Son et al. (2017) also found that the MJO period is increased and the propagation speed is reduced during the easterly phase of the QBO, although, following the analysis by Seo and Kumar (2008), they speculated that this changes may be a direct consequence of the increased amplitude. While there is a strong correlation between the DJF MJO activity and the QBO, the correlation in other seasons is weak and not statistically significant. The mechanisms

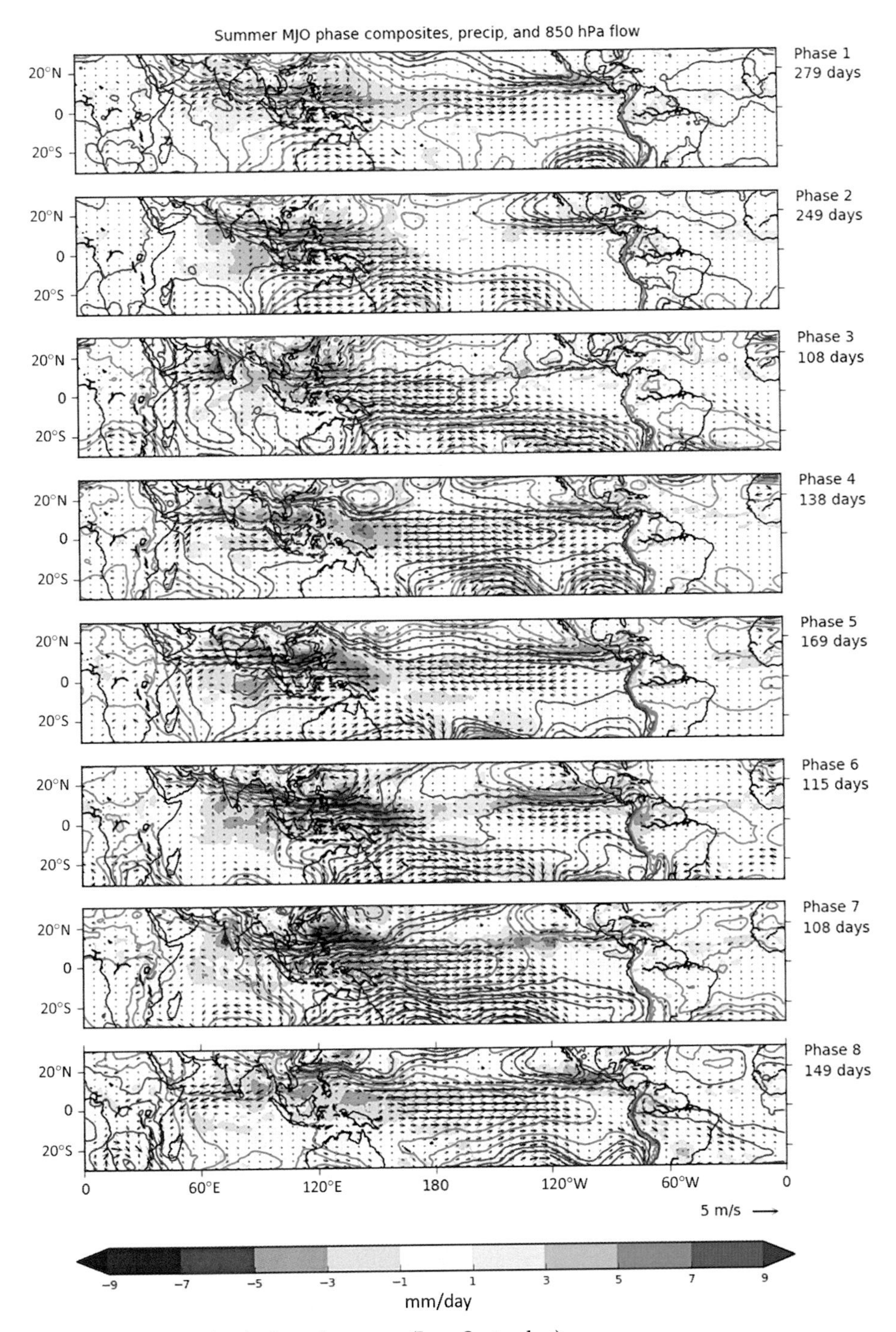

FIG. 8 As in Fig. 6 except for the boreal summer (June-September).

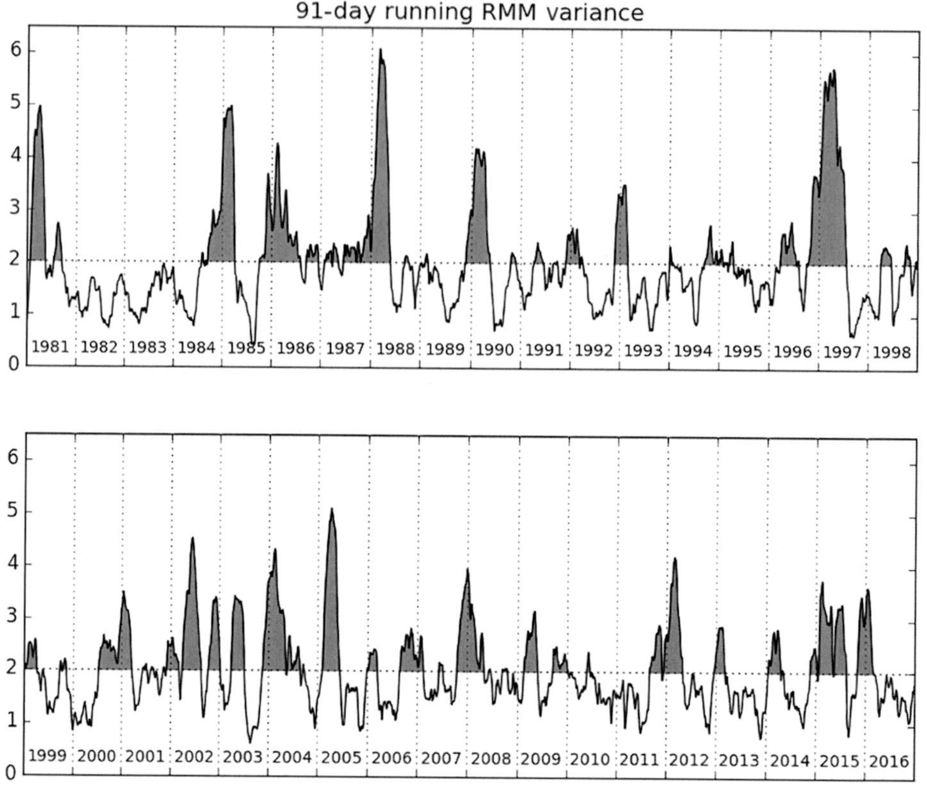

FIG. 9 Running 91-day RMM variance ($RMM_1^2 + RMM_2^2$) showing the interannual modulation of MJO activity. *After Wheeler, M., Hendon, H., 2004. An all-season real-time multivariate MJO index: development of an index for monitoring and prediction. Mon. Weather Rev. 132, 1917–1932.*

underlying this relationship are not clear, but during the easterly phase of the QBO, the static stability in the upper troposphere and lower stratosphere between about 150 and 70 hPa is reduced as a result of cooler temperatures at about 70 hPa, and Son et al. (2017) speculated that this change in static stability or increased radiative effects from increased cloudiness associated with the lower temperatures could enhance the MJO activity.

4 THE RELATIONSHIP BETWEEN THE MJO AND TROPICAL AND EXTRATROPICAL WEATHER

As the dominant mode of sub-seasonal variability in the tropical climate system, the MJO exerts a considerable influence on a range of weather and climate phenomena within both tropics and extratropics, leading to a global influence of the MJO in modulating temperature and precipitation, including extremes (e.g., Jones et al., 2004; Donald et al., 2006; Matsueda and Takaya, 2015). A comprehensive discussion of these impacts is outside the scope of this chapter but can be found, for example, in the relevant chapters of Lau and Waliser (2012).

Within the Indo-Pacific Warm Pool, the MJO (or its boreal summer counterpart) is found to influence the onset and active break cycles of the major monsoon systems (e.g., Wheeler and Hendon, 2004; Evans et al., 2014; Annamalai and Slingo, 2001; Hung and Hsu, 2008). For example, Wheeler and Hendon (2004) found that 80% of onset dates for the Australian monsoon occur during phases 4–7, when the MJO is active over the Maritime Continent, and Evans et al. (2014) found that the active-break cycles of the Australian monsoon are closely related to the phase of the MJO. During boreal winter, the MJO modulates the formation and intensity of cold surges and the Borneo Vortex within the South China Sea (e.g., Chang et al., 2005; Lim et al., 2017) and the frequency of occurrence of heavy precipitation across the region (e.g., Aldrian, 2008; Wheeler et al., 2009; Xavier et al., 2014).

Outside of the Indo-Pacific Warm pool, region the MJO is found to modulate rainfall in Africa (e.g., Pohl and Camberlin, 2006a,b; Pohl et al., 2009; Zaitchik, 2017) and tropical South America and Central America (e.g., Barlow and Salstein, 2006; Souza and Ambrizzi, 2006). Away from the Indo-Pacific Warm pool, these tropical MJO influences can arise through a mixture of local and remote responses to the MJO, including through extratropical pathways. The MJO diabatic heating and associated divergent circulation can act as a Rossby wave source, leading to extratropical circulation anomalies (see Stan et al., 2018, for a review), including in the North Pacific (e.g., Mori and Watanabe, 2008), North Atlantic (e.g., Cassou, 2008; Lin et al., 2009), and Southern Hemisphere (e.g., Berbery and Nogués-Paegle, 1993; Carvalho et al., 2005).

The MJO is also known to influence the formation of tropical cyclones across the tropics (e.g., Klotzbach, 2014), with enhanced formation and rapid intensification during and following the convectively active phase, both as a result of changes in the large-scale environmental conditions and the presence of preexisting cyclonic circulations associated with the westerlies to the west of the active convection (e.g., Carmago, 2009). On longer timescales, the westerly wind events to the west of the active convection as the MJO passes into the West Pacific can trigger downwelling oceanic equatorial Kelvin waves, which subsequently propagate across the Pacific and can trigger or intensify El Niño events. These westerlies can also drive surface currents that advect the eastern edge of the Warm pool toward the central Pacific, further supporting the developing El Niño (e.g., Lengaigne et al., 2004).

These wide-ranging impacts of the MJO mean that the MJO is an important source of predictability on the sub-seasonal to seasonal (S2S) timescales across the globe.

5 THEORIES AND MECHANISMS FOR MJO INITIATION, MAINTENANCE, AND PROPAGATION

Since the discovery of the MJO, a number of theories and mechanisms for the maintenance and propagation of it have been developed (see Wang, 2012, for a review). The essential components for a theory for the MJO are a model for the large-scale dynamical response to the anomalous heating, primarily the latent heat release in the convection, but also anomalous radiative and surface fluxes; and a model for how this anomalous heating evolves in response to the anomalous large-scale flow.

Unlike the equatorially trapped wave modes of Matsuno (1966), the MJO is not a normal mode of the tropical atmosphere. Theories for the large-scale dynamical structure of the MJO are commonly based on a linearized set of equations of motion for the tropical dynamics, with a simplified vertical structure with a limited number of vertical modes for the free troposphere, with or without a representation of the boundary layer. Some of the elements of this dynamical model are a critical component of the subsequent evolution of the convection; for example, the frictional convergence mechanism of Wang and Rui (1990a) depends on the inclusion of a boundary layer in the dynamical model, and the sensitivity to heating profile depends on the retention of a greater number of vertical modes (e.g., Mapes, 2000). However, the major differences between these theories are in how the heating evolves in response to the anomalous MJO circulation.

One class of theories relates the heating to the (moisture) convergence in the wave, including, for example, free tropospheric wave-induced convergence (e.g., Lau and Peng, 1987) and frictionally induced boundary-layer convergence (e.g., Wang, 1988; Wang and Rui, 1990a). Wang (1988) found that this frictionally induced convergence to the east of the convection can couple the Kelvin wave and Rossby wave, leading to an eastward-propagating disturbance with circulation anomalies that resembles the response to stationary heating of Gill (1980) and the observed circulation anomalies of the MJO. In these theories, the heating is related to the dynamical convergence through a prescribed humidity field (often related to an underlying SST distribution), and precipitation efficiency, which controls the fraction of moisture convergence and in turn produces latent heating, but in which the moisture field does not evolve with the flow.

More recently, attention has focused on the role of the moisture variations in modulating the precipitation intensity (e.g., Raymond and Fuchs, 2009; Adames and Kim, 2016), in part motivated by observed variations in moisture associated with the MJO (e.g., Sperber, 2003) and in part by results from numerical modeling experiments (e.g., Kim et al., 2014b), which show a relationship between MJO simulation fidelity and the sensitivity of precipitation to the relative humidity. In these theories, precipitation is determined by a measure of the column humidity rather than by the moisture convergence. While precipitation removes moisture from the column, it is recharged by the large-scale circulation and surface fluxes. Jiang (2017) analyzed the moist static energy (MSE) budget in reanalyses and found that the horizontal advection of the mean basic state moisture gradient by the MJO anomalous winds is an important component of the positive MSE (and moisture) tendency to the east of the active phase of the MJO, suggesting an important role for the background moisture field in the characteristics of the MJO. An MSE analysis of the MJO (e.g., Sobel et al., 2014) also identifies an important role for cloud radiative effects in maintaining the MSE anomalies associated with the MJO. As noted by Raymond (2001), this can be interpreted as the cloud radiative warming driving ascent and moistening by the large-scale circulation.

In a further class of theories under the heading of multiscale interactions (e.g., Biello and Majda, 2005; Majda and Stechmann, 2009, 2012; Liu and Wang, 2013), the interaction between the convection and the large-scale dynamics occurs through modulation of the heating by the MJO's modulation of the synoptic scale disturbances, motivated by observations of the multiscale structure in the MJO (e.g., Nakazawa, 1998). These models typically include a prognostic equation for "wave activity," which depends on the evolution of the moisture field, and this wave activity determines the heating.

A number of observational, modeling, and theoretical studies have considered the role of air-sea interaction in the MJO (see DeMott et al., 2015, for a review), particularly in the Indian Ocean and West Pacific. In these theories, the cloud and wind anomalies associated with the MJO modify the surface fluxes that drive intraseasonal SST variability. To the east of the convection, suppressed convection increases the surface shortwave flux and easterly wind anomalies on a mean westerly wind reduce the surface wind stress, which reduces the latent heat flux and causes a shoaling of the mixed layer. In phase with the active convection and to the west, the SST cools due to the enhanced cloud cover and strengthened surface winds. This forcing drives intraseasonal SST anomalies in a quadrature with the convection, with warmer SSTs to the east of the convection and cooler SSTs to the west of the convection. While the mechanisms by which the MJO drives the SST anomalies are well understood, the role of the ocean feedback in the atmosphere and the mechanisms by which this occurs are less so. DeMott et al. (2016) showed that the effect of the SST anomalies is to increase the surface heat fluxes (and moist state energy tendency) ahead of the convection and reduce them behind it. SST-induced boundary layer convergence (e.g., Hsu and Li, 2012) may also lead to additional moisture (and MSE) convergence ahead of the convection.

While many of these theories have some success in explaining the observed characteristics of the MJO, our theoretical understanding of it remains incomplete, and it is likely that any complete theory of the MJO will need to account for many of the mechanisms that have been identified in individual theories or classes of theories. Underlying all these theories is the importance of the relationship between the anomalous large-scale circulation associated with the MJO and anomalous diabatic heating, including latent heat release by convection, radiative heating, and surface fluxes, and consequently the critical importance of these largely parameterized processes for the simulation of the MJO in weather and climate models.

6 THE REPRESENTATION OF THE MJO IN WEATHER AND CLIMATE MODELS

Simulation of the MJO in weather and climate models remains a challenge. In two recent intercomparison studies, Jiang et al. (2015) and Ahn et al. (2017) found that, in agreement with many earlier intercomparison studies (e.g., Slingo et al., 1996; Lin et al., 2006), few climate models are able to produce realistic simulations of the MJO with the observed amplitude and systematic eastward propagation, although Ahn et al. (2017) noted some modest improvement since these earlier intercomparisons. Models with good simulations of the amplitude and propagation of the MJO also tend to capture the observed westward tilt of convection with height, and the boundary layer and low-level moistening ahead of the convection.

Over the years, a number of studies have identified a large sensitivity of the MJO simulation fidelity to the representation of convection in the model, including convective triggering (e.g., Wang and Schlesinger, 1999); modifying the sensitivity of the convective plume to environmental humidity (e.g., Tokioka et al., 1988; Bechtold et al., 2008; Klingaman and Woolnough, 2014b); enhancing the shallow convective heating and moistening (e.g., Zhang and Mu, 2005); or producing a more top-heavy heating profile in deep convection (e.g., Seo and Wang, 2010). Models with an explicit representation of the convection, either through replacing the parameterization scheme with a cloud-resolving model ("super-parametrization"; e.g., Benedict and Randall, 2009), or in global (e.g., Miyakawa et al., 2015) or regional (e.g., Holloway et al., 2013, 2015) convection permitting models, have shown improved representations of the MJO. The sensitivity of the MJO simulation to the representation of convection highlights the important role of convection in the maintenance and propagation of the MJO. Many of the modifications to the representation of convection have the effect of increasing the suppression of deep convection in the suppressed phase of the MJO, either by making it harder to trigger convection or by making it harder for convection to penetrate the atmosphere in dry conditions.

In addition to the configuration of the atmosphere model, a number of studies have assessed the impact of coupling on the MJO simulation (see DeMott et al., 2015, for a review). While there has been a mixed impact, most studies have shown an improved simulation of the MJO in coupled models over their atmosphere-only counterparts, although the impact is found to be sensitive to the fidelity of the atmosphere-only simulation of the MJO (e.g., Klingaman and Woolnough, 2014a), the vertical resolution of the upper ocean (e.g., Bernie et al., 2008), and the mean basic state of the model, particularly due to the dependence of the MJO-driven latent heat flux variability on the presence of mean westerly surface winds in the Indian Ocean and West Pacific (e.g., Inness et al., 2003).

Both the sensitivity studies and the model intercomparison studies have often tried to develop so-called process-oriented diagnostics of the model behavior that are not direct measures of MJO performance, but are well correlated with it, as a guide to improving model simulations. These metrics can often be linked to theoretical ideas regarding the MJO and may include measures of the basic state of the model (e.g., the background moisture field, Jiang, 2017); the coupling between convection and the large-scale circulation (e.g., the gross moist stability, Benedict et al., 2014); or the relationship between convection and the environment (e.g., the relationship between the precipitation and environmental humidity, Kim et al., 2014b).

7 MJO PREDICTION

Despite the difficulties in simulating the MJO in climate models, considerable progress has been made in MJO prediction over the past 10–15 years. This may in part be due to smaller errors in the basic state of forecast models compared to climate models, or differences in the metric by which we are measuring skill. Waliser (2005) reviewed the status of MJO prediction for both empirical and dynamical forecast models and found that a number of empirical MJO prediction systems had "useful" skill out to about 15 days; at that time, there were few operational dynamical prediction systems that extended to this time range, but a few studies had attempted to assess the skill for the MJO (and intraseasonal variability more generally) in dynamical forecast systems and found some useful skill out to about 6–9 days. Jones et al.

(2000) analyzed 5 years of daily 50-day forecasts and found that the anomaly correlation for intraseasonally filtered 200-hPa equatorial zonal wind reached 0.5 after about 5 days, with a small increase during active MJO events and for particular phases of the MJO. Hendon et al. (2000) showed from the same dataset that the forecast skill for the tropical and Northern Hemisphere circulation more generally was worse when there was an active MJO in the initial conditions than when there was not; the poor forecasts of the large circulation anomalies associated with the MJO introduced larger-forecast errors than weaker anomalies associated with nonactive MJO periods.

In 2006 the US Climate Variability and Predictability Program (CLIVAR) supported the formation of an MJO Working Group, which proposed and developed an MJO prediction activity (Gottschalck et al., 2010), to evaluate the MJO forecast skill across a number of operational forecast activities. This project proposed the use of the bivariate correlation (*COR*, shown in Eq. (1)) and root-mean-square error (*RMSE*, shown in Eq. (2)) of the forecast RMM indices as standard measures of MJO forecast skill, following Lin et al. (2008) and Rashid et al. (2011):

$$COR(\tau) = \frac{\sum_{i=1}^{N}[R_1^F(t,\tau)R_1^O(t) + R_2^F(t,\tau)R_2^O(t)]}{\sqrt{\sum_{i=1}^{N}[R_1^O(t)^2 + R_2^O(t)^2]}\sqrt{\sum_{i=1}^{N}[R_1^F(t,\tau)^2 + R_2^F(t,\tau)^2]}} \tag{1}$$

$$RMSE(\tau) = \sum_{i=1}^{N}[R_1^F(t,\tau) - R_1^O(t)]^2 + [R_2^F(t,\tau) - R_2^O(t)]^2, \tag{2}$$

where $COR(\tau)$ and $RMSE(\tau)$ are the bivariate correlation and RMSE at lead time τ, respectively; $R_{1,2}^O(t)$ are the observed RMM indices at time t; $R_{1,2}^F(t,\tau)$ are the lead time τ forecast RMM indices valid at time t; and N is the number of forecasts.

Two recent studies have assessed the forecast skill of a number of models from large, subseasonal hindcast (or reforecast) datasets. In contrast to numerical weather prediction (NWP) and medium-range (to about 15 days) forecasting for operational sub-seasonal (i.e., beyond about 15 days) prediction, the forecast anomalies are computed relative to a lead-time-dependent model climatology to account for the development of the systematic biases in the forecast model.

Neena et al. (2014) and Lee et al. (2016) analyzed the prediction skill from seven operational and research models run as part of the Intraseasonal Variability Hindcast Experiment (ISVHE). The hindcast dataset has a limited number of start dates, ranging from 1 to 3 per month, over 11–26 years depending on the model. Using the time at which the bivariate correlation for the ensemble mean reaches 0.5, Lee et al. (2016) reported skill for between 8 and 26 days for the participating models. Zhang et al. (2013) computed the correlation skill for RMM1 and RMM2 for multimodel ensembles (MME) from the ISVHE. Using all the models, correlation reaches 0.5 at about (21, 23) days for (RMM1, RMM2), which is no better than the best individual model; but using just the best two or three models, the MME skill extends to about (26, 27) days, which exceeds that of any individual model. Using a different metric, specifically the time at which the RMSE reaches the mean amplitude of the observed MJO, Neena et al. (2014) found skill for individual ensemble members for between 6 and 18 days depending on the model and for the ensemble mean (5–11 members depending on the model), there is skill for between 8 and 28 days, with the ensemble mean adding between 1 and 10 days over the individual ensemble member skill.

Vitart (2017) analyzed the skill of 10 operational models submitted to the Sub-seasonal to Seasonal Prediction Project (S2S) database (Vitart et al., 2017) for the reforecast period (1999–2010). Over the whole year, the skill (when the bivariate correlation reaches 0.6) of the ensemble mean ranges from 5 to 28 days depending on the model (Fig. 10), with most models having skill out to about 12–18 days and for individual members, this range reduces to 6–21 days. For the extended winter (December-March) period, there is skill for 10–25 days depending on the model. The ECMWF model has significantly greater skill than the other models, although this gap is reduced for the extended winter period, in part due to higher skill in the other models, but also due to a reduction in the skill for the ECMWF model during the extended winter compared to the whole year. While providing a useful overall metric of MJO forecast skill, the bivariate correlation is unable to distinguish between phase errors and amplitude errors. For the S2S reforecasts, most models underestimate the amplitude of the MJO by 10% to 40%; however, the CNRM model shows an MJO amplitude that grows through the forecast. Early in the forecast period (out to about 10 days), the models tend to have a positive phase error (too fast a propagation), although over the lead time of 10–15 days, this phase error reduces to the extent that from about day 15 onward, the phase errors tend to be negative (too slow a propagation).

Vitart (2014) shows the improvement of the MJO skill scores in the ECMWF monthly forecast system over the period 2002–12. During that period, the MJO skill as measured by the time at which the bivariate correlation reaches (0.5, 0.6, 0.8) has increased from (22, 15, 9) days to (31, 27, 14) days. Although there is steady improvement in forecast skill through the period, when analyzing the skill in terms of phase and amplitude error, it is clear that the largest improvements came during the period 2006–08, and in particular, both the amplitude error and phase error reduce. Hirons et al. (2013) were able to attribute these improvements to changes in the formulation of entrainment in the convection scheme, which made the entrainment rate sensitive to the environmental humidity and suppressed deep convection in dry environments. These improvements in amplitude error, with forecast amplitudes similar to these observations, had significant impacts on the forecasts of MJO-driven variability in the NAO. Prior to this version of the model, the ECMWF forecast model had less skill for the NAO when there was a strong MJO in the initial conditions, as the weak MJO in the model was unable to provide a sufficiently strong teleconnection to the North Atlantic and the forecasts were missing this important driver of North Atlantic variability. Beyond 2007, the increase in the amplitude of the forecast MJO provided a stronger teleconnection to the NAO, and the NAO forecast model was able to exploit this source of predictability for the North Atlantic and the skill became greater for initial conditions with an active MJO than for those without.

Vitart (2017) also reported that the models contributing to the S2S database are also typically underdispersive in their MJO forecasts. While there is some relationship between spread and skill at longer lead times, the standard deviation in ensemble spread is much less than in the RMSE suggesting that there is significant room for improvement in ensemble generation techniques for the MJO, in order to provide better information on the expected skill of

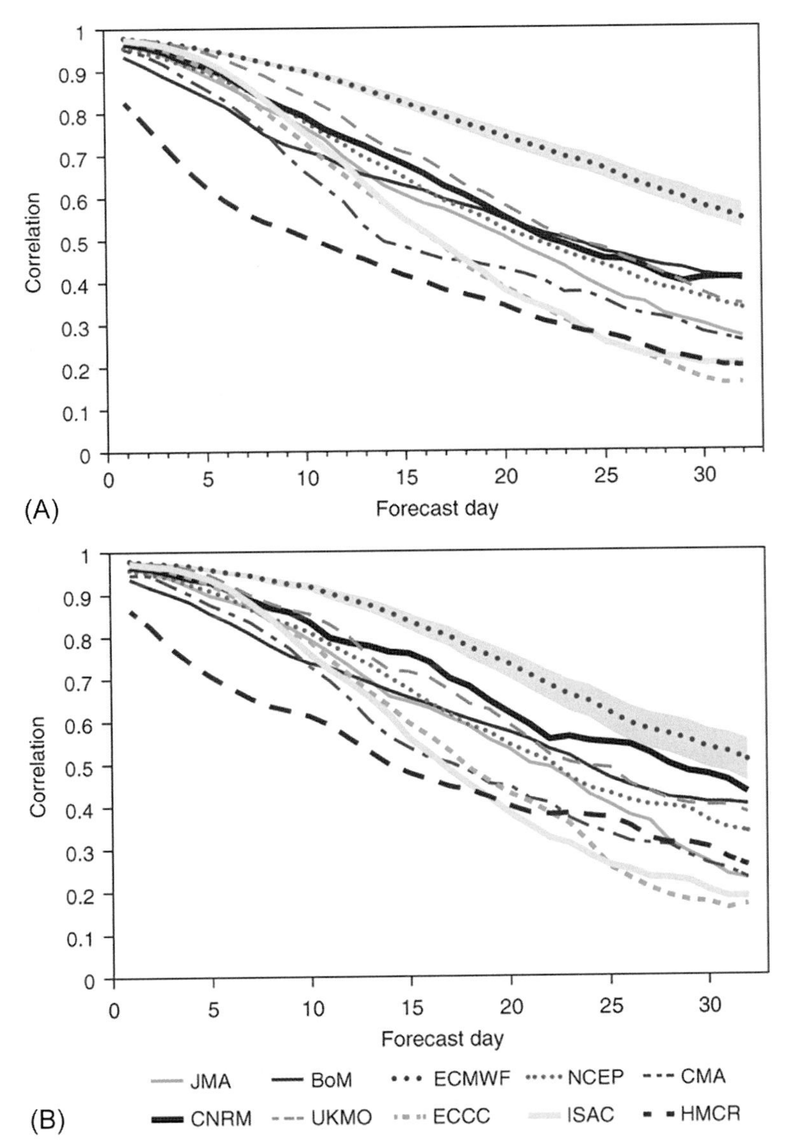

FIG. 10 Evolution of the bivariate correlation between the model ensemble means and ERA-Interim as a function of lead time for 10 S2S models. The MJO bivariate correlation has been computed over the period 1999–2010 for (A) all seasons and (B) extended winters (December-March). The *shaded area* represents the 95% level of confidence computed from a 10,000 bootstrap resampling procedure. *Reproduced from Vitart, F., 2017. Madden-Julian Oscillation prediction and teleconnection in the S2S database. Q. J. R. Meteorol. Soc. 143, 2210–2220.*

a particular forecast. While the spread skill relationship within a model is weak, Neena et al. (2014) showed a significant correlation between the ensemble spread and the improvement in skill of the ensemble mean forecast compared to the single-member forecast for a given model in the ISVHE dataset.

The forecast skill metrics given above provide a deterministic measure of skill, but the ensemble nature of the forecasting systems allows a probabilistic forecast, which requires evaluation. Relatively few studies have provided such a probabilistic evaluation. Hamill and Kildadis (2014) and Marshall et al. (2016) used the Continuous Ranked Probability Skill Score (CRPSS) (e.g., Wilks, 2011) to provide a probabilistic evaluation of forecast skill for the bivariate RMM index for the National Centers for Environmental Prediction Global Ensemble Forecast System (NCEP GEFS) and the Bureau of Meteorology's Predictive Ocean Atmosphere Model for Australia (POAMA), respectively, and found that the forecasts outpeform climatological forecasts to lead times of 14 days for the GEFS and 37 days for the POAMA. Marshall et al. (2016) also found skill for the MJO amplitude out 18 days and, using the Ranked Probability Skill Score (RPSS) for the MJO phase, found skill relative to climatology for 25 days for the MJO phase. The probabilistic skills for the POAMA are broadly consistent with the deterministic forecast skill reported in Rashid et al. (2011) and Vitart (2017).

Consistent with theoretical and general circulation model (GCM) modeling experiments on the role of air-sea coupling on the MJO, a number of studies have reported improved MJO forecast skill in coupled models compared to atmosphere-only models (e.g., Woolnough et al., 2007; Vitart et al., 2007; Seo et al., 2009; Fu et al., 2013; Shelly et al., 2014). These studies considered relatively short hindcast periods or few MJO events, but all found improved forecast skill for the MJO over atmosphere-only counterparts, with persisted SSTs of between 2 and 7 days, with, typically, larger improvements for the active phase of the MJO in the Indian Ocean or West Pacific rather than over the Maritime Continent. Woolnough et al. (2007) showed the sensitivity to the upper-ocean resolution, and in particular the representation of the diurnal cycle during the suppressed phase of the MJO, by comparing the skill of a coupled model with a full dynamical ocean and 10-m resolution in the upper layer with a one-dimensional (1D) mixed-layer ocean model with 1-m resolution near the surface.

Following on from theoretical arguments and climate model studies on the role of the basic state moisture field in MJO propagation and maintenance, Lim et al. (2018) examined the relationship between basic state errors in the mean moisture field and MJO prediction skill in the S2S database, and found that those models with better basic state meridional and zonal moisture gradients in the Indian Ocean and Maritime Continent region have better MJO forecast skill. Further, Kim (2017) showed that basic state errors in the moisture field in the eastern Indian Ocean and Maritime Continent region in the ECMWF model lead to errors in the MSE budget of the MJO, as well as errors in the representation of the propagation of the MJO across the Maritime Continent.

Over the last decade, there has been considerable improvement in MJO forecast skill, but there is still a large variation in skill between models. The relatively poor skill of the dynamical forecast skill compared to statistical models in Waliser (2005) led to the question of what the potential limits of predictability for the MJO are. The 15-days skill of the statistical methods suggests that this figure is a lower limit for the predictability. Neena et al. (2014) assessed the predictability of the MJO using "the perfect model" approach in the ISHVE hindcast by assessing the ability of the model to predict itself. Using this approach, they found predictability estimates of between 20 and 30 days for single members and 35 to in excess of 45 days for the ensemble mean, which is between 5 and 20 days greater than the respective skill of the models. These predictability estimates, however, should be treated with caution, not least because, they are based on the ability of a model to predict itself, but also because as documented by Vitart (2017), these prediction systems are typically underdispersive.

7.1 Sub-seasonal and Interannual Variations in Forecast Skill

The analysis given in this chapter provides a measure of the overall skill of the forecast models for the MJO, but this skill itself may depend on the initial state of the atmosphere, in terms of the MJO variability itself—for example, whether there is an active MJO in the initial conditions, or the phase of the MJO; the time of year; or the interannual variability, such as, ENSO or the QBO. While there has not been systematic analysis of the MJO forecast skill as a function of MJO phase across models, a number of studies have addressed aspects of it, with varying conclusions. Lin et al. (2008) found that for GCM3 of the Canadian Center for Climate Modeling and Analysis and Global Environmental Multiscale Model (GEM) of the Recherche en Prévision Numérique, there is higher skill for forecasts initialized in strong MJOs (RMM amplitude >1) in phases 1–3 than in phases 4–8, although this difference is smaller for GEM than for GCM3, while Rashid et al. (2011) found a much smaller phase dependence in the POAMA. Kim et al. (2014c) found that for the ECMWF varEPS, beyond about day 10, the highest skill is for forecasts initialized in phases 4 and 7, with the lowest skill in phases 2 and 5, but that the NCEP CFS2 shows much weaker variation in skill as a function of MJO phase. Kim et al. (2016) reported that for a later version of the ECMWF forecast model, the skill of the forecasts in phase 2 is much improved.

Vitart et al. (2017) considered the ability of the forecast models in the S2S database to propagate the MJO across the Maritime Continent by considering the percentage of cases initialized with a strong MJO in phases 2–3 that never propagate into the Western Pacific (phases 6–7) in the next 30 days, even as a weak event. In the ERA-Interim reanalysis, only 10% of events never reach phases 6–7, compared to between 19% and 46% of events in the forecast models, suggesting that there are systematic biases in all the models in modeling the propagation of the MJO across the Maritime Continent. Lin et al. (2008), Rashid et al. (2011), and Kim et al. (2014c) also considered the dependence on the initial amplitude of the MJO and found that for initial conditions with active MJO events (RMM amplitude >1), there is improved skill compared to all forecasts.

As described in Section 3, the QBO influences the strength of the MJO during DJF. Marshall et al. (2017) assessed the impact of the QBO on MJO forecast skill in the POAMA. They reported increased forecast skill in the easterly phase of the QBO when the MJO is stronger than in the westerly phase of the MJO, with bivariate correlations > 0.5 out to lead times of 31 days for the easterly phases of the QBO, compared to 23 days for the westerly phase of the QBO. As noted previously, a number of studies have shown improved MJO forecast skill for stronger MJO events than for weaker MJO events. To rule out the possibility that it is simply the stronger MJO events during the easterly phase of the QBO that lead to improved forecast skill, Marshall et al. (2017) stratified the forecasts by initial MJO amplitude and showed that while the improvement in MJO forecast skill in the easterly phase of the QBO is larger for strong MJO events, it remains for all initial MJO amplitudes.

8 FUTURE PRIORITIES FOR MJO RESEARCH FOR S2S PREDICTION

The MJO is the dominant mode of intraseasonal variability in the tropics, and as such is a considerable source of potential predictability on sub-seasonal timescales. While considerable progress has been made in our understanding of the MJO since its discovery in the 1970s and in its prediction since the review by Waliser (2005), there is still much to learn.

8.1 Linking Theory and Modeling

Theories for the maintenance and propagation have developed substantially since the early wave-CISK models (e.g., Lau and Peng, 1987) to include elements of the basic state, the role of humidity variations, and radiative feedback, and to account for the multiscale nature of convection; however, there is still no consensus theory. Furthermore, while the representation of the planetary-scale dynamics of the MJO by simplified equation sets is probably not limiting our ability to relate these theories to numerical models, MJO-related heating is usually represented by a bulk relationship to the dynamical fields or moisture convergence, and its free parameters are not easily related to the details of convection parameterizations used in numerical models, but more closely to diagnostics of their behavior. This disconnect between the specification of the heating in our theories and its representation of our models hinders our ability to use these theories in the development of our numerical models, or to use our numerical models to test our theories.

8.2 MJO Initiation

Diagnosing processes leading to MJO initiation remains a challenge, not least because of the difficulty in identifying MJO initiation highlighted by Straub (2013), but also because of the divergence in MJO events and the paucity of observations of the small-scale structures likely to be important in identifying whether similar large-scale conditions do or do not evolve into an MJO. Forecasting systems still have less skill for MJO prediction for primary MJO events compared to secondary events, and improving our ability to predict MJO initiation is likely to depend on an improved understanding of the processes that control it.

8.3 Predicting the Impacts of the MJO

Considerable progress has been made in MJO prediction over the last decade, but there have been relatively few studies that have assessed the skill of models at predicting the impact of the MJO on the weather (e.g., Marshall et al., 2011). To realize the full potential of the MJO as a source of predictability on sub-seasonal timescales, our models must be able to predict the evolution of the MJO itself and its impact on the weather variables of interest. There is a need for more studies that can link the predictability and predictive skill for the MJO itself to either predictive skill for the weather and its impact or an understanding of the conditional skill associated with the MJO (e.g., see Vitart, 2014, about the skill of the NAO conditional on the MJO).

Acknowledgments

SJW was supported by the National Centre for Atmospheric Science, a NERC collaborative center under contract R8/H12/83/001. The updated RMM indices of Wheeler and Hendon (2004) used to produce Figs. 3 and 4 were obtained from the Australian Bureau of Meteorology at http://www.bom.gov.au/climate/mjo/. Version 1.2 of the GPCP 1 Degree Daily (1DD) Precipitation dataset was obtained from ftp://ftp.gcd.ucar.edu/archive/PRECIP.

CHAPTER

6

Extratropical Sub-seasonal to Seasonal Oscillations and Multiple Regimes: The Dynamical Systems View

Michael Ghil,†, Andreas Groth*, Dmitri Kondrashov*, Andrew W. Robertson‡*

*Department of Atmospheric and Oceanic Sciences, University of California at Los Angeles, Los Angeles, CA, United States †Geosciences Department and Laboratoire de Météorologie Dynamique (CNRS and IPSL), Ecole Normale Supérieure and PSL Research University, Paris, France ‡International Research Institute for Climate and Society (IRI), Columbia University, Palisades, NY, United States

OUTLINE

https://doi.org/10.1016/B978-0-12-811714-9.00006-1

1 INTRODUCTION AND MOTIVATION

The John von Neumann point of view (Von Neumann, 1960):

> Short-term numerical weather prediction (NWP) is the easiest—that is, it is a pure initial-value problem; long-term climate prediction is next easiest—it corresponds to studying the system's asymptotic behavior; intermediate-term prediction is hardest—both initial & boundary values are important. The modeling hierarchy & successive bifurcations as Ariadne's thread through the rungs of the hierarchy.

Intraseasonal timescales, more recently called sub-seasonal to seasonal (S2S), range from the deterministic limit of atmospheric predictability (about 10 days) up to a season (say, 100 days). These timescales occupy a window of overlap between low-frequency variability (LFV) intrinsic to the atmosphere and short-climatic timescales that also involve the upper-ocean and land-surface features, as well as the stratosphere. These timescales are of particular importance to sub-seasonal prediction. Theoretical and observational studies of LFV over the past half-century have used two complementary ways of describing atmospheric LFV in the extratropics: (i) episodic, by means of multiple weather (Reinhold and Pierrehumbert, 1982) or flow (Charney and DeVore, 1979; Legras and Ghil, 1985) regimes; and (ii) oscillatory, by means of broad-peak, slowly modulated oscillations (Ghil and Robertson, 2002, and references therein).

We pursue throughout this chapter the dynamical systems point of view first formulated by Von Neumann (1955), according to which there are three levels of difficulty in understanding and predicting atmospheric phenomena: (a) short-term numerical weather prediction (NWP) is the easiest, because it represents a pure initial-value problem, as formulated by Bjerknes (1904) and Richardson (1922); (b) long-term climate prediction is next easiest because it corresponds to studying the system's asymptotic behavior—that is, the possible attractors, such as fixed points, limit cycles, and strange attractors, and the statistical properties thereof (Ghil and Childress, 1987; Dijkstra and Ghil, 2005; Dijkstra, 2013); and (c) intermediate-term prediction is hardest because both the initial and the parameter values are important. We deal with the latter problem by pursuing a full hierarchy of models, from the simplest to the most detailed (Schneider and Dickinson, 1974; Held, 2005), as well as the tool of successive bifurcations as Ariadne's thread through the rungs of this hierarchy (Ghil and Robertson, 2000; Ghil, 2001).

This chapter starts in Section 2 with a brief summary of the main characteristics of the oscillatory versus intermittent viewpoints from a dynamical systems perspective, based on Ghil and Robertson (2002) and updated using studies published since then. Section 3 presents a new analysis of the observational characteristics of oscillatory LFV, using a recently developed multivariate spectral analysis tool. Based on this background and results, Section 4 discusses low-order modeling (LOM) approaches and their application to LFV, while outlining how these approaches have developed over the years from linear inverse models (LIMs) or principal oscillation patterns (POPs) to empirical model reduction (EMR) and multilayer stochastic models (MSMs). The chapter concludes in Section 5 by discussing the prospects for improving S2S prediction in the extratropics based on the theoretical LFV framework sketched herein and on data-driven LOMs.

2 MULTIPLE MIDLATITUDE REGIMES AND LOW-FREQUENCY OSCILLATIONS

Blocking, zonal flow, and teleconnection patterns; phenomenological description of extratropical flows in the S2S band (10–100 days). Mechanical- and thermal-topography effects, i.e., mountains & land–sea contrast. Bifurcations & symmetry breaking as the unifying point of view for the theoretical interpretation of the phenomenology. Markov chains of regimes & predictability.

2.1 The Case for Multiple Regimes and Their Classification

Persistent LFV anomalies, in which the flow patterns differ significantly from the normal climatological circulation and remain stationary for more than a week, have been objectively identified over the North Pacific and North Atlantic in the 1980s (Wallace and Gutzler, 1981; Dole and Gordon, 1983; Mo and Ghil, 1988). Their onsets and breaks, on the other hand, are rather abrupt. The most familiar of these persistent patterns, or "Grosswetterlagen" (Namias, 1968)—albeit not the only ones—are blocked and zonal atmospheric flows. Within the last few decades, it has been demonstrated that these patterns can be identified by examining the probability distribution function (PDF) of the corresponding large-scale maps in the atmosphere's phase space (Cheng and Wallace, 1993; Kimoto and Ghil, 1993a,b; Smyth et al., 1999). The resulting patterns resemble those found previously by using correlation analysis.

Many of the methods used to classify weather maps are summarized in Table 1. In such a classification, an individual atmospheric map is thought of as a point in phase space. To achieve a reliable, statistically significant classification, it is necessary to consider a low-dimensional subspace of this phase space that still captures most of the variance in the dataset. The usual choice is to compute the analyzed record's empirical orthogonal functions (EOFs)—that is, the eigenvectors of the covariance (or correlation) matrix—and to select a subspace spanned by a few leading eigenvectors (Mo and Ghil, 1988; Cheng and Wallace, 1993; Kimoto and Ghil, 1993a,b; Smyth et al., 1999).

Many of the classification methods define the regimes as classes of distinct atmospheric states that have a high probability of occurrence and are separated by regions of lower probability. Some of these methods seek the maxima of the PDF by using kernel density estimation (Kimoto and Ghil, 1993b; Corti et al., 1999) or more ad hoc methods (Molteni et al., 1990). Each regime is then formed by the points, or maps, that exceed a given probability threshold in the neighborhood of a PDF maximum. The number of PDF peaks depends on the kernel-smoothing parameter used, which can be determined objectively by using a least-squares cross-validation procedure (Silverman, 1986).

Smyth et al. (1999) used a mixture model that approximates the PDF by the sum of a small number of multivariate Gaussians. In this case, the regimes are "fuzzy" in the sense that they overlap, and that each particular daily weather map can be assigned a probability of belonging to one regime or another. This was also the case in Mo and Ghil (1988), which used a different classification algorithm and obtained therewith a different number of regimes. This number as well as the spatial patterns associated with the regimes are discussed in Section 4.

Cluster analysis is a less ambitious approach to classifying atmospheric states: It localizes high concentrations of points, called clusters, but does not pretend to estimate the PDF. There

TABLE 1 Classification Methods for Weather Maps

Approach	Method	Datasets	References	Comments
REGIME CLASSIFICATION BY POSITION				
Cluster analysis	Categorical	NH	Mo and Ghil (1988)	Fuzzy
		NH + sectorial	Michelangeli et al. (1995)	Hard (*k*-means)
		Model	Dawson and Palmer (2014)	Hard (*k*-means)
		Model + NH + sectorial	Muñoz et al. (2017)	Hard (*k*-means)
		Model + sectorial	Straus and Molteni (2004), Straus et al. (2007, 2017)	Hard (*k*-means)
	Hierarchical	NH + sectorial	Cheng and Wallace (1993)	3 NH clusters
PDF estimation	Univariate	NH	Benzi et al. (1986), Hansen and Sutera (1995)	Bimodality
	Multivariate	NH	Kimoto and Ghil (1993a)	3 modes
		NH + sectorial	Kimoto and Ghil (1993b)	Multimodal
			Smyth et al. (1999)	3 NH clusters
REGIME CLASSIFICATION BY PERSISTENCE				
Pattern correlations		NH	Horel (1985)	
		SH	Mo and Ghil (1987)	3 regimes
Minima of tendencies		Models	Legras and Ghil (1985), Mukougawa (1988), Vautard and Legras (1988)	
		Atlantic-European sector	Vautard (1990)	4 regimes
TRANSITION PROBABILITIES				
Counts		Model + NH	Mo and Ghil (1988)	Elementary
Monte Carlo		NH + SH	Vautard et al. (1990)	Advanced
		NH + sectorial	Kimoto and Ghil (1993b)	Advanced

Notes: NH, *Northern Hemisphere;* SH, *Southern Hemisphere;* PDF, *probability density function.*
Source: Based on Ghil and Robertson (2002) and updated.

are two main types of clustering algorithm: hierarchical and partitioning. In hierarchical algorithms, one builds a classification tree iteratively, starting from single data points and merging them into clusters according to a similarity criterion. Cheng and Wallace (1993) used Ward's method to do this. In partitioning algorithms, a prescribed number of clusters is chosen, and data points are agglomerated around kernels initially picked as random seeds. The kernels are iteratively modified so as to globally minimize the data scatter about them, as done in the *k*-means method of Michelangeli et al. (1995).

Using low-pass-filtered data, based on running means of 5–10 days, introduces some measure of persistence into these methods, which are based on frequency of occurrence. A second broad class of methods uses quasistationarity explicitly. Here, the regimes are defined as comprising states for which large-scale motion is slow in the statistical sense. More precisely, one seeks the large-scale patterns that have, on average, a small time derivative (Legras and Ghil, 1985). This phase-space speed can be computed for maps that include synoptic-scale motions by a nonlinear equilibration technique (Vautard and Legras, 1988; Vautard, 1990; Michelangeli et al., 1995).

Huth et al. (2008) provided a more recent review of classification schemes of atmospheric circulation patterns and included a table that brought up to date the one from Ghil and Robertson (2002); an updated version of the Ghil and Robertson (2002) table is also included here as Table 1 (above). Straus et al. (2007) compared the circulation regimes in the National Centers for Environmental Prediction (NCEP) reanalyses with the general circulation model (GCM) simulations of the Center for Ocean-Land-Atmosphere Studies (COLA), showing that three out of the four observed clusters have identifiable counterparts in the model simulations. These authors, as well as Straus and Molteni (2004), also studied possible changes in the structure of the COLA GCM's regimes in response to changes in tropical sea surface temperature (SST) forcing. Christensen et al. (2014), Dawson and Palmer (2014), and Muñoz et al. (2017) used weather regimes to evaluate more broadly the simulation skill of climate models, with Dawson and Palmer (2014) in particular showing that higher horizontal resolutions in NWP models improve regime simulation.

Most recently, Hannachi et al. (2017) reviewed the state of knowledge on the low-frequency fluctuations of the extratropical troposphere. These authors showed that retaining periods longer than 10 days yield statistically significant multiple regimes, while Stephenson et al. (2004), to the contrary, found no strong evidence in monthly mean reanalysis data for rejecting the single-regime multinormal hypothesis. The obvious reason for the latter result seems to be that the number of maps retained by these authors was too small or that the subspace that they retained was too large for a satisfactory estimation of the PDF. Another way of looking at the difference between the results of Hannachi et al. (2017) and those of Stephenson et al. (2004) is the fact that the persistent anomalies associated with passage through the distinct regimes in phase space rarely lasts longer than 10 days (Dole and Gordon, 1983; Kimoto and Ghil, 1993a,b). Thus, Hannachi et al. (2017) were separating such passages from the more diffuse flow in phase space outside the regimes, while Stephenson et al. (2004) were not.

Besides the fairly low degree of controversy that persists on the very existence of multiple regimes, there is still some lack of agreement as to their causes. Thus, Majda et al. (2006) found in a Hidden Markov Model analysis that metastable regime transitions can occur despite a simple model's nearly Gaussian PDF, while Sura et al. (2005) showed that deviations from Gaussianity also can result from linear, stochastically perturbed dynamics with multiplicative noise statistics. Deremble et al. (2012) showed that thermal forcing by a time-independent

SST front can affect the atmospheric flow patterns and bifurcation sequence obtained in atmospheric models with likewise time-independent but mechanical (i.e., topographic) forcing (Legras and Ghil, 1985; Jin and Ghil, 1990).

One may conclude that the number and variety of methods that have been used to identify and describe LFV regimes are leading up to a tentative consensus on their existence, robustness, and characteristics. We turn therewith to reviewing some of the theoretical foundations for the explanation of multiple regimes.

2.2 Theoretical Basis of Multiple Regimes

Rossby Wave Propagation and Interference

The slowly traveling, large-scale wave patterns that were first associated with weather phenomena in the 1930s are solutions of the partial differential equation for the conservation of potential vorticity q along a particle trajectory (Gill, 1982; Pedlosky, 1987). For the purposes of this expository review, q can be defined as the vorticity ζ of a column of fluid divided by its height h (i.e., $q = \zeta/h$). Conservation of q thus means, for instance, that a column of fluid's counterclockwise rotation (defined as $\zeta > 0$) will slow down (i.e., ζ decreases to smaller positive values) as the column moves over a mountain range (i.e., $h > 0$ decreases). This type of vorticity balance leads to slow Rossby waves (Rossby et al., 1939; Haurwitz, 1940) that propagate westward with respect to the mean westerly jet. In Chapter 4, Brunet and Methven present a more complete view of the role of such waves in S2S variability.

One view of persistent anomalies in midlatitude atmospheric flows is that they result simply from the coincidental slowing or linear interference of such Rossby waves (Lindzen et al., 1982; Lindzen, 1986). Another view is that a standing wave induced by topography can lead to a resonant interaction with two separate Rossby waves of distinct wave numbers, and thus produce a long lived, resonant wave triad (Egger, 1978; Ghil and Childress, 1987, Section 6.2). Neither of these views provides an explanation of the observed clustering of persistent anomalies into distinct flow regimes. But the second one does suggest the more radically nonlinear theory described next.

Charney and DeVore (1979) took a major step in formulating a self-consistent atmospheric model for multiple equilibria and connecting it to observations of blocked and zonal flow. They used a highly idealized barotropic model to study the interaction between a zonal flow and simple topography with zonal wave number 2. Their model exhibits two stable equilibria for the same-strength ψ_A^* of the prescribed zonal forcing, which represents here the strength of the pole-to-equator temperature contrast. Charney et al. (1981) confronted the barotropic theory of Charney and DeVore (1979) with observations, while Charney and Straus (1980) extended it to baroclinic flows. In a somewhat complementary vein, Mitchell and Derome (1983) followed up on the suggestion (e.g., McWilliams, 1980, and references therein) that steady-state solutions of the inviscid potential vorticity equation, written as $q = G(\psi, p)$, can simulate blocking patterns and may be resonantly excited by periodic forcing; here, $\psi = \psi(x, y, p)$ is the stream function, p is pressure, and (x, y) are horizontal coordinates pointing eastward and northward.

Fig. 1 shows the Charney and DeVore (1979) model's bifurcation diagram, with the strength ψ_A of the zonal jet in the model's steady-state solutions plotted against the corresponding

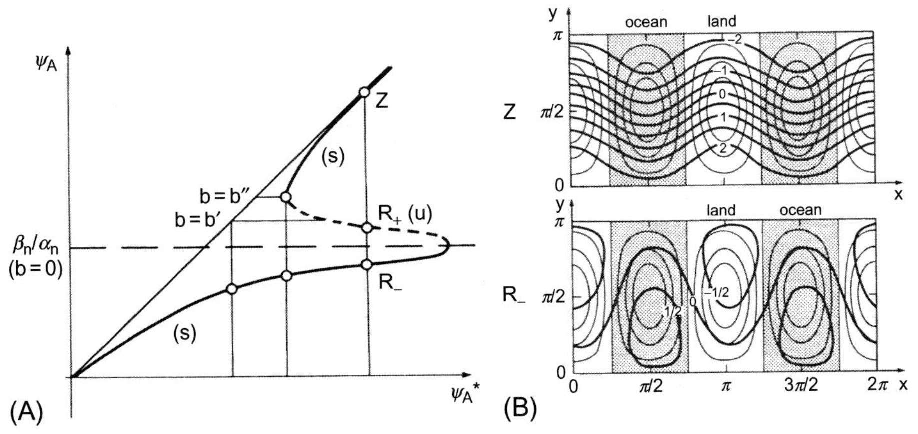

FIG. 1 Multiple equilibria of a three-mode, quasigeostrophic model with simplified forcing and topography. (A) Bifurcation diagram showing the model's response to changes in forcing. The *"S"-shaped bifurcation curve* is typical of two back-to-back, saddle-node bifurcations that give rise to two stable solution branches (*solid line*) separated by an unstable one (*dashed line*). (B) Flow patterns of the zonal (*upper*) and blocked (*lower*) branches, corresponding to the two stable equilibria Z and R_ (after Charney and DeVore, 1979). *Reprinted with permission from Ghil, M., Childress, S., 1987. Topics in Geophysical Fluid Dynamics: Atmospheric Dynamics, Dynamo Theory, and Climate Dynamics. Springer, New York. Copyright 1987, Springer Nature.*

strength ψ_A^* of the forcing. The two stable equilibria—marked Z and R_-—are associated with zonal and blocked flow, respectively, as illustrated in Fig. 1B. The near-zonal solution is close in amplitude and spatial pattern to the forcing jet and is influenced very little by the topography, whereas the blocked solution is strongly affected by it. In the blocked-flow solution, a high-amplitude ridge is located upstream of the model's highly idealized mountains, a situation that is highly similar to the one that prevails during typical observed blocks off the west coast of North America. This configuration, with a negative zonal pressure gradient on the windward slope of the mountains, corresponds to a negative mountain torque on the atmosphere.

Benzi et al. (1986) and Hansen and Sutera (1995) found evidence of bimodality in a composite index of wave amplitude in the Northern Hemisphere (NH) midlatitude flow. Although the statistical significance and robustness of their findings have been subject to criticism (Nitsche et al., 1994), direct confrontation of theoretical bimodality with observations has clearly stimulated LFV research during the 1980s. Further comparisons between the various approaches to LFV and to S2S prediction are drawn in Section 5.

A reasonable classification of low-pass-filtered flow maps into discrete regimes provides only a static view of LFV. The next step is to study the transitions between these regimes over time. A matrix of probabilities for transitions from regime *i* to regime *j* is constructed simply by counting the transitions occurring in the dataset. This yields an estimated set of conditional probabilities, in line with long-range forecasting experience (Namias, 1968; Kalnay and Livezey, 1985), and the physical intuition that certain pathways of transition are more probable than others.

One kinematic approach to LFV is based on the Markov chain of these transitions. In this approach, knowledge of the system's present state is put to use to make a forecast, rather than using only unconditional probabilities. The Markov-chain view of LFV (and, hence,

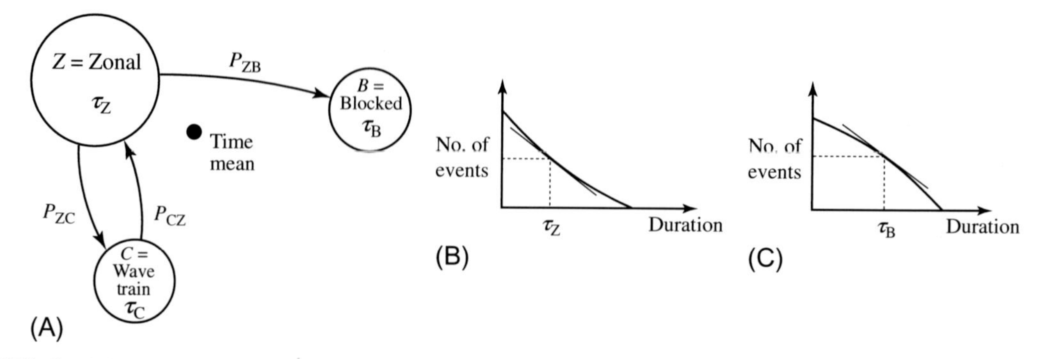

FIG. 2 Schematic Markov chain with three regimes, B, C, and Z, for "Blocked," "Wave Train," and "Zonal." (A) Some preferential paths between pairs of regimes are shown, along with the corresponding transition probabilities, such as p_{ZB}. (B and C) The distribution of residence times in log-linear coordinates differs from the *straight line associated with a red-noise process*; the mean residence time for each regime is denoted by τ. *(Reprinted with permission from Ghil, M., 1987. Dynamics, statistics and predictability of planetary flow regimes. In: Nicolis C., Nicolis G. (Eds.) Irreversible Phenomena and Dynamical Systems Analysis in the Geosciences. D. Reidel, Dordrecht, Boston, Lancaster, pp. 241–283. Copyright 1987, Springer Nature.*

long-range forecasting) is based on the existence of multiple regimes, the expected time of residence in each regime, and the probabilities of transitioning from one regime to another (Fig. 2).

Finally, several simple and intermediate model studies have addressed the issue of early warning indicators for regime transitions. For instance, Kondrashov et al. (2004) used the PDF of the exit angle from a given regime in a global baroclinic, quasigeostrophic, three-level (QG3) model with topography, originally due to Marshall and Molteni (1993), while Deloncle et al. (2007) used for the same model a fairly recent statistical learning method called random forests (Breiman, 2001). Kondrashov et al. (2007) then applied the random forests algorithm for studying regime transitions in NH datasets, while Tantet et al. (2015) used transfer operator theory for the study of transitions from zonal to blocking events in a hemispheric barotropic model. Methods for the study and prediction of regime transitions are discussed in greater detail in Section 4.2, after having provided additional insight into LFV dynamics in Section 3.2.

3 EXTRATROPICAL OSCILLATIONS IN THE S2S BAND

Phenomenological description of the extratropical oscillations in the S2S band. Variations of atmospheric angular momentum (AAM). Topographic instability and Hopf bifurcation; transition to irregular, chaotic behavior. Extratropical oscillations in a GCM with no Madden-Julian Oscillation (MJO) in the tropics.

3.1 Phenomenological Description

Variations of Geopotential Height

In this section, we analyze the geopotential height field at 500 hPa (Z500) from the European Reanalysis (ERA)-Interim reanalysis of Dee et al. (2011), which spans 37 years

(January 1979 to December 2016). We focus on the region 20°S–90°N, so as to include the tropics and the NH extratropics. This choice, along with that of the Z500 level, will become clearer as we proceed to the results.

To identify spectral components of spatiotemporal behavior in the extensive ERA-Interim reanalysis datasets, we apply here the multichannel singular spectrum analysis (M-SSA) methodology. For a comprehensive overview of the SSA and M-SSA methodology and of the related literature, see Ghil et al. (2002) and Alessio (2016, Chapter 12). M-SSA essentially diagonalizes the lag-covariance matrix of the multivariate dataset to yield a set of EOFs and the corresponding eigenvalues, which describe the variance captured by each EOF. In contrast to classical principal component analysis (PCA), M-SSA is able to capture oscillatory behavior in pairs of EOFs with approximately equal eigenvalues and dominant frequencies (Vautard and Ghil, 1989; Plaut and Vautard, 1994). To improve the separability of distinct frequencies, we rely here on a subsequent varimax rotation of the EOFs (cf. Groth and Ghil, 2011).

Prior to the M-SSA, the composite seasonal cycle is first subtracted from the daily time series. This cycle is computed by averaging the time series at each grid point over all years for each calendar day, and by smoothing this average using a 15-day running mean. Next, following Feliks et al. (2010), we apply a Chebyshev type I low-pass filter to the time series of anomaly maps (i.e., of the differences between the raw maps and the seasonal cycle so obtained) that removes high-frequency oscillations with a period shorter than 20 days. The low-pass-filtered Z500 anomalies are subsampled with a sampling rate of 10 days, which yields time series of roughly 1300 samples. Finally, the subsampled anomalies are projected onto the 40 leading spatial EOFs of a classical PCA, capturing 90% of the total variance. The corresponding principal components, finally, give the $D = 40$ input channels for the M-SSA; see Groth et al. (2017) for additional details of M-SSA analyzing extensive reanalysis datasets.

Oscillatory Features in Time and Space

Fig. 3A shows the eigenvalue spectrum of the subsampled Z500 anomalies. The spectrum exhibits several highly significant spectral peaks in the intraseasonal band; their frequencies agree very well with those identified and described in detail in previous studies (Ghil and Mo, 1991; Plaut and Vautard, 1994). In particular, two eigenvalue pairs stand out clearly above the red-noise spectrum at periods of 50 and 43 days, along with two shorter-period oscillations at 28 and 26 days.

While Weickmann et al. (1985) discussed overall oscillatory behavior in the 30- to 60-day band, Dickey et al. (1991) showed that variations in global AAM exhibit spectral peaks with distinct periods near 40 and 50 days. The latter authors demonstrated that the 50-day peak is largely associated with AAM fluctuations in the tropics, while the 40-day peak is associated primarily with variations in the strength of the midlatitude westerlies, particularly in the NH. Indeed, the amplitude of the 40-day oscillation in zonal winds is known to be largest during boreal winter, when the winds are strongest in the NH (Weickmann et al., 1985; Ghil and Mo, 1991; Strong et al., 1993, 1995)

Our analysis of the Z500 anomalies goes into greater spatial detail than the zonally averaged one of Dickey et al. (1991). Herein, both oscillations, at 50 and 43 days, make significant contributions to extratropical variability, with a Rossby wave train-like pattern across the

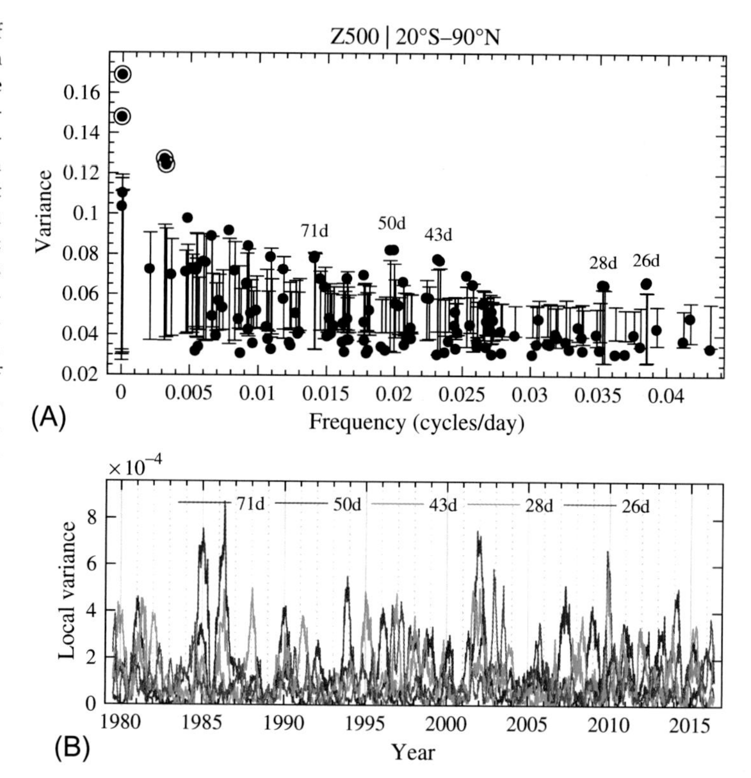

FIG. 3 Intraseasonal oscillations of 500-hPa geopotential height (Z500) in the ERA-Interim reanalysis for the time interval 1979–2016 and the region 20°S–90°N. (A) Spectrum of eigenvalues (*black dots*) plotted as a function of the associated dominant frequency from an M-SSA with a window length of $M = 35$ (350 days); the subsequent varimax rotation uses EOFs 1–40. The corresponding period length (in days) for the key oscillatory pairs in the S2S band is indicated. *Lower and upper limits of the error bars* correspond to the 2.5% and 97.5% quantiles of a Monte Carlo test against the composite null hypothesis of EOFs 1–4 plus AR(1) noise; the test ensemble has 1000 members. EOFs 1–4 (*target dots*) correspond to the trend and to a peak that lies below the S2S range of interest herein. (B) Local variance captured by the oscillatory pairs that are indicated in the *upper panel*, as a function of time.

North Atlantic extending into Eurasia, cf. Fig. 4A and B. Fig. 4A, however, also shows a larger contribution of the 50-day oscillation to the dynamics in the tropics, while a pattern that resembles the Pacific-North American (PNA) teleconnection extends northeastward from the tropical Central Pacific; the 43-day oscillation, cf. Fig. 4B, is mainly limited to the NH and contributes little to the tropical LFV, except in the Atlantic sector.

Furthermore, the 28- and 26-day oscillations that show up as significant in Fig. 3A have dynamical patterns (not shown) that resemble a westward-traveling wave of this periodicity (Branstator, 1987; Kushnir, 1987; Dickey et al., 1991; Ghil and Mo, 1991). Fig. 3A also points to a 71-day oscillation, whose period associates it with the 70-day oscillation emphasized by Plaut and Vautard (1994). The 71-day oscillation found in Z500, however, falls below the stringent 95% confidence interval level.

To visualize the dynamics underlying the variability pattern in Fig. 4A and B, we also have calculated phase composites of the 43- and 50-day mode. The phase composites of the 43-day mode are shown in Fig. 5. This mode's spatiotemporal structure is characterized by a retroregressing wave train, and it bears a remarkable resemblance to the 30- to 35-day oscillation of Plaut and Vautard (1994). In phase category 1, we observe an amplification of the North Atlantic dipole structure, while in category 3, we observe a blocking pattern over Scandinavia. The overall pattern also shows greater variability in the Atlantic sector, with very little participation by the Pacific one.

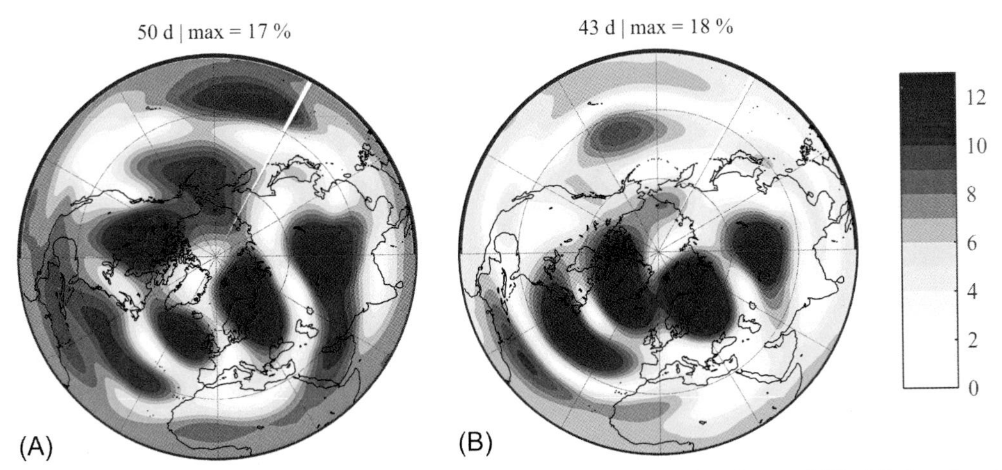

FIG. 4 Spatial structure of intraseasonal Z500 variability with (A) periods of 50 days and (B) of 43 days; based on the M-SSA in Fig. 3. Shown is the relative standard deviation (in %) at each grid point, that is, relative to the total standard deviation of overall Z500 variability.

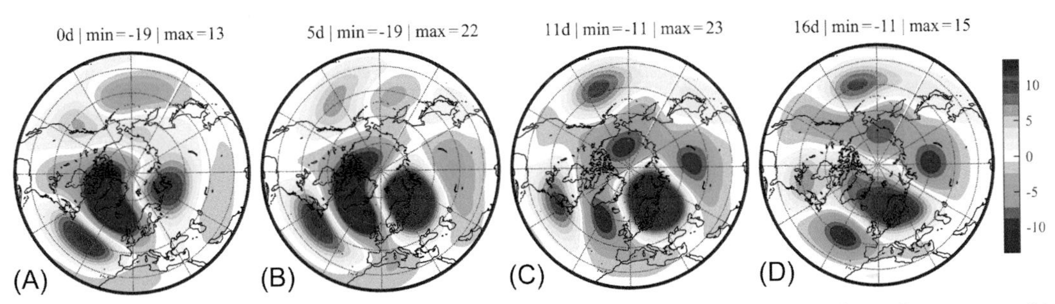

FIG. 5 Phase composites of the 43-day oscillation from the M-SSA of Z500 in Fig. 3. Four phases that correspond to one-half of the cycle are shown in *color* (in meters). The phase categories are labeled in the panel legends by their midpoint (in days); also shown are the extreme values (min and max, in meters).

The phase composites of the 50-day oscillation in Fig. 6 share certain features with the 43-day oscillations in the Atlantic sector, but they exhibit much stronger variations in the Pacific sector. The 50-day tropical amplitude, although quite weak, suggests that the Pacific sector is much more involved in the tropical-extratropical interaction, whereas the 43-day mode in Fig. 5 is more focused on the Atlantic extratropical sector, in good agreement with the results of Plaut and Vautard (1994).

The present results thus concur, overall, with the conclusion of Plaut and Vautard (1994) that tropical-extratropical interactions in the intraseasonal band are much more active in the Pacific sector and with a 50-day periodicity, whereas the 43-day mode is more active in the Atlantic sector's extratropics. A similar distinction between the MJO and a separate 50-day mode in the Indian monsoon was made by Krishnamurthy and Achuthavarier (2012) and was found to help S2S predictability there by Krishnamurthy and Sharma (2017). Even so, Cassou (2008) showed that the MJO does affect two of the four weather regimes defined

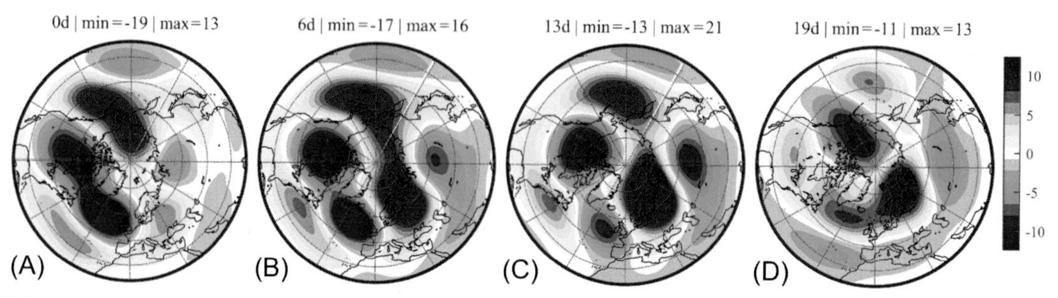

FIG. 6 Same as Fig. 5, but showing the phase composites of the 50-day oscillation.

by Vautard (1990) in the North Atlantic sector and Western Europe. Cassou's result is still consistent with Fig. 4, in which we observe a significant contribution of the 50-day oscillation in the NH.

In Fig. 3B, we have furthermore calculated the local variance (Plaut and Vautard, 1994; Groth et al., 2015) for some of the intraseasonal oscillations, which shows a strong episodic modulation. This modulation illustrates one of the interesting relationships between the episodic, multiple-regime description of atmospheric LFV and the oscillatory one, based on intraseasonal oscillations (Kimoto and Ghil, 1993b; Plaut and Vautard, 1994; Koo et al., 2002; Crommelin, 2003, 2004; Kondrashov et al., 2004). The complementarity of these two points of view, episodic and oscillatory (Ghil and Robertson, 2002), will be discussed in Section 5.

3.2 Topographic Instability and Hopf Bifurcation

We outline in this section how a hierarchy of models (Ghil, 2001; Ghil and Robertson, 2000, 2002) can be used to formulate and test the hypothesis that the 40-day oscillation is an intrinsic mode of the NH extratropics, associated with the interaction between the jet stream and midlatitude mountain ranges. The rudiments of this hypothesis originate in the highly idealized barotropic model of Charney and DeVore (1979), which was discussed in Section 2.2.

More complex models—both barotropic (i.e., single-layer) and baroclinic (i.e., multilayer), with more spatial degrees of freedom than the Charney and DeVore (1979) model—exhibit multiple flow patterns, some of which are similar to the blocked and zonal ones found in Fig. 1, even for fairly realistic values of the forcing. The crucial differences are that the equilibria found in the more complex models are no longer stable, and the system oscillates around the blocked solution or fluctuates between the zonal and blocked solutions in an irregular, chaotic way (Legras and Ghil, 1985; Ghil and Childress, 1987, Chapter 6).

The Legras and Ghil (1985) model, for instance, has 25 spherical harmonics, and its bifurcation diagram also appears as Fig. 8 in Ghil and Robertson (2002). In this diagram, the branch of blocked equilibria is destabilized by an oscillatory instability as the intensity of the forcing jet increases. The transition from a stable stationary solution—called a fixed point in dynamical systems theory—to a stable periodic solution, called a limit cycle in this theory, is termed a Hopf bifurcation. The limit cycle that arises from this bifurcation increases in amplitude as the forcing jet becomes stronger, and it has a period of roughly 40 days. This limit cycle loses its stability in turn as forcing is increased further and the flow becomes chaotic, but the 40-day periodicity survives in the power spectrum of the chaotic flow.

Jin and Ghil (1990) showed that when a sufficiently realistic meridional structure of a solution's zonal jet is allowed, the back-to-back saddle-node bifurcations of Fig. 1A are indeed replaced by a Hopf bifurcation, and thus transition to finite-amplitude periodic solutions can occur. Eigen analyses of the unstable equilibria in a barotropic model with higher horizontal resolution, as well as its time-dependent solutions, indicate oscillatory instabilities with intraseasonal (35–50 days) and biweekly (10–15 days) timescales (Strong et al., 1993). Floquet analysis of this model's limit cycles (Strong et al., 1995) confirms that the 40-day oscillations that arise in it by oscillatory topographic instability, like the observed NH oscillations (Ghil and Mo, 1991; Knutson and Weickmann, 1987), are stronger in winter than in summer.

To test the theory of NH extratropical oscillations developed in simpler models, Marcus et al. (1994, 1996) studied a 3-year perpetual-January simulation that had been performed with a version of the UCLA GCM, in which no self-sustained MJO was apparent in the tropics. A robust 40-day oscillation in AAM is found to arise in the model's NH extratropics when standard topography is present. Three shorter runs with no topography produced no intraseasonal oscillation; this result is consistent with a topographic origin for the NH extratropical oscillation in the standard model. The spatial structure of the circulation anomalies associated with the model's extratropical oscillation is shown in an 8-minute video, which can be seen at https://doi.org/10.7916/D8X36F5B, as displayed in several model-simulated fields (namely, 500-hPa geopotential heights, 250-hPa stream function, and surface pressure torques).

The oscillation is dominated by a standing wave number-2 pattern, which undergoes a predominantly barotropic, tilted-trough vacillation. High values of AAM are associated with low 500-hPa heights over the northeast Pacific and the North Atlantic oceans, and vice versa. These flow patterns resemble the configurations seen in the Charney and DeVore (1979) simple model (see Fig. 1B). The NE-SW-tilting and NW-SE-tilting phases in the video's 500-hPa fields are strongly reminiscent of the extremes and intermediate phases of the 40-day oscillation that arises by Hopf bifurcation from the blocked equilibrium in the Legras and Ghil (1985) model; see the discussion of Ghil et al. (2003, Fig. 3).

4 LOW-ORDER, DATA-DRIVEN MODELING, DYNAMICAL ANALYSIS, AND PREDICTION

Linear and nonlinear LOMs and the role of memory effects. Empirical model reduction (EMR) and EMR-based prediction. Partial observations and stochastic closure in LOMs. Slow changes in the forcing and random effects: nonautonomous and random dynamical systems.

4.1 Background and Methodological LOM Developments

State-of-the-art, highly resolved GCMs, while able to simulate detailed interactions within the climate system over a wide range of scales, generate detailed four-dimensional (4D) climate variability that is visually as complex as currently available observational datasets, and is hence no less challenging to interpret. Dynamical analysis of climatic phenomena typically involves a set of multiple GCM simulations that are designed to isolate physical processes governing the simulated, and by inference, observed climate variability. These simulations are computationally expensive, however, and their interpretation is hindered by the presence of model biases due to incomplete or imperfect parameterizations of the unresolved physical processes.

Moreover, GCMs represent a broad range of time and space scales and use a state vector that has many millions of scalar variables. While detailed weather prediction out to a few days does require such high numerical resolution, considerable work, both theoretical (Ghil and Childress, 1987, Section 6.5) and data-based (Toth, 1995), has shown that important aspects of observed atmospheric LFV can be represented and predicted with a substantially smaller number of degrees of freedom, as encapsulated by LOMs.

This statement extends beyond the issues of simulating and understanding LFV: it clearly includes prediction as well, and it is also consistent with the fact that climate concerns statistical time and space aggregates of weather. In fact, while there has been a recent surge of interest in S2S prediction through coordinated international modeling activity, most of this activity involves GCM-based ensemble prediction systems. Empirical, data-driven prediction, though, has a long history in forecasting the El Niño-Southern Oscillation (ENSO), and the best LOMs (e.g., Kondrashov et al., 2005) were still outperforming—no longer than 6 years ago—most GCM-based ENSO forecasts, even for lead times beyond the S2S range (Barnston et al., 2012). Unfortunately, since the Barnston et al. (2012) review, the plume of real-time Niño-3.4 predictions—which includes those of stochastic-dynamic models, as well as of GCMs—that is continuously monitored at the International Research Institute (IRI) for climate and society no longer allows one to isolate the skill of individual LOMs or GCMs. Still, inspection of the existing documentation at https://iri.columbia.edu/our-expertise/climate/forecasts/enso/2017-October-quick-look/?enso_tab=enso-sst_table does indicate that—over the last seven boreal winters, 2012–13 to 2017–18—ENSO prediction has remained a tough problem for both GCMs and simpler models. Hence, the contribution of the latter to real-time S2S forecasting can still be quite useful in and of itself, as well as in stimulating improvements in GCM predictions.

A theoretical approach to LOM development relies on timescale separation in the full governing equations to derive a reduced set of differential equations for the LFV and to parameterize the fast, unresolved variability—such as atmospheric convection on synoptic and mesoscales—by stochastic forcing (e.g., Epstein, 1969b; Fleming, 1971; Majda et al., 1999). An alternative approach to climate diagnosis and prediction is based on empirical modeling of the LFV, as observed in selected fields, under the assumption that it can be described as a spatiotemporal, nonlinear, and stochastic process. This statistical data-driven approach does not require knowledge of the governing equations: even though it lacks the immediate dynamical interpretability of climate models that are derived from first principles, it does allow one to reproduce detailed aspects of the observed statistics by inverse modeling techniques, albeit in the subset of fields being simulated by the LOM. The available datasets—whether given by direct observations, reanalyses, or GCM simulations—are used to estimate both the model's low-order, deterministic part and its driving noise.

Development of data-driven LOMs can thus be cast as a closure problem: These models can simulate and predict LFV, provided that they are able to account properly for (a) linear and nonlinear interactions between the resolved components (i.e., the high-variance LFV modes); and (b) the interactions among the resolved components and the huge number of unresolved ones that represent the unobserved small-scale processes and are not explicitly included in the LOM. The key steps, thus, are estimating interactions among the macroscopic variables, identifying the hidden variables, and modeling the cross-interactions among the macroscopic and hidden variables.

LIMs and POP analysis (Penland, 1989, 1996; Penland and Ghil, 1993; Penland and Sardeshmukh, 1995) assume a linear dynamical model for the macroscopic variables, while their interactions with the small-scale, hidden processes are approximated by spatially correlated white noise. The EMR approach (Kravtsov et al., 2005, 2009; Kondrashov et al., 2013) generalizes LIMs by allowing (i) quadratic interactions among macroscopic variables; (ii) time-delayed dynamical effects via memory terms; and (iii) a richer temporal structure of the noise. The memory effects that may appear in both the stochastic forcing and the dynamical operator are conveyed by hidden variables arranged into a stacked system of levels. Each additional level includes a new hidden variable, which is less auto-correlated than the one included on the previous level, until the last-level variable can be approximated by spatially correlated white noise.

Chen et al. (2016) conducted a comprehensive suite of EMR experiments using a 4000-year preindustrial control simulation dataset from the CM2.1 coupled model of the Geophysical Fluid Dynamics Laboratory (GFDL) at the National Oceanic and Atmospheric Administration (NOAA), with the goal of better understanding ENSO diversity, nonlinearity, seasonality, and the effects of memory in the simulation and prediction of the Tropical Pacific's SST anomalies. The results show that multilevel nonlinear EMR models that account for SST history improve ENSO prediction skill dramatically.

The EMR model coefficients introduced on each level are estimated by multilevel regression techniques; see Kravtsov et al. (2009) for a comprehensive review of the EMR methodology. Strounine et al. (2010) systematically compared various model reduction methods, as applied to wintertime geopotential height anomalies, and demonstrated that, in the absence of clear scale separation, EMR methodology's success is rooted in its multilayer structure, which accounts for the memory effects needed to achieve optimal closure (Kravtsov et al., 2005).

Newman et al. (2003) showed that at the beginning of the past decade, the forecast skill of a LIM model of NH extratropical, weekly averaged circulation is comparable to that of NCEP's medium-range forecast (MRF) model at week 2 (days 8–14) and week 3 (days 15–21). Zhang et al. (2013) and Vitart (2017) have documented the progress in MJO simulation and prediction by GCMs since then. Thus, the best GCM predictions have achieved a bivariate correlation skill near 0.6 at 4 weeks, as measured between the two observed and forecast real-time multivariate MJO indices known as RMM1 and RMM2.

At the same time, LOMs have also progressed beyond LIMs, as outlined earlier. Kondrashov et al. (2013) improved upon the EMR-based predictions used in real-time ENSO forecasting (Kondrashov et al., 2005) by including in the forecast ensemble information on the "weather noise" that prevailed at particular phases of the MJO cycle, via the past-noise forecasting (PNF) methodology of Chekroun et al. (2011a). These authors demonstrated that in retrospective forecasting, PNF-enhanced EMR methodology yields bivariate-RMM correlation results that are quite comparable to those of the best GCMs (Kondrashov et al., 2013, Fig. 2).

More broadly, EMR-based climate predictions have proved to be highly useful and competitive as statistical benchmarks against physics-based models for prediction on the S2S timescales associated with the MJO (Kondrashov et al., 2013), as well as with ENSO (Kondrashov et al., 2005).

The multilayer stochastic model (MSM) framework introduced by Kondrashov et al. (2015) establishes, in a data-driven context, a clear connection between the EMR methodology and the Mori-Zwanzig (MZ) formalism of statistical physics. It is this sound physical basis that helps derive a closed system of stochastic-dynamic equations for a subset of the original system's variables (Kondrashov et al., 2015, Sections 4 and 5). This closed system of equations includes Markovian and non-Markovian deterministic terms that are not necessarily quadratic, as well as stochastic noise terms (Chorin et al., 2002; Chorin and Hald, 2006).

The major methodological problem is that in the highly nonlinear climate system, there is typically a continuum of scales, from the fastest and shortest to the slowest and longest. In this case, the MZ formalism allows one to treat the interactions between LFV and smaller scales as memory effects. These memory effects represent a significant deviation from the Markovianity of the dynamical models that are most often used in modeling atmospheric, oceanic, and climate phenomena and processes. Markovianity means dependence on the initial state alone, as opposed to states that preceded it. Thus, ordinary and partial differential equations, as well as the Markov chains mentioned in Section 2, are Markovian: delay-differential and other functional differential equations (Hale and Verduyn Lunel, 1993) are not. Bhattacharya et al. (1982) introduced the latter into the climate sciences, and they have been used extensively since then in ENSO modeling, with memory terms that account for travel times of Kelvin and Rossby waves across the Tropical Pacific (see also Ghil et al., 2015, and references therein).

The non-Markovian terms become particularly important in the absence of scale separation, whereas Stinis (2006) showed that, for relatively large-scale separation, the results of the MZ methodology are similar to those of the model reduction formalism of Majda et al. (1999). The MSM framework thus improves upon EMR by its greater generality of the deterministic terms, as well as by the proper understanding of the memory effects. The improvements in simulation and prediction achieved by nonlinear (EMR and MSM) versus linear (LIM and POP) data-driven models (e.g., Barnston et al., 2012) seem to further buttress the credibility of the highly nonlinear Devil's Staircase scenario for ENSO (Jin et al., 1994; Tziperman et al., 1994) versus stable noise-forced dynamics (Neelin et al., 1998).

Kondrashov and Berloff (2015) showed that a change of basis by M-SSA can help in applying the MSM methodology to capture the LFV in multiscale oceanic turbulence because M-SSA utilizes time-lagged information and therefore implicitly conveys the memory effects. Data-adaptive harmonic decomposition (DAHD; Chekroun and Kondrashov, 2017) further advances the dynamic interpretation of nonlinear data-driven models. DAHD provides, for a broad class of time-evolving datasets, reduction coordinates that can be efficiently modeled by systems of paired stochastic differential equations (SDEs), while using a fixed set of predictor functions. The methodology has proven itself already in the modeling, simulation, and prediction of the Arctic's sea ice extent and concentration field (Kondrashov et al., 2017, 2018), but it remains to be tested for other climatic time series.

In summary, the field of low-order, data-driven stochastic-dynamic models has undergone remarkable development in recent years. These models offer, therewith, new opportunities to further our understanding and prediction on S2S timescales. The following two sections present examples of their successful application to midlatitude LFV variability and to ENSO phenomena, respectively.

4.2 Dynamical Diagnostics and Empirical Prediction on S2S Scales

LOMs can be very useful for the dynamical interpretation of atmospheric oscillations and multiple regimes. Such models not only help compact the dataset's information content, but also can provide insights into the dynamics and predictability of climatic LFV via the analysis of the reduced model's mathematical structure. Thus, Kondrashov et al. (2006, 2011) constructed and analyzed an LOM of extratropical atmospheric LFV by applying EMR to the output of a long simulation of the QG3 model mentioned in Section 2.2. This model has a fairly realistic climatology and variability, and it has been used quite extensively to study NH midlatitude flows (e.g., D'Andrea and Vautard, 2001; D'Andrea, 2002; Deloncle et al., 2007, and references therein).

The full QG3 model's phase space has a dimension of $\mathcal{O}(10^4)$, while the derived EMR model has only 45 variables. The regimes of both the EMR model and the QG3 model were computed using a Gaussian mixture model by following Smyth et al. (1999) and references therein. The centroids of the four regimes for the full QG3 model are plotted in Fig. 7, and they correspond to opposite phases of the Arctic Oscillation (AO^+ and AO^-) and of the North Atlantic Oscillation (NAO^+ and NAO^-), in good agreement with the earlier results of Molteni and Corti (1998). The EMR model captures very well the non-Gaussian features of the full QG3 model's PDF and shares with the latter its four anomalously persistent flow patterns, as well as the Markov chain of transitions between these regimes (not shown).

The spatial correlations between the anomaly patterns in Fig. 7 and those of the EMR model (not shown) all exceed 0.9. This good match between the regime centroids in the two models, full and reduced, as well as between the corresponding Markov chains, emphasizes the intrinsic dynamical nature of both the NAO and the AO and strongly suggests that the NAO is not merely a regionalization of the AO, at least not in the fairly realistic QG3 model.

The four regimes that are best supported by synoptic experience, as well as by the statistical analysis of the upper-air data for the past half-century, are the zonal and blocked phases of westerly flow in the Atlantic-Eurasian and Pacific-North-American sector, respectively (Cheng and Wallace, 1993; Smyth et al., 1999; Ghil and Robertson, 2002). The AO, also called the NH annular mode (Wallace, 2000), seems to be only statistically annular: it represents a redistribution of mass between the poles and subtropics, but actual "sloshing" events are sectorial, by and large. Hence, the correlation between the subtropical highs over the North Atlantic and the North Pacific in observational and reanalysis data is quite low at sub-monthly timescales (e.g., Kimoto and Ghil, 1993b).

Both the AO and NAO indices are dominated by Arctic sea-level pressure variations, however, and hence the NAO and AO are highly correlated in observations. Moreover, both the QG3 model (Kondrashov et al., 2006) and the no-MJO version of the UCLA GCM (Marcus et al., 1994, 1996) seem to possess some form of 40-day sloshing mode that originates in midlatitudes. Seeing the hemispheric AO^+ and AO^- patterns of Fig. 7C and D replace the PNA-sector patterns of Ghil and Robertson (2002) and references therein is thus not too surprising: the hemispheric $AO^\pm$ centroids overlap largely in the Atlantic-Eurasian sector with $NAO^\pm$,

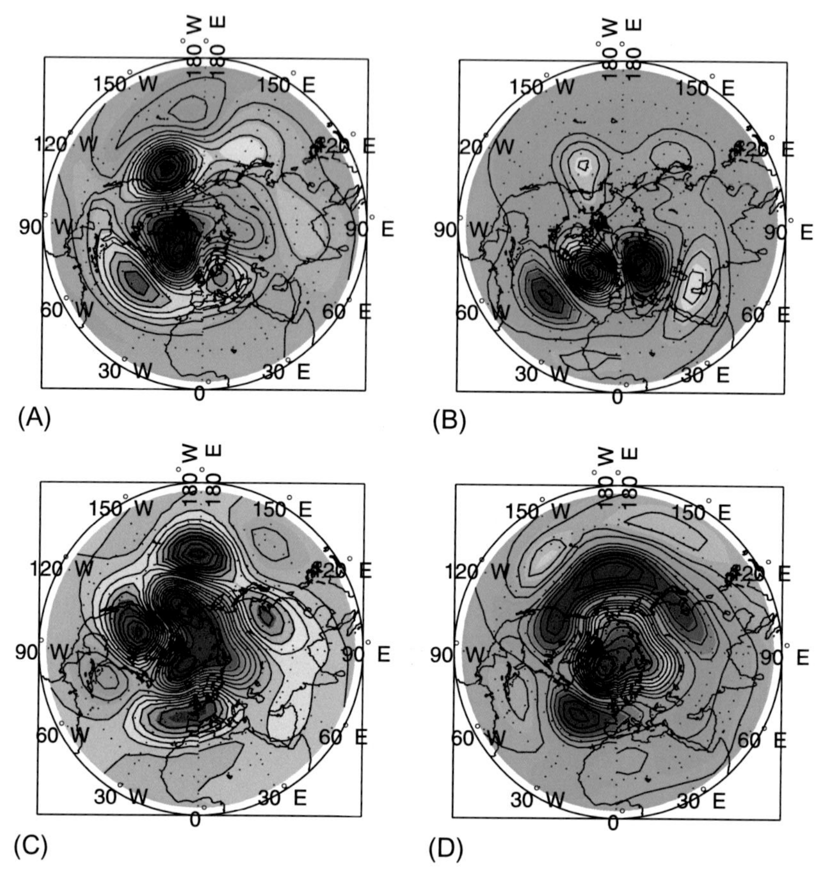

FIG. 7 Gaussian mixture-model centroids, showing stream function anomaly maps at 500 hPa, for the QG3 model: (A) NAO$^+$; (B) NAO$^-$; (C) AO$^+$; and (D) AO$^-$. Positive contours are *solid* and landmasses are *shaded*; 20 contour levels between maximum and minimum values are used, with the following intervals (in 10^6 m^2/s): (A) 1.1; (B) 0.8; (C) 0.8; and (D) 1.1. *After Kondrashov, D., Kravtsov, S., Ghil, M., 2006. Empirical mode reduction in a model of extratropical low-frequency variability. J. Atmos. Sci. 63 (7), 1859–1877. https://doi.org/10.1175/jas3719.1. American Meteorological Society. Reproduced with permission.*

respectively, from which they differ mainly by much stronger centers of action in the PNA sector.

In addition, M-SSA identifies intraseasonal oscillations with a period of 35–37 days and of 20 days in the data generated by both the full QG3 model and its EMR version. These oscillatory modes are similar to those seen in the observations, as plotted in Fig. 3, and in previous studies (e.g., Ghil and Robertson, 2000, 2002, and references therein). The former one is clearly these models' version of the extratropical, 40-day mode found by Legras and Ghil (1985), while the QG3 model is expected to lack the MJO-related 50-day mode seen in Fig. 3. Lott et al. (2001, 2004a,b) showed that observational NH data confirm the importance of mountain torques in an extratropical oscillation that has a shorter periodicity than the MJO. Mountain torque anomalies were also found to anticipate the phases of the AAM oscillation, as well as the onsets and breaks of certain flow regimes.

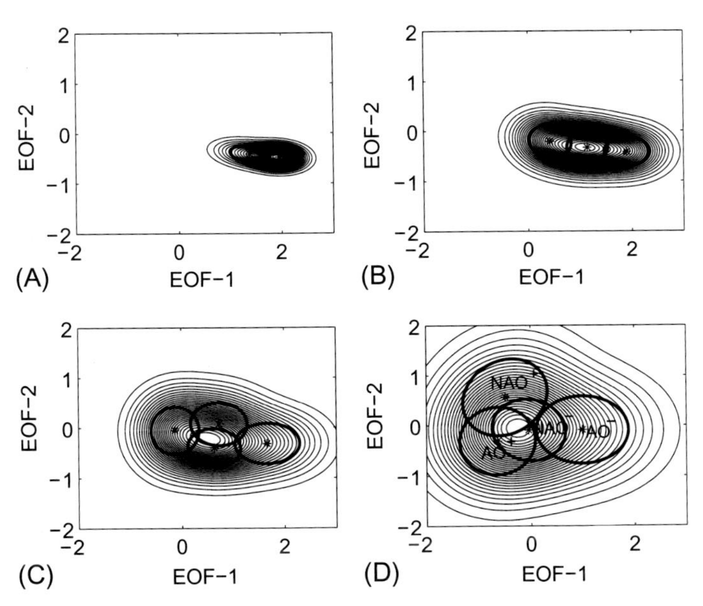

FIG. 8 Mixture-model clusters for EMR model simulations of the full QG3 atmospheric model. The four panels differ by the amplitude of the EMR's stochastic forcing, which is multiplied by a factor ϵ: (A) $\epsilon = 0.2$; (B) $\epsilon = 0.4$; (C) $\epsilon = 0.6$; and (D) $\epsilon = 1$ (optimal case). Note that the number of the clusters and their separation in phase space increase with ϵ. The four regimes are only labeled in panel D. *(After Kondrashov, D., Kravtsov, S., Ghil, M., 2006. Empirical mode reduction in a model of extratropical low-frequency variability. J. Atmos. Sci. 63 (7), 1859–1877. https://doi.org/10.1175/jas3719.1. American Meteorological Society. Reproduced with permission.)*

The QG3 model's LFV features can be interpreted via dynamical analysis of the reduced model. In particular, the AO^- regime arises from the unique steady state of the EMR model's three-level deterministic operator, at the larger positive values of PC-1, while the PDF ridge that is present in the QG3 model (cf. Kondrashov et al., 2006, Fig. 6), coincides with the location of a plateau of quasi-stationary states of the reduced model; cf. Fig. 8B–D. The increase in the extent of the PDF plateau in both the full and the reduced models—first along the EOF-1 direction and then at lower negative values of PC-1, along the EOF-2 direction—appears to be due to preferential instabilities in the subspace of the QG3 model's leading modes. These instabilities are to be equilibrated by the full model's transient feedback that is captured in the EMR by the noise and memory effects.

Changes in the number and amplitude of the PDF's modes were also observed by Molteni and Corti (1998) when varying the potential vorticity forcing in the QG3 model to represent a subtropical Rossby wave source appropriate for cold ENSO events. The changes plotted in Fig. 8 can be interpreted as changes that might arise in a GCM subject either to stochastic parametrizations with varying noise intensities (e.g., Palmer and Williams, 2009, and references therein) or to fluctuating changes in external forcing. The rather surprising success of EMRs, in their various guises, at simulating and predicting S2S variability may be largely attributed to the presence and proper intensity of the noise processes they include.

The dominant intraseasonal oscillation in both the QG3 and the optimal EMR model has a period of about 35–37 days, and it is associated with the least-damped eigenmode of the latter when linearized about its climatological state. While there is no clear scale separation in the QG3 simulations, it is the interactions between the model's largest-scale modes—whose variance is concentrated in the subspace of its four leading EOFs—and the intermediate scales, captured by EOFs 5–15, that appear to be responsible for the slowdown of the intraseasonal oscillation's trajectory in the full QG3 model, as well as in the optimal EMR model. This slowdown is associated with the emergence of the AO^+ and NAO^- regimes.

The EMR model's stochastic forcing represents the unresolved smaller-scale processes in the QG3 model. The effect of this forcing on the full dynamics is revealed by a bifurcation analysis with respect to the noise's amplitude ϵ. In particular, Fig. 8 shows that as ϵ increases from zero to its optimal value, the model trajectory is initially confined to a small region near the EMR model's unique steady state (panel A) and that it gradually fills up the quasistationary ridge along the EOF-1 axis (panels B–D). When ϵ becomes large enough, the intraseasonal oscillatory mode is excited, resulting in the model's PDF expanding in the EOF-2 and EOF-3 directions, while additional quasistationary states appear. The sequence of panels in Fig. 8 can be taken to imply that the stochastic forcing needs to be sufficiently strong to obtain realistic LFV regimes. The role of stochastic processes in nonlinear dynamics will be discussed more fully at the end of this chapter, in Section 5.

Kondrashov et al. (2011) also analyzed this EMR model's nonlinear dynamical operator and showed that it gives rise to "swirls" in the time-mean tendencies of the model's state-averaged trajectories when projected onto a low-dimensional subspace. These swirls are dominated by the deterministic interactions between the large-scale modes. More recently, Tantet et al. (2015) associated similar swirl patterns with persistent weather regimes in the output of a barotropic atmospheric model with an intermediate resolution of 231 spherical modes. These authors applied transfer operator methods (Chekroun et al., 2014, and references therein) to develop early warning indicators for regime transitions, in particular from zonal to blocked flows, as mentioned in Section 2.2.

The idea of considering preferred directions of instability of a regime's centroid goes back to the "ghost equilibria" of Legras and Ghil (1985). In particular, Deloncle et al. (2007) and Kondrashov et al. (2007) have applied the random forests algorithm (Breiman, 2001) to learn the predictors connected to these preferred directions in a low-dimensional space. This statistical approach worked rather well in medium- to long-range prediction of regime breaks in both the QG3 model (Deloncle et al., 2007) and NH atmospheric observational data (Kondrashov et al., 2007). The results of Tantet et al. (2015) provide further dynamical understanding of why that appears to be the case. In the context of S2S prediction, the dynamically motivated predictors might turn out to correspond to preferred instability directions of either fixed points that anchor multiple regimes, such as El Niño and La Niña (Dijkstra, 2005, Chapter 7, and references therein), or to similar instabilities of limit cycles that are associated with oscillatory modes, such as the MJO or the extratropical, 40-day mode discussed in this section and in Section 3.1. Instability of limit cycles in atmospheric LFV was studied, for instance, by Strong et al. (1995), and is mentioned in Section 3.2.

4.3 LFV and Multilayer Stochastic Closure: A Simple Illustration

In this section, we provide an illustrative example that will help the reader understand the reasons for the effectiveness of MSM modeling, and hence of EMR models, in simulating and potentially predicting extratropical S2S variability. This example illustrates intraseasonal ENSO prediction.

Consider the following nonlinear, periodically and stochastically forced, two-variable (x_1, x_2) model from Chekroun et al. (2011a) Supplementary Information:

$$\begin{cases} \mathrm{d}x_1 = \{(r+\sigma \mathrm{d}W_t)x_1(\alpha+x_1)(1-x_1) - cx_1x_2 + a\sin(2\pi f_0 t)\}\mathrm{d}t, \\ \mathrm{d}x_2 = \{-m\alpha x_2 + (c-m)x_1x_2\}\mathrm{d}t. \end{cases} \tag{1}$$

Here, the variable x_1 is subject to deterministic, additive forcing $a \sin(2\pi f_0 t)$, while the stochastic term σdW_t represents a random perturbation of the parameter r by white noise, and we assume that only the variable x_2 is observed. The default values of the model parameters are $\sigma = 0.3$, $m = r = 1$, $c = 1.5$, and $\alpha = 0.3$. The model is integrated from time $t = 0$ to $t = T_f = 2000$ (in dimensionless units), with the initial state $(x_1(0), x_2(0)) = (0.5, 0.5)$, by using a classical stochastic Euler-Maruyama scheme with step size $\Delta t = 0.1$. When both periodic forcing and noise are absent, the model has only one globally stable equilibrium.

When turning on the periodic forcing with amplitude $a = 0.05$ and frequency $f_0 = 0.25$, the system exhibits only one periodic orbit of period 4, which is globally stable. In the presence of noise, with $\sigma = 0.3$, an LFV mode with a period equal to approximately 25 units—i.e., a frequency of $f = 0.04 \neq f_0$—becomes dominant, as seen in Fig. 9. This mode (shown in blue in the figure) is captured by SSA reconstruction, and it can be attributed to a damped nonnormal mode that is sustained by the noise (Chekroun et al., 2011a). In the present ENSO-related illustration, the periodic forcing $T_0 = 1/f_0 = 4$ corresponds to the seasonal cycle, while the intrinsic periodicity of $T \simeq 25$ is associated with ENSO's low-frequency mode of roughly 6 years.

By applying EMR solely to a time series of the observed variable x_2 (cf. Kravtsov et al., 2005), a univariate, two-level EMR model is obtained for x_2 alone. This scalar model includes a cubic polynomial, as well as multiplicative, interaction with the periodic forcing, on its main level (not shown). Note that this parametric form is entirely different from the correct form of

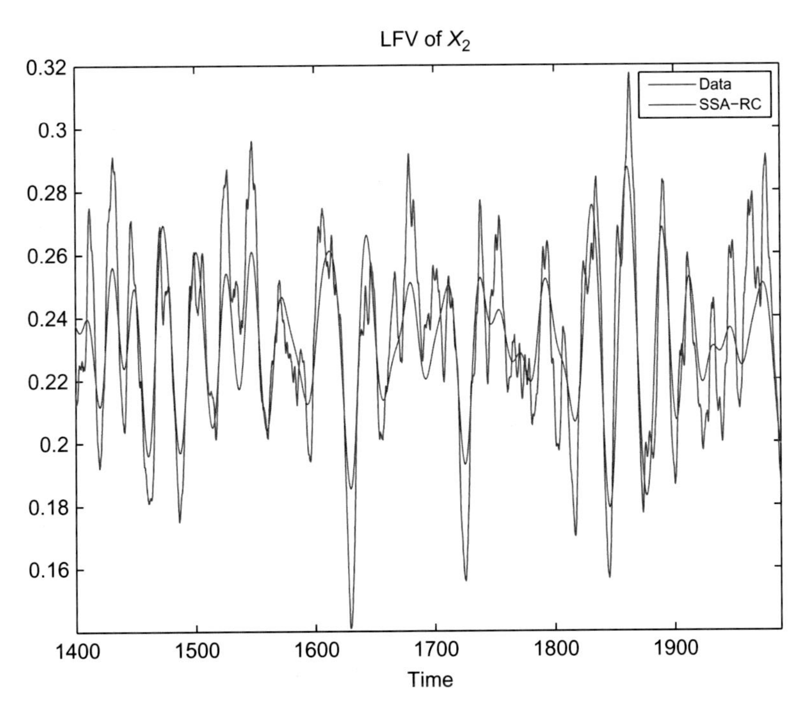

FIG. 9 Time series of the component x_2 (*red curve*) and its LFV mode obtained by SSA reconstruction (*blue curve*). The latter has a dominant period of 25 nondimensional units, and it captures 36% of the total variance.

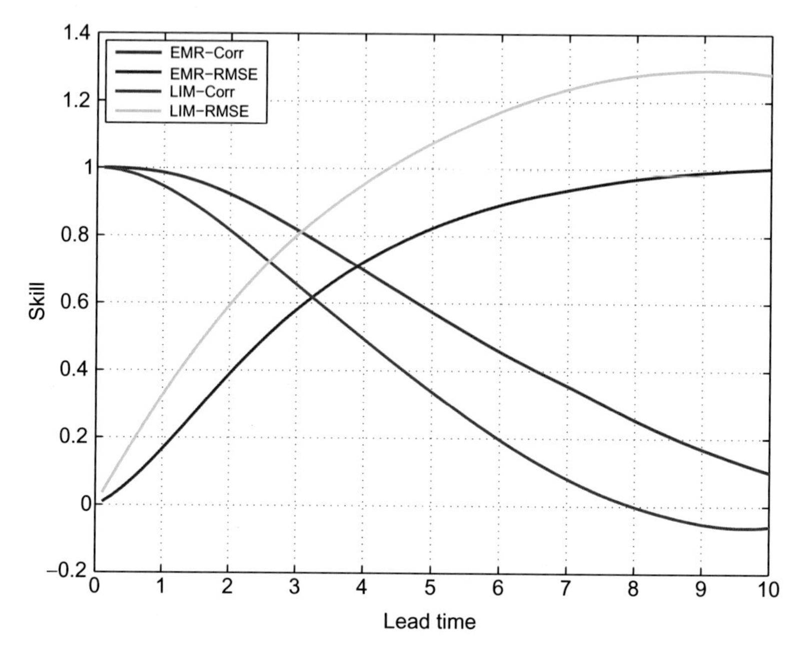

FIG. 10 Prediction skill for the optimal EMR model obtained solely from the $x_2(t)$ time series. The best cubic two-level EMR model clearly outperforms LIM both in terms of anomaly correlations (*red curve vs magenta*) and root-mean-square (RMS) errors (*blue vs cyan*).

the right-hand side of the x_2-equation in Eq. (1), as it does not explicitly include interactions with the periodically and stochastically forced unobserved variable x_1. These interactions are parameterized by the memory effects conveyed in the two-level EMR model.

Prediction skill is plotted in Fig. 10 in terms of both root-mean-square errors (RMSE) and anomaly correlations (Corr). The figure shows that the out-of-sample prediction skill of the ensemble mean for the estimated EMR model is significantly better than that of a linear, single-level LIM model, with the EMR maintaining useful prediction skill (i.e., Corr ≥ 0.5), up to a lead time of roughly 6 time units, which represents roughly a quarter of a period of the LFV mode.

This example illustrates the benefits for prediction of including nonlinear and memory terms in empirical stochastic models, as demonstrated by comparing the performance of the UCLA EMR model in real-time ENSO forecasting with other statistical forecasts; for more information, see Barnston et al. (2012).

5 CONCLUDING REMARKS

Considerable progress has been made in the 15 years since Ghil and Robertson (2002) in low-order predictive modeling of what was called at that time extended or long-range forecasting and is now more specifically referred to as S2S prediction. Four of the approaches that were considered in this modeling are summarized in Fig. 11.

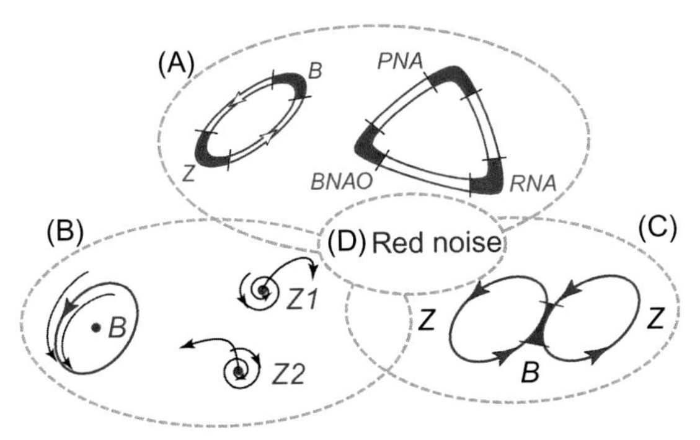

FIG. 11 Schematic overview of atmospheric LFV mechanisms.

One approach to persistent anomalies in midlatitude atmospheric flows on 10- to 100-day timescales is to consider them simply as due to the slowdown of Rossby waves or to their linear interference (Lindzen et al., 1982; Lindzen, 1986). This approach is illustrated in the sketch labeled (c) within the figure: zonal flow Z and blocked flow B are simply slow phases of an harmonic oscillation, like the neighborhood of $t = \pi/2$ or $t = 3\pi/2$ for a sine wave $\sin(t)$; or else they are due to an interference like that occurring for a sum $A\sin(t) + B\sin(3t)$ near $t = (2k + 1)\pi/2$. A more thorough, quasilinear version of this approach is to study long-lived resonant wave triads between a topographic Rossby wave and two free Rossby waves (Egger, 1978; Trevisan and Buzzi, 1980; Ghil and Childress, 1987, Section 6.2). Neither version of this approach, though, explains the anomalies' organizing into distinct flow regimes.

Rossby et al. (1939) initiated a different, genuinely nonlinear approach by raising the possibility of multiple equilibria as an explanation of preferred atmospheric flow patterns. These authors drew an analogy between such equilibria and hydraulic jumps, and formulated simple models in which similar transitions between faster and slower atmospheric flows could occur. This multiple-equilibria approach was pursued quite aggressively in the 1980s (Charney and DeVore, 1979; Charney et al., 1981; Legras and Ghil, 1985; Ghil and Childress, 1987, Sections 6.3–6.6) and it is illustrated in Fig. 11 by the sketch labeled (a): one version of the sketch illustrates models that concentrate on the B–Z dichotomy (Charney and DeVore, 1979; Charney et al., 1981; Benzi et al., 1986), and the other on models (e.g., Legras and Ghil, 1985) that allow the presence of additional clusters, like those found by Kimoto and Ghil (1993a) and Smyth et al. (1999), viz. opposite phases of the NAO and PNA anomalies (*PNA*, *RNA*, and $BNAO \simeq NAO^{-}$ in sketch (a) of Fig. 11). The LFV dynamics in this approach is given by the preferred transition paths between the two or more regimes; look again at Table 1 in Section 2.1 and references therein.

A third approach is associated with the idea of oscillatory instabilities of one or more of the multiple fixed points that can play the role of regime centroids. Thus, Legras and Ghil (1985) found a 40-day oscillation arising by Hopf bifurcation off their blocked regime B, as illustrated in sketch (b) of Fig. 11. An ambiguity arises, though, between this point of view and a complementary possibility—namely, that the regimes are just slow phases of such an oscillation, caused itself by the interaction of the midlatitude jet with topography (cf. Section 3.2). Thus, Kimoto and Ghil (1993b) found, in their observational data, closed paths within a Markov chain whose states resemble well-known phases of an intraseasonal oscillation. Such a possibility was confirmed in the QG3 model by Kondrashov et al. (2004). Furthermore, multiple regimes and intraseasonal oscillations can coexist in a two-layer model on the sphere within the scenario of "chaotic itinerancy" (Itoh and Kimoto, 1996, 1997).

Finally, sketch (d) in Fig. 11 refers to the role of stochastic processes in S2S variability and prediction, whether it be noise that is white in time, as in Hasselmann (1976) or in LIMs (Penland, 1989, 1996; Penland and Ghil, 1993; Penland and Sardeshmukh, 1995), or red in time, as in EMRs and MSMs (Kravtsov et al., 2005, 2009; Kondrashov et al., 2006, 2013, 2015), or even non-Gaussian (Sardeshmukh and Penland, 2015). Stochastic processes may enter into models situated on various rungs of the modeling hierarchy, from the simplest conceptual models to high-resolution GCMs. In the latter, they may enter via stochastic parametrizations of subgrid-scale processes (e.g., Palmer and Williams, 2009, and references therein), while in the former, they may enter via stochastic forcing, whether additive or multiplicative, Gaussian or not (e.g., Kondrashov et al., 2015, and references therein).

Fig. 11 simply summarizes some of the key dynamical mechanisms of midlatitude S2S variability, as discussed in this chapter, without providing a definitive answer to which approach to low-order modeling and prediction is most productive or will be so in the near future. Such an answer is likely to be long in coming, if ever. On the more practical aspects of S2S prediction, though, it is clear that the ideas presented herein—on the role of LOMs versus more detailed and highly resolved models, and on extratropical sources of variability and hence of predictability—are certain to be useful. The most promising LOMs appear to be those that do include nonlinear dynamics, memory effects, and colored noise. EMRs and MSMs that do this have proved quite competitive in ENSO forecasting, according to Barnston et al. (2012), and are likely to provide important prediction benchmarks for S2S forecasting, as well as important tools for a more complete understanding of the dynamics of extratropical S2S variability and teleconnections.

We conclude with a few comments about the role of climate change in the S2S variability and predictability problem. As mentioned in Section 1, the difficulty of the S2S problem is largely due to its being midway between the short-term weather problem and the long-term climate problem (Von Neumann, 1955). A mathematically self-consistent way out of this difficulty is provided by the theory of nonautonomous and random dynamical systems, as introduced recently into the climate sciences (e.g., Ghil et al., 2008; Chekroun et al., 2011b; Drótos et al., 2015; Ghil, 2017, and references therein). This theory—as opposed to the classical theory of autonomous dynamical systems, in which coefficients and forcing do not depend on time—allows one to include very fast forcing, by stochastic processes, as well as much slower forcing. Examples of the latter include not only changes in radiative forcing due to anthropogenic changes in aerosol and greenhouse gas concentrations, but also forcing by ocean-atmosphere (e.g., Ghil, 2001, 2017, and references therein) or stratosphere-troposphere interactions (e.g., Holton et al., 1995, and references therein). There is no room here to either provide a more detailed account of the appropriate concepts and methods or to actually apply it to NH LFV. Fairly readable accounts for the climate scientist do exist, though (Ghil, 2014; Moron et al., 2015; Chang et al., 2015, Chapter 2), and might encourage the interested reader to use them in the S2S context.

Acknowledgments

It is a pleasure to acknowledge detailed and constructive comments by Fréderic Vitart and an anonymous reviewer. This work was partially supported by US Office of Naval Research under MURI grants N00014-12-1-0911 and N00014-16-1-2073, and by National Science Foundation (NSF) under grants NSF OCE-1243175 and OCE-1658357. Results of Section 4.3 were developed with the support of Russian Science Foundation (grant No. 18-12-00231).

CHAPTER

7

Tropical-Extratropical Interactions and Teleconnections

Hai Lin, Jorgen Frederiksen†, David Straus‡, Cristiana Stan‡*

*Environment and Climate Change Canada, Dorval, Quebec, Canada †CSIRO Oceans and Atmosphere, Aspendale, Victoria, Australia ‡George Mason University, Fairfax, VA, United States

OUTLINE

1 INTRODUCTION

Tropical-extratropical interactions occur on different timescales, in different forms. For example, on decadal timescales, the thermohaline circulation provides an effective oceanic teleconnection for interhemispheric climate interactions. Interactions between the tropics and subtropics can take place through the atmospheric Hadley circulation and the oceanic

Sub-seasonal to Seasonal Prediction
https://doi.org/10.1016/B978-0-12-811714-9.00007-3

subtropical cell (e.g., Liu and Alexander, 2007). A tropical cyclone often can transform into an extratropical cyclone as it moves poleward, a process known as *extratropical transition (ET;* e.g., Harr and Elsberry, 2000; Hart and Evans, 2001). Occasionally, an extratropical cyclone can penetrate the tropics, lose its frontal features, develop convection near the center of the storm, and transform into a tropical cyclone (e.g., Davis and Bosart, 2004).

Of particular interest in this chapter are the tropical-extratropical interactions involving atmospheric teleconnection patterns. Extratropical weather is frequently influenced by recurring and persistent large-scale circulation patterns, usually referred to as *teleconnection patterns,* that connect widely separated locations of hemispheric or even global scales (e.g., Wallace and Gutzler, 1981). Typically, these teleconnection patterns last for weeks to months. By connecting atmospheric variability from a source region to remote locations, teleconnections contribute greatly to the atmospheric predictability on the sub-seasonal to seasonal (S2S) timescale.

A significant part of the extratropical atmospheric variability has a tropical origin. The sea surface temperature (SST) variability in the equatorial eastern Pacific associated with El Niño–Southern Oscillation (ENSO), for example, was found to be related to a teleconnection pattern across the North Pacific and North America on a seasonal-to-interannual timescale (e.g., Horel and Wallace, 1981). Tropical convection anomalies induce Rossby wave trains that extend eastward and poleward across the middle latitude (Hoskins and Karoly, 1981; Jin and Hoskins, 1995; Bladé and Hartmann, 1995). Such an extratropical atmospheric response to tropical thermal forcing usually takes the form of teleconnection patterns.

On seasonal and longer timescales, atmospheric teleconnection patterns often can be regarded as stationary or as responses to persistent anomalies in the lower boundary. On the other hand, the sub-seasonal teleconnections tend to evolve in time and space (e.g., Blackmon et al., 1984). For example, the atmospheric response to the eastward-propagating tropical convection of the Madden-Julian Oscillation (MJO) results in a lagged association between the extratropical teleconnections and the tropical forcing (e.g., Cassou, 2008; Lin et al., 2009). The perturbations of S2S timescales interact with not only the climatological mean flow, but also synoptic-scale transients.

There is also considerable extratropical influence on the tropics. Tropical convection is influenced by extratropical waves (e.g., Liebmann and Hartmann, 1984; Webster and Holton, 1982; Hoskins and Yang, 2000). Tropical MJO variability also can be induced by extratropical disturbances (e.g., Lin et al., 2007; Ray and Zhang, 2010; Vitart and Jung, 2010).

The association between the extratropical atmosphere and the organized tropical convection is not just a one-way influence from the tropics to the extratropics or vice versa. Instead, it is a two-way interaction. Some earlier studies found coherent circulation anomalies across the tropical and extratropical regions (e.g., Lau and Phillips, 1986) and suggested a global view of intraseasonal variability (e.g., Hsu, 1996). This is supported by the instability theory of Frederiksen (2002), who found that some of the unstable modes couple the extratropics with a tropical 40–60-day disturbance, which is similar to the MJO.

In this chapter, we summarize some observational characteristics of tropical-extratropical teleconnections on the S2S timescale and discuss basic dynamical processes related to such interactions. The discussion focuses on the teleconnection associated with the MJO, which is a major source of S2S predictability (e.g., Waliser et al., 2003). The tropical influence on the extratropics is discussed in Section 2, with a focus on the impact of the MJO, and the various dynamical processes associated with extratropical atmospheric response to tropical thermal forcing are also covered. The influence of extratropics on the tropics is documented in Section 3. In Section 4, tropical-extratropical, two-way interactions, as well as an instability theory, are reviewed. Finally a summary and discussion are given in Section 5.

2 TROPICAL INFLUENCE ON THE EXTRATROPICAL ATMOSPHERE

2.1 Observed MJO Influences

The MJO is the dominant mode of sub-seasonal variability in the tropics, which has a direct impact on the weather in the tropics by organizing convection and precipitation. An increasing number of studies have provided evidence that the variability of tropical convection associated with the MJO has a considerable influence on the extratropical weather and climate. This global impact can provide an important signal for sub-seasonal climate prediction (e.g., Waliser et al., 2003).

The MJO has a significant impact on the extratropical atmospheric teleconnection patterns. Through a composite analysis of multiyear (1985–93), global reanalyses produced by the National Centers for Environmental Prediction/National Center for Atmospheric Research (NCEP/NCAR) and the National Aeronautics and Space Administration/Data Assimilation Office (NASA/DAO), Higgins and Mo (1997) demonstrated that the development of persistent North Pacific circulation anomalies during the Northern Hemisphere winter is linked to tropical Intraseasonal Oscillations (ISOs). Mori and Watanabe (2008) found that the development of the Pacific-North American (PNA) pattern can be triggered by the MJO convection activity in the tropical Indian Ocean and western Pacific. The eastward progression of the convectively active phase of the MJO is associated with a corresponding shift in the tendency and sign of the Arctic Oscillation (AO) index (L'Heureux and Higgins, 2008). Observational studies show a robust lagged connection between the MJO and North Atlantic Oscillation (NAO; Cassou, 2008; Lin et al., 2009). A significant increase in the probability of a positive (negative) NAO happens about 10–15 days after the occurrence of MJO phase 3 (phase 7), which corresponds to a dipole structure of tropical convection anomalies with enhanced (reduced) convection in the equatorial Indian Ocean and reduced (enhanced) precipitation in the western Pacific. Here, the MJO phase is defined according to the real-time, multivariate MJO index as proposed by Wheeler and Hendon (2004). Shown in Fig. 1 are lagged composites of Northern Hemisphere 500-hPa geopotential height anomalies with respect to MJO phase 3. The wave train in the North PNA region associated with the tropical forcing of MJO phase 3 leads to the development of a positive phase of NAO (Lin et al., 2009).

Through atmospheric teleconnections, the MJO has a profound impact on a wide range of extratropical weather and climate events. Vecchi and Bond (2004) found that the phase of the MJO has a substantial systematic and spatially coherent effect on sub-seasonal variability in wintertime surface air temperature in the Arctic region. Using lagged composites and projections with the thermodynamic energy equation with respect to different MJO phases, Yoo et al. (2012) suggested that the Arctic surface air temperature change is associated with poleward-propagating Rossby waves, and the adiabatic heating, eddy heat flux, and radiative effects are also important. The phase of the MJO has a substantial systematic and spatially coherent effect on sub-seasonal variability in wintertime surface air temperature in North

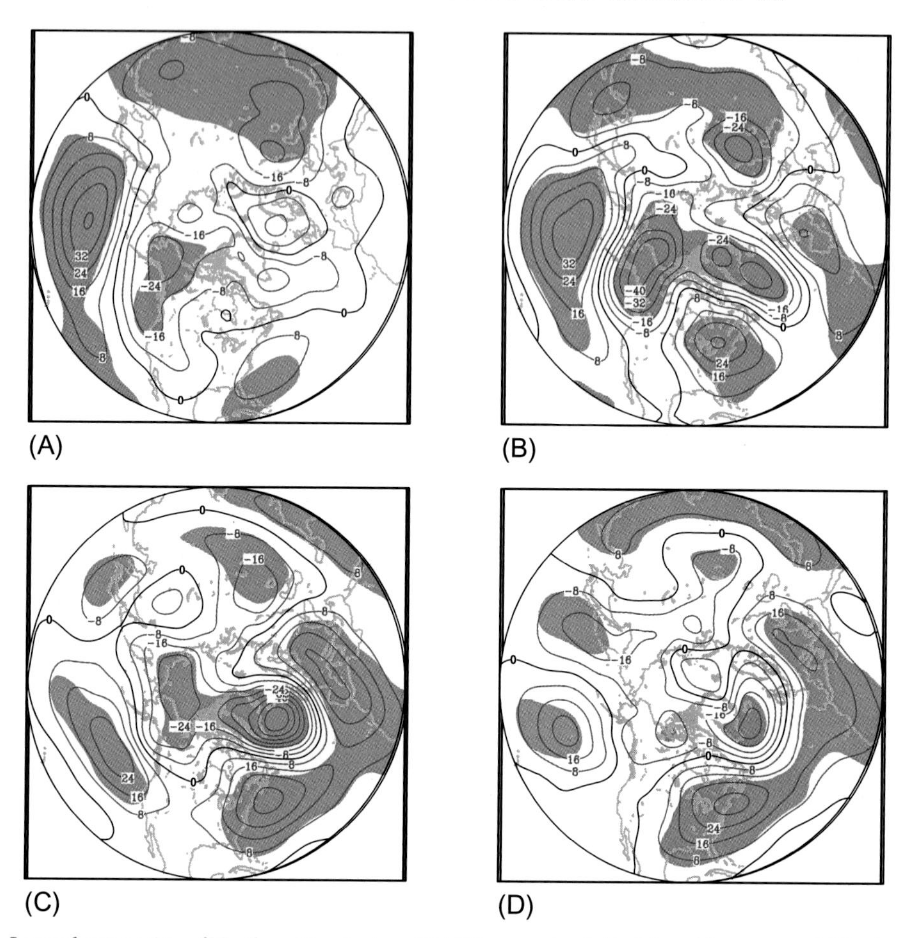

FIG. 1 Lagged composites of Northern Hemisphere 500-hPa geopotential height with respect to MJO phase 3. Contour interval is 8 m. Lag *n* means that the temperature anomaly lags MJO phase 3 by *n* pentads (5 days). (A) lag = 0 pentad; (B) lag = 1 pentad; (C) lag = 2 pentads; (D) lag = 3 pentads. Contours with positive and negative values are *red* and *blue*, respectively, and the zero contour line is *black*. Areas in orange are those where the composite anomaly is statistically significant at the 0.05 level according to a Student-t test. The analysis was performed for the 37 extended winters (November–April) from 1979/80 to 2015/16 using the NCEP/NCAR reanalysis data.

America (Lin and Brunet, 2009; Zhou et al., 2012; Baxter et al., 2014). As shown in Fig. 2, the surface air temperature over Canada and the eastern United States in winter tends to be anomalously warm 10–20 days after the MJO phase 3 (Lin and Brunet, 2009). Such a lagged association implies that North American temperature anomalies in winter are likely predictable up to about 3 weeks, given knowledge of the initial state of the MJO. Studies using statistical models with the MJO as a predictor have shown that there is indeed some skill in predicting North American temperature anomalies beyond 20 days, especially in strong MJO cases (e.g., Yao et al., 2011; Rodney et al., 2013; Johnson et al., 2014). The benefit of MJO in predicting North American surface air temperature also has been seen in operational monthly forecasting systems (e.g., Lin et al., 2016).

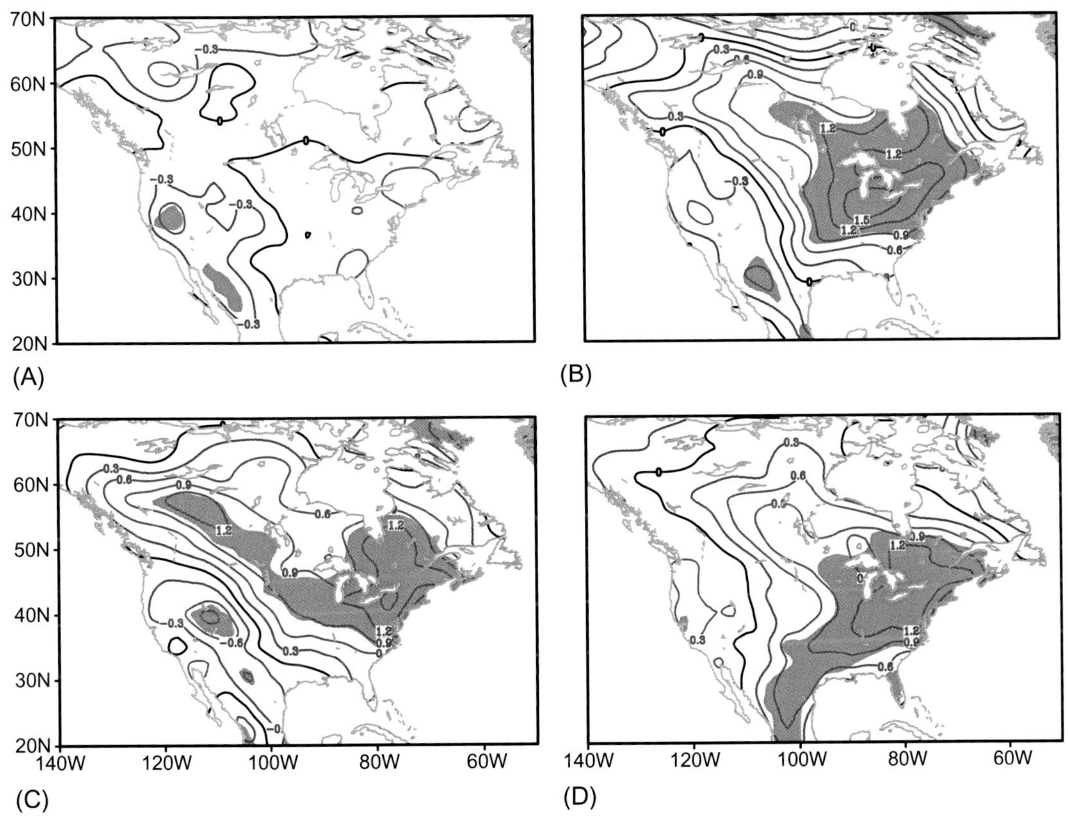

FIG. 2 Lagged composites of 2-m air temperature anomaly in North America with respect to MJO phase 3. Contour interval is 0.3°C. Lag n means that the temperature anomaly lags MJO phase 3 by n pentads (5 days). (A) lag = 0 pentad; (B) lag = 1 pentad; (C) lag = 2 pentads; (D) lag = 3 pentads. Contours with positive and negative values are *red* and *blue*, respectively, and the zero contour line is *black*. Areas in orange are those where the composite anomaly is statistically significant at the 0.05 level according to a Student t-test. The analysis was performed for the 37 extended winters (November–April) from 1979/80 to 2015/16 using the NCEP/NCAR reanalysis data.

There have been many studies on the impact of the MJO and the Boreal Summer Intraseasonal Oscillation (BSISO) in the South and East Asian regions. In boreal summer, the BSISO disturbances tend to propagate northeastward and significantly influence the "active" and "break" monsoon rainfall fluctuations (e.g., Yasunari, 1979; Murakami et al., 1984; Wang et al., 2006). The disturbances associated with the MJO, the BSISO, or both directly modulate the rainfall over the Asian continent through their influence on the genesis of higher-frequency monsoon lows and depressions. As revealed by Goswami et al. (2003), a majority of such lows and depressions develop during the wet phase of the MJO. Zhang et al. (2009) reported a significant impact of the MJO on summer rainfall in southeast China. The impact of the MJO on wintertime weather in the midlatitude East Asian region also has been reported. Jeong et al. (2005) and Jeong et al. (2008) studied the influence of the MJO on wintertime surface air temperature and precipitation in East Asia, respectively. In boreal winter, the eastward-propagating MJO convective activity is largely confined to the tropical region. Its influences on the midlatitude region of East Asia was found to be related to a local Hadley cell overturning and Rossby wave response (He et al., 2011).

Beyond East Asia, the MJO influence on precipitation through extratropical waves and teleconnections has been detected as well. The tropical convection associated with the MJO was found to be correlated with precipitation on the U.S. West Coast (e.g., Higgins et al., 2000; Mo and Higgins, 1998; Bond and Vecchi, 2003; Becker et al., 2011). Based on 30-year station precipitation observations, a lagged association was found between the sub-seasonal precipitation variability over the West Coast of Canada and northeast regions of Canada and the tropical convection dipole anomaly of MJO phase 3 or phase 7 in boreal winter (Lin et al., 2010a). Extreme rainfall over the contiguous United States was found to be more likely to happen when the MJO is active than inactive, and most frequently when the MJO convection center is over the Indian Ocean (Jones and Carvalho, 2012; Barrett and Gensini, 2013).

The MJO influence on the extratropics also has been found in the Southern Hemisphere. A consistent signal on winter temperature variability and precipitation in different MJO phases was found in southeastern South America (Naumann and Vargas, 2010). The Pacific-South American (PSA) patterns (Mo and White, 1985) were observed to be associated with enhanced convective activity in different phases of the MJO (Mo and Higgins, 1998). Several Southern Hemisphere teleconnection patterns in June–August exhibit oscillatory behavior on timescales of 20–30 days and with the frequency of occurrence modulated by the MJO phases (Chang and Johnson, 2015). Flatau and Kim (2013) demonstrated that on the sub-seasonal timescale, enhanced MJO convection in the Indian Ocean precedes changes in the Antarctic Oscillation (AAO). It was suggested that the MJO directly affects regional circulation and climate in the New Zealand region, potentially through extratropical Rossby wave response to tropical diabatic heating (Fauchereau et al., 2016). Whelan and Frederiksen (2017) found that tropical-extratropical interactions associated with the MJO contributed to the extreme rainfall and flooding in northern Australia during January 1974 and January 2011.

In summary, observational studies have shown that the MJO has a profound impact on a wide range of extratropical weather and climate events (e.g., Stan et al. 2017; Zhang, 2013). It was found to influence the sub-seasonal forecast skill of the NAO (Lin et al., 2010b). There is a strong seasonality of the MJO-related extratropical teleconnections. Except for the direct influence of BSISO on the South and East Asian regions, the extratropical teleconnections of MJO are stronger in the boreal winter season. The midlatitude westerlies, which are favorable for Rossby wave propagation (as will be discussed in Section 3), are stronger in the winter season. Operational sub-seasonal forecasting models are in general able to capture the MJO-related teleconnections, and their forecasts beyond 2 weeks in North America and Europe depend strongly on the presence of an MJO event in the initial conditions (e.g., Vitart and Molteni, 2010; Lin et al., 2016). An improved understanding of the MJO teleconnections and a better representation of the associated processes in numerical models are important to improving S2S predictions.

2.2 Extratropical Atmospheric Response to Tropical Thermal Forcing

The latent heat released by tropical convection provides a major energy source that drives the global atmospheric circulation. Deep tropical convection produces upward motion, leading to divergence in the upper troposphere. Such upper divergent flow converges in the subtropical westerly jet regions, generating a source of extratropical Rossby waves (Sardeshmukh and Hoskins, 1988), which propagate in the midlatitude westerlies bounded by the pole and the critical latitude where the climatological zonal wind becomes easterlies (e.g., Webster and Holton, 1982; Hoskins and Ambrizzi, 1993). In addition to the Rossby wave propagation, at least two other dynamical processes in the extratropics influence the atmospheric response to tropical heating. The first is that the disturbances tend to grow in preferred locations (e.g., the eastern North Pacific and North Atlantic) by extracting kinetic energy from the zonally asymmetric climatological flow through barotropic conversion (e.g., Simmons et al., 1983; Branstator, 1985) or baroclinic-barotropic conversion (Frederiksen, 1982, 1983; Frederiksen and Webster, 1988). The second process is that the midlatitude synoptic-scale transients, which are displaced or influenced by the large-scale response anomalies to tropical heating, feed back to reinforce the response through transient eddy vorticity flux convergence (e.g., Lau, 1988; Klasa et al., 1992).

The extratropical response to a tropical forcing is influenced by the barotropic instability of the climatological mean flow. As noted in Simmons et al. (1983), the barotropic kinetic energy conversion from the climatological mean flow can be expressed as

$$C = -\left(\overline{u'^2} - \overline{v'^2}\right)\frac{\partial \overline{u}}{\partial x} - \overline{u'v'}\frac{\partial \overline{u}}{\partial y} \tag{1}$$

The overbar is the time mean, and the prime is the departure, which here refers to the low-frequency (10–100 days) anomaly that responds to tropical forcing. The first term on the right side represents energy transfer between the basic state and the perturbation in places where the climatological mean zonal wind varies along the east-west direction. In the midlatitude westerly jet exit regions of the North Pacific and Atlantic, where $\frac{\partial \overline{u}}{\partial x} < 0$, zonally elongated eddies ($\overline{u'^2} > \overline{v'^2}$), which are like most low-frequency disturbances in those regions, would develop by extracting energy from the basic flow. Through the mechanism of barotropic instability, the zonal distribution of climatological flow determines the preferred location of low-frequency disturbances. The Rossby wave response to the MJO diabatic heating was found to be intensified in the Pacific jet exit region by extracting kinetic energy from the climatological mean flow (Bao and Hartmann, 2014; Adames and Wallace, 2014).

Another process that affects the extratropical atmospheric response to a tropical forcing is the interaction with the synoptic-scale, high-frequency, transient eddies. The barotropic vorticity equation takes the following form:

$$\frac{\partial \zeta}{\partial t} = -\nabla \cdot \left[\vec{V}\,(\zeta + f)\right] + F \tag{2}$$

where ζ is the vorticity, f the Coriolis parameter, $\vec{V}$ the horizontal velocity, and F the forcing.

Each variable X can be decomposed into three parts corresponding to different frequency bands:

$$X = \overline{X} + X_l + X_h$$

where the overbar represents the time mean (e.g., of a winter season); and the subscript l represents the low-frequency (i.e., sub-seasonal timescale) and h the high-frequency (i.e., synoptic timescale) parts of the time series. Then, the equation for the vorticity tendency of the low-frequency disturbances can be expressed as

$$\frac{\partial \zeta_l}{\partial t} = -\nabla \cdot \left(\vec{V}_h \zeta_h\right)_l - \nabla \cdot \left(\vec{V}_l \zeta_l\right)_l + R_l \tag{3}$$

The first and second terms on the right side represent the low-frequency vorticity tendencies associated with the vorticity flux convergence by the synoptic-scale transients and low-frequency eddies, respectively. R_l includes all remaining components in the low-frequency vorticity equation, such as horizontal advection of low-frequency vorticity by time-mean flow, interactions between low- and high-frequency eddies, and the low-frequency part of the forcing. In the extratropics, the vorticity tendency can be expressed as a geopotential height tendency by taking an inverse Laplacian, which changes the sign and has the effect of spatial smoothing (e.g., Lau, 1988). Observational studies indicate that the geopotential height tendency caused by synoptic-scale transient vorticity flux convergence is positively correlated with the low-frequency height anomaly itself, helping to maintain and reinforce the variability (e.g., Lau, 1988; Lin and Derome, 1997; Feldstein, 2003). As discussed in Held et al. (1989), the linear Rossby wave propagation as the direct response to anomalous diabatic heating is relatively small in the extratropics, and the response to the anomalous transients, particularly the anomalous upper-tropospheric transients, is important for the extratropical response.

Although the extratropical response to a tropical thermal forcing tends to occur in preferred geographical locations, as discussed previously, its sign and amplitude are sensitive to the longitudinal location of the tropical forcing (Simmons et al., 1983). To assess the sensitivity of the extratropical response to the longitudinal location of MJO heating, several numerical experiments were conducted in Lin et al. (2010a), with equatorial thermal forcings at different longitudes. They found that strongest extratropical response is produced when the heating is located in either the Indian Ocean or the western Pacific, whereas a very weak response is induced when the thermal forcing is near the Maritime continent longitudes. The response to a tropical Indian Ocean forcing is out of phase with that caused by a heating in the western Pacific (Fig. 3). Tropical thermal forcing with heating in the equatorial Indian Ocean and cooling in the western Pacific (or vice versa) is the most effective in exciting extratropical circulation anomalies (Fig. 4). Such a tropical, west-east, dipole thermal forcing structure is similar to that of MJO phases 2–3 and 6–7, which is consistent with observational studies (e.g., Cassou, 2008; Lin et al., 2009).

The global atmospheric circulation anomalies associated with the tropical forcing of MJO can in broad terms be simulated by numerical models (e.g., Matthews et al., 2004; Lin et al., 2010a,b; Vitart and Molteni, 2010; Seo and Son, 2012). Vitart (2017) found that all the models in the World Weather Research Program (WWRP)/World Climate Research Program (WCRP) Sub-seasonal to Seasonal Prediction Project (S2S) database produce realistic patterns of MJO teleconnections at 500 hPa, with an increased probability of positive NAO following MJO phase 3 and of negative NAO following MJO phase 7. The amplitude of the MJO teleconnection patterns, however, was found to be significantly weaker than the observations over the Euro-Atlantic sector, and often were too strong over the western North Pacific. Using the CMIP5 database, Henderson et al. (2017) assessed the impact of the model MJO and basic state quality on MJO teleconnection pattern quality, suggesting that both the basic state and

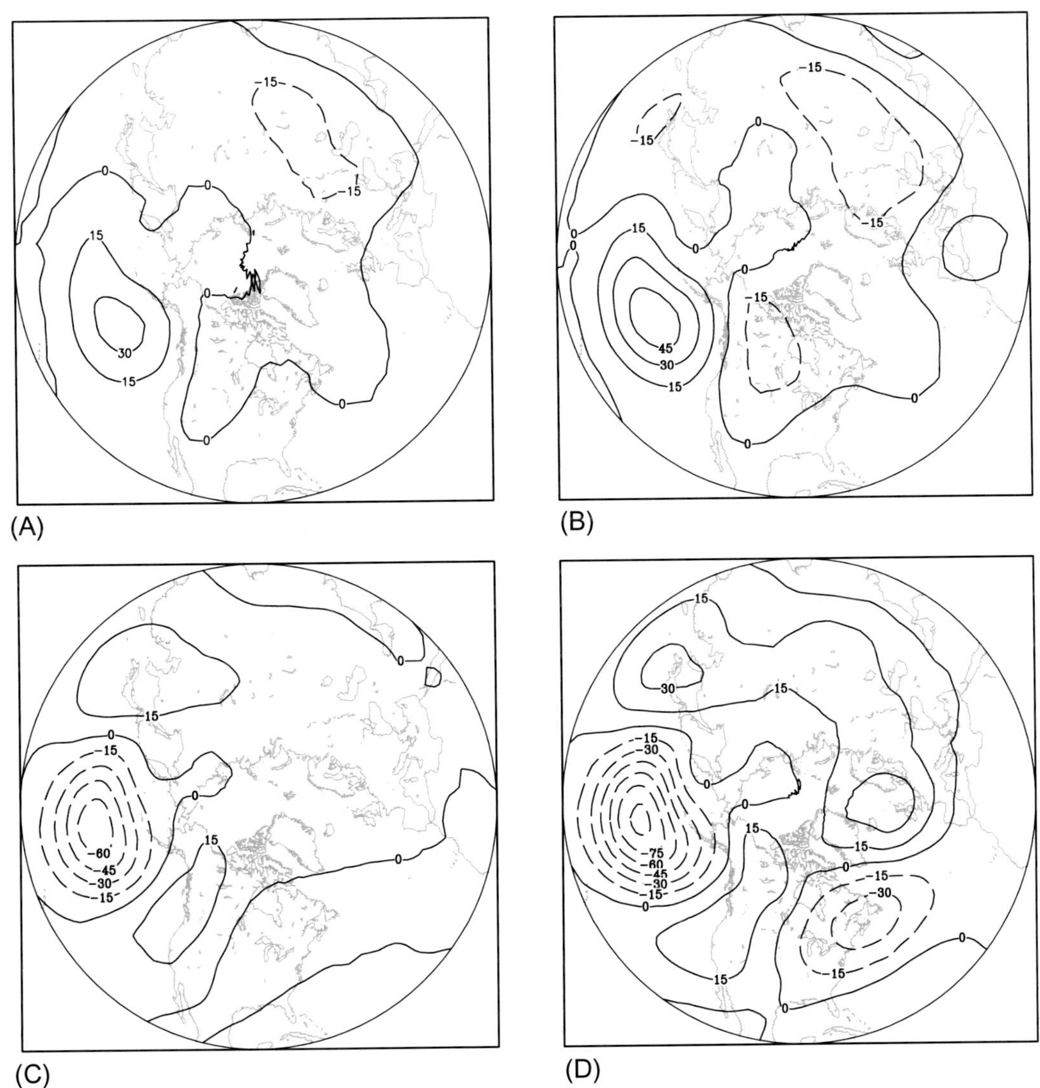

FIG. 3 500-hPa geopotential height response averaged between days 6 and 10 (left) and between days 11 and 15 (right) for a 80°E heating run (top) and 160°E heating run (bottom). (A) 80°E: days 6–10; (B) 80°E: days 11–15; (C) 160°E: days 6–10; (D) 180°E: days 11–15. A linear model was used with a basic state of Northern Hemisphere winter climatological flow, forced by a thermal forcing at the equator. The contour interval is 15 m. Contours with negative values are dashed. *From Lin, H., G. Brunet, and R. Mo, 2010a: Impact of the Madden-Julian Oscillation on wintertime precipitation in Canada, Mon. Wea. Rev., 138, 3822–3839 © Her Majesty the Queen in Right of Canada, as represented by the Minister of the Environment, [2010].*

the MJO must be well represented to reproduce the correct teleconnection patterns. Lin and Brunet (2017) investigated the nonlinearity of the extratropical response to the MJO and found that the response to the MJO at phases 6–7 is not a mirror image of that to the MJO at phases 2–3. Instead, there is a phase shift of the extratropical wave train in the PNA region between these two cases. The negative NAO following MJO phases 6–7 is stronger than the

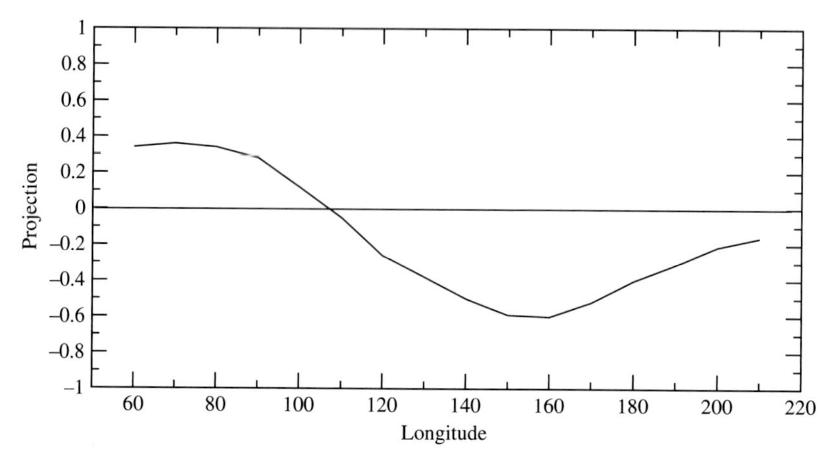

FIG. 4 Normalized projection of Northern Hemisphere, extratropical, 500-hPa geopotential height response averaged between days 11 and 15 as a function of the longitude of the equatorial heating center location. The projection was made onto the response to the tropical dipole heating similar to MJO phase 3. *From Lin, H., G. Brunet, and R. Mo, 2010a: Impact of the Madden-Julian Oscillation on wintertime precipitation in Canada, Mon. Wea. Rev., 138, 3822–3839 © Her Majesty the Queen in Right of Canada, as represented by the Minister of the Environment, [2010].*

positive NAO after MJO phases 2–3. The strength and location of the response were found to be dependent on the north-south location and amplitude of the East Asian subtropical westerly jet, respectively. Yoo et al. (2015) examined the impacts of cumulus parameterizations on the simulation of the boreal wintertime MJO teleconnection in the Northern Hemisphere and found that the unified convection scheme (UNICON), which is a process-based parameterization, substantially improves the MJO teleconnection. It is likely that the improved MJO teleconnection simulation benefits from a more realistic climatological flow and MJO structure.

In summary, the extratropical atmospheric response to tropical forcing is influenced by the climatological mean flow and synoptic-scale transients. The action centers of a response pattern tend to occur at preferred locations, although the amplitude is sensitive to the location of tropical heating relative to the extratropical westerly jet. For a climate model to have a reasonable response to tropical forcing and teleconnection patterns, the model should have a realistic representation of (1) distribution of the tropical diabatic heating; (2) climatological mean flow, and thus small systematic errors, for Rossby wave propagation and wave-mean flow interactions; and (3) extratropical storm tracks and associated transient eddies.

3 EXTRATROPICAL INFLUENCE ON THE TROPICS

3.1 Extratropical Influences on Tropical Convection and the MJO

Considerable extratropical influence on the tropics has been found in many previous studies. For example, during East Asian winter monsoons, cold surges from the midlatitudes are found to be connected to variations in tropical convection (e.g., Chang and Lau, 1980).

Through lead and lag correlation analysis between 5- and 10-day averaged midlatitude 500-mb heights and tropical outgoing longwave radiation (OLR), Liebmann and Hartmann (1984) found that energy predominantly propagates from the middle latitudes to the tropics, especially over the eastern Pacific. Extratropical waves propagate into the tropics through regions of westerly zonal wind (e.g., Webster and Holton, 1982), and influence tropical convections (e.g., Kiladis and Weickmann, 1992; Matthews and Kiladis, 1999).

Examples of wave energy impinging on the tropical atmosphere from the extratropics include the convectively coupled equatorial waves in the Pacific Inter Tropical Convergence Zone (ITCZ) (Zangvil and Yanai, 1980; Yanai and Lu, 1983; Zhang and Webster, 1989; Zhang and Webster, 1992; Hoskins and Yang, 2000; Dickinson and Molinari, 2002; Straub and Kiladis, 2003; Yang et al., 2007; Roundy, 2012; Kiladis et al., 2009; Liebmann et al., 2009; Fukutomi and Yasunari, 2009, 2014; Roundy, 2014; Maloney and Zhang, 2016). Straub and Kiladis (2003) and Yang et al. (2007) provided observational evidence of extratropical Rossby wave-train activity as a precursor of convectively coupled Kelvin waves emanating over the western Pacific. Zangvil and Yanai (1980) and Yanai and Lu (1983) found that mixed-Rossby gravity waves could be forced by lateral forcing from middle latitudes in regions where no critical latitude exists.

Zhang and Webster (1992) applied a linear model to study the equatorial waves generated by a midlatitude forcing. They found that the amplitude of the tropical wave response depends on the tropical mean zonal wind. The Rossby-wave mode is stronger in mean westerlies than in mean easterlies, while the Kelvin-wave response exhibits a larger amplitude in mean easterlies than in mean westerlies. The Doppler-shifting effect of a mean zonal flow on equatorial waves was analyzed in Hoskins and Yang (2000), where the tropical atmosphere was forced by a moving wavelike extratropical forcing. They showed that even if extratropical waves do not propagate directly into the tropics through regions of westerly wind, extratropical, zonally propagating waves can nevertheless disturb the equatorial wave guide and help organize convection in the tropics.

The extratropical waves modulate tropical convection over a range of timescales, including the MJO timescale, when they propagate into the tropics (e.g., Matthews and Kiladis, 1999). Lin et al. (2007) showed that a tropical MJO-like wave can be generated in a long integration of a dry atmospheric model with time-independent forcing. The variability in the tropics comes from middle latitudes, and there is coherent variation between the tropical and extratropical regions in the model atmosphere. Using a tropical channel model, Ray and Zhang (2010) investigated the initialization of two MJO events and found that the only factor critical to the reproduction of the MJO initiation is time-varying lateral boundary conditions from the reanalysis. When such lateral boundary conditions are replaced by time-independent conditions, the model fails to reproduce the MJO initiation. These results support the idea that extratropical influences can be an efficient mechanism for MJO initiation. It was also found that the latitudinal momentum transport is important for MJO initialization (Ray and Zhang, 2010).

Lin et al. (2009) observed that MJO phases 6–7 (phases 2–3) tend to occur about 20 days after a positive (negative) phase of the NAO. This is associated with equatorward wave activity flux from the North Atlantic into the tropics following the amplification of the NAO. Hong et al. (2017), analyzed the impact of extratropical perturbation on the onset of the 2015 MJO-El Niño event and found that the southward penetration of northerly wind anomalies associated with extratropical disturbances in the extratropical western North Pacific triggered the tropical convective instability that led to the onset of the MJO to the west of the dateline. The critical effect of the extratropical disturbances on the MJO onset was confirmed by numerical experiments in an atmospheric general circulation model (GCM) coupled with an ocean mixed-layer model.

Hall et al. (2017) conducted several experiments with various lateral boundary conditions in a tropical channel model. Comparison between the experiments indicates that about half the intraseasonal variance in the tropics can be attributed to the boundary influence of the middle latitudes, and specifically to the presence of an intraseasonal extratropical signal.

3.2 Diagnosing Intraseasonal Extratropical Influences on the Tropics

Extratropical, transient, upper-level troughs that propagate eastward and equatorward during boreal winter have been linked to "tropical plumes," extensive elongated cloud bands with a length scale of 4000–16,000 km and lifetime of 3–9 days, as reviewed by Knippertz (2007). This interaction takes place in the active portions of the ITCZ, where there is a westerly duct allowing the upper-level extratropical disturbances to penetrate the tropics. These disturbances can be described as tilted troughs, or equivalently as elongated potential vorticity (PV) "streamers" transporting stratospheric PV equatorward, leading to enhanced momentum flux and strong subtropical jet streaks. As pictured in Fig. 6, they can be associated with Rossby wave breaking. Diagnosing wave breaking can be an involved process (Swenson and Straus, 2017), so it is useful to explore a simpler approach to identifying episodes of extratropical forcing of tropical convection.

From the streamlines indicated in Fig. 5, it can be seen that positive meridional flow in the subtropics (near 30°N) is associated with positive zonal flow, while negative meridional flow is associated with weak or negative zonal flow. Thus, this type of anticyclonic wave-breaking incursion can be associated with a poleward flux of zonal momentum on synoptic timescales (as in, e.g., Cassou, 2008). From a more general point of view, positive momentum flux is associated with equatorward Eliassen-Palm (EP) flux in the Northern Hemisphere, indicating an influence of higher-latitude processes on lower latitudes (Hoskins et al., 1983).

This association suggests a method for identifying instances in which this extratropical incursion is associated with tropical heating. To set the stage for this, we should point out that the boreal winter climatology of the synoptic-scale momentum flux $\overline{u'v'}$ is quite substantial (and positive) in the northern subtropics down to about 10°N. This is seen in Fig. 6, which shows the high-pass (i.e., with periods greater than 10 days filtered out) covariance $\overline{u'v'}$ at 200 hPa, averaged over many winters from the ERA-Interim reanalysis (Dee et al., 2011) (Here, u is the zonal wind, v is the meridional wind, and the primes denote the time filtering.) Apparently, there is generally equatorward propagation of EP flux in the northern tropics.

To see if such fluxes are associated with large-scale, intraseasonal, tropical, diabatic heating, we estimate the diabatic heating from the ERA-Interim reanalysis, using an approach similar to that of Hagos et al. (2009), but with some differences in detail. Taking the daily temperature, horizontal winds, and vertical motion as an accurate estimate of the truth from the reanalysis, we compute the total diabatic heating as a residual in the thermodynamic equation. The estimate of daily diabatic heating was initially carried out at the original resolution

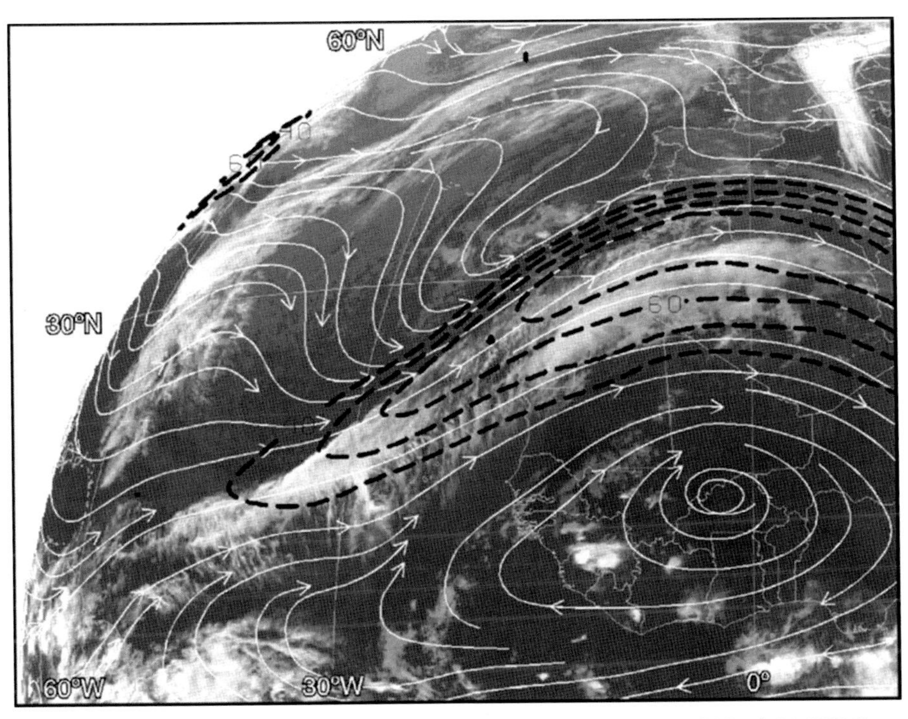

FIG. 5 Meteosat infrared image of a tropical plume over northwest Africa at 00 UTC March 31, 2002. Superimposed are streamlines and isotachs on the 345K isentropic level (dashed contours at 40, 50, 60, and 70 m/s) from the European Centre for Medium-Range Weather Forecasts (ECMWF) TOGA analysis. The 345K level is close to 200 hPa in the tropics. Streamlines indicate extratropical wave incursion into the tropics. *Reprinted from Knippertz, P., 2007. Tropical-extratropical interactions related to upper-level troughs at low latitudes. Dyn. Atmos. Ocean. 43, 36–62. ©Elsevier. Used with permission.*

(equivalent to T255 spectral truncation) for all 37 vertical levels in the reanalysis, and then consistently truncated to a T42 equivalent grid to focus on the large-scale component (see the Appendix of this chapter for full details). The heating rates were vertically integrated over a number of layers; the one used here is the 700–300-hPa layer. The resulting diabatic heating was filtered further to retain only periods greater than 20 days, and an estimate of the seasonal cycle for *each winter* was removed (again, see the Appendix). This yields Q, the large-scale deep heating on timescales of roughly 20–90 days.

The key diagnostic quantity computed is the intraseasonal covariance between the momentum flux $\overline{u'v'}$ and Q for each boreal winter (taken as December 1–March 16), in which the seasonal cycle of both quantities for each winter is removed. Positive covariance in the tropics would indicate at least an association of equatorward-propagating disturbances with tropical, large-scale heating. To identify those winters in which there is strong extratropical forcing of tropical convection, we have examined plots of this covariance (averaged between 15°S and 15°N) as a function of winter (not shown). Large values are seen in the mid-Pacific and eastern Pacific only during specific winters—generally ones in which the MJO was active,

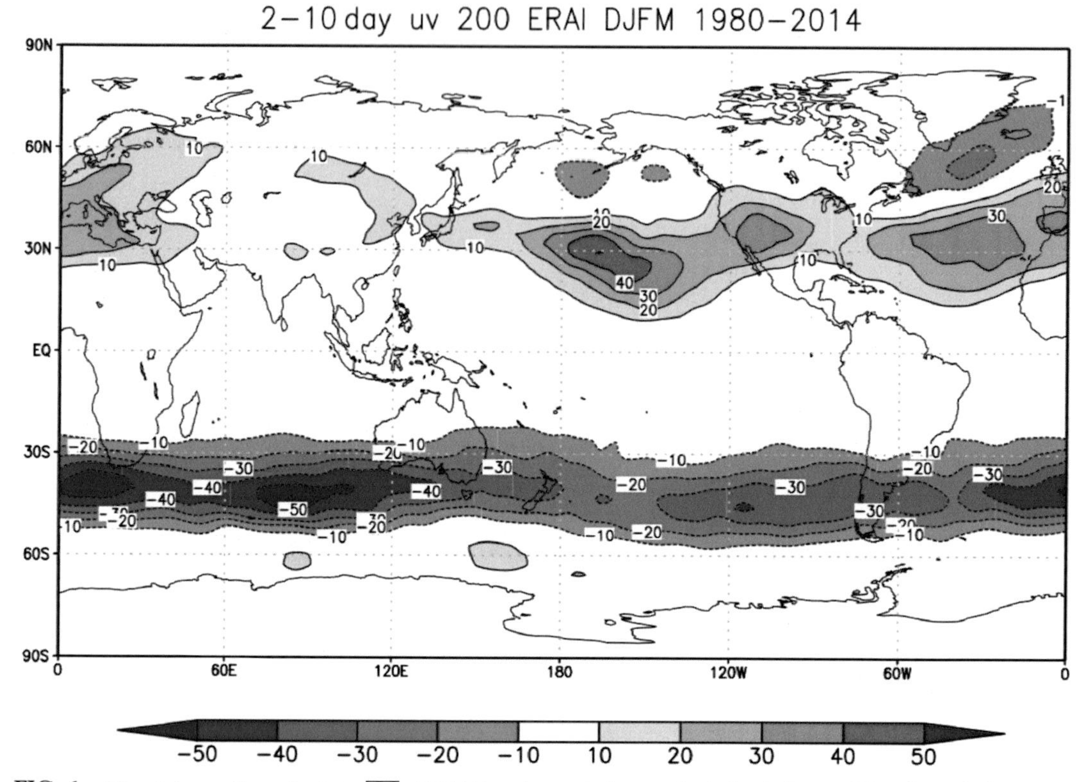

FIG. 6 Climatology of boreal winter $\overline{u'v'}$ at 200 hPa, where u is the zonal wind, v is the meridional wind, the primes denote a high-pass filter (retaining periods less than 10 days), and the overbar denotes an average from December 1 to March 16. Computed from ERA-Interim reanalysis, averaged over the 35 winters 1980/81–2014/15.

as indicated by the MJO amplitude shown by the Meteorological Society of Japan (http://ds.data.jma.go.jp/tcc/tcc/products/clisys/mjo/rmm.html).

We show a map of the climatological covariance for the Pacific Ocean averaged over the winters of 1980/81–2014/15 in the left panel of Fig. 7A, and the covariance for the winter 1989/90 only in Fig. 7B. Climatologically, the covariance between the moment flux and diabatic heating is quite weak equatorward of 30°N in both hemispheres. In some winters, however, the covariance is much stronger in the tropics, as indicated, for example, in Fig. 8B, which shows positive values of the covariance near 20°N of the same order of magnitude as in midlatitudes.

Fig. 8 shows another example, taken from the winter of 2012–13. Fig. 8A shows the high-pass meridional wind v' on February 19 (during an active MJO period), along with the full, low-pass, midlayer (700–300 hPa) heating. There is a clear indication of a wave train propagating from midlatitudes into the tropics in the eastern Pacific, where it is associated with diabatic heating (shown in shading). The high-pass momentum flux $\overline{u'v'}$ associated with this wave train is shown in Fig. 8B), which also shows penetration into the tropics and association with heating. This extratropical interaction is similar to those reported by Matthews and Kiladis (1999), who used OLR as an indicator of tropical heating.

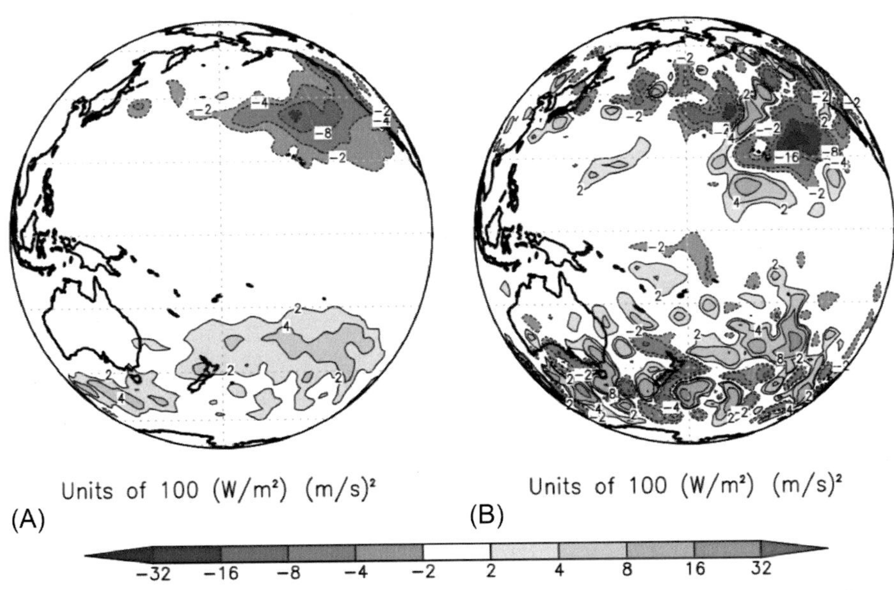

FIG. 7 Daily boreal winter (December 1–March 16) intraseasonal covariance between 200-hPa, high-pass momentum flux $u'v'$ and low-pass, layer-integrated (700–300-hPa) diabatic heating Q, averaged over all winters 1980/81–2014/15 (left panel), and for 1989/90 winter (right panel). (A) DJFM ISCov(uv,Q) Clim; (B) DJFM ISCov(uv,Q) 1989. Here, u indicates zonal wind, v is meridional wind, the high-pass filter retains periods of less than 10 days, and the low-pass filter retains periods of greater than 20 days. The interval is 100 (W/m^2) (m^2/s^2). Map projections are orthographic with equatorial aspect. The central longitude is 180°E.

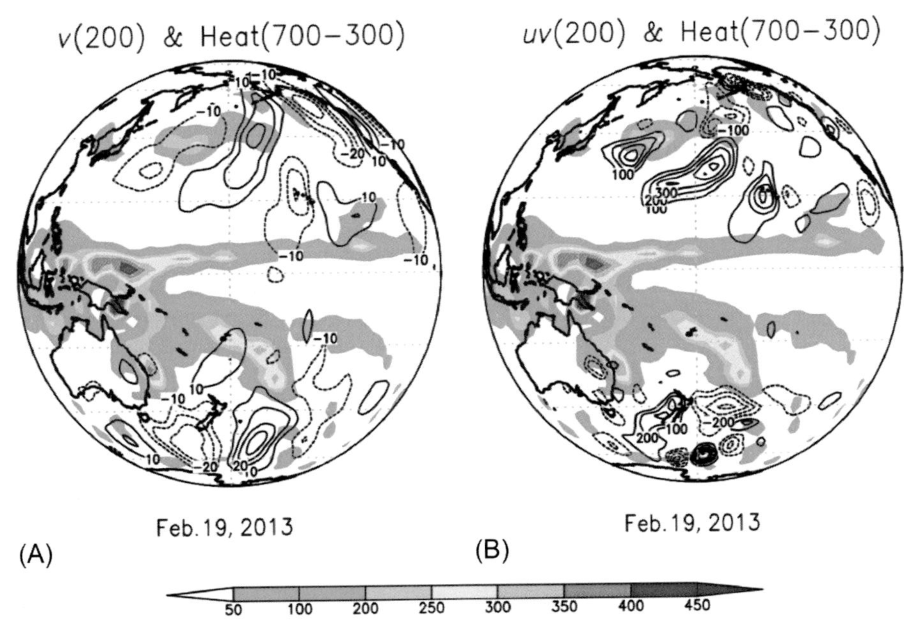

FIG. 8 (A) High-pass meridional wind v' on February 19, 2013 (contours), and low-pass, layer-integrated (700–300 hPa) diabatic heating Q (shading). (B) Product of high-pass meridional and zonal wind $u'v'$ (contours) and low-pass, layer-integrated (700–300-hPa) diabatic heating Q (shading). The high-pass filter retains a period of less than 10 days, and the low-pass filter retains periods greater than 20 days. Interval is 10 m s^{-1} in (A), 100 m^2 s^{-2} in (B). Heating is given in (W/m^2). Map projections are orthographic with equatorial aspect. The central longitude is 180°E.

We should point out here that the diagnosis of large-scale, low-pass, and intraseasonal diabatic heating can be applied to newer reanalysis products, and it also can be modified to provide more detail regarding the vertical structure of heating. The covariance between intraseasonal heating and the high-pass momentum flux should be a useful indicator of extratropical forcing, easily applied to both reanalyses and model output.

4 TROPICAL-EXTRATROPICAL, TWO-WAY INTERACTIONS

4.1 Forcing of Extratropical Waves Through Two-Way Interactions

In addition to the propagation of wave energy from the extratropics into the tropics through the westerly duct (Webster and Holton, 1982), extratropical forcing can project directly onto the equatorial waves in the regions where equatorial waves can extend into the extratropics (Zhang and Webster, 1989; Hoskins and Yang, 2000). In this framework, the interaction between the tropics and extratropics can be interpreted as a two-way interaction that is modulated by the zonal mean flow. Zhang and Webster (1989) showed this by expending Matsuno's (1966) shallow-water-equation model to include a zonal basic state with meridional shear. This basic state is in geostrophic balance with the surface pressure gradient. The impact of the basic state on the equatorial waves manifests as a Doppler shift induced by the horizontal advection through the zonal mean wind. The solutions of the dispersion equation for the equatorial waves become dependent on the meridional structure of the zonal basic state.

Zhang and Webster (1989) also showed that the basic zonal flow affects the meridional extension of the oscillatory regions. The Rossby wave extends farther poleward from the equator in westerly flow and remains confined to the equator in easterly flow. The wave amplitude increases and the position of the maximum moves poleward with the increase in the meridional wave number. Fig. 9 shows the meridional dependence of Rossby-wave structures for two meridional wave numbers ($n = 1, 2$) and three values of -10, 0, and 10 m/s for the basic-state zonal wind, similar to the results of Zhang and Webster (1989; see their Fig. 6). The dispersion relationship and the associated solutions of this model are not exact, and the order of approximation also has an impact on the amplitude of the wave and the meridional position of the maximum. The solutions shown in Fig. 9 were computed using a higher-order approximation than in Zhang and Webster (1989). In the higher-order approximation, the position of maximum wave amplitude moves farther poleward, especially for the geopotential height. This meridional extension of Rossby waves into the middle latitudes allows the extratropical forcing located within the turning latitudes of the Rossby wave to interact with the equatorially trapped waves.

4.2 Three-Dimensional Instability Theory

Coupled instability modes of tropical-extratropical dynamics were found to have properties very similar to 30–60-day ISOs in Frederiksen and Frederiksen (1993, 1997) and Frederiksen (2002). They used a linearized, two-level, primitive-equation model with three-dimensional (3D) basic states for northern winter. They employed the cumulus

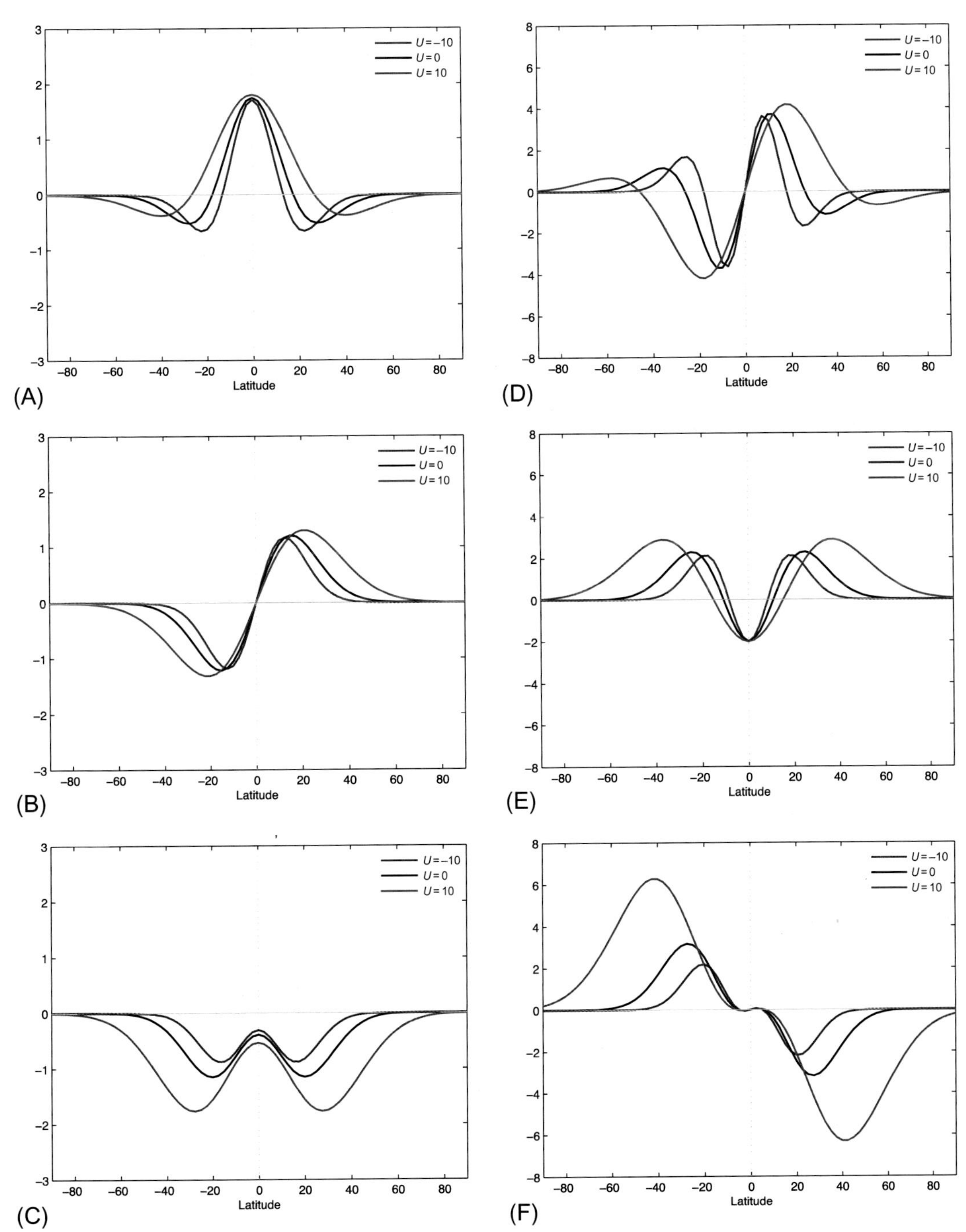

FIG. 9 Meridional distribution of Rossby wave with zonal wave number $k = 5$ and meridional wave number 1 (left side) and 2 (right side) in the zonal wind, meridional wind, and geopotential height. (A) Zonal wind, $n = 1$; (B) zonal wind, $n = 2$; (C) meridional wind, $n = 1$; (D) meridional wind, $n = 2$; (E) geopotential, $n = 1$; (F) geopotential, $n = 2$.

convection parameterization of Kuo (1974), and an evaporation-wind-feedback parameterization (Emanuel, 1987; Neelin et al., 1987) was employed by Frederiksen (2002). The theoretical model encapsulated the combination of extratropical and tropical processes that have been studied to some extent in isolation. These include the role of the baroclinic-barotropic instability of the 3D basic state in the development of extratropical teleconnection patterns, the role of convection in reducing the moist static stability of the tropical atmosphere, and the role of evaporation-wind feedback in improving the propagation characteristics of the ISO. These mechanisms have been studied separately, as detailed next, but it is only in combination that ISOs, with realistic tropical and extratropical structures, emerge.

As reviewed by Frederiksen and Webster (1988) and Frederiksen (2007), 3D instability has been widely employed for understanding many aspects of large-scale atmospheric disturbances. The first studies with observed climatological basic states, but without convection or evaporation-wind feedback (Frederiksen, 1982, 1983), elucidated the generation mechanisms of localized storm tracks, blocking, and teleconnection patterns. In particular, patterns like the PNA and NAO were derived as growing modes on Northern Hemisphere winter basic states within two-level, quasi-geostrophic models, with the NAO having an intraseasonal period of about 40 days. Similar teleconnection pattern modes were found in multilevel, quasi-geostrophic models (Frederiksen and Bell, 1987) and in primitive-equation models (Frederiksen and Frederiksen, 1992), with barotropic processes playing an important role (Frederiksen, 1983; Simmons et al., 1983; Schubert, 1985) but with baroclinic processes able to increase growth rates by as much as 65%, as in the case of the NAO (Frederiksen, 1983).

Frederiksen and Frederiksen (1993, 1997) and Frederiksen (2002) revealed a wide variety of disturbances, including growing ISO modes, which couple the tropics and extratropics, with properties similar to the MJO. The first internal mode, or baroclinic, structure in the tropics of their coupled ISOs was shown to be due to convection reducing the moist static stability there. They also found that in order to obtain realistic ISO tropical structures, the tropical moist static stability needs to be relatively small compared with the dry static stability. They compared the differences between moist ISO modes and dry ISO modes and the role of evaporation-wind feedback. The moist static stability can be negative for moist ISO modes, as is characteristic of wave—Convective Instability of the Second Kind (CISK), but this is not a requirement because the generation mechanism is primarily moist baroclinic-barotropic instability. An advantage of a positive moist static stability, due to convective interaction with dynamics (Neelin and Yu, 1994), is that it avoids the ultraviolet catastrophe of wave-CISK theories in which the smallest resolved scale grows fastest. Instead, scale selection for ISOs is governed by moist baroclinic-barotropic instability, and their velocity potentials and vertical shear stream functions have a dominant zonal wave number of $m^* = 1$.

Frederiksen (2002) also examined the properties of the convectively coupled and equatorially trapped waves and found his theoretical modes to compare closely the results of Takayabu (1994) and Wheeler et al. (2000) based on observations. As in the observations, the theoretical ISO modes and the $m^* = 1$ Kelvin wave were distinctly different, indicating that the MJO is not a Kelvin wave, as earlier theories had suggested.

Detailed comparisons between the properties and tropical-extratropical interactions of the observed MJO based on global observations during northern winter and the leading theoretical ISO mode of Frederiksen (2002) were made by Frederiksen and Lin (2013). The upper-tropospheric velocity potential of the MJO evolves in a broadly similar way to that of the theoretical ISO (Frederiksen and Lin, 2013, Figs. 2 and 3). Also, the observed and theoretical upper-tropospheric stream functions have PNA-like and NAO teleconnection patterns at the corresponding phases, which is an important test of the theory (their Figs. 4 and 5). To diagnose wave-activity flux associated with teleconnections, the **W** vector proposed by Takaya and Nakamura (2001) is used, which is based on the conservation of wave-activity pseudomomentum. The **W** vector is independent of phase, which allows snapshots of wave dispersion to be taken at each stage of the circulation evolution. We consider the horizontal components of the **W** vector based on the 300-hPa, basic-state vector winds $\mathbf{U} = (U, V)$ and the 300-hPa perturbation stream function ψ for the theoretical intraseasonal mode. These components are given by

$$\mathbf{W} = \frac{1}{2|\mathbf{U}|}\begin{bmatrix} U\left(\psi_x^2 - \psi\psi_{xx}\right) + V\left(\psi_x\psi_y - \psi\psi_{xy}\right) \\ U\left(\psi_x\psi_y - \psi\psi_{xy}\right) + V\left(\psi_y^2 - \psi\psi_{yy}\right) \end{bmatrix} \tag{4}$$

where the subscripts denote partial derivatives.

Shown in Fig. 10 are the wave-activity flux vectors overlaid on the contours of Northern Hemisphere 300-hPa stream function for the leading theoretical intraseasonal mode, with the EVAP basic state, at phases 3–5. They exhibit structures very similar to those of Lin et al. (2009) for observed anomalies of the extratropics linked to MJO convection. Theoretical ISOs for other basic states were also studied by Frederiksen and Lin (2013) and shown to have similar relationships between the tropical ISO signal and PNA-like and NAO teleconnection patterns. The close comparisons of these theoretical modes with observations of the tropical *and* extratropical circulations associated with the MJO, made by Frederiksen and Lin (2013), suggest that the theory is verified by these observations.

Frederiksen and Frederiksen (2011) performed theoretical studies, with a primitive-equation-instability model, of the changes during the 20th century in austral winter synoptic weather modes, including ISOs, with a Southern Hemisphere focus. In particular, they found increases in the latter part of the 20th century in the growth rate of ISOs. Interestingly, as shown in the Fig. 7 in Frederiksen and Frederiksen (2011), at a particular phase, there are wave trains in both hemispheres associated with the tropical signal, including one from over Africa and the Indian Ocean across Australia to the Southern Ocean and South America.

Whelan and Frederiksen (2017) studied the properties of dynamical modes and their tropical-extratropical interactions during the 1974 and 2011 La Niña events using a primitive-equation-instability model with Kuo convection and evaporation-wind feedback. They related these interactions on the intraseasonal timescale to the observed extreme austral summer rainfall and flooding over Australia. Through spectral analysis of observational data, they found that the extremes were associated with rapid growth of the MJO and the constructive interference with Kelvin waves. Moreover, the leading theoretical ISO and Kelvin wave modes were rapidly growing with extratropical responses over Australia in broad agreement with the observations, including those in the Fig. 4 of Whelan and Frederiksen (2017).

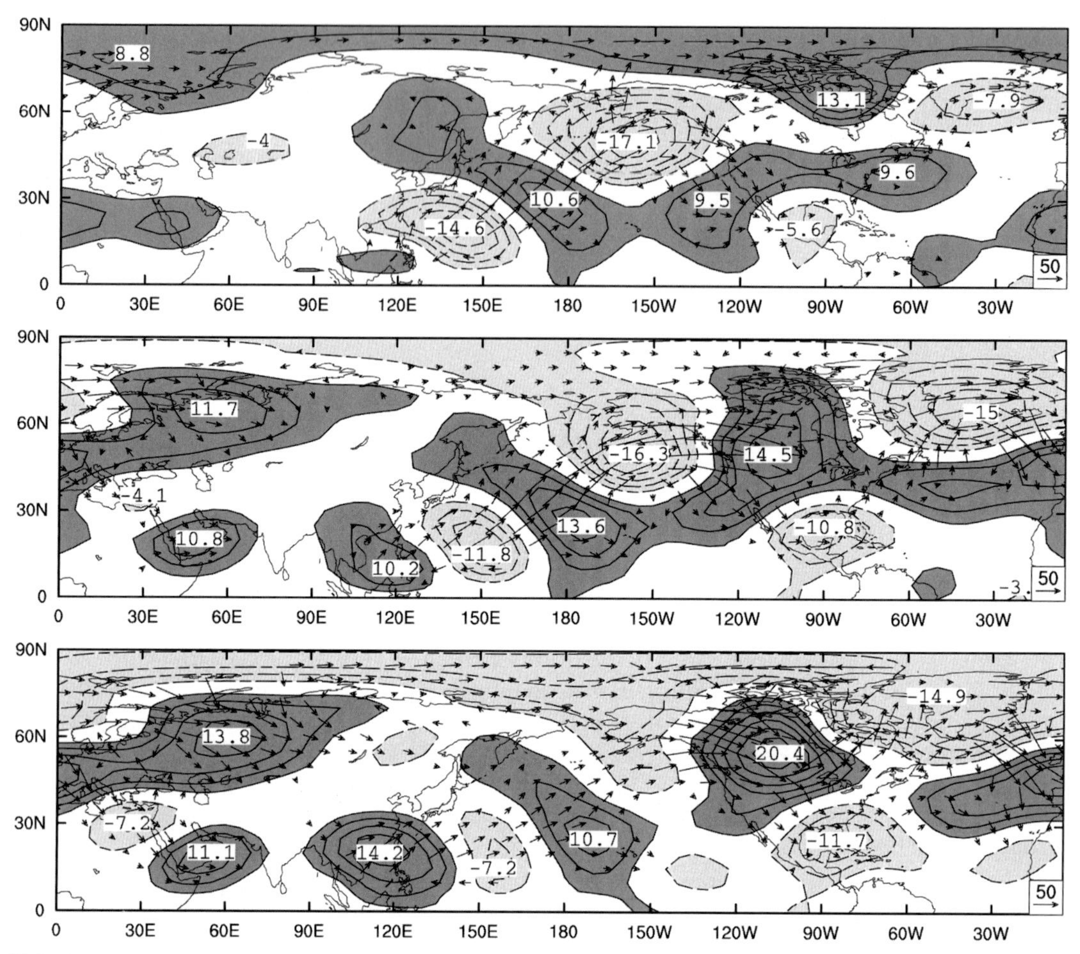

FIG. 10 Wave-activity flux vectors overlaid on the contours of Northern Hemisphere, 300-hPa stream function for the leading theoretical intraseasonal mode, with the EVAP basic state, at phases 3–5. The stream function contour interval is 3; an arrow length corresponding to 50 is shown in the bottom-right corners of each panel.

5 SUMMARY AND DISCUSSION

In this chapter, tropical-extratropical interactions and teleconnections on the sub-seasonal timescale were reviewed. As the dominant mode of sub-seasonal variability in the tropics, the MJO has a significant influence on the extratropical weather and climate. The influence is stronger in the winter hemisphere due to stronger westerlies. The MJO phases of 2–3 and 6–7, according to the phase definition of Wheeler and Hendon (2004), which correspond to a west-east dipole structure of convection anomaly in the tropical Indian Ocean and western Pacific, tend to be more effective at forcing the extratropical atmosphere. Dynamical processes associated with extratropical atmospheric response to tropical thermal forcing include Rossby-wave propagation and interactions with the basic flow and with synoptic-scale transients. Extratropical waves can propagate into the tropics through regions of westerly zonal wind and influence tropical convections and the MJO. The global view of sub-seasonal variability with tropical-extratropical, two-way interactions is supported by an instability theory of 3D basic flow.

There is great potential of gain in sub-seasonal forecast skill if the model can capture the tropical-extratropical interactions and atmospheric teleconnections. However, many scientific questions remain to be answered. For example, what is the relative importance of tropical convection in generating teleconnections in comparison to other dynamical processes, such as interactions with synoptic-scale eddies? What are the processes involved in the initiation of tropical convection by Rossby wave trains propagating from the extratropics into the tropics? There has not been a systematic assessment of how the current models perform in simulating global teleconnections on the sub-seasonal timescale, especially for those related to tropical-extratropical interactions. The question of which processes determine the model's ability to capture teleconnections is also unclear. Further studies are required to understand the role of the ocean in tropical-extratropical interactions on a sub-seasonal timescale.

APPENDIX. TECHNICAL MATTERS RELATING TO SECTION 4.2

(a) *Time filtering*. The fields of zonal wind (u) and meridional wind (v) were filtered to retain only periods of 10 days and less using a digital filter, as described in Blackmon (1976) and Duchon (1979). These high-pass winds (u', v') were used in Fig. 8. The diabatic heating was likewise filtered to retain periods of 20 days and more (i.e., low-pass). Note that the low-pass components still retain the seasonal cycle. In particular, the low-pass heating shown in Fig. 8 contains the climatological heating. To obtain intraseasonal components of the diabatic heating and the momentum flux ($u'v'$), the method of Straus (1983) was used: For each winter season (consisting of 105 days from December 1–March 15), a single parabola was fitted to the time series at each grid point and then removed. This is an efficient way of estimating variability on intraseasonal timescales. The intraseasonal components of momentum flux and heating were used in Fig. 7.

(b) *Estimate of diabatic heating*. We start from the following thermodynamic equation:

$$T\frac{ds}{dt}=Q$$

where s is the entropy per unit mass, T is the temperature, Q is the diabatic heating, and the time derivative is the material derivative. Next, use the entropy for an ideal gas:

$$s=c_p \ln\theta$$

where c_p is the specific heat per unit mass at constant pressure and θ the potential temperature, calculated as $\theta=\left(\frac{p_0}{p}\right)^{\kappa}T$, with p being the pressure, $p_0=1000$ hPa; and $\kappa=\frac{R}{c_p}$, with R being the ideal gas constant.

The thermodynamic equation then can be written as

$$\left(\frac{p}{p_0}\right)^{\kappa} c_p \frac{d\theta}{dt} = Q$$

$$\left(\frac{p}{p_0}\right)^{\kappa} c_p \left(\frac{\partial\theta}{\partial t} + \vec{v} \cdot \vec{\nabla}\theta + \omega\frac{\partial\theta}{\partial p}\right) = Q$$

$$\left(\frac{p}{p_0}\right)^{\kappa} c_p \left(\frac{\partial\theta}{\partial t} + \vec{\nabla} \cdot \left(\vec{v}\theta\right) - \theta\vec{\nabla} \cdot \vec{v} + \omega\frac{\partial\theta}{\partial p}\right) = Q$$

In the second step given here, the material derivative has been written in pressure coordinates, with $\vec{v} = (u, v)$ referring to the horizontal velocity and $\omega = \frac{dp}{dt}$. The final equation is convenient for evaluation using spherical harmonics.

The ERA-Interim reanalysis was read on the original (512 × 256) (lon × lat) Gaussian grid at all 37 available pressure levels for the 35 winters 1980/81–2014/15. The spherical harmonic coefficients of temperature, vertical velocity, vorticity, and divergence up to truncation T255 were evaluated once per day. These spectral coefficients were truncated to T42, and these coefficients used to estimate each of the terms on the right side of the last equation given here. The temporal derivative of θ at day $(d + {}^1/_2)$ was estimated by $\frac{(\theta_{d+1} - \theta_d)}{\tau}$, where τ is 24 hours, and the other (advection) terms were estimated as the average of their values at day d and day (d + 1). These level values of heating were integrated vertically (with mass weighting) into layers. The 700–300-hPa layer vertical integral is used.

CHAPTER

8

Land Surface Processes Relevant to Sub-seasonal to Seasonal (S2S) Prediction

Paul A. Dirmeyer, Pierre Gentine†, Michael B. Ek‡, Gianpaolo Balsamo§*

*Center for Ocean-Land-Atmosphere Studies, George Mason University, Fairfax, VA, United States †Columbia University, New York, NY, United States ‡National Center for Atmospheric Research, Boulder, CO, United States §European Centre for Medium-Range Weather Forecasts, Reading, United Kingdom

OUTLINE

Sub-seasonal to Seasonal Prediction
https://doi.org/10.1016/B978-0-12-811714-9.00008-5

1 INTRODUCTION

There has been a recognition of the prospects for sub-seasonal to seasonal (S2S) prediction for at least 15 years, including the potential role of the land surface (Schubert et al., 2002). Recent recommendations on the next generation of Earth-system prediction (NAS, 2016) point to soil moisture as a source of S2S prediction with "benefits likely in the short term." The relatively slowly varying aspects of the land surface, including soil moisture and snowpack, are noted as forming the basis of a large portion of S2S predictability; the peak impact of the land on prediction is thought to be between about 1 week and 2 months after forecast initialization, but it also affected shorter (weather) and longer (seasonal) timescales to some degree (Dirmeyer et al., 2015). Better representation of the land surface and hydrological processes in models, assimilation and prediction of soil states, and seasonal vegetation growth are seen as key components of S2S prediction. A report by the National Academy of Sciences (2016) also champions research to better understand the significant interactions with the weather and climate system, as well as "the need for dynamic integration into operational forecasting systems."

In this chapter, we provide specific background on the theories underpinning the relevance of land-surface states to weather and climate, including some history of the evolution of land-surface models for operational prediction. In addition, we describe basic physical processes in the interactions between land and atmosphere and key land-surface states relevant to S2S prediction. Atmospheric boundary layer (BL) responses to land-surface variations are explained, followed by an outline of the necessary conditions for S2S predictability and prediction skill derived from the land surface. Specifically, there are three "ingredients" necessary in order for the land surface to contribute to sub-seasonal prediction, which are described in Section 4. We close with a discussion of how to improve land-driven S2S prediction in the future.

2 PROCESS OF LAND-ATMOSPHERE INTERACTION

Interactions between the land surface and atmosphere, which can provide an important source of predictability to weather and climate on S2S timescales, occur via the fluxes between them, These include fluxes of radiant energy (primarily in visible and infrared wavelengths), heat, water, and momentum. Thus, these interactions directly affect atmospheric temperature, humidity, density, and winds near the surface, which can be conveyed vertically and horizontally in the atmosphere by dynamical and thermodynamic processes (see Fig. 1). In this section, we provide a brief primer on the physical processes involved in land-atmosphere feedback.

2.1 Surface Fluxes

Shortwave radiation originates mainly as light from the sun. The light that makes it through the atmosphere to the land surface is mostly absorbed, but some is reflected. Open water absorbs nearly all the sunlight falling upon it. Verdant vegetation absorbs strongly in

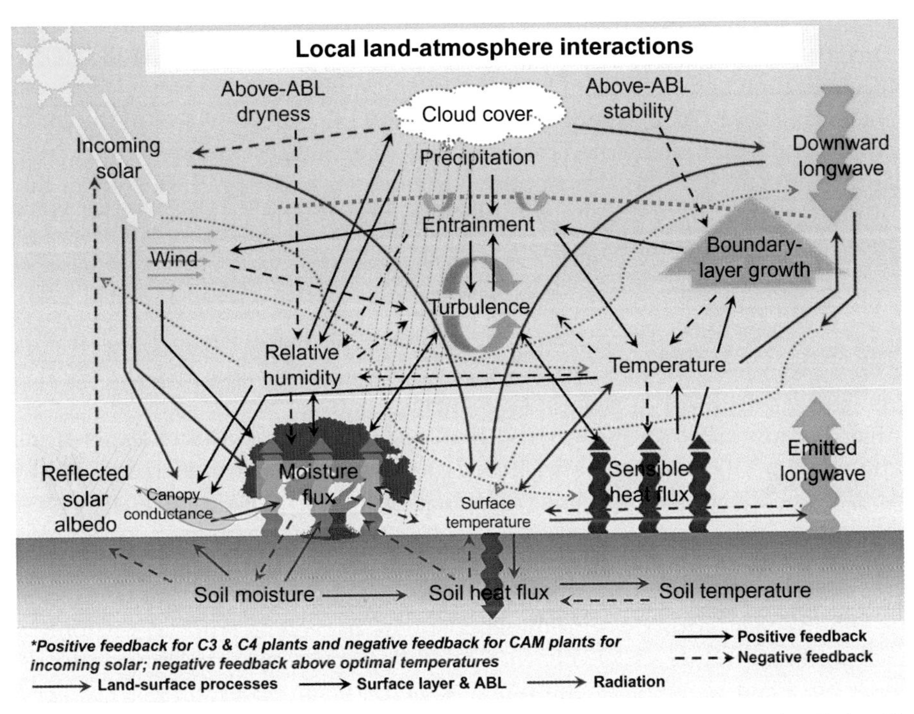

FIG. 1 Schematic of the complex interactions among the land surface, atmospheric BL, and radiation. These interactions are not all well understood and are often poorly represented in numerical models.

the visible spectrum, but it is highly reflective in the near-infrared range, where more than 40% of solar energy resides. Dormant or senescent vegetation reflects up to a quarter of visible light, and desert sands can reflect as much as 40% of all incoming shortwave radiation (Bonan, 2008). Fresh snow is the most highly reflective natural substance on the land surface, reflecting over 80% of solar radiation, but snow progressively absorbs more radiation as it ages and compacts. These spatial and temporal variations in land-surface reflectivity have huge consequences for the surface energy balance and the distribution of surface heating.

Absorbed solar radiation heats the land surface—this heat may be conducted deep into the soil, radiated back to the atmosphere as thermal radiation, or returned to the atmosphere by sensible heating (direct contact between air and the surface) or latent heating via the evaporation of water or sublimation of snow and ice. Latent heating also links the surface energy balance to the water balance, allowing the land a degree of control over atmospheric humidity as well as temperature (Shuttleworth, 2012).

The primary exchanges of water between land and atmosphere occur by the processes of precipitation (namely, rain and snow) and evapotranspiration (ET), which has several components (Shuttleworth, 2012). The more complex processes of land-atmosphere interaction involve the role of vegetation in regulating *transpiration,* the conduction of water from soil to air via the vascular system of plants. Transpiration is strongly connected to photosynthesis. Plants lose water vapor during the process of taking carbon dioxide from the atmosphere. Vegetation is constantly optimizing the production of biomass against environmental stresses of temperature, sunlight, and water availability (Bonan, 2008). This has direct consequences for the surface energy budget, as plant regulation of transpiration affects the partitioning of net surface radiation between latent heat fluxes and other heating terms (e.g., Teuling et al., 2010; Seneviratne et al., 2010; Gentine et al., 2007, 2011a). Surface water in the form of lakes, rivers, ponds, and water intercepted by vegetation or artificial structures can evaporate at the potential rate directly into the atmosphere. Large water bodies can thus affect the regional Bowen ratio, contributing extra moisture to the atmosphere; provide evaporative cooling; and induce mesoscale circulations.

2.2 Land-Surface States

Soil moisture is the most important land-surface state for S2S prediction over most continental areas because of its strong variability in space and time, as well as its crucial role in modulating ET. Specifically, anomalies in soil moisture can contribute to deviation from normal conditions on S2S timescales. For example, dry soil can exacerbate heat waves (Mueller and Seneviratne, 2012; Miralles et al., 2014), while unusually moist soil in semiarid regions can keep temperatures anomalously low. Moisture in the top few centimeters of the soil can interact directly with the atmosphere via evaporation to the air. Below the surface, soil moisture still interacts strongly with the atmosphere in regions with vegetation via transpiration. S2S prediction models typically consider this connection as reaching only 1–3 m below the surface, but plant roots may penetrate tens of meters into the soil (Jackson et al., 1996).

At higher latitudes, and in many mountainous areas, snow is an important surface variable. Snow is unique in that it has two distinct processes of interaction with the atmosphere, one radiative and one hydrologic, each of which operates on a different timescale (Xu and Dirmeyer, 2013). Snow provides both a much more reflective surface than the underlying ground and a saturated surface over which sublimation can occur at the potential rate. When substantial solar energy is present, such as during the transition from winter to spring, the presence or absence of snow cover can cause drastic contrasts in absorbed shortwave radiation, and thus the surface energy budget. Snow also represents another water reservoir at the land surface. Thus, anomalies in winter snow cover can translate into anomalies in spring or summer soil moisture after the snow melts, exhibiting all the controls on S2S predictability discussed here (Xu and Dirmeyer, 2013).

Vegetation regulates the transpiration component of ET, as mentioned earlier, but it also affects exchanges of heat and momentum with the atmosphere via its coloration and density. Human land management is another kind of land-surface perturbation. Agriculture replaces native vegetation with crops, which often have different physical characteristics and seasonality. Cropping in particular can introduce drastic changes on S2S timescales, as cultivars typically green up very quickly and harvest drastically changes land-surface properties on a scale of days to weeks (cf. Mahmood et al., 2014).

Lakes are important surface elements that influence the local weather and climate. Over midlatitude regions, they foster mild "microclimate" conditions acting as thermal inertial bodies. Lakes also can trigger locally higher precipitation rates through enhanced evaporation and mesoscale circulations driven by differential daytime heating relative to adjacent land. At high latitudes, lakes typically freeze in winter; freezing changes the surface albedo and thermal capacity, with consequent effects on the surface fluxes exchanged with the atmosphere. The freeze/thaw state of lake water can make the difference between a light or a major snowfall event (via so-called lake effect snow enhancement) downwind from a lake. Lake effects are also important in temperate and tropical areas, where they are often linked with high-impact weather (HIW), contributing to the formation of nighttime convective cells due to moisture convergence and breeze effects.

2.3 Boundary Layer (BL) Response

A major difference between continents and oceans is the presence of a strong diurnal cycle over land (Gentine et al., 2007, 2011a,b, 2013a; D'Andrea et al., 2014). This cycle is mostly due to the small heat capacity of land surfaces compared to open water—vegetation and soil contain less moisture that limits heat storage and conductance. Thus, land exchanges energy with the atmosphere through sensible heating proportional to the temperature difference between the solid surfaces and the overlying air. This leads to the diurnal growth of the BL (Deardorff, 1979; Deardorff and Willis, 1980; Stull, 1988; van Heerwaarden et al., 2009; Gentine et al., 2013a, 2016), and it modifies the atmospheric stability near the surface (Businger et al., 1971; Wyngaard and Moeng, 1992).

In addition, the diurnal cycle of the BL is tightly connected to the generation of forced (i.e., negatively buoyant thermals) and active (i.e., shallow and deep) convection (Brown et al., 2002; Khairoutdinov and Randall, 2003, 2006; Ek and Holstlag, 2004; Gentine et al., 2013b, c; D'Andrea et al., 2014). The development of shallow or deep convection depends on atmospheric stability and its interplay with the surface Bowen ratio (Gentine et al., 2013c). Under relatively weak stratification in a deep BL, convection occurs preferably over drier surfaces, whereas under stronger stratification in a shallow BL, it tends to occur over wetter soil (Findell and Eltahir, 2003; Findell et al., 2011; Guillod et al., 2015).

A consequence of this linkage is that feedback can develop. A growing BL over land necessarily gains heat from the surface, but it also can gain moisture from ET, while gaining heat and losing moisture from the entrainment of air from the free atmosphere above the growing BL, which enhances ET rates when turbulence mixes that air down to the surface (Betts et al., 1996; Lhomme, 1997). Positive feedback includes the direct water cycle loop, wherein precipitation increases soil moisture, enabling ET and adding water vapor to the air that can lower the lifting condensation level and reinforce precipitation (Dirmeyer, 2006; Santanello et al., 2011a). Such positive hydrologic feedback may have played a role in the seasonal-scale 1993 floods over the midwestern United States (Beljaars et al., 1996), and synoptic-scale flooding events as well (Milrad et al., 2015; Teufel et al., 2016). Droughts can be maintained by the same processes acting to reinforce negative anomalies in water states and fluxes (Mo et al., 1997). However, daytime BLs grow as a result of surface heating, which is favored over dry surfaces (Betts, 1994). Therefore, in circumstances where growing a deeper BL is more efficacious to produce clouds and precipitation than adding water vapor to the atmosphere (which lowers the base of potential clouds), negative hydrologic feedback exists (Guillod et al., 2015). The relative cost of generating clouds via surface-sensible heat fluxes versus latent heat fluxes can be quantified in a single framework (Ek and Holstlag, 2004; Santanello et al., 2009, 2011a; Gentine et al., 2013c; Tawfik and Dirmeyer, 2014).

2.4 Timescales

Variations in the interactions between land and atmosphere occur on a range of timescales and space scales. The diurnal cycle is particularly strong in many locations because solar forcing oscillates strongly on the daily scale, and the land surface has relatively low heat capacity. Because soil moisture is an integrating reservoir term in the surface water budget, it has a red-noise spectrum, as opposed to the rather white spectrum of precipitation (Rodrı´guez-Iturbe et al., 1991a,b). As a result, soil moisture extends the hydrologic timescale at the land surface to days or weeks, providing a potential source of S2S predictability via extended memory. Spatially, horizontal gradients in land-surface cover, albedo, soil wetness, and elevation can induce gradients in surface fluxes. Atmospheric properties also can play a role in determining the preferred scales to maximize the impact of such gradients; a horizontal scale of variations about 4–9 times the depth of the BL is optimal for establishing mesoscale circulation (van Heerwaarden et al., 2014), but the scale of impact of mesoscale effects and the development of deep convection induced by mesoscale heterogeneity depend on the BL and free tropospheric conditions (Rotunno, 1983; Gentine et al., 2013a,b,c; Rieck et al., 2015).

3 A BRIEF HISTORY OF LAND-SURFACE MODELS

Plausible ideas about the potential for land-surface impact on weather and climate go back at least to Leonardo da Vinci, who accurately imagined many aspects of the surface water and energy cycles (Pfister et al., 2009). During the westward expansion of European settlement in North America, agronomist Cyrus Thomas is credited with promulgating the notion in the 1870s that "rain follows the plow." Land speculators picked up on the notion that agriculture could affect weather, suggesting that "after deep plowing ... a reduction of temperature must at once occur, accompanied by the usual phenomena of showers" (Wilber, 1881). Similar notions were popularized during the settlement of the interior of Australia (Diamond, 2005), although such ideas were based on avarice rather than scientific research.

Statistical analysis by Namias (1962, 1963) showed a lagged relationship between spring and summer temperatures and rainfall over the central United States, which he attributed to land acting as a reservoir for heat and moisture, communicating anomalies across seasons. However, early numerical models used for weather prediction and climate research did not include an interactive land-surface component. The land was thought to be unimportant compared to the role of atmospheric dynamics.

3.1 Origin and Evolution of Land-Surface Models

The potential effect of the land on the atmosphere was first suspected at seasonal timescales, whereby controls could occur via access to deep soil moisture by tree roots, seasonal variations in vegetation phenology that affect transpiration, seasonally frozen soils, snow cover, rains, and dry periods. The original implementation of a parameterization of land-surface hydrology in a global numerical model was the so-called bucket model of Manabe (1969), consisting of a 15-cm reservoir for soil moisture underlying each atmospheric grid box over continents. The bucket was filled by precipitation, overflowed into runoff, and was emptied by evaporation, represented as a simple linear function of soil moisture. A "leaky bucket" (e.g., Huang et al., 1996) was developed to account for the gravitational loss of soil moisture into the vadose zone and water table.

In the early days of numerical weather prediction (NWP), the effect of land (and thus the inclusion of land models) was initially ignored. On weather timescales, the focus was on solving the Primitive Equations for the atmosphere, as well as the parameterization of Physics, such as radiative transfer, convection, and turbulent energy dissipation. Throughout the 1960s and 1970s, the only role for the land surface was the influence of varying terrain heights on air flow and surface friction effects. However, as the diurnal cycle of downwelling shortwave radiation was explicitly included in atmospheric models, the role of daily terrestrial thermal and hydrologic cycling became recognized, and land-surface models (LSMs) arose.

First-generation LSMs maintain complete surface water and energy budgets. One popular parameterization was the Force-Restore method, which recognized that there are two primary cycles of variability at the land surface: the diurnal cycle and the seasonal cycle (Bhumralkar, 1975). This approach solves two wave functions whose penetration into the soil, and thus the volume of soil under consideration for heat and water content estimation, decreases with the square root of the frequency. Such an approach was mathematically elegant but proved difficult to calibrate or validate. Multilayer soil models became the standard, as heat and moisture transfer can be simulated by conduction equations and directly validated against direct subsurface measurements.

Second-generation LSMs include the effects of vegetation, both in terms of the spatial heterogeneity of plant coverage across the globe and their impact on the surface budgets through processes like transpiration and radiative transfer within plant canopies. They followed the "big leaf" approach, treating vegetation properties homogeneously within each grid box as separate circuits for heat and moisture transport with resistances, potentials, and fluxes analogous to electric current. Third-generation LSMs explicitly model photosynthesis and the carbon cycle, along with the water and energy cycles, as all three cycles are closely connected (Sellers et al., 1997).

Yet, in some ways, the science of land-atmosphere interactions has proceeded backward to the traditional scientific progression from observation of natural phenomena, formulation of hypotheses, development of experiments, and construction of models. As implied previously, LSMs were developed initially to provide lower boundary conditions for existing atmospheric models and applied largely by scientists outside the disciplines of hydrology, ecology, soil science, and biogeochemistry. This was done before there were wide-ranging observations of the land surface or land-atmosphere interactions relevant to model development. Early LSMs were calibrated on a handful of locations (Sellers and Dorman, 1987) and then applied globally out of necessity.

3.2 LSMs at Operational Forecast Centers

Here, we detail a number of examples of LSM evolution at several operational forecasting centers. Numerical weather forecast models from the 1960s through much of the 1980s neglected the land surface entirely, focusing on deterministic atmospheric forecasts determined solely by evolution from the initial atmospheric state. Energy and water budget closure at the surface were gradually recognized as important, especially as the diurnal cycle of radiative (solar) forcing was incorporated into models. The U.S. National Weather Service (NWS) Nested Grid Model (Hoke et al., 1989) included a single-layer soil slab with a bucket-style hydrology having a specified wetness factor with separate winter and summer settings, diurnal land-surface albedo effects, surface energy balance and surface skin temperature calculations, and inclusion of snow cover, but not depth. In the early 1990s, the Oregon State University (OSU) LSM (Mahrt and Pan, 1984) was introduced into the NWS global model, having two soil layers with soil heat diffusion equations and soil hydraulic properties (Clapp and Hornberger, 1978), a specified annual cycle of vegetation accounting for plant stomatal control on transpiration (Mahrt and Ek, 1984), and a simple snow physics treatment. The OSU LSM was upgraded for implementation in the Eta regional forecast model (Chen et al., 1996a, 1997), applying the emerging big leaf approach to modeling vegetation (Jarvis, 1976; Noilhan and Planton, 1989).

Upgrades in the early 2000s at the National Centers for Environmental Prediction (NCEP)—notably new infiltration and runoff formulations, improved soil thermal conductivity, snowpack physics and surface fluxes that accounted for patchy snow cover, and inclusion of frozen soil physics (Ek et al., 2003)—led to the renaming of the LSM to Noah, with the name acknowledging contributions from **N**CEP, **O**SU, the U.S. **A**ir Force, and the Office of **H**ydrology at the National Oceanic and Atmospheric Administration (NOAA). Noah was implemented in the NCEP global model in 2005, and into the new Weather Research and Forecasting (WRF) mesoscale model (Chen and Dudhia, 2001).

Noah has evolved toward a third-generation configuration that includes options to choose among multiple parameterizations (Noah-MP; Niu et al., 2011). Noah-MP uses an efficient but more accurate iterative solution of the surface energy budget and has a physically based CO_2-photosynthesis formulation for canopy conductance, dynamically predicted vegetation phenology, variable multilayer snowpack layering, snow water melt-refreeze processes, and an explicit groundwater module. Noah-MP is now an option in the community WRF model and is being tested in the NCEP seasonal forecast system with the aim to implement it for medium-range and NWP forecasting.

The European Centre for Medium-range Weather Forecasts (ECMWF) introduced treatment of the land surface via a climatology for deep soil states and two interactive layers to represent slow and fast processes (Blondin, 1991). Viterbo and Beljaars (1995) introduced a fully interactive, four-layer soil scheme, but globally uniform vegetation. The Tiled ECMWF Scheme for Surface Exchanges over Land (TESSEL) introduced the capacity to represent up to seven subgrid surface types (van den Hurk et al., 2000), but soil texture remained globally fixed. HTESSEL added revised soil hydrology (thus the added H) with variable soil texture, revised formulations of hydrological conductivity and diffusivity, and a new treatment of surface runoff (van den Hurk and Viterbo, 2003). Balsamo et al. (2009) verified the impact of the soil hydrological revisions from field sites to global scales, in both atmospheric coupled experiments and data assimilation. The HTESSEL snow hydrology was revised to replace the previous scheme based on Douville et al. (1995). It introduced rainfall interception, liquid water retention in the snowpack, melting/refreezing, and the main processes of compaction that affect density. Snow albedo and snow cover fraction parameterizations were revised, and forest albedo in the presence of snow was retuned based on Moderate Resolution Imaging Spectroradiometer (MODIS) satellite estimates (Dutra et al., 2010a).

CHTESSEL introduced CO_2 and vegetation seasonality (thus the additional C). The leaf area index (LAI), which expresses the vegetation density, was temporally constant in ERA-Interim, assigned by a lookup table depending on the vegetation type. An LAI monthly climatology based on MODIS was implemented in November 2010 (Boussetta et al., 2013a). CHTESSEL can also simulate CO_2 exchanges with the biosphere (Boussetta et al., 2013b). In 2012, a further revision allowed evaporation from bare soil for soil moisture content below the wilting point (Balsamo et al., 2011), which results in more realistic soil moisture for dry land. The changes reported in Balsamo et al. (2011) have been extensively evaluated by Albergel et al. (2012) over the United States. The introduction of a subgrid thermodynamic lake model followed preparatory work by Dutra et al. (2010b) and Balsamo et al. (2010, 2012).

Early on, the Met Office in the United Kingdom (UKMO) used a force-restore method to solve surface energy and water budgets (Davies and Warrilow, 1986), but later switched to a multilevel soil model (Warrilow et al., 1986). LSM development in the United Kingdom was driven primarily from the climate modeling interests of the Hadley Centre, which produced one of the first third-generation schemes, the Met Office Surface Exchange Scheme (MOSES; Cox et al., 1999; Gedney and Cox, 2003). The current LSM, which evolved from MOSES, is the Joint UK Land Environment Simulator (JULES; Blyth et al., 2006). Similarly, Météo-France has evolved from a force-restore parameterization (Noilhan and Planton, 1989) to ever-more-physically realistic representations of the land surface (Noilhan and Mahfouf, 1996; Boone et al., 1999; Calvet et al., 2004; Decharme and Douville, 2006).

Often, regional interests drive aspects of LSM development, such as the historical emphasis on snow processes in the Canadian Land-Surface scheme (CLASS; Verseghy, 2000). The Japan Meteorological Agency (JMA) LSM is derived from the Simple Biosphere (SiB) model (Sellers et al., 1986), as its first implementation in a global atmospheric model was accomplished by a visiting JMA scientist (Sato et al., 1989), but it has evolved further with an emphasis on simulating "urban canopies" becoming prevalent across Asia. In other operational centers, the heritage of the LSM may be linked to origins of the atmospheric model.

3.3 LSM Initialization and Data Assimilation

In the absence of land observations, forecast models that include LSMs would typically be initialized with climatological states of land temperature, snow cover, or soil moisture (Mintz and Serafini, 1981). An alternative method was developed in the early 1990s, whereby LSMs were driven directly by gridded, observationally based analyses of near-surface meteorology (Liston et al., 1993). This approach was developed, standardized, and applied in a multimodel framework by the Global Soil Wetness Project (Dirmeyer et al., 1999, 2006a) in a research context, and then further developed into a real-time system appropriate for forecast initialization, the Land Data Assimilation System (LDAS; Mitchell et al., 2004; Rodell et al., 2004). This approach was first applied to seasonal forecast models, but it has worked its way into operational weather forecast systems, such as Global LDAS (GLDAS; Meng et al., 2012) and European LDAS (ELDAS; Jacobs et al., 2008).

The original LDAS approach did not actually assimilate land-surface observations; rather, it relied on input from observationally constrained atmospheric analyses using the assimilation of meteorological data. Among the first attempts to use near-real-time observed land-surface data was by the Environmental Modeling Center of NCEP, which used the analysis of Northern Hemisphere snow coverage produced by the U.S. Air Force (Kopp and Kiess, 1996).

Another earlier approach was introduced by Bouttier et al. (1993a,b), who realized that the soil moisture state is critical to controlling near-surface air temperature and humidity. They updated soil moisture not based on actual soil moisture observations, but rather to minimize model air temperature and relative humidity errors. This approach assumes soil moisture is the main source of such errors, and is still used operationally. One consequence is that errors of the atmospheric model such as in precipitation or downward shortwave radiation are essentially transferred to the soil; short-term forecasts improve, but at the expense of correct characterization of soil moisture dynamics. This hampers the use of additional surface observations to constrain data assimilation, as they tend to degrade the model forecast.

Error and information characterizations of satellite soil moisture products may help bridge the gap to successful applications in S2S forecast models (Kumar et al., 2017). In recent years, a wealth of new remotely sensed observations have become available that can be used to constrain the state of the land surface. Here are some of the major ones (but the list is by no means exhaustive):

(a) *Soil moisture content* in the top few centimeters of the soil can be assessed from space using microwave remote sensing, especially L-band passive measurements (Kerr et al., 2010; Entekhabi et al., 2010; Kolassa et al., 2016; Wigneron et al., 2017). Surface soil moisture may provide important information to constrain the surface Bowen ratio, vegetation phenology, and hydrology (e.g., McColl et al., 2017). It is an ideal candidate for land-surface data assimilation (Entekhabi et al., 1994; Wigneron et al., 2002; Reichle et al., 2007, 2008; Reichle, 2008; De Lannoy and Reichle, 2016), possibly adding other constraints for S2S prediction (Draper and Reichle, 2015).

(b) *Surface albedo* is critical to constraining the surface available radiation (Oleson et al., 2003) and the surface energy budget. Long-term and near-real-time records of surface albedo are now available (Salomon et al., 2006; Liu et al., 2009a), which can be used to constrain historical simulations and forecasts. Many models still use a specified climatology of phenology and albedo, not allowing anomalous variations that could be extremely impactful. For example, during the spring in midlatitude and cold regions, green-up, the new cycle of seasonal plant growth, can be delayed or accelerated by cold or warm sub-seasonal anomalies affecting albedo, as well as transpiration, surface drag, and turbulence (Fitzjarrald et al., 2001). Another example is in monsoonal regions, where phenology is strongly related to the timing of the monsoon onset (Dahlin et al., 2015).

(c) *Vegetation indexes* such as the Normalized Difference Vegetation Index (NDVI) or Enhanced Vegetation Index (EVI), which provide crucial information on vegetation phenology and greenness, have been available for several decades. Such indexes provide crucial information about vegetation phenology and response to stresses or global change (Konings et al., 2017). Vegetation indexes have been used in LSMs (Jarlan et al., 2008) and coupled climate models (Lu and Shuttleworth, 2002) in order to understand their impact on the hydrology and ecosystem functioning (Quaife et al., 2008) and regional climate. Assimilation of vegetation indexes has not been implemented for S2S prediction, but it could provide substantial improvements.

(d) *Solar-induced fluorescence (SIF)* occurs from photosynthetically active radiation lost by chlorophyll as an inefficiency in the process of photosynthesis. Recently, observation of

SIF has been made from remote-sensing platforms (e.g., GOSAT, GOME-2, and OCO-2; Frankenberg et al., 2011, 2012, 2014). On sub-seasonal timescales, SIF is nearly linearly related to the rate of photosynthesis (and thus transpiration), providing information about both the carbon and water cycles.

The advent of routine, reliable satellite observations of soil moisture, vegetation states, and snow cover has allowed genuine land data assimilation in LSMs that were either uncoupled or coupled to an atmospheric model (e.g., Dee et al., 2011; de Rosnay et al., 2014). Nonetheless, few examples of operational land-surface data assimilation examples currently exist (Barbu et al., 2014; Balsamo et al., 2015). Reasons may include the poor performance of LSMs (Nearing and Gupta, 2015; Haughton et al., 2016), the difficulty in prescribing model and observational errors for accurate assimilation updates, and the limited memory of surface soil moisture in many regions (Dirmeyer et al., 2016; McColl et al., 2017). As such, the assimilation of vegetation products such as LAI, NDVI, or SIF may lead to more rapid success for S2S forecasting (Balsamo et al., 2015; Norton et al., 2017), as vegetation activity appears to strongly constrain S2S land-atmosphere feedback in observations (Green et al., 2017). Also, future P-band missions could help one infer deeper soil moisture and therefore add substantial memory (Konings et al., 2013; Tabatabaeenejad et al., 2014).

4 PREDICTABILITY AND PREDICTION

As mentioned previously, land-surface states generally evolve more slowly than atmospheric states, so the land surface can maintain anomalies and serve as a source of memory affecting the atmosphere beyond typical weather timescales (Koster and Suarez, 2001; Dirmeyer et al., 2009). Nevertheless, there are natural modes of variability of the atmosphere on S2S timescales, as described in other chapters in Part II of this book. Anomalous planetary waves in the extratropical troposphere, such as blocking episodes, can have natural life cycles lasting weeks to months. Land-surface anomalies can cause such waves to amplify and persist, exacerbating associated extremes (Koster et al., 2014). This positive feedback mechanism has been implicated in the persistence of major droughts (Vautard et al., 2007; Roundy and Wood, 2015) and heat waves (Miralles et al., 2012; Ford and Quiring, 2014).

However, atmospheric conditions need not be extreme for the land surface to play a role in S2S climate variations. Broadly speaking, the land surface can affect the atmospheric state on S2S timescales significantly if three conditions are strong enough (Dirmeyer et al., 2015). The first is sensitivity of the atmosphere to variations at the land surface. Sensitivity exists when surface fluxes are primarily controlled by land-surface states and the lower troposphere responds significantly to those fluxes. An example is the relationship between soil moisture and ET. At times when there is a positive correlation between day-to-day variations of soil moisture and ET, there is a demonstrable sensitivity of surface heat fluxes to soil water content. This occurs when the limitation on the evaporation rate is due to the availability of water in the soil, and not energy (net radiation). When and where the correlation is negative, ET variations control soil moisture variations. This situation occurs when energy is the limiting factor for evaporation and the feedback becomes negative.

The second necessary condition is variability in land-surface states. For instance, we find very strong sensitivity of surface fluxes to soil moisture over deserts. In a severely dry location, such as over the central Sahara or the Atacama Desert, rainfall is exceedingly rare, and soil moisture is almost always negligible. Therefore, although there is high sensitivity, soil moisture rarely changes, and surface fluxes are not driven by terrestrial moisture content in practice. In regions such as the Great Plains of North America, the Sahel of Africa, and the Indus River basin, we find both strong sensitivity and sufficient variability—these are among the classical hot spots of land-atmosphere coupling (Koster et al., 2004).

To foment predictability on S2S timescales, a third condition is necessary: memory or persistence of land-surface anomalies. Where the timescale of persistence of soil moisture anomalies is no longer than the deterministic range of atmospheric prediction, initial land-surface states will have only a marginal impact on S2S predictions. For example, this can occur in middle and high latitudes where, even when net radiation is abundant and moisture is available, baroclinic systems still dominate the precipitation events. Northern Europe in summer is one such location. In such regions, land-surface initialization in dynamical prediction models may have an impact on weather forecasts out to a few days, but the impact on S2S predictions may be minimal. The regions where the atmosphere appears most reliably responsive to land-surface states are the hot spots between arid and humid regions in both the middle latitudes and subtropics—these regions largely correspond to major areas of agricultural production. Table 1 broadly summarizes the land-atmosphere coupling

TABLE 1 Variation of Land Impact on Atmosphere Across Various Climate Regimes, Including How the Necessary Ingredients for Land-Atmosphere Coupling Vary

Region	Land-Atmosphere Coupling Strength	Soil Moisture Memory	Land Impact on Atmosphere
Semihumid middle latitudes (summer)	High	Moderate	*High* (in major agricultural areas)
Monsoonal subtropics	High	Long in dry season; short in wet season	*High* in dry season, not in wet season
Arid middle latitudes (summer)	Medium (mainly terrestrial leg)	Long (dry)	*Medium*, during wet spells
Humid middle latitudes (summer)	Medium (mainly atmospheric leg)	Moderate-short	*Medium*, during dry spells
Arid subtropics	Low	Long (dry)	*Low* (generally little variability)
Humid tropics	Low	Short	*Low* (not moisture-limited), except in severe droughts
High latitudes (winter)	Low	Long (frozen)	*Low* until spring melt (lack of sunlight)
High latitudes (summer) and middle latitudes (winter)	Low	Short	*Low* (not moisture limited; atmospheric dynamics dominate)

characteristics of various climate zones of the world and how they affect the ability of the land surface to contribute to predictability.

Finally, it is important to keep in mind the difference between predictability in its idealized sense and realizable predictability. Fig. 2 shows that in version 2 of the NCEP Climate Forecast System (CFSv2), there is a clear relationship between forecast skill, the strength of land-atmosphere coupling, and soil moisture memory. However, taking the soil moisture memory timescale as a limit on the duration of extended skill derived from land-surface initialization,

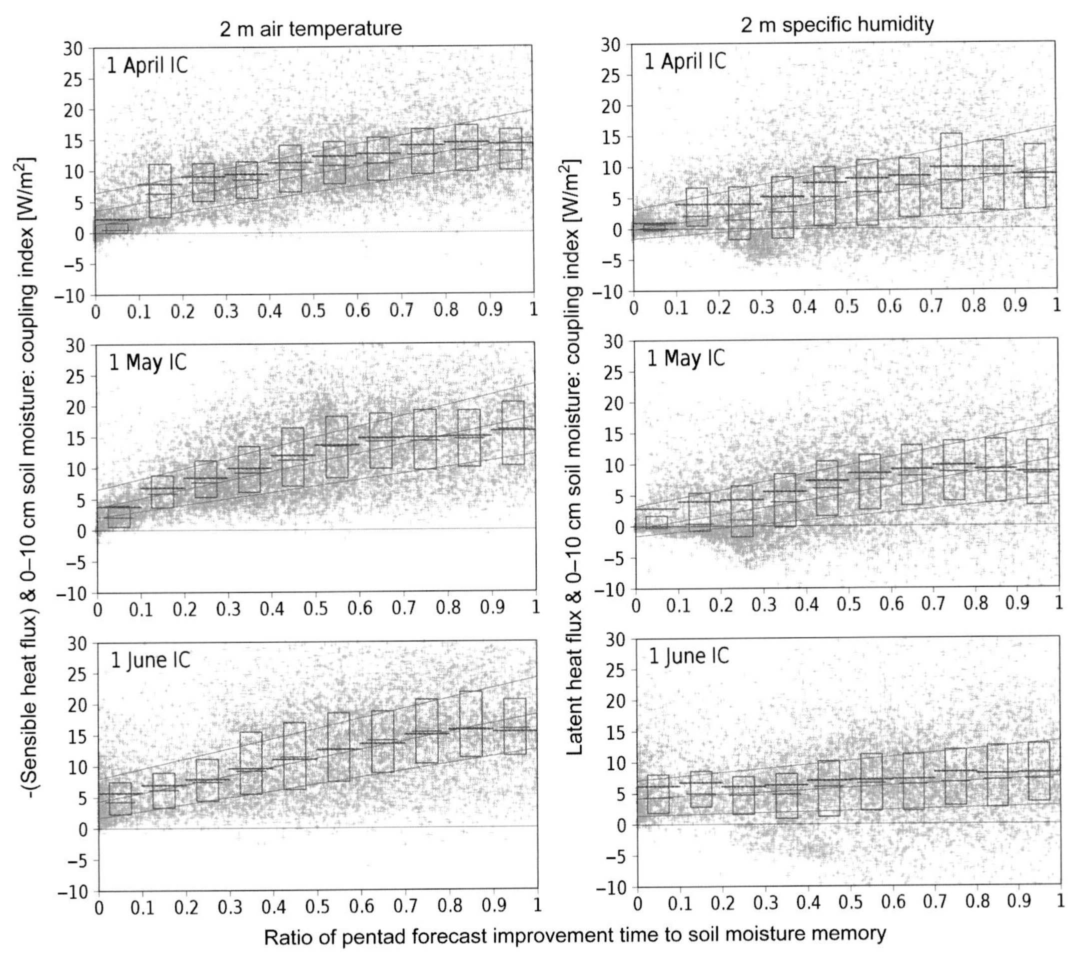

FIG. 2 Scatter distributions of terrestrial coupling indexes (y-axes; sensible heat flux linkage to soil moisture on the left, ET linkage to soil moisture on the right) as a function of the fraction of soil moisture memory duration realized as forecast skill improvement (x-axis) in CFSv2 for (left) 2-m air temperature and (right) specific humidity for ensembles, with the indicated start dates over all global land points that are ice free and do not have frozen soil. *Red bars* show mean values; blue boxes mark median, first quartile, and third quartile; straight lines are linear regressions though the medians and quartiles. *IC* refers to the date of the initial conditions of the forecasts. *From Dirmeyer, P.A., Halder, S., 2017. Application of the land-atmosphere coupling paradigm to the operational Coupled Forecast System (CFSv2). J. Hydrometeorol., 18, 85–108. ©American Meteorological Society; used with permission.*

we see that many locations fall short. There is a significant correlation between the ability of S2S forecasts to realize potential predictability as actual skill and the strength of the coupling between land and atmosphere via both the water and energy cycles (Dirmeyer and Halder, 2017).

5 IMPROVING LAND-DRIVEN PREDICTION

5.1 Validation

Demonstration of the role of the land surface in driving atmospheric anomalies within global numerical models began decades ago. Shukla and Mintz (1982) showed the sensitivity of a climate model to globally wet versus dry land-surface conditions. Subsequent studies (e.g., Rowntree and Bolton, 1983; Oglesby and Erickson III, 1989; Koster and Suarez, 1995; Dirmeyer, 1999; Douville et al., 2001) used specific models to show that sensitivity extended to less severe, more realistic land-surface anomalies. It was not until the 21st century that systematic, multimodel experiments illustrated the role of land-atmosphere coupling in determining temperature and precipitation at S2S timescales (Koster et al., 2002, 2006; Guo et al., 2006), and that realistic land-surface initialization improves sub-seasonal multimodel prediction (Koster et al., 2010a). However, models used for S2S prediction are demonstrably deficient in simulating some aspects of coupled land-atmosphere behavior (Dirmeyer et al., 2006b; Dirmeyer, 2013), while in many ways, they are wholly unvalidated.

The problem lies with a lack of observationally based verification of the behavior of forecast models, stemming largely from a lack of necessary land observations. There have been field campaigns dating back to the late 1980s that were designed to measure and elucidate important land-atmosphere processes (e.g., Sellers et al., 1992, 1995; Famiglietti et al., 1999; Jackson and Hsu, 2001; Andreae et al., 2002; Weckwerth et al., 2004; Redelsperger et al., 2006; Miller et al., 2007; Wulfmeyer and Turner, 2016). Such campaigns have been crucial to improving process understanding and encouraging improvements in model parameterizations. However, field campaigns are very limited in duration and regional in scope, whereas model parameterizations are applied globally and continuously. Very little of the phase space of coupled land-atmosphere model behavior has been calibrated. While thermometers, barometers, and rain gauges are nearly universal, most of the world has never had an in situ measurement of heat flux, surface radiation, or BL. Satellite remote sensing provides global coverage with multiple sensors, but many surface variables, particularly fluxes, are not easily measured from space (Mueller et al., 2013), and the vertical resolution of orbiting atmospheric sounders is inadequate to resolve the critical structure of humidity and temperature in the BL (Santanello et al., 2015).

A valuable uncoupled benchmarking project to examine biases in land models is led by the Global Land/Atmosphere System Study (GLASS), part of the Global Energy and Water Exchanges (GEWEX) project (Best et al., 2015; Haughton et al., 2016). Other activities under GEWEX/GLASS extend model testing into the coupled realm, such as the Local Land-Atmosphere Coupling (LoCo; Santanello et al., 2011b) project and the DIurnal land/atmosphere Coupling Experiment (DICE; Best et al., 2013). The goals of LoCo and DICE are to understand, model, and predict the role of local land-atmosphere coupling in the evolution of surface fluxes and state variables, including BL clouds.

5.2 Initialization

Better-quality land-surface initialization for S2S forecasts holds great promise. Improved forecasts depend on models having the best possible initial conditions. Fig. 3 shows that realistic land-surface initialization in CFSv2 extends prediction skill by 1–2 pentads over most of the globe, and longer in some locations (Dirmeyer and Halder, 2017). NCEP has an uncoupled North American Land Data Assimilation System (NLDAS; Xia et al., 2014), as well as GLDAS, a part of operational CFSv2 (Meng et al., 2012) that provides evolving/cycled land states. ECMWF land data assimilation has a two-dimensional (2D) optimal interpolation

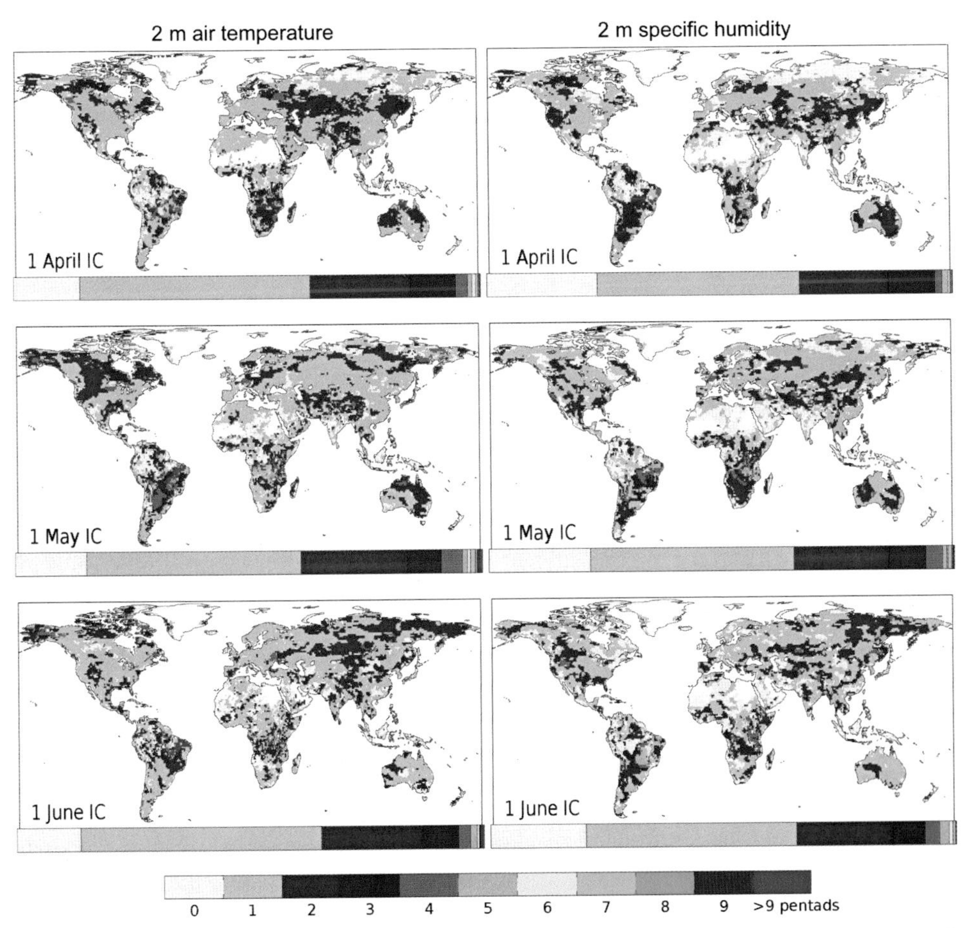

FIG. 3 Number of pentads where the monthly mean forecast of the indicated variables in 28 years of CFSv2 hindcasts with realistic land-surface initialization for the indicated dates remains significantly better than forecasts with randomized initial land states; skill is quantified by anomaly correlation coefficient. *IC* refers to the date of the initial conditions of the forecasts. The *colored bars* beneath each map reflect the fraction of land area occupied by each skill duration category. *From Dirmeyer, P.A., Halder, S., 2017. Application of the land-atmosphere coupling paradigm to the operational Coupled Forecast System (CFSv2). J. Hydrometeorol., 18, 85–108, ©American Meteorological Society; used with permission.*

analysis for snow depth, but soil moisture nudging is still based on minimization of near-surface temperature and humidity errors (Balsamo et al., 2014). The near-real-time green vegetation fraction becomes an assimilable quantity for LSMs predicting vegetation phenology. Other satellite and ground-based observations in an LDAS may yield optimal estimates of the current land states and surface fluxes, such as the use by the National Aeronautics and Space Administration (NASA) of the Land Information System (LIS; Peters-Lidard et al., 2007; Kumar et al., 2008) to assimilate soil moisture, snow, and other land data.

Necessary parameter sets for LSMs include distributions of vegetation type and coverage, soil properties, and related surface parameters (e.g., roughness or snow-free surface albedo) on the model's grid. These may be fixed global data sets, such as the International Geosphere-Biosphere Programme's global map of the land-use/vegetation type (Loveland et al., 2000) or STATSGO/FAO soil texture (Wolock, 1997; FAO, 1988). The quality and resolution of soil data are highly variable around the world, relying heavily on national-level data sets. Time-varying parameters may include mean annual cycles or near-real-time values of green vegetation fraction, LAI, or net surface albedo monitored by MODIS satellite (Hansen et al., 2003). Near-real-time parameters can be especially useful for S2S prediction since in any given year, vegetation green-up or senescence may be ahead of or behind the climatological values of vegetation parameters.

5.3 Unconsidered Elements

Operational LSMs used for S2S forecasting are often less complex than LSMs used in Earth-system models for climate predictions and projections. There are two main reasons. First, the longer the timescale under consideration, the more processes become significant in affecting the evolution of land and atmosphere states. The second is that operational numerical modeling is inherently conservative; only model changes that improve forecast skill are made, or more relevantly, no changes are made that might degrade forecast skill, even temporarily. As a result, operational forecast model development proceeds cautiously and slowly.

Groundwater hydrology and river-routing (completing the water cycle linking freshwater flow to the ocean), lakes, urban areas, ecosystem phenology, biogeochemical cycles (namely carbon, but also nutrients), and human land and water management are all elements that have become common components of LSMs in climate-projection models, but they remain rare in weather and S2S forecast systems. The hydrology option in WRF, WRF-Hydro, along with the Noah-MP land option from WRF, form the basis of the NWS Office of Water Prediction (OWP) National Water Model (NWM; Gochis et al., 2015). This model operates over the continental United States at 1-km LSM resolution and 250 m for surface and subsurface hydrology. The ECMWF modeling system now employs a computationally efficient freshwater lake model (FLake; Mironov et al., 2010) and CO_2 exchanges (Boussetta et al., 2013a; Agustí-Panareda et al., 2014).

As model resolutions increase, more processes become locally significant and the need for concomitant, high-resolution parameter data sets becomes urgent. Remote sensing can provide some information at resolutions of 0.1–1 km^2. However, soil properties that are crucial for determining soil moisture availability, memory, and vegetation stress response are measured in situ unevenly across the globe, with more complete and accurate data for developed nations as well as in agricultural areas. These are crucial parameter sets for LSMs that are failing to meet the needs of high-resolution models. Also unaccounted for in LSMs is the impact of subsurface geology on the drainage of soil moisture. Regions underlain by karst or other porous and fractured substrates experience soil drying much more rapidly that other regions (Dreybrodt, 1988). These variations manifest in the surface water budget primarily beyond the weather timescales. About one-quarter of the conterminous United States is underlain by karst formations (Weary and Doctor, 2014).

5.4 Coupled Land-Atmosphere Model Development

Traditionally, LSM and atmospheric model development are conducted separately by personnel with different scientific backgrounds and expertise. Uncoupled LSM tests are computationally inexpensive, isolating evaluation of the land model without allowing land-atmosphere interactions. Some LSM testing and validation is done in a coupled mode, which is valuable for assessing how models will perform in an operational setting, but it is much more computationally costly. The model development process, however, continues to proceed separately for land and atmospheric models, with coupling done only at the final stage of model upgrade cycles.

We contend that as the potential value of land states for S2S forecasting is emerging, coupled processes are becoming better measured and understood, and model development needs to proceed in a coupled mode throughout the process. Soil, vegetation, and the lower atmosphere behave on a continuum, particularly in terms of the hydrologic cycle, but also for heat and carbon cycles (Dirmeyer et al., 2014). After a long history of dissociated observing networks, meteorological, hydrological, and ecological interests are coordinating to measure consistently from "bedrock to boundary layer" (Duffy et al., 2006). This makes possible the construction of better-performing coupled land-atmosphere models, but only if the interaction of their constituent components is considered throughout the development cycle.

CHAPTER

9

Midlatitude Mesoscale Ocean-Atmosphere Interaction and Its Relevance to S2S Prediction

R. Saravanan, P. Chang

Department of Atmospheric Sciences and Department of Oceanography, Texas A&M University, College Station, TX, United States

OUTLINE

1 INTRODUCTION

The ability to forecast using numerical models has been termed a "quiet revolution" (Bauer et al., 2015). Using a computer to solve the equations governing atmospheric motions, one can deterministically predict the occurrence of individual weather events with some skill for periods of up to 2 weeks. Accurate knowledge of the initial state of the atmosphere is needed to achieve this predictive skill. Beyond about 2 weeks, the chaotic nature of atmospheric flows begins to dominate, and deterministic predictions are not very skillful. Any predictions

Sub-seasonal to Seasonal Prediction
https://doi.org/10.1016/B978-0-12-811714-9.00009-7

beyond this timescale can describe only the statistical properties of weather, which defines climate, rather than the exact timing and location of individual weather events. Furthermore, the role of atmospheric initial conditions in providing predictive skill diminishes beyond 2 weeks, and the skill increasingly arises from the knowledge of atmospheric surface boundary conditions, such as the state of the land, ocean, and sea ice (Lorenz, 1975). Predictions on timescales of 2–12 weeks are referred to as *sub-seasonal to seasonal (S2S)* predictions and have received increasing attention in recent years (Brunet et al., 2010; White et al., 2017). On seasonal and longer timescales, the atmospheric initial state is no longer relevant, the transformation of numerical forecasting from an initial value problem to a boundary value problem is complete, and predictions are referred to as *climate predictions*.

In this chapter, we focus on the S2S predictability, specifically on the influence of the oceanic surface conditions on the atmosphere in the middle latitudes. To achieve predictable skill in the atmosphere through this influence, one needs to be able to predict the state of the ocean on timescales longer than 2 weeks. This is indeed possible because the oceanic state evolves on timescales much longer than atmospheric weather timescales due to the ocean's large thermal inertia. Mesoscale eddies in the ocean—the dynamical analogs of atmospheric weather—can persist for months (e.g., Chelton et al., 2011). This means that even a persistence forecast for the ocean state can provide significant skill on these timescales. Slowly evolving sea surface temperature (SST) anomalies can act as heat sources or sinks that alter atmospheric flow patterns.

In the middle latitudes, atmospheric and oceanic flows are approximately quasi-geostrophic—that is, characterized by small Rossby numbers (Gill, 1982). These flows are also baroclinically unstable, exhibiting intrinsic variability on spatial scales corresponding to the Rossby radius of deformation, which is of the order of 1000 km in the atmosphere and 100 km or less in the ocean. These roughly correspond to the scales of atmospheric weather systems and oceanic mesoscale eddies, respectively, thus explaining the spatial scales of observed variability in the middle latitudes. Typically, the basin-scale SST anomalies (order of 1000 km) are forced by atmospheric variability, whereas the mesoscale SST anomalies tend to be intrinsically generated by oceanic processes.

A simple model for explaining the oceanic response to atmospheric variability is the *stochastic climate model* (Frankignoul and Hasselmann, 1977). Intrinsically generated atmospheric modes of low-frequency variability (LFV) occur on timescales relevant to S2S prediction (10–100 days). In the stochastic climate model, the LFV modes randomly force an oceanic SST response that is essentially an imprint of the atmospheric surface flux anomaly. By associating the spatial structure of this stochastic variability with the spatial patterns of atmospheric low-frequency modes, such as the North Atlantic Oscillation (NAO) or the Pacific-North American (PNA) pattern, one can largely explain the spatial structure of basin-scale SST anomalies. The stochastic atmospheric variability can be approximated as having a white-noise temporal power spectrum, but with preferred spatial structures (e.g., Saravanan, 1998; Saravanan and McWilliams, 1998). Due to the greater heat capacity of the ocean, the SST response exhibits a red-noise temporal spectrum. In this stochastic red-noise model, there is no feedback from the oceanic SST anomalies to the atmosphere, and thus the ocean does not contribute to atmospheric predictability. A refinement to this model was proposed by Barsugli and Battisti (1998), where the oceanic SST response to atmospheric forcing weakens the air-sea temperature difference and leads to increased persistence of atmospheric thermal anomalies due to ocean feedback. This thermodynamic feedback may contribute to increased predictability on S2S timescales, but this effect is likely to be restricted to shorter lead times (2–4 weeks) because the LFV modes of atmospheric variability typically have *e*-folding times of less than 2 weeks (e.g., Feldstein, 2003).

Midlatitude SST anomalies can also affect the atmosphere through dynamic processes (i.e., by altering flow patterns). This effect has been studied extensively at basin scales of O(1000 km) using observational data and coarse-resolution global climate models (GCMs). The modeling studies find that midlatitude SST anomalies have only a modest influence on the midlatitude atmospheric flow (e.g., Kushnir et al., 2002). Observationally, it is difficult to distinguish between atmospheric forcing of the ocean and the oceanic influence on the atmosphere due to the coupled nature of the system. Nevertheless, lagged correlation analyses suggest that the dominant interaction is one of atmospheric forcing of oceanic SST anomalies on basin scales (e.g., Davis, 1976; Deser and Timlin, 1997). An important signature of such interaction is a negative correlation between the net heat flux out of the ocean and SST anomalies (i.e., positive heat flux out of the ocean being associated with negative SST anomalies and vice versa). Typically, this positive upward heat flux is dominated by the latent cooling associated with increased evaporation driven by increased surface wind speeds. Thus, the correlation between SST and surface wind speed also tends to be negative when the dominant interaction is atmospheric forcing of the ocean, which is consistent with the stochastic climate model mechanisms (Barsugli and Battisti, 1998; Saravanan and McWilliams, 1998).

A positive correlation between SST and surface wind speed is indicative of oceanic forcing of the atmosphere because positive SST anomalies would be associated with net upward heat flux. With the advent of satellite-based scatterometer measurements, it has become possible to analyze surface wind at fine spatial scales over the oceans in conjunction with satellite SST measurements. When filtered spatially to emphasize oceanic mesoscale variability, the scatterometer winds show positive correlations between surface wind speed and SST over oceanic regions that exhibit significant mesoscale eddy activity (Chelton and Xie, 2010; Chelton et al., 2004). These include the Gulf Stream region in the North Atlantic and the Kuroshio Extension Region (KER) in the North Pacific. This suggests that oceanic fronts and mesoscale eddies affect atmospheric flow over these regions. In addition, the Gulf Stream region and KER are located near the baroclinic source regions for the Atlantic and Pacific storm tracks (e.g., Chang et al., 2002), which means that any local influence of oceanic fronts and eddies on the atmosphere could affect the downstream development of the storms. Because the oceanic surface features can persist for a month or longer, this can act as a source of predictive skill on S2S timescales.

Small-scale oceanic surface features that potentially can influence the atmosphere can be divided into two types: (1) oceanic fronts and (2) mesoscale eddies. (The two are not completely distinct because the eddies can be associated with meanders of the front; we use the term *front* to refer to SST gradients in a coarse-grained view of the flow, where the eddies have been smoothed out.) Western boundary currents such as the Gulf Stream or the Kuroshio transport warm tropical waters toward the pole. This generates a strong frontal region in SST along the path of these currents when they separate from the coast. These oceanic fronts are anisotropic, characterized by short spatial scales *across* the front but long spatial scales *along* the front. Mesoscale eddies, on the other hand, are characterized by closed circulations and are far more isotropic spatially. The temporal scales associated with the fronts and eddies are also different. The fronts persist across seasons, although their location shifts slowly with the annual cycle of heat transport. Most ocean eddies do not persist across seasons and move much faster, with propagation speeds determined by a combination of baroclinic Rossby wave speeds and background currents. Because SST fronts move rather slowly, with the seasonal cycle, their impact on the atmosphere becomes important on seasonal-to-decadal timescales, but it is less interesting on timescales of a few weeks. The oceanic mesoscale eddies evolve significantly on timescales of several weeks, and their evolution could affect predictability on S2S timescales.

Representing the evolution of SST fronts and eddies remains a challenge for coupled ocean-atmosphere models, even at relatively high resolution. Errors in the location of simulated SST fronts can affect the large-scale flow and degrade S2S forecasts. Numerous observational and modeling studies have considered the influence of midlatitude SST fronts associated with western boundary currents on atmospheric variability (e.g., Minobe et al., 2008; Feliks et al., 2016, and the reviews by Kelly et al., 2010; Kwon et al., 2010). This chapter reviews some recent studies of the role of ocean mesoscale eddy-atmosphere (OME-A) interaction in the midlatitudes (Ma et al., 2015, 2016, 2017), with a focus on the potential implications for S2S predictability in the North Pacific region.

The outline of this chapter is as follows: Section 2 describes the data and models used. Sections 3 and 4 discuss the local impact of OME-A interaction in the atmospheric boundary layer and above the atmospheric boundary layer, respectively. Remote tropospheric impacts of OME-A interaction are discussed in Section 5, and the feedback on the ocean is discussed in Section 6. Implications for S2S predictability are considered in Section 7, followed by a summary and conclusions in Section 8.

2 DATA AND MODELS

2.1 Uncoupled Integrations

Studying the OME-A interaction requires the use of an atmospheric model with high spatial resolution (Bryan et al., 2010). Because the focus of this discussion is on the atmospheric response to mesoscale eddies both in the vicinity of the eddies themselves and farther downstream, a regional atmospheric model can be used to reduce computational costs. We use the Weather Research and Forecasting (WRF) developed by the National Center for Atmospheric Research (NCAR) as the atmospheric model (Skamarock et al., 2008). WRF is a freely available community model that has been used for a variety of weather and climate-related applications. The WRF computational domain covers the entire North Pacific Ocean from 3.6°N to 66°N, 99°E to 270°E. The horizontal grid is set at 27 km based on a Mercator projection. The model atmosphere is divided into 30 vertical levels. The model configuration includes the Lin et al. (1983) microphysics scheme, the RRTMG and Goddard scheme for longwave and shortwave radiation, a Noah land surface scheme, the Yonsei University (YSU) boundary layer scheme, Kain-Fritsch (KF; Kain, 2004) cumulus parameterization, and a horizontal, first-order closure Smagorinsky scheme for calculating eddy diffusion.

Integrations were carried out for a six-month period, from October 1, 2007, to March 31, 2008. This corresponds to a period of strong eddy activity in the KER, but with neutral conditions for El Niño–Southern Oscillation (ENSO) and the Pacific Decadal Oscillation (PDO). This minimizes the influence of either of these two modes of climate variability on our analysis. Further, 6-hourly SST interpolated from the satellite-derived Medium Wavelength-Infrared (MW-IR) daily values at 0.09° resolution was used as the lower boundary condition. The initial condition and lateral boundary conditions for the same period were derived from the 6-hourly NCEP/DOE AMIP-II (NCEP2) reanalysis data (Kanamitsu et al., 2002; Saha et al., 2010).

To address the atmospheric response to mesoscale SST variability, two ensembles of 27-km-resolution WRF integrations were carried out. The two ensembles, each with 10 members, only differed in the SST forcing field: a control (CTRL) simulation forced with the 10-km MW-IR SST (Fig. 1A), and a mesoscale eddy-filtered simulation (MEFS) forced with spatially lowpass-filtered SST to remove mesoscale eddies (Fig. 1B). A Loess filter with a 15° (longitude) × 5° (latitude) cutoff wavelength was used to remove mesoscale SST variability under approximately 800 km, following previous studies. As expected, the SST difference between CTRL and MEFS simulations exhibits features mostly confined to the eddy-active regions along the KER (Fig. 1C).

To address the sensitivity of the atmospheric response to model resolution, two additional ensembles of WRF integrations were carried out at a coarser horizontal resolution of 162 km. These ensembles, referred to as *LR-CTRL* and *LR-MEFS*, are identical to CTRL and MEFS,

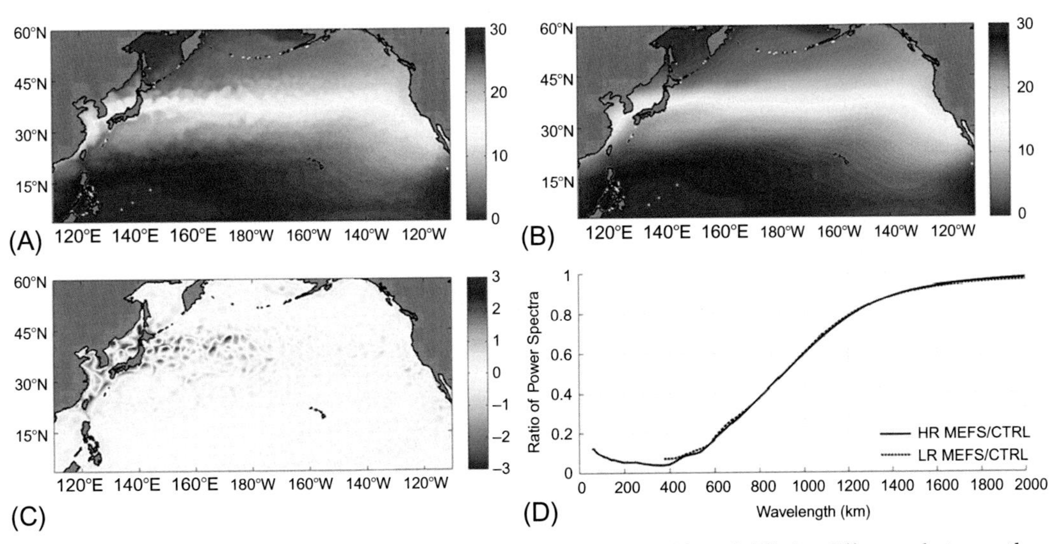

FIG. 1 Winter season (NDJFM) mean SST in (A) CTRL and (B) MEFS; and (C) the difference between them. (D) Ratio (MEFS/CTRL) of SST power spectra computed in the KER (278–428N, 1558E–1808) for high-resolution *(solid)* and low-resolution *(dash)* simulations. *Reproduced from Ma, X.H., Chang, P., Saravanan, R., Montuoro, R., Nakamura, H., Wu, D.X., Lin, X.P., Wu, L.X., 2017. Importance of resolving Kuroshio Front and Eddy influence in simulating the North Pacific storm track. J. Clim. 30, 1861–1880, © American Meteorological Society. Used with permission.*

except for the atmospheric resolution. The high-resolution SST was regridded to a coarser grid for these integrations. The half-power wavelength of the Loess filter is approximately 900 km (Fig. 1D), which suppresses mesoscale SST variance in both MEFS and LR-MEFS integrations.

The 10 ensemble members were initialized with independent realizations of the atmospheric state corresponding to October 1 which was obtained from NCEP2 reanalysis, but for 10 different years (2002–11). Note that the lateral and surface boundary conditions for all ensemble members were from October 1, 2007, to March 31, 2008. The first month of integration (October) was omitted from all the analyses to account for model spin-up, and averages were computed over the remaining five months, November–March (NDJFM). The different initial conditions allow us to assess the uncertainty associated with intrinsically generated variability. The use of identical lateral boundaries between CTRL and MEFS simulations in our regional modeling approach yields a higher signal-to-noise ratio compared to similar global model experiments because variability propagating into the domain of interest from other regions is suppressed.

2.2 Coupled Integrations

To study two-way coupled feedbacks associated with OME-A interactions, a five-member ensemble of integrations was carried out using a high-resolution coupled regional climate model (CRCM) developed at Texas A&M University. The atmospheric component of CRCM is the WRF model described previously and the oceanic component is the Regional Ocean Modeling System (ROMS) developed by Rutgers University and the University of California, Los Angeles (UCLA) (Haidvogel et al., 2008; Shchepetkin and McWilliams, 2005; http://myroms.org). The two components interact via a custom coupling software framework that allows frequent mass, momentum, and energy exchanges at the ocean-atmosphere interface every hour. Both WRF and ROMS are configured on a common Arakawa-C grid, with 9-km horizontal resolution, over the same North Pacific domain as the uncoupled WRF model. Apart from the finer horizontal resolution, WRF was configured in the same manner as in the uncoupled integrations. ROMS was configured using 50 levels for the vertical terrain-following coordinate system and included a K-profile parameterization turbulent mixing closure scheme for vertical mixing and a biharmonic horizontal Smagorinsky-like mixing for the momentum.

The coupled CRCM-CTRL simulation is comprised of an ensemble of five 6-month integrations, initialized on October 1 in 2003, 2004, 2005, 2006, and 2007, respectively. Each of the initial conditions for ROMS was derived from a 6-year spin-up run, carried out using version 2 of the Coordinated Ocean-ice Reference Experiments (CORE-II; Large and Yeager, 2009) as surface forcing and 5-day averaged Simple Ocean Data Assimilation (SODA; Carton and Giese, 2008) output as lateral boundary conditions. Initial and lateral boundary conditions for WRF were derived from NCEP2 reanalysis, as in the uncoupled integrations. The setup of the filtered CRCM-MEFS ensemble was identical to CRCM-CTRL, except that the ROMS-simulated SST was processed by an embedded low-pass Loess filter with a 15° (longitude) × 5° (latitude) cutoff wavelength before being provided to WRF at each coupling step. This prevented the atmospheric model from seeing the oceanic mesoscale eddies, even though they were present in the ocean model.

3 MESOSCALE OCEAN-ATMOSPHERE INTERACTION IN THE ATMOSPHERIC BOUNDARY LAYER

Small et al. (2008) examine many aspects of the observed features of OME-A interaction in the atmospheric boundary layer. Two different mechanisms have been proposed to explain the influence of fine-scale SST features. Atmospheric pressure gradients in the boundary layer response adjust to the SST gradients, driving surface wind convergence in regions with warmer SST (Lindzen and Nigam, 1987). This so-called pressure adjustment mechanism (PAM) can explain the occurrence of the precipitation band along the warmer flank of the Gulf Stream front, as described in Minobe et al. (2008). An alternative mechanism was proposed by Wallace et al. (1989), where warmer SST leads to a more unstable boundary with deeper vertical mixing. This causes momentum associated with stronger winds in the free troposphere to be brought down to the surface. In this vertical mixing mechanism (VMM), warm SST anomalies would be colocated with positive surface wind-speed anomalies. In contrast, for the PAM, the strongest wind speeds would occur over regions with the strongest SST gradients.

Recent observational and modeling studies have shed further light on OME-A interaction.

Frenger et al. (2013) used a Lagrangian approach to analyze the atmospheric response to oceanic mesoscale eddies in the Southern Ocean by tracking individual oceanic eddies and creating composites of satellite-observed atmospheric properties over the eddies. The composites indicated that over warm eddies, not only does the surface wind speed increase, as expected from the VMM, but there are also increased cloudiness and precipitation. Masunaga et al. (2016) further confirmed that these atmospheric boundary layer responses to SST anomalies were associated with the interannual variability of the KER using satellite and reanalysis data. The coupled modeling study of Putrasahan et al. (2013) was able to reproduce many aspects of the boundary layer OME-A interaction, suggesting that both the PAM and VMM mechanisms may play an important role.

The most robust aspect of OME-A interaction in the middle latitudes is the positive correlation between mesoscale SST and the surface wind speed that is observed in eddy-rich regions of the ocean. This correlation is clearly evident in the KER during the winter of 2007/2008, when satellite analyses of SST and surface wind-speed anomalies superimposed on each other, after high-pass filtering to retain only the mesoscales (Fig. 2A). The CTRL simulation using WRF reproduces this correlation remarkably well, with an almost one-to-one correspondence between observed and simulated features (Fig. 2B). This serves to validate the use of the WRF model for further analysis and sensitivity studies.

The CTRL simulations using WRF allow us to analyze the atmospheric response above the ocean surface, and we note a strong positive correlation between anomalies in SST and anomalies in the height of the planetary boundary layer (PBL; Fig. 2C). This demonstrates that the simulated PBL height is greater over positive SST anomalies, and vice versa. Such a relationship is consistent with the VMM for OME-A interaction, where positive SST anomalies destabilize the PBL and deepen it, thus mixing the momentum downward from the free troposphere and increasing the surface wind speed. The anomalies in convective available potential energy (CAPE) also exhibit strong positive correlations with SST anomalies, confirming the occurrence of increased destabilization (Fig. 2D).

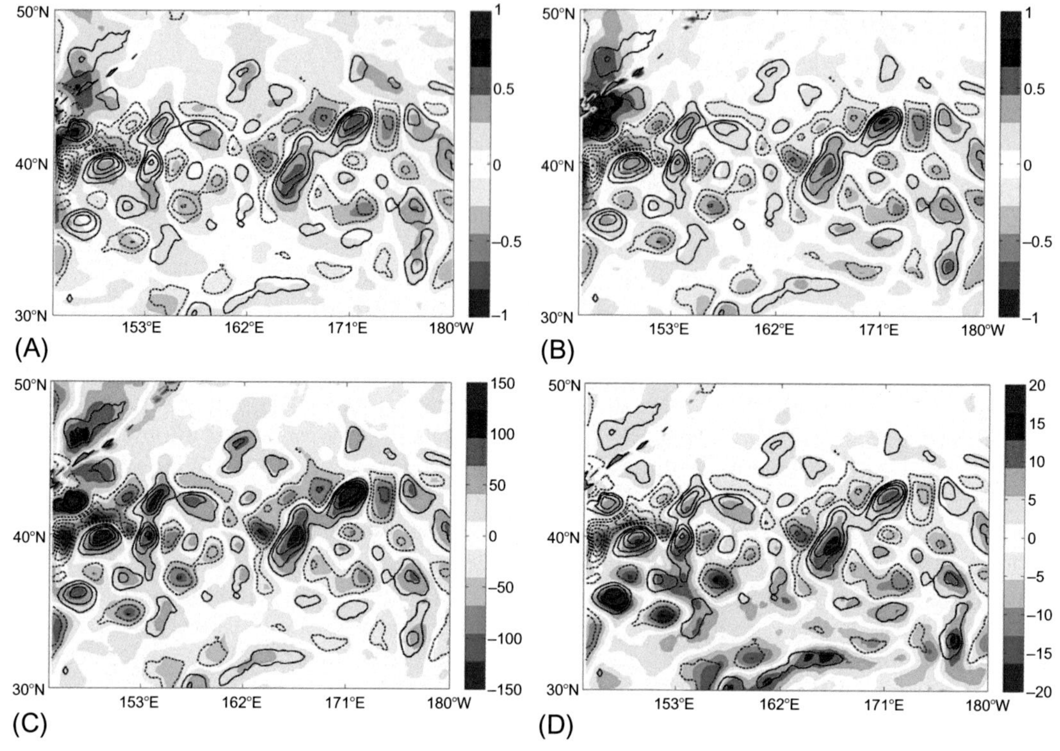

FIG. 2 Satellite-observed and model-simulated mesoscale air-sea interactions in KER. (A) 2007/08 winter season mean (NDJFM) spatially high-pass-filtered SST (contour with interval of 0.5°C) and 10 m wind speed (color in m s^{-1}) derived from MW-IR and CCMP satellite observations and (B) from the ensemble mean of 27-km, uncoupled WRF CTRL simulations; (C) high-pass-filtered SST (contour with interval of 0.5°C) and PBL (color in m); (D) high-pass-filtered SST (contour with interval of 0.5°C) and CAPE (color in J kg^{-1}) derived from the ensemble mean of 27-km, uncoupled WRF CTRL simulations. *Reproduced from Ma, X.H., Chang, P., Saravanan, R., Montuoro, R., Hsieh, J.S., Wu, D.X., Lin, X.P., Wu, L.X., Jing, Z., 2015. Distant influence of Kuroshio Eddies on North Pacific weather patterns? Sci. Rep. 5, 17785. https://doi.org/10.1038/srep17785.*

4 LOCAL TROPOSPHERIC RESPONSE

Analyses of observations and the CTRL simulations discussed earlier in this chapter clearly demonstrate the robust, near-surface atmospheric response to mesoscale SST forcing in surface, boundary layer height, and other elements. However, it becomes more difficult to identify the impact of OME-A interaction above the atmospheric boundary layer in these analyses. Atmospheric variability above the boundary layer is dominated by propagating synoptic-scale weather systems. The small-scale response correlated with oceanic eddies is expected to weaken with altitude, making discerning it against the background of weather "noise" more difficult.

Comparison of the CTRL simulation to the MEFS simulations, where the influence of oceanic eddies is artificially suppressed, provides a better estimate of the impact of OME-A above the boundary layer and away from the source regions. Additional analysis of the LR-CTRL and

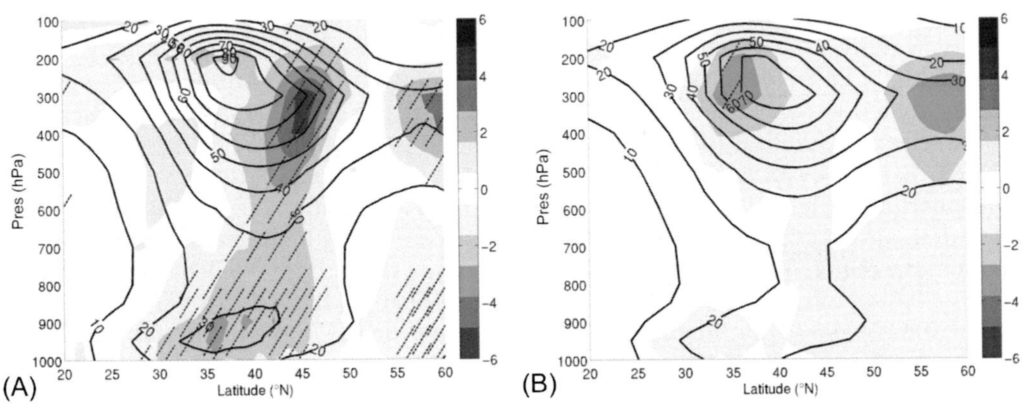

FIG. 3 (A) Vertical profile of winter-season mean (contours) storm track $\langle v'v' \rangle$ in the KER averaged from 150°E to 180°E and simulated in CTRL and difference *(color shaded)* of that between MEFS and CTRL ($m^2\ s^{-2}$). (B) As in (A), but for LR-CTRL and LR-MEFS. The difference significant at the 95% confidence interval (CI) level based on a two-sided Wilcoxon rank sum test is marked by hatching. *Reproduced from Ma, X.H., Chang, P., Saravanan, R., Montuoro, R., Nakamura, H., Wu, D.X., Lin, X.P., Wu, L.X., 2017. Importance of resolving Kuroshio Front and Eddy influence in simulating the North Pacific storm track. J. Clim. 30, 1861–1880, © American Meteorological Society. Used with permission.*

LR-MEFS simulations will allow us to assess the impact of model resolution. The KER is located near the entrance of the Pacific storm track, a region of enhanced midlatitude cyclogenesis. Therefore, we assess the possible local impact of OME-A interaction on storm activity. To isolate synoptic storm activity, a 2–8-day bandpass filtering is applied to the data.

Fig. 3A shows the vertical section of the eddy variance $\langle v'v' \rangle$ of the meridional wind v', zonally averaged over the western North Pacific (150°E–180°E), where the prime denotes bandpass-filtered data. The CTRL simulation captures the northward tilt of the eddy variance with increasing height, as in observations (Booth et al., 2010). The difference in variance between MEFS and CTRL shows that eddy variance decreases locally when mesoscale eddies are suppressed, both at the surface and aloft in the storm track region. (There is also a slight increase aloft north of the storm track, indicating some northward shift in addition to the overall weakening of the storm track.) In the low-resolution simulations, the local weakening of the eddy variance is much less pronounced and not statistically significant (Fig. 3B). This demonstrates that fine atmospheric spatial resolution is required to represent OME-A interaction properly. (The mean storm structure is also less realistic in the low-resolution simulations, with the lack of a northward tilt in the vertical.)

To explain the local strengthening of the atmospheric eddy variance due to OME-A interactions (e.g., as seen in Fig. 3A), one needs to invoke a nonlinear mechanism. Any linear, large-scale atmospheric response to mesoscale ocean eddies would tend to cancel out between the positive and negative SST anomalies associated with cyclonic and anticyclonic eddies. Indeed, the time-averaged SST gradients in the MEFS and CTRL simulations are nearly identical in the KER, indicating that the baroclinicity associated with the mean state is not significantly altered by the presence of oceanic eddies. One plausible nonlinear mechanism is the surface moisture flux response. Due to the strong nonlinearity of the Clausius-Clapeyron relation between saturation humidity and temperatures, the reduced moisture flux associated with negative SST anomalies would not cancel out the positive moisture flux associated with positive SST anomalies (Ma et al., 2015). The resulting net positive moisture flux associated with OME-A interaction would increase the diabatic heating and reduce static stability, leading to stronger baroclinic growth and increased eddy variance.

To highlight the potential influence of OME-A interactions on storms, we construct a storm index using the surface turbulent heat flux (THF). We use a THF threshold criterion in which we first select all periods during which the daily mean THF averaged over the KER (32°–42°N, 140°–170°E) exceeds its 80th percentile and then for each of these periods we choose the day corresponding to the maximum THF as a storm day. Physically, this selection identifies the days when strong cyclones occur slightly downstream of the KER, leading to extreme THF values in the region (see Ma et al., 2015). The resultant storm days in each ensemble account for about 20% of total winter season days from November to March. Fig. 4A shows the vertical section of the water vapor mixing ratio averaged over the western North Pacific, differenced between MEFS and CTRL simulations. We find a clear signature of OME-A in the moisture content above the boundary layer, with MEFS runs showing a moisture deficit compared to the CTRL simulation. The deficit is strongest in the boundary layer but extends to the midtroposphere (up to about 500 hPa). In the low-resolution simulations, this signal almost disappears (Fig. 4B), once again underscoring the need for fine resolution to simulate the impact of OME-A interactions.

Decreased moisture availability in the MEFS simulations can be expected to lead to decreased moist diabatic heating in the source region of the storm track during storm days. The Pacific storm track is characterized by a localized maximum in diabatic heating centered near the dateline (Fig. 5A). In the MEFS runs, the diabatic heating is weakened in the source region of the storm track (west of the dateline), as result of the lack of OME-A interactions (Fig. 5B). The vertical cross section of geopotential height and diabatic heating along the western portion of the storm track during storm days shows that the CTRL simulation has a 25% higher baroclinic wave amplitude and double the strength of diabatic heating compared to the MEFS integrations (Fig. 5C and D).

In summary, the comparison of CTRL and MEFS simulations shows clear evidence for a local impact of OME-A interaction over the KER regions that extends above the atmospheric boundary layer. Oceanic mesoscale eddies have a nonlinear rectified effect that leads to increased moisture flux and diabatic heating that results in stronger storm development (Fig. 6). This effect is essentially absent in the lower-resolution simulations, confirming the need to resolve atmospheric processes properly on the scale of oceanic mesoscale eddies.

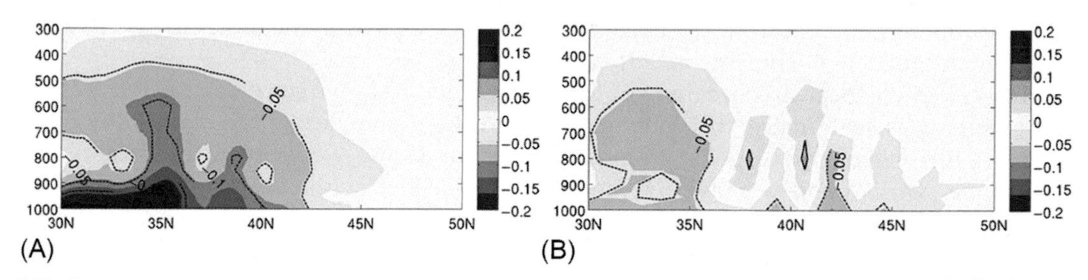

FIG. 4 Vertical section of zonally averaged (from 140°E to 180°E) water vapor mixing ratio Q (kg m^{-2} s^{-2}) averaged over storm days (A) differenced between MEFS and CTRL and (B) between LR-MEFS and LR-CTRL. *Reproduced from Ma, X.H., Chang, P., Saravanan, R., Montuoro, R., Nakamura, H., Wu, D.X., Lin, X.P., Wu, L.X., 2017. Importance of resolving Kuroshio Front and Eddy influence in simulating the North Pacific storm track. J. Clim. 30, 1861–1880, © American Meteorological Society. Used with permission.*

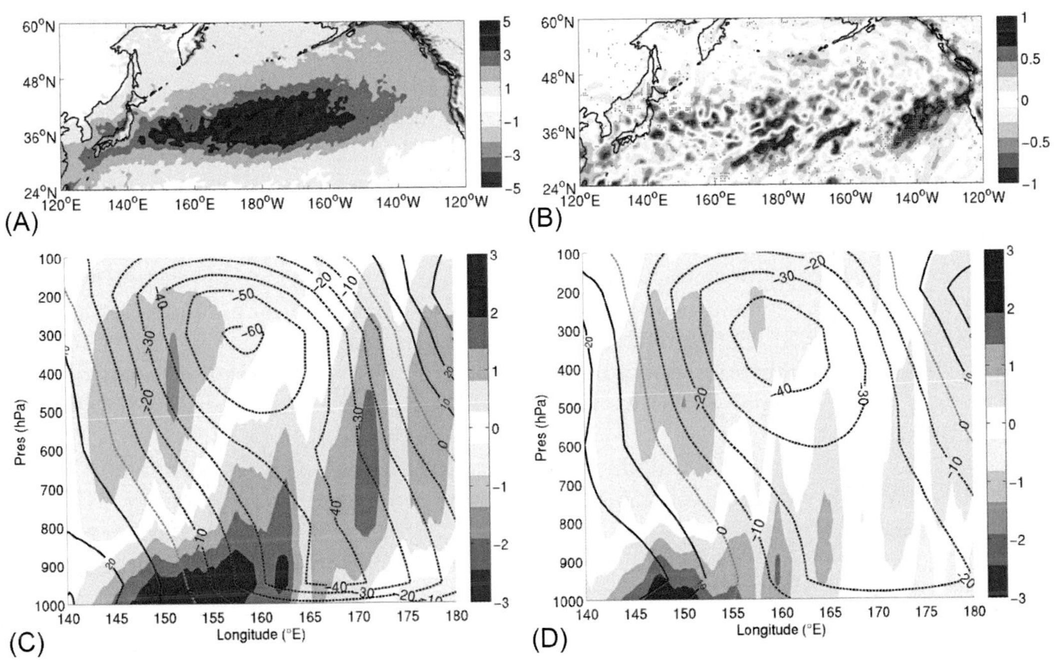

FIG. 5 Simulated diabatic heating difference and synoptic baroclinic waves over the KER during cyclone development periods in MEFS and CTRL. (A) Vertically integrated (from 1000 to 300 hPa) storm-day diabatic heating (Pa K/s) in CTRL. (B) Difference of vertically integrated (from 1000 to 300 hPa) storm-day diabatic heating (Pa K/s) (MEFS-CTRL). The difference significant at 95% CI level based on a two-sided Wilcoxon rank sum test is shaded by gray dots. (C) Vertical structure of geopotential height (contours, m) and diabatic heating (shaded, Pa K/s) along the storm path (denoted by *yellow dotted lines* in Fig. 7) for all the developing synoptic (2–8-day) storms in CTRL, composited during the simulated storm. (D) Same as (C), but for MEFS. *Reproduced from Ma, X.H., Chang, P., Saravanan, R., Montuoro, R., Hsieh, J.S., Wu, D.X., Lin, X.P., Wu, L.X., Jing, Z., 2015. Distant influence of Kuroshio Eddies on North Pacific weather patterns? Sci. Rep. 5, 17785. https://doi.org/10.1038/srep17785.*

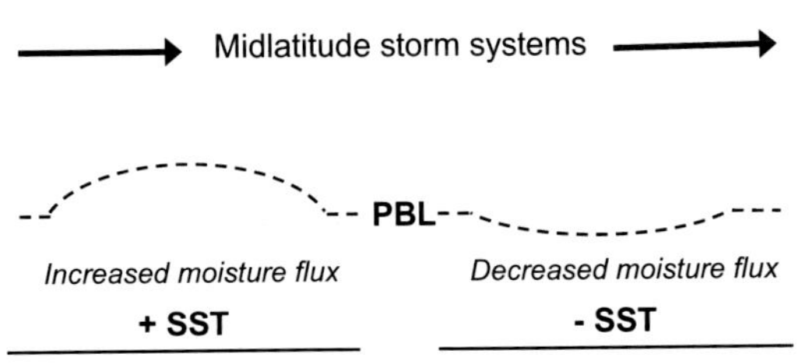

FIG. 6 Schematic illustrating the proposed mechanism of local OME-A influence on the storm track: the atmospheric PBL is deeper over the warm SST anomalies and is associated with increased vertical moisture flux. The opposite is true over cold SST anomalies, but the amplitude is weaker due to the nonlinearity of the response. Therefore, there is a net rectified effect of increased moisture flux as storm systems pass over alternating warm and cold SST anomalies, leading to stronger diabatic heating and intensified moist baroclinic cyclogenesis.

5 REMOTE TROPOSPHERIC RESPONSE

Although the local atmospheric response to ocean eddies is interesting, the local effect can contribute to S2S prediction skill only over the ocean. More important is its effect on S2S prediction skill over continental regions. In this section, we address the question of whether the local tropospheric response associated with OME-A interactions leads to remote impacts downstream of the KER. It is difficult to address this question using observational analysis on S2S timescales because the ocean eddies evolve very slowly and the atmospheric weather noise dominates. Several mechanistic modeling studies have addressed this issue. By filtering out the mesoscale ocean eddies in high-resolution global model simulations, Small et al. (2014) and Piazza et al. (2016) demonstrated a shift in the storm tracks associated with OME-A interaction in the Gulf Stream region. Zhou et al. (2015) showed that the introduction of high-frequency (i.e., daily) SST variability also can affect atmospheric variability in a global model.

In our mechanistic model simulations, evidence for a remote impact is shown in Fig. 5C, which illustrates the difference in diabatic heating during storm days between MEFS and CTRL simulations. East of the dateline, in the exit region of the storm track, there is increased diabatic heating south of the storm track and decreased diabatic heating to the north. This suggests that the lack of OME-A interaction has caused a southward shift of the storm track. To confirm this, a lag-composite analysis of 2–8-day bandpass filtered meridional wind v at 850 hPa was carried out for CTRL and MEFS simulations, from lag −2 days to lag 2 days, with lag 0 corresponding to peak values of the THF index when storms pass over the KER (Fig. 7). All the composited v fields were normalized by the maximum northerly v amplitude near the storm center at lag −3 days in CTRL and MEFS, respectively, for clearer comparison. The composite reveals a clear eastward-propagating baroclinic wave pattern, and it is evident that the wave intensifies more rapidly over the KER in CTRL than in MEFS. Note also that at lag 2 days, the storms appear to propagate farther northward in the CTRL composites, as compared to the MEFS composites (Fig. 7E and F), which is consistent with the diabatic heating signal in Fig. 5B.

Changes in the tracks of storms not only will affect the downstream weather, but also can affect the mean flow conditions through nonlinear wave-mean flow interaction associated with changes in eddy momentum and heat fluxes. Associated with the Pacific storm track, there is a maximum in the zonal winds (known as the *Pacific jet* and shown in Fig. 8A) in the upper troposphere that is maintained in part by the convergence of eddy momentum flux. In the MEFS simulation, the jet stream shifts southward east of the dateline as the path of the storms shifts southward (Fig. 8B). This dipolelike, upper-tropospheric response is associated with a surface low-pressure anomaly, suggesting an equivalent barotropic response (Fig. 8C). Thus, we see enhanced storm activity over California and decreased storm activity north of California (Fig. 8D).

6 IMPACT ON OCEAN CIRCULATION

Thus far, our analysis has focused on the impact of oceanic mesoscale eddies on the atmosphere. However, the atmospheric boundary layer response to oceanic eddies also will alter the surface fluxes of heat and momentum, which in turn can affect oceanic eddies themselves.

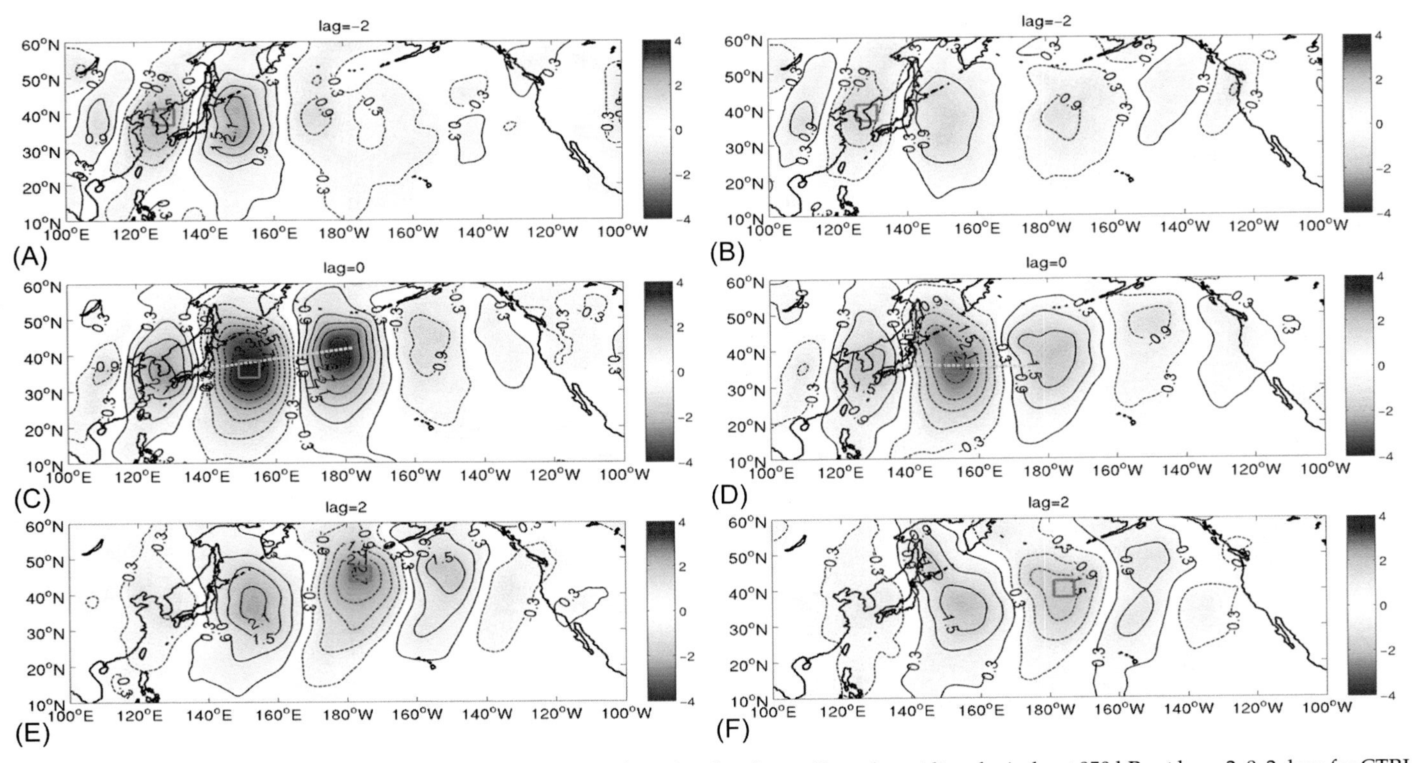

FIG. 7 Simulated synoptic storm development. Lag-composite of 2–8-day, bandpass-filtered meridional wind v at 850 hPa at lag −2, 0, 2 days for CTRL (A, C, E) and MEFS (B, D, F), respectively. In each case, the composites are normalized by the maximum v at lag −3 days at which the storm center is about to enter the ocean from the Eurasian continent. The *dotted lines* in lag 0 composites indicate the storm path along which the vertical structure of the composited storm is shown in Fig. 5C and D. *Reproduced from Ma, X.H., Chang, P., Saravanan, R., Montuoro, R., Hsieh, J.S., Wu, D.X., Lin, X.P., Wu, L.X., Jing, Z., 2015. Distant influence of Kuroshio Eddies on North Pacific weather patterns? Sci. Rep. 5, 17785. https://doi.org/10.1038/srep17785.*

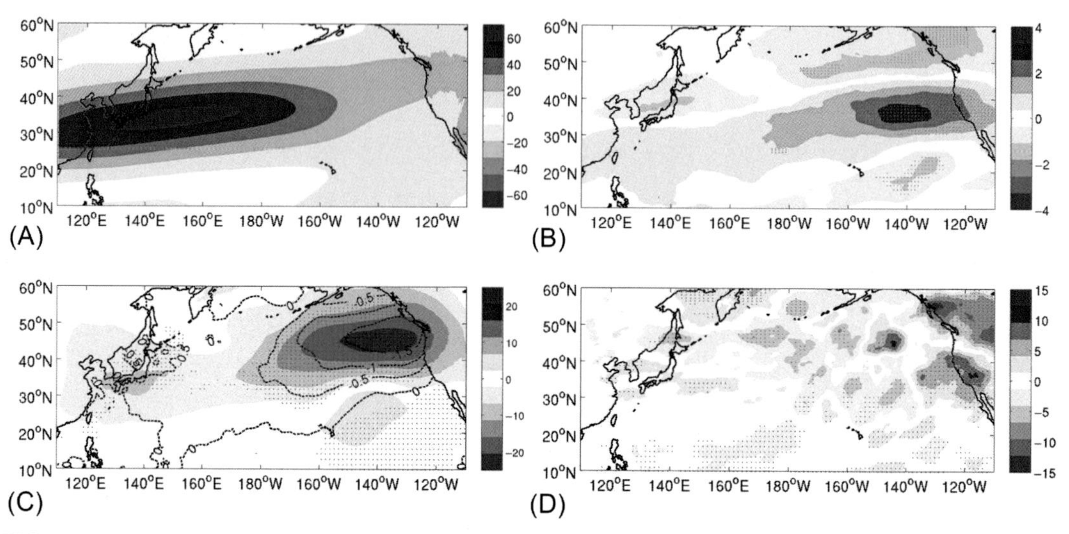

FIG. 8 Upper atmospheric response to mesoscale SST forcing. (A) Winter season (NDJFM) mean zonal wind U at 300 hPa, U300 (m s^{-1}), in CTRL. (B) Difference of U300 (m s^{-1}), (C) sea-level-pressure (mb) (contour) and geopotential height at 500 hPa (Z500) (m) (color), and (D) transient eddy kinetic energy at 300 hPa (m^2 s^{-2}) between MEFS and CTRL (MEFS-CTRL). The transient kinetic energy was derived using 2–8-day, bandpass-filtered variables. In (B)–(D), the difference significant at the 95% CI level based on a two-sided Wilcoxon rank sum test is shaded by *gray dots. Reproduced from Ma, X.H., Chang, P., Saravanan, R., Montuoro, R., Hsieh, J.S., Wu, D.X., Lin, X.P., Wu, L.X., Jing, Z., 2015. Distant influence of Kuroshio Eddies on North Pacific weather patterns? Sci. Rep. 5, 17785. https://doi.org/10.1038/srep17785.*

For example, one may expect the increased surface wind-speed response over warm oceanic eddies to cool the eddies more strongly than in the absence of OME-A interactions. To study such feedback, ensembles of coupled ocean-atmosphere integrations were carried out *with* (CRCM-CTRL) and *without* (CRCM-MEFS) the OME-A feedback. These are analogous to the uncoupled simulations, except that the horizontal resolution is finer (9 km) and the ensemble size is smaller (five members).

To assess the impact of OME-A interaction on the strength of oceanic eddies, we compute the ratio of the wavenumber spectra of eddy kinetic energy (EKE) and eddy enstrophy (ENS) between the CRCM-MEFS and CRCM-CTRL simulations in the KER (Fig. 9). The spectra of EKE and ENS show increased values in the MEFS simulations for wavelengths less than 500 km, with the largest increase of about 30% occurring for the smaller eddies of 100-km scale or less. The increase in eddy strength in the MEFS simulation is explained using an energy budget analysis (Ma et al., 2016). The mean available potential energy (MAPE) associated with the thermal gradients across the Kuroshio front is converted to eddy potential energy (EPE) when the eddies are generated. The destruction of the eddies may occur through one of two mechanisms: (1) conversion of EPE to EKE to energize the eddy circulation or (2) dissipation of the EPE through oceanic mixing or heat loss to the atmosphere. It is the latter process that is amplified by the OME-A feedback, leading to increased eddy dissipation in the CRCM-CTRL simulation. Decreased eddy dissipation in the MEFS simulations leads to increased conversion of EPE to EKE, as well as a weaker and more meandering Kuroshio jet.

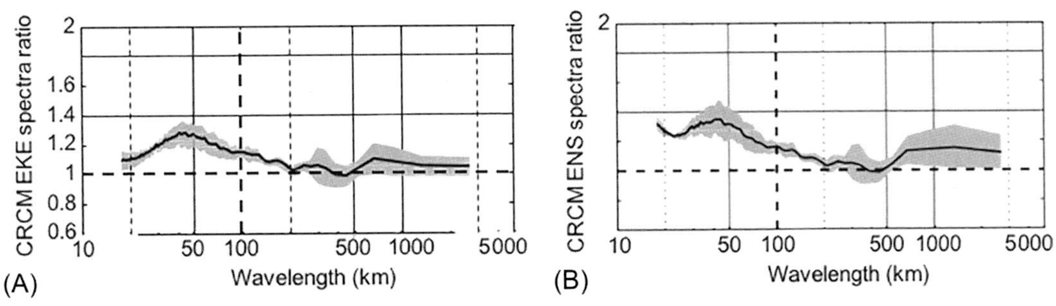

FIG. 9 (A) Simulated ratio (CRCM-MEFS/CRCM-CTRL) of winter-season (ONDJFM) mean EKE spectra in the KER (145°E–160°E, 30°N–42°N) based on the CRCM simulations. (B) As in (A), but for the simulated ratio of eddy enstrophy (ENS; half the vorticity variance). The eddy velocity is defined as the departure from the zonal mean between 145°E and 160°E. The shading indicates the geometric standard deviation of the EKE and the enstrophy ratio in individual years of CRCM. The vertical dashed line represents eddies with a 100-km wavelength. The horizontal dashed line indicates the MEFS/CTRL ratio of unity. *Reproduced from Ma, X., Jing, Z., Chang, P., Liu, X., Montuoro, R., Small, R.J., Bryan, F.O., Greatbatch, R.J., Brandt, P., Wu, D., Lin, X., Wu, L.., 2016. Western boundary currents regulated by interaction between ocean eddies and the atmosphere. Nature 535, 533–537. https://doi.org/10.1038/nature18640.*

Fig. 10 shows the meridional cross section of the strength of the Kuroshio jet in the CRCM-CTRL and CRCM-MEFS simulations. The Kuroshio jet is weaker in the MEFS simulation, which is associated with increased oceanic eddy activity.

7 IMPLICATIONS FOR S2S PREDICTION

S2S prediction attempts to fill the gap between deterministic weather forecasts and probabilistic seasonal outlooks. Atmospheric initial conditions play an important role in the former but are ignored in the latter. Atmospheric boundary conditions, such as SSTs or sea ice, are generally assumed to be constant in the former, but need to be explicitly predicted for the latter. For S2S prediction, both atmospheric initial and boundary conditions can play a significant role, which makes it more challenging in some respects. This also could explain why S2S prediction has begun receiving attention only recently, unlike weather forecasting or seasonal-to-interannual prediction.

SST serves as an important boundary condition for the atmosphere. The genesis regions for the two dominant storm tracks in the Northern Hemisphere, in the Pacific and the Atlantic, are colocated with the SST fronts and eddies associated with oceanic currents in the KER and the Gulf Stream region, respectively. As discussed previously, variations in SST associated with mesoscale eddies can influence the atmosphere both locally and remotely. Therefore, any impact of OME-A on atmospheric circulation can contribute to improving the skill of S2S forecasts in the storm-track regions.

One advantage of S2S prediction over seasonal prediction is that one can use fairly simple models to predict the evolution of SST over the first few weeks of an S2S forecast. Ocean mesoscale eddies can persist for several months and propagate very slowly, at a rate of about a

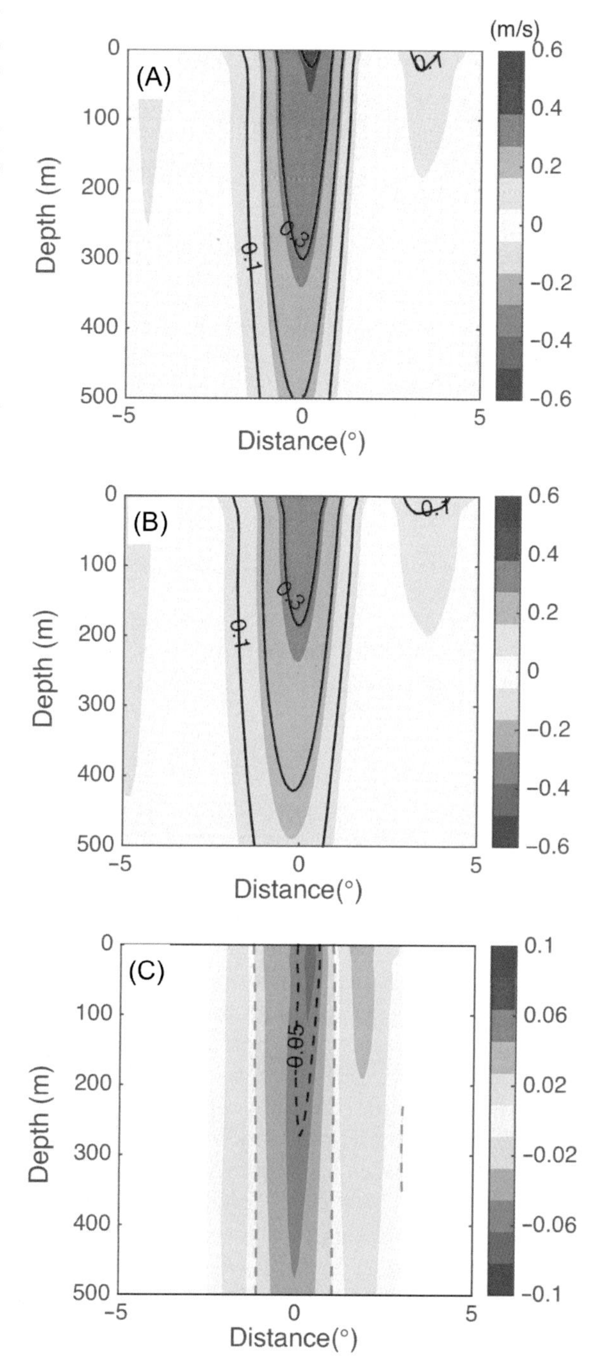

FIG. 10 Vertical section of zonally averaged (145° E–160°E) oceanic zonal current U response across the Kuroshio extension front simulated by eddy-resolving coupled regional climate model. (A) CRCM-CTRL, (B) CRCM-MEFS, and (C) CRCM-MEFS—CRCM-CTRL, based on the winter-season (ONDJFM) mean of the ensemble of CRCM simulations. The x-axis is the distance from the axis of the Kuroshio jet at 0, which is defined by the latitude of the winter-season mean maximum U, and the y-axis is depth (m). *Reproduced from Ma, X., Jing, Z., Chang, P., Liu, X., Montuoro, R., Small, R.J., Bryan, F.O., Greatbatch, R.J., Brandt, P., Wu, D., Lin, X., Wu, L.., 2016. Western boundary currents regulated by interaction between ocean eddies and the atmosphere. Nature 535, 533–537. https://doi.org/10.1038/nature18640.*

few kilometers per day (Chelton et al., 2011). Therefore, a persistence model for SST may be adequate for the first 2 weeks, and simple statistical models may work well for a few more weeks.

The advantage of using statistical models is that they will not suffer from the systematic drift and bias problems associated with a full dynamical ocean model. Of course, for timescales longer than a month or so, a full dynamical ocean model may be necessary to predict the evolution of SST in the coupled system. For such a coupled model, it may be important to take into account the impact of OME-A on surface fluxes, as discussed in Section 6. The dynamical predictability of oceanic mesoscale eddies in the vicinity of western boundary currents on S2S timescales remains to be explored. Predictability studies focusing on other eddy-active regions, such as the Loop Current eddy in the Gulf of Mexico region, suggest that oceanic eddies may be dynamically predictable on 4–6-week timescales. Predictability of oceanic eddies near western boundary currents drops significantly beyond the S2S timescales, with intrinsic oceanic variability causing a degradation of skill over interannual and longer timescales (Nonaka et al., 2016).

8 SUMMARY AND CONCLUSIONS

Ample evidence exists from high-resolution satellite observations that ocean mesoscale eddy–induced SSTs can influence the atmospheric PBL. This PBL response to ocean mesoscale eddies can be faithfully simulated by atmospheric models, provided that model resolutions are sufficiently high to resolve eddy forcing. The work of Bryan et al. (2010) suggests that a minimum horizontal grid resolution of 25 km may be necessary in coupled ocean-atmosphere models in order to capture OME-A interactions. Evidence is also mounting that the eddy-forced atmospheric response can extend beyond the PBL into the lower troposphere to the midtroposphere, producing coherent rainfall and cloudiness anomalies directly over ocean eddies. High-resolution climate model simulations further suggest that resolving ocean eddy forcing can lead to a local strengthening of the atmospheric eddy variance in the lower troposphere to the midtroposphere. This strengthened atmospheric eddy variance cannot be explained simply by changes in atmospheric baroclinicity associated with the mean state, but rather is more likely attributable to changes in moist diabatic processes in the atmosphere caused by the rectified effect of ocean eddies on surface moisture fluxes. More important, recent modeling studies present evidence that ocean eddies along the western boundary current regimes can force remote atmospheric responses downstream of eddy-active regions. Although the detailed mechanisms behind the local and remote tropospheric response to ocean eddies still need to be worked out, the evidence at hand does show that OME-A interaction is the strongest during winter season, particularly over periods when extratropical cyclones develop in eddy-active, western boundary current regimes. As such, the potential that OME-A interaction can have an impact on winter storm-track predictability on S2S timescales is worth exploring.

The OME-A interaction mechanism is fundamentally different from the proposed mechanisms for the influence of SST fronts on the atmosphere. The latter act on scales of the atmospheric Rossby radius of deformation, and they can be modeled as a linear effect.

The influence of small-scale oceanic eddies above the atmospheric boundary layer can be explained only by invoking nonlinear mechanisms, such as the asymmetric amplitudes of response to positive versus negative SST anomalies (Fig. 6). This asymmetry can result in a rectified vertical moisture flux that can act to amplify the genesis and growth of storms. The MEFS simulation is analogous to a coarse-resolution atmospheric model that fails to capture the impact of oceanic mesoscale eddies. Our results suggest that poor representation of OME-A interactions in models can result in systematic errors in large-scale atmospheric circulation far from the eddy-rich oceanic regions.

OME-A interaction is also shown to have an impact on the maintenance of western boundary currents, such as the Kuroshio and Gulf Stream, through its influence on ocean eddy energetics. Inaccurate representation of OME-A interaction can contribute to the severe climate model biases in simulating western boundary current regimes. Therefore, improving the OME-A interaction representation in climate models can have important implications for S2S prediction, both in terms of providing a potential source of predictability and in mitigating climate model biases in western boundary regimes. The current generation of climate prediction models do not take OME-A interaction into full consideration. We submit that the role of OME-A interaction in S2S prediction may be significant and should be explored by future studies.

Acknowledgments

This review summarizes research carried out by numerous collaborators and coauthors, most notably the doctoral and postdoctoral research of Xiaohui Ma, along with Raffaele Montuoro, Jen-Shan Hsieh, Dexing Wu, Xiaopei Lin, Lixin Wu, Zhao Jing, R. J. Small, F. O. Bryan, R. J. Greatbatch, P. Brandt, H. Nakamura, and Jesse Steinweg-Woods. Support from NSF Grant AGS-1462127 and NOAA Grant NA160AR4310082 is also acknowledged.

CHAPTER

10

The Role of Sea Ice in Sub-seasonal Predictability

Matthieu Chevallier[*,†], *François Massonnet*[‡,§], *Helge Goessling*[¶], *Virginie Guémas*[*,§], *Thomas Jung*[¶]

[*]CNRM, Université de Toulouse, Météo France, CNRS, Toulouse, France [†]Division of Marine Forecasting and Oceanography, Météo France, Toulouse, France [‡]Georges Lemaître Centre for Earth and Climate Research, Earth and Life Institute, Université catholique de Louvain, Louvain-la-Neuve, Belgium [§]Barcelona Supercomputing Centre, Barcelona, Spain [¶]Alfred Wegener Institute for Polar and Marine Research, Bremerhaven, Germany

OUTLINE

Sub-seasonal to Seasonal Prediction
https://doi.org/10.1016/B978-0-12-811714-9.00010-3

1 INTRODUCTION

The polar regions, the Arctic and the Antarctic, are the scene of a number of unique weather and climate phenomena and conditions, such as polar lows, stable boundary layers, tip jets, katabatic winds, sea and land ice, snow, and mixed-phase clouds. Polar regions are also characterized by their remoteness, which makes them naturally undersampled by conventional observation systems, although they are currently well covered by polar-orbiting satellites. The Arctic is home to >4 million people, including an increasing majority of nonindigenous settlers with growing economic activity. The Antarctic has no permanent human habitation, but it does host a number of permanent research stations.

Sea ice is arguably one of the most iconic features of the polar system. Due to its insulating properties, it regulates exchanges between the atmosphere and ocean. Its high albedo has a direct impact on the Earth's energy balance. Sea-ice growth and melting affect the upper-ocean stratification, possibly feeding back on the thermohaline circulation. It is also a central player in the process of polar amplification, which affects the meridional structure of the atmosphere. Therefore, the recent trends observed in many sea ice parameters, which are often interpreted as early-warning signals of climate change, also may have far-reaching consequences for the whole climate system.

Along with its key role in the Earth's climate system, sea ice is at the center of several socioeconomic, geopolitical, and operational concerns. Rapid sea ice changes unlock invaluable opportunities for the shipping, tourism, and offshore industries (Smith and Stephenson, 2013; Lloyd's, 2012; COMNAP, 2015). At the same time, sea ice inevitably poses a serious threat to all types of marine operations, regardless of their purpose. In this context, the need for sea ice (and, more generally, polar environmental) prediction on timescales ranging from hours (tactical) to months (operational) and years (strategic) has become pressing (e.g., Jung et al., 2016).

The scientific community has eagerly responded to the need for sea ice predictions, focusing mainly on seasonal to decadal timescales (e.g., Guémas et al., 2016). After initiating research for operational purposes, especially for predicting marine accessibility in the Beaufort Sea along the Alaskian coast (Barnett, 1980), this area has become well established for research. Since 2008, interests in seasonal sea ice predictability and predictions also have benefited from the momentum generated by the Sea Ice Outlook (SIO; Stroeve et al., 2014; https://www.arcus.org/sipn/seaice-outlook), a forum of Arctic sea ice prediction providers with the purpose of liaising with a growing stakeholder community. However, research on seasonal sea ice predictability in the Southern Ocean is far less advanced.

Sea ice lies between the atmosphere, which has a memory of about 1–4 weeks, and the ocean, which is known to have persistence at timescales longer than a few seasons (Frankignoul and Hasselmann, 1977), especially in the tropical oceans (e.g., Latif et al., 1998). The oceanic origin of sea ice provided the hope that it could hold a considerable memory, like many components of the Earth's cryosphere (e.g., snow and land ice). However, its relative thinness (only a few meters thick, compared to a few hundreds of meters or kilometers for land ice) and the fact that it is strongly driven by the atmosphere, especially on synoptic timescales, drastically limit its predictability beyond months compared to the ocean. By acting as an effective insulator, the presence of sea ice (and snow on top) transforms the surface such that from an atmospheric perspective, it behaves less like an open ocean and more like land, allowing atmospheric near-surface temperatures to vary more strongly and rapidly. The location of the sea ice edge, therefore, is not only of particular interest to potential forecast users, but it also has a pronounced impact on the overlying atmosphere. The marginal ice zone and adjacent regions, hence, are places where the predictability of the second kind of the atmosphere—that is, the influence of slow variations in ocean-surface characteristics on weather statistics (discussed further in Chapter 1)—can be particularly pronounced. Combined with growing evidence that a substantial fraction of nonseasonal variations in the ice-edge location are potentially predictable at sub-seasonal to seasonal (S2S) timescales, there is the potential for trustworthy predictions of the atmosphere on S2S timescales as well.

The critical role of sea ice in forcing atmospheric variability, as will be discussed later in this chapter, is one of the main reasons why prediction systems used for shorter-term forecasts increasingly account for it, either to improve the forcing of weather forecasting models or to provide dedicated forecasts (Pellerin et al., 2002). The focus has been primarily on sea ice concentration—the fraction of an area covered with sea ice, which is the quantity that (1) primarily determines how air-sea fluxes are modulated by sea ice, and (2) provides the basic information on the presence of sea ice in an area. Stakeholders, however, require information on other variables, including sea ice thickness.

The goal of this chapter is to present sea ice in relation to S2S prediction. The field of S2S prediction attempts to bridge the gap between numerical weather prediction (NWP) and climate prediction. Interestingly, thus far, the role of sea ice on sub-seasonal climate predictability is not well understood. From the existing literature, which mostly deals with seasonal-to-interannual timescales, we will review the sources of sea ice predictability at timescales from 2 weeks to 1 year. Based on this analysis, we will characterize the predictability of the second kind as related to sea ice and provide an overview of our understanding of the possible role of sea ice as a source of S2S atmospheric predictability, in the polar regions and beyond.

2 SEA ICE IN THE COUPLED ATMOSPHERE-OCEAN SYSTEM

2.1 Sea Ice Physics

The process of sea ice formation and subsequent growth is a complex topic that is beyond the scope of this chapter (for a comprehensive review, see Petrich and Eicken, 2017). Sea ice initially forms through the freezing of the surface ocean. While sea ice forms, it captures only a fraction of sea salt (from 2 to about 10 g/kg), which is dissolved in brine pockets. The remaining salt is released to the ocean, thereby modifying the density profile of the water column. As sea ice grows, its salinity keeps decreasing though drainage. After the initial ice formation, thermodynamic ice growth is sustained by thermal imbalance as a consequence of air-sea temperature differences. The thermodynamical growth rate is rapidly damped as sea ice thickens; thicker ice grows more slowly than thin ice (e.g., Bitz and Roe, 2004). Sea ice can reach up to 2–3 m only through thermodynamical growth (e.g., Maykut and Untersteiner, 1971). During sea ice growth, snow plays a significant role. It is a powerful insulator, limiting the loss of heat in the ocean to the atmosphere during winter (e.g., Semtner Jr, 1976). Furthermore, the thermal conductivity of snow is lower than that of sea ice by one order of magnitude. The presence of a thick layer of snow is one of the reasons why Antarctic sea ice is on average thinner than its Arctic counterpart. Note that when the snow load is sufficient to depress the ice-snow interface below the sea surface, snow ice may form. Again, this process is mostly observed to occur in the Southern Hemisphere.

Dynamical processes play an important role in shaping the space-time variability of sea ice thickness. Sea ice floes—discrete elements of the sea ice cover—drift in response to atmospheric winds and ocean currents. Sea ice motion is not a free drift, and part of the kinetic energy input is dissipated in the sea ice interior. Sea ice deformation occurs under ridging (e.g., collision and accumulation of ice along a ridge) and rafting (e.g., sliding of ice floes above one another) in areas of convergence. Under mechanical deformation, sea ice can form a pile of up to 10–20 m. In areas of divergence, leads open within the sea ice pack.

As temperatures rise in the spring, sea ice undergoes top, bottom, and lateral melting. Similar to its central role during the growth season, snow plays a key role in modulating surface melt. If present, snow delays the ice melt onset due to its relatively high albedo. And when snow melts, ponds of liquid water—known as *melt ponds*—form at the surface of the ice, significantly lower the surface albedo, and deepen to reach the bottom of the ice eventually.

2.2 Sea Ice Observations

Satellite-based retrieval of sea ice concentration has been done on a regular basis since the early 1970s using passive microwave sensors. Brightness temperatures from Scanning Multichannel Microwave Radiometer (SMMR), Special Sensor Microwave Imagers (SSMI), and Special Sensor Microwave Imager/Sounder (SSMI/S) are used to estimate sea ice concentration using a variety of retrieval algorithms. For instance, the Bootstrap algorithm (Comiso, 1995) is used to determine sea ice concentration at a resolution of 25 km on a daily basis since 1987 (every other day since 1979), as well as to calculate various sea ice indices, such as the total sea ice extent provided by the National Snow and Ice Data Centre (NSIDC; Fetterer et al., 2002).

Direct or indirect measurements of sea ice thickness are more scarce. In the Arctic, sources of in situ data include measurements from in situ drillings and submarine and moored, upward-looking sonar and airborne electromagnetic freeboard measurements (e.g., Lindsay, 2010). The only comprehensive and long-term, in situ data set on ice thickness in the Southern Ocean is provided by the Antarctic Sea-ice Processes and Climate (ASPeCt) group (Worby et al., 2008), a compilation of visual, ship-based observations of ice thickness. More recently, spatial altimetry provides enhanced coverage of both poles (ICESAT, CryoSat-2).

Other parameters are monitored from space or buoy data, such as sea ice drift, deformation, snow, surface temperature, albedo, or melt pond fraction. Since the remainder of this chapter focuses on sea ice concentration and thickness, the interested reader is referred to more detailed studies (e.g., Leppäranta, 2011; Kwok, 2011; Heygster et al., 2012; Meier and Markus, 2015) for more information on how these other sea ice parameters are retrieved.

2.3 Sea Ice in Models and Reanalyses

Due to the relatively short observational records on sea ice concentration, and the scarcity of sea ice thickness observations, our knowledge of sea ice variability and predictability relies mostly on modeling studies. Most sea ice models now include all the dynamics and thermodynamics processes described thus far, as well as reasonable formulations for coupling with the atmosphere and the ocean (e.g., Notz and Bitz, 2017). State-of-the-art sea ice models include representations of subgrid-scale physics (Hunke et al., 2010). Historically, sea ice models have been developed in the context of climate modeling, and their formulations are based on assumptions that are believed not necessarily valid at fine space scales and timescales (i.e., a few kilometers and a few hours), which are usually of interest in operational short-term prediction. For example, sea ice is assumed to act as a continuous viscous-plastic medium (i.e., a viscous behavior under low-stress forcing and a plastic behavior under intense-stress forcing) in most sea ice models, an assumption that is not supported by careful examination of observations from drifting buoys (Rampal et al., 2008). Most of these models also lack a proper representation of wave-ice interactions.

In general, ocean-sea ice models forced by atmospheric reanalyses provide reasonable estimates of the position of the sea ice edge, especially in winter. This is largely due to the strong constraints imposed by prescribed atmospheric forcing; atmospheric reanalyses are produced with atmospheric models using prescribed sea ice concentrations as lower boundary conditions (e.g., Lindsay et al., 2014). However, even under the same atmospheric conditions, forced ocean-sea ice models differ significantly in their simulated ice-thickness distribution in the Arctic (e.g., Danabasoglu et al., 2014; Wang et al., 2016) and the Antarctic (Downes et al., 2015), due to a combination of factors.

Sea ice mean state and variability in fully coupled atmosphere-ocean-sea ice climate models differ considerably from one model to another, and there are substantial biases (e.g., Flato et al., 2013; Day et al., 2016). Model shortcomings in atmospheric and oceanic physics, as well as inaccurate formulation of atmosphere-sea ice-ocean couplings, contribute to biases in sea ice models (e.g., Notz and Bitz, 2017).

A good representation of sea ice in ocean reanalyses is key, especially for S2S prediction purposes. First, for poorly observed properties like sea ice thickness, reanalyses provide unique sources of information regarding long-term trends and variability. Second, reanalyses are widely used for predictions, such as boundary conditions for atmosphere-only or regional simulations and predictions, or as initial states of sub-seasonal and seasonal hindcasts (Guémas et al., 2016). Up to now, none of the current operational reanalyses assimilates sea ice thickness data directly, and most ocean-sea ice reanalyses have shown significant biases in the sea ice thickness fields, at least in the Arctic Ocean (Chevallier et al., 2017).

Among all the various reanalyses, the Panarctic Ice-Ocean Model Assimilation System (PIOMAS) is an ocean-sea ice reanalysis dedicated to the Arctic Ocean; it uses an ocean-sea ice model driven by atmospheric reanalyses, with assimilation of sea surface temperature and sea ice concentration. PIOMAS estimates of sea ice thickness and volume have been evaluated through comparisons with observations from U.S. Navy submarines, oceanographic moorings, and satellites (Schweiger et al., 2011). A few global ocean-sea ice reanalyses show similar performance in their Arctic sea ice simulations as PIOMAS, such as ORAP5 (Zuo et al., 2015), from the European Centre for Medium-range Weather Forecasts (ECMWF). Recent intercomparisons suggest that ORAP5 provides reasonable solutions for Arctic (Chevallier et al., 2017) and Antarctic sea ice. The PIOMAS and ORAP5 reanalyses will be used in the following discussion to assess sea ice volume variability.

3 SEA ICE DISTRIBUTION, SEASONALITY, AND VARIABILITY

In both hemispheres, sea ice cover has its maximal extension at the end of the winter and its minimal extension at the end of the summer. The amplitude of the seasonal cycle however, is different in each hemisphere.

The mean seasonal cycle of Arctic sea ice extent (i.e., the area covered with sea ice having a concentration higher than 15%) has an amplitude of about 9.2 million km^2, with a climatological maximum of 15.5 million km^2 and a minimum of 6.3 million km^2 over the reference period 1979–2015. The spatial distribution of sea ice concentration is primarily constrained by the presence of land. In winter, sea ice expands in the northern North Atlantic and Pacific oceans roughly to the mean position of the ocean thermal front, as reflected in Fig. 1. Ocean heat advection plays an important role in shaping the sea ice edge in the winter (Bitz et al., 2005), as evidenced in the east-west asymmetry in both the Atlantic and Pacific (e.g., approximately 45°N offshore North America versus northward 80°N in the Barents Sea). Sea ice motion is also responsible for the sea ice presence along the eastern coast of Greenland. Within the interior Arctic basin, the ice motion is constrained by coastlines. As a result, ice piles up to thicknesses well above 10 m (Thorndike, 1992). A large fraction of this ice can survive the melt season in the Arctic Ocean, this latter fraction being the basis of the ice cover that would grow over the next year. The spatial distribution of sea ice thickness in the Arctic Ocean, with the thickest ice located against the northern coast of Greenland and the Canadian Arctic archipelago, partially reflects large-scale atmospheric and oceanic circulation in the Arctic (e.g., Bourke and Garrett, 1987).

The seasonal cycle of sea ice extent in the Southern Ocean is larger than in the Arctic Ocean, which is due to both a larger maximum sea ice extent and a smaller minimum sea ice extent than in the Arctic. The position of the winter sea ice edge is mostly thermodynamically driven (Bitz et al., 2005). In the winter, the sea ice extent is limited by the westerly winds and the position of the Antarctic Circumpolar Current, which has surface temperatures above the freezing point and acts as a permanent heat source. In the summer, sea ice survives only in the Weddell and Ross seas. As a consequence, the Antarctic winter sea ice cover is mostly ice that forms during the same freezing season, with an average thickness below 2 m. Although mechanical deformation seems less possible there than in the Arctic Ocean, instances of sea ice thickness larger than 10 m can be found (Williams et al., 2015).

Interannual variability of sea ice extent and sea ice volume is different in the two hemispheres as well. In the Arctic Ocean, the variability of sea ice extent is much greater in the summer than in the winter. In the Southern Ocean, sea ice extent variability is greater during the transition seasons (spring and fall) and lower at the February minimum and September maximum. The summer Arctic sea ice extent variability is 2–3 times larger than the summer Antarctic sea ice extent variability. According to a reanalysis, the interannual variability of sea ice volume is larger in the Arctic than in the Antarctic at all months.

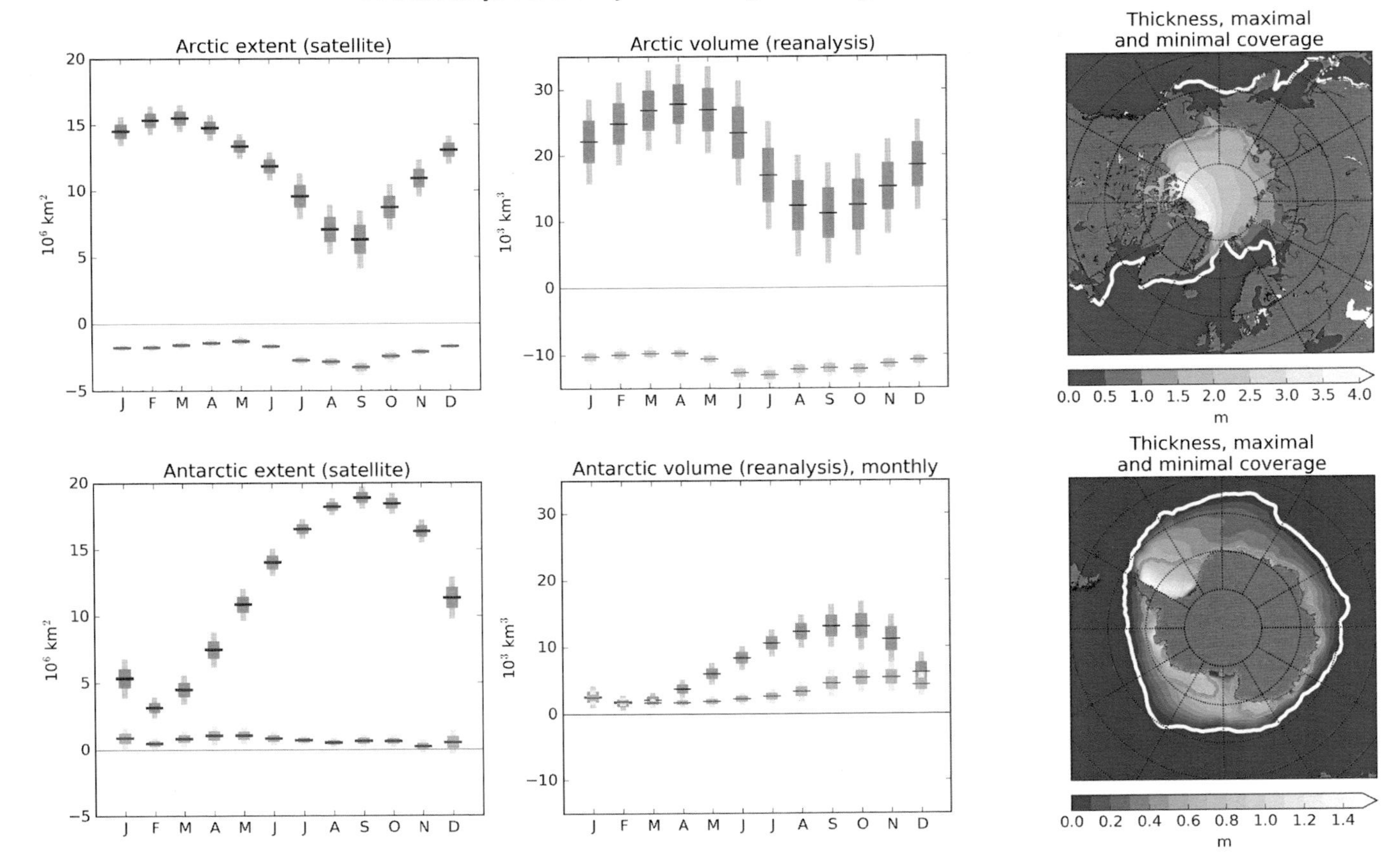

FIG. 1 Arctic (top) and Antarctic (bottom) sea ice mean state and variability. The left column displays annual cycles of 1979–2015 monthly sea ice extent (gray; data: NSIDC sea ice index) and changes in sea ice extent estimated from a quadratic fit over 1979–2015 (the *shadings* denote one and two standard deviations around the estimated quantities, respectively). The central column displays the same statistics for sea ice volume (from the PIOMAS reanalysis for the Arctic, 1979–2015, and the ORAP5 reanalysis for the Antarctic, 1979–2012). The right column displays the annual mean sea ice thickness from PIOMAS (1979–2015) and ORAP5 (1979–2012), together with the positions of the ice edge at the annual minimum and maximum (September and March in the Arctic, February and September for the Antarctic; data: NASA Bootstrap).

Over the recent decades, sea ice has undergone significant changes. Fig. 1 shows estimates of long-term trends over the last four decades (1979–2012 or 1979–2015) for the Arctic and Antarctic sea ice extent and volume in all months. The picture is remarkably different in each of the two hemispheres. There has been a strong and significant decrease (negative trend) in Arctic sea ice extent and volume in every month of the year. The negative trends reached their maximum in September for sea ice extent (Stroeve et al., 2012), and June–July for sea ice volume. The rate of thinning, however, seems more consistent year round. Kwok and Rothrock (2009) documented a 40% reduction of sea ice thickness in an interior Arctic basin based on submarine measurements and recent altimetry data.

Trends for total Antarctic sea ice extent are positive in all months, though they are small and not always significant; in addition, regional trends may differ substantially from each other (Holland, 2014). The origin of the positive trend in Antarctic sea ice extent is a topic of particular interest. Recent studies highlight regional discrepancies, as well as a possible role played by natural variability and feedback (e.g., Polvani and Smith, 2013; Goosse and Zunz, 2014). However, there is some evidence, based on early proxy, satellite, or whaling records, of an abrupt decline in sea ice extent between the 1930s and the beginning of the satellite era in the late 1970s (Curran et al., 2003; Edinburgh and Day, 2016; Gagné et al., 2015).

4 SOURCES OF SEA ICE PREDICTABILITY AT THE SUB-SEASONAL TO SEASONAL TIMESCALE

4.1 Persistence

The term *persistence* was originally introduced in meteorology to describe a series of several days with similar weather characteristics. The extension of that concept to climate scales and other components than the atmosphere has been natural since then. In this regard, the case of sea ice is of particular interest. Dynamically and thermodynamically forced by the relatively fast atmosphere (i.e., short persistence) and the relatively slow ocean (i.e., long persistence), a complex situation can be expected, especially because sea ice evolution also is governed by internal processes with their own characteristic timescales. This section reviews the typical timescales at which sea ice exhibits significant memory associated with persistence, as well as some physical processes that provide the source of this memory.

Persistence is loosely defined as the time necessary for a time series to decorrelate from itself. While seemingly simple in its formulation, this definition is more complicated in its implementation. First, *decorrelation* must be properly defined, and various approaches exist for doing so (e.g., Flato, 1995). Second, as shown in the previous section, contemporary sea ice signals are highly seasonal and bear the imprint of background climate change. The imprint of these two external drivers must be accounted for (and possibly removed) if the goal is to study the inherent persistence of the system itself, not the persistence offered by these external drivers. This implies the adequate estimation and removal of these forced contributions, which are far from trivial (Mudelsee, 2014). Third, there is evidence of nonstationarity in sea ice properties (e.g., Holland and Stroeve, 2011; Goosse et al., 2009), which makes the choice of the baseline period important to estimate autocorrelations. Similarly, memory properties are likely to be season dependent (Chevallier and Salas-Mélia, 2012; Day et al., 2014). Finally, most sea ice parameters are difficult to monitor accurately, further challenging the idea that robust estimates of persistence can be retrieved accurately from observationally based estimates only.

For all these reasons, it is not surprising that estimates of sea ice anomaly persistence vary from study to study. In the following discussion, we refer to persistence as *persistence of anomalies* with respect to the long-term linear trend. Arctic sea ice areal properties are found to exhibit persistence from 1 to 5 months, depending on the product, the methods, and the season in question (Walsh and Johnson, 1979; Lemke et al., 1980; Blanchard-Wrigglesworth et al., 2011a; Guémas et al., 2016). This range of 1–5 months tends to be generally overestimated by climate models (Day et al., 2014; Blanchard-Wrigglesworth et al., 2011a), possibly due to the lack of representation of important physical processes in these models. Estimates of sea ice volume persistence are more uncertain, owing to the lack of reliable observational thickness estimates. Early modeling studies by Flato (1995) and Bitz et al. (1996) estimated the total Arctic sea ice volume as persisting for up to 6–7 years, although more recent estimates point to values closer to 2–4 years (Bushuk et al., 2017; Blanchard-Wrigglesworth et al., 2011b; Day et al., 2014), possibly as a consequence of using more advanced and fully coupled models in those latter studies. Antarctic sea ice persistence has largely been disregarded in the literature.

Fig. 2 offers an updated estimation (following a consistent definition) of the persistence of various dynamic and thermodynamic sea ice parameters in the Arctic and the Antarctic. Three remarkable points must be noted. First, persistence ranges from synoptic (about 1 day) to annual and even interannual timescales. This gives full justification for considering sea ice as a key source of S2S predictability in the polar regions and even beyond (see Section 6). Second, Antarctic sea ice generally exhibits less persistence than Arctic sea ice. This is likely due to the difference in geographical configurations (see Section 3 and Fig. 1, earlier in this chapter). Antarctic sea ice variability has a strong regional component (e.g., Parkinson and Cavalieri, 2012) and hence is characterized by strong decouplings (e.g., Lemke et al., 1980) that reduce the persistence of the hemispheric quantities. Besides, Antarctic sea ice is on average much thinner than the Arctic and almost entirely seasonal (Fig. 1). In an Arctic study based on coupled climate models, Blanchard-Wrigglesworth and Bitz (2014) suggested that thinner ice is generally associated with shorter-lived anomalies. Finally, and as expected, persistence is shorter at the local than the global scale. Still, the persistence of sea ice concentration and thickness, on the order of weeks and seasons, respectively (see Fig. 2, but also Lukovich and Barber, 2007, and Blanchard-Wrigglesworth and Bitz, 2014), indicate the potential for climate information based solely on the intrinsic memory of the ice, provided that the initial state is estimated accurately.

Persistence of sea ice area depends on the season. Using monthly mean observational and model data of Arctic sea ice area, Blanchard-Wrigglesworth et al. (2011a) noticed that correlations were lower between successive months when the initial sea ice is most rapidly advancing or retreating. Fig. 3 presents these results differently for both polar oceans based on daily data. The longest persistence (between 5 and 60 days) is found when the correlation varies the least, which happens in the summer of both hemispheres, before the annual minimum. It coincides with a slowing of seasonal ice loss during the melt season. Shorter persistence can be found during the spring in both hemispheres, shortly after the annual maximum, and during the fall in the Arctic. During these seasons, anomalies of sea ice extent are anticorrelated with those 1–2 months later.

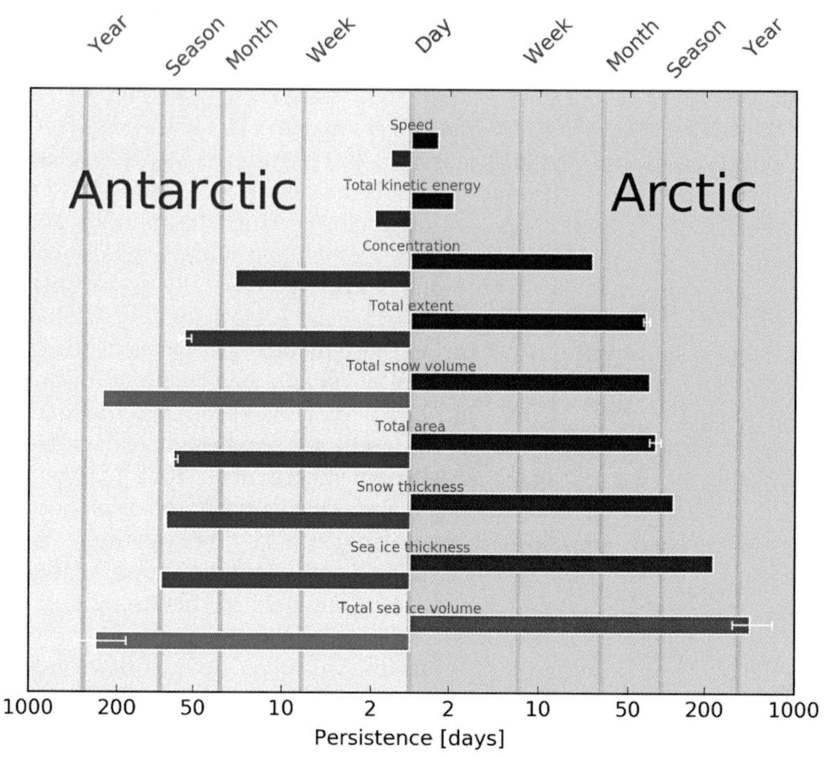

FIG. 2 Persistence of common sea ice parameters. From top to bottom: Sea ice speed at a point in the Beaufort Sea (P1: 81 N, 145°W) and at a point in the Ross Sea (P2: 74°S, 165°W); hemispheric average of kinetic energy per unit mass; concentration at points P1 and P2; hemispheric sea ice extent; hemispheric snow-on–sea ice volume; hemispheric sea ice area; snow-on–sea ice thickness at points P1 and P2; sea ice thickness at points P1 and P2; hemispheric sea ice volume. Persistence is primarily estimated from a 1979–2015 ocean-sea integration. When possible, other data sources are used (e.g., sea ice reanalyses and satellite-based retrievals), and whiskers are displayed to reflect the range between these sources. Persistence is estimated as the time necessary for the autocorrelation function of the daily mean, quadratically detrended time series to reach $1/e$.

4.2 Other Mechanisms

Several physical mechanisms can offer other sources of predictability besides persistence. These mechanisms can be split into two types: those related to the sea ice itself and those related to other agents (e.g., the atmosphere or the ocean). We provide a few examples in the following discussion.

The sea ice cover is far from uniform. At horizontal scales of about 10 m, sea ice thickness can vary substantially, by as much as a few meters (Thorndike et al., 1975). The way that sea ice thickness is distributed in a given area (e.g., a grid box typically employed in climate models) has a significant impact on the amount of energy, mass, and momentum exchanged with the atmosphere and the underlying ocean. This is because many fluxes depend nonlinearly on the ice thickness. Therefore, it has been hypothesized that the wintertime ice thickness distribution (ITD) could carry predictability over to the area during the next

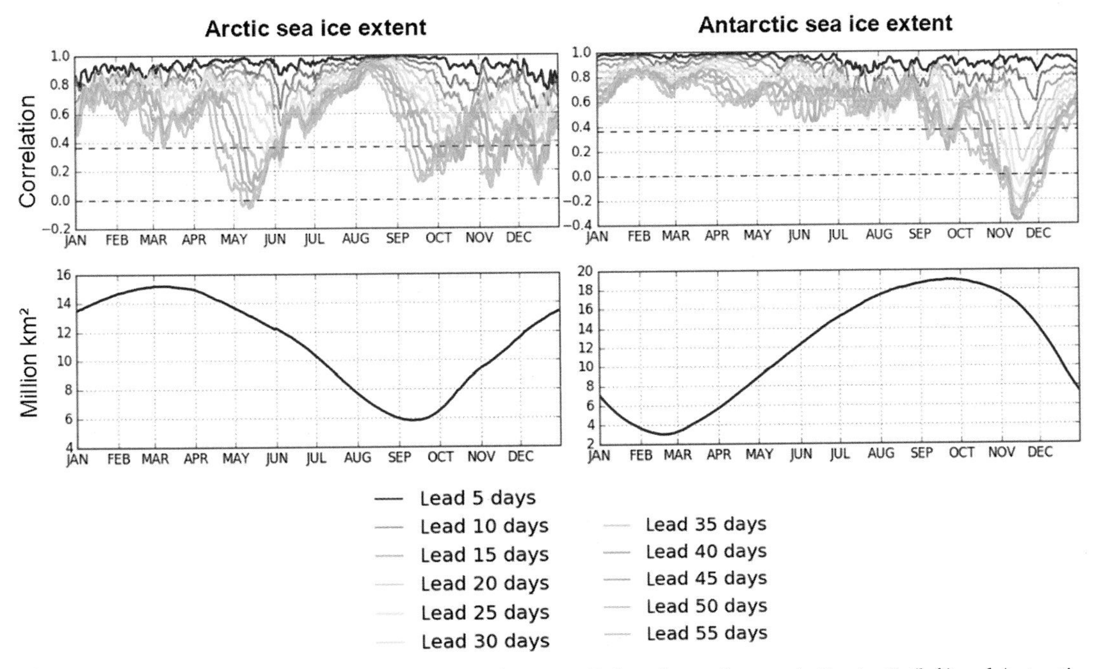

FIG. 3 Top row: lagged correlation for lead times from 5 to 60 days for sea ice area in the Arctic (left) and Antarctic (right) oceans. Bottom row: annual cycle of sea ice area. The dotted lines in the top row indicate 95% significance and null correlation. *Source: NSIDC Bootstrap sea ice concentration (daily data).*

summer because thicker ice has less propensity to melt away than thin ice. Blanchard-Wrigglesworth et al. (2011a) first identified a mechanism of summer-to-summer reemergence of Arctic sea ice area as follows: An anomaly of sea ice area in summer causes ice thickness to become anomalous in winter, which eventually causes a summer sea ice area anomaly; that is, sea ice partly "remembers" its summer area from year to year, even though the summer-to-winter link is nonsignificant (hence the term *reemergence*). Chevallier and Salas-Mélia (2012) explored this mechanism further and determined that the anomalous summer ice area is significantly determined by the area of ice thicker than 0.9–1.5 m up to 6 months earlier. The tight coupling between wintertime sea ice thickness and summertime sea ice area has been confirmed in other studies in various setups (Guémas et al., 2016; Day et al., 2014; Massonnet et al., 2015) and appears to be a robust feature. Owing to the lack of long-term and reliable sea ice thickness observational data, this proposed mechanism has not yet been assessed with observational data.

The low albedo of melt ponds in spring is thought to be a source of predictability for summer sea ice extent, although modeling and observational estimates differ in the range at which this process operates. Using melt pond fractions reconstructed with a stand-alone sea ice model, Schroeder et al. (2014) showed that the amount of melt ponds in May explained up to 64% of the variance of the September Arctic sea ice extent over the period 1979–2013. However, Liu et al. (2015) found no evidence of such predictive skill in May using satellite retrievals of melt ponds over the period 2000–2010. Nevertheless, greater predictability arises when the amount of melt ponds is integrated from May to late July, suggesting that the relationships may be robust for sub-seasonal prediction. The possible role of melt pond fraction as a precondition of summer melting can be explained by a positive feedback mechanism and a larger melt pond fraction, leading to more absorbed solar radiation and thus more melting.

Because sea ice is interacting with the atmosphere and the ocean, these two media may have an imprint on sea ice predictability. A number of studies suggest that the ocean is the main source of sea ice predictability on interannual and longer timescales (e.g., Guémas et al., 2016, for a comprehensive review). It also plays a role at S2S timescales. Woodgate et al. (2010), for example, emphasized the role played by anomalous warm water inflow through the Bering Strait in the summer of 2007 and showed that it accounted for one-third of sea ice melt in that year, leading to the then-record-breaking September sea ice minimum.

Sea ice drift results from the integration by sea ice of forcing from the atmosphere and the ocean. As shown in Fig. 2, the memory of sea ice speed is very low, which is consistent with the predominance of wind forcing, which has a very limited memory. However, sea ice advection has long been considered as a potential source of predictability at various timescales (e.g., Koenigk and Mikolajewicz, 2009). More recently, Holland et al. (2013) showed an eastward-propagating signal of potential predictability arising in the fall and continuing into the winter following the initial January, providing predictability beyond initial persistence. In the Arctic Ocean, advection during May–June acts as a key driver of September sea ice distribution through a redistribution of winter sea ice thickness anomalies (Kauker et al., 2009). It is thus likely that sea ice advection may play a role in S2S timescales at a regional scale, especially in the marginal ice zone.

Past studies have documented a modulation of the sea ice extent by the El Niño–Southern Oscillation (ENSO) in the Arctic (Liu et al., 2004), and possibly in both hemispheres (Gloersen, 1995). Guémas et al. (2016) showed the major role played by the atmosphere in the Arctic Ocean. The Arctic surface circulation is primarily wind-driven (Gudkovich, 1961). The North Atlantic Oscillation (NAO) is thought to drive a seesaw in sea ice conditions between the Labrador Sea and the Greenland and Barents seas, with sea ice area in the former positively correlated with the NAO index (e.g., Deser et al., 2000) and a maximum correlation at a lag of about 2 weeks in the sea ice response to NAO. In the Southern Ocean, the Antarctic Dipole, characterized by out-of-phase sea ice concentration anomalies between the South Atlantic and the South Pacific, persists 3–4 months after being triggered by ENSO (e.g., Yuan, 2004). Fluctuations in the polarity of the Southern Annular Mode (SAM) are also thought to affect sea ice area. Positive Southern Annular Mode induces an overall sea ice expansion (though with regional differences) of sea ice at the short timescale through enhanced northward surface Ekman drift (Hall and Visbeck, 2002; Lefebvre et al., 2004), but sustained positive Southern Annular Mode conditions eventually may lead to a decrease of sea ice area due to the upwelling of warm water from below the mixed layer (Ferreira et al., 2015; Holland et al., 2016).

Henderson et al. (2014) studied the response of winter and summer Arctic sea ice concentration to specific phases of the Madden-Julian Oscillation (MJO), which is the leading mode of atmospheric intraseasonal variability. Building on previous studies showing the modulation of high-latitude climate by the MJO (e.g., Cassou, 2008; Lin and Brunet, 2009), the authors show coherent regions of ice concentration variability in the Atlantic (phases 4 and 7), and in the Pacific sectors (phases 2 and 6) during January, and for the North Atlantic (phases 2 and 6) and Siberian sectors (phases 1 and 5) during July. These active regions are coherent with corresponding anomalies in surface wind. The authors argued that the MJO still could project onto the Arctic ice margins in the future, while specific phase relationships may change.

5 SEA ICE SUB-SEASONAL TO SEASONAL PREDICTABILITY AND PREDICTION SKILL IN MODELS

5.1 Potential Sea Ice Predictability

The term *predictability* is sometimes used loosely as a synonym for *predictive skill*. In the following section, we use it specifically to refer to the inherent or potential predictability of sea ice; that is, the theoretical limit for the predictive skill of a sea ice forecast system that would be achieved with a model that perfectly resembles reality and with close-to-perfect knowledge of the initial state of the atmosphere-sea ice-ocean system. For classical weather forecasts, this definition is somewhat vague, as the obtained predictability limits depend strongly on how exactly the initial state is known. The definition is better constrained when it comes to assessing the predictability of the ocean and sea ice (and the associated predictability of the second kind) on S2S and longer timescales. The increasingly uncorrelated atmospheric forcing is the main factor driving ice/ocean states apart. However, whether atmospheric states diverge within 5 or 10 days does not much affect the subsequent speed of ice/ocean state divergence on longer S2S timescales (e.g., Juricke et al., 2014); therefore, the obtained estimates of potential predictability should be more robust.

Obviously there will never be a model perfectly resembling reality, so estimates of potential predictability can be obtained only in the so-called perfect-model world, using a model itself as a surrogate reality. Boer (2000) termed this type of predictability "prognostic potential predictability" (in contrast to diagnostic approaches that estimate predictability based on the temporal characteristics of a time series, as described in Section 4). In this framework, one can conduct pseudoforecast experiments, where one just slightly perturbs the initial state and investigates how quickly different system trajectories diverge. Following numerous studies of this kind with individual models (Koenigk and Mikolajewicz, 2009; Blanchard-Wrigglesworth et al., 2011b; Holland et al., 2011; Tietsche et al., 2013; Day et al., 2014), recently a number of global climate modeling groups contributed to the Arctic Predictability and Prediction on Seasonal to Inter-annual Timescales (APPOSITE) project, following a common perfect-model-type experimental protocol. More specifically, initial conditions for the forecast ensembles were taken from a long control simulation under constant present-day forcing (greenhouse gas concentrations, etc.), with minute perturbations added to the SSTs (Tietsche et al., 2014; Day et al., 2016).

Studies following the perfect-model approach under constant conditions have the advantage that there are no uncertainties in observations or reanalyses because the model's surrogate reality is perfectly known. Even more so, there are no secular trends that are hard to separate from long-term variability in the real world. While limitations accompanying this strongly idealized approach, such as neglecting model biases and unrealistic forcing, need to be kept in mind, these circumstances considerably facilitate the clean quantification of the potential predictability of a given system.

To quantify the predictability of sea ice also requires the specification of what characteristic of the sea ice precisely is to be assessed (see also Fig. 2). For Arctic sea ice, the most commonly addressed quantities are simple scalar quantities such as the pan-Arctic sea ice extent and volume (also compare Fig. 1). For such scalars, one meaningful way to quantify potential predictability is to compute a root-mean-squared error (RMSE) based on all possible pairs of ensemble members, and to derive a normalized root-mean-squared error (NRMSE) by dividing this error by a climatological error that is obtained when the procedure is repeated with pairs of sea ice states from different years (but from the same time of year). The NRMSE quantifies the degree to which ensemble members are more similar to each other than states randomly chosen from a long time series. To give an example, in the APPOSITE simulations, initialized July 1, the NRMSE in September ranges from about 0.3–0.6 for pan-Arctic sea ice extent, and from about 0.1–0.3 for volume. Here, a value of zero implies perfect predictability, whereas a value of 1 implies a complete loss of predictability (see Day et al., 2016, Fig. 5). Thus, the APPOSITE results imply that roughly half the interannual variations in September sea ice extent are potentially predictable from July 1, and even about 80% of the variations in sea ice volume, consistent with the longer memory associated with volume compared to extent as mentioned previously.

Potential predictability could be estimated locally (e.g., to sea ice concentration or thickness). For sea ice concentration, however, this is not very meaningful, as the climatological RMSE will be close to zero in most places, which is trivial because significant variability occurs only around the marginal ice zone. A reasonable alternative is to sum up all areas where a forecast disagrees with the observations on whether sea ice is present (e.g., using the typical 15% sea ice concentration threshold); such an approach, called the Integrated Ice Edge Error (IIEE), has been proposed by Goessling et al. (2016).

Fig. 4 illustrates the predictability of the Arctic sea ice cover at S2S timescales. The ice edges in the four forecast ensembles (Fig. 4A–D) exhibit a clear coherency even after 2.5 months when atmospheric states have diverged completely due to internal atmospheric variability. For example, at 150°E, all ice edges are far north of 80°N in the ensemble depicted in Fig. 4B, but all edges are south of 80°N in Fig. 4D, meaning that this information resided already in the difference in initial states on July 1. This qualitative assessment is confirmed by the quantitative estimates provided in Fig. 4E (where the gray dashed vertical line corresponds to the situation shown in Fig. 4A–D); measured by the IIEE, about 50% of the ice-edge variations in September can potentially be predicted from July 1.

While such perfect-model estimates give rise to optimism that sea ice can be predicted at S2S timescales with some useful skill, we cannot rule out that coupled climate models might systematically overestimate potential predictability compared to the real world. Indeed, there is some indication that sea ice may be less predictable in the real world (Day et al., 2014). The abovementioned limitations, however, make it difficult to diagnose the potential predictability in the real world. In summary, previous work on potential predictability provides room for optimism when it comes to predicting sea ice on S2S timescales.

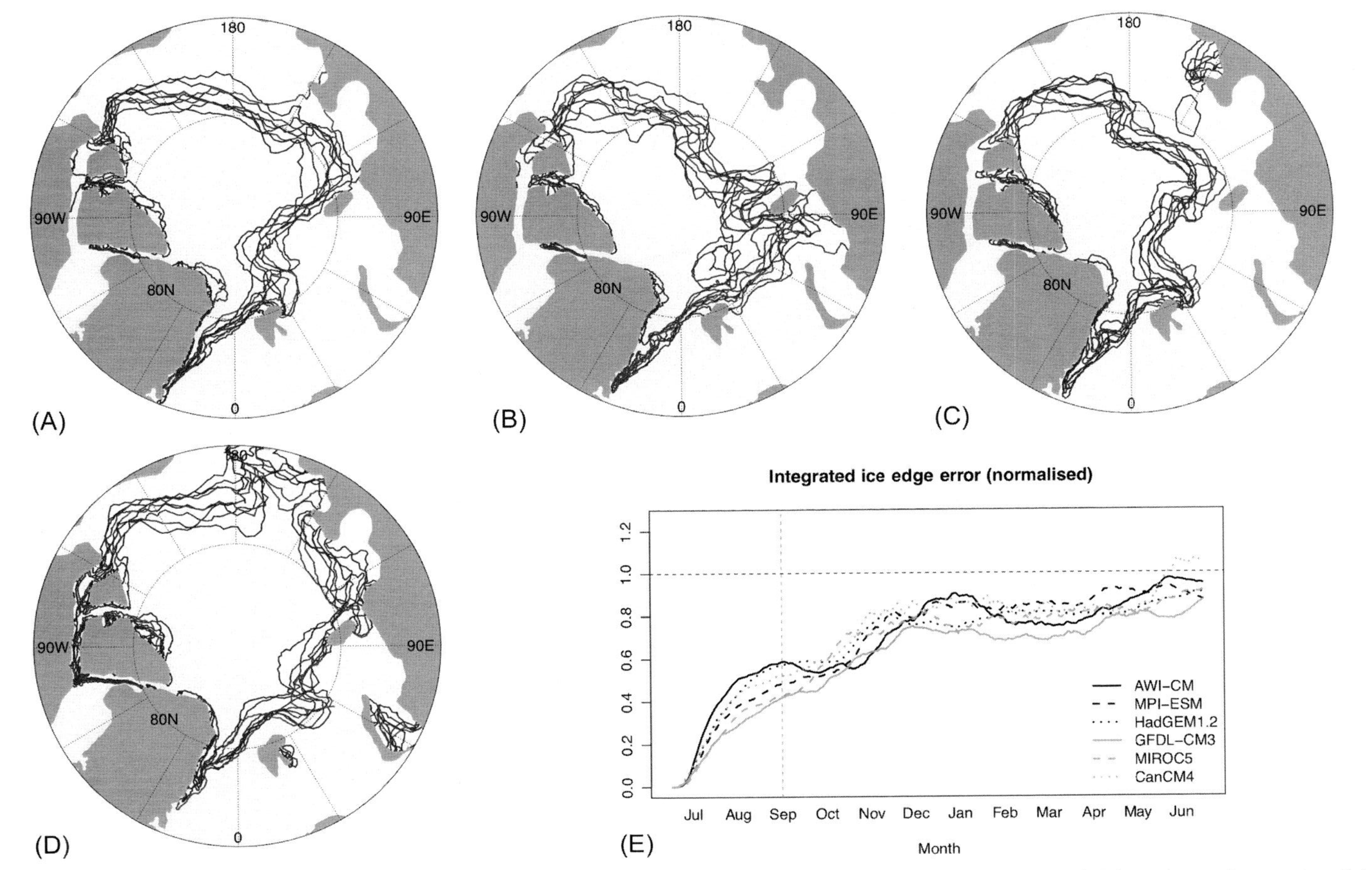

FIG. 4 Potential predictability of the Arctic sea ice edge in coupled climate models. (A)–(D) Ice edge locations (15% concentration contours) in 9-member idealized (perfect-model) forecast ensembles on September 14 in four different years from one of the models (AWI-CM). Ensemble members are shown in different colors. The ensembles were started 2.5 months earlier on July 1, with initial conditions taken from a long control simulation with minute perturbations added to the SSTs. (E) The IIEE, the total area of mismatch, averaged over all possible pairs of ensemble members and over a number of cases for six climate models following the same protocol. Errors are normalized by a climatological error, such that IIEE = 0 indicates perfect predictability and IIEE = 1 indicates the complete loss of predictability (for details, see Goessling et al., 2016).

5.2 Skill of Sea Ice Prediction Systems at Sub-seasonal Timescales

5.2.1 Short-Term Predictions

Sea ice models were originally developed as part of climate models. While climate models now all include a dynamic-thermodynamic sea ice model, most forecasting systems used for short- to medium-range predictions still use a persisted sea ice cover (Jung et al., 2016). Recently, ECMWF upgraded its operational forecasting system by incorporating a fully dynamic-thermodynamic sea ice model. Such a choice was motivated by the belief that accounting for sea ice in prediction matters, both for prediction skill itself and for users of forecast products.

Accounting for interactive sea ice cover, rather than persisting sea ice throughout the forecast, has been shown to improve predictions on short timescales (e.g., Jung et al., 2016, their Fig. 7). Sea ice predictions at lead times from a few days to 2 weeks are typically done using ocean-sea ice models forced with operational weather forecasts. This two-tier approach has some inconsistencies because operational atmospheric forecasts often used a persisting sea ice cover. Since Van Woert et al. (2004), the added value of model-based forecasts of sea ice concentration in the Arctic Ocean has been assessed relative to the persistence forecasts[1] in areas where sea ice concentration changed by >5% over the forecast. In the Polar Ice Prediction System, Van Woert et al. (2004) showed that the skill of 24-h forecasts of sea ice concentration is higher than that of persistence in all months except during the freeze-up period, when a combination of persistence and climatology offers a better estimate.

More recently, in the Canadian Global Ice-Ocean Prediction System (GIOPS; Smith et al., 2014), improved skill of sea ice concentration forecasts up to 7 days into the forecast compared to persistence were reported for the Arctic and Antarctic oceans. Errors present at such lead times arise from inconsistencies between GIOPS forecasts and atmospheric fields in areas where sea ice concentration evolves rapidly. Using pseudoforecasts in which the ice-ocean model is forced with a new forecast every day (instead of a single, 7-day-long forecast), Smith et al. (2014) showed that the error relative to persistence is even more reduced in both polar oceans. This suggests that more accurate atmosphere fluxes could provide better forecasts, which could be the case in fully coupled atmosphere-sea ice-ocean forecasting systems (e.g., Faucher et al., 2010).

5.2.2 Sub-seasonal to Seasonal Predictions

Sea ice forecasts aiming at timescales longer than a few days have classically used empirical models that tried to exploit statistical relations between the state of the sea ice at the target time and the state of the sea ice and other physical quantities at earlier times. For example, Barnett (1980) used the intensity of the Siberian High to predict ice conditions off Alaskian coasts in August. Walsh (1980) used similar methods to forecast the sea ice extent north of Alaska in all months based on sea level pressure, air temperature, and sea ice concentration. Along similar lines, Johnson et al. (1985) generalized the use of periodic regression coefficients to predict sea ice anomalies in various sectors of the Arctic and Antarctic oceans. These early empirical models suggested that using "internal" sea ice predictors such as sea ice concentration or lateral advection improve forecasts compared to persistence, whereas using

[1]This technique assumes that the information on an initial sea ice anomaly is maintained during the forecast.

"external" predictors such as sea level pressure, air temperature, and SST can even deteriorate forecasts. Overall, significant forecast skill up to 2 months in advance was reported.

Drobot and Maslanik (2002) found for the period 1979–2001 that 85% of the variance in Beaufort Sea summer ice extent is explained jointly by spring total ice concentration, winter multiyear ice concentration, the October East Atlantic Index, and the March North Atlantic Oscillation Index. In a follow-up study, Drobot et al. (2006) showed that February sea ice concentration, surface skin temperature, surface albedo, and downwelling longwave radiation jointly explain 46% of September Arctic sea ice extent variance over the 1984–2004 period. However, once again, most of the skill originates from the sea ice concentration—that is, through persistence—which is also the case for regional forecasts (Drobot, 2007). While the latter studies were based on observational data, Lindsay et al. (2008) followed a similar statistical approach to predict pan-Arctic sea ice extent based on a variety of atmospheric, oceanic, and sea ice predictors derived from atmosphere and ocean-sea ice reanalyses, finding that apparent skill for lead times of 3 months or more derives only from the long-term trend.

Similar statistical approaches have been used to predict sea ice in the Antarctic, such as by Chen and Yuan (2004), who conducted principal component analysis (PCA) with seven atmospheric and sea ice variables. Remarkably, significant skill for Antarctic winter sea ice conditions has been achieved up to 1 year in advance. This forecast skill seems to be due to linkages between the variability of the Bellingshausen/Weddell seas dipole and tropical modes of variability (Yuan and Martinson, 2001; Yuan, 2004; Holland et al., 2005).

Coming back to the Arctic, for which more studies exist but where there is also a much stronger long-term trend to cope with, Lindsay et al. (2008) and Holland and Stroeve (2011) highlighted the limitation of statistical approaches in the presence of nonstationarity. This is one motivation to move from statistical approaches toward dynamical forecasts for seasonal timescales. Another argument in favor of dynamical systems is that any statistical approach condenses the information contained in the initial state (and earlier states) to a reduced set of state variables. To provide a simple example, forecasts based solely on current (and past) sea ice anomalies neglect any information that other state variables, such as ocean temperature anomalies or modes of atmospheric circulation, may contain about the future sea ice evolution. Forecasts based on dynamical models that capture the physics of the climate system as comprehensively as possible largely overcome this limitation. Nevertheless, statistical methods are still useful benchmarks for forecast skill (e.g., Chevallier et al., 2013).

Dynamical forecasts of the sea ice extent include the use of (1) ocean-sea ice models forced by atmospheric reanalyses (e.g., Zhang et al., 2008); and (2) fully coupled atmosphere-ocean-sea ice models. Skill assessment of these techniques has been made in hindcast mode, which means a set of reforecasts over the past, typically from the 1990s to the 2010s, initialized with sea ice reanalyses (Peterson et al., 2015) or reconstruction (Chevallier et al., 2013). Recent studies have shown that the forecast skill varies significantly among systems. Guémas et al. (2016) showed that anomaly correlation for the reforecast of September sea ice area initialized in May can vary from 0.2 to 0.7. Forecast skill seems to depend on the initialization of sea ice thickness (Dirkson et al., 2017) or sea ice concentration (Bunzel et al., 2016; Msadek et al., 2014). Combining forecasts by several models seems to improve forecast skill (Merryfield et al., 2013).

A community effort that since 2008 has coordinated S2S predictions of the September Arctic sea ice extent, based on statistical as well as dynamical methods and allowing even heuristic estimates, is the SIO of the Sea Ice Prediction Network (SIPN). Stroeve et al. (2014) concluded for the SIO predictions, which are initialized at the beginning of the months of June, July, and August, that (1) the prediction error for the September sea ice extent is only slightly better than predictions based on linear trends, and (2) there is no indication yet of fully coupled models yielding better sea ice predictions compared to other methods. The apparent gap between real-world forecast skill and potential predictability estimates from perfect-model studies suggests that there is strong potential for improved sea ice predictions, although one cannot preclude that the reality might be less predictable. Blanchard-Wrigglesworth et al. (2015) also showed that the skill of forecasts submitted to SIO is lower than in the hindcasts, which suggest that summer sea ice extent could have been even less predictable in recent years compared to previous decades.

While the SIO will continue to provide interesting data to document and advance our ability to forecast sea ice, a new data set of high relevance in this context, rooted in the operational NWP community, is the S2S data set (Vitart et al., 2017). While some of the systems still prescribe the sea ice cover (e.g., by persisting the ice edge at the beginning and relaxing toward climatology after some time), seven of the contributing forecast systems include a dynamic sea ice component, recently including the abovementioned ECMWF system. A systematic assessment of the sea ice forecast skill of these systems will allow the approach to the sea ice prediction problem to be from a really seamless perspective.

6 IMPACT OF SEA ICE ON SUB-SEASONAL PREDICTABILITY

Skillful predictions of sea ice, discussed in previous sections of this chapter, are also potentially important for predicting the atmosphere and ocean, both in the polar regions and in midlatitudes. Over its lifetime, sea ice acts as a unique boundary condition for the atmosphere. From an atmospheric perspective, it is a highly variable surface, both in time and space (Persson and Vihma, 2017). Sea ice surface properties (e.g., albedo, temperature, and roughness) can change rapidly in time as a result of dynamic and thermodynamic processes. The presence of a mixture of sea ice (possibly snow-covered) and open waters has a strong impact on surface-air turbulent heat fluxes. As a consequence, near-surface temperatures may vary by 30 K (or even more) across relatively short distances depending on whether sea ice is present. For instance, in a large-eddy simulation model, Lüpkes et al. (2008) showed that a 1% variation in sea ice concentration resulting from opening leads could change the surface air temperature by 3.5 K in winter. The presence of sea ice also affects the stratification in the lower atmosphere: the atmospheric boundary layer over sea ice has a stable or near-neutral stratification over most of the year, while over open waters (leads, polynyas), localized convection takes place. Furthermore, sea ice can influence the stratification of the upper ocean through changes in salinity and heat, with implications for deep convection in the ocean. Therefore, there are strong reasons to believe that skillful forecasts of sea ice are a source of predictability for atmospheric and oceanic parameters on the timescales considered here.

The rapid decline of Arctic sea ice and its possible impact on Northern Hemisphere weather and climate have triggered a large number of scientific studies aimed at understanding the influence of sea ice on the atmosphere. Although most of these studies target the climate change problem, it can be argued that the lessons learned apply equally well to shorter S2S timescales, given the relatively fast atmospheric response to an external forcing (e.g., Semmler et al., 2016a,b).

6.1 Impacts in the Polar Regions

For the Arctic, there is consensus that reduced sea ice leads to a warming of the lower atmosphere due to increased heating from the ocean. This low-level warming goes along with a large-scale baroclinic atmospheric response, which is reflected by reduced sea level pressure and increased geopotential height at the 500-hPa level (e.g., Semmler et al., 2016a). It can be expected that a similar robust baroclinic response can be found in the Antarctic, should similar ice retreat occur.

Changes in sea ice also accompany oceanic changes, most notably the heat content of the upper ocean. A reduction of Arctic sea ice, for example, leads to lower turbulent heat fluxes out of the ocean in the vicinity of the ice edge, which results in positive SST anomalies. It has been argued that these anomalies south of the Arctic sea ice edge are instrumental in triggering an midlatitude response (Blackport and Kushner, 2017). Furthermore, it is plausible that thinner sea ice leads to larger momentum transport from the atmosphere into the ocean, thereby influencing the strength of the wind-driven ocean circulation in sea ice–covered regions (e.g., Roy et al., 2015).

6.2 Impacts Outside Polar Regions

The impact of sea ice anomalies on the atmosphere outside of the polar regions is much more controversial. In fact, numerous recent workshops have concluded that we are in a preconsensus state, not unlike with ENSO global impacts in the 1980s (Overland et al., 2015; Jung et al., 2015). The controversy arises from the fact that various modeling studies have found different atmospheric responses to similar imposed sea ice perturbations. Several possible explanations for these differences have been proposed, such as a weak atmospheric response, which leaves the results prone to sampling variability, and the importance of nonlinearity, which makes the results sensitive to small details in the imposed forcing and used numerical protocol (e.g., coupled versus atmosphere-only experiments).

It is accepted, however, that in principle, there are physical mechanisms through which sea ice *can* influence midlatitude weather. A good summary of these mechanisms is given by Barnes and Screen (2015). It turns out that the most robust midlatitude response to sea ice change is thermodynamic in nature, owing to advection of warmer (colder) air masses for low (high) sea ice years; this effect may even offset dynamical temperature changes associated with atmospheric circulation regimes (Screen, 2017). The atmospheric circulation response over the Northern Hemisphere is less certain, although there is some agreement among modeling studies, suggesting that low sea ice winters go along with a strengthened Siberian high pressure system (Deser et al., 2010), which may be explained by an external forcing of the atmosphere in the Barents Sea (Petoukhov and Semenov, 2010). For the wintertime NAO, observational studies suggest a strong link to sea ice anomalies in the Arctic (Cohen et al., 2014). This link involves the impact of SST anomalies in the Barents-Kara seas, snow anomalies in

Siberia, and stratosphere-troposphere interaction. Modeling studies strongly disagree on the degree (and even the sign) to which the NAO is influenced by Arctic sea ice anomalies. A similar situation is found for the Aleutian low-pressure system in the North Pacific.

A completely different approach to studying the impact of the Arctic on midlatitude weather and climate was employed by Jung et al. (2014). They carried out sub-seasonal prediction experiments with the ECMWF model (atmosphere-only) with and without relaxation of the Arctic troposphere toward ERA-Interim reanalysis data (see also Semmler et al., 2017). By studying linkages from a prediction perspective, Jung et al. (2014) identified two main pathways out of the Arctic—one over Eurasia and one over North America. A relatively small impact was found over the North Pacific and North Atlantic, where it was argued that midlatitude dynamics and tropical forcing are more important. They also highlighted the fact that midlatitude prediction skill may benefit only intermittently from Arctic processes due to the strongly flow-dependent nature of these linkages. Strong flow dependence calls for using ensemble prediction systems to exploit the potential of linkages between the Arctic and midlatitudes in operational S2S prediction.

A similar study, employing the relaxation approach for the Southern Hemisphere, has found a smaller impact of the Antarctic troposphere on midlatitude weather; and it was argued that this has to do with the fact that planetary waves in the Southern Hemisphere are weaker than those in the Northern Hemisphere (Semmler et al., 2016c).

7 CONCLUDING REMARKS

This chapter has reviewed various aspects of sea ice, following mainly two directions: (1) the sources and mechanisms of sea ice S2S predictability and (2) the role played by sea ice on atmospheric S2S predictability.

The key concepts to remember about sea ice predictability may be summarized as follows:

(1) Sea ice is part of a *coupled* system. Sea ice physics is driven by forcing from the atmosphere and the ocean. Thus, sea ice predictability is influenced (enhanced or damped) by the two media. Additionally, from a prediction perspective, sea ice exemplifies the added value of using coupled models even for short timescales, showing the importance of coupled processes right from day 1.

(2) Sea ice is present in *both* polar regions. However, different geographical configurations, climates, and physical processes (at the large and small scales) lead to differences in mean state, seasonality, and variability (see Fig. 1). As a result, predictability mechanisms differ in both polar oceans. Noteworthy, research on seasonal sea ice predictability in the Southern Ocean is far less advanced than in the Arctic.

(3) Sea ice has *memory* at the S2S timescale. In observations and models, persistence is identified as the primarily source of predictability for sea ice area properties at the S2S timescale (see Fig. 2). This is a promising result that provides hope for skillful predictions based on the knowledge of sea ice concentration, but also physical grounds on an impact of sea ice on atmospheric predictability in the polar regions. Other sea ice properties have memory at longer (sea ice thickness) and shorter (sea ice speed) timescales, and reemergence mechanisms provide further predictability at longer timescales.

(4) Sea ice has a strong *seasonal* component. As a result, persistence properties are season dependent (see Fig. 3). Sea ice extent anomalies observed in the summer have longer memory than in the spring and fall. This, along with the intrinsic coupled nature of sea ice, argues for the use of fully coupled models for S2S predictions.

(5) Like other components of polar climate, sea ice exhibits *transiency*. In the relatively short observational records, it is difficult to disentangle actual predictability from the signal due to the negative or positive trends. Model studies can partially help, for instance in providing information on unobserved variables, in examining predictability under stable forcing (e.g., preindustrial conditions) or in extrapolating the climate system behavior under future conditions.

This chapter is an invitation to explore the world of sea ice predictability further. We would like to stress the value of coordination in advancing knowledge on sea ice predictability and improving sea ice predictions in the Arctic and Antarctic. Such coordination has been successful for the Arctic, as exemplified by the momentum gained by SIO since 2008. There is a strong demand for a similar initiative in the Southern Ocean. As already mentioned, a systematic assessment of the hindcasts and forecasts available from the S2S database would allow significant progress in many areas, with a seamless perspective. We also encourage further investigations using the Year Of Polar Prediction (YOPP) data portal (Jung et al., 2016; https://yopp.met.no/), which will host coordinated model experiments designed to explore sea ice predictability and polar to lower-latitude connections in weather and climate, run during the YOPP core phase (2017–19) and consolidation phase (2019–22).

Acknowledgments

The authors would like to thank Edward Blanchard-Wrigglesworth, Mitch Bushuk, and David Salas y Mélia for their useful comments and suggestions.

CHAPTER

11

Sub-seasonal Predictability and the Stratosphere

Amy Butler, Andrew Charlton-Perez†, Daniela I.V. Domeisen‡, Chaim Garfinkel§, Edwin P. Gerber¶, Peter Hitchcock||, Alexey Yu. Karpechko#, Amanda C. Maycock**, Michael Sigmond††, Isla Simpson‡‡, Seok-Woo Son§§*

*Cooperative Institute for Research in Environmental Sciences (CIRES)/National Oceanic and Atmospheric Administration (NOAA), Boulder, CO, United States †Department of Meteorology, University of Reading, Reading, United Kingdom ‡Institute for Atmospheric and Climate Science, ETH Zürich, Zürich, Switzerland §Freddy and Nadine Hermann Institute of Earth Science, Hebrew University of Jerusalem, Jerusalem, Israel ¶Courant Institute of Mathematical Sciences, New York University, New York, NY, United States ||École Polytechniqe, Saclay, Paris, France #Finnish Meteorological Institute, Helsinki, Finland **School of Earth and Environment, University of Leeds, Leeds, United Kingdom ††Canadian Centre for Climate Modelling and Analysis, Environment and Climate Change Canada, Victoria, BC, Canada ‡‡Climate and Global Dynamics Laboratory, National Center for Atmospheric Research, Boulder, CO, United States §§School of Earth and Environmental Sciences, Seoul National University, Seoul, Republic of Korea

OUTLINE

Sub-seasonal to Seasonal Prediction
https://doi.org/10.1016/B978-0-12-811714-9.00011-5

1 INTRODUCTION

The stratosphere is the layer of highly stratified air that extends for roughly 40 km above the tropopause and contains approximately 20% of the mass of the atmosphere. The climatology, seasonal evolution, and variability of the stratospheric circulation are strongly governed by the combined influences of solar and infrared radiation, ozone chemistry, and transport and momentum transport by Rossby and gravity waves that propagate upward from the troposphere below. While it contains a smaller fraction of atmospheric mass than the troposphere, the stratosphere is far from being a passive bystander to tropospheric influences. It exhibits a diverse range of variability on a spectrum of timescales with, in many cases, a well-established influence on the tropospheric circulation below. As a result, knowledge of the state of the stratosphere has the potential to enhance the predictability of the troposphere on sub-seasonal to seasonal (S2S) timescales and beyond.

This chapter reviews our knowledge of the coupling between the stratosphere and troposphere in the tropics (Section 2) and the extratropics (Section 3) to provide a clear understanding of where and when coupling is important. In Section 4, we review the progress to date in trying to harness stratosphere-troposphere coupling to enhance predictability on the S2S timescale, a key focus of the World Climate Research Programme/Stratosphere-Troposphere Processes And Their Role in Climate (WCRP/SPARC) Stratospheric Network for the Assessment of Predictability (SNAP) project. Finally, in Section 5, we examine a number of open questions and provide some perspective on where and how improved understanding and simulation of stratosphere-troposphere coupling are most likely to lead to improved skill. Throughout the chapter, it is important to emphasize that one of the significant difficulties in assessing and understanding stratosphere-troposphere coupling (in common with other low-frequency phenomena, such as Deser et al., 2017) is the relatively short observational record that exists for the stratosphere.

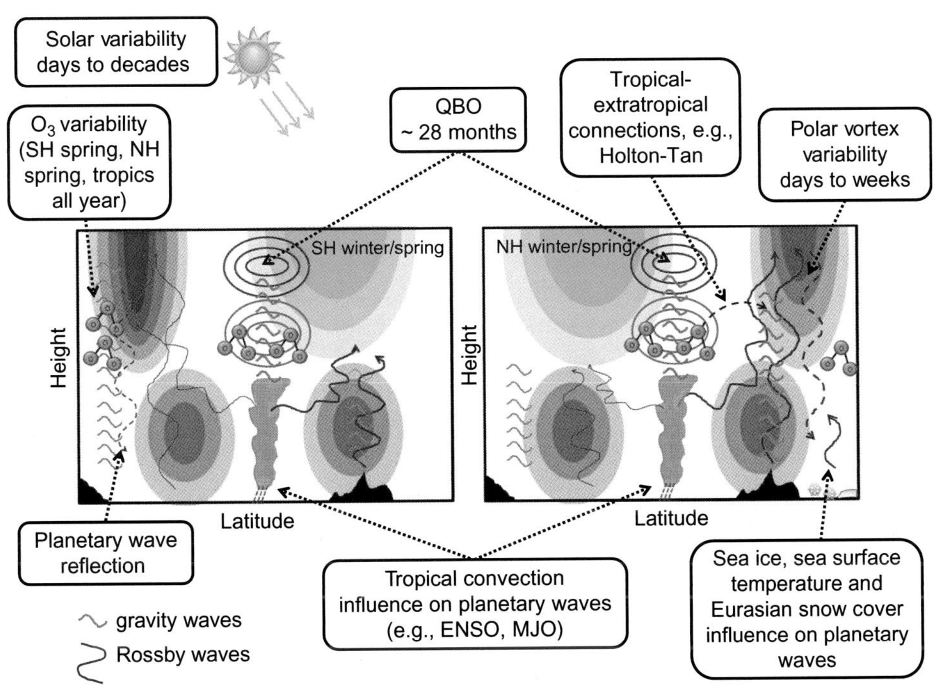

FIG. 1 Schematic showing phenomena of known relevance to stratosphere-troposphere coupling. Contours show the zonal mean zonal wind schematically (*red*-filled contours indicate mean westerly winds and *blue*-filled indicate easterly winds). Left panel shows Southern Hemisphere (SH) winter and spring and right panel shows Northern Hemisphere (NH) winter and spring. *Red*- and *blue*-unfilled contours indicate zonal mean, zonal wind anomalies associated with the QBO.

Fig. 1 highlights the major phenomena relevant to coupling between the stratosphere and troposphere, including the Quasi-Biennial Oscillation (QBO), solar variability, ozone, and the role of tropospheric planetary-scale waves.

2 STRATOSPHERE-TROPOSPHERE COUPLING IN THE TROPICS

In the tropics, the dominant feature of stratospheric variability is a remarkably regular succession of downward-migrating easterly and westerly zonal jets known as the QBO. Over a period of roughly 28 months, the equatorial winds transition between westerly QBO (WQBO) and easterly QBO (EQBO) as a result of the selective absorption of tropical waves propagating upward from the troposphere below (Lindzen and Holton, 1968; Holton and Lindzen, 1972; Baldwin et al., 2001).

In addition to the effects of the QBO on polar vortex variability discussed later in Section 4.3, recent work has emphasized the important role that the QBO may play in determining tropical, tropospheric variability and predictability. This section reviews the understanding of QBO-troposphere coupling in the tropics and the potential to exploit these links for improved predictability.

2.1 How Does the QBO Influence the Tropical Troposphere?

The QBO can affect the characteristics of tropical deep convection (e.g., Collimore et al., 2003; Liess and Geller, 2012). Satellite observations and numerical model simulations indicate that tropical deep convection across the western Pacific is stronger during EQBO winters than during WQBO winters (Collimore et al., 2003). Additionally, sub-seasonal Madden Julian Oscillation (MJO)-like convective activity is significantly modulated by the QBO, with stronger and more organized MJO convection during 50-hPa EQBO winters (Liu et al., 2014b; Yoo and Son, 2016; Son et al., 2017; Nishimoto and Yoden, 2017).

The mechanism for QBO-induced changes in tropical convection is not well understood. The possible impact of the QBO on tropical deep convection often has been explained by local instability and tropopause property changes (Giorgetta et al., 1999; Collimore et al., 2003; Yoo and Son, 2016). Recently, radiative feedback and the associated large-scale vertical motion change have been proposed as possible mechanisms (Nie and Sobel, 2015; Son et al., 2017). Next, these hypotheses are briefly introduced.

The change in vertical wind shear in the upper troposphere associated with the downward propagation of QBO wind anomalies could modify tropical deep convection (Gray et al., 1992). For example, over the Indo-Western Pacific warm pool region, absolute vertical wind shear across the tropopause becomes anomalously strong under WQBO (see Fig. 3 of Gray et al., 1992). This could disrupt convective organization, especially by shearing off deep convection that overshoots into the stratosphere. This may result in less-organized deep convection in WQBO, but more-organized convection in EQBO.

The QBO modifies not only the vertical wind shear, but also the thermal stratification. It is well documented that the secondary circulation induced by the QBO effectively changes the tropical temperature profile (e.g., Baldwin et al., 2001). For instance, 50-hPa easterlies accompany cold anomalies centered at 70 hPa that extend into the upper troposphere. These temperature anomalies can act to destabilize the upper troposphere. If deep convection is influenced by the upper-tropospheric thermal stratification, this destabilization may allow more-organized deep convection during EQBO (Gray et al., 1992; Giorgetta et al., 1999; Collimore et al., 2003; Yoo and Son, 2016).

The two mechanisms described here are essentially based on local instability, which could modify very deep convection. However, in the tropics, cloud tops are typically located a few kilometers below the tropopause (Gettleman and Forster, 2002). Convection that crosses the tropopause is relatively rare. As such, it is questionable whether these mechanisms are really acting in the atmosphere.

Another possible mechanism for QBO influence on tropical convection (Reid and Gage, 1985; Gray et al., 1992) suggests that QBO-induced tropopause changes can modify deep convection. During EQBO, when the lower stratosphere is anomalously cold, tropopause height is slightly increased (Collimore et al., 2003; Son et al., 2017). A higher tropopause may provide a favorable condition for deep convection through enhanced organization. It is also possible that the cold tropopause itself can directly change tropical deep convection (e.g., Emanuel et al., 2013).

A further alternative is that radiative processes could play a role. Son et al. (2017) showed that tropical cirrus clouds are significantly modulated by the QBO. For example, near-tropopause cirrus clouds increase during EQBO winters due to an anomalously cold tropopause. This cirrus cloud change then could cause additional longwave radiative heating in the troposphere (Hartmann et al., 2001; Yang et al., 2010; Hong et al., 2016), as simulated by a cloud-resolving model (Nie and Sobel, 2015). This radiative process might be particularly important in QBO-related MJO convection changes because the MJO is partly organized by cloud-radiative feedback (e.g., Andersen and Kuang, 2012).

Finally, the QBO could influence variability directly in the subtropical troposphere. QBO-related equatorial wind anomalies must be accompanied by a meridional circulation that extends to the subtropical tropopause to maintain thermal wind balance, and this circulation appears to affect tropospheric eddies (Garfinkel and Hartmann, 2011a,b). The effect is particularly strong over East Asia (Inoue et al., 2011; Seo et al., 2013).

2.2 Predictability Related to Tropical Stratosphere-Troposphere Coupling

Predictability of the tropical stratosphere is strongly related to the predictability of the QBO and therefore exceeds sub-seasonal timescales. Many modern numerical prediction systems are capable of internally generating the QBO; however, model-generated QBOs often exhibit biases in amplitude and period (Schenzinger et al., 2016). Model forecasts of the QBO have significant skill beyond 12 months (Scaife et al., 2014a), but similar skill can be achieved with a simple statistical model representing a cosine with a period of 28 months. The predictive capabilities of forecast systems were recently tested by an interruption of the regular QBO behavior when an easterly jet unexpectedly appeared within a descending westerly phase in the lower stratosphere in early 2016 (Newman et al., 2016; Osprey et al., 2016). Seasonal forecasts initialized in November 2015 were not able to predict this event, instead predicting a regular descent of the westerly phase. Although unusual in observational records, such interruptions are occasionally seen in long climate model simulations (Osprey et al., 2016). The predictability limits of similar deviations from the regular QBO have not yet been examined.

Predictive skill in the QBO might be translated to tropospheric skill via a direct influence of the QBO on the MJO phase. As described in the previous section, QBO modulates interannual variations of MJO convection and its teleconnection (Son et al., 2017). A series of studies have shown that MJO-like, sub-seasonal convective activities become anomalously strong during EQBO winters (Liu et al., 2014b; Yoo and Son, 2016; Marshall et al., 2016b; Son et al., 2017; Nishimoto and Yoden, 2017). Such enhancement is observed in all phases of MJO from the Indian Ocean to the central Pacific (e.g., Yoo and Son, 2016). In addition, during EQBO winters, MJO convections tend to propagate more slowly and its period becomes longer (Son et al., 2017; Nishimoto and Yoden, 2017). Consistent with these changes, the MJO power spectrum is sharply peaked in the 40–50-day band during EQBO winters (Marshall et al., 2016b).

The fact that the MJO is generally stronger and better organized during EQBO winters could be translated into improved MJO prediction in EQBO winters. Marshall et al. (2016b), looking at 30 years of retrospective forecasts from the Bureau of Meteorology (BoM) seasonal prediction model, showed that the MJO is indeed better predicted when the equatorial lower-stratosphere is in the EQBO phase. In this model, the MJO prediction skill increases by up to 8 days between WQBO and EQBO winters (Fig. 2). Interestingly, this

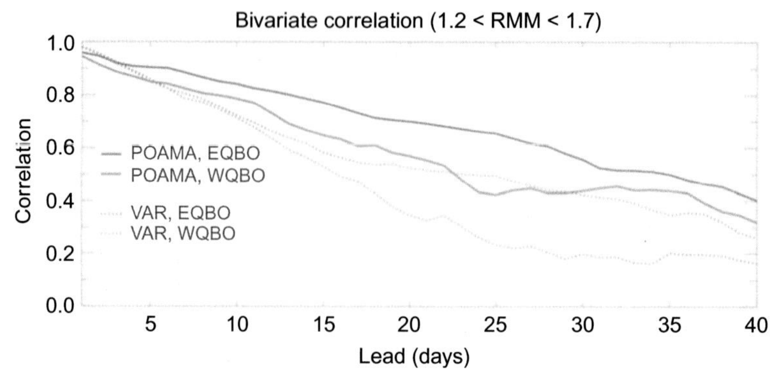

FIG. 2 Bivariate correlation skill for the ensemble-mean Real-time Multivariate MJO (RMM) index, predicted by the BoM coupled seasonal prediction model, as a function of lead time during EQBO *(blue)* and WQBO *(red)* winters. For reference, the statistical forecast, based on vector autoregression, is also shown by *dotted lines*. Only MJO events that have initial amplitudes of 1.2–1.7 at the initial time are considered over the period of 1981–2010. *Reproduced with permission from Marshall, A.G., Hendon, H.H., Son, S.-W., Lim, Y., 2016. Impact of the quasi-biennial oscillation on predictability of the Madden–Julian oscillation. Clim. Dyn., doi:10.1007/s00382-016-3392-0.*

model simulated an increase in skill from the QBO-MJO connection, even though it does not have a highly resolved stratosphere (Marshall et al., 2016b).

A higher MJO prediction skill is not simply caused by the fact that MJO events initialized in EQBO winters are stronger than those in WQBO winters. In fact, MJO events of similar amplitude at the initial time showed essentially the same result—that is, a higher MJO prediction skill in EQBO winters (Fig. 2). This suggests that a structural change of MJO convection or upper-tropospheric circulation by the QBO may play a role in modulating MJO prediction skill. A more persistent MJO, which is typically found during EQBO winters, may also contribute to the extended MJO prediction. Although further analyses are required, especially using stratosphere-resolving models, this result suggests that the QBO is an untapped source of MJO predictability in boreal winter.

3 STRATOSPHERE-TROPOSPHERE COUPLING IN THE EXTRATROPICS

Many of the proposed pathways for a stratospheric influence on near-surface weather and climate have emphasized the role of the polar vortex, particularly the dynamic variability of the vortex in the Northern Hemisphere. This is reinforced by robust evidence from observations and a wide variety of modeling studies. Accordingly, much of the effort toward understanding the mechanisms of stratosphere-troposphere dynamical coupling has focused on this pathway. This section reviews our understanding of these extratropical links between the stratosphere and troposphere.

3.1 An Overview of Polar Vortex Variability

Due to the annual variation of solar heating over the poles, the stratosphere undergoes a strong seasonal cycle. In the extratropical winter hemisphere, the stark contrast in stratospheric temperatures between the cold polar night and the warmer low latitudes leads to the development of a strong westerly stratospheric polar vortex (Waugh et al., 2017), as shown in Fig. 3 for the Northern Hemisphere (NH; left panel) and Southern Hemisphere (SH; right panel). In the Northern Hemisphere, the stratospheric vortex exhibits maximum variability in January and February. The seasonal reversal of the climatological stratospheric winds from westerly to easterly as sunlight returns to the pole in the spring (the so-called final warming) occurs on average in mid-April, but it is highly variable due to the presence of significant dynamical variability. The Southern Hemisphere exhibits considerably weaker interannual variability (see shading in Fig. 3), both in midwinter and in the onset of the final warming, due to its weaker wave forcing; see Andrews et al. (1987) for a comprehensive review of stratospheric climate and dynamics.

There is active coupling between the stratosphere and troposphere during periods when significant stratospheric variability occurs (winter and spring in the Northern Hemisphere and spring in the Southern Hemisphere) (Thompson and Wallace, 2000). Variability in the position and strength of the stratospheric polar vortex is largely driven by planetary-scale Rossby waves (whose sources lie within the troposphere), which vertically amplify into the stratosphere and break (see Section 3.2 for more detail). When the polar stratospheric winds become easterly in spring-summer, downward-coupling mechanisms that involve vertically propagating Rossby waves from the troposphere to the stratosphere no longer will be in effect (Charney and Drazin, 1961).

In extreme cases, the wintertime polar vortex is so perturbed by the effects of Rossby wave breaking that the climatological westerly winds become temporarily easterly in events known as major Sudden Stratospheric Warmings (SSWs). Although there is a variety of criteria and terminology used to define these events (Butler et al., 2015), they are typically associated with

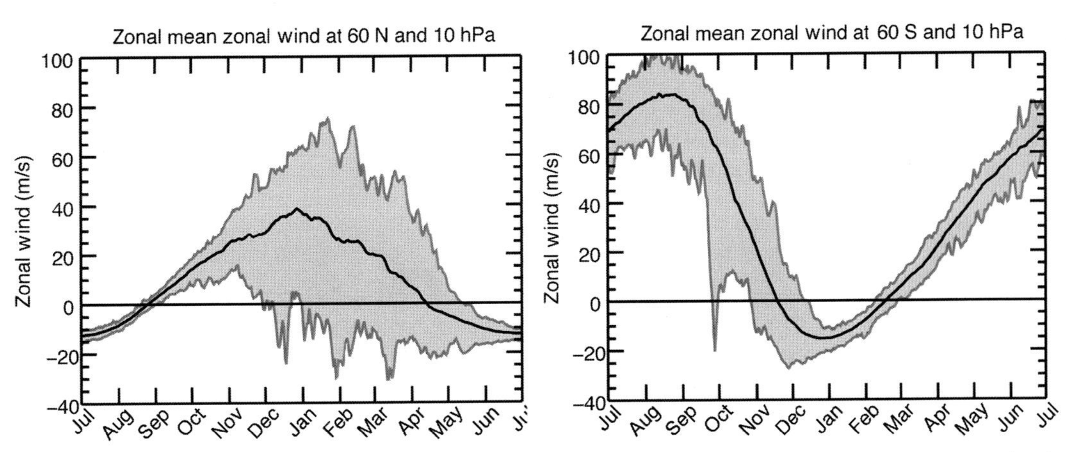

FIG. 3 Zonal-mean zonal winds at 10 hPa and 60°N (left, for the Northern Hemisphere) and 60°S (right, for the Southern Hemisphere). The *solid black line* is the daily mean value, and the *gray shading* shows the range of values between the daily maximum and minimum values, using JRA-55 reanalysis data from 1958 to 2016.

a rapid adiabatic warming of the polar stratosphere of up to 70K over only a few days (Labitzke, 1977; Limpasuvan et al., 2004). SSWs are often classified by whether the vortex is displaced off the pole or whether it splits into two (Charlton and Polvani, 2007; Mitchell et al., 2011); these permutations correspond roughly to the zonal wave number of the waves responsible for the disruption of the vortex. While some SSW events are followed by a fairly fast recovery of the vortex over a period of a week or two, roughly half of SSWs undergo an extended-timescale recovery that lasts up to 2 months (Hitchcock et al., 2013), producing a stratospheric circulation pattern often termed the *Polar night Jet Oscillation (PJO)* (Kuroda and Kodera, 2001) and resulting in an extended influence on the troposphere below.

In the Northern Hemisphere, midwinter SSWs occur roughly six times per decade (Charlton and Polvani, 2007), most frequently in January and February, though they can occur any time from November through March. SSWs are much less common in the more quiescent Southern Hemisphere; only one event has been observed in September 2002 (Newman and Nash, 2005), although observational records in the Southern Hemisphere are short.

3.2 What Drives Polar Vortex Variability?

It has long been recognized that SSWs are associated with rapid amplifications of planetary waves from the troposphere (Finger and Teweles, 1964; Julian and Labitzke, 1965; Matsuno, 1971). Just how and why this amplification occurs remains a matter of longstanding debate. Some authors have argued that this is predominantly controlled by changes in the stratospheric basic state on which the waves are propagating; others say that the amplification arises from changes in the tropospheric source of the waves.

On the one hand, SSWs can occur in models that lack any explicit representation of tropospheric variability (Holton and Mass, 1976; Scott and Haynes, 2000; Scott, 2016) or in which the tropospheric variability has been strongly suppressed (Scott and Polvani, 2004). In these models, the internal state of the stratosphere itself determines the wave fluxes at the lower boundary. This control is often understood to occur through a resonance effect: when the phase speed of a free, traveling wave mode of the stratosphere approaches zero, it comes into resonance with the topographic forcing (Tung and Lindzen, 1979). Nonlinear effects can act to tune this resonance, leading to the rapid growth of the wave mode throughout the column (Plumb, 1981; Matthewman and Esler, 2011). Evidence for this behavior has been found in a number of case studies, particularly involving vortex-split events (Smith, 1989; Esler and Scott, 2005; Albers and Birner, 2014). However, just because idealized representations of the stratosphere can produce variability independent of tropospheric sources of variability does not mean that observed SSWs are unaffected by these sources. Many studies have pointed out various tropospheric precursors to SSWs (Nishii et al., 2009; Coy et al., 2009; Colucci and Kelleher, 2015; O'Neill et al., 2017). Constructive interference between climatological stationary waves and the anomalous waves generated by such precursors has been proposed as the relevant mechanism (Garfinkel and Hartmann, 2008; Garfinkel et al., 2010; Cohen and Jones, 2011; Smith and Kushner, 2012; Watt-Meyer and Kushner, 2015; Martineau and Son, 2015).

What implications do these alternate paradigms of the causes of stratospheric variability have for its predictability? Idealized models of vortex variability that show strong stratospheric control over the wave field sometimes exhibit bifurcations in their behavior that are extremely sensitive to initial conditions and external parameters (Yoden, 1987; Matthewman and Esler, 2011). This would imply that SSWs are extremely difficult to predict beyond the predictability horizon of weather systems, at least in a deterministic sense. Evidence for such bifurcations in forecast models has been found (Noguchi et al., 2016). However, sensitivity to the stratospheric basic state suggests that certain processes may provide skill in predicting the likelihood of the occurrence of SSWs, as has been proposed in various contexts, including solar variability and the QBO (Holton and Tan, 1980; Kodera, 1995; Haigh, 1996). On the other hand, if the wave field is primarily controlled by their sources, then tropospheric processes ranging from ENSO to Eurasian snow cover to Arctic sea ice should provide greater skill in predicting SSWs (e.g., Cohen et al., 2007; Ineson and Scaife, 2009; Kim et al., 2014). Tropospheric precursors themselves are also likely subject to both predictable and chaotic influences. The truth likely lies somewhere between the two paradigms; a recent study demonstrated explicitly that both the stratospheric and tropospheric states are essential for reproducing the amplification of the waves in an idealized model (Hitchcock and Haynes, 2016). Much future work remains in order to fully understand the extent to which polar stratospheric variability is predictable.

3.3 How Does Stratospheric Polar Vortex Variability Influence Surface Climate?

Much of the interest in stratosphere-troposphere coupling has arisen from studies of annular modes, which are the dominant structures of large-scale extratropical atmospheric variability in each hemisphere (Thompson and Wallace, 2000). To leading order, the annular modes in the stratosphere represent variations in the strength of the stratospheric polar vortex; by convention, positive indices correspond to a stronger, colder vortex. In the troposphere, the annular mode represents a latitudinal shifting of the eddy-driven, midlatitude westerlies; here, positive indices correspond to a poleward excursion.

Although not the first study to propose a downward influence from the stratosphere (e.g., Boville, 1984; Perlwitz and Graf, 1995; Hartley et al., 1998), the Northern Annular Mode (NAM) composites presented by Baldwin and Dunkerton (2001) have become an iconic visualization of stratospheric influence on the troposphere. They demonstrate that the tropospheric eddy-driven jet is, on average, shifted systematically equatorward for days to weeks following weak polar vortex events (essentially SSWs), associated with a negative phase of the NAM, and thus colder surface temperatures over much of the Northern Hemisphere midlatitudes and warmer surface temperatures over the Arctic. Likewise, strong vortex events are usually followed by a poleward shift of the eddy-driven jet (or positive phase of the NAM).

Given that the variability in the Northern Hemisphere polar vortex is primarily driven by planetary-scale Rossby waves whose sources lie within the troposphere, it has been questioned whether the apparent downward descent of annular mode index anomalies represents a true downward propagation of information from the stratosphere to the troposphere (Plumb and Semeniuk, 2003). However, controlled experiments with models of various degrees of complexity, in which stratospheric perturbations are imposed into a tropospheric model state that has no memory of the conditions that contributed to the stratospheric perturbation (Polvani and Kushner, 2002; Jung and Barkmeijer, 2006; Douville, 2009; Gerber and Polvani, 2009; Hitchcock and Simpson, 2014), have shown clearly that the state of the stratosphere does affect the troposphere.

Early attempts to explain the downward influence from the stratosphere to the troposphere focused on the consequences of the large-scale dynamical balance between winds and temperatures. This balance implies that the forcings responsible for inducing the stratospheric anomalies will have a direct (albeit weak) impact on the troposphere as well (Robinson, 1988; Hartley et al., 1998; Ambaum and Hoskins, 2002). Moreover, diabatic processes tend to strengthen this influence in an effect termed "downward control" (Haynes et al., 1991). Efforts to quantify this effect, however, have generally found it to be too weak to explain the full surface response (Charlton et al., 2005; Thompson et al., 2006a; Hitchcock and Simpson, 2016).

Strong feedback between tropospheric eddies and the jet is recognized as an essential component of internal variability of the tropospheric annular modes (Robinson, 1991, 1996; Hartmann and Lo, 1998; Limpasuvan and Hartmann, 2000). The idea that this tropospheric eddy feedback could play a role in the response to anomalous stratospheric conditions was suggested as early as Hartmann et al. (2001) and was confirmed by a series of studies demonstrating that the strength of the stratospheric influence on the troposphere in a given model was closely related to the strength of the tropospheric eddy feedback (Chan and Plumb, 2009; Gerber and Polvani, 2009; Garfinkel et al., 2013). The feedback is often measured by the decorrelation timescale of the annular modes: the stronger the feedback, the more persistent the annular modes (Ring and Plumb, 2008). The relationship between the decorrelation timescale and the response to an external forcing for a given dynamical mode is expected on the basis of a general result known as the *Fluctuation-Dissipation Theorem* (Leith, 1975). However, recent studies have shown that the correspondence between eddy feedback strength and annular mode decorrelation timescales is not always reliable (Simpson and Polvani, 2016), and alternative methods for quantifying this feedback continue to be explored (Lorenz and Hartmann, 2001, 2003; Simpson et al., 2013; Nie et al., 2014).

Recognition of the importance of tropospheric eddy feedback in determining the response to a stratospheric perturbation does not clarify how the stratosphere triggers this feedback in the first place. This question was first clearly articulated by Song and Robinson (2004), who proposed that the downward control response, while weak in and of itself, may serve as a trigger for tropospheric feedback between the eddies and the mean flow. However, their numerical experiments also suggested that planetary wave feedback might play a role. The possibility of some direct influence by the stratosphere on the synoptic scale eddies also has been raised (Tanaka and Tokinaga, 2002; Wittman et al., 2007). However, more recently, several modeling studies have clearly identified planetary waves as the key coupling pathway (Martineau and Son, 2015; Smith and Scott, 2016; Hitchcock and Simpson, 2016), at least for the Northern Hemisphere winter.

Although our understanding of stratosphere-troposphere coupling processes has advanced substantially in the last two decades, major open questions remain. Many theoretical and modeling studies to date have focused on the zonally symmetric component of the tropospheric response, despite the clear agreement in observations and models that the Northern Hemisphere response to stratospheric variability is strongest within the storm tracks, particularly in the Atlantic basin (Charlton and Polvani, 2007; Garfinkel et al., 2013; Hitchcock and Simpson, 2014). Additionally, it has been tacitly assumed that the same mechanisms are at play in Northern Hemisphere and Southern Hemisphere stratosphere-troposphere coupling; yet the tropospheric circulation is significantly different, and in particular, planetary wave activity is far weaker in the Southern Hemisphere.

3.4 Other Manifestations of Extratropical Stratosphere-Troposphere Coupling

SSWs are certainly the most dramatic form of stratospheric vortex variability, but they are not the only form. A full spectrum of other vortex variability exists, ranging from anomalously strong vortex events, to less dramatic weak vortex events that do not pass the threshold criteria for major SSWs, to individual planetary wave reflection events (Limpasuvan et al., 2005; Dunn-Sigouin and Shaw, 2015; Maury et al., 2016).

Planetary wave reflection events are an example of coupling through wave motions, as opposed to through the zonal mean circulation. The state of the stratosphere can affect the propagation of planetary waves such that they are reflected back down toward the troposphere, with subsequent tropospheric impacts (Perlwitz and Harnik, 2003). Wave coupling events occur on timescales of a few days to weeks and tend to be followed by a positive sign of the NAM or a poleward shift of the North Atlantic storm track (Shaw and Perlwitz, 2013). In the Southern Hemisphere, wave reflection plays a stronger role in stratosphere-troposphere variability, whereas in the Northern Hemisphere, it is found to be of comparable importance to the zonal mean coupling (Shaw et al., 2010). Less is currently known about the predictability of stratospheric wave reflection events (Harnik and Lindzen, 2001; Harnik, 2009; Shaw et al., 2010), although they tend to be sensitive to the QBO and sea surface temperature (SST) variability (Lubis et al., 2016a).

Chemistry-climate feedback is another important factor for stratosphere-troposphere coupling. Over the latter half of the 20th century, anthropogenic emissions of chlorofluorocarbons (CFCs) into the atmosphere have led to the chemical destruction of ozone (O_3) within the Southern Hemisphere polar vortex in springtime. Temperatures in the Southern Hemisphere polar stratosphere routinely drop below 195K in winter due to weaker dynamic variability compared to the Northern Hemisphere, which allows significant amounts of polar stratospheric clouds (PSCs) to form, upon which catalytic chemical reactions that destroy ozone can occur. The radiative cooling associated with chemical depletion of ozone at high latitudes results in a stronger polar vortex, and often a delayed seasonal breakup as well, and is thus a key driver of stratosphere-troposphere coupling on interannual (Son et al., 2013) and multidecadal (Thompson and Solomon, 2002; McLandress and Shepherd, 2011; Polvani et al., 2011) timescales in the Southern Hemisphere. While chemical ozone destruction within the Northern Hemisphere polar vortex is much lower than in the Southern Hemisphere, as temperatures are not typically cold enough to form large amounts of PSCs, substantial springtime Arctic ozone loss also can occur (Manney et al., 2011) and may be linked to interannual Northern Hemisphere tropospheric variability in the spring (Karpechko et al., 2014; Smith and Polvani, 2014; Calvo et al., 2015; Xie et al., 2016; Ivy et al., 2017). Ozone layer recovery, due to the Montreal Protocol and its amendments, may reverse these effects in the future, particularly if greenhouse gases continue to increase (Eyring et al., 2013).

4 PREDICTABILITY RELATED TO EXTRATROPICAL STRATOSPHERE-TROPOSPHERE COUPLING

The fact that variability in the stratospheric polar vortices has a substantial impact on the tropospheric circulation in both hemispheres is now well established. Because the stratospheric anomalies associated with SSWs can persist for several weeks, this fact alone is of considerable value for S2S forecasts in the extratropics (Sigmond et al., 2013). However, if the onset of stratospheric polar vortex anomalies can themselves be forecast, the value for forecasting is even greater, potentially leading to higher skill of extratropical surface climate at longer lead times. In order to exploit stratosphere-troposphere coupling, sub-seasonal prediction models need to be able to:

1. Skillfully forecast stratospheric variability
2. Accurately simulate the dynamical coupling between the stratosphere and troposphere

These issues are discussed in turn in the following subsections.

4.1 How Accurately Can the Polar Stratosphere be Predicted?

The smaller role of baroclinic instabilities and the strongly reduced Rossby wave spectrum in the stratosphere suggest that, in general, predictability timescales in the stratosphere should be longer than in the troposphere. One way to demonstrate the intrinsic, enhanced predictability of the stratosphere is to look at the decorrelation timescales of the annular modes (Baldwin et al., 2003; Gerber et al., 2010). In the Northern Hemisphere extratropical stratosphere (Fig. 4C), the characteristic timescale of NAM anomalies is about 1 month during winter. Even longer timescales exceeding 2 months can be seen in the Southern Hemisphere extratropical stratosphere during late winter and early spring (Fig. 4D). This contrasts sharply with decorrelation timescales in the troposphere, which are typically less than 10 days and peak at about 2 weeks during December–January in the Northern Hemisphere and November–December in the Southern Hemisphere. Note that extended persistence in the troposphere tends to coincide with enhanced variance in the stratosphere (Fig. 4A and B).

Numerical Weather Prediction (NWP) models have long been able to reproduce extended predictability in the stratosphere compared to the troposphere. For example, Waugh et al. (1998, in the Southern Hemisphere), Jung and Leutbecher (2007, in the Northern Hemisphere) and Zhang et al. (2013b, in both hemispheres) showed that forecast skill in the stratosphere is roughly twice that of the troposphere for the same forecast lead time. This skill is mostly linked to the ability to capture and maintain anomalies in the zonal mean circulation, even if models are unable to skillfully forecast planetary waves.

Recent studies have demonstrated stratospheric predictability in operational forecast models at sub-seasonal timescales in the Northern Hemisphere (Zhang et al., 2013b; Taguchi, 2014; Vitart, 2014). Correlation skill scores for stratospheric parameters can be higher than 0.6 for forecasts with a lead time of more than 20 days. During SSWs or periods of anomalously strong polar vortex, correlation skill scores can be as high as 0.8 for forecasts at 4 weeks lead time (Tripathi et al., 2015). There is also some evidence for modest skill in predicting the probability of an SSW or strong vortex event in seasonal forecasts initialized on November 1, based

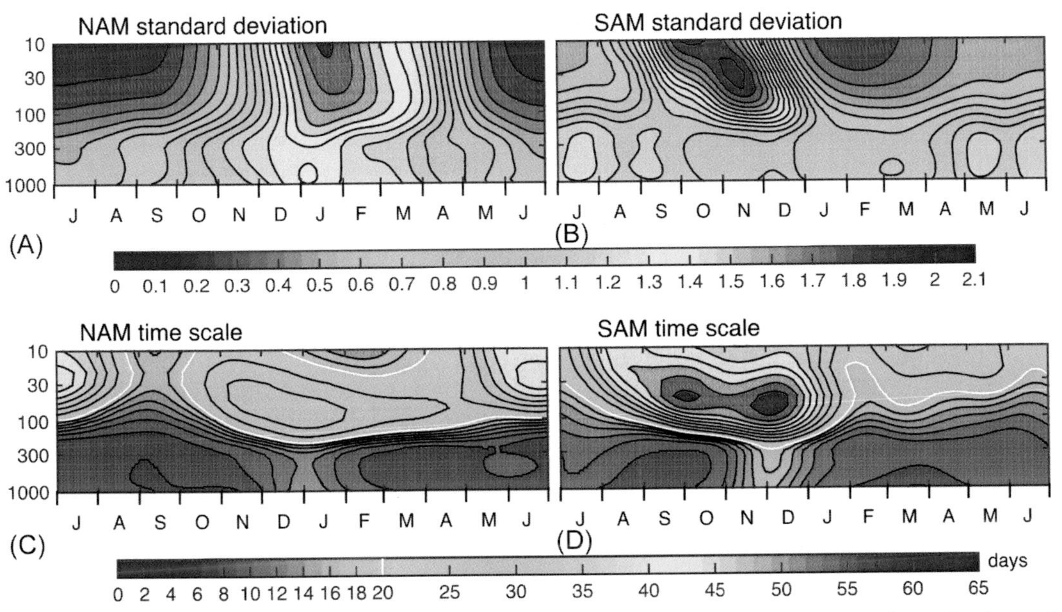

FIG. 4 Standard deviation of the (A) Northern Hemisphere and (B) Southern Hemisphere annular modes (normalized such that indices at each pressure level have unit variance in the annual mean), and the estimated decorrelation timescale of the (C) Northern Hemisphere and (D) Southern Hemisphere annular modes, as a function of pressure and month, using JRA-55 reanalysis data (after Gerber et al., 2010). The NAM is based on data from 1960 to 2009, and the Southern Annular Mode (SAM) from 1980 to 2009.

on ensemble spread (Scaife et al., 2016). To date, fewer studies have assessed stratospheric predictability in the Southern Hemisphere, but similar or higher levels of skill have been found on sub-seasonal timescales (Roff et al., 2011), with some indications of even higher skill on seasonal timescales (Seviour et al., 2014).

As might be expected intuitively, stratospheric predictability is lowest just prior to SSW events when pulses of planetary wave activity reach the stratosphere, leading to nonlinear interactions between waves and the mean flow and weakening of the polar vortex (Taguchi, 2014; Noguchi et al., 2016). The predictability of SSW events is typically between 5 and 15 days (Tripathi et al., 2015), comparable to that of tropospheric weather systems. Tripathi et al. (2016) found high predictability for the onset of the January 2013 SSW in initialized numerical prediction systems for lead times of up to 10 days, but diminished predictability for longer lead times. This was partly attributed to decreased predictability of the amplified wave number-2 activity in the troposphere that induced the SSW. In general, vortex-weakening cases are thought to be less predictable than vortex-strengthening cases, even when the wave activity anomalies leading to these events—either wave amplification or attenuation—were of comparable magnitude (Taguchi, 2015). The largest stratospheric forecast errors are associated with cases where models fail to correctly predict wave activity fluxes over western Siberia and northern Europe, which is likely linked to an underestimation of tropospheric blocking (Lehtonen and Karpechko, 2016). In some cases, accounting for a mismatch of a few days between forecast and observed dynamical events (which is comparable to considering time-averaged forecasts) may lead to improved predictability (Cai et al., 2016).

4.2 S2S Extratropical Forecast Skill Associated With Strong and Weak Polar Vortex Events

In general, initializing forecast models with information about the state of the stratosphere does improve tropospheric prediction skill on S2S timescales. For example, Baldwin et al. (2003), Charlton et al. (2003), and Christiansen (2005) used simple statistical models and found modestly improved Northern Hemisphere extratropical surface skill at 10–45 days when using a stratospheric predictor as opposed to tropospheric predictors. More recent studies have combined information about the state of the stratospheric polar vortex with other tropospheric predictors to show significant skill in forecasting the winter North Atlantic Oscillation (NAO), the Atlantic-sector manifestation of the NAM. For example, Dunstone et al. (2016) used a linear regression model based on four November predictors (tropical Pacific SSTs, the Atlantic SST tripole pattern, Barents-Kara sea ice, and the strength of the stratospheric polar vortex) and found significant skill in the wintertime NAO ($r = 0.60$). Even higher skill ($r \sim 0.7$) using statistical forecasts with stratospheric predictors is found in other recent studies (Wang et al., 2017).

Following early experiments that examined the tropospheric response to a significant diminution in the stratospheric representation in models (Norton, 2003; Kuroda, 2008), two main modeling approaches have been used to demonstrate and quantify the impact of stratospheric perturbations on tropospheric predictability:

- Imposing perturbations to the stratospheric state through artificial nudging or damping to bring the stratospheric state closer to observations can produce model forecasts with substantially increased skill in the extratropical troposphere (Charlton et al., 2004; Scaife and Knight, 2008; Douville, 2009; Hansen et al., 2017; Jia et al., 2017), although not for all cases and models (e.g., Jung et al., 2011).
- Splitting a large set of hindcasts into groups initialized during strong, weak, and neutral stratospheric vortex conditions. S2S forecast skill of atmospheric circulation (including the NAO), surface temperature (particularly in eastern Canada and northern Russia), and North Atlantic precipitation is enhanced for both weak-vortex (Mukougawa et al., 2009; Sigmond et al., 2013) and strong-vortex (Tripathi et al., 2015) cases.

These studies show that S2S predictability associated with weak and strong stratospheric vortex conditions can be realized in dynamical forecast systems, with generally higher skill of surface climate predictions when forecasts are initialized during periods when the stratospheric state is significantly disturbed from its climatology.

4.3 S2S Extratropical Forecast Skill Associated With Stratosphere-Troposphere Pathways

As shown previously in Fig. 1, the stratospheric circulation is sensitive to a number of different processes in the Earth system. Certain relationships, or pathways, between the troposphere and the stratosphere persist for weeks, or even over the course of a season or longer, and can be exploited to improve probabilistic forecasts of surface variables. In the context of sub-seasonal predictability, these relationships can contribute to the overall likelihood of significant variability in the polar vortex. These relationships, and related studies examining associated forecast skill, are briefly reviewed here, arranged in order from shorter to longer timescales:

- *Blocking*: Tropospheric blocking exerts an influence on wave propagation into the stratosphere and can act as a precursor to SSW events (Quiroz, 1986) in terms of their spatial structure (Martius et al., 2009; Castanheira and Barriopedro, 2010), and also in terms of their characteristic anomalies in heat fluxes (Ayarzagüena et al., 2015; Colucci and Kelleher, 2015).
- *The Madden-Julian Oscillation (MJO)*: Both Garfinkel et al. (2012b) and Kang and Tziperman (2017) have demonstrated that the likelihood of SSW events increases during MJO events. Further, Garfinkel and Schwartz (2017) showed that there is a tight relationship between tropical convection in the West Pacific and polar stratospheric variability.
- *Snow cover and sea ice*: Extratropical surface conditions, such as snow cover and sea ice extent, can modulate the tropospheric wave field, and therefore, through promotion of a large-scale wave pattern that linearly interferes with the climatological Rossby wave field, can affect wave amplification and propagation into the stratosphere (Cohen and Entekhabi, 1999; Smith et al., 2010). Arctic sea ice variability also has been found to influence Northern Hemisphere polar vortex variability (Peings et al., 2013; Kim et al., 2014; Kretschmer et al., 2016), although the linkages seem to depend on the region of sea ice change (Sun et al., 2015; Screen, 2017a), with loss of sea ice in the Barents and Kara seas tied to a weakening of the polar vortex in late winter and spring (Kim et al., 2014; King et al., 2016; Yang et al., 2016). In some studies, these relationships have been tied to improved seasonal forecast skill (Cohen and Jones, 2011; Riddle et al., 2013; Orsolini et al., 2013, 2016; Kretschmer et al., 2016).
- *El Niño–Southern Oscillation (ENSO)*: On seasonal timescales, the ENSO tends to affect the midlatitudes through Rossby wave trains that propagate poleward on timescales of days to weeks (Hoskins and Karoly, 1981). El Niño events tend to strengthen the Aleutian low in the North Pacific (e.g., Barnston and Livezey, 1987), which in turn increases the wave flux into the Northern Hemisphere stratosphere through linear interference with the climatological stationary wave pattern (Garfinkel and Hartmann, 2008; Fletcher and Kushner, 2011; Smith and Kushner, 2012). The increased wave flux from the troposphere tends to weaken the Northern Hemisphere polar vortex and increase the probability of a negative NAO. This stratospheric pathway has been found to constitute a significant influence of ENSO on Eurasian climate (Ineson and Scaife, 2009; Bell et al., 2009; Cagnazzo and Manzini, 2009; Manzini, 2009; Li and Lau, 2013; Butler et al., 2015; Polvani et al., 2017). A few studies have indicated improved seasonal prediction skill for Eurasian climate during El Niño winters, when the stratospheric pathway is active (Domeisen et al., 2015; Butler et al., 2016).
- *The QBO*: The QBO exerts an influence on the polar vortex via the Holton-Tan effect (Baldwin et al., 2001), whereby the EQBO is typically associated with a weaker and more variable polar vortex in Northern Hemisphere winter, through its influence on planetary wave propagation (Holton and Tan, 1980; Naito et al., 2003). The communication between the tropical stratosphere and the polar vortex occurs through the altered characteristics of Rossby wave propagation in the subtropical stratosphere between the WQBO and EQBO (Garfinkel et al., 2012c; Anstey and Shepherd, 2014). Prediction skill

based on the phase of the QBO can be translated into an enhanced or reduced likelihood of polar stratospheric variability and coupling to the extratropical tropospheric jet. The QBO has been shown to enhance skill over the North Atlantic (Boer and Hamilton, 2008; Marshall and Scaife, 2009; Scaife et al., 2014a), although models appear to underestimate the magnitude of the effect apparent in observational or reanalysis data (Scaife et al., 2014b; Butler et al., 2016).

- *Decadal variability*: On decadal timescales, the 11-year solar cycle can affect the stratospheric temperature structure in the tropics (Crooks and Gray, 2005), and it has been proposed that this has subsequent effects on the stratospheric polar vortex (Bates, 1981; Kodera, 1995; Camp and Tung, 2007). In addition, tropical lower stratospheric temperature anomalies associated with the solar cycle may influence the tropospheric eddies and jet streams directly (Haigh et al., 2005).
- *The Pacific Decadal Oscillation (PDO)*: The PDO also may influence polar stratospheric variability (Woo et al., 2015; Kren et al., 2016) and any future large volcanic eruption likely would influence the stratospheric polar vortex for at least 1–2 years (Timmreck et al., 2016). Decadal changes in the polar vortex strength (Garfinkel et al., 2017) or position (Zhang et al., 2016) have been found to influence the extratropical tropospheric circulation; whether these changes are internally generated or forced via a tropospheric or surface driver is unclear (e.g., Kim et al., 2014; McCusker et al., 2016).

Nonlinear interactions among the various factors that can influence the polar vortex strength described here also may be important. For example, the QBO can affect the magnitude of the ENSO-stratosphere teleconnection (Richter et al., 2011, 2015), yielding a strengthening of the teleconnection during the QBO westerly phase (Calvo et al., 2009; Garfinkel and Hartmann, 2011a). The Indian Ocean Dipole (IOD) has been found to alter the teleconnection of ENSO into the stratosphere (Fletcher and Cassou, 2015). The QBO relationship to the extratropical polar stratosphere may be altered by the 11-year solar cycle (Labitzke and van Loon, 1992), although the relationship is less clear in some long climate simulations (Kren et al., 2014). The QBO also may modulate the NAO's response to snow forcing (Peings et al., 2013), and ENSO's influence on the stratosphere may be modulated by the solar cycle (Calvo and Marsh, 2011). Understanding these complex interactions may improve our ability to simulate these processes and ultimately improve extratropical predictive skill. Taken together, multiple forcings may provide windows of opportunity for forecasts in which sub-seasonal forecasts are and could be expected to be more skillful, but a great deal more work is required to clearly establish the dynamical basis for such periods.

5 SUMMARY AND OUTLOOK

The previous sections of this chapter have provided an overview of the wide-ranging scientific literature demonstrating stratosphere-troposphere dynamical coupling and its effects on S2S predictions. However, a number of outstanding research questions remain regarding the mechanisms underpinning stratosphere-troposphere coupling and its representation in numerical models, which we discuss next. In the sub-seasonal context, the ultimate aim of any research targeted at stratosphere-troposphere coupling should be to improve its representation in models so that it can be exploited to improve tropospheric predictability. The recent widespread availability of sub-seasonal prediction hindcast data sets presents a unique and unprecedented opportunity to study the predictability of the stratosphere-troposphere system.

5.1 What Determines How Well a Model Represents Stratosphere-Troposphere Coupling?

Broadly, a minimum requirement in order for numerical prediction systems to exploit the potential predictability associated with stratosphere-troposphere coupling is that they are capable of simulating the range of atmospheric and climate phenomena that induce the coupling described in Sections 2–4. In reality, there is likely to be a complex array of factors that determine how stratosphere-troposphere coupling and its impacts on tropospheric predictability are manifest in individual modeling systems.

5.1.1 Role of Model Lid Height and Vertical Resolution

On sub-seasonal timescales, adequately simulating the Northern Hemisphere polar vortex and its variability during winter and spring (including the planetary Rossby waves that drive this variability) is essential for reproducing the observed connections between the polar vortex and surface climate. Numerical models with an upper boundary below the stratopause consistently underestimate the frequency of SSWs compared to "high-top" stratosphere-resolving models (Marshall and Scaife, 2010; Maycock et al., 2011; Charlton-Perez et al., 2013), showing that model lid height may be an important limitation in some models. Biases in Northern Hemisphere vortex variability are also related to the ability of models to capture the relative occurrence of wave number-1 and wave number-2 type disturbances (Seviour et al., 2016). Low model lid height also has been connected to biases in the occurrence of extreme eddy heat flux events in models, which may have a causal influence on biases in the midlatitude tropospheric circulation (Shaw et al., 2014).

Coarse vertical resolution in the stratosphere also may affect a model's ability to simulate stratosphere-troposphere coupling, including the evolution of stratospheric wind anomalies during an SSW event and the spring breakup of the polar vortex in the Southern Hemisphere (Kuroda, 2008; Wilcox and Charlton-Perez, 2013). Realistic simulation of the QBO is also highly dependent on model lid height (Osprey et al., 2013) and model vertical resolution (Geller et al., 2016; Anstey et al., 2016). However, as yet, there is no theoretical basis for determining what is adequate vertical resolution, and this is likely to depend on several other factors, such as the representation of parameterized processes (Sigmond et al., 2008), and thus will vary from model to model.

5.1.2 Influence of the Tropospheric State and Biases

Simulating the low-frequency climate phenomena that are known to influence stratospheric variability (including ENSO, QBO, sea ice, and snow cover) is likely to be important for producing skillful probabilistic forecasts of polar vortex variability. However, simply simulating all relevant phenomena alone is unlikely to be sufficient because in many cases, the impact of stratosphere-troposphere coupling on predictability occurs via complex pathways, along which a model may fail to resolve key processes at multiple stages. Thus, the tropospheric state and biases in its representation in models could influence stratosphere-troposphere coupling through the following:

1. Modulation of drivers and precursors to the tropospheric processes forcing stratospheric variability
2. Modulation of the tropospheric response to stratospheric variability

For example, in relation to point 1, simulation of ENSO effects on the Northern Hemisphere polar vortex requires a realistic representation of the Rossby wave train response to anomalous tropical convection and the impact of tropospheric Rossby waves on stratospheric dynamics (Garfinkel et al., 2013). In this regard, some models simulate more linear ENSO teleconnections—particularly over the North Pacific, an important precursor region for stratospheric variability—than those observed (Garfinkel et al., 2012a).

In relation to point 2, a model's ability to simulate the tropospheric response to stratosphere-troposphere coupling also can be affected by systematic biases in the representation of the tropospheric circulation and its response to external forcing (e.g., Kidston and Gerber, 2010; Son et al., 2010). This may include the representation of tropospheric jet streams and storm tracks; tropospheric stationary and transient waves; feedback between eddies and the mean flow; the effects of parameterized processes (e.g., surface drag) on the tropospheric flow; and tropical circulation and convection. However, the representations of these factors may not be independent of the stratosphere itself (e.g., Shaw et al., 2014), posing the potential for complex, interdependent relationships to exist.

5.1.3 Influence of Different Drivers on Stratosphere-Troposphere Coupling Efficacy

As outlined in Sections 2–4, stratosphere-troposphere coupling is associated with a wide range of climate phenomena. At present, the equivalence of these different drivers for inducing stratosphere-troposphere coupling events is not well understood. Many of the drivers discussed in this chapter occur in tandem, and their combined effects may not be linearly additive. Thus, there is a need to study the combined effects of various phenomena on the coupled stratosphere-troposphere system.

Furthermore, there is a lack of quantitative understanding of the comparability of stratosphere-troposphere coupling induced by different phenomena. For example, is the coupling efficacy associated with a midwinter SSW comparable to that associated with Arctic springtime ozone depletion? Are the dynamical mechanisms underlying stratosphere-troposphere coupling in these two cases similar? How sensitive is the efficacy of stratosphere-troposphere coupling to the initial state of the troposphere and stratosphere; the type of stratospheric event, such as whether the vortex is displaced or split in two; and the amplitude of stratospheric anomalies (e.g., Son et al., 2010; Maycock and Hitchcock, 2015; Karpechko et al., 2017)?

Addressing these questions requires a set of quantitative dynamical metrics that can be applied consistently to study stratosphere-troposphere coupling related to different phenomena and its representation in models (Son et al., 2010; Shaw et al., 2014; Maycock and Hitchcock, 2015; Lubis et al., 2016b,a).

5.2 How Can We Use Sub-seasonal Prediction Data in New Ways to Study Stratospheric Dynamics and Stratosphere-Troposphere Coupling?

Following its renaissance in the early 2000s, the study of stratosphere-troposphere dynamical coupling has advanced rapidly, as described in Sections 2–4. Evidence for the importance of this coupling for extratropical climate in winter and spring is now clear, and a number of mechanisms for how this coupling works have been developed and refined. Much of the recent improvement in prediction on the S2S timescale is thought to have been related to the improvements that modeling centers have made in their representation of the stratosphere and its coupling to the troposphere.

However, challenges remain in arriving at a set of general unifying principles that can provide a quantitative description of the role of stratosphere-troposphere coupling on an event-by-event basis (Gerber and Polvani, 2009; Butler et al., 2017). It is, therefore, desirable to shed new light on the characteristics that determine the efficacy of stratosphere-troposphere coupling and its influence on weather and climate prediction. The large S2S hindcast data set (Vitart et al., 2017) offers a tremendous resource for pursuing such inquiries. At present, we see four main opportunities by which the study of stratosphere-troposphere coupling can benefit from the increased availability of high-quality sub-seasonal hindcast data sets:

1. Examination of the growth of model errors in the troposphere and stratosphere and their impact on coupling
2. Separating competing drivers of stratospheric variability and coupling and examining how these interact either linearly or nonlinearly
3. Determining what sets the efficacy of stratosphere-troposphere coupling on an event-by-event basis
4. Developing a probabilistic understanding of the likelihood of significant stratospheric variability

Improving our understanding in these areas may allow us to further exploit the enhancements in predictability that the stratosphere has to offer.

PART III

S2S MODELING AND FORECASTING

CHAPTER

12

Forecast System Design, Configuration, and Complexity

Yuhei Takaya

Meteorological Research Institute, Japan Meteorological Agency, Ibaraki, Japan

OUTLINE

1 INTRODUCTION

An operational forecast aims to provide future weather and climate forecast information to a wide range of users on a routine basis. For this purpose, the operational forecast system requires a high level of forecast quality, timeliness, and cost-efficiency under various competing resource requirements, where the forecast quality embraces accuracy (forecast skill) and benefit (usability). Therefore, it is vitally important for operational weather services to design and configure their forecast systems, including sub-seasonal forecast systems, to maximize the accuracy and benefit of the forecast information. The forecast system design is one of the important research and practical issues for the sub-seasonal forecast community (National Academies of Sciences, Engineering, and Medicine, 2016; World Meteorological Organization, 2013).

Sub-seasonal to Seasonal Prediction
https://doi.org/10.1016/B978-0-12-811714-9.00012-7

Operational ensemble forecast systems, which are utilized for predictions beyond a deterministic forecast limit (about 10 days), consist of several elements: real-time analysis and reanalysis, generation of perturbed initial conditions, integration of numerical models, a posteriori calibration, generation and dissemination of various products. This chapter mainly focuses on the first three elements. The analysis of real-time climate states is obviously essential to make a prediction with numerical models. This analysis is usually conducted using a data assimilation system, which synthesizes various observations and forecast model simulations to make the best estimate of climate states. The sub-seasonal forecast is produced based on ensemble forecasts, which are numerous realizations of forecasts to estimate the forecast uncertainty. To represent the uncertainty, ensemble forecasts may start from slightly perturbed ensemble initial conditions.

The sub-seasonal forecast is different from that of numerical weather prediction (NWP) models (for up to about 10 days) in many ways. Besides the use of the ensemble prediction approach mentioned here, S2S prediction systems often employ atmosphere-ocean coupled models. In such prediction systems, oceanic states are analyzed with an ocean data assimilation system. When a sea ice component is incorporated into a forecast system, sea ice conditions also may be analyzed in the assimilation system. In addition, sub-seasonal predictions require so-called reforecasts, as discussed later in this chapter, and carrying out the reforecasts requires reanalysis (analysis over the past long period). As briefly described here, sub-seasonal forecast systems are different from NWP systems in many aspects. The specifications of the sub-seasonal forecast system are elaborated on in this chapter.

To design and configure sub-seasonal forecast systems, several choices need to be made, including resolution of models, ensemble size (the number of ensemble members), forecast frequency, and reforecast period. The choice of these configurations demands careful considerations because each choice affects the forecast cost and quality. Under the constraint of fixed resources, these choices compete against one other. This is sometimes also described as the tradespace for allocating fixed computational resources. Despite their important impact on forecast quality, these aspects are rarely discussed in the research literature, and experience across the operational forecasting community is seldom shared. To address this gap, this chapter aims to cover a wide range of topics related to the design and configuration of sub-seasonal forecast systems.

However, as will be discussed later in this chapter, the practical decisions related to the forecast system design and configuration are subject to a grand design of a forecast service (and, ultimately, a strategy of model development) at each operational weather service, and currently not much is known about their impact on forecast performance. As such, it is hard to determine the optimal configuration. Therefore, this chapter will review the current practices at operational weather services and discuss the pros and cons of their approaches. The U.S. National Academies of Sciences, Engineering, and Medicine (2016) also gave a comprehensive review of the current status of sub-seasonal prediction and discussed strategies and recommendations, to which interested readers may refer for information on any topics not addressed in this chapter (e.g., the design of the multimodel ensemble system, observation, and assimilation).

Although poorly documented, the current sub-seasonal forecast systems have been built on knowledge and experience from the long history of operational weather and seasonal climate forecasting. After the European Centre for Medium-range Weather Forecasts (ECMWF) and the National Meteorological Center (NMC) introduced operational ensemble prediction

TABLE 1 Upgrades of JMA 1-Month Ensemble Prediction System

Version	Date	Resolution	Ensemble Size
GSM9603	March 1996	T63L30 (180 km)	10
GSM0103	March 2001	T106L40 (110 km)	26 (13 × 2, LAF)
GSM0603C	March 2006	TL159L40 (110 km)	50 (25 × 2, LAF)
GSM0803C	March 2008	TL159L60 (110 km)	50 (25 × 2, LAF)
GSM1304	March 2014	TL319L60 (55 km)	50 (25 × 2, LAF)
GSM1603E	March 2017	TL479L100 (40 km, day 0-18) TL319L100 (55 km, day 18-34)	52 (13 × 3 + 12, LAF)

Note: Only major upgrades are listed.

systems for the medium-range forecast in 1992 (Toth and Kalnay, 1993; Palmer et al., 1993), the Japan Meteorological Agency (JMA) was the first center to produce an operational sub-seasonal (1-month) forecast in March 1996 (Tokioka, 2000). During the more than 20 years of operations of sub-seasonal forecast systems at JMA, many changes have been made to the systems (Table 1). All these changes have justifications and to some degree have influenced the forecast quality and cost. This chapter also discusses the choices made to configure the JMA 1-month forecast systems as an example of operational practices, with the hope that the author's experience at JMA will offer a basis for further discussion.

2 REQUIREMENTS AND CONSTRAINTS OF THE OPERATIONAL SUB-SEASONAL FORECAST

Before digging into the design of sub-seasonal forecast systems, some background information on operational systems should be reviewed. Operational centers providing sub-seasonal forecasts with lead times of about 2 weeks to 1–2 months, which approximately coincides with the extended-range (10–30-day) weather forecast of the World Meteorological Organization (WMO) definition, usually also provide forecasts for the short and medium range (up to 10–14 days ahead of time), as well as long-range or seasonal forecasts for several months in advance. In such centers, the operational forecast system is part of a high-level design for providing seamless forecast information across all time ranges from minutes to seasons (World Meteorological Organization, 2015). As part of the operational enterprise, the sub-seasonal forecast system is designed to meet requirements and to fill the gap between medium-range (issued every day) and seasonal forecasts (issued typically every month). Some centers use a unified system for medium-range and sub-seasonal forecasts, while others use a unified system for sub-seasonal and seasonal (S2S) forecasts. The former design allows the use of systems with higher resolution and the provision of seamless and consistent forecast information between medium-range and sub-seasonal forecasts compared to the latter. The latter approach may be possible if a seasonal forecast system is operated in the Lagged Averaged Forecasting (LAF; Hoffman and Kalnay, 1983; also see Section 3) mode in a continuous way (i.e., every day), and one can produce sub-seasonal ensemble forecasts with relatively short lags before the initial time.

The design and configuration of an operational sub-seasonal forecast system depend on various important and obvious factors: specific requirements, priority, and resources in an operational weather service, as well as the inherent predictability of sub-seasonal climate variability of specific phenomena and regions, which determines the benefits. For instance, in countries exposed to the risks of high-impact weather events such as tropical cyclones and flash floods, their national weather services may place the highest priority on providing forecast information about these types of events, and invest more resources to predict them with greater accuracy a few days in advance. On the contrary, in countries that benefit from sub-seasonal forecasting because of certain predictability and predictive skill several weeks ahead of time, national weather services may invest more resources on sub-seasonal forecast systems. As this example shows, resources for sub-seasonal forecasting are allotted in accordance to the priority of each operational center and cannot be indiscriminately determined.

With this background, this chapter now will discuss the optimization and compromise in configuring the sub-seasonal forecast systems under the constraint of finite resources. The next section briefly reviews the design of ensemble forecasts, with a special focus on ensemble method because this design is a fundamental and key element of the forecast system configuration. Then configurations of reforecasts and real-time forecasts will be discussed in Sections 4 and 5.

3 EFFECT OF ENSEMBLE SIZE AND LAGGED ENSEMBLE

3.1 Effect of Ensemble Size

Forecasts with a lead time longer than typically a week should essentially be regarded as probabilistic forecasts due to a chaotic behavior of the climate system. This means that these forecasts need to represent numerous possible future scenarios for a given target period by applying an ensemble forecast approach.

For the later discussion, we examine the required ensemble size of reforecasts and real-time forecasts to meet practical requirements of operational forecasts. First, we investigate the relationship between the ensemble size and expected forecast skill scores. To this end, a simple simulation experiment was carried out to illustrate several essential features. Here, the forecast skill dependency on the ensemble size was assessed by carrying out an idealized simple simulation following the method used by Kumar (2009). In the simulations, the expected scores were estimated using 10,000 sets of Monte Carlo simulations of 50 independent samples (i.e., cases), which is the typical verification sample size for seasonal and sub-seasonal problems and is a reasonable choice. The samples were generated by a random number generator and the Box-Muller method by assuming that samples belong to a Gaussian distribution and have a signal-to-noise ratio (SNR) corresponding to a specified correlation skill. The variances in the simulation were configured following Tippett et al. (2010).

Fig. 1 illustrates the expected correlations of M-member ensemble mean predictions (c_M) for a given averaged single-member correlation (c_1) or equivalent SNR (Kumar, 2009; Tippett et al., 2010). The correlation coefficients of the ensemble mean increase with the ensemble size, but skill gains by increasing the ensemble size gradually level off for a sufficiently large ensemble size. Another important result is that the skill gains are largest for moderate

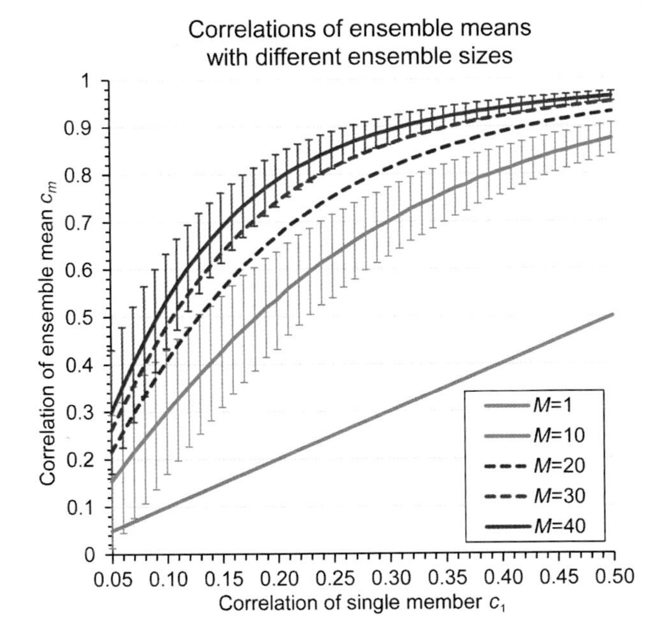

FIG. 1 Dependency of correlation coefficients on the ensemble size of M. The whiskers indicate the intervals of 15.9%–84.1%, corresponding to a 1-standard-deviation range of the Gaussian distribution. For simplicity, the whiskers are shown only for $M = 10$ and 40.

correlation skill ($0.1 < c_1 < 0.3$ or $0.3 < c_{10} < 0.7$). These results are consistent with Kumar and Hoerling (2000), and a theoretical relationship between ensemble size and correlations derived by Murphy (1988b) and Kharin et al. (2001). Because the typical correlations in sub-seasonal forecasts in the middle latitudes are often moderate, this result suggests that the skill of sub-seasonal forecasts benefits from a larger ensemble size. Fig. 2 displays the actual skill of a 10-member ensemble forecasts of 2-m temperature from reforecasts of the ECMWF forecast systems. The 10-member ensemble reforecasts display relatively high skill in the tropics and moderate skill in many parts of the middle latitudes for week 3, with a lead time of 14 days.

Several studies have been performed to estimate the required ensemble size for seasonal forecasts. Although few studies have dealt with this topic in the context of the sub-seasonal forecast, research on seasonal forecasts can provide helpful guidance for the forecast skill dependency on the ensemble size for the sub-seasonal forecast with a similar level of skill. However, because the forecast skill depends on a number of factors, including variables, regions, lead time, averaging period, seasons, and low-frequency variability such as El Niño–Southern Oscillation (ENSO) and Intraseasonal Oscillation (ISO) (e.g., Li and Robertson, 2015; Weigel et al., 2008), the required ensemble size for seasonal forecasts may not necessarily be the same as for sub-seasonal forecasts. Branković and Palmer (1997) analyzed 5-year atmospheric ensemble simulation and concluded that relatively large ensembles (i.e., about 20 members) are needed for the seasonal forecast of surface temperatures in the extratropics. Déqué (1997) evaluated the required ensemble size with multimodel seasonal forecast experiments and suggested an ensemble size of 20 for predicting midlatitude geopotential height, and 40 for temperatures over Europe. Kumar et al. (2001) investigated the relationship between ensemble size and ranked probability skill score and indicated that the ensemble size of 10–20 is sufficient to ensure an average skill level that is close to that of a large ensemble size

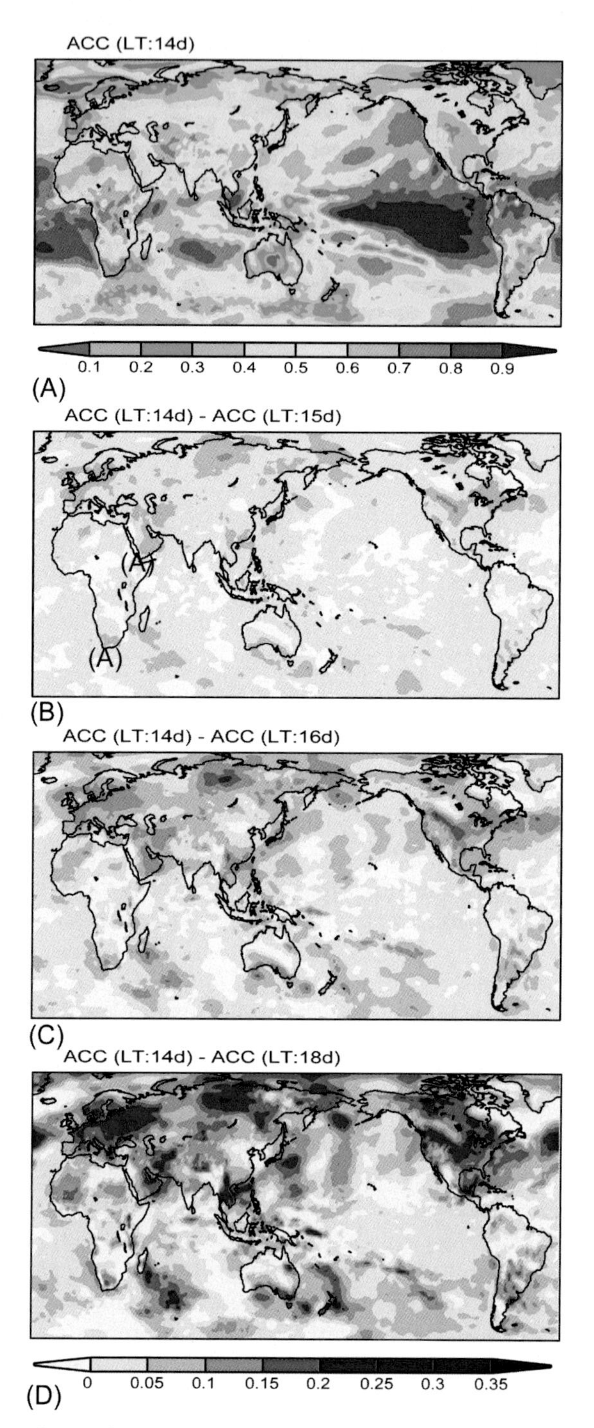

FIG. 2 The temporal anomaly correlation coefficient (ACC) of week-averaged 2-m temperatures between analyses and 10-member ensemble mean forecasts with a lead time of 14 days (A) and ACC difference between a lead time of 14 days and 15 days (B), 16 days (C), and 18 days (D). The 10-member (excluding the control member) means of reforecasts of ECMWF model (CY43R1) started from January 2, 9, 16, 23, and 30 were verified against ERA-Interim reanalysis during 1997–2016.

for SNRs of 0.4–0.5. All the studies cited here indicated that an ensemble size of 20-40 is acceptable for making operational seasonal forecasts and effectively enhancing predictive skills. For predictions with low inherent predictability and low actual skill, more ensemble members are required to extract the skillful forecast information (Fig. 1; Kumar, 2009; Kumar and Chen, 2015). From the results of early studies and Figs. 1 and 2, an ensemble size of roughly 20–40 seems to be reasonable and recommended for the purpose of operational forecasting. These results support the choice of the ensemble size in the current operational subseasonal forecasting systems (Table 2).

TABLE 2 Configurations of Sub-seasonal Prediction Systems Participating the S2S Project (as of May 2017)

	Real-Time Forecast					Reforecast		
Center	**Resolution**	**LAF/ Burst**	**Ensemble Size**	**Start Days**	**Frequency**	**Period**	**Ensemble Size**	**Fixed/on the Fly**
BoM	T47L17	Burst	3 × 11	Sun. Thu.	1, 6, 11, 16, 21, and 26 of every month	1981–2013	3 × 11	On the fly
CMA	T106L40	LAF	4/day	Every day	Every day	1994–2014	4	Fixed
CNR-ISAC	0.75 × 0.56L54	Burst	41	Thu.	Every 5 days	1981–2010	1	Fixed
CNRM	T255L91	Burst	51	Thu.	1, 15 of every month	1993–2014	15	Fixed
ECCC	0.45 × 0.45L40	Burst	21	Thu.	Every Thu.	1995–2014	4	On the fly
ECMWF	Tco639/ 319L91	Burst	51	Mon. Thu.	Every Mon. Thu.	Last 20 years	11	On the fly
HMCR	1.1 × 1.4L28	Burst	20	Wed.	Every Wed.	1985–2010	10	On the fly
JMA	T479/319L100	LAF	13/ 6 hours	Tue. Wed.	10, 20, and the last	1981–2010	5	Fixed
KMA	N216L85	LAF	4/day	Every day	1, 9, 17, 25 of every month	1991–2010	3	On the fly
NCEP	T126L64	LAF	16/day	Every day	Every day 1, 9, 17, 25 of every month	1999–2010	4	Fixed
UKMO	N216L85	LAF	4/day	Every day		1993–2015	7	On the fly

Abbreviations: BoM, Bureau of Meteorology; CMA, China Meteorological Administration; CNR-ISAC, Institute of Atmospheric Sciences and Climate; CNRM, Centre National de Recherches Météorologiques; ECCC, Environment and Climate Change Canada; HMCR, Hydrometeorological Centre of Russia; KMA, Korea Meteorological Administration; UKMO, Met Office.

3.2 Uncertainty of Skill Estimate

We will now look at the uncertainty of the estimated ensemble correlation by Monte Carlo simulations. In Fig. 1, the whiskers indicate the uncertainty range corresponding to one standard deviation in the Monte Carlo simulations. The uncertainty of the skill estimate decreases as the correlation and ensemble size increase, which is consistent with results of Kumar (2009). The graph shows that the uncertainty is relatively large for the 10-member ensemble, which is a typical ensemble size in the operational reforecast configuration (5–15), compared with that of the 40-member ensemble. This result shows the limitations of verifying sub-seasonal reforecasts and also indicates that the skill estimated in the verification of reforecasts is often expected to be significantly lower than the skill of real-time forecasts, which have a much larger ensemble size than reforecasts. The skill dependency on the ensemble size suggests that care is to be taken in skill comparison among reforecast sets from systems with different ensemble sizes. As seen here, the small ensemble size limits the evaluation of the forecast performance with reforecasts. However, some scores have been proposed to alleviate the dependency of scores on the ensemble size (e.g., Müller et al., 2005; Weigel et al., 2007a). Moreover, the statistical significance of the skill's estimate changes when using predictions from multiple initial dates because the autocorrelation affects the statistical tests (Geer, 2016), and this is also the case for the sub-seasonal prediction. For all these reasons, the current configurations of the sub-seasonal reforecasts at operational centers may not be ideal for accurate skill assessment.

The same approach could be easily adapted for evaluating other metrics than correlations (cf. Kumar, 2009). For example, probabilistic verification measures such as the Brier Skill Score (BSS) and the Ranked Probability Skill Score (RPSS) are also determined by the SNR (cf. Kumar, 2009). However, the gains in the skill scores may differ for various scores (see, e.g., Murphy, 1988b; Kumar, 2009; Richardson, 2001; Ferro, 2007). One of the advantages of the Monte Carlo approach applied here is that the uncertainty estimates are obtained along with the score estimates. On the other hand, a weakness of the idealized simulation is the assumption of the stationary potential predictability for all samples. In reality, this is not the case because the inherent potential predictability is state dependent. Because the predictability depends on the climate conditions such as ENSO and Madden-Julian Oscillation (MJO; Li and Robertson, 2015; Weigel et al., 2008; Johnson et al., 2014), an insufficiently short reforecast length (i.e., small number of reforecast years) inhibits accurate estimates of a state-dependent skill (also referred to as a *conditional skill*). In NWP verification, usually a sample size (i.e., number of forecast cases) is sufficiently large to assess the forecast performance accurately, even though it is also subject to autocorrelation (Geer, 2016). Bear in mind that the state dependency of the predictive skill is inevitable in the verification of operational prediction systems from sub-seasonal to seasonal (S2S) time ranges because ENSO events happen once over several years. As discussed here, several inherent difficulties of sub-seasonal prediction challenge the verification of the sub-seasonal prediction.

3.3 Effect of LAF Ensemble

In this section, we discuss the next issue in the design of sub-seasonal forecast systems: namely, the approaches for generating ensembles. Here, we compare two ensemble

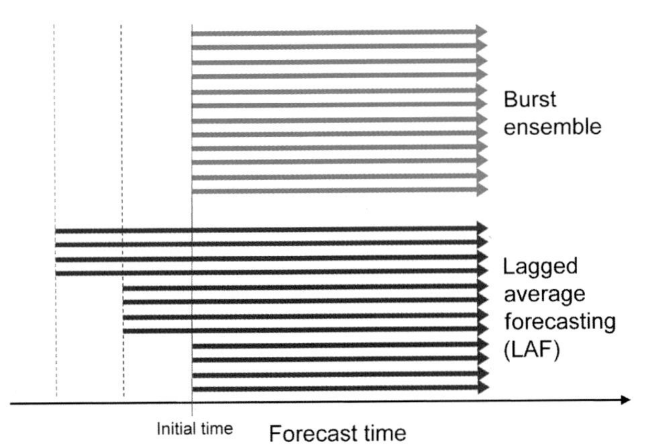

FIG. 3 Schematic diagram of the ensemble configurations of the burst and LAF ensembles.

generation approaches currently adopted at major operational centers (Fig. 3). The first approach is often referred to as a *burst ensemble,* in which the members of a large ensemble (e.g., 50 members) are initialized from the same initial time (Vitart et al., 2017). The second approach is LAF (Hoffman and Kalnay, 1983), in which a smaller set of ensemble integrations is generally started from consecutive initial times with a certain time interval (e.g., 4 members every 6 hours over a period of 2 days preceding the start of the forecast). The burst ensemble approach would ideally provide the best forecast skill because there is no lag before the initial time, but it requires more computer resources at one time than does the LAF. On the other hand, the LAF ensemble system divides the computation of required ensemble forecasts into several chunks and distributes and balances computer loads; therefore, it is more suitable when computer resources at a given time are limited. Another possible advantage of the LAF approach is that it can better sample the rapidly changing initial conditions, as well as their influence on forecasts. Therefore, these merits and disadvantages should be quantitatively assessed in some way to optimize the forecast system, and these are discussed in this section.

The LAF is one of the most widely used ensemble generation methods in operational S2S forecasts (Table 2; Chen et al., 2013; Takaya et al., 2017; Arribas et al., 2011). It uses ensemble forecasts starting from multiple lagged initial times to compose a full ensemble. In the LAF ensemble, forecasts starting from old initial conditions should have slightly degraded skill compared with those starting from the latest initial conditions (Figs. 2 and 3). Thus, the LAF method has two opposite effects on forecast skill: Adding more ensemble members contributes to increasing the forecast quality while a longer lead time degrades the forecast's quality (Chen et al., 2013). Therefore, finding the optimal configuration is crucial when using the LAF method.

Now, consider the optimal configuration of the LAF ensemble. For simplicity, suppose that an ensemble forecast is made of forecasts starting from two consecutive initial times. To quantify the effect of the LAF ensemble, the ensemble size of the burst ensemble, which displays the same forecast skill as the LAF ensemble, has been assessed. Fig. 4 shows the equivalent

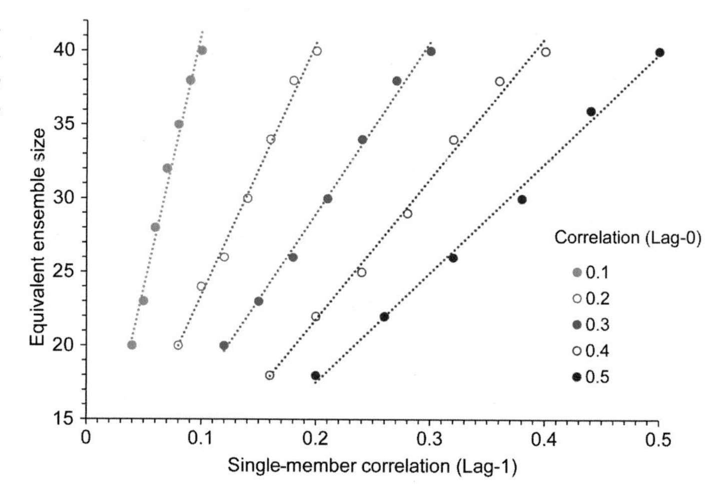

FIG. 4 Equivalent sizes of LAF ensembles to corresponding burst ensembles when two sets of 20-member ensembles with different skills compose one large (40-member) ensemble.

ensemble sizes for various combinations of the ensembles. The estimated scores with the LAF method were obtained by combining the two groups of ensembles (20 members each) with the forecast skill of one group slightly degraded to simulate a LAF ensemble. In this experiment, if the two groups have the same skill, the equivalent ensemble size is 40. However, combining two 20-member ensembles with $c_1 = 0.2$ and 0.3 (approximately $c_{10} \sim 0.5$ and 0.7) leads to a 40-member LAF ensemble with the same skill as a 30-member burst ensemble (Fig. 4).

In sub-seasonal forecasts, we are interested in forecast properties averaged over a specific time period (e.g., 1 week) because this time-averaging produces more skillful forecasts due to reduced uncertainty (Leith, 1973). The skill scores of the averaged sub-seasonal forecast decline more slowly than the short-range forecast based on instantaneous values or averages over a short period (typically 1 day). Because of the slow decline of the skill as a function of lead time, the lagged ensemble approach works reasonably well for S2S forecasts. As expected, a longer LAF period, which combines less skillful lagged forecasts, produces less skillful forecasts (Fig. 2). However, the results (Figs. 1, 2, and 4) suggest that the LAF ensembles with a lag of 1–2 days improve extended-range forecast skill. This result also implies that care should be taken to design a multimodel ensemble (National Academies of Sciences, Engineering, and Medicine, 2016) because multimodel systems are prone to deterioration due to the difference in initial time if they are not well coordinated and have start dates of more than a few days apart.

It should be noted that the simulation presented here neglects the error covariance in the LAF ensemble and therefore indicates the maximum potential benefit of the LAF ensemble. The effect of the LAF ensemble has been evaluated using operational forecast data by Chen et al. (2013), who pointed out that the most effective time range of the LAF method varies depending on the forecasted properties, regions, and seasons. More recently, Trenary et al. (2017) proposed a novel method for determining the optimal lagged ensemble by taking into account the error covariance of the lagged ensemble. These results validate the LAF ensemble methodology and demonstrate its benefits.

4 REAL-TIME FORECAST CONFIGURATION

In this section, the configurations of the real-time forecast (in particular, ensemble size, forecast frequency, ensemble generation, and initialization) are discussed.

Real-time forecast products are generally delivered from ensemble forecasts with large ensemble sizes. In general, the required ensemble size depends on the forecast skill of the variables, which is related to the inherent (potential) predictability, and the type of forecast (ensemble means; tercile, quintile, or decile probabilities; or full probabilistic distributions). A larger ensemble size is required to produce more detailed probability information and detect risks of extreme (rare) events. It also depends on the potential predictability or predictive skill. A larger ensemble size will be needed to improve the forecast skill if the predictable signal (SNR) is relatively small, as discussed in the previous section, although the skill remains low because of the low SNR (Kumar and Chen, 2015). Based on these considerations, the current operational sub-seasonal forecasts are carried out with 15–50 ensemble members (Table 2).

Another important aspect of the forecast system design is resolution versus ensemble size. According to previous studies and experience at operational centers, the sub-seasonal forecast skill increases with resolution until it saturates for relatively fine resolution. Increasing the resolution requires a good deal of computational resources; doubling the horizontal resolution of the atmospheric model (which halves the grid spacing) may increase the computational cost to run the model by a factor of 8. Therefore, it may be more beneficial to increase relatively small ensembles (10–15 members) first, rather than increasing the resolution. If resources are limited, a choice must be made. In the early phase of the JMA 1-month forecasting, computational resources were preferentially allotted to increase the ensemble size rather than the model resolution (Table 1) because a system with larger ensemble size can produce more skillful and reliable probabilistic forecast information for general use.

Another approach to maximize the benefits of resolution when computing resources are limited is to adopt the variable resolution approach, where a higher-resolution model is employed for the first several integrations, and a lower-resolution model for the rest of the integrations (Tracton and Kalnay, 1993; Szunyogh and Toth, 2002; Buizza et al., 2007a). It offers a more efficient way to optimize the forecast skill with finite resources (Buizza, 2010) and is adopted in the National Centers for Environmental Prediction (NCEP) Global Ensemble Forecast System (GEFS) and ECMWF ensemble forecast systems. This approach also has been introduced at JMA in 2017 (Table 1)

The ensemble configuration is an essential component for the operational sub-seasonal forecast system. As reviewed in Section 3, Chen et al. (2013) and Trenary et al. (2017) specifically discussed the effect and optimal configuration of the LAF ensembles. Recently, JMA changed the frequency of the lagged ensemble in the 1-month forecast system from the LAF ensemble with two 25-member ensembles with a 1-day interval to that with three 13-member ensembles and one 12-member ensemble with a 6-hour interval. This more frequent LAF ensemble configuration slightly improved the predictive skill in a feasible operational schedule (not shown).

The forecast frequency and schedule still may be contentious in current operational sub-seasonal forecasts. In contrast to the deterministic short-range forecasts, which are usually run from the same initial time (00 UTC, 06 UTC, 12 UTC, 18 UTC) at all operational centers, the S2S forecasts are run with varying schedules. Many centers issue the sub-seasonal forecast once or twice a week to fill the gap of forecast periods between extended-range (several times a day) and seasonal forecasts (usually once a month). The majority of centers participating to the S2S project start their forecasts on Thursday and issue the forecast to the public on Thursday or Friday. Meanwhile, in 2014, JMA changed the forecast issue time from Friday afternoon to Thursday afternoon and accordingly changed the LAF initial days from Wednesday and Thursday to Tuesday and Wednesday. The change was intended to improve the practical usability of these sub-seasonal forecasts. If an operational center issues its official sub-seasonal forecasts on Thursday afternoon, users have enough time to make better use of the sub-seasonal forecasts for their decision-making for the next calendar week. This change in schedule may seem like a trivial matter, but it actually has a large impact on the usability and benefit of the operational forecasts.

Recently, a few centers have increased the release frequency of their sub-seasonal forecasts. For instance, JMA introduced 2-week forecasts in 2008 (see the description of this system in Chapter 13) to supplement the 1-month forecast once a week to support early warnings of extreme weather events, and ECMWF started to issue the sub-seasonal forecasts twice a week in 2011 instead of once a week. Some centers adopting the LAF approach (e.g., NCEP) also provide more frequent sub-seasonal forecast products. The increase of the forecast frequency also contributes to improving the usability of forecast information. Therefore, the operational schedule is an important element of the forecast system design. The coordination of the forecast schedule among the various operational centers is also important for multimodel ensemble forecasts. Currently, operational multimodel ensembles for seasonal forecasting are coordinated and produced in various organizations, such as the WMO Lead Centre for Long-Range Forecasts Multimodel Ensemble (LC-LRFMME; Graham et al., 2011) and EUROpean Seasonal to Inter-annual Prediction (EUROSIP). Similar multimodel ensemble systems also may be beneficial for sub-seasonal forecasting. Because large differences in initial forecast dates (i.e., the schedule) among the contributing centers would make sub-seasonal multimodel ensemble systems difficult to set up, as discussed in Section 3 earlier in this chapter; coordination across operational centers in scheduling the sub-seasonal forecasts is crucial for climate services that issue multimodel products.

The ensemble generation and initialization are other important topics in the real-time forecast system design. Initial conditions for the ensemble are often slightly perturbed to represent uncertainty in them. Several techniques have been proposed to generate perturbed initial conditions—namely, a singular vector (Buizza et al., 1993), breeding vector (breeding method; Toth and Kalnay, 1997), LAF, Ensemble Transform Kalman Filter (Evensen, 1994; Anderson, 2001; Wang and Bishop, 2003), and Ensemble Transform (Wei et al., 2006) (see Chapter 13 for more details). Furthermore, initial perturbation schemes have been designed specifically for sub-seasonal coupled model forecasts at the Australian Bureau of Meteorology (Hudson et al., 2013).

Some studies have compared forecast performance using various ensemble perturbations (e.g., Houtekamer and Derome, 1995; Hamill et al., 2000; Magnusson et al., 2008; Wang et al., 2007). However, only a few studies have focused on the sub-seasonal forecast. It is still unclear which method is preferable for sub-seasonal forecasting and what their influence is at this time scale. In addition, operational centers tend to combine multiple approaches to make a desired number of ensemble members. For example, JMA concurrently use the LAF and singular vectors, which makes such comparative studies more difficult. An in-depth comparison is beyond the scope of this article, but it is also worth noting that in practice, the choice of the ensemble generation technique affects not only the forecast quality, but also the design of the real-time forecast and the reforecast. For example, the singular vectors require instantaneous initial conditions only, while the breeding vector requires continuous cycles to generate perturbations and suits the forecast systems adopting the LAF approach. Furthermore, a recent trend in development of the ensemble technique is to combine a flow-dependent assimilation system and ensemble forecast system like the Ensemble Transform Kalman Filter (Evensen, 2003) or hybrid ensemble 4D-Var (Bonavita et al., 2016). This means that the initial ensemble perturbations are provided through a data assimilation process. Therefore, the strategy for initial perturbation generation is strongly linked to the general design of the forecast system. More work is needed to develop and implement improved initial perturbation schemes for the sub-seasonal forecasts.

5 REFORECAST CONFIGURATION

The reforecast, which is also known as a *hindcast*, is a retrospective forecast with a fixed forecast system for a long period in the past (Hamill et al., 2006). The reforecast has two main purposes. The first is to estimate the model climatology for some a posteriori forecast bias correction and calibration. In general, numerical models depart (drift) from the observed climate at a long lead time due to imperfect model physics and insufficient resolution. Therefore, sub-seasonal forecasts require bias correction and calibration using outputs of the reforecasts to enhance forecast skills. This forecast postprocessing is another important component of the forecast system design (discussed in Chapter 16). The second purpose is to assess the forecast performance expected in the real-time prediction to support the use of real-time forecast products (discussed in Chapter 17). Reforecast configurations include the reforecast length (number of years), reforecast ensemble size (which is generally smaller than the ensemble size of the real-time forecasts), frequency, initialization method, and other elements. The rest of this section will assess how the reforecast configurations influence the quality of the real-time predictions and reforecast verifications.

We begin with a discussion on the frequency and length of reforecasts. The sub-seasonal variability is modulated by relatively large interannual and intraseasonal variability. Moreover, the observed climatology of sub-seasonal variation sometimes exhibits rapid seasonal changes, as seen during abrupt monsoon onsets. This suggests that the observed and model climatologies, which are estimated with long-term analyses and reforecasts, require accurate calculations. The Commission of Climatology (CCl) of the WMO recommends defining the climatological normals (climatology) from at least 30 years of data (Arguez and Vose, 2011). The WMO Manual on the Global Data-Processing and Forecasting System (GDPFS) recommends more than 15 years for the seasonal (long-range) forecast (WMO, 2010).

The minimum required period to define the climatology also may vary between regions and seasons. In practice, the reforecast periods chosen by the operational centers are the most recent consecutive periods of more than 15 years, or fixed 30 years following the CCl recommendation. The frequency of the reforecasts is another important factor to define the climatology. Some centers have reforecasts produced on a weekly basis, while others have reforecasts starting on specific days of the month. For instance, JMA currently conducts its reforecasts over more than 30 years to cover most recent climatological normal period (1981–2010), with initial dates roughly 10 days apart. Other centers may have longer or shorter intervals (Table 2).

The ensemble size is another important element of the reforecast configuration. A larger ensemble size allows for a better estimation of the model climatology (not only climatological means, but also probabilistic distributions), and a better assessment of the performance of forecast systems. A reasonably accurate model climatology can be obtained with a relatively small ensemble size if properties such as means and tercile thresholds are considered. However, we need to bear in mind that the climatology computed from the analysis, which has a single realization, contains larger uncertainty. The accurate verification requires a large ensemble size, as described in Section 3, especially in the low-skill regions/elements. As seen in Table 2, the choices of the reforecast ensemble size at operational centers differ from one another, presumably reflecting the priority of each center. It is also noted that the diversity of the reforecast configuration makes it hard to compare the predictive skills of operational centers and to standardize operational sub-seasonal forecast systems among operational centers.

As discussed previously, there is a trade-off between the configurations of the ensemble size, reforecast frequency, and reforecast length. A large ensemble size allows for more accurately verifying the skill of the real-time system and assessing probabilistic forecast skill (Weigel et al., 2008). Higher frequency and longer reforecast length are beneficial for obtaining more accurate climatologies, and longer training and verification samples for forecast postprocessing. With finite resources, operational centers have to decide the configuration that fits best with their priorities. Some tests done at JMA showed that the longer reforecast period and longer and more frequent training sample is beneficial for forecast postprocessing; thus, the center decided to carry out more than 30 years of reforecasts for operational systems. The best configurations may depend on the postprocessing training.

Apart from the specifications described here, the way to execute reforecasts is another important aspect in the operational reforecast configuration. There are two types of execution adopted at operational centers: namely, fixed reforecasts and reforecasts that are done on the fly (spontaneously; Table 2). In the fixed-reforecast configuration, a set of reforecasts is produced before the implementation of a new system. This enables model developers to conduct holistic evaluation of forecast performance, and users to understand characteristics of the new forecast system and to better calibrate their own application models by using a full set of reforecasts. In the on-the-fly reforecast configuration, the reforecasts are carried out shortly before or simultaneously with calculation of the real-time forecasts. This approach is favorable for frequent model upgrades and expedites model development and implementation (Arribas et al., 2011). NWP systems can be replaced in a relatively short time because they do not require reforecasts, but usually a parallel suite test is carried out for a few months to years.

Keeping the consistency of initial conditions in reforecasts and real-time forecasts is also a challenge for the operational sub-seasonal forecast because sub-seasonal forecast systems require long-term analyses of various model components for initialization. This situation is not the case for NWP systems, which require real-time analyses only. In operational sub-seasonal forecast systems, the atmospheric initial conditions are usually obtained from existing atmospheric reanalysis, which may use an old forecast model or a model from another operational center, except for the NCEP system (Saha et al., 2014), whose reanalysis component uses a consistent model as that of real-time seasonal and sub-seasonal forecasts. Because Earth-system components such as land, ocean, and sea ice potentially add to the forecast skill of the sub-seasonal forecast (Koster et al., 2010a), the consistent initial conditions of these components are crucial for operational sub-seasonal forecasts. Inconsistent initial conditions may have potentially detrimental effects (e.g., bias errors) on the real-time forecast products (Kumar et al., 2012). Therefore, care should be taken to ensure good consistency of the initial conditions between reforecast and real-time forecast configurations.

6 SUMMARY AND CONCLUDING REMARKS

This chapter reviewed the design and configuration of sub-seasonal forecast systems. In contrast to the weather forecast systems, sub-seasonal forecast systems have a large degree of freedom and complexity in terms of system configuration. As discussed, the configuration of real-time forecast and reforecast configurations at operational centers are diverse (Table 2), presumably reflecting the strategy and priority at each modeling center. This also implies that the consensus for the optimal forecast configuration and design has not been reached yet. Currently, the configurations in operational centers are very different, which makes coordination in exchanging the forecasts across the operational centers for the development of climate services complicated.

It would be fair to say that the design of the operational sub-seasonal forecast system is still in an early stage, and there is much room for improvement. There are still many open questions regarding the optimal forecast system design for sub-seasonal prediction. In particular, topics like (1) benefits and deficiencies of burst and LAF ensemble approaches for operational forecasts; (2) optimal configurations of real-time forecasts, such as frequency and ensemble size; (3) optimal configurations of reforecasts in terms of ensemble size, length, and frequency; and (4) techniques of ensemble generation and data assimilation are all emergent and practical matters to improve the operational sub-seasonal forecasting. Furthermore, the system design may require additional consideration as the system complexity increases (with use of Earth-system models). For instance, consistent initialization of submodels (e.g., ocean, ocean wave, sea ice, and land) for real-time forecasts and reforecasts is likely to benefit sub-seasonal forecasts with Earth-system models. Future developments for the seamless forecasts with Earth-system models will include more advanced analysis and initialization with coupled models (Brassington et al., 2015).

Forecast design and configuration will remain fundamental issues for operational forecast centers, no matter how much the forecast system evolves. Hopefully, this chapter will spawn a discussion on the practical issues relevant to forecast system design and configuration and contribute to the improvement of operational sub-seasonal systems in the future.

Acknowledgments

The author would like to thank Dr. Arun Kumar, at CPC/NOAA, for his valuable comments and suggestions; S. Hirahara and T. Kanehama at CPD/JMA and Y. Takatsuki and M. Sugi at MRI; and two editors, Drs. Frederic Vitart and Andrew Robertson, for their help to improve the manuscript. This work was supported by JSPS KAKENHI Grant Numbers JP17K14395 and JP17K01223.

CHAPTER

13

Ensemble Generation: The TIGGE and S2S Ensembles

Roberto Buizza

Scuola Superiore Sant'Anna, Pisa, Italy and ECMWF, Reading, United Kingdom

OUTLINE

1 GLOBAL SUB-SEASONAL AND SEASONAL PREDICTION IS AN INITIAL VALUE PROBLEM

Before discussing sub-seasonal and seasonal (S2S) prediction, and how ensembles can help to provide reliable and accurate forecasts, let us first define the meaning of a few key terms that are going to be used in this chapter.

https://doi.org/10.1016/B978-0-12-811714-9.00013-9

We are talking about *S2S forecasts of variables that matter for human lives and activities*. These include atmospheric variables for the lower part of the troposphere, where humans live and most human activities occur (temperature, wind, precipitation, clouds, sea-level pressure); land surface variables (temperature, water cycle, soil humidity concentration); and ocean variables (waves and currents, temperature and salinity, carbon cycle). Concerning the time range, by *sub-seasonal*, we mean forecasts valid for between 2 weeks and 2 months, and by *seasonal*, we mean forecasts valid for between 3 and 24 months.

By the term *Earth-system model*, we mean a model that simulates the processes of the atmosphere, land, ocean, and sea ice components that are relevant for S2S prediction. *Relevant* here means that removing or adding these processes to the model has a detectable impact on the quality of the forecasts, or that these processes are necessary to generate forecasts for certain variables (e.g., if you want to predict how the sea state varies in the future, you have to include the simulation of the sea state in your model). To date (see, e.g., Vitart et al., 2014; Molteni et al., 2011; Arribas et al., 2011; Saha et al., 2014), S2S predictions (say, forecasts up to 1 year ahead) are produced using Earth-system models that include the atmosphere (up to a height of about 80 km), the land surface, the ocean waves and dynamical currents, and the sea ice.

The generation of S2S prediction involves solving a *global* (as compared to regional in space) problem. On average, signals and errors propagate throughout the globe following the atmospheric flow, with velocities that vary depending on the flow of the day. On average, errors and signals move between 15° and 30° of longitude every day. This zonal propagation speed was confirmed by the targeted-observation work of Buizza et al. (2007) and Kelly et al. (2007), who investigated the propagation of initial condition errors in the Northern Hemisphere ocean basins. They showed that removing observations in the Northern Pacific (Atlantic) Ocean increases that forecast error over the downstream landmass of North America (Europe) after about 2 days. They also showed that removing observations over the Northern Pacific can have an impact on the prediction of the atmospheric flow over the Atlantic after about 5–7 days. In other words, information coming from observations can propagate throughout the globe in about 2–3 weeks. This means that predictions on a monthly to seasonal timescale have to be global, if we want to extract predictable signals coming from a proper and correct initialization of the relevant Earth systems' components.

We are talking about predictions made by solving the *dynamical equations* that describe the time evolution of the Earth-system components that are relevant for the predictions themselves. The dynamical equations are deduced from Newton's law of physics, written for a fluid on a rotating sphere, adding terms that simulate the impact of physical processes (e.g., convection) on the tendencies of the Earth-system state vector.

S2S prediction is an *initial value problem* because its solution (i.e., the time integration of the dynamical equations that describe the system evolution) depends on the initial state of the system. This is why we have to compute a reliable and accurate estimate of the initial state of the system in order to generate a forecast.

It is worth bearing in mind the meanings of four additional terms:

- *A numerical forecast* is a prediction generated by solving numerically the physics equations that describe the evolution of the Earth system, or of some of its components.
- *A single forecast* is one forecast that describes the possible future state of a system.

- *An ensemble of forecasts* is a group of *N* forecasts (also called *ensemble members*) that describe the possible distribution of future states of a system.
- *A probabilistic forecast* is a forecast expressed in terms of probability (e.g., computed by counting the number of ensemble members predicting a specific phenomenon).

It is also worth clarifying that in terms of forecast ranges, these are the meanings of the terms that we are going to use throughout this chapter [see the definitions by the World Meteorological Organization (WMO): http://www.wmo.int/pages/prog/www/DPS/GDPS-Supplement5-AppI-4.html]:

- *Short-range forecast*: A prediction that is valid for a few days (say, up to 2–3 days)
- *Medium-range forecast*: A prediction that is valid for up to 2 weeks
- *Extended-range forecast*: A prediction that is valid for a period of 10–30 days
- *Long-range*: A prediction valid for a period longer than 30 days
- *Monthly and sub-seasonal forecast*: A prediction that is valid for up to 1–2 months
- *Seasonal forecast*: A prediction that is valid for a period longer than 2 months

In our discussion, by *weather*, we mean the state of the Earth system at a specific point and time, and/or averaged over a three-dimensional (3D) volume and over a short time period (say, up to a few days), expressed in terms of the state variables. By contrast, with *climate* we mean the state of the Earth system averaged over a large, 3D volume (say, with the size of a few hundreds of kilometers) and over a long time period (say, longer than a few days).

2 ENSEMBLES PROVIDE MORE COMPLETE AND VALUABLE INFORMATION THAN SINGLE STATES

Since the early 1990s, when we have started using ensembles in operational weather prediction, forecasters and other users have been able to estimate the probability distribution function (PDF) of forecast states rather than just one possible outcome, as opposed to what was happening in the 1970s–1980s, when only single forecasts were available. This means that forecasters and users can compute the probability of occurrence of any event of interest, such as events linked with maximum acceptable losses.

This paradigm shift from a single forecast to an ensemble of forecasts started in 1992, with the implementation of global ensemble predictions systems at the European Centre for Medium-range Weather Forecasts (ECMWF; Buizza and Palmer, 1995; Molteni et al., 1996) and at the National Centers for Environmental Predictions (NCEP, Tracton and Kalnay, 1993; Toth and Kalnay, 1993). These centers were followed, in 1996, by the Meteorological Service of Canada (MSC, Houtekamer et al., 1996), and in the subsequent years by many others (see Section 5 of this chapter). Ensembles are more valuable than single forecasts (see, e.g., Buizza, 2008), provided that they offer reliable and accurate probabilistic forecasts.

In the next two subsections, we will introduce the concepts of reliability and accuracy and discuss how they can be measured (see also Chapter 16). Then we will show that an ensemble of forecasts (e.g., in the form of an ensemble-based probabilistic forecast) is more valuable than a single forecast.

2.1 Reliability and Accuracy of an Ensemble

An ensemble of N forecasts can be used to compute the probability of occurrence of any event of interest. The simplest way to compute such a probability is to count how many ensemble members predict that the event will occur, as well as to define $p = n/N$. More sophisticated ways to compute such a probability would involve giving different weights to the ensemble members (e.g., if the ensemble includes members with different start dates, and younger forecasts are on average more accurate than older ones, a relatively higher weight could be given to the younger forecasts).

A probabilistic forecast p that a certain event will occur (e.g., the 2-m temperature anomaly at a specific point) is reliable if, on average, the predicted event will occur with an o frequency. A way to measure the reliability of an ensemble is to construct so-called reliability diagrams, which contrast the average forecast probability p against the average observed frequency [i.e., it shows the position of the points $(p;o)$] over many cases, taking into account all grid points within an area. If an ensemble is reliable, the points $(p;o)$ lie on the diagonal, and the Brier Score (BS), which measures the distance of these points from the diagonal (Brier, 1950; Wilks, 2005), is zero. The corresponding skill score, the Brier Skill Score (BSS), which is defined as the relative difference between the BS of the forecasts compared with a reference (e.g., a statistical forecast based on climatology), would be 1 for a perfect forecast and 0 for a forecast that does not provide any more valuable information than the reference.

Fig. 1 shows four examples of reliability diagrams for the ECMWF operational (May 2017) seasonal System-4 (S4; see Molteni et al., 2011) forecasts, starting on November 1 and valid for December–January–February (DJF, +2–4 months), computed considering the forecasts valid over all the S4 grid points over two regions, Tropics and Europe, issued over a 30-year period (1981–2010). These diagrams show that S4 is capable to provide reliable and accurate 2–4-month forecasts in November over the Tropics (Fig. 1A and B; positive BSS). By contrast, results indicate that, on average, the S4 forecasts for Europe started in November are not skillful (Fig. 1C and D; negative BSS).

Ensembles should be designed to simulate all possible sources of forecast error, and this can be achieved by simulating the effect of the uncertainty in the knowledge of the initial conditions (from which a forecast starts) and the effect of model approximations. Reliability can be seen as a measure of whether the sources of forecast errors have been simulated properly. If an ensemble has poor reliability, improvements in the simulation of the sources of forecast errors will lead to a better reliability.

Reliability is important because in a perfectly reliable ensemble, there is a one-to-one correspondence between forecast probabilities and observed frequencies. Another way to understand why this is important is to think about the hypersphere (in the phase space of the system) spanned by the ensemble members. In a reliable ensemble, on average the mean distance between the ensemble members and the ensemble-mean (the ensemble spread, measured by the ensemble standard deviation) should be equal to the average error of the ensemble-mean. When this happens, on average the hypersphere spanned by the ensemble members includes the true state of the system as one possible solution. This means that when the ensemble has a smaller-than-average spread, we should expect the forecast error to be smaller than average as well. This is represented in a schematic way in Fig. 2. This is why reliability is a key property that should be looked at when designing ensemble systems.

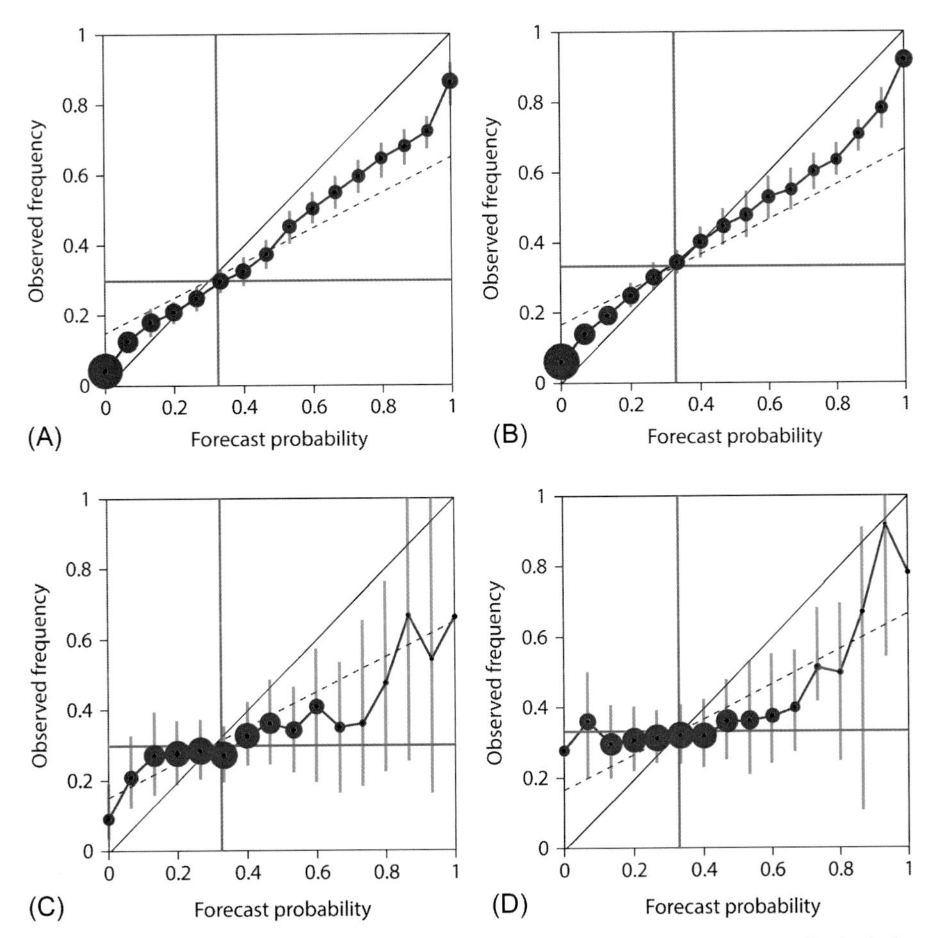

FIG. 1 Reliability diagram for the ECMWF seasonal S4 prediction of 2-m temperature anomalies in the lower tercile (A–C, left) and 2-m temperature anomalies in the upper tercile (B–D, right), over the tropics (A and B, top) and Europe (C and D, bottom). The diagrams refer to the 2–4-month forecasts started on November 1 and valid for DJF, and have been computed considering the 30-year S4 reforecasts (i.e., the forecasts started from November 1, 1981, to November 1, 2010, included). The size of each circle depends on the number of events in that specific category (the largest, the highest the number). BSSs are (A) 0.248, (B) 0.279, (C) −0.053 and (D) −0.081.

Reliability alone does not guarantee that ensemble-based, probabilistic forecasts provide accurate information (i.e., information that is more accurate than that based on climatology). For example, a climatological ensemble defined by the past states would be reliable on average, but it would not provide very accurate forecasts on each single day.

Accuracy can be measured using various metrics. The four most commonly used metrics (see, e.g., Wilks, 2005 for an overview) are the following:

- *The root-mean-square error (RMSE) of the ensemble-mean forecast*: It provides information on how close, on average, the mean of the ensemble PDF is to the observed state.
- *The BS*: It measures the correspondence between forecast probabilities and observed frequencies for specific events, such as the probability that the 2-m temperature anomaly will exceed 4°.

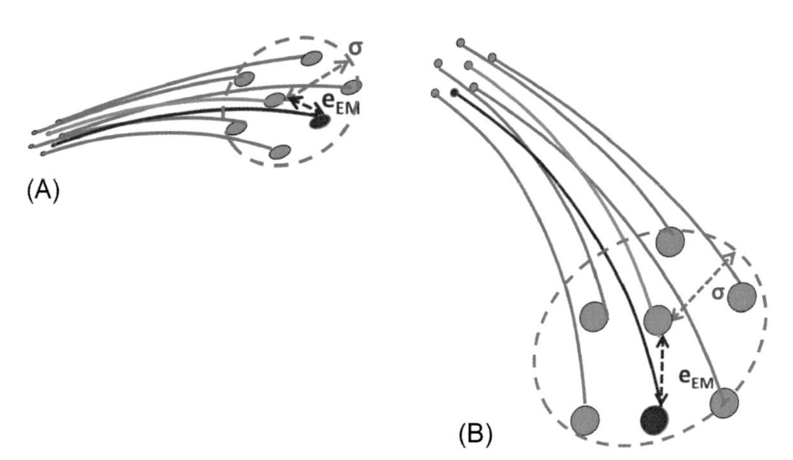

FIG. 2 Schematic of the relationship between the ensemble spread, measured by the ensemble standard deviation, and the error of the ensemble-mean forecast. The *blue circles* identity ensemble members, the *green circles* identify the ensemble-mean forecast, and the *red circles* identify the true state of the system. (A) shows that when the ensemble spread is smaller than average, one should expect that the ensemble-mean error is smaller-than-average; and (B) shows the opposite situation.

- *The area under a Relative Operating Characteristic (ROC) curve*: It measures the capacity of a probabilistic system to discriminate between the occurrence and the nonoccurrence of events.
- *The Continuous Ranked Probability Score (CRPS)*: It measures the average distance between the forecast PDF and the observation (i.e., a narrow distribution centered on the observed value, and with a width that represents the observation uncertainty); this score is the equivalent of the RMSE, which is often used to evaluate single forecasts, for probabilistic forecasts.

For any forecast *f*, and given a reference forecast *ref*, one can define the skill score *sk*(*f*) for each of these scores *sc*(*f*) as

$$sk(f) = \frac{sc(ref) - sc(f)}{sc(ref)} = 1 - \frac{sc(f)}{sc(ref)} \tag{1}$$

Fig. 1 has shown examples of the reliability diagram for four probabilistic forecasts, with the corresponding BS and BSS reported in the title of each panel. A zero BS and a 1 BSS would indicate a perfect forecast; a zero BSS would indicate that the forecast is as good as the reference forecast used to define the skill score (usually a climatological forecast). Fig. 3 shows the corresponding ROC curves for the four probabilistic forecasts whose reliability was shown in Fig. 1. There is clearly a correspondence between the two metrics, although the relative merit of a forecast with respect to the reference might change depending on the metric. For example, both figures indicate that the seasonal probabilistic forecasts are more accurate over the tropics than over Europe, with ROC area values around 0.8 for the tropics (Fig. 3A and B) and just above 0.5 for Europe (Fig. 3C and D).

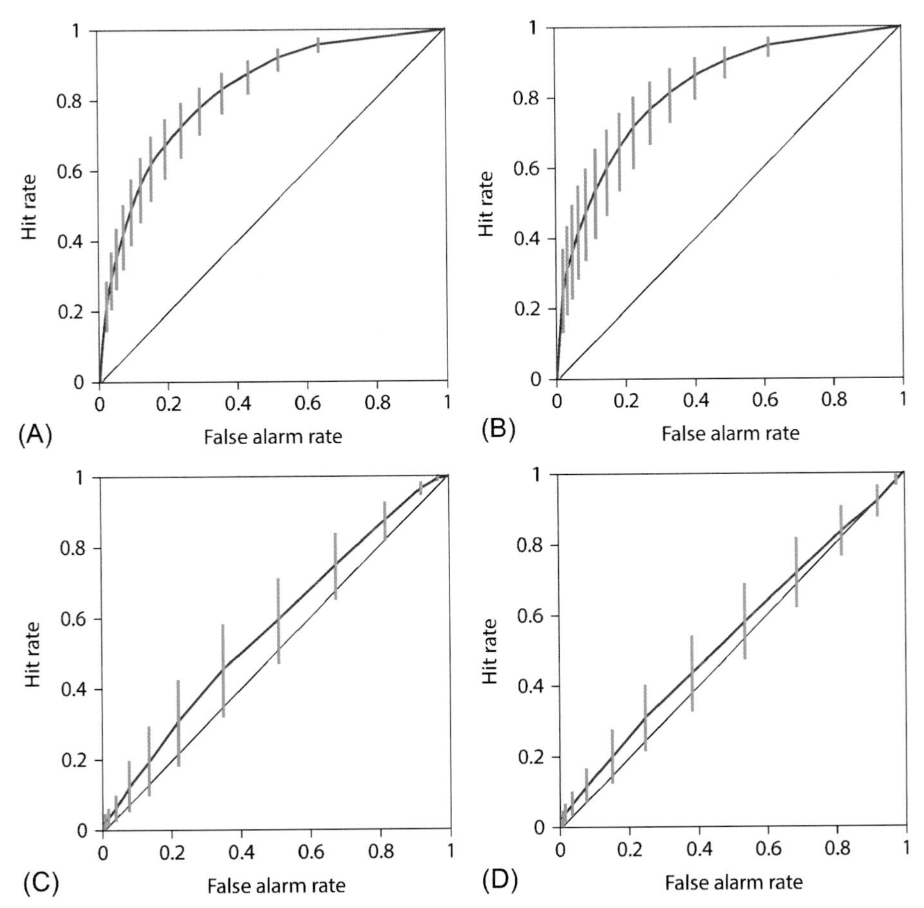

FIG. 3 ROC curves for the ECMWF seasonal S4 prediction of 2-m temperature anomalies in the lower tercile (A–C; left panels) and 2-m temperature anomalies in the upper tercile (B–D; left panels), over the tropics (A and B; top panels) and Europe (C and D; bottom panels). The diagrams refer to the 2–4-month forecasts started on November 1 and valid for DJF, and have been computed considering the 30-year S4 reforecasts (i.e., the forecasts started from November 1, 1981, to November 1, 2010, included). ROC-area scores are (A) 0.820, (B) 0.820, (C) 0.570, and (D) 0.540.

An example of the Continuous Rank Probability Skill Score (CRPSS), applied to ECMWF medium-range ensemble forecasts, is given by Fig. 2.6 in Chapter 2, which shows the time evolution of the CRPSS from 1994 to 2016, for forecasts valid from days 2–15. This diagram shows that, for the prediction of synoptic-scale features represented by the 500-hPa geopotential height, the forecast quality of the ECMWF probabilistic forecasts have been continuously improving at a rate of about 2 days per decade. Because the CRPSS is sensitive to both the reliability and the accuracy of the ensemble, the fact that it has been improving is an indication that there has been a continuous improvement in the whole system (model, definition of the initial conditions and simulation of the initial and model uncertainties).

2.2 Value of Single and Ensemble-Based Probabilistic Forecasts

Section 2.1 discussed two key properties of ensemble-based probabilistic forecasts, reliability, and accuracy. Let us now briefly present two reasons why ensemble-based probabilistic forecasts are more valuable than single forecasts.

The first reason is that they make it possible not only to predict the most likely scenario, but also to estimate the probability that any event can occur. In other words, ensembles provide users with more complete information, with extra pieces of information about the future weather scenario. One way to measure such a difference is to apply simple cost-loss (C/L) models and apply a measure called the Potential Economic Value (PEV; Richardson, 2000; Buizza, 2001) of a forecasting system.

The PEV is based on a simple C/L model, whereby a user can decide to pay an amount (cost) C to protect against a loss L, linked to a specific weather event. Forecasts then can be assessed by considering users with different C/L ratios, and by constructing a curve that shows the savings that users can make if they used the forecasts. Clearly, PEV is a function of the reliability and accuracy of the forecasts: reliable and accurate forecasts have a ROC curve that draws closer to the top ($y = 1$) of the diagram. As an example, Fig. 4 shows the average PEV for the ECMWF single high-resolution forecast and the medium-range/monthly ensemble (ENS) probabilistic forecast of four events:

- 2-m temperature cold anomaly (with respect to climatology) lower than 4°C
- 2-m temperature warm anomaly (with respect to climatology) greater than 4°C
- 10-m wind speed stronger than 10 m/s
- Total precipitation larger than 1 mm

The $t + 144$-hour forecasts have been verified against SYNOP observations over Europe and Northern Africa, and the PEV have been computed for the 3-month period October–December 2016. Fig. 4 shows that the ENS-based probabilistic forecasts have a higher PEV for all ranges of users.

The second reason why ensemble-based, probabilistic forecasts are more valuable is that an ensemble system provides forecasters with more consistent (i.e., less changeable) successive forecasts. This was studied (e.g., by Zsoter et al., 2009), who compared the consistency of ENS control and the ensemble-mean forecasts. Ensemble-mean forecasts issued 24 hours apart and valid for the same verification time jump less (i.e., are more consistent) than the corresponding single forecasts. In other words, ensemble-based, dynamical averaging makes successive forecasts more consistent.

3 A BRIEF INTRODUCTION TO DATA ASSIMILATION

By the term *data assimilation,* we mean the process followed to estimate the initial state of the Earth system. In this section, we briefly review the key stages of the process so that readers can understand better the methods used to generate the ensembles that are discussed in Section 5, later in this chapter. For more information on this subject, the reader is referred to more topical books, such as Daley (1991), Ghil et al. (1997), and Kalnay (2012).

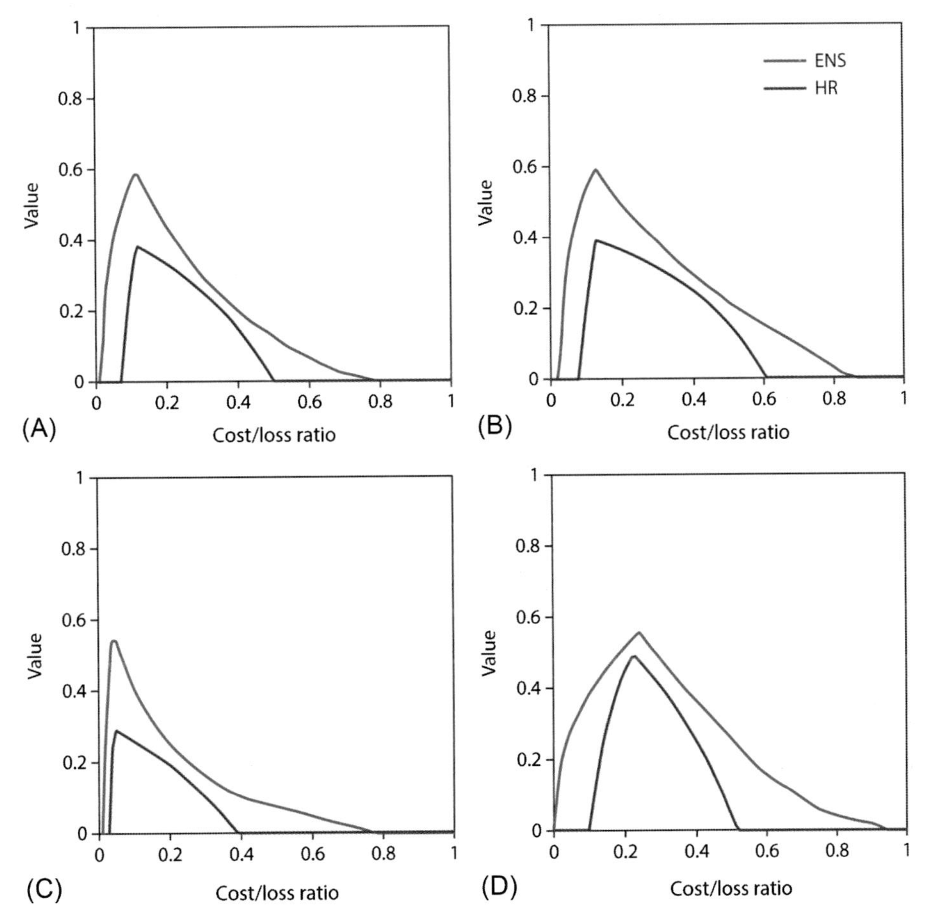

FIG. 4 PEV of ECMWF single high-resolution forecasts *(blue lines)* and ENS-based probabilistic forecasts *(red lines)*, for C/L ratios ranging from 0 to 1, for four different forecasts: 2-m temperature cold anomaly lower than 4° (A), 2-m temperature warm anomaly greater than 4° (B), 10-m wind speed stronger than 10 m/s (C) and total precipitation larger than 1 mm (D). PEV average values have been computed considering the ECMWF operational forecasts for October–November–December 2016, verified against surface synoptic (SYNOP) observations.

Fig. 5 shows a schematic of the numerical weather prediction (NWP) process, followed every day by ECMWF to generate a global forecast:

1. As many observations as possible are collected and exchanged using a global telecommunication network, so that weather prediction centers have timely access to them.
2. A few times a day (for global prediction, this happens every 6 hours, at times coinciding with what are called the synoptic times: 00, 06, 12, and 18 UTC, where UTC stands for Coordinated Universal Time), a data-assimilation procedure is performed to estimate the state of the system at a specific time T. This procedure merges observations collected a few hours before or centered on time T, with a short-range forecast that provides an estimate of the state of the atmosphere.

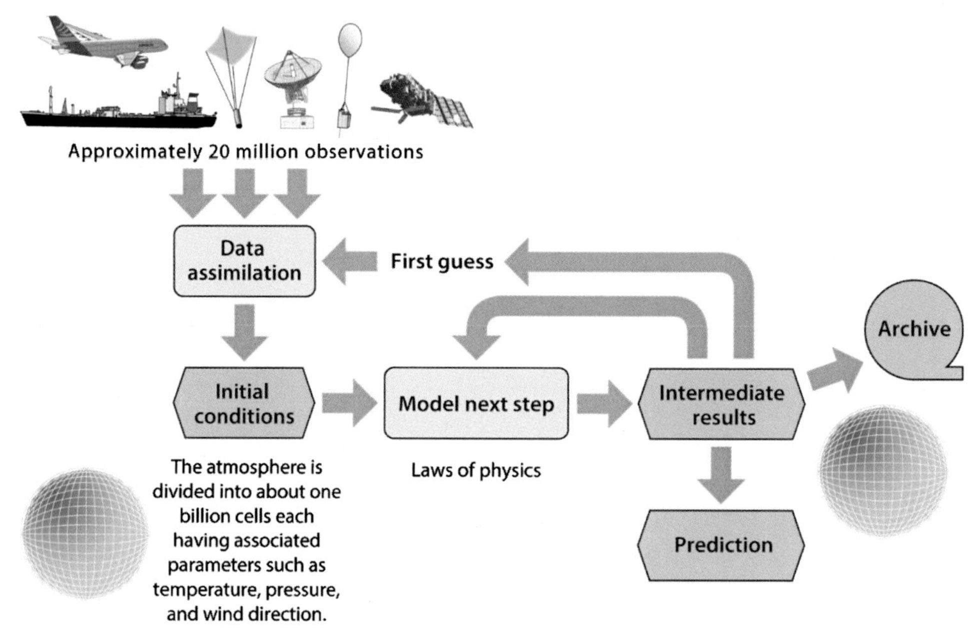

FIG. 5 Schematic of the NWP process followed at the ECMWF to generate a global forecast.

3. At the end of the data-assimilation procedure, initial conditions are available to launch the next forecast; at ECMWF, for example, forecasts with different forecast lengths are launched at the four synoptic times every day.
4. The forecasts then are used as input to the next data-assimilation procedure, and to generate forecast products.
5. All analysis and forecast data is disseminated to the users and is also copied to an archive system, so that users can go back and revisit each case for diagnostic and verification purposes.

At the time of writing, every day, about 200–250 million observations are collected worldwide and exchanged. Of these, about 20–25 million (and thus about 10% of the received ones) are used to estimate the initial conditions required to start a numerical integration of the equation of motions. These 10% observations are selected so that they provide good and uniform coverage of the whole globe. Some of the collected observations are discarded because they do not satisfy quality-control checks, while some are discarded because they are redundant. A thinning process is also applied to make sure that the spatial resolution of the observations is similar to the model grid. Of these 25 million observations, about 95% comes from instruments aboard satellites. Fig. 6 shows a few examples of the data coverage of four different types of data used at ECMWF to generate the initial conditions at 12 UTC of May 27.

Observation quality is very important because it influences the accuracy of the best estimate of the true state of the atmosphere that is generated using data-assimilation procedures. In general, observation quality depends on the instrument, and for satellites, also on

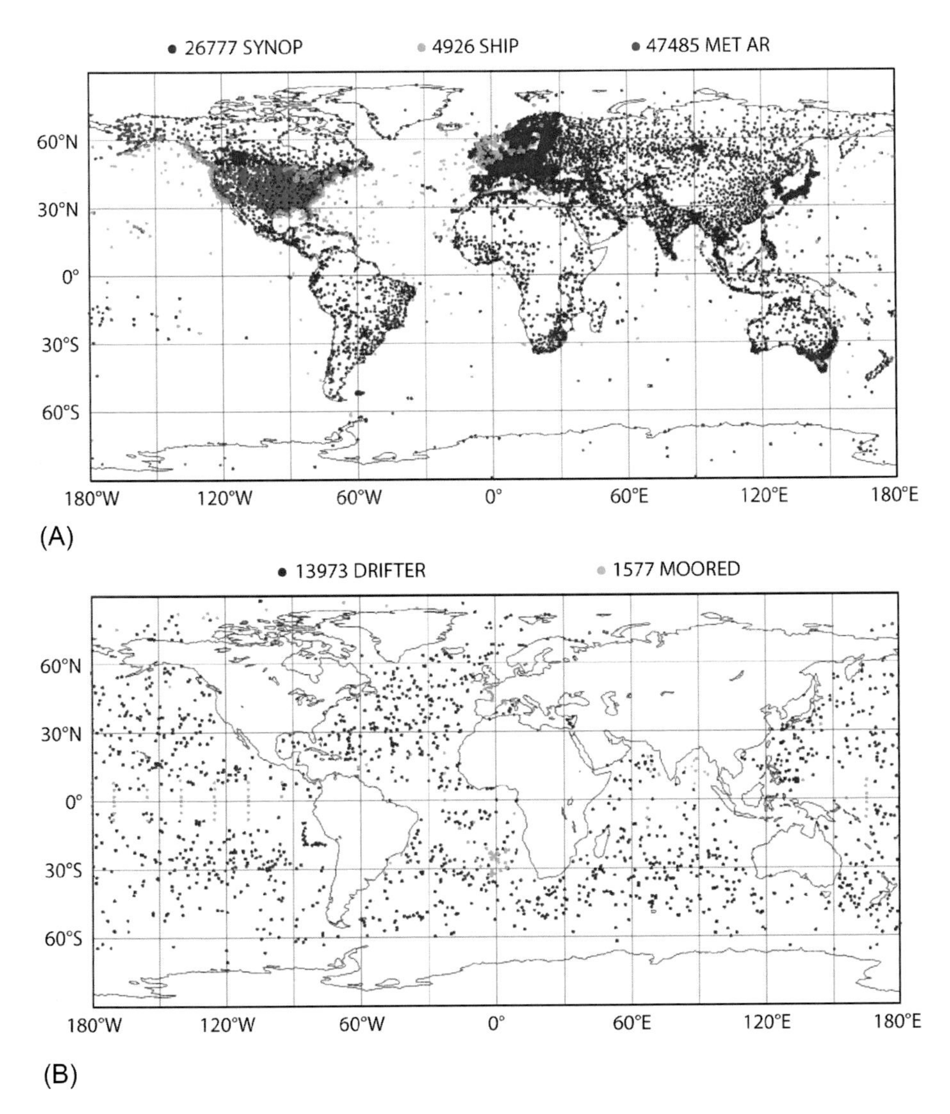

FIG. 6 Coverage of the data received at ECMWF and used to generate the analysis of May 27, 2017, at 12 UTC: (A) station (synop and metar) and ship data; (B) buoys;

(Continued)

the instrument position with respect to the area under observation. Furthermore, satellite observation quality is influenced by the state of the atmosphere, with lower (higher) observation errors in cases of clear-sky (cloudy) conditions. For example, because water vapor concentration is higher close to the surface, satellite observations are less accurate closer to the Earth surface. The fact that observations are affected by errors that depend on the type

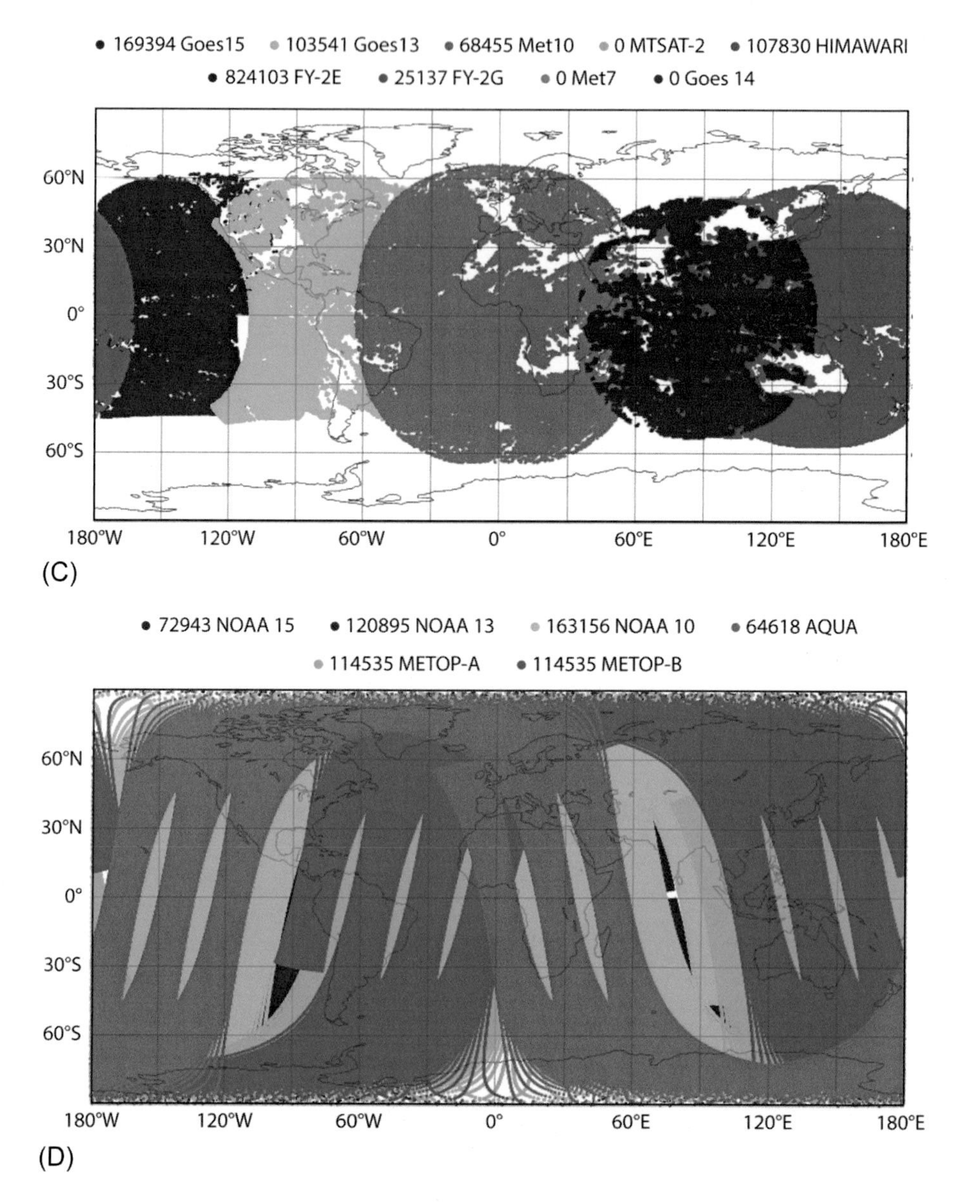

FIG. 6—Cont'd (C) atmospheric motion vectors from stationary satellites; and (D) Advanced Microwave Sounding Unit-A (AMSU-A) data aboard polar-orbiting satellites.

and on the region of the atmosphere that is observed is taken into account in the data-assimilation procedures. Table 1, for example, lists the prescribed RMSEs for observations of the two horizontal wind components (u and v) and temperature at different vertical levels (from A. Geer, ECMWF, personal communication). By using various RMSE estimates for each

TABLE 1 Example of observation errors used in the ECMWF assimilation procedure for some variables

	U and V Wind Components' Observations (m/s)			Height Observations (m)			Temperature Observations (K)		
Vertical Level (hPa)	Temp/ Pilot	SATOB	SYNOP	Temp/ Pilot	SYNOP (Manual Land)	SYNOP (Auto Land)	Temp	AIREP	SYNOP (Land)
1000	1.80	2.00	3.00	4.30	5.60	4.20	1.40	1.40	2.00
850	1.80	2.00	3.00	4.40	7.20	5.40	1.25	1.18	1.50
700	1.90	2.00	3.00	5.20	8.60	6.45	1.10	1.00	1.30
500	2.10	3.50	3.40	8.40	12.10	9.07	0.95	0.98	1.20
250	2.50	5.00	3.20	11.80	25.40	19.05	1.15	0.95	1.80
100	2.20	5.00	2.20	18.10	39.40	29.55	1.30	1.30	2.00
50	2.00	5.00	2.00	22.50	59.30	44.47	1.40	1.50	2.40

Note: Columns 2–4: prescribed RMSEs for the U and V wind components for three types of observations at seven different heights (from the surface at 1000–50 hPa), used in the ECMWF data-assimilation procedure (see ECMWF IFS documentation). Columns 5–7: as with columns 2–4, but for height observations. Columns 8–10: as with columns 2–4, but for temperature. Synop, Temp, and Pilot are types of reports of station observations, SATOB is satellite observations, and AIREP is reports of airplane observations.
Data from A. Geer, ECMWF, personal communication.

observation, the data assimilation can take into account their accuracy and give more or less weight to the more or less accurate observations. Complex techniques also can take into account the fact that the observations taken closely to each other can have correlated observation errors.

When observations are merged with a first-guess estimate of the state of the atmosphere to generate the initial conditions of the next forecasts, different weights are given to them, depending on the relative quality of the observations and the first guess. For the observations, the quality is a function of the instrument: for example, SYNOP observations have a higher quality (smaller observation error) than satellite data (Table 1). For the first guess, the quality depends on the model accuracy (the first guess is simply a short-range forecast), which is a function of the model resolution, for example.

Complications arise from several factors. For example, the model grid and the location where observations are taken do not always coincide, or the time when observations are taken does not coincide with an available model state, and the variables used by the model to define the state of the atmosphere (say, temperature, wind, humidity, surface pressure, and cloud concentration) do not coincide with the observed variables. Because of this, so-called observation operators need to be developed to map a model state onto an observation state. These operators have to be developed for each observation type, and because they simulate the true physical phenomena only in an approximate way (as does the model of the atmosphere), they can introduce uncertainties in the way that the initial conditions are estimated.

A further complication arises from the chaotic nature of the Earth system and the fact that the relative observation and first-guess errors depend on the state of the Earth system itself.

This flow-dependency can be taken into account in data assimilation, such as by using an ensemble of first guesses (i.e., an ensemble of short-range forecasts) to provide an estimate of the expected first-guess error at each grid point.

Data assimilation is an area of constant research and innovation. Most of the main meteorological operational centers have developed their own assimilation systems to generate the best estimates of the Earth-system initial state. The methods used to initialize the ensembles that are going to be discussed in Section 5 can be grouped into three main categories: variational methods, either 3D or four-dimensional (4D); ensemble methods (EnKF, ETKF); and/or hybrid methods, which combine an ensemble component, used to provide flow-dependent statistics, and a variational component.

4 A BRIEF INTRODUCTION TO MODEL UNCERTAINTY SIMULATION

Numerical models have limitations in the way that they describe the dynamical and physical evolution of the Earth system, arising from both numerical approximations and the assumptions involved in the parametrizations of subgrid physical processes. Therefore, tendencies from various models for the same state in a given column will differ, and any numerical forecast is bound to develop errors due to differences between the model tendencies and the real-world tendencies, even if the initial conditions were perfect.

A properly designed, reliable ensemble forecasting system should aim at representing the random errors in the tendencies in order to predict reliable uncertainty estimates. This can be achieved by using alternative numerical and physical formulations in each integration, by including a stochastic component designed to represent the difference between alternative, physically plausible representations of a given process, or by doing both. At present, four main approaches are followed in ensemble prediction to represent model uncertainties:

- A multimodel approach, in which different models are used in each ensemble members; models can differ entirely, or only in some components (e.g., in the convection scheme).
- A perturbed parameter approach, in which all ensemble integrations are made with the same model, but with different parameters defining the settings of the model components; one example is the Canadian ensemble (see Section 5.1.7).
- A perturbed-tendency approach, in which stochastic schemes designed to simulate the random model error component are used to simulate the fact that tendencies are known only approximately: one example is the ECMWF Stochastically Perturbed Parameterization Tendency scheme (SPPT; see Section 5.1.4).
- A stochastic backscatter approach, in which a Stochastic Kinetic Energy Backscatter (SKEB) scheme is used to simulate processes that the model cannot resolve (e.g., the upscale energy transfer from the scales below the model resolution to the resolved scales): another example is the SKEB scheme used in the ECMWF-ENS (see Section 5.1.4).

Given that forecast models will never be perfect, due to the necessary numerical approximations and approximations in the representation of physical processes, model uncertainty must be taken into account.

Indeed, introducing an adequate representation of model uncertainties has been shown to be beneficial in forecasts and data assimilation. The first group that detected an advantage of simulating model uncertainties was the Meteorological Service of Canada (MSC; Houtekamer et al., 1996), as documented by the fact that the first version of the MSC ensemble prediction system included a scheme to simulate model uncertainty. The MSC scheme was based on a multimodel approach: Two dynamical cores with different orographies were used, and each ensemble member was integrated by calling different parameterization schemes of horizontal diffusion, convection, radiation, and gravity wave drag.

At ECMWF, the advantage of including model error was detected first in prediction mode, and this led to the implementation of the first version of the stochastic scheme in 1999, and later on in assimilation, as follows:

- In forecast mode, Buizza et al. (1999), who designed the first version of the stochastic scheme implemented in the ensemble system, showed that "stochastic physics increases the spread of the ensemble and improves its performance, in particular for the probabilistic prediction of precipitation." Since then, the ECMWF ensemble ENS has been running with stochastic model error simulation schemes. See Palmer et al. (2009) and Leutbecher et al. (2016) for reviews of the schemes used currently in the operational ENS, and for some recent results that confirm that without model error schemes, the ECMWF medium-range/monthly ensemble (ENS) would be underdispersive and the probabilistic skill lower, and the seasonal system would be less reliable.
- In assimilation mode, since its first implementation in 2008, the Ensemble of Data Assimilations (EDA; Isaksen et al., 2007; Buizza et al., 2008) has been using a one-time level version of the SPPT. The stochastic scheme is essential to have an EDA spread that is in line with analysis error estimates.

Today, schemes to represent model uncertainty have been introduced in most of the operational ensembles s, as will be discussed in Section 5. Among them are the following:

- At the NCEP, a stochastic parameterization scheme with total model tendencies perturbed with stochastic forcing sampled from the differences in the conventional tendencies between the ensemble perturbed members, and the control is used in their ensemble system (Hou et al., 2008).
- At the Met Office in the United Kingdom (UKMO), two schemes are used to simulate the effect of model errors in its ensemble system (Bowler et al., 2008): a random parameter scheme, to simulate the fact that the parameters that control the physical processes' schemes are uncertain; and a stochastic convective vorticity scheme, to simulate uncertainties due to subgrid-scale processes, particularly the ones linked to convection, that are not represented by the model.
- At the MSC, they are still using different model parameterizations (but only one dynamical core), and since 2007, they also use both a perturbed tendency perturbation scheme and a stochastic backscatter scheme (Gagnon et al., 2007; Houtekamer et al., 2009; Charron et al., 2010). They are also using a fourth scheme, called the *parameterized system error scheme* (Houtekamer and Mitchell, 2005), which simulates the effect of model error sources that degrade the quality of the initial conditions by adding a random perturbation field to the initial conditions.

TABLE 2 Key Characteristics of the Nine TIGGE Operational, Global, Medium-Range Ensembles

Center	Initial Uncertainty Method (area)	Model Uncertainty	Truncation (°, km)	# Vert Level (TOA, hPa)	Forecast Length (d)	No. of Members (Perturbed + Unperturbed)	No. of Runs per Day (UTC)
BMRC	SV (NH,SH)	No	TL119 (1.5°; 210 km)	19 (10.0)	10	32 + 1	2 (00/12) (Until 2011)
CMA	BV (globe)	No	T213 (0.56°; 70 km)	31 (10.0)	10	14 + 1	2 (00/12)
CPTEC	EOF (40° S:30°N)	No	T126 (0.94°, 120 km)	28 (0.1)	15	14 + 1	2 (00/12)
ECMWF	SV (NH, SH, TC (tropics)) +EDA (globe)	Yes	Tco639 (0.14°; 16 km)	91 (0.01)	0–15	50 + 1	4 (00/06/ 12/18)
			Tco319 (0.28°; 32 km)		15/46		
JMA	LETKF and SV (NH, TR, SH)	Yes	TL479 (0.35°; 40 km)	100 (0.01)	11	25 + 1	2 (00/12)
KMA	ETKF (globe)	Yes	N400 (0.28°; 32 km)	70 (0.1)	12	24 + 1	2 (00/12)
MSC	EnKF (globe)	Yes	800 × 400 (0.45°, 50 km)	40 (2.0)	16/32	20 + 1	2 (00/12)
NCEP	ETR(globe)	Yes	T574 (0.30°; 34 km)	64 (2.7)	0–8	20 + 1	4 (00/06/ 12/18)
			T372 (0.60°; 55 km)		8–16		
UKMO	ETKF(globe)	Yes	N216 (0.45°; 60 km)	70 (0.1)	15	23 + 1	2 (00/12) (Until 2014)

Note: Listed in alphabetical order (column 1): initial uncertainty method (column 2), model uncertainty simulation (column 3), truncation and approximate horizontal resolution (column 4), number of vertical levels and top of the atmosphere in hPa (column 5), forecast length in days (column 6), number of members for each run (column 7), and number of runs per day (column 8). For columns 3–8, blue (brown) color identifies the ensemble with the finest (coarsest) characteristics; for column 8, red identifies ensembles that stopped producing medium-range (say, up to 10 days) forecasts. NH, Northern Hemisphere; SH, Southern Hemisphere. See Buizza (2014) for more details.

- At the U.S. Navy Operational Global Atmospheric Prediction System (NOGAPS) ensemble system, a "stochastic convection" approach (Teixeira and Reynolds, 2008) is used to simulate model error uncertainties mainly caused by convection parameterization. With this approach, parameterization schemes are used with control parameters sampled from a distribution of probable values instead of with a unique, most likely value.

The challenge in the development and evaluation of model error schemes is that the random component of the tendency error of the forecast model cannot be observed directly, and this makes it extremely difficult to assess their realism and effectiveness. This lack of a verification target is one of the reasons why pragmatic approaches often are followed, as one can see from the brief overview presented here.

5 AN OVERVIEW OF OPERATIONAL, GLOBAL, SUB-SEASONAL, AND SEASONAL ENSEMBLES, AND THEIR INITIALIZATION AND GENERATION METHODS

Meteorological centers who have been producing global ensemble forecasts for the medium range have been exchanging data since 2007 as part of the THORPEX Interactive Grand Global Ensemble (TIGGE), a project of the WMO. Since 2007, as part of TIGGE, about 600 global forecasts are exchanged and archived at ECMWF every day: these forecasts can be accessed in quasi-real time (i.e., with a 48-hour delay) from the TIGGE data portal (http://apps.ecmwf.int/datasets/data/tigge/levtype=sfc/type=cf/). The TIGGE archive includes forecasts from nine operational centers from about 2004, although today only seven of the original nine ensembles are generating global, medium-range forecasts (say, forecasts up to at least 10 days) on a daily basis. The key characteristics of the original nine TIGGE ensembles will be discussed in Section 5.1.

In 2010, TIGGE was extended to TIGGE-LAM, which includes weather forecasts from European limited-area model (LAM) ensembles. These forecasts are produced at higher resolution than the TIGGE medium-range global ensembles (say, between 12 and 2 km grid spacing) and provide detailed information for the short range (normally up to 3 days, and only in a few cases, up to 7 days ahead). Data from the TIGGE-LAM ensembles can be accessed from the ECMWF data portal: http://apps.ecmwf.int/datasets/data/tigge-lam/expver=prod/type=pf/. The characteristics of these ensembles are not going to be discussed in this chapter.

TIGGE can be considered as a precursor of the WMO Sub-seasonal to Seasonal (S2S)project. Within S2S, about 650 sub-seasonal forecasts are generated and exchanged every week. The S2S database contains ensemble forecasts and reforecasts from 11 operational centers; their key characteristics will be discussed in Section 5.2.

5.1 The TIGGE Global, Medium-Range Operational Ensembles

Today, thanks to the TIGGE project, 588 medium-range forecasts are produced and exchanged every day. These forecasts are produced by eight centers (note that of the nine production centers listed in this section, the Bureau of Meteorology Research Center, Australia (BMRC) stopped producing medium-range ensemble forecasts in 2010), with a horizontal resolution that ranges from about 16 km (ECMWF) to about 210 km (BMRC). Five of the eight ensembles simulate both initial and model uncertainties, and one ensemble (ECMWF) uses an atmosphere/land/ocean coupled model with an interactive sea ice model from the initial time.

Table 2 lists the key characteristics of the nine TIGGE global, medium-range (i.e., with a forecast length beyond 7 days) operational ensembles as of May 2017, generated by the following institutions:

- BMRC
- The China Meteorological Administration (CMA)
- The Brazilian Center for Weather Prediction and Climate Studies (Centro de Previsao de Tempo e Estudos Climatico, or CPTEC)
- ECMWF
- The Japanese Meteorological Administration (JMA)
- The Korean Meteorological Administration (KMA)
- MSC
- NCEP
- UKMO

TIGGE ensemble data is available from the TIGGE website hosted at ECMWF (http://www.ecmwf.int/en/research/projects/tigge).

Each TIGGE ensemble member is defined by the numerical integration of the model equations adopted by the production center in question to simulate the Earth system. In other words, each single T-hour forecast starting at day d is given by the time integration of a set of model equations from initial time 0 to time T:

$$e_j(d;T) = e_j(d;0) + \int_0^T [A_j(t) + P_j(t) + dP_j(t)]dt \tag{2}$$

where A_j is the tendency due to the adiabatic processes (say, advection, Coriolis force, and pressure gradient force); P_j is the tendency due to the parameterized physical processes (say, convection, radiation, turbulence, ...); and dP_j represents the tendency due to stochastic model errors and unresolved processes.

In the MSC ensemble, each numerical integration starts from initial conditions defined by an independent data-assimilation procedure:

$$e_j(d,0) = F[e_j(d - T_A; T_A), o_j(d - T_A, d)] \tag{3}$$

where $F[..,..]$ denotes the data-assimilation process of merging the model first guess $e_j(d - T_A, T_A)$ and the observations spanning the time period T_A (the window covered by the data-assimilation process), from $(d - T_A)$ to d. The first guess $e_j(d - T_A; T_A)$ is given by the T_A-hour integration of the model equations from $(d - T_A)$ to d. The data-assimilation process uses the observations $o_j(d - T_A, d)$, taken between $(d - T_A)$ and d, and available at the time when the data-assimilation procedure starts. More precisely, each member's initial conditions are selected among the members of its Ensemble Kalman Filter (EKF).

In the other eight TIGGE ensembles, each numerical integration starts from the initial conditions defined by adding a perturbation to an unperturbed initial state:

$$e_j(d,0) = e_0(d,0) + de_j(d,0) \tag{4}$$

$$e_0(d,0) = F[e_0(d - T_A; T_A), o_0(d - T_A, d)] \tag{5}$$

where the unperturbed initial conditions are defined by the data-assimilation process spanning the time period T_A. The initial perturbations $de_j(d,0)$ are defined in different ways in each ensemble. Eqs. (2)–(5) provide a simple, unified framework to describe how the jth member of the TIGGE forecasts is produced every day. Buizza (2014) summarizes the way that initial and model uncertainties are simulated in each ensemble and lists the main references that describe each ensemble.

5.1.1 BMRC-ENS

The BMRC-ENS started production in July 2001 and stopped in July 2010, when BMRC decided to adopt the United Kingdom's data-assimilation and forecasting system. BMRC data is available in TIGGE from July 2007 to July 2010. BMRC is planning to restart its medium-range ensemble production in 2018 (M. Naughton, BMRC, personal communication).

In 2010, when BMRC stopped production, the BMRC-ENS had the same configuration as when it started production in 2001. It comprised 33 members, 1 unperturbed and 32 perturbed (Bourke et al., 1995, 2004), and forecasts were run twice a day for up to 10 days. The system had a spectral triangular truncation $T_L119L19$ (about 1.5°, 160 km in physical space and 19 vertical levels), with the top of the atmosphere at 10 hPa. The forecast model included only a description of land and atmospheric processes (no wave or ocean model component was used).

In terms of Eqs. (2)–(5), each BMRC-ENS member is given by the following time integration:

$$e_j(d;T) = e_j(d;0) + \int_0^T [A_0(t) + P_0(t)]dt \tag{6}$$

where A_0 and P_0 represented the unperturbed model dynamical and physical tendencies (i.e., there was only one dynamical core and one set of parameterizations called with the same parameters, and no model error scheme), and the forecast length T was 10 days. The ensemble did not simulate model uncertainties.

The initial conditions were defined by adding perturbations to the unperturbed initial conditions:

$$e_j(d,0) = e_0(d,0) + de_j(d,0) \tag{7}$$

The unperturbed initial conditions were produced by interpolating to the ensemble resolution the initial conditions defined by the BMRC 3D multivariate statistical interpolation scheme. The initial perturbations were defined by T42L19 singular vectors (SVs), computed as in the first version of the ECMWF-ENS (see Section 5.1.4 for more details).

The individual perturbations were defined by linear combinations of the SVs through an orthogonal phase space rotation, followed by an amplitude-scaling factor, as was done in the original ECMWF-ENS (Molteni et al., 1996):

$$de_j(d,0) = \sum_{k=1}^{32} \alpha_{j,k} SV_k \tag{8}$$

The SVs were computed only over the Southern Hemisphere (for all grid points with latitude $\lambda < 20°S$), and were scaled to have amplitudes, locally, that are comparable to analysis error estimates.

5.1.2 CMA-ENS

The CMA-ENS started production in 2001, with a T213L31 resolution (about 0.56°, 70 km in physical space, and 31 vertical levels). The CMA-ENS data has been available in the TIGGE archive since May 2007.

In summer 2017, the CMA-ENS comprised 15 members, 1 unperturbed and 14 perturbed (Su et al., 2014). Forecasts are run twice a day, at 00 and 12 UTC, for up to 10 days. The forecast model included only a description of land and atmospheric processes (no wave or ocean model component is used). The ensemble did not simulate model uncertainties.

In terms of Eqs. (2)–(5), each CMA-ENS member is given by the following time integration:

$$e_j(d;T) = e_j(d;0) + \int_0^T [A_0(t) + P_0(t)]dt \tag{9}$$

where A_0 and P_0 represents the unperturbed model dynamical and physical tendencies (i.e., there is only one dynamical core and one set of parameterizations called with the same parameters, and no model error scheme), and the forecast length T is 10 days.

The initial conditions are defined by adding perturbations to the unperturbed initial conditions:

$$e_j(d,0) = e_0(d,0) + de_j(d,0) \tag{10}$$

$$de_j(d,0) = BV_j(d,0) \tag{11}$$

The unperturbed initial conditions are produced by interpolating the T213L31 analysis to the ensemble resolution. The initial perturbations are defined by bred vectors (BVs), computed as in the original NCEP-ENS (Toth and Kalnay, 1997; see Section 5.1.8).

5.1.3 The CPTEC-ENS

The CPTEC-ENS has been producing ensemble forecasts since 2001. Its data has been available in the TIGGE archive since March 2008.

In summer 2017, the CPTEC-ENS had a T126L28 resolution, with a corresponding 0.94° grid (about 120 km). In the vertical, it has 28 vertical levels, up to 0.1 hPa. It used a spectral model, has 15 members (1 unperturbed and 14 perturbed), runs twice a day (at 00 and 12 UTC), and produced forecasts up to 15 days. The forecast model included only a description of land and atmospheric processes (no wave or ocean model component is used). The ensemble did not simulate model uncertainties.

In terms of Eqs. (2)–(5), each CPTEC-ENS member is given by the following time integration:

$$e_j(d;T) = e_j(d;0) + \int_0^T [A_0(t) + P_0(t)]dt \tag{12}$$

where A_0 and P_0 represent the unperturbed model dynamical and physical tendencies (i.e., there is only one dynamical core and one set of parameterizations called with the same parameters, and no model error scheme), and the forecast length T is 15 days.

The initial conditions are defined by adding perturbations to the unperturbed initial conditions:

$$e_j(d,0) = e_0(d,0) + de_j(d,0) \tag{13}$$

$$de_j(d,0) = EOF_j(d,0) \tag{14}$$

The unperturbed initial conditions are defined by the NCEP operational analysis (see Section 5.1.8), interpolated to the CPTEC resolution.

The initial perturbations are defined using empirical orthogonal functions (EOFs; Coutinho, 1999; Zhang and Krishnamurti, 1999). The method consists of (1) computing 36-hour BVs (by adding random perturbations to unperturbed initial conditions and running pairs of 36-hour forecasts), (2) constructing a time series of these BVs, and (3) performing an EOF analysis of this time series to obtain the fastest growing perturbations. These EOF-based perturbations are computed for the region 45°S–30°N.

5.1.4 The ECMWF-ENS

The ECMWF-ENS has been producing ensemble forecasts from November 1992. Its data has been available in the TIGGE archive since October 2006.

In summer 2017, the ECMWF-ENS comprised 51 members, 1 unperturbed and 50 perturbed. Forecasts were run with a variable resolution (Buizza et al., 2007): Tco639L91 (spectral triangular truncation T639 with a cubic-octahedral grid, which corresponds to about 18 km spacing in the horizontal dimension of the physical space, and 91 vertical levels) during the first 15 days, and Tco319L91 (i.e., about 35 km spacing in the horizontal dimension) thereafter. Forecasts were run twice a day, with initial times at 00 and 12 UTC, up to 15 days; at 00 UTC on Mondays and Thursdays, the forecasts were extended to 46 days (Vitart et al., 2008). Since summer 2016, ENS also has been run at 06 and 18 UTC up to forecast day 6.5.

The forecasts were coupled to an ocean wave and a dynamical ocean circulation model. The ocean wave model was WAM (Janssen et al., 2005, 2013; The WAMDI Group, 1988): it had a 20-km resolution and 24 directions and 30 frequencies up to day 10, and 12 directions and 25 frequencies afterward, and it was coupled to the atmosphere every time-step. The ocean model was the Nucleus for European Ocean Modelling (NEMO; Madec, 2008), with the ORCA025z75 grid, which had a 0.25° horizontal resolution and 75 vertical layers. NEMO is a state-of-the-art modeling framework for oceanographic research, operational oceanography, seasonal forecast, and climate studies, developed by the NEMO Consortium (http://www.nemo-ocean.eu/). See Mogensen et al. (2012a,b, and references therein) for a description of how NEMO and NEMOVAR have been coupled and implemented at ECMWF.

In terms of Eqs. (2)–(5), each ECMWF-ENS member is given by the following time integration:

$$e_j(d;T) = e_j(d;0) + \int_0^T \left[A_0(t) + P_0(t) + dP_j(t)\right] dt \tag{15}$$

where A_0 and P_0 represent the unperturbed model dynamical and physical tendencies (i.e., there is only one dynamical core and one set of parameterizations called with the same parameters); dP_j represents the model uncertainty simulated using two model error schemes, the SPPT (Buizza et al., 1999; Palmer et al., 2009) and SKEB (Berner et al., 2008; Palmer et al., 2009; Leutbecher et al., 2016); and the forecast length T is 15 or 46 days.

For the atmosphere, the initial conditions are defined by adding perturbations to the unperturbed initial conditions:

$$e_j(d,0) = e_0(d,0) + de_j(d,0) \tag{16}$$

The unperturbed initial conditions are given by the ECMWF high-resolution 4D variational assimilations (4DVAR), run at Tco1279L137 resolution and with a 12-hour assimilation window, interpolated from the Tco1279L137 resolution to the Tco639L91 ensemble resolution.

The perturbations are defined by a linear combination of SVs (Buizza and Palmer, 1995) and perturbations defined by the ECMWF (EDA; Buizza et al., 2008; Isaksen et al., 2010):

$$de_j(d,0) = \sum_{a=1}^{8}\sum_{k_a=1}^{50} \alpha_{j,k_a} SV_{k_a} + \left[f_{m(j)}(d-6,6) - \langle f_{m=1,25}(d-6,6)\rangle\right] \tag{17}$$

The SVs are computed over up to eight areas (Northern Hemisphere: all grid points with points with latitude $\lambda > 30°N$; Southern Hemisphere: all grid points with latitude $\lambda < 30°S$; tropics: up to six areas where tropical depressions have been reported). The SVs, the fastest-growing perturbations over a 48-hour interval, are computed at T42L91 resolution. SVs optimized to have maximum total-energy growth over the different areas are linearly combined and scaled to have amplitudes, locally, comparable to analysis error estimates provided by the ECMWF high-resolution 4DVAR.

Each EDA member is generated by an independent 4DVAR with the same resolution as ENS, but with 137 vertical levels (Tco639L137). Each EDA member uses perturbed observations, with their perturbations sampled from a Gaussian distribution with zero mean and the observation error standard deviation. Each EDA member's nonlinear trajectory is generated by using the SPPT scheme to simulate model uncertainties (a description of the SPPT scheme will be presented later in this chapter). Since November 2013, the EDA has been including 25 independent 4DVARs run with a 12-hour assimilation window (the EDA had only 11 members before November 2013).

SV- and EDA-based perturbations are combined as in Eq. (17). The EDA-based perturbations are defined by differences among 6-hour forecasts started from the most recent available EDA analyses (these analyses are valid for 6 hours earlier than the ENS initial time). Differences are computed between each of the 25 perturbed forecasts and their ensemble-mean, and the 25 perturbations are added to and subtracted from the unperturbed analysis. SV- and EDA-based perturbations are defined such that full symmetry is maintained in the ENS initial perturbations [i.e., the even member ($2n$) has minus the total perturbation of the odd member ($2n-1$), for $n = 1, 25$].

The initial conditions for the ocean are different: these are defined by the five-member ensemble of ocean analyses, produced by NEMOVAR, the NEMO 3D variational assimilation system (Mogensen et al., 2012a). Each ocean analysis is generated using all available in situ temperature and salinity data, an estimate of the surface forcing from ECMWF short-range atmospheric forecasts, sea surface temperature (SST) analysis, and satellite altimetry measurements. One member (the unperturbed control) is generated using unperturbed wind forcing provided by the high-resolution 4DVAR, while the other four are generated using perturbed versions of the unperturbed wind forcing.

Model uncertainties are simulated only in the free atmosphere (i.e., not in the land surface, nor in the ocean), using two stochastic schemes: the SPPT (Buizza et al., 1999; Palmer et al., 2009) and the backscatter (SKEB; Shutts, 2005; Berner et al., 2008)) schemes. SPPT is designed to simulate random model errors due to parameterized physical processes; the current version uses three spatial and time-level perturbations. SKEB is designed to simulate the upscale energy transfer induced by the unresolved scales on the resolved scales.

Since March 2008, when the ECMWF medium-range and monthly ensembles were joined into a seamless system, a key component added to the ECMWF-ENS used to generate some bias-corrected and/or calibrated products, has been the reforecast suite (Vitart et al., 2008; Leutbecher and Palmer, 2008). This suite included a 5-member ensemble that has been run once a week with the operational configuration (resolution, model cycle, ..) for the past 20 years. Since March 2016, the reforecast suite configuration has been upgraded from 5 to 22 members a week, with an 11-member ensemble run twice a week. These reforecasts are used to estimate the model climate required to generate some ensemble products [e.g., the Extreme Forecast Index (EFI), or weekly average anomaly maps] and to calibrate the ENS forecasts.

This is how the reforecasts are used to generate some of the operational ENS products. For each date (e.g., December 15, 2012), ensemble reforecasts are generated by running 11-member ensembles, twice a week and for 5 weeks for the past 20 years. That is, the initial conditions of these reforecasts are defined by the same day of the week of the 5 weeks centered around the current date (in this case, December 1, 4, 8, 11, 15, 18, 22, 25, and 29). These reforecasts start from ECMWF ERA-Interim reanalyses instead of the operational analyses because for older dates, ERA-Interim provides more accurate initial conditions. In the reforecast suite, the simulation of the initial uncertainties is slightly different from the one in the operational suite: the reforecast ensembles use SVs of the day (which are recomputed, as in the operational suite), but use EDA-based perturbations of the current year instead of the correct year. This is because the EDA has not been available for the past 20 years (the EDA started running in 2010); see Buizza et al. (2008) for more details. Buizza et al. (2008) described the rationale behind this choice and showed that despite this, the reforecast and the forecast ensembles perform very similarly. Once the 51-member forecasts and the few hundred reforecasts are completed, products such as probabilities of anomalies (computed with respect to the model climate) or the EFI, which needs the model climatological distribution function to be computed (see Lalaurette, 2003), are generated.

The most recent change in the ECMWF-ENS configuration was implemented in November 2016, when NEMO was upgraded, the resolution was increased from 1° to 0.25°, and the number of vertical layers increased from 42 to 75 (Zuo et al., 2014). Furthermore, a dynamical sea ice model, the Louvain-la-Neuve Sea Ice Model version 2 (LIM2; Fichefet and Morales Maqueda, 1997; Bouillon et al., 2009), was included. LIM2 is a two-level, thermodynamic-dynamic sea ice model, with sensible heat storage and vertical heat conduction within snow and ice determined by a three-layer model (one layer for snow and two for ice).

Since March 2015, ENS has been extended to 46 days (as opposed to up to 32 days previously) twice a week (at 00 UTC of Mondays and Thursdays); see Section 5.2.4 and Vitart et al. (2014) for more details.

5.1.5 The JMA-ENS

The JMA-ENS has been producing ensemble forecasts since March 2001. Its data has been available in the TIGGE archive since August 2011.

In summer 2017, the JMA-ENS (Yamaguchi and Majumdar, 2010) included 25 forecasts with a resolution of T_L479L100 (spectral triangular truncation with linear grid, 0.35° spacing that corresponds to about 40 km in physical space), with the top of the atmosphere at 0.01 hPa. The most recent change was introduced in March 2017, when the number of vertical levels was increased from 60 to 100 and the top of the model was raised from 0.1 to 0.01 hPa. This change followed an upgrade that was implemented in March 2014, when the resolution was increased from T_L319 to T_L479, and the daily configuration was changed from producing 51 forecasts once a day (at 12 UTC), to producing 26 twice a day (at 00 and 12 UTC).

Initial uncertainties are simulated using a combination of SVs and perturbations generated using a Local Ensemble Transform Kalman Filter (LETKF). The SVs are computed at T63L40 resolution over the Northern Hemisphere and Southern Hemisphere extratropics (north and south of 30°) with a 48-hour optimization time interval, and over the tropics (30°S–30°N) with a 24-hour optimization time interval. The JMA ENS perturbations are generated using perturbations from the JMA LETKF (Miyoshi and Sato, 2007), which were developed following the example of NCEP (Szunyogh et al., 2008; Whitaker et al., 2008).

Model uncertainties are simulated using a stochastic scheme similar to the original ECMWF SPPT (Buizza et al., 1999). The forecast model includes only a description of land and atmospheric processes (no wave or ocean model component is used).

In terms of Eqs. (2)–(5), each JMA-ENS member is given by the following time integration:

$$e_j(d;T) = e_j(d;0) + \int_0^T \left[A_0(t) + P_0(t) + dP_j(t)\right]dt \tag{18}$$

where A_0 and P_0 represent the unperturbed model dynamical and physical tendencies (i.e., there is only one dynamical core and one set of parameterizations called with the same parameters); dP_j represents the model uncertainty simulated using the JMA stochastic scheme; and the forecast length T is 11 days. The initial conditions are defined by adding perturbations to the unperturbed initial conditions:

$$e_j(d,0) = e_0(d,0) + de_j(d,0) \tag{19}$$

The unperturbed initial conditions are given by the JMA high-resolution 4DVAR, which has a resolution of T_L959L100 (about 20 km in physical space), interpolated at the ensemble resolution.

The perturbations are defined by a linear combination of SVs computed over three regions (Northern Hemisphere, Southern Hemisphere, and the tropics and LETKF analyses):

$$de_j(d,0) = \sum_{a=1}^{3} \sum_{k_a=1}^{25} \alpha_{j,k_a} \mathrm{SV}_{k_a} + \mathrm{LETKF}_j(d,0) \tag{20}$$

It is worth mentioning that JMA also produces monthly forecasts with a 50-member lagged ensemble, which has a $T_L319L60$ resolution (about 70 km in grid-point space). This monthly ensemble uses BVs (Toth and Kalnay, 1993, 1997) instead of SVs to define the initial perturbations. Monthly products are issued once a week, on Fridays, and are based on the 25 forecasts started the previous Wednesday and Thursday at 12 UTC. The forecast model includes only a description of land and atmospheric processes (no wave or ocean model component is used) and does not simulate model uncertainties. The JMA monthly forecasts are expected to become available in the S2S archive.

5.1.6 The KMA-ENS

The KMA-ENS has been producing ensemble forecasts since March 2000. The data has been available in the TIGGE archive since October 2007.

The original KMA-ENS used initial perturbations defined by a breeding method (Goo et al., 2003). It included 16 perturbed members, run with a T106L21 resolution, with the 16 initial perturbations defined by rotated BVs. The forecasts were run once a day (at 12 UTC) for up to 10 days. It did not simulate model uncertainties and included only a description of land and atmospheric processes (no wave or ocean model component is used).

Since 2011, KMA has been generating its operational forecasts using the Unified Model (UM) and related preprocessing/postprocessing system imported from the UKMO (Kai and Kim, 2014). Thus, since then, the KMA-ENS has been practically the same as the UKMO-ENS (see Section 5.1.9).

In summer 2017, the KMA-ENS was based on 24 members, 1 control and 23 perturbed ones, with the initial perturbations generated using an Ensemble Transformed Kalman Filter (ETKF) with localization (Bowler et al., 2008; Kai and Kim, 2014). It has a horizontal resolution of approximately 40 km and 70 vertical levels (N400L70).

Model uncertainties are simulated using a stochastic-physics schemes that consist of random parameters and stochastic convective vorticity schemes (Bowler et al., 2008).

If we now reconsider Eqs. (2)–(5), this is how they can be written for the KMA-ENS:

$$e_j(d;T) = e_j(d;0) + \int_0^T \left[A_0(t) + P_0(t) + dP_j(t)\right] dt \tag{21}$$

where A_0 and P_0 represent the unperturbed model dynamical and physical tendencies (i.e., there is only one dynamical core and one set of parameterizations called with the same parameters); dP_j represents the model uncertainty simulated using the KMA stochastic scheme; and the forecast length T is 10 days.

The initial conditions are defined by adding perturbations to the unperturbed initial conditions:

$$e_j(d,0) = e_0(d,0) + \mathrm{ETKF}_j(d,0) \tag{22}$$

The unperturbed initial conditions are given by the KMA version of the UKMO 4DVAR system. The perturbations are defined by ETKF perturbations (see Section 5.1.9).

5.1.7 The MSC-ENS

The MSC-ENS has been producing ensemble forecasts since February 1998. Its data has been available in the TIGGE archive since October 2007. Since December 2015, the resolution of MSC-ENS has been N800 × N400 (800 grid points in longitude and 400 in latitude), which corresponds to about 0.45°, 50 km (Gagnon et al., 2014a,b, 2015).

The MSC perturbed-observation approach attempts to represent all sources of uncertainty by adding random perturbations to as many system components as possible. The Canadian Ensemble Kalman Filter (EnKF) has been used to supply initial conditions (Houtekamer et al., 2009, 2014), with sources of uncertainty simulated by different sets of random perturbation schemes (Houtekamer et al., 1996; Houtekamer and Lefaivre, 1997; Anderson, 1997).

Compared to the original 1995 configuration, the number of the MSC ensemble members has increased from the original 8 to 16, and stands now at 20. The number of members of the MSC EnKF that are used to define the initial conditions also has increased from 48 to 96, 192 and now stands at 256.

The most recent change in the MSC-ENS was implemented in December 2015, when the resolution was increased to 50 km. In 2013, the treatment of the SST was changed so that the SST evolves while persisting the anomaly (deviation from the climatology). Since December 2013, monthly forecasts have been produced on Thursdays by extending the forecast length to 32 days. Furthermore, reforecasts based on four-member ensembles have been run once a week for the past 20 years, so that the model climate can be estimated and calibrated products can be generated (Gagnon et al., 2014b).

In summer 2017, the MSC-ENS included 21 members, 1 unperturbed and 20 perturbed, and was run twice a day (at 00 and 12 UTC) up to forecast day 16. Once a week, at 00 UTC on each Thursday, the ensemble was extended to 32 days. The initial conditions are obtained directly from the Canadian EnKF. Model uncertainties are sampled using four schemes: isotropic perturbations at the initial time, different physical parameterizations, stochastic physical tendency perturbations (Charron et al., 2010), and stochastic kinetic energy backscatter (Houtekamer et al., 2009). The forecast model includes only a description of land and atmospheric processes (no wave or ocean model component is used).

In terms of Eqs. (2)–(5), each MSC-ENS member is given by the following time integration:

$$e_j(d;T) = e_j(d;0) + \int_0^T \left[A_0(t) + P_j(t) + dP_j(t)\right] dt \tag{23}$$

where A_0 represents the unperturbed model dynamical core; P_j represents the physical tendencies, which varies in each member because different parameterization schemes and/or different parameters are used; and dP_j represents the model uncertainty simulated using various stochastic schemes.

Each member's initial conditions are defined by one of the EnKF members as follows:

$$e_j(d,0) = \mathrm{ENKF}_{k(j)}(d,0) \tag{24}$$

It is worth mentioning that the MSC-ENS and the NCEP-ENS forecasts are exchanged in real time to generate the multimodel products of the North American Ensemble Forecast System (NAEFS; Candille, 2009); see also the Canadian government's website (http://weather.gc.ca/ensemble/naefs/index_e.html). NAEFS is a joint project involving MSC, the U.S. National Weather Service (NWS) and the National Meteorological Service of Mexico (NMSM). NAEFS, launched in November 2004, provides users with operational products generated by blending the MSC-ENS and the NCEP-ENS. The research, development, and operational costs of the NAEFS system are shared among the three partners.

5.1.8 *The NCEP-ENS*

The NCEP-ENS has been producing ensemble forecasts since December 1992. Its data has been available in the TIGGE archive since March 2007.

The original version of the NCEP-ENS simulated only initial uncertainties using BVs (Toth and Kalnay, 1993, 1997). The breeding method involves the maintenance and cycling of perturbation fields that develop between two numerical model integrations. These fields, once rescaled, define the initial perturbations. In its original form with a single global rescaling factor, the BVs represented a nonlinear extension of the Lyapunov vectors (Boffetta et al., 1998). In the operational NCEP-ENS, multiple breeding cycles were used, each initialized at the time of implementation with independent perturbation fields ("seeds"). The original system was based on 10 perturbed ensemble members, run both at 00 and 12 UTC every day, out to 16 days of lead time. For both times, the generation of the initial perturbations was done in five independent breeding cycles, originally started with different perturbations and using a regional rescaling algorithm. Since then, the method used to define the initial perturbations has been upgraded several times.

In summer 2017, the NCEP-ENS consisted of four runs a day (at 00, 06, 12, and 18 UTC), with a T574L64 resolution up to forecast day 8, and a T372L64 resolution from days 8 to 16 (Zhou et al., 2016). Each run included 1 unperturbed and 20 perturbed forecasts. The forecast model included only a description of the land and atmospheric processes (no wave or ocean model component is used).

The initial perturbations were now generated using the Ensemble Transform with Rescaling (ETR; Wei et al., 2006, 2008) technique. ETR is an extension of the original breeding approach (in an ensemble with only two members, both methods should produce the same perturbations). To improve the simulation of initial uncertainties in cases of tropical storms, the perturbed initial conditions are generated using a tropical storm relocation method (Liu et al., 2006; Snyder et al., 2010).

In the current NCEP-ENS, model uncertainties are represented using the STTP (Hou et al., 2008), designed to represent model uncertainty by adding a stochastic forcing term to the total tendency. For each 6-hour forecast period, the forcing term is defined by a linear combination of the past ensemble tendencies. In the linear combination, the total tendencies are rescaled so that, on average, the ensemble standard deviation matches the error of the ensemble-mean.

In terms of Eqs. (2)–(5), each NCEP-ENS member is given by the following time integration:

$$e_j(d;T) = e_j(d;0) + \sum_{k=1}^{N} \Delta T_{j,k} + \Delta S_{j,k} \tag{25}$$

$$\Delta T_{j,k} = \int_{T_k}^{T_k+6} \left[A_{0,j}(t) + P_{0,j}(t)\right] dt \tag{26}$$

$$\Delta S_{j,k} = \sum_{m=1}^{N} w_{m,k} \Delta T_{m,k} \tag{27}$$

where $A_{0,j}$ and $P_{0,j}$ represent the unperturbed model dynamical and physical tendencies (i.e., there is only one dynamical core and one set of parameterizations called with the same parameters). The jth subscript indicates that each finite-time tendency is different for each ensemble member because it is computed starting from a different initial state.

For each member, the 6-hour tendency $\Delta T_{j,\,k}$ is computed by integrating ahead in time the model equations for 6 hours. Once the 6-hour tendency $\Delta T_{j,\,k}$ has been computed, the stochastic perturbation $\Delta S_{j,\,k}$ is defined by a linear combination of all 6-hour tendencies. It should be evident now why the scheme is called STTP, where T stands for total, because the original tendencies also include the dynamical tendencies (by contrast, the ECMWF SPPT method does not perturb the tendencies due to the dynamics). Once the STTP term has been computed, the initial states are then advanced by 6-hour by adding ($\Delta T_{j,\,k} + \Delta S_{j,\,k}$). Then the process is repeated.

The initial conditions are defined by adding perturbations to the unperturbed initial conditions:

$$e_j(d,0) = e_0(d,0) + \mathrm{ETR}_j(d,0) \tag{28}$$

The unperturbed initial conditions are given by the NCEP T382L64 4D variational assimilation system.

As mentioned also in Section 5.1.7, note that the NCEP-ENS and the MSC-ENS forecasts are exchanged in real time to generate the multimodel products of the NAEFS (http://weather.gc.ca/ensemble/naefs/index_e.html).

5.1.9 The UKMO-ENS

The UKMO-ENS started production in August 2005 and stopped in July 2014. Its data has been available in the TIGGE archive since October 2006. UKMO is investigating the possibility to restart generating medium-range ensembles using a lagged approach.

The UKMO-ENS used an ETKF (Wei et al., 2006; Bishop et al., 2001; Bowler et al., 2007, 2008), to generate the initial perturbations. The ETKF is a simplified version of the EKF, a data-assimilation scheme that updates the mean state of the atmosphere and the error covariance in that estimate using background information obtained from an ensemble. The ETKF can be viewed as a transformation of the NCEP error-breeding scheme.

The initial perturbations were defined by a linear combination of the forecast perturbations from the previous cycle of the ensemble. The weights were calculated by considering the spread of the ensemble in the space of the observations (Wang et al., 2004), ensuring that the perturbations are centered on the control analysis, and that they are orthogonal. The perturbations were inflated to ensure that the ensemble has the correct spread for the next analysis time (corresponding to $t + 12$ hours), with the inflation factor calculated on the fly, so that the system automatically retuned itself to model changes.

Two stochastic physics schemes were used to represent the effects of structural and subgrid-scale model uncertainties: the Random Parameters (RP) scheme and the Stochastic Convective Vorticity (SCV) scheme. The RP scheme treated a selected group of parameters as stochastic variables (Lin and Neelin, 2000; Bright and Mullen, 2002) and involved perturbing a total of eight parameters from four physical parameterizations (large-scale precipitation, convection, boundary layer, and gravity wave drag). The main aim of the SCV scheme (Gray and Shutts, 2002) was to represent potential vorticity anomaly dipoles similar to the one typically associated with a mesoscale convective system.

In July 2014, when production was stopped, the UKMO-ENS included 1 unperturbed and 24 perturbed members, run twice a day with a 60-km resolution and 70 vertical levels, and up to a forecast range of 15 days. The forecast model included only a description of land and atmospheric processes (no wave or ocean model component is used).

In terms of Eqs. (2)–(5), each UKMO-ENS member is given by the following time integration:

$$e_j(d;T) = e_j(d;0) + \int_0^T \left[A_0(t) + P_j(t) + dP_j(t)\right] dt \qquad (29)$$

where A_0 represented the unperturbed model dynamical tendencies (i.e., there was only one dynamical core); P_j represented the fact that the physical parameterizations were integrated by perturbing some key parameters as defined by the RP scheme; dP_j represented the model uncertainty simulated using the SCV scheme; and the forecast length T was 15 days.

The initial conditions were defined by adding perturbations to the unperturbed initial conditions:

$$e_j(d,0) = e_0(d,0) + \mathrm{ETKF}_j(d,0) \qquad (30)$$

The unperturbed initial conditions were given by the UKMO high-resolution 4D variational assimilations, and the perturbations were defined by ETKF perturbations.

5.2 The S2S Global, Monthly Ensembles

Today, thanks to the S2S project, about 650 monthly forecasts are produced and exchanged every week. They are produced by 11 centers, with a horizontal resolution that ranges from about 32 km (ECMWF) to about 250 km (BMRC). A total of 7 of the 11 ensembles use an atmosphere/land/ocean coupled model, with some of them using also an interactive sea ice model.

Table 3 lists the key characteristics of the 11 S2S monthly (i.e., with a forecast length beyond 30 days) operational ensembles as of May 2017, generated by the following:

- BMRC
- CMA-BCC, the Beijing Climate Center (BCC) Climate Prediction System of the CMA
- The Environment and Climate Change Canada (ECCC)
- ECMWF
- The Hydrometeorological Research Center (HMRC) of the Russian Federation
- The Italian Istituto Scienze dell'Atmosfera e del Clima (ISAC-CNR)
- JMA
- KMA
- Météo France (MF)
- MSC
- NCEP
- UKMO

Because a very important component of extended-range ensembles is a reforecast suite, Table 4 lists the key characteristics of the reforecast suites of the 11 S2S ensembles. A reforecast suite is an ensemble of forecasts of past cases used to estimate the model climate. Knowledge about the model climate, and ensemble forecast errors, can be used to correct, a posteriori, the real-time forecasts.

Data from these ensembles can be downloaded from the S2S website hosted at ECMWF (http://apps.ecmwf.int/datasets/data/s2s/expver=prod/type=pf/).

In the following sections, we will briefly review the methodologies used to generate the S2S ensembles. Because the methods used to define the initial conditions and simulate model uncertainties are similar to the ones used in the medium-range ensembles, see Section 5.1 for more details about them. Hereafter, attention will focus more on whether the ensembles are coupled to a dynamical ocean, as well as whether they also include a reforecast suite. More information about these ensembles can be found on the ECMWF S2S website (https://software.ecmwf.int/wiki/display/S2S/Models).

5.2.1 The BMRC Monthly Ensemble

In summer 2017, the BMRC global, monthly ensemble simulated initial uncertainties using coupled BVs and perturbations generated using an ensemble data assimilation. Model uncertainties due to model errors were simulated using three slightly different model versions. It included 33 members (11 from each of three model versions), run twice a week (Sunday and Thursday at 00 UTC) for up to 9 months. The forecasts were generated using a coupled ocean-atmosphere model (Hudson et al., 2013). The ocean model is ACOM2 (Schiller and Godfrey, 2003), which is based on the Geophysical Fluid Dynamics Laboratory (GFDL) MOM2 ocean model with a 2.0° zonal resolution, a 0.5 meridional resolution, and 25 vertical levels. Forecasts are initialized from the POAMA Ensemble Ocean Data Assimilation System (PEODAS; Yin et al., 2011) that assimilates subsurface temperatures and salinity with an approximate Kalman filter. Surface-forcing fields (stress and evaporations minus precipitation) during the ocean assimilation cycle for the reforecast (hindcast) reanalyses are taken from the

TABLE 3 Key Characteristics of the 11 S2S Monthly Operational Ensembles

Center	Initial Perturbation Method	Model Uncertainty	Horizontal Resolution (km)—Vertical Levels (TOA, hPa)	Dynamical Ocean (model)	Forecast Length	No. of Members (Perturbed + Unperturbed)	No. of Runs per Week (UTC)
BMRC, monthly	BV (globe)	Yes	T42 (250 km)—L17 (10.0)	Yes (ACOM2)	62 days	33 + 1	Twice a week (00, Thu and Sun)
CMA-BCC, monthly	LAF method	No	T106 (110 km)—L40 (0.5)	Yes (MOM4)	60 days	4 + 1	Daily
ECCC, monthly	EnKF (globe)	Yes	0.45° × 0.45°—L40 (2.0)	No	32 days	20 + 1	Once a week (00, Thu)
ECMWF ENS, monthly extension	SV (NH, SH, TC) + EDA (globe)	Yes	Tco639 (0.14°; 16 km)—L91 (0.01) Tco319 (0.28°; 32 km)—L91 (0.01)	Yes (NEMO)	0–15 days 46 days	50 + 1	Twice per week (00, Mon and Thu)
HMCR, monthly	BV (globe)	No	1.1° × 1.4° (120 km)—L28 (5.0)	No	61 days	20 + 1	Once a week (00, Wed)
ISAC-CNR, monthly	LAF method + BV (globe)	No	0.8° × 0.56° (80 km)—L54 (6.8)	No	31 days	40 + 1	Once a week (00, Mon)
JMA, monthly	LAF method + BV (globe)	Yes	TL319 (0.70°; 60 km)—L60 (0.1)	No	34 days	24 + 1	Twice a week (12 UTC, Tue and Wed)
KMA	LAF + ETKF (globe)	Yes	N216 (0.8° × 0.56°; 60 km)—L85 (0.1)	Yes (NEMO)	60 days	4 + 1	Daily
MF, monthly	No	Yes	TL255 (80 km)—L91 (0.01)	Yes (NEMO)	32 days	50 + 1	Once a week (00, Thu)
NCEP	LAF + BV (globe)	No	T126 (100 km)—L64 (0.02)	Yes (MOM4P0)	45 days	16	Daily
UKMO	LAF + ETKF (globe)	Yes	N216 (0.8° × 0.56°; 60 km)—L85 (0.1)	Yes (NEMO)	60 days	4	Daily

Note: Listed in alphabetical order (column 1): initial uncertainty method (column 2), model uncertainty simulation (column 3), horizontal resolution and number of vertical levels (including the top of the atmosphere, in hPa; column 4), coupling to a dynamical ocean (with model name; column 5), forecast length in days (column 6), number of members for each run (column 7), and frequency (number of runs per week; column 8). For columns 3–5, blue (brown) color identifies the ensemble with the finest (coarsest) characteristics. NH, Northern Hemisphere; SH, Southern Hemisphere.

TABLE 4 Key Characteristics of the Reforecast Suite of the 11 S2S Operational, Global, and Monthly Ensembles

Center	No. of Years (From–To)	No. of Members Run Each Week	Initial Conditions for Atmosphere/Land	Initial Conditions for Ocean/Sea Ice
BMRC, monthly	34 (1981–2013)	33	ECMWF ERA-Interim	BMRC PEODAS
CMA-BCC, monthly	21 (1994–2014)	28 (4 × 7)	NCEP reanalysis	BCC GODAS
ECCC, monthly	20 (1995–2014)	4	ECMWF ERA-Interim for atmosphere and ECCC land scheme	–
ECMWF ENS, monthly extension	20 (most recent)	22 (11 × 2)	ECMWF ERA-Interim	ECMWF ORAS5
HMCR, monthly	26 (1985–2010)	10	ECMWF ERA-Interim	–
ISAC-CNR, monthly	30 (1981–2010)	1	ECMWF ERA-Interim	–
JMA, monthly	30 (1981–2010)	5	JMA reanalysis JRA-55	–
KMA	20 (1991–2010)	3	ECMWF ERA-Interim	UKMO reanalysis
MF, monthly	22 (1993–2014)	15 (every 2 weeks)	ECMWF ERA-Interim	MERCATOR reanalysis
NCEP	22 (1999–2010)	28 (4 × 7)	NCEP CFSR reanalysis	NCEP CFSR reanalysis
UKMO	23 (1993–2015)	7	ECMWF ERA-Interim	UKMO ocean

Note: Listed in alphabetical order (column 1): number of year (column 2), number of ensemble members (column 3), initial conditions for the atmosphere/land component (column 4) and initial conditions for the ocean/sea ice components (column 5). For columns 2–3, blue (brown) color identifies the ensemble with the highest numbers.

ECMWF ERA-interim reanalyses for a real-time system from the Bureau of Meteorology (BoM) global NWP system. Frequency of coupling is every 24 hours. The reforecast suite includes 33-member ensembles that are run once a week, and it covers a 34-year period (1981–2013). The reforecasts' initial conditions for the atmosphere and land are given by nudged fields constructed using the ECMWF ERA-Interim reanalysis, while the ocean initial conditions are given by the PEODAS reanalysis. Forecasts have been copied to the S2S database since January 2015.

5.2.2 The CMA-BCC Monthly Ensemble

In summer 2017, the CMA-BCC ensemble used a lagged average forecasting (LAF) method to simulate initial uncertainties, whereby forecasts initialized every 6 hours were combined to provide monthly forecasts (Liu et al., 2017). Each forecast consisted of four LAF ensemble members, which were initialized at 00 UTC of the first forecast day and 18, 12, and 06 UTC of the previous day, respectively. The atmosphere/land component had a T106L40 resolution, with the top of the model set to 0.5 hPa. Model uncertainties are not currently simulated.

The model is the fully coupled BCC Climate System Model version 1.2 (BCC-CSM1.2), which uses the GFDL MOM4 model with a 1°/3–1° horizontal resolution, 40 vertical levels, initialized from BCC Global Ocean Data. The reforecast suite includes four members, run in lagged mode as in the operational suite, and covers 21 years (1994–2014). The reforecasts' initial conditions for the atmosphere and the land are given by the NCEP reanalysis, while the ocean's initial conditions are given by the BCC GODAS. Forecasts have been copied to the S2S database since January 2015.

5.2.3 *The ECCC Monthly Ensemble*

In summer 2017, the ECCC ensemble (Lin et al., 2016) initial conditions were generated with an EKF, as was done for the medium-range ensemble (MSC-ENS; see Section 5.1.7 for more details on this method of simulating initial uncertainties). The atmosphere/land component has a 0.45° × 0.45° resolution with 40 vertical levels, with the top of the model set to 2.0 hPa. Model uncertainty was simulated by using different model configurations (multiparametrization physics), and stochastics schemes, as with the MSC-ENS.

The ECCC monthly ensemble is practically based on 21 MSC-ENS members (20 perturbed and 1 control) that are extended to day 32 once a week (Thursday at 00 UTC). The atmosphere/land model is not coupled to a dynamical ocean. The reforecast suite includes four members, run in lagged mode as in the operational suite, and covers 20 years (1995–2014). The reforecasts' initial conditions for the atmosphere are given by the ECMWF ERA-Interim, while the land-surface initial conditions are given by an offline run of the ECCC surface prediction system (SPS) cycle driven by near-surface atmospheric ERA-interim reanalysis and its associated precipitation. Forecasts have been copied to the S2S database since January 2015.

5.2.4 *The ECMWF Monthly Ensemble*

In summer 2017, the ECMWF monthly forecasts were generated with the medium-range/monthly ENS twice a week (00 UTC on Thursdays and Sundays), by extending the medium-range forecasts from 15 to 46 days with a lower resolution (Tco319, equivalent to about 36 km, instead of about 18 km). The atmosphere/land component was coupled to a dynamical ocean and sea ice model. See Section 5.1.4 for a description of the methods used to simulate initial and model uncertainties (see also Vitart et al., 2014). Reforecasts were based on 11-member ensembles run twice a week for the past 20 years.

The reforecasts' initial conditions for the atmosphere and the land are given by the ECMWF ERA-Interim reanalysis, and for the ocean, initial conditions are given by the ocean reanalysis ORAS5. Forecasts have been copied to the S2S database since January 2015.

5.2.5 *The HMRC Monthly Ensemble*

The HMRC global ensemble system (Tolstykh et al., 2014), includes 20 members and has a resolution of 1.125° × 1.40° with 28 vertical levels (top of the atmosphere at 5 hPa). It includes a simulation of initial uncertainties using a breeding method (for more information about the method, see Section 5.1.8). It does not simulate model uncertainties, and it is not coupled to a dynamical ocean model. It is based on 20 members, run weekly (Wednesday at 00 UTC) up to day 61. The ensemble reforecasts consist of a 10-member ensemble starting the same day and month as a Wednesday real-time forecast and cover 26 years (1985–2010). The reforecasts' initial conditions are given by the ECMWF ERA-Interim reanalysis. Forecasts are available in the S2S archive from April 2015.

5.2.6 The ISAC-CNR Monthly Ensemble

The ISAC-CNR ensemble is based on the grid-point, hydrostatic, and atmospheric general circulation model (GCM) GLOBO (Malguzzi et al., 2011; Mastrangelo et al., 2012). It is coupled to a slab ocean. The ensemble includes 40 perturbed members plus the control, and it is run once a week (Monday at 00Z) for up to 31 days. Initial perturbations are taken from the GEFS-NCEP operational ensemble with a mixed lagged-ensemble technique: 10 perturbed initial conditions are taken every 6 hours (on Sunday at 00, 06, 12, 18 UTC), with the unperturbed (control) analysis taken on Monday at 00Z. It does not simulate model uncertainties, and it is not coupled to a dynamical ocean. The ensemble reforecasts include 1-member run every 5 days for 30 past years, from 1981 to 2010. The reforecasts' initial conditions are given by the ECMWF ERA-Interim reanalysis. Forecasts are available in the S2S archive from April 2015.

5.2.7 The JMA Monthly Ensemble

The JMA global, monthly ensemble system simulates initial uncertainties using BVs and a lagged averaging scheme. Model uncertainties due to physical parameterizations are simulated using a stochastic scheme. The model does not include an ocean model. It includes 50 members (24 perturbed plus the control run on Tuesdays, and 24 plus the control run on Wednesdays), and it has a resolution of about 60 km and 60 vertical levels. The 50-member products are generated once a week (12 UTC on Wednesdays) by merging the lagged-forecasts generated on the previous Wednesday and Tuesday, up to day 34. The JMA reforecasts use a fixed set of dates, which means that the reforecasts are produced once from a frozen version of the model and are used for a number of years to calibrate real-time forecast. The JMA reforecasts consist of a five-member ensemble running three times a month from 1981 to 2010. The start dates correspond to the 1st, 11th, and 21st of each month at 00Z minus 12 hours (February 28 instead of February 29). The reforecasts' initial conditions for the atmosphere and the land are given by the JMA reanalysis JRA-55 (Kobayashi et al., 2015). Forecasts are available in the S2S archive from January 2015.

5.2.8 The KMA Monthly Ensemble

The KMA global monthly forecasts are generated using the KMA seasonal ensemble, which is developed at the UKMO (see Sections 5.2.11 and 5.1.9). Forecasts are available in the S2S archive from April 2016. It simulates initial uncertainties using an ETKF method, and model uncertainties using a SKEB scheme. Monthly forecasts are generated by combining lagged forecasts generated over a number of days. Each day, four ensemble members are initialized and run—two for 75 days and two for 240 days. The reforecasts include three members initialized on frozen dates (the 1st, 9th, 17th, and 25th of each month) for 20 years

(1991–2010). The reforecasts' initial conditions for the atmosphere and the land are given by the ECMWF ERA-Interim, and the ocean initial conditions are given by the UKMO ocean reanalysis. Forecasts are available in the S2S archive from April 2016.

5.2.9 *The MF Monthly Ensemble*

The MF ensemble has an 80-km resolution and 91 vertical levels, includes 51 members, and is run once a week (at 00UTC on Thursdays) for up to 32 days. It simulates initial uncertainties using a stochastic scheme (Batté and Déqué, 2012). The forecasts are generated using a coupled model, with the NEMO3.2 ocean model with a 1° horizontal resolution, and 42 vertical levels, initialized from unperturbed MERCATOR-OCEAN Ocean and Sea Ice Analysis. The reforecasts include 15-member ensembles starting the 1st and the 15th calendar day of each month for the period 1993–2014 (22 years). The reforecasts' initial conditions for the atmosphere and land are given by ERA-Interim, and for the ocean, the initial conditions are given by the MERCATOR-OCEAN ocean reanalysis.

5.2.10 *The NCEP Monthly Ensemble*

The NCEP monthly ensemble includes 16 lagged forecasts run every day for up to 45 days, with a T126 (about 100-km) resolution and 64 vertical levels, up to 0.02 hPa. The model is CFSv2, the NCEP global ensemble developed for monthly and seasonal applications (Saha et al., 2014). The 16-member ensemble includes lagged forecasts, with 3 perturbed members and 1 unperturbed member, which are run every day at 00, 06, 12, and 18 UTC. Perturbations are defined using BVs (see Section 5.1.8). It does not simulate model uncertainties. It uses a coupled model, with the MOM4P0 ocean model. The reforecasts include four lagged forecasts, run every day at 00, 06, 12, and 18 UTC, for 22 years (1999–2010). The reforecasts' initial conditions are generated using the NCEP coupled reanalysis CFSR. Forecasts are available in the S2S archive from January 2015.

5.2.11 *The UKMO Monthly Ensemble*

The UKMO monthly forecasts are generated by using some of the individual members of their seasonal ensemble, GloSEA5 (MacLachlan et al., 2014). GloSEA5 uses the coupled HadGEM3 model, with the following components:

- Atmosphere: MetUM (Walters et al., 2011), Global Atmosphere 3.0
- Land surface: Joint UK Land Environment Simulator (JULES; Best et al., 2011), Global Land 3.0
- Ocean: NEMO (Madec, 2008), Global Ocean 3.0
- Sea ice: The Los Alamos Sea Ice Model (Community Ice CodE (CICE); Hunke and Lipscomb, 2010), Global Sea ice 3.0

The GloSEA5 atmosphere and land component has a 0.833° × 0.556° grid, and the ocean and sea ice models have a 0.25° resolution. It simulates initial uncertainties using a combination of lagged forecasts and ETKF perturbations (see Section 5.1.9). Model uncertainties are simulated using a random parameter perturbation scheme (Bowler et al., 2008) to simulate, using a single model framework, a range of representations of physical processes, thus sampling a large part of model uncertainty. Each day, four forecasts are run, for up to 60 days.

The reforecasts include seven members initialized on frozen dates (the 1st, 9th, 17th, and 25th of each month) for 23 years (1993–2015). The reforecasts' initial conditions for the atmosphere and the land are given by the ECMWF ERA-Interim, and the ocean's initial conditions are given by the UKMO ocean reanalysis. Forecasts are available in the S2S archive from February 2015.

5.3 Does an Ensemble Performance Depend on its Configuration?

It is evident from Sections 5.1 and 5.2 that there is not a unique recipe for generating reliable and accurate ensembles, and indeed the nine TIGGE ENS and the 11 S2S ensembles use different methods to simulate the effects' initial condition and model errors. It could be interesting to look at some average scores and assess whether we can link their performance with their configuration. To investigate this aspect, let us compare the performance of the TIGGE ensembles for one season: winter 2016–17.

To understand the degree of sensitivity of ensemble performance to the quality of the model and the initial conditions, resolution and membership, and perturbation methods followed to simulate initial and model uncertainties, ensemble forecasts generated by the different centers have been compared in the past. Buizza et al. (2005) compared the performance of the ECMWF, MSC, and NCEP ensembles for one season (summer 2002). They concluded that the performance of ensemble prediction systems strongly depends on the quality of the data-assimilation system used to create the unperturbed (central) initial condition, as well as the numerical model used to generate the forecasts. They also stated that a successful ensemble prediction system should simulate the effect of both initial and model-related uncertainties on forecast errors.

Park et al. (2008) compared the first set of TIGGE forecasts and concluded that there was a large difference between the performances of the single ensembles. For the 500-hPa geopotential height over the Northern Hemisphere in the medium range (say, around forecast day 5), the difference in predictability between the worst and the best control or ensemble-mean forecasts was about 2 days, while the difference between the worst and the best probabilistic predictions was larger, about 3–4 days.

More recently, Hagedorn et al. (2012) not only discussed the performance of the TIGGE ensembles, but also investigated the possibility of combining the TIGGE ensembles into a multimodel one. The skill of this multimodel ensemble was compared with the skill of an ensemble defined by calibrated ECMWF-ENS, with the calibration based on the ECMWF ensemble reforecast suite. Considering the statistical performance of global probabilistic forecasts of 850-hPa and 2-m temperatures, they concluded that a multimodel ensemble containing the nine TIGGE ensembles did not improve on the performance of the best single model, the ECMWF-ENS. However, a reduced multimodel system, consisting of only the four best ensembles (ECMWF, MSC, NCEP, and UKMO) showed an improved performance. They also concluded that the ECMWF-ENS was the main contributor for the improved performance of the multimodel ensemble; that is, if the multimodel system did not include the ECMWF contribution, it was not able to improve on the performance of the ECMWF-ENS alone. These results were shown to be only marginally sensitive to the verification data set used.

Yamaguchi et al. (2012) looked at the tropics and compared tropical cyclone track predictions from each TIGGE ensemble, with tracks generated by what they called a multicenter grand ensemble (MCGE) that included all of them. Their work considered 58 tropical cyclones in the western North Pacific from 2008 to 2010. In the verification of tropical cyclone strike probabilities, the BSS of the MCGE was larger than that of the best single ensemble, which was the ECMWF-ENS in the medium range. By contrast, the reliability was improved by the MCGE, especially in the high-probability range.

Buizza (2014) compared maps of the initial time ensemble-mean (i.e., the central point of the distribution spanned by an ensemble) and the ensemble spread. He showed that the ensemble-means at the initial time are very similar over the extratropics, while at $t + 48$ hour, some small differences start to appear. By contrast, they are more different over the tropics, even at the initial time. This reflects the fact that the TIGGE centers' analyses are very similar in the extratropics, while they differ substantially over the tropical region—for instance, between 20°S–20°N (see also Park et al., 2008). By contrast, initial time differences are much larger in terms of ensemble spread. The KMA, MSC, and UKMO ENSs have the largest initial spread, to compensate for their initial perturbations' slower growth rate. As shown, for instance, in Buizza et al. (2005), initial perturbations generated using the EnKF and EKTF methods grow slower than forecast error, and thus to get the right level of spread in the medium range, the initial perturbations must be set with rather large amplitudes. At forecast time $t + 48$ hour, the spread of all ensembles are more similar than at the initial time, in terms of both coverage and local maxima. Broadly speaking, they are also rather similar in terms of structure, with maxima localized in the same areas, where the jet-stream is stronger or where cyclonic developments occur. This is not surprising because all these ensembles have been designed for the medium range (say, 3–10 forecasts), and to achieve this, they have all been configured to have, on average, the right level of spread from about forecast days 2–3.

Fig. 7 gives a brief, updated view, as of May 2017, of the performance of five of the TIGGE ensembles for the probabilistic prediction the 850-hPa temperature. The ensemble performance has been measured using the CRPS, which is the equivalent of the RMSE for probabilistic forecasts. The CRPS measures the average distance between the forecast probability density function and the observed density function, which is a delta function if observation errors are not taken into account (in this particular case), or a very narrow distribution if they are taken into account. The CRPS is zero for a perfect forecast. The corresponding skill score, the CRPSS, has been defined using a climatological forecast as reference.

Figs. 8 and 9 show the reliability of four of the TIGGE ensembles: they contrast the RMSE of the ensemble-mean against the ensemble spread for the 850-hPa temperature over the Northern Hemisphere and the tropics at four forecast times. In a reliable ensemble, the two values should be similar, and the reliability curve on the scatter diagram should lie as close as possible to the diagonal.

These results confirm earlier indications (see the previous section and related references) that for all ensembles, the performance is better over the extratropics than the tropics, and synoptic-scale features are easier to predict than small-scale, surface phenomena. Considering the individual ensembles, these results confirm that overall, the ECMWF ensemble continues to provide the most accurate and reliable forecasts. As pointed out in earlier studies (see, e.g., Buizza et al., 2005), this is due to a combination of reasons:

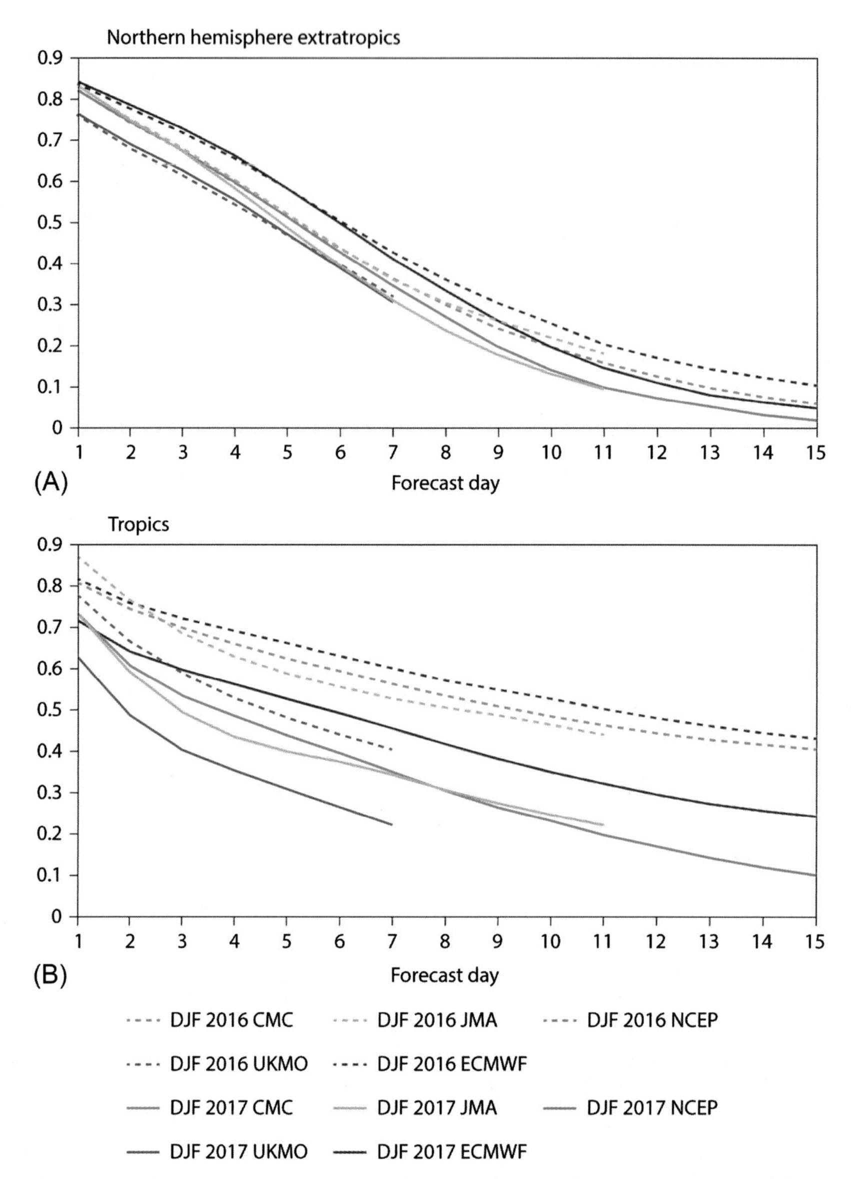

FIG. 7 Winter 2016 (D15–JF16, *dotted lines*) and winter 2017 (D16–JF17, *solid lines*) average CRPSS of the probabilistic prediction of the 850-hPa temperature over the Northern Hemisphere (20°N–90°N;A) and the tropics (20°S–20°N; B) of five TIGGE ensembles: ECMWF *(red)*, CMC Canada *(cyan)*, JMA Japan *(orange)*, NCEP *(green)* and UKMO (*blue*, only up to forecast day 7). Each ensemble has been verified against its own analysis.

- ECMWF has developed a very accurate model, and its data-assimilation system generates very accurate good initial conditions.
- The ECMWF ensemble uses the finest resolution and the largest ensemble size.
- The ECMWF ensemble uses very good methods to simulate the impact of initial and model uncertainties on forecast accuracy.

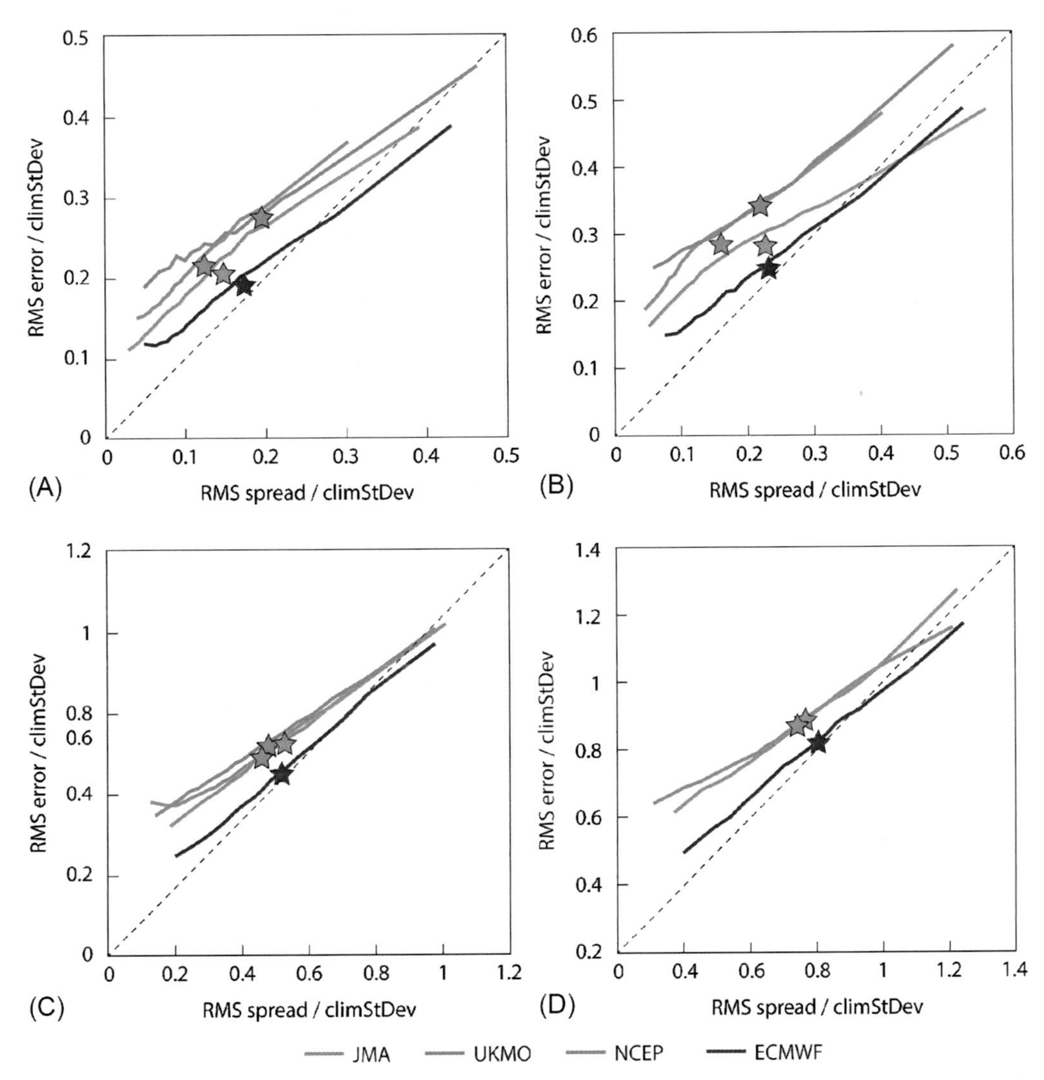

FIG. 8 Winter 2017 (D16–JF17, *solid lines*) average scatterplot of the RMSE of the ensemble-mean (y-axis) and the ensemble standard deviation (x-axis) of the prediction of the 850-hPa temperature over the Northern Hemisphere (20° N–90°Nl) of four TIGGE ensembles: ECMWF *(red)*, JMA Japan *(orange)*, NCEP US *(green)* and UKMO *(blue*, only up to forecast day 7), at four forecast times: (A) day 1, (B) day 2, (C) day 6, and (D) day 10. Each ensemble has been verified against its own analysis.

A very good model helps to slow the growth rate of the forecast error and makes it easier to assimilate the observations. A good data-assimilation system, based on a good model, makes the initial conditions being more accurate. Using a high resolution makes it possible for the model to simulate more accurately the small scales and their interaction with the larger (synoptic, planetary waves) scales. Adopting a large ensemble size increases the ensemble reliability, makes the resolution in probability space finer, and has a positive impact on probabilistic predictions, especially of rare events. Adopting good methods to simulate initial and model uncertainties improves both the ensemble reliability and its accuracy.

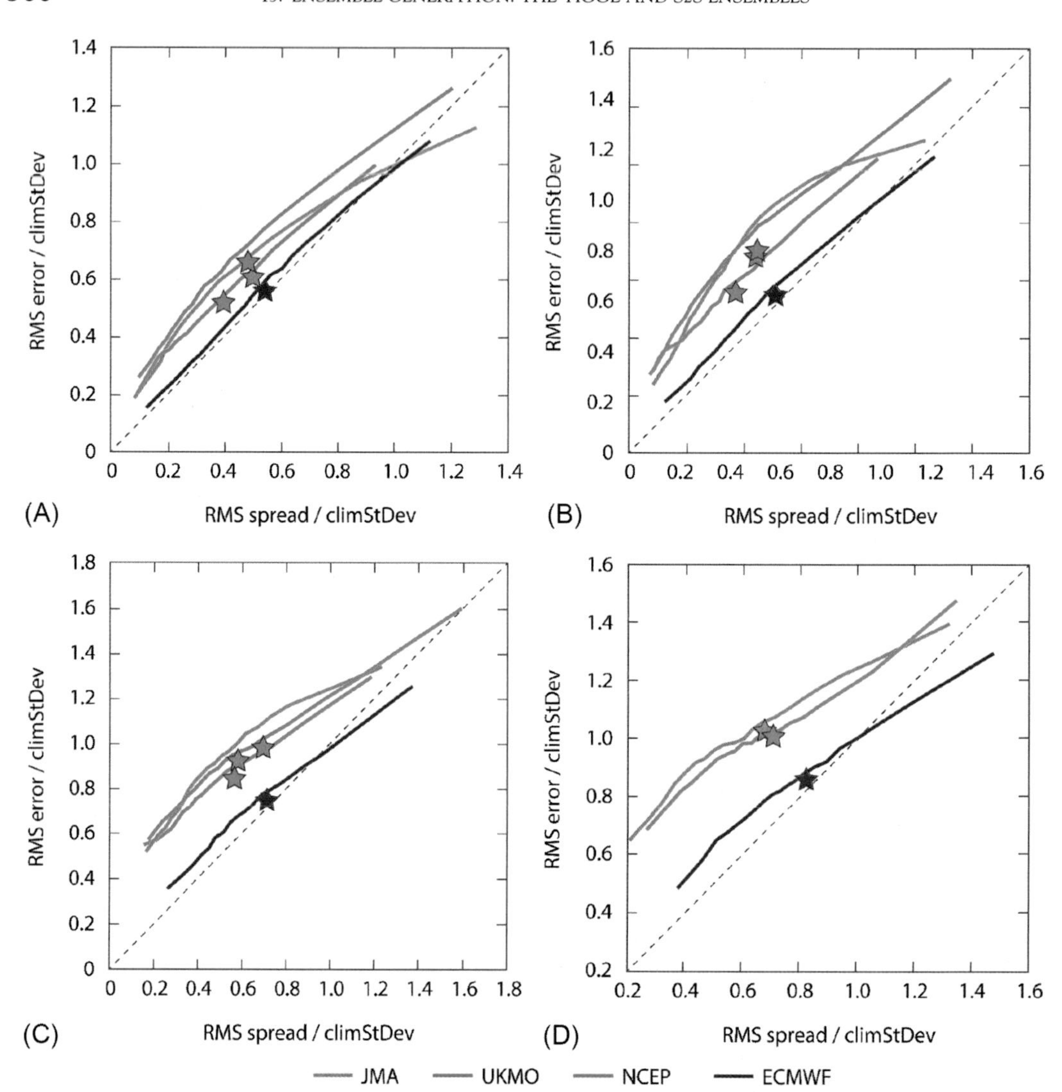

FIG. 9 Winter 2017 (D16–JF17, *solid lines*) average scatterplot of the RMSE of the ensemble-mean (y-axis) and the ensemble standard deviation (x-axis) of the prediction of the 850-hPa temperature over the tropics (20°S–20°N) of four TIGGE ensembles: ECMWF *(red)*, JMA Japan *(orange)*, NCEP US *(green)* and UKMO (*blue*, only up to forecast day 7), at four forecast times: (A) day 1, (B) day 2, (C) day 6, and (D) day 10. Each ensemble has been verified against its own analysis.

6 ENSEMBLES: CONSIDERATIONS ABOUT THEIR FUTURE

In the last 25 years, following the pioneering work in such studies as Thompson (1957), Epstein (1969a), Lorenz (1969a,b), and Leith (1965), we have witnessed a paradigm shift in operational NWP from a deterministic approach, based on a single forecast, to a probabilistic

one, whereby multiple ensembles are used to estimate the probability density function of initial and forecast states. The year 1992 saw the implementation of the first two operational ensemble systems, at ECMWF in Europe and at NCEP in the United States. They were followed by MSC in Canada in 1995 and other centers a few years later.

Today, it is widely accepted that forecasts have to include uncertainty estimations, confidence indicators that allow forecasters to estimate how predictable the future situations are. These estimates can be expressed in various ways, as a range of possible scenarios or as probabilities that events of interest can occur. Today, short- and medium-range forecasts, monthly and seasonal forecasts, and even decadal forecasts and climate projections are based on ensembles, so not only the most likely scenario, but also its uncertainty can be estimated. Furthermore, ensembles are widely used to provide an estimate of the initial state uncertainty, in order to estimate the analysis error more accurately.

The discussion in this chapter has illustrated that although all ensembles have been designed with the same goal (to estimate the probability density function of forecast states), different techniques have been used to simulate initial and model uncertainties. Thus, a first conclusion that we can draw is that there is no unique recipe to generate reliable and skillful ensembles. Considering the performance of the TIGGE ensembles, a second conclusion that we can draw is that the system design affects performance. A third possible conclusion is that TIGGE, S2S, and similar projects designed to give data access to the scientific community and the users of ensemble forecasts are essential and fundamental resources that can help our understanding of how best to design these ensembles, so that we can extract predictable signals from our forecasts.

Looking into the future, two trends can be detected in the way that ensembles are being upgraded:

- A move toward an Earth-system approach to modeling and assimilation
- A move toward a seamless approach in the design of the analysis and forecast ensembles

The first trend is linked to results obtained in the past two decades that showed that by adding relevant processes, we can further improve the quality of the existing forecasts, and we can further extend the forecast skill horizon at which dynamical forecasts lose their value. Buizza and Leutbecher (2015), for example, looked at the evolution of the skill of the ECMWF ensemble from 1994 to date, and concluded, that "Forecast skill horizons beyond 2 weeks are now achievable thanks to major advances in numerical weather prediction. More specifically, they are made possible by the synergies of better and more complete models, which include more accurate simulation of relevant physical *processes* (e.g. the coupling to a dynamical ocean and ocean waves), improved data-assimilation methods that allowed a more accurate estimation of the initial conditions, and advances in ensemble techniques."

The second trend comes partly from scientific reasons and partly from technical reasons. From the scientific point of view, for example, there is evidence that processes that were thought to be relevant for the extended range are also relevant for the short range. An example comes from the introduction of a dynamical ocean in the ECMWF ensembles. We started using a coupled ocean-land-atmosphere more for the seasonal and the monthly timescales (see, e.g., Vitart et al., 2014; Molteni et al., 2011; Anderson et al., 2007), and then we introduced it in the medium-range ensemble once we realized that it could contribute to improving its reliability and accuracy (Janssen et al., 2013). From the technical point of view, having an integrated approach whereby the same model is used in analysis and prediction mode, from day 0 to year 1, simplifies maintenance and the implementation of upgrades. Furthermore, it helps the diagnostic and evaluation of a model version because tests carried out over different timescales can help identifying undesirable behaviors that could lead to forecast errors. The ECMWF strategy for the next 10 years (ECMWF Strategy 2016–2025) has these two aspects as two key pillars that should help us continue to advance our science and improve the quality of our forecasts.

I think that the future ensembles will have a higher degree of coupling in the way their initial conditions are generated, and the ensembles of analyses and forecasts will be more seamless than today.

Consider, for example, ECMWF: Today, we use four different ensembles to provide estimates of the PDF of analysis (i.e., initial time) and forecast states:

- The 25-member EDA and the high-resolution single analysis, to generate the initial conditions for the atmosphere and land components of its ensembles
- The 5-member ORAS5, to generate the initial conditions for the dynamical ocean and the sea ice
- The 51-member medium-range/monthly ENS, to generate sub-seasonal forecasts
- The 51-member SEAS4 (upgraded to SEAS5 in November 2017), to generate seasonal forecasts

Both ENS and SEAS4 use an Earth-system model based on a dynamical ocean (currents, waves, and sea ice) and a dynamical land/atmosphere model, but they use different model versions and slightly different initialization methods, and different resolutions as well. A step forward has been made in November 2017, when the seasonal ensemble was upgraded to SEAS5, which has the same resolution and the same initialization as ENS. In terms of the settings of the atmosphere model component, SEAS5 is still slightly different than ENS.

Considering the initial conditions, we are still generating them in uncoupled mode. The (atmosphere/land) EDA includes only 25 members, and the (ocean/sea ice) ORAS5 has only 5 members instead of 51. Each ORAS5 ocean member is not driven by one EDA member, but rather by the unperturbed analysis fields, perturbed in a way that is not consistent with any EDA member. Furthermore, considering the atmosphere, because the EDA is very costly, we do not have enough computing resources to complete the 25 assimilation cycles in time to initialize the forecasts. Therefore, we use EDA short-range forecasts (started from the previous EDA cycle) to generate the atmosphere/land initial perturbations. This means that we are not yet generating the initial conditions with the same degree of coupling as in the forecast model. Although it is not clear what level of coupling is needed in data assimilation, we think that we could make the ensemble of analyses and of forecast more seamless, and we could increase the degree of coupling in the assimilation. An example of this latter aspect is given by CERA-20C, the first coupled ocean/land/atmosphere reanalysis of the 20th century to be generated at ECMWF as part of the ERA-CLIM2 project.

Thus, we are using suboptimal initial conditions (they are generated in uncoupled mode), suboptimal initial perturbations (they are also generated in uncoupled mode and with different techniques in the atmosphere, land and ocean), and suboptimal model error schemes (we perturb only atmospheric processes). In other words, we are not yet using a seamless approach.

If enough computing resources are made available, in the next 5–10 years, we could evolve the existing four ensembles into one seamless coupled ensemble, with N (51, or possibly 101) coupled forecasts starting directly from N coupled initial conditions. The open questions that we will have to address to get there are the level of coupling required in assimilation (Do we need a strong coupling formulation in a coupled atmosphere/land/ocean data-assimilation scheme?), and the level of complexity that we will need to insert in the coupled models (How many other processes do we need to include? For example, do we need to include an interactive aerosol scheme to further extend the forecast skill horizon?).

7 SUMMARY AND KEY LESSONS

In this chapter, first we discussed in general terms how dynamical models can be used to produce S2S predictions. Then we introduced the concept of ensemble prediction, discussed how each ensemble's reliability and accuracy can be measured, and illustrated why ensembles provide more valuable information than single forecasts. Next, we briefly reviewed how data assimilation works and generates initial conditions by merging observations and estimates of the state of the system using data-assimilation procedures. We briefly reviewed how model uncertainties can be simulated in the ensembles. And we reviewed the methods used to initialize global, operational S2S ensembles and illustrated how the nine TIGGE ensembles and the 11 S2S ensembles are generated. We compared their key characteristics, highlighted the rationale behind their design, and discussed how their designs influence their performance. Finally, we discussed some considerations on how we expect the ensembles to evolve in the near future.

What are the key lessons that we can draw from this chapter?

- Initialization is important and influences performance in the early forecast ranges.
- Ensembles provide more complete and more valuable information than single forecasts.
- There is no unique way to generate ensemble forecasts, either in the way that initial and model uncertainties are simulated, or in the ensemble configuration (e.g., number of members, forecast length, resolution, and frequency).
- An ensemble's reliability and accuracy depend on the quality of the model and the initial conditions, on the resolution and number of members, and on the methods used to simulate initial and model uncertainties.
- In the future, we expect ensembles to include more and more relevant processes and move toward an Earth-system approach to modeling and assimilation. We also expect an evolution toward a seamless approach in the design of the analysis, medium-range, S2S ensembles.

Ensembles are here to stay, and in the future, they will be used even more, both at initial time and at all forecast ranges, in order to help us take into account initial and model uncertainties in our predictions.

CHAPTER

14

GCMs With Full Representation of Cloud Microphysics and Their MJO Simulations

In-Sik Kang[*,†], *Min-Seop Ahn*[‡], *Hiroaki Miura*[§], *Aneesh Subramanian*[¶,||]

[*]Indian Ocean Operational Oceanographic Research Center, SOED/Second Institute of Oceanography, Hangzhou, China [†]Center of Excellence of Climate Change Research, King Abdulaziz University, Jeddah, Saudi Arabia [‡]School of Earth Environment Sciences, Seoul National University, Seoul, Republic of Korea [§]Department of Earth and Planetary Science, University of Tokyo, Tokyo, Japan [¶]AOPP, Department of Physics, University of Oxford, Oxford, United Kingdom [||]Scripps Institution of Oceanography, UCSD, San Diego, CA, United States

OUTLINE

1 INTRODUCTION

Although moist physical parameterizations have improved substantially in recent years, most recent general circulation models (GCMs) still have problems with simulating the precipitation statistics (Kang et al., 2015) and the Madden-Julian Oscillation (MJO), as evaluated

https://doi.org/10.1016/B978-0-12-811714-9.00014-0

by Hung et al. (2013) and Ahn et al. (2017) with the CMIP5 models. To overcome the limitation of parameterized convection, several recent studies have attempted to include full representation of cloud microphysical processes, so-called explicit convection, in regional and global models (Miura et al., 2007b; Benedict and Randall, 2009; Kang et al., 2016). Moncrieff and Klinker (1997) showed that explicit convection produces a more realistic simulation of superclusters than parameterized convection does. More recently, Holloway et al. (2013, 2015) performed MJO simulations with parameterized and explicit convection with varying horizontal mesh sizes and found better performance with explicit moist physics.

The global cloud-resolving model (global CRM), which includes the cloud microphysical processes explicitly expressed by GCM state variables, has been tested by a Japanese group since early 2000 (Tomita et al., 2005; Satoh et al., 2014; and many others). This model is a global version of the CRM in a nonhydrostatic framework, with various horizontal resolutions from a few kilometers to 14 km. With several modifications and improvements of their earlier version of the global CRM, they recently reported that the model is able to reproduce the eastward propagation of the observed MJO and typhoon genesis reasonably well (Miura et al., 2007b; Oouchi et al., 2009; Miyakawa et al., 2014; Kodama et al., 2015).

One benefit of using CRM is to permit turbulent motions associated with clouds, although it is questionable if smaller-scale turbulence is essential for large behaviors of synoptic- or planetary-scale phenomena. The other benefit of using CRMs is to permit direct coupling between the fluid dynamics and the cloud microphysics. This provides a pathway to avoid difficulties of the "cumulus parameterization deadlock" (Randall et al., 2003) by computing transports and phase changes of water substances directly. However, this approach requires very heavy computing resources due to an ultrahigh horizontal resolution of an order of kilometers, and it is not possible at present to use this kind of model for sub-seasonal to seasonal (S2S) predictions. Moreover, their climatological behaviors have not been well described, and many aspects of cloud physical processes are not known yet.

The so-called superparameterization is a promising alternative strategy for representing the effects of moist convection explicitly through a CRM embedded within each grid of the GCM with a horizontal resolution of 100 km or more (Iorio et al., 2004; DeMott et al., 2007). The superparameterized GCM has been shown to simulate the MJO reasonably well with computational efficiency (Benedict and Randall, 2009; Zhu et al., 2009), although the MJO intensity is somewhat exaggerated. It is also pointed out that the superparameterization has some deficiencies, in that it does not consider the interaction between clouds in neighboring GCM grids and the cloud properties of each grid point must be quickly adjusted to the boundary conditions prescribed by GCM state variables.

Recently, Kang et al. (2015, 2016) presented a GCM with full representation of cloud microphysics at 50-km horizontal resolution. For this GCM, the cloud microphysics, which is expressed by GCM state variables, was modified to be suitable to 50-km horizontal resolution. Kang et al. (2015) demonstrated that this GCM is able to simulate the heavy and extreme precipitation statistics, which cannot be simulated with a GCM with convective parameterization, and better simulates the MJO compared to the conventional GCM (Kang et al., 2016). In particular, Kang et al. (2016) demonstrated that the cloud microphysics is not enough by itself to simulate the vertical profile of the moisture field, particularly in the lower troposphere, in a 50-km resolution GCM, and a shallow convection is required to simulate the mean low-level moisture and its anomalies with MJO timescales. Recently, they found that the

TABLE 1 Three Types of Global Models with Cloud Microphysics

	Global CRM	Superparameterized GCM	Gray-Zone GCM
Horizontal resolution	About 1 km	About 100 km	About 10 km
Model framework	Expansion of CRM to the global domain	CRM embedded in each GCM grid	Cloud microphysics expressed by GCM state variables
Cloud and convection	Explicitly resolved and computed by CRM physics	Explicitly resolved, but only within the GCM grid box	Modified cloud microphysics, plus convective parameterization
Cloud motion	Fully expressed	Only within the GCM grid (no interaction between GCM grids)	Only large-scale interaction (expressed by the GCM physics)
Computational cost	Very high	Relatively low	Moderate
Possibility of S2S prediction	No, in the near future	Yes	Yes, in the near future

GCM used by Kang et al. (2016) still produces dry bias in the middle and upper levels. For a further improvement of their GCM, they added a scale-adaptive deep convection to enhance the vertical transport of moisture and temperature, which helps improve the dry bias in middle- and upper-troposphere and MJO simulation.

In this chapter, we review three types of GCMs with cloud microphysics as mentioned here: an ultrahigh-resolution global CRM, a low-resolution, superparameterized GCM, and a medium-range-resolution GCM with cloud microphysics and a scale-adaptive cumulus parameterization. Characteristic differences among the three types of GCMs are summarized in Table 1. The three types of models are still being developed, and more time will be needed to create an operational system to adapt the models. Therefore, parameterized convections are still used in current operational systems, particularly for S2S predictions.

In this chapter, the development strategies of those GCMs and the simulation qualities of their precipitation climatology and MJO are described. The global CRM with an order of 1-km horizontal resolution is discussed, and its simulation quality along the equator is covered in Section 2. The superparameterized GCM with an order of 100-km horizontal resolution is described in Section 3, and the GCM with full representation of cloud microphysics and scale-adaptive convection at 50-km horizontal resolution is described in Section 4. A summary and concluding remarks are given in Section 5.

2 GLOBAL CRM

The global CRM, described here, is the Nonhydrostatic Icosahedral Atmosphere Model (NICAM; Satoh et al., 2008, 2014), which uses the finite-volume discretization of the fully compressible nonhydrostatic Euler equations on an optimized geodesic grid. The standard terrain-following height coordinate system is used in the vertical (Tomita and Satoh, 2004).

The Arakawa A-grid staggering is used in the horizontal discretization (Tomita et al., 2001). The current standard package of the physics schemes is a single-moment microphysics of Tomita (2008), a modified version of the Mellor-Yamada turbulence (Noda et al., 2010), a two-stream radiative transfer model of Sekiguchi and Nakajima (2008), an improved version of the Louis scheme for the surface fluxes (Uno et al., 1995), the MATSIRO land surface model (Takata et al., 2003), and the mixed-layer ocean. Here, we describe some of the development history of this NICAM, with a particular focus on the MJO and related tropical phenomena.

In the first trial of the global CRM simulation with NICAM (Miura et al., 2007a), tropical cyclones were exaggerated, but individual convective clouds were muted. The cause of this unrealistic result was a significant overestimation of the upward transport of moisture by a turbulence scheme. Holloway et al. (2013) found that their 4-km simulations were sensitive to the choice of a turbulence scheme. It was also found from sensitivity tests of Miura et al. (2007a) that the extent of cloud organizations depended strongly on the subgrid-scale turbulence; weak and strong vertical mixings induced scattered and organized convections, respectively.

In a subsequent study, Miura et al. (2007b) updated the boundary layer scheme from a dry Mellor-Yamada to a moist Mellor-Yamada (Noda et al., 2010), with a global, quasi-uniform, 7-km mesh. The updated model reproduced realistic eastward movement of the convectively active region of the MJO event initiated over the Indian Ocean in mid-December 2006. Liu et al. (2009b) examined the fidelity of this 7-km simulation further. The 1-week simulation using a global 3.5-km mesh replicated the distribution of clouds and its evolution over the maritime continents realistically. Recently, this model was used to simulate the first MJO event during CINDY2011/DYNAMO (Fig. 1A). As seen in Fig. 1A and B, the model appears to reproduce the event to some extent. Fig. 1 shows the longitude-time cross section of the Outgoing Longwave Radiation (OLR) averaged between 10°S and 10°N.

For the project called Athena (Kinter et al., 2013), which addressed a comparison between NICAM and the Integrated Forecast System (IFS) of the European Centre for Medium-range Weather Forecasts (ECMWF), some physics packages of NICAM had been updated from those of the MJO simulation by Miura et al. (2007b). One significant change was in the cloud microphysics scheme. A simple scheme of Grabowski et al. (1998) was replaced with a one-moment bulk microphysics scheme (NSW6) of Tomita (2008).

The parameters of the NSW6 scheme were chosen to increase ice clouds, so that the energy budget at the top of the atmosphere (TOA) was approximately balanced. The motivation of using those parameters was to address the issue of future cloud changes under global warming (Satoh et al., 2012; Tsushima et al., 2014). Although the energy balance at the TOA was improved, this tuning caused severe biases in the fields of temperature, clouds, and the precipitation in the tropics. The atmospheric stability became fairly strong, particularly in the upper troposphere because of the positive biases of upper-tropospheric moisture and clouds. As a result, convective activity was suppressed over the ocean and was exaggerated over land areas because the turbulence was more strongly forced by the surface-sensible heat flux over land.

When the first MJO event during CINDY2011/DYNAMO was simulated with the same configuration as in the Athena project, the NICAM model almost failed to reproduce the event (Fig. 1C), while the same configuration as Miura et al. (2007b) reproduced it to some extent (Fig. 1B), as mentioned previously. From these tests, it turned out that the set of the

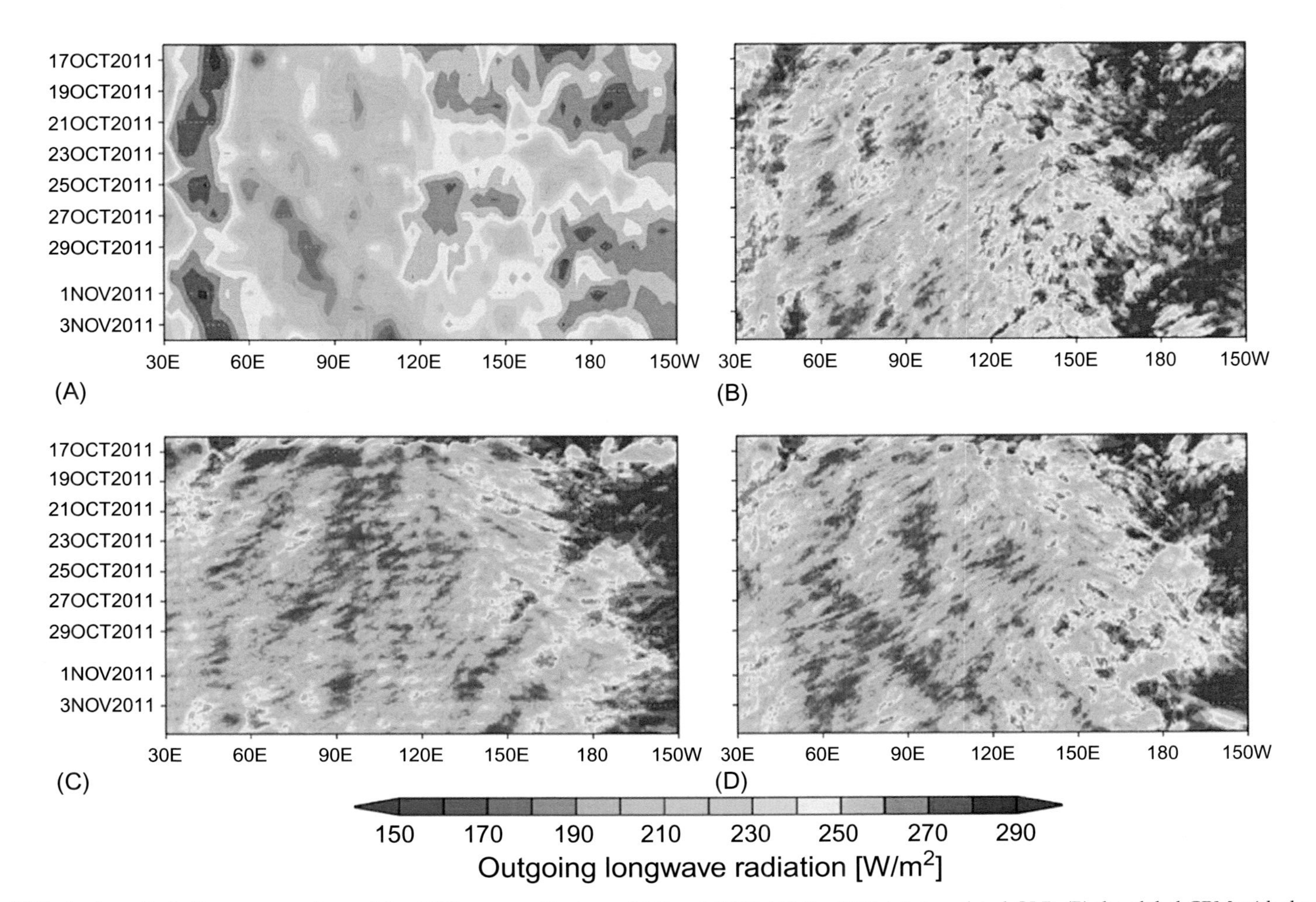

FIG. 1 Longitude-time cross sections of the OLR, averaged between 10°S and 10°N. (A) the NOAA Interpolated OLR, (B) the global CRM with the Grabowski et al. (1998) scheme, (C) with the Tomita (2008) scheme with the Athena parameters, and (D) with the Tomita (2008) scheme with the retuned parameters.

NSW6 parameters chosen to increase ice clouds (Satoh et al., 2012; Tsushima et al., 2014) was unsuitable for the MJO simulations. Next, the parameters were retuned (Miura et al., 2012) to recover the MJO event (Fig. 1D). Note that the important parameters tuned were the autoconversion rate of the cloud ice and the falling speeds (terminal velocities) of rain and snow.

It is obvious that these parameters can strongly influence the vertical distribution of water substances and water vapor. The better performance of the new set of the NSW6 parameters is not limited to the first event of CINDY/DYNAMO, but it seems robust for various MJO events. Miyakawa et al. (2014) used the NICAM on the K computer with the retuned NSW6 scheme and showed that the skill score of MJO prediction was maintained for 26–28 days, depending on the MJO phase at the time of model initialization.

This experience of the loss and the recovery of the MJO in NICAM with the NSW6 scheme informs us that the global or near-global CRMs are still sensitive to the vertical redistribution of moisture that is inevitably represented by the subgrid-scale unresolved processes. It is noted that the MJO simulations of NICAM are also sensitive to the settings of the surface latent heat flux. These results might disappoint us that a "cloud microphysics deadlock" will emerge, even if we go "beyond the deadlock" (Randall, 2013) of the cumulus parameterization. It is also noted that all failures listed in this section are, to a greater or lesser extent, associated with an unintended use or artificial tuning of physical parameterizations.

Hopefully, we may not need to be afraid of tuning the subgrid-scale processes unnecessarily if global or near-global CRMs are appropriately constrained by a variety of observations. Although more than a decade has passed since Tomita et al. (2005) performed the first global CRM simulation using a quasi-uniform, 3.5-km mesh under aqua-planet conditions, only limited knowledge has been obtained about the characteristics of global CRMs. We need to continue to develop cloud microphysics schemes (Seiki et al., 2015) and to improve them through comparisons with satellite data and in situ observations (Masunaga et al., 2008; Inoue et al., 2010; Dodson et al., 2013; Hashino et al., 2013; Roh et al., 2017).

3 SUPERPARAMETERIZED GCM

The GCM used in the present superparameterized GCM is from the ECMWF IFS (Wedi et al., 2013). The model is triangularly truncated at a spherical harmonics total wave number 159, which is equivalent to a horizontal resolution of about 112.5 km in the tropics. In the vertical, the atmosphere is discretized into 91 layers between the surface and the 5-hPa level (about 35-km altitude). The integration time-step is 3600 s. The CRM used for the superparameterized IFS is a three-dimensional (3D) model developed initially at the Colorado State University, as described in detail by Khairoutdinov and Randall (2003). The CRM is based largely on a large eddy simulation (LES) model by Khairoutdinov and Kogan (1999).

The prognostic thermodynamic variables include the liquid/ice-water moist static energy, the total nonprecipitating water, and the total precipitating water. The model is run at a high horizontal resolution of 4 km. A detailed description of the model dynamic and thermodynamic equations can be found in Khairoutdinov and Randall (2003). This CRM has been coupled to global models in a multiscale modeling framework (Khairoutdinov et al., 2005;

Randall, 2013), in which the CRM is embedded in each GCM grid and is constrained by large-scale forcing and surface forcing fields from GCM grid-mean fields. In the present superparameterization, the CRM is embedded in each GCM grid of the IFS global model. In this discussion, the model will be referred to as *SPIFS*.

The two atmospheric models, one with superparameterization (SPIFS) and the other with conventional convective parameterization (IFS), were run for 4 years, and the mean climate and intraseasonal variability were compared to corresponding observations. The mean precipitation along the Inter Tropical Convergence Zone (ITCZ) in the IFS experiment is comparable to that in the Global Precipitation Climatology Project (GPCP) observations, while the SPIFS experiment has too much precipitation along the ITCZ band. The wave number-frequency diagram for precipitation, as diagnosed by Wheeler and Kiladis (1999), shows the tropical variability in the convectively coupled equatorial waves (not shown). The IFS experiment shows excessive power in the Kelvin wave band and reduced power in the MJO band (wave numbers 1–5 at frequencies of 30–90 days). This bias is reduced in the SPIFS experiments, with improved Kelvin waves and an increase in MJO power. Similar improvements in the MJO representation from the superparameterization also have been shown in other models, such as the NCAR Community Atmosphere Model (Kim et al., 2009) and in the National Centers for Environmental Prediction (NCEP) CFSv2 (Goswami et al., 2011).

We further examine the usefulness of the superparameterized IFS in a forecast mode, in particular in the generation of ensemble perturbations of the initial condition for the ensemble forecast. The ECMWF extended-range ensemble prediction system (up to 46 days) uses initial perturbations from ensemble data assimilation and singular vectors operationally. We conducted two sets of hindcast experiments. The first one with the ensemble superparameterization (ESP), where we use no initial condition perturbations on the large-scale GCM grid, but the CRM embedded in the SPIFS was perturbed at the initial condition. In the current set of experiments, we perturbed the CRM, with perturbations given to the temperature field in the boundary layer. The bottom five layers of the CRM grid cells are perturbed with $\pm 0.5°C$ multiplied by a Gaussian random number. Therefore, the IFS grid variables are unperturbed in the ensemble setup at the initial time step. Hence, every ensemble member has exactly the same initial conditions, while the subgrid CRM has perturbed different initial conditions in each ensemble member. In this vein, we propose modeling uncertainty in deep convection using the superparameterization approach as a process-level uncertainty modeling framework (Palmer, 2012). In the second set of experiments, we used a convection parameterized IFS ensemble with initial perturbations (IniPert) on the GCM grids. The model runs were initialized from February 1 every year from 1989 until 2009. The MJO skill scores were computed for these 20 years for comparison. Thus, we have tested the paradigm to represent convective error growth in our current experiments by comparing it to the IFS forecasts, with perturbations only on the GCM initial conditions.

Fig. 2A shows the root-mean-square error (RMSE; continuous lines) and the ensemble spread (dashed lines) for the 20 years of wintertime forecasts. The RMSE for the first two principal components of the multivariate empirical orthogonal function (EOF) analysis is computed as defined in Gottschalck et al. (2010). The ESP experiments in the first week have a higher RMSE, but also a larger ensemble spread. The IniPert experiment has a lower spread than the ESP experiments, and hence it is a less reliable ensemble. Yet, both experiments have

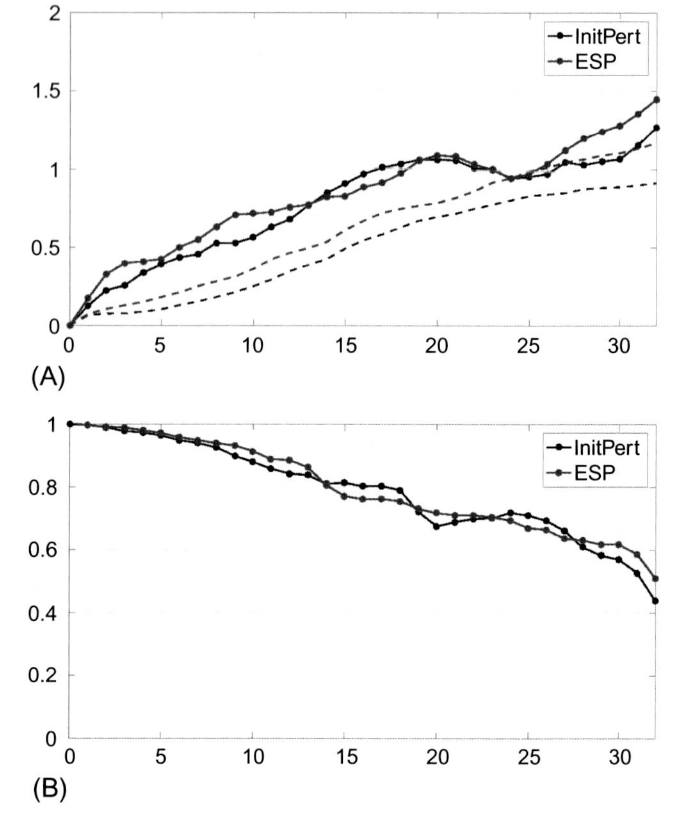

FIG. 2 (A) RMSE (continuous lines) and ensemble spread (dashed lines) and (B) bivariate correlations of the SPIFS forecasts for different lead times in days (*x*-axes). The RMSE and bivariate correlations are computed using the first two principal components (RMM1 and RMM2) of the multivariate EOF analysis of MJO, as in Gottschalck et al. (2010). The black and red lines represent the IniPert *(black)* and ESP *(red)* experiments, respectively.

a lower ensemble spread than the RMSE at all lead times, and thus are under dispersive and unreliable EPSs for the MJO. The IniPert ensemble has the lower ensemble spread at all lead times than the ESP ensemble. This implies that the IniPert ensemble is less reliable for forecasting the MJO in the second and third weeks. Fig. 2B shows the bivariate correlation coefficient of the first two principal components of the multivariate EOF analysis for the MJO over the 20 years of wintertime forecasts. The bivariate correlation coefficient is computed as defined in Gottschalck et al. (2010). The RMSE skill score for each experiment shows similar relative skill, as does the bivariate correlation metric. The skill in MJO forecasts is very comparable for the first week in both experiments. In the second week, the ensemble with initial perturbations on GCM scales has the least skill, and the forecasts made in the ESP case have higher correlation and increased skill than the other experiments. As we move into the third and fourth weeks, the skills of both experiments drop and become comparable. Predictability at S2S timescales is a key challenge for the forecasting community (Vitart, 2014), and having a reliable forecasting system for the MJO is one of the key aspects of this challenge. The result shown here demonstrates the usefulness of the superparameterized GCM for S2S prediction, particularly for generating initial perturbations, but further studies may be needed to improve the prediction for longer lead times.

4 GCM WITH FULL REPRESENTATION OF CLOUD MICROPHYSICS AND SCALE-ADAPTIVE CONVECTION

This section describes an atmospheric GCM (AGCM) and a coupled GCM (CGCM) at a medium-range horizontal resolution of 50 km, with a comprehensive cloud microphysics and a scale-adaptive convective parameterization. The AGCM used in this section is a Seoul National University (SNU) model. The SNU AGCM has a finite volume dynamical core, with a hybrid sigma-pressure vertical coordinate developed by Lin (2004), represented by a 50-km horizontal resolution and 20 vertical levels.

The convective parameterizations include a deep convection scheme based on a simplified version of a Relaxed Arakawa-Schubert cumulus convection scheme (Moorthi and Suarez, 1992), a large-scale condensation scheme based on Le Trent and Li (1991), and a diffusion-type shallow convection scheme as described by Tiedtke (1984). Radiation processes are parameterized by the two-stream k-distribution scheme developed by Nakajima et al. (1995). A detailed description of the physical parameterizations of the AGCM can be found in Lee et al. (2001) and Kim and Kang (2012). The SNU CGCM is a coupled version of the SNU AGCM and the MOM2.2 Ocean GCM developed at the Geophysical Fluid Dynamics Laboratory (GFDL). The CGCM includes the mixed-layer model developed by Noh and Kim (1999). The ocean zonal resolution of the CGCM is 1.0°, and the meridional grid spacing is 1/3° between 8°S and 8°N and gradually increases to 3.0° between 30°S and 30°N, and 3.0° poleward. A detailed description of the CGCM can be found in Ham et al. (2010).

The cloud microphysics adopted in the present study is obtained from the Goddard Cumulus Ensemble (GCE) model, developed at the Goddard Space Flight Center of the National Aeronautics and Space Administration (NASA; Tao et al., 2003). The cloud microphysics includes the Kessler-type, two-category liquid water scheme and the three-category ice-phase scheme, developed by Lin et al. (1983) and Rutledge and Hobbs (1983, 1984). Based on the sensitivity experiments of microphysical processes to the horizontal resolutions, Kang et al. (2015) developed a modified cloud microphysics suitable for the 50-km resolution in order to overcome a resolution dependency of cloud microphysics (Weisman et al., 1997; Grabowski et al., 1998; Bryan et al., 2003; Jung and Arakawa, 2004; Pauluis and Garner, 2006; Arakawa et al., 2011; Bryan and Morrison, 2012).

The major parts of the modification involve the condensation process and the terminal velocity. The original CRM condensation formula is replaced by the large-scale condensation formula of Le Trent and Li (1991), except that the relative humidity criterion for condensation is 90%. The coefficient in the terminal velocity formula adapted is half the original value. The details are described in Kang et al. (2015). Fig. 3 shows the vertical profiles of cloud hydrometeers and various microphysical processes simulated by the CRMs with the original cloud microphysics of 1- and 50-km resolutions, and that with the modified microphysics at 50-km resolution. The modified cloud microphysics at 50-km resolution (red lines) produces the vertical profiles, close to those of 1-km resolution (black lines). Fig. 3A–D shows the vertical profiles of cloud water, graupel/hail, rainwater, and cloud ice from the model of 1-km resolution and the models of 50-km resolution, both with and without modification (blue dotted lines). Fig. 3E–H shows the vertical profiles of various microphysical processes: accretion of cloud water by rainwater (Fig. 3E), accretion of cloud water by graupel (Fig. 3F), melting of

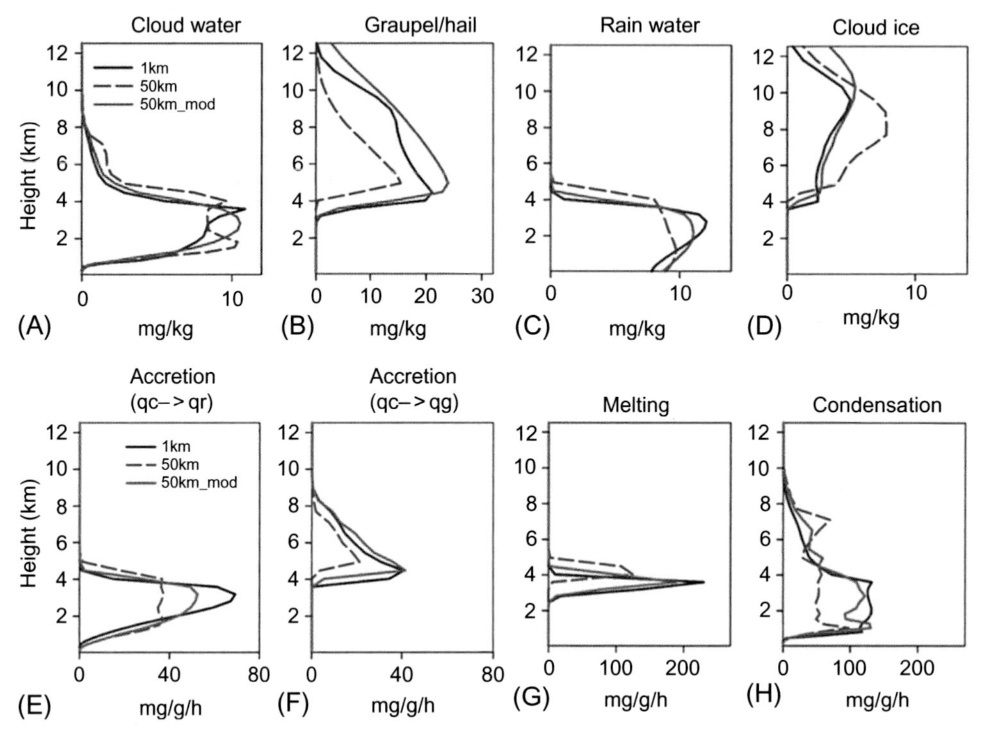

FIG. 3 Vertical profiles of domain-averaged (A) cloud liquid water, (B) graupel/hail, (C) rainwater, (D) cloud ice water, (E) accretion of cloud water by rainwater, (F) accretion of cloud water by graupel, (G) melting of graupel, and (H) condensation from the CRM with 1-km horizontal resolution *(black line)*, the CRM with 50-km horizontal resolution *(blue dashed line)*, and the modified CRM with 50-km horizontal resolution *(red line)*.

graupel (Fig. 3G), and condensation (Fig. 3H) from the 1- and 50-km-resolution models. It is noted that the water species (cloud water and rainwater) are less sensitive to the horizontal resolution compared to those of ice species (graupel and cloud ice). As the horizontal resolution becomes coarse, most cloud microphysics processes weaken and become underestimated, especially in the accretion and condensation processes (Fig. 3E–H). But, as mentioned previously, the modification improves the simulation of the cloud microphysics at 50-km resolution.

The modified cloud microphysics was implemented in a SNU AGCM at 50-km resolution, where the conventional parameterizations (both convection and large-scale condensation schemes) were replaced by the modified cloud microphysics of the CRM described previously. In this GCM, the cloud hydrometers are treated as prognostic variables, and the cloud microphysics are computed explicitly by using GCM state variables. As expected and shown by Kang et al. (2015), the GCM with the modified cloud microphysics produces a large bias in low-level moisture due to the insufficient vertical transport of moisture from the surface to the free atmosphere, and thus they added a diffusion-type shallow convection in the model.

The model with cloud microphysics and shallow convection was shown to simulate the climatological-mean precipitation distribution and 3-hourly precipitation statistics reasonably well. In contrast to the conventional GCM with parameterized convection schemes, which produces too much light rain and relatively less heavy precipitation, the model with the modified cloud microphysics produces precipitation frequencies of light and heavy precipitations close to those of Tropical Rainfall Measuring Mission (TRMM) satellite observation (Kang et al., 2015).

In a subsequent study, Kang et al. (2016) examined the MJO structure of the GCM with the cloud microphysics and shallow convection, used by Kang et al. (2015). The model MJO was much improved compared to that of the conventional SNUGCM with convective parameterizations. However, they found that the moisture anomalies associated with the MJO has a large bias in the upper troposphere. This problem motivated the inclusion of a certain portion of the deep convection generated by the convection scheme in the original AGCM. The subgrid-scale convective mixing should be dependent on the horizontal grid size.

The convection scheme, such as a Simplified Arakawa-Schubert (SAS) scheme, could be developed for a coarse-resolution GCM with a horizontal resolution of an order of several hundred kilometers, and therefore modifying the convection scheme suitable to the 50-km resolution model was considered. The resolution dependency of subgrid-scale vertical mixing is examined using a 3D CRM of 1-km horizontal resolution in a radiative convective equilibrium condition. Ratios of the subgrid-scale vertical transport to the total vertical transport of moist static energy at 850-hPa level as a function of grid size are examined. As expected, the ratio increases as the grid size increases. The value of the ratio is about 0.62 for 50-km resolution and close to 1 for 280 km, indicating that about 60% of the convective heating and condensation produced by the SAS scheme could be used for the model of 50-km resolution. In the present study, the deep convection from the SAS scheme is reduced to 60% by reducing the cloud base mass flux to about 10%. The scheme of resolution-dependent convection is referred to as a *scale-adaptive convection scheme*. Details of the scale-adaptive convection scheme are described in Ahn and Kang (2018).

The AGCMs and CGCMs with the modified cloud microphysics and the scale-adaptive convection scheme will be referred to in this chapter as the *MP-AGCM* and *MP-CGCM*, respectively. It is noted that these GCMs do not use a subtime interval for the cloud microphysics calculation, but the time interval of model integration is reduced to 600 s for all GCM and microphysics variables, except that the terms with the terminal velocity are computed every 20 s. The MP-AGCM with climatologically varying sea surface temperatures (SSTs) prescribed and the MP-CGCM are integrated for 5 years with 50-km horizontal resolution, and the climatological mean precipitations are shown in Fig. 4, along with the corresponding results obtained with the original SNU AGCM and CGCM with parameterized convection schemes.

A comparison of Fig. 4B and C indicates that the MP-AGCM appears to improve the precipitations over the extratropical storm track regions in the Pacific and Atlantic oceans and the distribution of the dry region over the eastern subtropical Pacific, compared to the conventional AGCM. However, the relatively large precipitations over the Indian Ocean and the tropical Atlantic Ocean are still not well simulated. It is interesting to note that the MP-CGCM (Fig. 4E) produces a very different distribution of precipitation from that of MP-AGCM (Fig. 4C) over the tropical oceans, particularly the western Pacific, indicating that the air-sea interaction significantly influences precipitation over the tropical oceans. Overall, the coupled model appears to simulate the precipitation intensity over the tropical oceans,

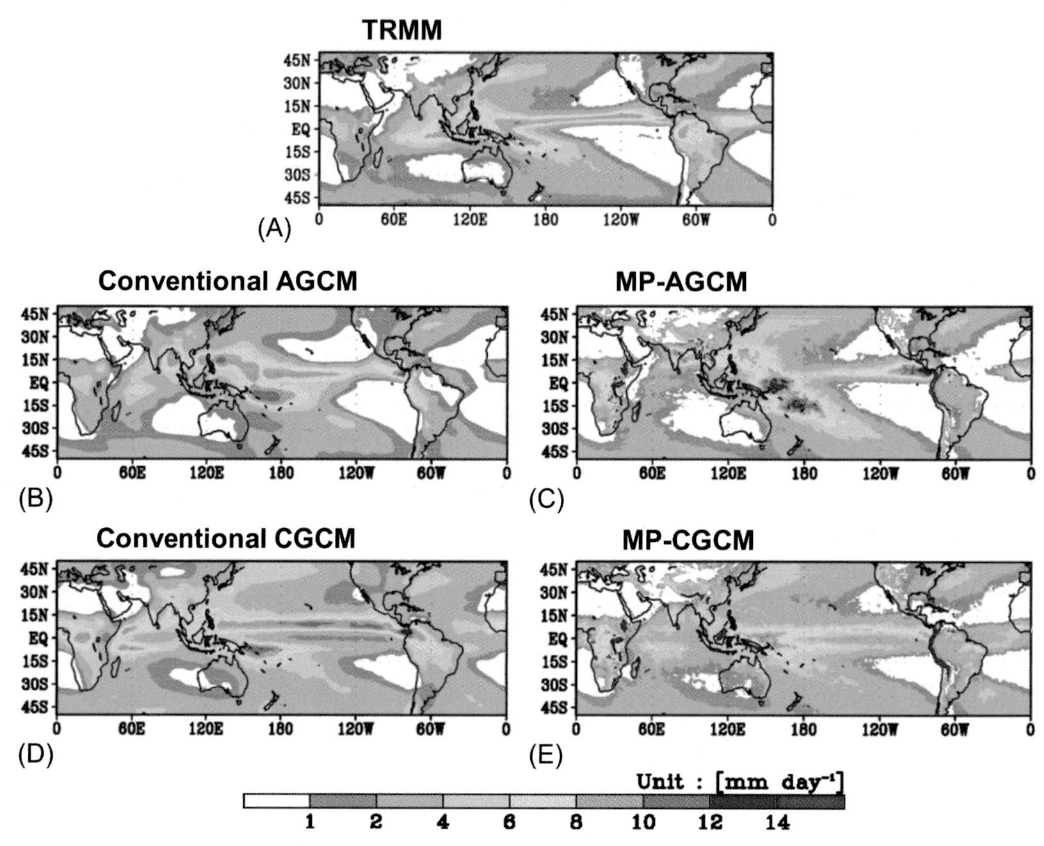

FIG. 4 Spatial distributions of annual-mean precipitation for (A) TRMM; (B) conventional AGCM; (C) AGCM with the modified cloud microphysics, shallow convection, and scale-adaptive deep convection (MP-AGCM); (D) conventional CGCM; and (E) CGCM with the modified cloud microphysics, shallow convection, and scale-adaptive deep convection (MP-CGCM). Here, 5-year simulations are used for the model cases, and a 10-year mean of 2000–09 is used for the TRMM. The TRMM data was interpolated to the 50-km-resolution model horizontal grid.

particularly in the western Pacific, better than the AGCM does. The double ITCZ, which is a common problem in most CGCMs, still appears in the MP-CGCM and the conventional CGCM.

The MJO simulated by the two models is examined in terms of the wave number-frequency power spectra of precipitation. As illustrated in Fig. 5A, the observation shows the strong eastward-propagating power in wave numbers 1–3 and periods of 30–100 days, whereas the conventional AGCM and CGCM show a lack of eastward-propagating power within the MJO wave number and period (Fig. 5B and D). The MP-AGCM (Fig. 5C) and MP-CGCM (Fig. 5E), on the other hand, produce strong eastward propagation in the wave numbers and frequencies close to the observed counterparts, although the MJO-like signals of both modes are somewhat stronger than observed.

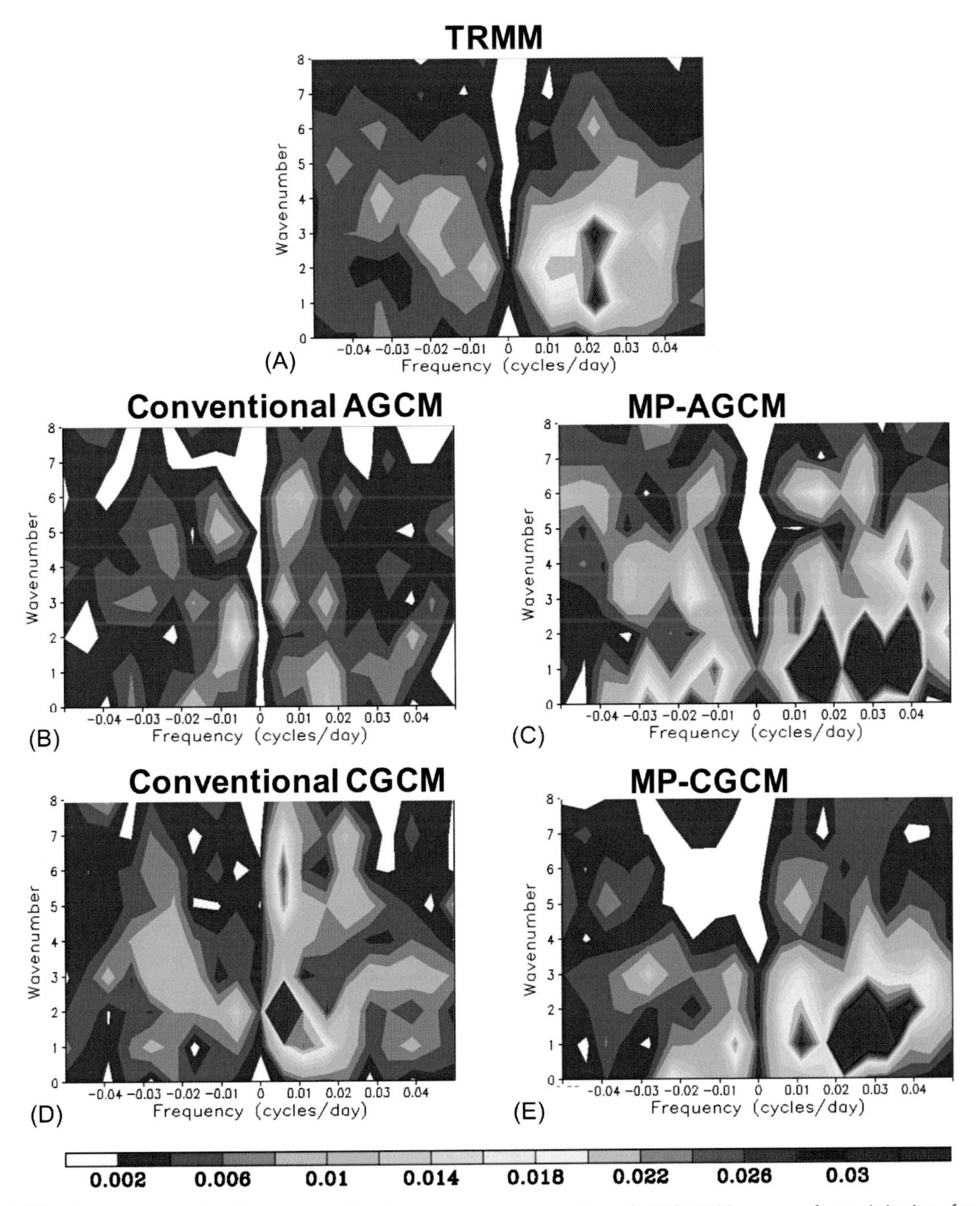

FIG. 5 November–April wave number-frequency power spectra of 10°S–10°N averaged precipitation for (A) TRMM, (B) conventional AGCM, (C) MP-AGCM, (D) conventional CGCM, (E) MP-CGCM. The *MP* stands for the model with the modified cloud microphysics and shallow and scale-adaptive deep convections.

As shown in Kang et al. (2016), the implementation of the cloud microphysics alone helps eastward wave propagation along the equator, but the propagation speed is relatively fast and precipitation is more or less scattered (not well organized) in the tropics. They also showed that adding the shallow convection makes the convection more organized, and eastward propagation becomes close to the observed MJO speed. Now, in the present study, the scale-adaptive convection is added to enhance the deep convective mixing, which results in the stronger, MJO-like, eastward propagation of precipitation. It is noted that the vertical profile of the moisture field is better simulated in the MP-GCMs in the upper troposphere compared to those of the models without the scale-adaptive deep convection (not shown). The better moisture field induces the enhancement of convection-moisture coupling, resulting in a stronger MJO, as demonstrated by Ahn et al. (2017) with the CMIP5 models. The results discussed here indicate that both cloud microphysics and ocean-atmosphere coupling contribute to improve the precipitation climatology and MJO. The ocean-atmosphere coupling strongly influences the precipitation climatology but plays a lesser role in the MJO, and the moisture physics associated with the cloud microphysics and parameterized convection appears to be a major contributor to the better simulation of MJO.

5 SUMMARY AND CONCLUSION

This chapter reviewed three types of GCMs, with full representation of cloud microphysics, and examined their simulations of MJO. The first model reviewed is a global CRM with an ultrahigh horizontal resolution of an order of a few kilometers, particularly NICAM models. The advantage of using the CRM is to permit turbulent motions associated with clouds and to permit a direct coupling between the fluid dynamics and the cloud microphysics. This provides a pathway to avoid difficulties of the "cumulus parameterization deadlock" (Randall et al., 2003) by computing transports and phase changes of water substances directly. However, this approach requires very heavy computing resources due to an ultrahigh horizontal resolution of an order of kilometers, and therefore their climatological behaviors are not well described and many aspects of cloud physical processes are not known yet. The experience of the loss and the recovery of the MJO in NICAM with some modification of physics, as described in this chapter, informs us that the NICAM is still sensitive to the vertical redistribution of moisture that may be inevitably represented by the subgrid-scale unresolved processes. These results might disappoint us that the cloud microphysics deadlock will emerge, even if we go beyond the deadlock of the cumulus parameterization.

The second model is the so-called superparameterized global model with CRMs embedded in each GCM grids. This kind of model has a horizontal resolution of an order of 100 km. Currently, there are several superparameterized models, first at Colorado State University, and subsequently at NASA and with an NCEP model. Here, we have described the SPIFS, a superparameterized model with the ECMWF IFS model. The SPIFS appears to improve the IFS (but not significantly) in simulating the Kelvin waves and MJO. Note that the IFS is already able to simulate the MJO reasonably well. This chapter also described the use of SPIFS in producing initial ensemble perturbations in the S2S-type prediction by perturbing the CRM variables, but without perturbing the GCM variables in the initial time.

The third model is an AGCM and CGCM with a full representation of microphysics at a medium-range horizontal resolution of 50 km. One issue of developing an order of

10-km-resolution GCM is to modify the cloud microphysics suitable to the horizontal resolution. For the present model, the modification was based on sensitivity experiments for the parameters of the important processes sensitive to the model resolution, particularly the condensation process and the terminal velocity. It was demonstrated that shallow convection and scale-adaptive deep convection are still needed in the present model of 50-km resolution with the cloud microphysics.

The present AGCM was shown to simulate the precipitation statistic, such as the light and heavy precipitation frequencies, reasonably well (Kang et al., 2015), and the MJO is simulated with a stronger intensity than the observed one. The MJO characteristics of the present CGCM is similar to that of the AGCM. However, the precipitation climatologies of the AGCM and CGCM are very different from each other in terms of both their distribution and intensity, indicating that the air-sea interaction plays an important role in determining the climatology, and this result suggests that we should tune the model physics and their parameters with CGCM rather than with AGCM.

The global CRM probably could be an ultimate framework of global weather and climate models. However, the global CRM with an ultrahigh horizontal resolution of an order of kilometers has not only a practical difficulty in use due to a huge computational resource required, but also a scientific difficulty to overcome the so-called cloud microphysics deadlock, as illustrated by the Japanese past experiences reviewed in this chapter. Huge efforts and resources for a long period may be required to develop such a model to use in weather and climate predictions.

The superparameterized GCM could be an alternative choice of a model with full representation of cloud microphysics. However, it also could be an intermediate choice for a period before a useful global CRM has been fully developed. The physics of the superparameterized GCM has a limitation in the representation of the large-scale cloud interaction between neighboring GCM girds. However, as reviewed in this chapter, the superparameterization also could provide a useful tool for a S2S-type prediction, particularly for generating many initial perturbations for the ensemble prediction.

Currently, the highest horizontal resolution of the S2S operational system is about 30 km at ECMWF. The order of 10-km horizontal resolution could be affordable in many operational centers in the near future. For such a model resolution, the cloud microphysics should be modified to be suitable to the resolution. However, as pointed out in this chapter, such a model still requires subgrid vertical mixing, which could be added by a scale-dependent cumulus parameterization. One of the major issues of the so-called gray-zone GCM, with a horizontal resolution of an order of 10 km, is to determine the ratio of the resolved and unresolved (parameterized) cloud and moisture physics. As reviewed in this chapter, there are many issues related to the three kinds of GCMs with cloud microphysics. Unlimited efforts may be needed in the modeling community to overcome these issues.

Acknowledgments

I.-S. Kang and M.-S. Ahn were supported by the National Research Foundation of Korea (NRF) grant, funded by the Korean government (MEST) (NRF-2009-C1AAA001-0093065) and by the Brain Korea 21 Plus. H. Miura was supported by Grant-in-Aid for Scientific Research (B-16H04048) from the Japan Society for the Promotion of Science. A. Subramanian was supported by an ERC grant (Towards the Prototype Probabilistic Earth-System Model for Climate Prediction, project reference 291406).

CHAPTER

15

Forecast Recalibration and Multimodel Combination

Stefan Siegert, David B. Stephenson

[a]Department of Mathematics, University of Exeter, Devon, United Kingdom

OUTLINE

1 INTRODUCTION

Computational models of the Earth's climate system are based on mathematical abstractions and numerical approximations. Not all physical processes of the real world are included in climate models. The chaotic nature of atmospheric dynamics leads to the forecast's sensitivity to the imprecisely known initial state of the system. Therefore, numerical model forecasts are imperfect representations of the real world. Discrepancies between the model forecast and the real world can be loosely classified into random and systematic. Random forecast errors are unpredictable, whereas systematic errors are (at least to some extent)

https://doi.org/10.1016/B978-0-12-811714-9.00015-2

predictable. The most illustrative example of a systematic forecast error is the mean bias of the forecast (i.e., a constant offset between the time mean of the forecast and the time mean of the real-world predictand). If it is known from past experience that, say, a temperature forecast consistently differs from the real-world temperature by +2 K, it is rational to adjust future forecast downward by 2 K to correct for the bias and thereby improve the forecast. Bias correction is a simple example of forecast recalibration.

There are two distinct uses of the term "calibration" in the literature, both of which are related to, but different from, the technical term "recalibration." Forecast calibration can refer to the *act of calibrating* a forecast by tuning parameters of the numerical model. Forecast calibration can also be used to characterize a forecast as being *reliably calibrated*. We will not be concerned with parameter tuning in this chapter, and use the term "calibration" only in the second sense, to refer to the degree of "reliability" of a forecast model. We will focus on forecast recalibration, which is the process of making a forecast model better calibrated by statistical postprocessing of its output.

It is often the case that there is not only a single forecast model, but also multiple forecast models for the same event. One way of viewing this collection of forecasts is that they are competitors, the best of which should be picked to issue the forecast, thereby discarding the information contained in the output of the other, "suboptimal" models. However, the decision for picking the best model is often ambiguous: Forecast models must be compared by calculating performance measures, such as the correlation between past forecasts and their verifying observations, or proper scoring rules. But these measures are uncertain due to sampling variability, so the forecast model that achieves the best performance measure over a few past cases is not necessarily the best model for future forecasts. Furthermore, there are many different measures of forecast performance, and the ordering of forecasts can depend on the measure used to evaluate them. This ambiguity gives rise to the idea of viewing the various forecasts as complementary sources of information that collectively contain more information about the real world than any one of them individually. When this view is adopted, the challenge changes from picking the best model to combining the various model predictions into a single forecast of the real world.

Fig. 1 provides an illustrative example of seasonal, multimodel ensemble forecast data. The dataset consists of seasonal forecasts of average surface temperatures over the Niño-3.4 region, which is an important indicator for the state of the El Niño-Southern Oscillation (ENSO), and for the occurrence of El Niño and La Niña events. ENSO is a dominant mode of climate predictability on seasonal timescales, and so the correlation coefficients between ensemble mean forecasts and verifying observations vary between 0.81 (CFSv2) and 0.91 (SYST4). Such high predictive skill is rather atypical for seasonal climate predictions. Furthermore, due to the high predictability of ENSO, the ensemble has a rather high signal-to-noise ratio (SNR); that is, the spread of the ensemble is small compared to the variance of the ensemble mean. However, other criteria such as sample size, ensemble size per model, between-model variability, systematic bias, and number of models, this hindcast dataset is representative for forecasts on seasonal timescales. Therefore, the hindcast dataset will be used throughout this chapter to demonstrate various concepts related to forecast recalibration and combination.

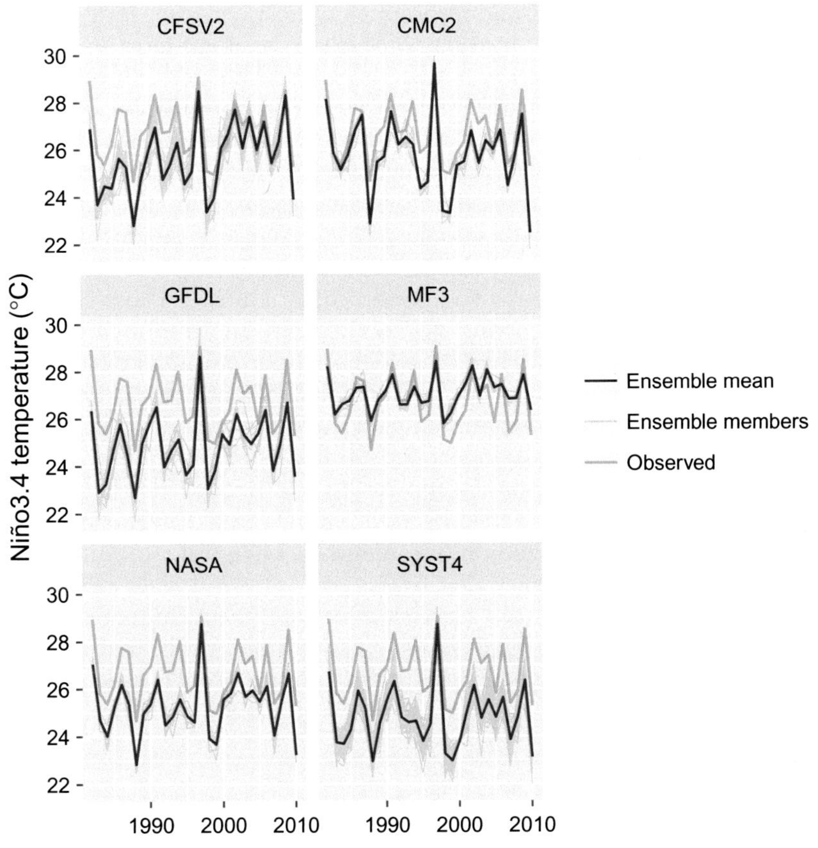

FIG. 1 Hindcast data of a seasonal multimodel ensemble forecasting system: average December temperature in the Niño-3.4 region 1982–2010, with forecasts initialized in August (lead time 5 months). *Light blue lines* represent the ensemble members, *dark blue lines* represent the ensemble mean forecasts, and *orange lines* represent the verifying observations (same in each panel).

Forecasts can be poorly calibrated in a variety of ways. Commonly observed types of forecast uncalibratedness include:

- *Constant bias of the mean*: A difference between forecast mean and observation mean.
- *Dispersion error*: The spread of an ensemble of forecasts does not correctly represent the uncertainty in the observations.
- *Lack of variability*: The year-to-year variability of the forecast is not representative of the variability in the observations.
- *Lack of association*: The correlation between forecasts and observations is low or zero.
- *Error in trends*: Slow average increases or decreases in the observations are not reproduced by the forecasts.

These violations of forecast calibration occur in atmosphere/ocean forecast products on all timescales, from short-term weather prediction to long-term climate projections. Reasons for lack of forecast calibration include initialization errors, structural model errors, model simplifications, numerical truncation errors, missing processes, and simply bugs in the code.

Statistical forecast recalibration is usually necessary for forecast products on all timescales. There are a number of challenges related to forecast recalibration and multimodel combination that are specific to seasonal to sub-seasonal (S2S) climate predictions. The training data to fit statistical recalibration models is often limited, and highly nonstationary. Formulations of the operational forecast model are revised periodically, which can change the statistical behavior of the data and require readjustment of recalibration and combination parameters. The internal variability of the forecasts is high due to the chaotic nature of the atmosphere, which decreases the SNR of the forecast. The correlation between forecast and observations is often low. Finally, multimodel hindcast experiments are not designed with model combinations in mind and are therefore often nonhomogeneous. Strategies to generate hindcast datasets can be loosely characterized as either "on the fly" or "fixed," depending on how changes to the forecast model are accounted for. In the S2S hindcast database (Vitart et al., 2017), for example, most forecasts that are initialized on different days have different hindcast periods and forecast out to different lead times.

By using statistical models to correct forecast errors of dynamical models, forecast calibration bridges the gap between empirical (statistical) and numerical forecasting. To issue reliable forecasts, we need robust statistical methods to issue probabilistic predictions, which take into account the correlation and error structure of multimodel ensemble forecasts.

2 STATISTICAL METHODS FOR FORECAST RECALIBRATION

Forecast calibration is an important diagnostic to differentiate good forecasts from bad forecasts. To characterize forecast calibration, Gneiting et al. (2007) introduced various *modes of calibration* (namely, probabilistic calibration, exceedance calibration, and marginal calibration). All modes of calibration characterize, in different ways, the agreement between the issued forecast distribution and the hypothetical distribution from which the real-world observation is drawn. Forecast recalibration is thus closely related to forecast verification, which is discussed in detail in Chapter 16 of this book.

In a similar spirit, Jolliffe and Stephenson (2012) define forecast calibration in terms of the equality between the forecast and the conditional mean of the observation, given the following forecast:

$$E_Y(Y|X=x)=x. \quad (1)$$

That is, if we collected all instances on which a particular value $X=x$ was forecast, the mean over all verifying observations should be equal to x if the forecast is calibrated. Calibration is thus a joint property of forecasts and observations that can be assessed by comparing several pairs of forecasts and observations. If a forecast is found to be uncalibrated, statistical recalibration methods can be used to correct for the lack of forecast calibration.

Eq. (1) suggests that to recalibrate a poorly calibrated forecast, we could replace the current forecast value x by the conditional mean of the observation, given that forecast value. The exact value of the conditional expectation is not known in general, and it has to be estimated from past forecast and observation data. More generally, one could estimate the conditional distribution of the observation, given the forecast. The conditional mean or distribution can be estimated by collecting all past forecasts that have a given value (or are sufficiently close to a given value) and averaging all past observations corresponding to these forecasts. This *nonparametric* way of recalibrating forecasts is appealing, as it can potentially account for complicated nonlinear relationships between forecasts and observations. However, it requires enough past forecasts that are close to the current forecast value in order to estimate the conditional mean robustly. In data-poor settings, where only a few pairs of past forecasts and observations are available, nonparametric estimation methods will suffer from large estimator variance. In these situations it is often useful (or even necessary) to assume a parametric relationship between forecasts and observations (i.e., to describe the conditional mean of the observation given the forecast by a function of the forecast that is parameterized by a small number of coefficients). Next we discuss two of the most commonly used parametric methods for forecast recalibration—namely, model output statistics (MOS) and nonhomogeneous Gaussian regression (NGR).

3 REGRESSION METHODS

3.1 Model Output Statistics

The most commonly used parametric methods for forecast recalibration are based on regression techniques. In the meteorological literature, using linear regression to recalibrate a forecast is also referred to as model output statistics (MOS; Glahn and Lowry, 1972; Glahn et al., 2009). In linear regression, the observation y_t at time t is modeled as a linear function of forecasted value (or forecasted values if several forecasts are available) $x_{1,\,t}, \ldots, x_{p,\,t}$, plus an independent, normally distributed error term:

$$y_t = \beta_0 + \beta_1 x_{1,\,t} + \cdots + \beta_p x_{p,t} + \sigma\epsilon_t, \tag{2}$$

where $\beta_0, \ldots, \beta_p$ and σ are unknown parameters and $\epsilon_t \sim \mathcal{N}(0,1)$. Eq. (2) can also be written in vector form as

$$y_t = \boldsymbol{x}_t'\boldsymbol{\beta} + \sigma\epsilon_t, \tag{3}$$

using the column vectors $\boldsymbol{x}_t = (1, x_{1,\,t}, \ldots, x_{p,\,t})'$ and $\boldsymbol{\beta} = (\beta_0, \ldots, \beta_p)'$. A common case is $p = 1$, where there is a single forecast, such as the ensemble mean taken from a single model, using the same variable and location as the predictand y. It is possible that multiple predictors are used (i.e., $p > 1$). These can be output from several forecast models, different variables than the predictand, different ensemble members started from perturbed initial conditions, or variables at different locations that are deemed informative about the predictand. The regression parameters $\boldsymbol{\beta}$ and σ can be estimated from previously observed pairs of forecasts and verifying observations.

It is useful to collect the verifying observations $y_1, \ldots, y_N$ into a column vector $\boldsymbol{y}$ and the row vectors $\boldsymbol{x}_1', \ldots, \boldsymbol{x}_N'$ into the rows of the design matrix $\mathbf{X}$. Then we can write

$$\boldsymbol{y} = \boldsymbol{X\beta} + \sigma\boldsymbol{\epsilon}, \tag{4}$$

where $\boldsymbol{\epsilon}$ is assumed to have a multivariate normal distribution with diagonal covariance matrix $\text{var}(\boldsymbol{\epsilon}) = \mathbf{1}$.

Under the assumption that the error term ϵ_t has a standard Gaussian distribution, and ϵ_t and $\epsilon_{t'}$ are uncorrelated for $t \neq t'$, the log-likelihood function of the linear regression model is proportional to

$$\ell(\boldsymbol{\beta}, \sigma^2; \boldsymbol{y}) \propto -\frac{N}{2}\log(\sigma^2) - \frac{(\boldsymbol{y} - \boldsymbol{X\beta})'(\boldsymbol{y} - \boldsymbol{X\beta})}{2\sigma^2}. \tag{5}$$

The maximum likelihood estimators of $\boldsymbol{\beta}$ and σ^2 are obtained by setting the partial derivatives of ℓ to zero:

$$\hat{\boldsymbol{\beta}} = (\mathbf{X}'\mathbf{X})^{-1}\mathbf{X}'\boldsymbol{y}, \tag{6}$$

$$\hat{\sigma}^2 = \frac{(\boldsymbol{y} - \boldsymbol{X}\hat{\boldsymbol{\beta}})'(\boldsymbol{y} - \boldsymbol{X}\hat{\boldsymbol{\beta}})}{N}. \tag{7}$$

The commonly used unbiased estimator of σ^2, denoted $\hat{\sigma}_u^2$, is given by subtracting the total number of estimated parameters from N in the denominator of Eq. (7), that is,

$$\hat{\sigma}_u^2 = \frac{(\boldsymbol{y} - \boldsymbol{X}\hat{\boldsymbol{\beta}})'(\boldsymbol{y} - \boldsymbol{X}\hat{\boldsymbol{\beta}})}{N - p - 1}. \tag{8}$$

After fitting the regression parameters by maximum likelihood, a future observation y^*, given a new forecast $\boldsymbol{x}^*$, is predicted by plugging $\boldsymbol{x}^*$ into Eq. (3), using the maximum likelihood estimators for the regression parameters. By transforming the forecast $\boldsymbol{x}^*$ by the regression relationship (3), some violations of calibration in the raw forecast $\boldsymbol{x}^*$ are corrected, namely constant bias, linear scaling, and ensemble dispersion errors.

It can be shown that the forecast distribution for the new observation y^* based on the new forecast vector $\boldsymbol{x}^*$ is a Student t-distribution:

$$y^* | \boldsymbol{x}^*, \hat{\boldsymbol{\beta}}, \hat{\sigma}^2 \sim t_{N-p-1}\left[(\boldsymbol{x}^*)'\hat{\boldsymbol{\beta}}, \hat{\sigma}_u^2\left(1 + (\boldsymbol{x}^*)'(\mathbf{X}'\mathbf{X})^{-1}(\boldsymbol{x}^*)\right)\right]. \tag{9}$$

So the forecast mean is at $(\boldsymbol{x}^*)'\hat{\boldsymbol{\beta}}$, and a 95% prediction interval for y^* is given by

$$(\boldsymbol{x}^*)'\hat{\boldsymbol{\beta}} \pm t_{0.975, N-p-1}\hat{\sigma}_u\sqrt{1 + (\boldsymbol{x}^*)'(\mathbf{X}'\mathbf{X})^{-1}\boldsymbol{x}^*}, \tag{10}$$

where $t_{\alpha, n}$ denotes the α-quantile of the Student t-distribution with n degrees of freedom. It is tempting to simply forecast a normal distribution with mean $(\boldsymbol{x}^*)'\hat{\boldsymbol{\beta}}$ and variance σ_u^2. But it has been shown that MOS forecasts issued using the predictive t-distributions are better calibrated than forecasts issued as normal distributions because the t-distribution accounts for the estimation uncertainty of the regression parameters (Siegert et al., 2016a).

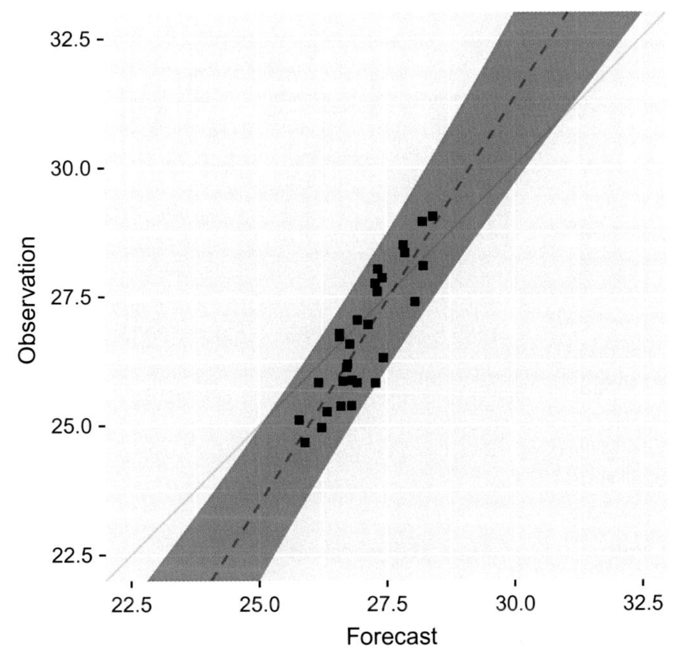

FIG. 2 Recalibration by linear regression applied to MF3 forecasts. The *dashed line* indicates the regression line, and the *blue ribbon* indicates the 95% prediction interval.

As an example, consider the MF3 Niño-3.4 ensemble mean seasonal forecasts at 5 months lead time, illustrated by a scatterplot in Fig. 2. The points do not lie along the diagonal, which indicates that mean and/or scale of the forecasts do not match the mean and scale of the observations. The forecasts are not well calibrated, and statistical recalibration is therefore necessary. The scatterplot further suggests that linear rescaling of the forecasts might be a good recalibration strategy, which makes MOS a suitable candidate. The maximum likelihood estimator of the coefficient vector $\boldsymbol{\beta}$ is $\hat{\boldsymbol{\beta}} = (-15.70, 1.57)'$ and the (unbiased) maximum likelihood estimator of σ^2 is $\hat{\sigma}_u^2 = 0.39$. For a new forecast value of 27.0°C that lies close to the mean of all previously observed forecast values, the recalibrated prediction equals 26.69°C, and a 95% prediction interval is given by (25.32°C, 27.91°C); that is, the width is 2.59. Likewise, for a new forecast value of 30.0°C, which is large compared to all previously observed forecasts, the prediction is 31.32°C and the 95% prediction interval is given by (29.67°C, 32.96°C) (i.e., the width is 3.30), which is considerably wider than for the intermediate forecast value. Informally, the widening of the prediction interval is caused by the extrapolation beyond previously observed forecast values, which increases uncertainty. Mathematically, the term $(\boldsymbol{x}^*)'(\boldsymbol{X}'\boldsymbol{X})^{-1}\boldsymbol{x}^*$ in Eq. (9) is responsible for widening of the prediction intervals.

Fig. 3 shows raw forecasts and MOS-recalibrated forecasts of the MF3 model, and their verifying observations. The effect of MOS is to bring the forecast means closer (in a mean-squared-error sense) to the verifying observations, and to increase the forecast variance compared to the ensemble spread. The result is better coverage of the observations by the prediction intervals, and therefore more reliable probability forecasts.

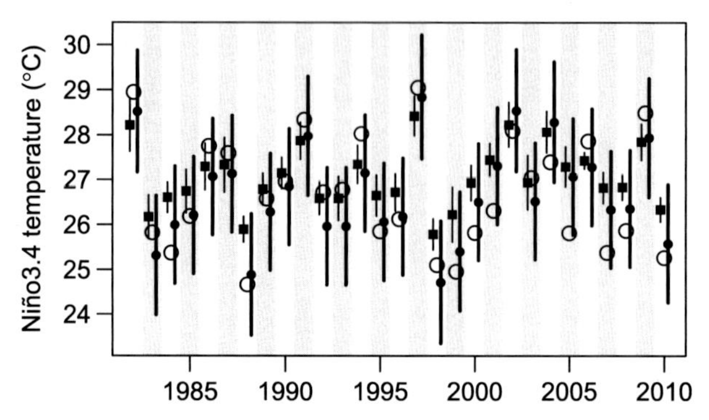

FIG. 3 Illustration of the effect of recalibration by linear regression. *Squares and thin lines* indicate ensemble mean forecasts (generated by the MF3 model) ± two ensemble standard deviations. *Filled circles* and *thick lines* indicate recalibrated ensemble means and 95% prediction intervals. *Open circles* represent observations. The recalibrated forecasts are on average closer to the observations, and the prediction intervals overlap the observations more often than the uncalibrated ensemble spread.

3.2 Nonhomogeneous Gaussian Regression

MOS can be extended to allow use of the ensemble spread information. A good ensemble forecasting system should be able to accurately represent the forecast uncertainty that results from imprecisely known initial conditions and model errors. Therefore, narrow ensembles, corresponding to high confidence in the forecast, should on average incur smaller forecast errors than very wide ensembles, which indicate low confidence in the forecast. Due to model errors and natural variability, the correspondence between ensemble spread and forecast error (the spread-skill relationship) cannot be expected to be perfect, but it is reasonable to assume that a linear relationship exists. A regression framework to recalibrate both ensemble mean and ensemble spread is nonhomogeneous Gaussian regression (NGR; Gneiting et al., 2005). NGR assumes that the observation has a normal distribution whose mean and variance depend linearly on ensemble mean and ensemble variance. In particular, let m_t denote the ensemble mean forecast at time t, and s_t^2 the ensemble sample variance at time t. The conditional distribution of the observations, given the ensemble forecast, is

$$\mathcal{N}(a + bm_t, c + d^2 s_t^2). \tag{11}$$

The recalibration parameters (a, b, c, d) are unknown and have to be estimated from historical forecast and observation data. Unlike linear regression (MOS), the maximum likelihood parameters cannot be determined analytically and therefore have to be estimated by numerical optimization. Given a series of ensemble mean forecasts $m_1, \ldots, m_N$, ensemble variances $s_1^2, \ldots, s_N^2$, and verifying observations $y_1, \ldots, y_N$, the log-likelihood function of the NGR model is proportional to

$$\ell(a,b,c,d;\{m_t,s_t^2,y_t\}_{t=1}^N) \propto -\frac{1}{2}\sum_{t=1}^{N}\left[\log(c + d^2 s_t^2) + \frac{(y_t - a - bm_t)^2}{c + d^2 s_t^2}\right]. \tag{12}$$

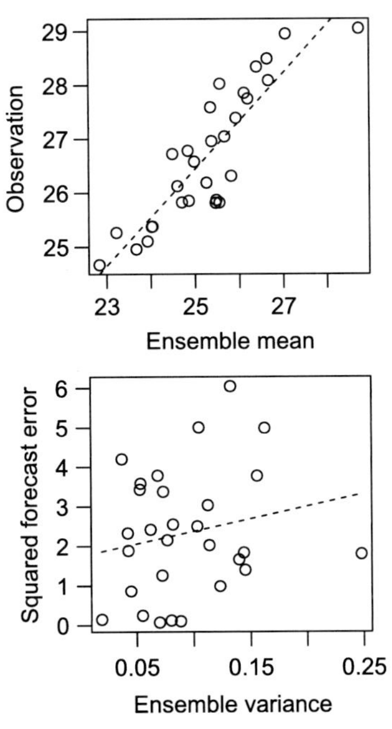

FIG. 4 Scatterplots of observation over ensemble mean and squared forecast error over ensemble variance for the Niño-3.4 ensemble forecasts issued by the NASA model. Least-squares linear fits have been added as guides.

To give a specific example, consider the Niño-34 temperature seasonal forecasts issued by the National Aeronautics and Space Administration (NASA) model at 5 months lead time. Scatterplots of verifying observations over ensemble means and squared forecast errors over ensemble variances are shown in Fig. 4. There is a strong linear relationship between ensemble means and observations (correlation 0.88). There is also a weak positive linear relationship between ensemble variances and squared forecast errors (correlation 0.19). The correlation between variance and error is not statistically significant, but it might still be beneficial for forecast recalibration.

To fit the NGR, we optimize the NGR log-likelihood (Eq. 12) using the Broyden-Fletcher-Goldfarb-Shanno (BFGS) algorithm, as implemented in the function `stats::optim` of the `R` statistical programming environment (R Core Team, 2017). The estimated values are given in Table 1. The parameter estimates suggest that the forecasts are not well calibrated, and there is scope for improvement by statistical recalibration of mean, scaling, and variance. However, the parameter d is very small, indicating very little relationship between ensemble spread and forecast variance.

It is worth noting that NGR was first proposed, and is mostly applied, in the context of numerical weather prediction (NWP). In NWP, atmospheric prediction is treated as an initial

TABLE 1 NGR Estimates

Parameter	a	b	c	$\lvert d \rvert$
Estimate	4.06	0.89	0.34	2.5×10^{-5}

value problem, so that varying levels of sensitivity to initial conditions can lead to a spread-skill relationship in ensemble forecasts. Seasonal climate forecasting, rather, is a boundary value problem, where long-term predictability is a result of slowly varying drivers of the climate system. A strong spread-skill relationship is therefore unlikely, so it is no surprise that NGR is not beneficial as a recalibration method for the seasonal forecasts shown here. For prediction on sub-seasonal timescales, lying between weather and seasonal climate forecasting; however, a systematic spread-skill relationship might be conceivable.

3.3 Comparing Recalibration Models

Forecast recalibration is a statistical modeling exercise. At any point in time, several recalibration models might be available, and the task of the forecaster is to choose one of them to make a prediction. The task of choosing the "best" among a number of candidate statistical models is called model selection. Here, we give an example to illustrate how to choose between MOS and NGR to recalibrate the NASA model. A good introduction to model selection and statistical modeling in general can be found in Hastie et al. (2009).

A commonly used model selection criterion is the Bayesian Information Criterion (BIC; Schwarz, 1978), defined as

$$\mathrm{BIC} = -2\hat{\ell} + k\log(n), \tag{13}$$

where $\hat{\ell}$ is the log-likelihood function evaluated at the mode (i.e., using the optimized parameter values); k is the number of parameters of the model; and n is the sample size. The model with lowest BIC is to be preferred. A low BIC is achieved by high values of $\hat{\ell}$ and low values of k. Thus, BIC reward models that fit the data well, while at the same time having a small number of free parameters. The BIC is closely related to the Akaike Information Criterion (AIC), which is calculated by replacing $\log(n)$ by 2 in Eq. (13).

We have seen in Table 1 that the optimal value for the parameter d is very small in the NASA ensemble, which suggests that taking the ensemble spread into account in the variance of the forecast distribution might be unnecessary. When d is zero, NGR is equivalent to MOS. The differences between the optimized log-likelihoods of NGR and MOS for this ensemble is on the order of 10^{-10} (i.e., the recalibration by NGR and MOS yields almost identical recalibrated forecasts). But since NGR has four free parameters, where MOS has only three, we get BIC = 64.5 for NGR and BIC = 61.1 for MOS, which suggest that MOS is the preferable recalibration model in this case. In other words, the hypothesized spread-skill relationship in the NASA ensemble cannot be considered useful for forecast recalibration.

Another widely used method for model comparison is cross-validation. In cross-validation the ability of a statistical recalibration model is assessed by evaluating its predictions on unknown data that were not part of the training dataset.

3.4 Further Remarks on Recalibration

Because forecast recalibration is a *statistical modeling* problem, all issues that apply to statistical modeling are relevant to forecast recalibration as well. We have discussed the important areas of parameter estimation and model selection in some detail. Here, we discuss a number of further problem areas that should be considered and refer the reader to the relevant literature.

If parameters are estimated from a finite number of training data, their estimation uncertainty must be taken into account. Siegert et al. (2016a) have shown that failing to account for parametric uncertainty can lead to degradation of the quality of the recalibrated forecasts. Accounting for estimation uncertainty in the recalibration parameters has the effect of inflating the tails of the forecast distribution, which leads to better calibrated and more skillful forecasts. It is often the case that prior information is available on recalibration parameters, in which case a Bayesian estimation framework is suitable. Siegert et al. (2016b) have shown that prior information on the correlation coefficient of the ensemble mean can improve the performance of recalibrated forecasts compared to standard methods. Furthermore, the Bayesian approach of Siegert et al. (2016b) allows one to address the problems of forecast verification and forecast recalibration in the same coherent statistical framework.

Delle Monache et al. (2011) and Obled et al. (2002) have used statistical analog techniques to improve forecast recalibration. The underlying idea is to construct the training dataset for parameter estimation by considering only past forecasts that are similar to the present one. A related technique is to use a sliding training window (e.g., Sweeney et al., 2011) to use only the most recent forecast and observation data to construct the training dataset for parameter estimation. A sliding window approach allows the recalibration strategy to adapt to changes in the forecasting system or the climate system.

Forecast data produced by climate models is usually high-dimensional, consisting of multiple climatological variables on a spatial grid and at many points in time for various ensemble members initialized at different times and initial conditions. Various techniques exist for multivariate recalibration, and especially the field of spatial recalibration has undergone rapid development in recent years.

Two important nonparametric methods for spatial recalibration are the *Schaake shuffle* (Clark et al., 2004) and *ensemble copula coupling* (Schefzik et al., 2013). These methods are based on the idea of reordering ensemble forecasts locally so as to better replicate the spatial correlation structure of the predictand (see also Schefzik, 2017; Vrac and Friederichs, 2015; Scheuerer et al., 2017). Parametric approaches for multivariate forecast recalibration have been proposed based on Gaussian random fields (Feldmann et al., 2015) and parametric copulas (Möller et al., 2012; Hemri et al., 2015). It can be noted that multivariate methods such as principal component regression (PCR) and canonical correlation analysis (CCA) have been used to recalibrate seasonal climate forecasts (e.g., Barnston and Tippett, 2017). However, recalibration based on explicit spatiotemporal statistical models is largely unexplored in the field of S2S predictions.

4 FORECAST COMBINATION

The development and maintenance of a climate forecasting system require considerable effort. It is therefore sensible to establish climate modeling centers, where scientists, developers, and administrators provide the necessary expertise and infrastructure. As a consequence, several climate modeling centers exist around the world, each one running its own forecast system. The multiplicity of modeling centers provides opportunities to share expertise and to compare various modeling strategies. But because each center provides its own climate forecasting products, using slightly various climate models, the user faces a conundrum of choice, having to make an informed decision as to which climate model to use. Better yet, the user might want to benefit from the "wisdom of crowds" and let the multiplicity of climate model forecasts act as a sort of committee that jointly provides a final, combined forecast product.

Various methods have been proposed to optimally combine forecasts from different numerical models. A key reference on forecast combinations in seasonal climate forecasting is DelSole (2007), who presents a unified Bayesian framework that accommodates a number of multimodel combination strategies. Sansom et al. (2013) discuss weighting strategies for multimodel ensembles in a climate change context. Further combination strategies are discussed in Stephenson et al. (2005), Doblas-Reyes et al. (2005), and Rajagopalan et al. (2002). The rest of this section follows the methodologies outlined in DelSole (2007), with particular focus on the hierarchical regression method of Lindley and Smith (1972).

4.1 Hierarchical Linear Regression

Assume, as before, that at times $t = 1, \ldots, N$, we have climate forecasts that were produced by p numerical models, $f_{1,t}, \ldots, f_{p,t}$. Each $f_{i,t}$ is assumed to be scalar, so it could be a spatial, temporal, and ensemble average produced from the output of a single climate model. We assume in this section that the vectors of forecasts $f_1, \ldots, f_p$ have been individually standardized to have zero mean and unit variance over time; DelSole (2007) reports that standardization of individual forecasts improved the quality of the combined forecast product. One possible method, motivated by the regression framework discussed in the previous section, is to combine the individual forecasts into a single forecast by a linear combination, and to assume the residual to be independently normally distributed:

$$y_t = \sum_{m=1}^{p} \beta_m f_{m,t} + \sigma \epsilon_t, \tag{14}$$

which can be collected into the matrix equation

$$\boldsymbol{y} = \boldsymbol{X}\boldsymbol{\beta} + \sigma \boldsymbol{\epsilon}, \tag{15}$$

where $\boldsymbol{X}$ is the $N \times p$ matrix of forecasts, $\boldsymbol{\beta}$ is the vector of combination weights, and $\boldsymbol{\epsilon}$ has a multivariate normal distribution with zero mean and identity covariance matrix. One then can estimate the vector $\boldsymbol{\beta}$ of forecast combination weights, as well as the residual variance σ^2.

There are two possible extremes that we could adopt when estimating the combination weights. On the one hand, we could assume that the combination weights can be completely different and are fully independent, such that we would not be surprised to learn that the weight of one model is orders of magnitude larger, and with a possibly different sign, than the combination weight of another model. On the other hand, we might judge that there should be no difference at all between the combination weights for different models because the individual models are judged to be exchangeable, and we do not expect any performance differences among them that would warrant upweighting one model forecast in favor of another one.

The framework of DelSole (2007) points out a middle way between those two extremes, using the results of Lindley and Smith (1972) on hierarchical regression. The framework essentially allows the shrinking of the combination weights β_m toward a common, but unknown, value β_0, thus reducing the variability of the combination weights. The underlying idea is that we are usually prepared to assign different weights to different model forecasts, but the weights are not expected to be very different from one another because we would generally not expect large differences between the quality of different forecasts. We will come back to the judgment of similar quality and its implications for forecast combination later in this chapter.

The notion of "different, but similar" combination weights can be modeled within a Bayesian statistical framework as follows. The result of a Bayesian computation is a posterior probability distribution of unknown model parameters, given observed data (i.e., $p(\boldsymbol{\beta}, \sigma^2|\boldsymbol{y})$), in the present context. The posterior distribution is computed by the Bayes' rule:

$$p(\boldsymbol{\beta},\sigma^2|\boldsymbol{y}) \propto p(\boldsymbol{y}|\boldsymbol{\beta},\sigma^2)p(\boldsymbol{\beta},\sigma^2), \tag{16}$$

where the likelihood $p(\boldsymbol{y}|\boldsymbol{\beta}, \sigma^2)$ derives from the linear model specification (Eq. 15) and the prior distribution $p(\boldsymbol{\beta}, \sigma^2)$ encodes prior knowledge about the model parameters. In the present context, we will be interested only in the maximum a posteriori (MAP) estimators of the model parameters (i.e., the values that maximize $p(\boldsymbol{\beta}, \sigma^2|\boldsymbol{y})$), and so the proportionality constant in Eq. (16) is unimportant.

The hierarchical regression framework developed by Lindley and Smith (1972) allows us to encode the notion that the combination weights β_m are different but similar in the prior distribution $p(\boldsymbol{\beta}, \sigma^2)$. The elements of $\boldsymbol{\beta}$ are assumed to be independently normally distributed around a common (unknown) mean β_0 and variance σ_β^2:

$$\beta_i \sim N(\beta_0, \sigma_\beta^2). \tag{17}$$

The normal distribution allows for the β_m to be different, but a small variance σ_β^2 will constrain them to be close to one another. We have to make further assumptions about β_0 and σ_β^2 to close the calculations. Either the values of β_0 and σ_β^2 must be specified, or if this is not possible, vague assumptions must be encoded as probability distributions over β_0 and σ_β^2. The following choices seem justified in the specific context of climate forecast combination and also will lead to a convenient and tractable method of estimating the MAP values of the combination weights. Users will probably not have strong prior beliefs about β_0, and therefore a very wide (uninformative) prior distribution for the central value β_0 is appropriate. A convenient choice of the prior for β_0 is therefore a normal distribution with zero mean and diverging variance. On the other hand, a user who wants to encode the idea of "not too different" combination weights will usually have an idea about what "too different" means quantitatively. For example, if we think that combination weights for our forecasts are unlikely to differ from their common value by more than 0.2, this can be encoded by specifying the variance $\sigma_\beta^2 = 0.1^2$. Finally, to complete the prior specifications, we choose an uninformative prior distribution for the residual variance σ^2—namely, an inverse χ^2 distribution with degrees of freedom $\nu = 0$, such that $p(\sigma^2) \propto 1/\sigma^2$.

Lindley and Smith (1972) show that under these prior assumptions, the MAP estimators of the combination parameters $\boldsymbol{\beta}$ and σ^2 can be obtained by solving the following system of equations:

$$\hat{\boldsymbol{\beta}} = \left[\mathbf{X}'\mathbf{X} + \frac{s^2}{\sigma_\beta^2}(\mathbf{1}_p - p^{-1}\boldsymbol{J}_p)\right]^{-1}\mathbf{X}'\boldsymbol{y}, \tag{18}$$

$$s^2 = \frac{(\boldsymbol{y} - \mathbf{X}\hat{\boldsymbol{\beta}})'(\boldsymbol{y} - \mathbf{X}\hat{\boldsymbol{\beta}})}{n+2}, \tag{19}$$

where $\mathbf{1}_p$ is the $p \times p$ identity matrix, and $\boldsymbol{J}_p$ is a $p \times p$ matrix with each element equal to 1. The equations cannot be solved analytically, but an approximate solution can easily be found iteratively by solving the two equations in turn, each time substituting the solution of one equation into the other. We found that this algorithm leads to convergence within a few (<10) iterations, with little dependence on the choice of initial values.

Note that the two extreme cases mentioned here (equal weighting and fully flexible unequal weighting) correspond to particular choices of the prior variance parameter σ_β^2: By setting $\sigma_\beta^2 \rightarrow \infty$, the additional term in the brackets in Eq. (18) vanishes and the estimate of $\boldsymbol{\beta}$ reduces to the least-squares estimator $(\mathbf{X}'\mathbf{X})^{-1}\mathbf{X}'\boldsymbol{y}$. Imposing no constraints on the elements of $\boldsymbol{\beta}$ by setting $\sigma_\beta^2 \rightarrow \infty$ thus amounts to ordinary multiple linear regression (MLR). On the other hand, by setting $\sigma_\beta^2 = 0$ (i.e., assuming that all combination weights are equal to β_0), the MAP estimate converges to the same estimate of β that would be obtained if we fitted a simple linear regression (SLR) to the multimodel ensemble mean; see Appendix of DelSole (2007) for a proof.

Fig. 5 shows the results of estimated combination weights for the six seasonal Niño-3.4 temperature forecasts shown in Fig. 1. The ensemble mean forecasts of all models were standardized before estimating the combination weights. Three distinct values of σ_β^2 were chosen:

- $\sigma_\beta^2 \rightarrow \infty$, corresponding to unconstrained MLR
- $\sigma_\beta^2 = 0.1^2$, corresponding to HLR
- $\sigma_\beta^2 = 0$, corresponding to SLR on the multimodel ensemble mean

Combination with equal weights yields $\beta_i = \beta = 0.16$. Using MLR with no constraints on the variability of the parameters, the combination weights vary wildly, between − 0.16 and 0.66.

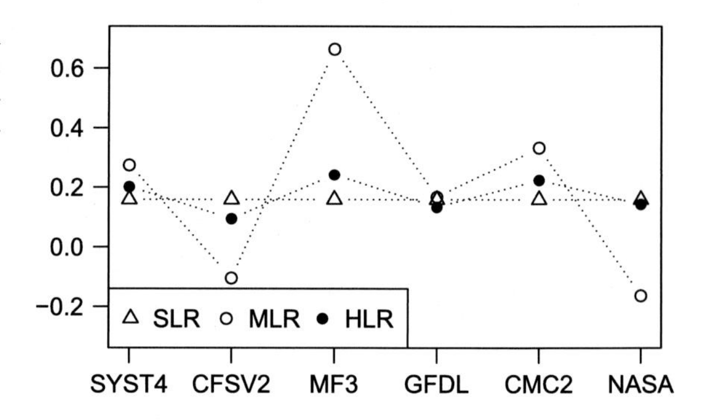

FIG. 5 Niño-3.4 combination weights assigned to numerical models by different methods: SLR = simple linear regression ($\sigma_\beta^2 = 0$); MLR = multiple linear regression ($\sigma_\beta^2 \rightarrow \infty$); HLR = hierarchical linear regression ($\sigma_\beta^2 = 0.1^2$).

This variability is somewhat damped using the hierarchical regression estimators with $\sigma_\beta^2 = 0.1^2$, which restricts the combination weights to a much more reasonable range of 0.09–0.24.

A relevant question to ask is which of the three combination method performs best. If we addressed this question simply by looking at the sum of squared residuals after fitting the regression models, we would find that $\sigma_\beta^2 \to \infty$ performs best. But this is, at least partly, due to the great flexibility of the unconstrained MLR model, which allows parameters to adapt to random variations in the data by setting one regression coefficient to a very large positive value and another coefficient to a very low negative value. But the sum of squared residuals is a measure of in-sample goodness of fit, which is not really relevant in practice. In practice, we would like to estimate how well the methods perform out of sample, on as-yet-unseen data that was not part of the training dataset.

To estimate out-of-sample performance, we conduct a leave-one-out cross-validation. We leave out 1 year of the N years in the hindcast archive and fit the combination weights using the $N-1$ remaining pairs of forecasts and observations. We then use the fitted combination weights on the left-out forecasts to predict the left-out observation. This process is repeated N times, each time leaving out a different year, which results in N out-of-sample predictions whose squared prediction errors can be used to assess out-of-sample performance. The equally weighted forecast combination ($\sigma_\beta^2 = 0$) obtains a leave-one-out mean-squared prediction error of 0.281. The forecast combination obtained by unconstrained multiple regression ($\sigma_\beta^2 \to \infty$) has a much larger mean-squared prediction error of 0.330. The constrained unequal weighting approach with $\sigma_\beta^2 = 0.1^2$ achieves a leave-one-out mean squared error of 0.277, which is a large improvement over multiple regression and a minor improvement over simple regression on the multimodel mean.

The choice of the prior parameter σ_β^2 can be guided by different principles. DelSole (2007) suggests using a nested cross-validation approach to estimate the optimal value σ_β^2. Lindley and Smith (1972) show how $\boldsymbol{\beta}$ and σ^2 can be estimated when an informative prior distribution (in the form of a scaled inverse-χ^2 distribution) is specified for σ_β^2, rather than setting a fixed value as we did earlier in this chapter. It also should be noted that the choice of the prior variance of the β_m should depend on the number of models. We might be more willing to accept larger differences between the weights of 2 models than between the weights of 10 models. It is also possible to specify a different prior distribution than a normal distribution for $\boldsymbol{\beta}$. In particular, a Laplace prior distribution, which leads to the so-called Lasso regression (Tibshirani, 1996), might be beneficial. The Laplace distribution has more probability mass close to the mode and more probability mass in the tails than the normal distribution. Therefore, it would set some of the weights to exactly equal values, while giving significantly higher or lower weight to only a few models. However, no closed-form solutions are available for the Lasso estimates of $\boldsymbol{\beta}$, and thus computationally more expensive numerical optimization methods would be required.

4.2 Why Is It So Hard to Beat the Recalibrated Multimodel Mean?

It is interesting to note the tiny difference between the leave-one-out prediction errors of constrained unequal weighting (0.277) and equal weighting (0.281). The improvement from unequal weighting compared to equal weighting is so tiny that it could well be simply due to chance, and even if it were a genuine improvement, its practical utility would be limited at best. Based on this data, we have no reason to believe that unequal weighting offers any considerable improvement over equal weighting. As a matter of fact, there are a number of reasons why we should not expect a large improvement of unequal weighting over equal weighting in the first place. All climate models in the multimodel ensemble have roughly the same complexity—they all run on supercomputers and are maintained and developed by government agencies. Obviously, all models contain the same basic physics—namely, a discretized and simplified version of the Navier-Stokes equations, thermodynamic closure relations, and parameterizations of unresolved processes. The models were not developed independently, but rather rely on the same body of knowledge about the practicalities of numerical climate modeling. Furthermore, from a statistical point of view, the small sample size of $N = 31$ years naturally limits the precision with which the combination weights can be estimated. The ensuing estimation variance of the combination weights will degrade the quality of out-of-sample predictions. Some authors (e.g., Weigel et al., 2010) have explicitly warned against using unequal weighting at all and recommend treating the different models as exchangeable, even though small differences are conceivable in principle. It should be noted, however, that there are cases where a single model is superior to all other models, and therefore, forecast combination with less skillful models is always detrimental (e.g., Vitart, 2017). We have shown that unconstrained MLR with a small training dataset can indeed degrade the performance of the combined forecast compared to equal weighting. But a suitable shrinkage strategy that limits the variability of the combination weights can reduce this problem and has the potential to gain slight improvements over equal weighting. However, for the reasons stated in this chapter, we should not expect the improvement to be large, even if we knew the "true" optimal combination weights.

5 CONCLUDING REMARKS

Forecasts of physical-dynamical models can suffer from various forecast biases that can be corrected by statistical methodology. Furthermore, the availability of several forecast models for the same predictand calls for statistical methods to optimally combine multiple forecasts into a single forecast. In this chapter, we have outlined various regression approaches and discussed relevant statistical concepts such as model selection, in-sample versus out-of-sample performance, and the incorporation of prior knowledge. The methods discussed are based on developments from short-term weather forecasting to longer-term seasonal climate forecasting, and thus they are fully applicable at the sub-seasonal scale.

Acknowledgments

The authors thank Caio Coelho for providing the seasonal Niño-3.4 hindcast dataset, and Thordis Thorarinsdottir for helpful input on multivariate recalibration. Andrew Robertson and Frédéric Vitart provided helpful feedback and comments on earlier drafts.

CHAPTER

16

Forecast Verification for S2S Timescales

*Caio A.S. Coelho**, Barbara Brown*†, Laurie Wilson*‡, Marion Mittermaier*§, Barbara Casati*‡*

*Centro de Previsão de Tempo e Estudos Climáticos, Instituto Nacional de Pesquisas Espaciais, Cachoeira Paulista, SP, Brazil †Research Applications Laboratory, National Center for Atmospheric Research, Boulder, CO, United States ‡Environment and Climate Change Canada, Montreal, QC, Canada §UK Met Office, Exeter, United Kingdom

OUTLINE

Sub-seasonal to Seasonal Prediction
https://doi.org/10.1016/B978-0-12-811714-9.00016-4

1 INTRODUCTION

Forecast verification (or evaluation) is a critical aspect of the forecast improvement process and is also fundamental to informing forecast users regarding their reliability, skill, accuracy, and other features in order to aid optimal use. The idea of evaluating forecasts and projections using quantitative methods dates back more than a century, and many measures commonly used today for assessing sub-seasonal and seasonal forecasts were developed early in the 20th century for weather forecasts (Murphy, 1996). However, several new measures and approaches have been developed over the last couple of decades in response to newly identified needs for different kinds of information, changes in forecast types, and the need to address certain forecast performance questions adequately. For example, spatial methods have become a part of the verification toolbox in only the past 20 years. Verification science continues as an active research area as new forecasts, such as sub-seasonal, are developed and new challenges are discovered. Sub-seasonal forecast verification, therefore, capitalizes on methodological developments on other timescales (e.g., weather and seasonal).

As defined by Murphy (1993), forecast "goodness" combines forecast quality, consistency, and value. Forecast verification, by definition, measures forecast quality through comparisons of forecasts to observations. Although forecast value (i.e., the value accrued to users by utilizing forecasts in decision-making) is typically related to forecast quality, its formulation is complex and dependent on other factors that affect the decision process (e.g., the cost assessment of action versus the losses due to missed action). Hence, quality is not equivalent to value. Nevertheless, it is possible to consider user perspectives in verification processes through the evaluation of meaningful variables and the impacts of specific thresholds, and by applying diagnostic verification approaches that examine forecast performance characteristics relevant to particular users or groups.

Forecast verification serves a number of purposes. The primary verification goals are categorized as follows:

- *Scientific*: To inform forecast system development and improvement
- *Administrative*: To monitor forecast performance over time or justify a new supercomputer acquisition
- *User-oriented*: To help users make better decisions

Each of these purposes may require a different verification approach. Administrative users may be interested only in simple measures that are easy to compute and follow through time, whereas for scientific purposes, a wider range of diagnostics is desirable to provide greater forecast performance understanding in different situations. Commonly, operational forecasting centers focus on administrative aspects, while scientists and developers focus on scientific aspects. However, incorporating information from the third aspect—forecast users' applications—in both administrative and scientific verification efforts often can lead to more meaningful information about forecast performance.

Relevant forecast quality attributes depend on the type of forecast (e.g., probabilistic, deterministic) and events (e.g., categorical, continuous) of interest. Examples of forecast performance attributes include the following:

- *Association*: The strength of the relationship between forecasts and observations
- *Accuracy*: The average difference (e.g., Euclidean distance) between forecasts and observations for deterministic predictions and between forecast probabilities and binary observations for probabilistic predictions
- *Bias*: The distance between the forecast and observation average values
- *Discrimination*: Conditioned on observed outcomes, the degree to which forecasts distinguish between different observations or events
- *Reliability (conditional bias)*: Conditioned on the forecast, correspondence between forecast probabilities and observed relative frequency (e.g., an event must occur on 30% of the occasions that the 30% forecast probability was issued to achieve perfect reliability)
- *Resolution*: Conditioned on the forecasts, the degree to which observed frequency of occurrence of an event differs as the forecast probability changes
- *Sharpness*: The degree to which forecasts deviate from the mean climatological value/category for deterministic forecasts or from the climatological mean probabilities for probabilistic forecasts; the unconditional variation in the forecasts

Because verification is a multidimensional problem, it is important to measure multiple attributes to obtain a meaningful forecast performance evaluation. That is, a single measure is unable to provide a meaningful evaluation of a forecast. Moreover, single measures can hide important information about forecast quality. For example, the root-mean-squared error (RMSE; Section 4.1) incorporates information about both bias and variance of errors; to avoid confusion about the source of a poor score, it is important to consider these two features individually.

Specific forecast types may require different treatment from other forecast types and also may create opportunities for novel evaluations. In particular, sub-seasonal to seasonal (S2S) forecast characteristics may lead to consideration of the S2S verification problem as being somewhat different from verification at other timescales. For example, as S2S models are tuned to represent meteorological phenomena on the sub-seasonal timescale (with a range that covers from day 15 to day 60 in some models), they are naturally suited to investigating seamless verification across the weather and seasonal timescales (Wheeler et al., 2017; Zhu et al., 2014). Another particular aspect of S2S verification is the special challenge of dealing with inhomogeneities in ensemble size between hindcasts and forecasts when evaluated together (Weigel et al., 2008), a challenge also faced in seasonal forecasting. Various studies investigated the effect of ensemble size on probabilistic forecast quality, including attempts to remove the dependence on ensemble size in some verification scores to allow comparisons (e.g., Richardson, 2001; Müller et al., 2005; Weigel et al., 2007a; Ferro, 2007; Ferro et al., 2008), and thus being relevant for S2S verification. An additional challenge for S2S verification is the need to evaluate more than one variable simultaneously for some forecast types, such as bivariate attributes of Madden-Julian Oscillation (MJO); for more discussion of this topic, see Section 5.3.

This chapter provides a brief overview of relevant methods for S2S forecast verification. Several additional resources exist that provide further details regarding forecast verification methods, including Wilks (2011a), Jolliffe and Stephenson (2012a), and a website coordinated by the Joint Working Group on Forecast Verification Research of the World Meteorological Organization (WMO; https://www.wmo.int/pages/prog/arep/wwrp/new/jwgfvr.html). Section 2 focuses on the initial steps in the verification process: factors affecting verification studies design. Section 3 considers the issues associated with identifying appropriate observations for use in verification, as well as some of the issues resulting from their uncertainties. Commonly used verification measures are introduced in Section 4, and current S2S verification practices are described in Section 5. Finally, Section 6 includes a summary and recommendations.

2 FACTORS AFFECTING THE DESIGN OF VERIFICATION STUDIES

Various factors need consideration prior to computing verification scores. The forecast type being verified is particularly important. S2S forecasts are often probabilistic rather than deterministic, and they are also multicategory (e.g., below normal, normal, and above normal). Dichotomous (yes/no) and continuous deterministic forecasts are less common, especially for user applications. This section lists the key factors and questions to be considered in the design of a verification framework or study, beginning with developing an understanding of what is required and the target user/audience.

2.1 Target Audience

The target audience is a key determinant of how a verification study should be designed, so it should be identified first. For example, is the verification aimed at a model developer or an end user? What are the forecast performance aspects that the target audience cares about? What forecast type will be evaluated? Are there user-specific thresholds to be considered? What is the scope of the verification for the user, and how will the verification results be used? Should the target audience influence how verification results are presented? What complexity of metrics is appropriate? Less scientific audiences require simpler, more intuitive metrics and graphics.

2.2 Forecast Type and Parameters

The verification methodology is tailored to the forecast type and characteristics of the parameter to be evaluated: for example, is the forecast deterministic or probabilistic? Is it a grid-point forecast or spatially defined? Is the variable smooth or episodic? S2S forecasts are often expressed as the likelihood of a particular weather regime, positive or negative anomalies, or multicategory (often tercile) probability forecasts. The use of anomalies is widespread and requires taking account of model climate drifts and biases. In this context, it is important to identify relevant thresholds for defining the events of interest to be verified. S2S forecasts are often area-based, but they also can be site-specific. The spatial and/or temporal resolution may require an analysis of representativeness (Section 3), a potential issue that arises when pairing gridded forecasts with observations or analyses.

2.3 Nature of Available Observations

Suitable and reliable observations are crucial for attaining informative verification results. It is fundamental to have observations able to capture the events that the forecasts attempt to predict. What observational resolution (temporal and spatial) is required to verify the forecasts adequately? This may depend on the parameter; for example, precipitation and temperature spatial and temporal variability are very different. The impacts of inadequate (inhomogeneous) spatial and temporal sampling and observation uncertainty on verification can be large; therefore, it is important to understand and take these known and unknown uncertainties into account. Questions such as the following are important to answer: Are the observations quality-controlled? Are faulty measurements corrected or disregarded? Is model information used in the quality control? How large is the observation uncertainty, and are its sources fully known? How can this uncertainty information (or the lack thereof) be included in the verification results and their interpretation?

2.4 Identification of Appropriate Methods and Metrics

Once the verification goals and purposes have been established according to user needs and the characteristics of the available data are established, then appropriate methods and metrics can be chosen. The objective is to identify multiple verification attributes to address the questions of interest, and find graphical presentations aligned with the requirements identified in Sections 2.1 and 2.2, taking account of the data issues discussed in Section 2.3. Section 4 provides a summary of the verification metrics used to assess the most common attributes.

3 OBSERVATIONAL REFERENCES

Observations are the cornerstone of verification. However, reliable, long-term and model-independent observations are difficult to find. This is particularly challenging for S2S, where daily resolution precipitation and near-surface temperature data is needed for user-oriented forecasts (such as for calculating weekly averages), as opposed to monthly values, while long time series are still required. Besides, accounting for observation uncertainty in verification practices is an unresolved challenge in verification research and practice. Verification practitioners need to recognize observational data sets uncertainty sources (e.g., measurement errors, remote-sensing retrieval algorithms, inhomogeneous and incomplete spatial and temporal sampling, and time series standardization and homogenization), and their effects on verification statistics. It is also important to acknowledge model dependencies of the verifying observation (e.g., calibration and quality control often use a model analysis as the reference) in order to interpret verification results correctly. This section reviews some of the challenges in the quest for appropriate observational references for verification purposes.

Long time series (i.e., around 30 years of measurements) are often required for climate forecast evaluation and to serve as climatological reference in S2S forecasting and verification. These time series are often affected by break points (e.g., due to instrument replacement) and therefore, complex procedures are needed to homogenize and standardize the data (e.g., Vincent and Mekis, 2006). These procedures enable producing temporally coherent time series, but they can affect the measured values (e.g., extremes) and introduce uncertainties into verification data sets.

Verification against point observations can suffer from representativeness issues: A pointwise measurement might differ substantially from a model value for the nearby grid cell simply because the model value is conceived to represent a grid-box average (e.g., precipitation), while the station measurement reports the value of a subgrid phenomena (e.g., a convective cell), which is not represented by the model. Typically, duc to the representativeness issue, coarse resolution models underestimate precipitation extremes; finer model spatial resolution results in better representation of intense precipitation. Similarly, coarser-resolution models predict trace amounts more often than finer-resolution models, leading to a positive bias for small precipitation quantities (see Fig. 1).

Gridded observations are usually obtained from remote-sensing instruments on satellites or ground-based radar networks. Satellite-based products can provide gridded measurements of temperature, humidity, cloud cover, soil moisture, and sea ice concentration and thickness. Radar-based products provide quantitative precipitation estimates. These physical variables are obtained from satellite retrieved radiances and radar-backscattered reflectivities, based on remote-sensing statistical and physical assumptions (e.g., the Marshall and Palmer (1948) Z-R relationship to convert reflectivity to precipitation rate). To mitigate the effects of these assumptions (and associated uncertainties), verification can be performed with a model-to-observation approach (e.g., by comparing model-simulated brightness temperature directly to satellite-retrieved radiances). Data assimilation algorithms are often used to harmonize, merge, and quality-control satellite and radar-based gridded observations, introducing model characteristics and dependence on these observations. Finally, gridding procedures (such as kriging) can introduce synthetic features and consequently affect verification statistics.

Verification practices require that forecast and observed values are matched in space (and time). Caution is needed when choosing appropriate interpolation procedures because interpolation can alter the forecast or observed values, affecting verification statistics. For example, bilinear precipitation interpolation often introduces small trace and lower extreme values; cubic interpolation often introduces small negative precipitation values. Area-conservative interpolation is best for precipitation upscaling (from high- to low-resolution grids) because model precipitation values usually represent a grid-box average, whereas a nearest-point interpolation is best for adjusting precipitation on two grids with similar resolution. Spatially smooth variables (e.g., temperature, geopotential height) are often interpolated using bilinear or bicubic schemes. Neighborhood verification approaches relax the exact spatiotemporal colocation for matching forecast and observations. These approaches (as well as some spatial verification distance metrics) do not require interpolation and therefore avoid the related issues.

Verification against a model-generated analysis is often performed because of a number of conveniences, including (1) representativeness issues, quality control, and gridding are addressed by data assimilation algorithms used for analysis generation; and (2) observations are spatially defined, with no spatiotemporal gaps. However, a forecast model verification assessment against its own model-based analysis is affected by interdependence, and the results must be interpreted with caution. Park et al. (2008) demonstrated that verification against model-based analyses strongly favors the model used to produce the analysis: therefore, for fairness in model intercomparisons, verification against one's own analysis is often adopted (at the cost of losing a single unique reference for all models). A best practice to

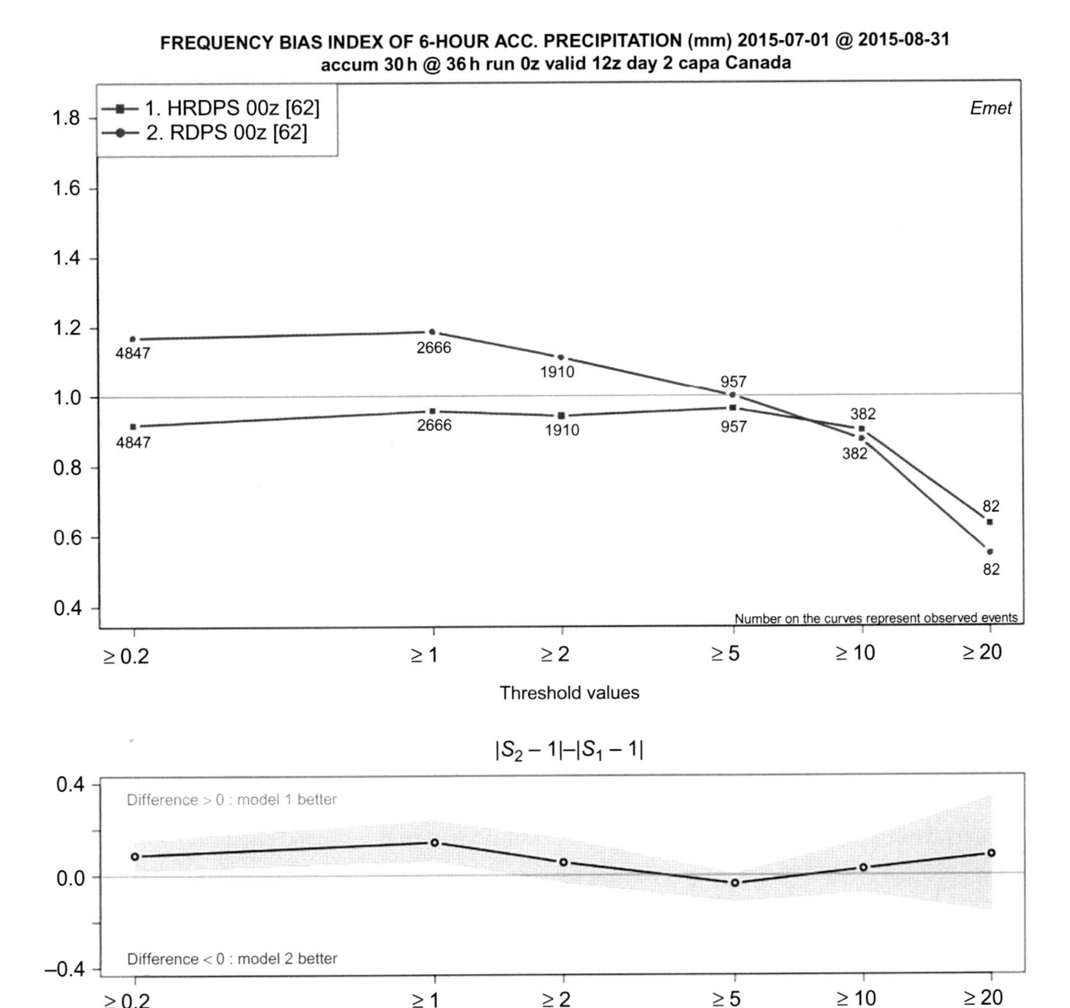

FIG. 1 Frequency bias for 6-h accumulated precipitation (from 30 to 36 UTC) for the Canadian RDPS (10-km resolution) and HRDPS (2.5-km resolution), for the summer 2015 against CaPA station measurements over Canada. The model with coarser resolution exhibits a larger and positive bias for smaller precipitation accumulations, as well as a more severe underestimation of high precipitation values, with respect to its higher resolution counterpart.

reduce the model-analysis dependence effect could include verifying against analysis at grid points where an observation has recently been assimilated (e.g., Lemieux et al., 2015). The use of the model background state in data assimilation algorithms nudges the observations toward the model climatology, which can affect scores based on climatology. Even if not used directly for verification, data assimilation can be exploited in verification practices to provide estimates of observation uncertainties and representativeness differences.

Finally, note that all verification procedures that use model-influenced observations via the analysis (i.e., observations are nudged toward the model climatology and upscaled to the model grid) and/or quality control (i.e., the filtering of observations which differ significantly from a short-range model forecast) reduce the verification results utility for all users outside the modeling community, and that generally leads to overestimating the model forecast quality. However, it is worth noting that S2S forecast verification mostly use reanalysis/analysis as reference data sets. Another important issue for S2S verification is the inconsistency in the computation of reference anomalies used to verify the forecasts (usually computed relative to the past 20–30 years) with reanalyses and operational (real-time) analyses, which are often based on different model versions. The difference is particularly large for surface parameters and can contaminate the reference anomalies used to verify the forecasts, which are computed by subtracting the operational (real-time) analysis from the reanalysis climatological mean over the past 20 or 30 years, because reanalyses are usually not available in real time and operational analyses are not available in the past.

4 REVIEW OF THE MOST COMMON VERIFICATION MEASURES

As outlined in Section 2, verification metrics selected for a specific application depend on several factors, the most important of which are the following:

a. The needs of the users of the verification results
b. The characteristics of the variable being verified
c. The nature of the forecast and available observations

Considering factors b and c leads to the classification of forecast variables and associated metrics into the following groups:

- *Deterministic variables (forecasts and observations)*—Characterized by specific variable values expressed in physical units (e.g., temperature in degrees Celsius). Deterministic variables are divided further into quasi-continuous, such as temperature, which takes any physically plausible value; and categorical, which is characterized by two or more ranges of values (categories) separated by one or more predetermined thresholds. Thresholds may have physical meaning. For example, the 0.5-mm daily rain threshold is often used to separate rainfall into "no rain" and "rain" categories. Thresholds also may be set to values particularly meaningful for forecast users (e.g., setting a 50-mm threshold in 24 h to indicate flooding risk). The set of categories thus defined is mutually exclusive (no overlap in values) and exhaustive (covers the whole range of possible values).
- *Probabilistic forecasts*—Forecasts indicating the probability of occurrence for predefined categorical variables or the forecast probability distribution for all possible variable values. Probability forecasts are usually verified with respect to deterministic observations, even though the observations may be subject to uncertainty. If observational uncertainty estimates are available, they can be used in probabilistic forecast verification (e.g., Candille et al., 2007). Most metrics described in this section can be generalized to incorporate observational uncertainty.

- *Spatial verification metrics*—These methods are designed to account for the spatial nature of the forecast variable and corresponding observations (e.g., how the shape of a forecast feature, such as a cloud band, compares with the observed shape). They can be applied to deterministic or probabilistic forecasts, but the former is more common and probably carries more physical meaning.

This section briefly describes and summarizes common metrics applicable to specific verification problems and data types. The forecast attributes (see Section 1) assessed by each metric are identified.

4.1 Metrics for Continuous Deterministic Forecasts

Table 1 summarizes the most common metrics used to verify deterministic continuous forecasts. The linear bias (B) identifies the average error for the verification sample, which is also implicitly included in both the mean absolute error (MAE) and the RMSE. Sometimes the bias is removed before computing the RMSE. Such removal not only reduces the RMSE but also implies that only the variable portion of the error is assessed. However, presentation of B and the bias-corrected RMSE together disentangles these two aspects of performance, which comprise the RMSE (see Murphy, 1988a for the decomposition of the MSE), and allows clearer understanding of the forecast errors. Comparing MAE and RMSE magnitudes for a

TABLE 1 Common Metrics (or Scores) for Verifying Continuous Deterministic Forecasts (F_i) Against the Observations (O_i)

Metric	Equation	Attribute Measured	Characteristics
Bias (linear bias, B)	$B=\frac{1}{N}\left[\sum_{i=1}^{N}(F_i-O_i)\right]$	Accuracy (average error)	Estimates the persistent or average error based on a specific data set; negative orientation (best when $B=0$)
MAE	$\text{MAE}=\frac{1}{N}\left[\sum_{i=1}^{N}\lvert F_i-O_i\rvert\right]$	Accuracy	Average error magnitude, negative orientation (best when MAE $=0$)
RMSE	$\text{RMSE}=\left[\frac{1}{N}\sum_{i=1}^{N}(F_i-O_i)^2\right]^{1/2}$	Accuracy	Average error magnitude weighted to larger errors; Negative orientation (best when RMSE $=0$)
SS (Skill Score)	$\text{SS}=\frac{S_f-S_r}{S_p-S_r}=\frac{S_r-S_f}{S_r}=1-\frac{S_f}{S_r}$	Skill (general format) For negatively oriented scores, perfect score $S_p=0$)	Fractional improvement of the forecast over an unskilled reference. Range: $-\infty$ to 1.
Pearson correlation coefficient (r)	$r=\frac{\sum_{i=1}^{N}(F_i-\overline{F})(O_i-\overline{O})}{\sqrt{\sum_{i=1}^{N}(F_i-\overline{F})^2}\sqrt{\sum_{i=1}^{N}(O_i-\overline{O})^2}}$	Association	Strength of the linear relationship between forecasts and observations Range: -1 to 1.

Note: Subscripts i refer to the ith case of the verification sample; the sample of forecast and observation pairs is of size N; the overbar indicates sample averaging. S_f (usually) refers to either the MAE or RMSE scores computed for the N pairs of F_i and O_i according to the equations in the table; S_r refers to the same score computed using a unskilled reference forecast such as the variable mean (climatology) or the latest observed value of the variable (persistence); S_p refers to the score for the perfect forecast. For perfect forecasts where $F_i=O_i$ for all N pairs, both MAE and RMSE equal zero (i.e., $S_p=0$).

particular sample gives an idea about the variability of the errors. The lower the variability, the smaller the difference between the two because large errors are more heavily penalized by the RMSE. Thus, the RMSE is favored when larger errors are considered relatively more important than smaller errors.

Skill scores (SSs) measure forecast accuracy relative to the accuracy of a reference forecast. Most SSs are in the general format shown in Table 1, which defines skill as the fractional improvement of the forecast score compared to the score for the reference-for-comparison. If the score for the forecast is worse than the reference, then the skill is negative. When the reference forecast accuracy is very high, when the sample size is small, or both, the SS can become unstable, with a small denominator. For this reason, SSs are always computed using the final summation score for a particular data set, not for the individual cases. SSs commonly use the MAE, the mean square error (MSE), or the RMSE as the score.

While climatology (sample mean, or long-term climatology if known), random chance and persistence (the last available observation) are the most commonly used reference forecasts, sometimes SSs are used to compare two competing forecasts, with the score for the poorer- or older-model version replacing the reference forecast. When the reference forecast is the climatology for the verification sample, the SS is the same as the reduction of variance or fraction of variance explained, which is the same as the square of the correlation coefficient between the forecasts and observations. This interpretation is more complicated when the reference is a second forecast.

The Pearson product moment correlation coefficient (r) is often used to measure the strength of the linear relationship between forecasts and observations (the association attribute). Perfect association (i.e., $r = 1$) is obtained when the forecasts and observations oscillate in exactly the same direction. This measure, however, provides only an indication of potential skill because correlation is insensitive to forecast biases, as well as differences in forecast versus observation variances. Several of the continuous measures can be displayed simultaneously using a Taylor diagram (Taylor, 2001). In particular, this diagram displays the correlation coefficient, RMS difference, and ratio of the standard deviations of the forecast and observed patterns.

4.2 Verification Methods for Categorical Deterministic Forecasts

Categorical deterministic forecasts are often synthesized using contingency tables. Table 2 shows the contingency table and scores for a 2×2 (2-category) variable, all of which are functions of the four table entries: hits (a), misses (c), false alarms (b), and correct negatives (d). Fundamental score characteristics are indicated in the table. Analogous table forms and scores exist for more than two categories, but multiple categories of a single variable are often treated as sequences of 2-category problems, with boundaries at each of the thresholds in turn. Murphy and Winkler (1987) related the contingency table and its entries to the forecast and observation joint probabilities, providing the statistical framework to interpret categorical scores as functions of joint, conditional, and marginal probabilities.

Several relationships exist between the categorical scores listed in Table 2, so a subset of those is often calculated, such as the frequency bias (FB) and the equitable threat score (ETS), or the Heidke skill score (HSS), used to assess bias and accuracy/skill. The FB is

TABLE 2 Contingency Table Format and Associated Scores

<table>
<tr><th>Measure</th><th>Equation/Format</th><th>Range-Orientation</th><th>Characteristics</th></tr>
<tr><td>Contingency table</td><td>
<table>
<tr><td></td><td></td><td colspan="2">Observed</td><td></td></tr>
<tr><td></td><td></td><td>Yes</td><td>No</td><td>Total fcst</td></tr>
<tr><td rowspan="2">Forecast</td><td>Yes</td><td>a
(Hits)</td><td>b
(False alarms)</td><td>$a+b$
(total events forecast)</td></tr>
<tr><td>No</td><td>c
(Missed events)</td><td>d
(Correct negatives)</td><td>$c+d$
(total non-events forecast)</td></tr>
<tr><td></td><td>Total obs</td><td>$a+c$
(total events obs)</td><td>$b+d$
(total non-events obs)</td><td>$N=a+b+c+d$
(sample size)</td></tr>
</table>
</td><td>Normally, as shown, columns are conditional observation totals, and rows are conditional forecast totals.</td><td>Equivalent to a scatterplot for categorized variable; 2 × 2 table most common—two categories, one threshold.</td></tr>
<tr><td>FB</td><td>$\mathrm{FB}=\frac{a+b}{a+c};\frac{c+d}{b+d}$
Ratio between the total number of events forecast (or not forecast) and the total number of events observed (or not observed).</td><td>0 to ∞</td><td>Best score = 1. Simple comparison of forecast frequency to observed frequency.</td></tr>
<tr><td>H (Probability of detection)</td><td>$H=\frac{a}{a+c}$</td><td>0 to 1</td><td>Best = 1. Incomplete score—does not account for false alarms.</td></tr>
<tr><td>F (probability of false detection)</td><td>$F=\frac{b}{b+d}$</td><td>1 to 0</td><td>Best = 0. Can be improved by forecasting the event less often to reduce false alarms.</td></tr>
<tr><td>FAR</td><td>$\mathrm{FAR}=\frac{b}{a+b}$</td><td>1 to 0</td><td>Best = 0. Sensitive to false alarms but ignores misses. Use with H.</td></tr>
<tr><td>TS (critical success index)</td><td>$\mathrm{TS}=\frac{a}{a+b+c}$</td><td>0 to 1</td><td>Best = 1. Sensitive to both false alarms and misses; ignores correct negatives.</td></tr>
</table>

Continued

TABLE 2 Contingency Table Format andAssociated Scores—cont'd

Measure	Equation/Format	Range-Orientation	Characteristics
ETS (Gilbert skill score)	$\text{ETS} = \frac{a - a_r}{a + b + c - a_r}$ where $a_r = \frac{(a+b)(a+c)}{N}$	−1/3 to 1; 0 indicates no skill over chance.	Best = 1. TS adjusted for the number correct by chance (guessing), a form of SS. Always <TS.
KSS (also true skill statistic TSS or Pierce skill score	$\text{KSS} = \frac{a}{a+c} - \frac{b}{b+d} = H - F$	−1 to 1; 0 indicates no discriminant ability	Best = 1. Related to the ROC area and EDI/SEDI scores. Indicates the ability of the forecast to discriminate between events and nonevents, as a basis for decision-making.
HSS	$\text{HSS} = \frac{(a+d) - E_r}{N - E_r}$ where $E_r = \frac{1}{N}[(a+c)(a+b) + (c+d)(b+d)]$	−∞ to 1	Best = 1. SS in the general format, with "chance" as the reference forecast.
EDI	$\text{EDI} = \frac{\ln F - \ln H}{\ln F + \ln H}$	−1 to 1; 0 indicates no accuracy.	Best = 1. Designed to avoid convergence to 0 or 1 for low frequency (rare) events. Most often used for verifying extreme event forecasts.
SEDI	$\text{SEDI} = \frac{\ln F - \ln H + \ln(1-H) - \ln(1-F)}{\ln F + \ln H + \ln(1-H) + \ln(1-F)}$	−1 to 1; 0 indicates no accuracy.	Best = 1. Similar to EDI, but approaches 1 only for unbiased forecasts.

Note: The letters a, b, c, and d refer to total counts of cases with the corresponding pairing of forecast and observation. Sample size is denoted as N.

not a verification score in the strict sense because it does not depend on matched forecasts and observation pairs. As a ratio of the forecast frequency to the observed frequency of each event category, it describes the forecast strategy as overforecasting if >1 and underforecasting if <1. Several categorical scores can be displayed simultaneously in a performance diagram (Roebber, 2009).

Often, the occurrence of one of the two categories is greater than the other, particularly in the case of extreme events, which are usually much less common than the corresponding nonevent. The extremal dependence index (EDI) and the symmetric extremal dependence index (SEDI) are specifically designed to score categories with low observed event frequency $(a + c)/N$ (called the *base rate* or *climatological frequency*). Under these conditions, scores such as the threat score (TS), the hit rate (H), the false alarm ratio (FAR), the false alarm rate (F), the ETS, and the Hanssen-Kuipers discriminant score (KSS) tend artificially toward their limit values (0 or 1), rendering the interpretation of the verification results challenging. The HSS also may become unstable for low base rates because the unskilled forecast accuracy is high.

When both categories are of similar interest, H, FAR, and TS can be computed for both categories (event and nonevent) separately. For example, for the nonevent category, H is $d/(b + d)$ and TS is $d/(b + c + d)$. In this situation, the proportion correct PC $= (a + d)/N$ is also an informative score. PC is not recommended otherwise because it becomes misleading when one category occurs more frequently than the other. See literature on the "Finley affair" (e.g., Murphy, 1996).

H and F refer to stratifying the verification data set in terms of (conditioned on) the observations. These scores are often used in pairs, and along with the relative operating characteristic curve (ROC) and relative operating characteristic area (ROCA) (discussed in more detail in Section 4.3), are useful for evaluating the a posteriori forecast quality as a basis for users' decision-making.

FAR is different from F (and sometimes is confused with it); it is the proportion of forecasts that are false alarms (i.e., it is conditioned on the forecasts). FAR is widely used and can be controlled by the forecaster by such actions as forecasting the event less often to reduce the number of false alarms. This strategy also would increase the number of missed events (c); therefore, FAR should be used in combination with H.

The correct negatives (d) sometimes are hard to determine for a contingency table because the nonevent may be spatially unbounded, temporally unbounded, or both; d also can be very large in the case of extreme events and overwhelm the contingency table computations. The H, FAR, and TS scores use only the other three entries of the contingency table, and therefore they can be computed without estimating or taking into account correct negatives. However, all SSs and scores useful for discrimination and decision-making need correct negative estimates. Some ways of estimating d for severe weather nonevents are suggested in Wilson (2014), and Wilson and Giles (2013).

4.3 Verification Measures for Probability Forecasts

Probability forecasts are estimates of the likelihood of occurrence of an event, which is usually defined as a category of a variable (e.g., the probability for the daily average temperature to be in the upper tercile of the temperature climatological distribution for a specific location or area). Categories are defined by thresholds as for categorical variables. Probability forecasts are difficult to verify meaningfully as single forecasts because the observation is usually treated as categorical (the event either occurred as forecast or not). Probability forecast verification thus proceeds after a sufficiently large sample of matched forecasts and observations is collected, allowing comparative assessment of the actual event occurrence frequencies with the forecast probabilities.

While probability forecasts are most often obtained from ensembles, it is worth noting that ensemble forecasts require postprocessing to generate probabilities, by calculating probabilities of the occurrence of events simply from the proportion of the ensemble members satisfying the threshold for the event, or by using the ensemble to estimate a full predicted distribution. *Ensembles* are collections of deterministic forecasts obtained from perturbed initial conditions, variations in the model formulation, or both (see Chapter 13). Raw ensembles are assumed to be a random selection from the unknown conditional probability density function (PDF) of possible forecast values, the associated cumulative distribution function (CDF), or both. The resulting forecast value distribution is inherently discrete given the relatively small ensemble sizes, but processing methods are available to estimate continuous PDFs. The verification methods summarized next are suitable for probability forecasts of specific events or for forecast PDFs.

Table 3 summarizes commonly used verification scores for probability forecasts. The Brier score (BS) is generally used for probability forecasts for a dichotomous (binary) variable, while the discrete ranked probability score (RPS) is preferred when there are more than two categories. The continuous ranked probability score (CRPS) is used to evaluate the full continuous or quasi-continuous forecast CDF. The BS, RPS, and CRPS all measure the attribute accuracy, while the corresponding scores shown in the bottom row of the table measure skill. The three scores can be partitioned into three components representing the attributes *reliability*, *resolution*, and *uncertainty*, the latter being a function of the observations only. The reliability (or attributes) diagram and the ROC curve offer concise and convenient graphical representations of most of the probability forecast attributes listed in Section 1.

Fig. 2A shows an example of an attributes diagram (which is based on a reliability diagram) produced by first binning the verification sample according to forecast probability, and next by computing the observed event frequency for all of the forecasts in each bin. The diagram is a plot of the observed frequency versus the forecast probability for each bin. Five bins are used (0%–20%, 20%–40%, 40%–60%, 60%–80%, 80%–100%). The points are plotted at the midpoints of the five bins, but it is possible (and perhaps more accurate) to plot the points at the actual mean forecast probability for each bin. Forecasts are considered perfectly reliable if the points lie along the 45° diagonal, indicating that, on average, the forecast probability equals the observed event frequency. Dashed lines are drawn horizontally at the climatological event frequency (the base rate), and vertically at the average forecast probability. Comparing these two lines indicates whether the event is, on average, overforecast or underforecast. In the example, there is no unconditional bias. The average forecast probability is about 33%, equal to the observed frequency.

The line that bisects the angle between the diagonal and the climatology line is known as the *no skill line*. On this line, the resolution of the forecasts equals the reliability component (and is of opposite sign), so that skill with respect to sample climatology along this line, as computed by the BSS, is zero. When the plotted curve lies within the shaded area, the

TABLE 3 Common Scores for Probabilistic Forecast Verification

Measure	Equation	Range-Orientation	Characteristics
BS	$BS=\frac{1}{N}\sum_{i=1}^{N}(p_i-o_i)^2$	1 to 0	Best = 0, negatively oriented, mean-squared error of the probability forecasts.
RPS (for discrete categories)	$RPS=\frac{1}{M-1}\sum_{m=1}^{M}\left[\left(\sum_{k=1}^{m}p_k\right)-\left(\sum_{k=1}^{m}o_k\right)\right]^2$	1 to 0	Best = 0, equals BS for 2 categories, for >2 categories, sensitive to distance between forecast and observed category.
CRPS	$CRPS=\int_{-\infty}^{\infty}\left[P_f(x)-P_o(x)\right]^2dx$	0 to ∞	Best = 0, Compares CDF for forecast with CDF of observation. Obs CDF is step function if deterministic; the result is in the units of the variable, reduces to MAE for deterministic forecast.
Brier Skill Score (BSS), Rank Probability Skill Score (RPSS), and Continuous Rank Probability Skill Score (CRPSS)	$SS=\frac{S_f-S_r}{S_p-S_r}=\frac{S_r-S_f}{S_r}=1-\frac{S_f}{S_r}$	$-\infty$ to 1	Best = 1. SSs in the standard format for negatively oriented scores. Caution: The reference forecast is defined by the sample over which the SS is computed (see Hamill and Juras, 2006).

The variables p_i and o_i refer to the ith forecast probability and ith observation in a sample of size N. The observation o_i is 0 (1) if the category predicted with probability p_i doesn't (does) occur. The subscript k refers to the kth category of a total of M categories, and $P_f(x)$ and $P_o(x)$ are the predicted and observed CDFs, respectively, with the latter taking the form of a step (heaviside) function, with the step at the observed value of the variable x. S_f, S_p, and S_r are defined exactly as in Table 1.

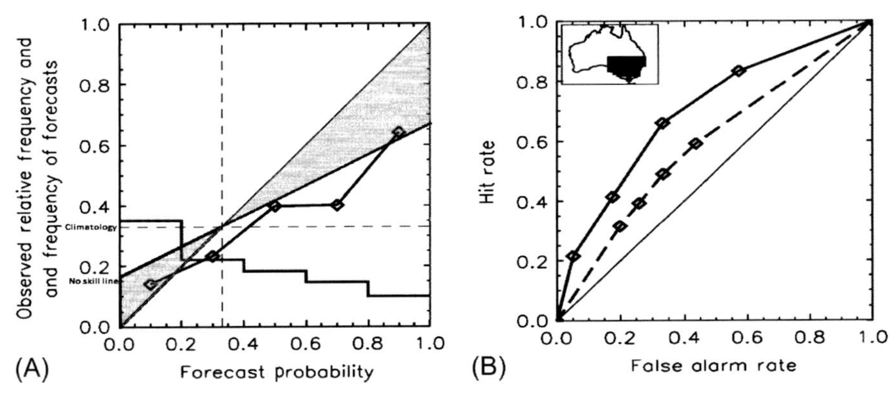

FIG. 2 Attributes/reliability diagram (A) and ROC plot (B) for the event average week 3–4 (second fortnight) maximum temperature forecasts in the upper tercile for Southeast Australia for spring start months. *Adapted from Fig. 7 of Hudson, D., Alves, O., Hendon, H.H., Marshall, A.G., 2011. Bridging the gap between weather and seasonal forecasting: intraseasonal forecasting for Australia. Q. J. R. Meteorol. Soc. 137, 673–689.*

forecasts have skill with respect to climatology. In the example, there is apparently little skill, but there is an indication of resolution in the forecasts because the plotted line is inclined with respect to the base rate line. The plotted curve presents a shallower than 45° angle, indicating that the forecasts are overconfident. The highest probabilities are overforecast, while the lowest probabilities are slightly underforecast. Curves lying at a steeper than 45° angle (in the area between the diagonal and the vertical dashed line) may be skillful, but they also are said to be overresolved, underconfident, or both. This situation does not often happen in practice. Finally, the histogram on the reliability diagram represents the percentage of the forecast probabilities sample falling into each bin, known as the *sharpness diagram*. Sharp forecasts have *u*-shaped histograms presenting high frequencies for near 0 and 100% forecast probabilities. In the example, forecasts are not particularly sharp, and they are skewed toward the lowest probabilities, with probabilities of <20% forecast nearly 40% of the time.

The ROC (Fig. 2B) and the ROCA measure the ability of the forecast to discriminate situations leading to events of impact from those that do not. Such discriminating ability is useful for those who must decide whether to take action to minimize adverse weather/climate impacts. The curve is obtained as follows:

1. Organize the verification sample in ascending order of the predictive variable (usually the forecast probability, but it also can be a physical variable such as the precipitation amount). If the forecasts are from ensembles, each set of ensemble forecasts can be ordered and pooled over all verification sample cases. The associated observation is binary (1 or 0), according to whether or not the event occurred for each case. For an ensemble forecast, the observation value 0 or 1 is assigned to all members of that particular ensemble.
2. For each unique prediction value, considered as a prediction threshold for the event of interest, compute the FAR and H for the resulting contingency table.
3. Plot H against the FAR. The result is a stepwise graph, approximating a curve. The more points that are possible (the more unique forecast values exist in the data set), the "smoother" the curve will be.
4. The area under the curve can be computed by triangulation using all plotted points.

Examples of the computation of the ROC by this method are shown in Mason and Graham (2002). It is common practice to bin the data into forecast categories, often forecast probability deciles, or in pentiles (as in Fig. 2B). This results in fewer points to estimate the curve, possibly leading to an underestimation of the ROCA. For example, if the frequency of occurrence of the event for a 5% forecast is lower than for 15% forecasts, then this discrimination information would be lost if the data were binned into 20% bins. However, the tradeoff is that there must be enough cases in each bin to support the plotted points, or else the plotted curve will be noisy and confidence in the location of the points will be low. Underestimates of the ROCA can be avoided by fitting a binormal model to the binned data (Wilson, 2000). The binormal model is described in Mason (1982).

Discrimination ability is indicated if the ROC curve lies in the upper left half of the diagram. The closer the curve lies to the upper left corner, the better the forecast discrimination ability. The diagonal is the no skill line (also known as the *no discrimination line*). This means that the forecast probabilities distribution when the event occurs is no different from the forecast probabilities distribution when the event does not occur, and therefore the user has no basis to decide whether to take action. The ROCA is the total area between the lower right corner and the ROC curve. ROCA values larger than 0.5 indicate discrimination ability. Perfect discrimination (ROCA = 1) occurs when there is no overlap at all between the conditional forecast distribution when the event occurred and the conditional forecast distribution when the event did not occur. Sometimes the ROC score is expressed as 2ROCA − 1 to give a positively oriented score varying between 0 and 1.

Fig. 2B shows two ROC curves. The solid line is for S2S model probability forecasts for the event "Second fortnight averaged maximum temperature in the upper tercile for southeastern Australia." The values on the dashed line are obtained by assuming that the first fortnight average temperatures persist for the second fortnight. The plot indicates modest discrimination, with some improvement of the S2S model (ROCA = 0.70) over the persistence forecast (ROCA = 0.59).

One cautionary note is needed with regard to application of ROC plots and SSs with respect to climatology. The relevant standard of comparison is always the mean observation of the sample used to compute the score. For the ROC, discrimination of all variation sources from the overall sample mean is credited. This effect is discussed in Hamill and Juras (2006).

4.4 Spatial Methods

Meteorological variables defined over spatial fields are characterized by spatial structures and features. Traditional point-by-point verification approaches do not account for the intrinsic spatial correlation existing between nearby grid points. This practice leads to double penalties (associated with small spatial displacements) and limited diagnostic power (traditional scores do not inform on displacements or error scale dependence). To address these issues, several spatial approaches have been developed and applied to weather forecasts in the past two decades. Spatial verification techniques aim to do the following:

- Account for spatial structure and features
- Provide information on the forecast error in physical terms (e.g., diagnose the location error as a distance in kilometers)
- Account for small time-space uncertainties

Spatial verification approaches are categorized into five classes:

1. Scale-separation approaches involve decomposition of the forecast and observation fields into scale components using a single-band spatial filter (Fourier transforms, wavelets, spherical harmonics), followed by a traditional verification of each spatial scale component. The rationale is to provide information on physical processes associated with weather phenomena on different scales (frontal systems versus convective precipitation; planetary, synoptic, and subsynoptic scales). These approaches enable the assessment of bias, error, and skill on each individual scale; are used to analyze predictability scale dependence (by determining the no skill-to-skill transition scale); and assess the forecast versus observation scale structure. Scale-separation techniques have been successfully applied to both weather and climate studies (e.g., Casati, 2010; Denis et al., 2002; Denis et al., 2003; Jung and Leutbecher, 2008; Livina et al., 2008) and can be useful in the S2S framework.

2. Neighborhood methods (Ebert, 2008) relax the requirement for an exact observation-forecast location match and define a neighborhood (both in space and time) where the forecast and observation are matched. Data treatment within the neighborhood differentiates the verification strategies, which include simple averaging (equivalent to upscaling, Yates et al., 2006); comparing the forecast versus observed event frequencies (Roberts and Lean, 2008); evaluating various attributes of the forecast versus observed PDF (Marsigli et al., 2005); and applying probabilistic and ensemble verification approaches to assess the forecast PDF within the observed neighborhood (Theis et al., 2005). Neighborhood approaches are suitable for comparing higher versus coarser resolution models. Moreover, they enable probabilistic evaluation of deterministic forecasts.
3. Field deformation techniques use a vector field to deform the forecast field toward the observed field until an optimal fit is found (by maximizing a likelihood function). A scalar (amplitude) field is then applied to correct the intensities of the deformed forecast field to those of the observed field. These morphing techniques were originally developed for data assimilation and nowcasting (Germann and Zawadzki, 2004; Nehrkorn et al., 2003) and have been used in verification only recently (Gilleland et al., 2010; Keil and Craig, 2007; Keil and Craig, 2009; Marzban and Sandgathe, 2010).
4. Feature-based verification techniques (Davis et al., 2006a,b; Ebert and McBride, 2000) first identify and isolate features in forecast and observation fields (by thresholding, image processing, using composites, or cluster analysis), and then assess different attributes (displacement, timing, extent, and intensity) for each pair of observed and forecast features.
5. Distance measures for binary images assess the distance between forecast and observation fields by evaluating the geographical distances between all the grid points exceeding a selected threshold. These metrics were developed in image processing for edge detection, pattern recognition, or both (Baddeley, 1992a,b; Dubuisson and Jain, 1994), and have been used for verification purposes only recently (Dukhovskoy et al., 2015; Gilleland, 2011; Schwedler and Baldwin, 2011). The distance measures are sensitive to differences in object shape and extent, in addition to the distance/displacement between forecast and observed features. Therefore, they are considered a hybrid between field-deformation and feature-based techniques.

5 TYPES OF S2S FORECASTS AND CURRENT VERIFICATION PRACTICES

5.1 Deterministic S2S Forecast Verification Practices

Sub-seasonal forecasts are often presented as weekly averages for the forthcoming 4 weeks, either defined as averages over days 1–7 (week 1), days 8–14 (week 2), days 15–21 (week 3), and days 22–28 (week 4), as in Li and Robertson (2015); or as averages over days 5–11 (week 1), days 12–18 (week 2), days 19–25 (week 3), and days 26–32 (week 4), as in Weigel et al. (2008). Some studies (Hudson et al., 2011, 2013) investigate averages over days 1–14 (first fortnight) and days 15–28 (second fortnight).

As in weather and seasonal forecasting practices, the mean of the available ensemble members is commonly used as an estimate of the forecast distribution central value. Deterministic forecasts are expressed as ensemble mean anomalies, computed by subtracting the ensemble mean forecast from the model long-term mean (climatology), estimated using retrospective forecasts produced for a number of previous years for a first-order model bias correction. This procedure used for computing ensemble mean anomalies is typically lead time dependent.

The simplest S2S verification practice is eyeball (visual) comparison of the forecast ensemble mean anomaly with the corresponding observed anomaly. As shown in Fig. 1 of Vitart et al. (2017), for example, one can visually compare 2-m temperature ensemble mean forecast anomaly maps for different models with the observed anomaly. Eyeball comparison is useful for an initial qualitative assessment of specific forecasts, but it also is prone to subjective interpretation biases and therefore must be used with caution. Quantitative assessment obtained by computing verification metrics based on a collection of past forecasts and observations provides a more complete view of forecast quality.

A common verification practice for deterministic (ensemble mean) S2S forecasts is to compute the linear correlation between the forecast and observed anomalies at each grid point (over the available retrospective forecasts) and produce a map with the obtained values (see Fig. 1 of Hudson et al., 2011 and Li and Robertson, 2015, respectively; and Fig. 10 of Weigel et al., 2008). The Pearson product moment correlation coefficient is often used for this purpose, providing an association measure (as discussed in Section 4.1). However, due to its insensitivity to forecast biases, complementary accuracy metrics are required to quantify forecast errors. A standard S2S metric used for this purpose is linear bias (see Section 4.1). Fig. 3 of Weigel et al. (2008) provides an example of 2-m temperature ensemble mean bias for weeks 1–4 retrospective forecasts.

Deterministic S2S forecast skill can be estimated using the mean squared error skill score (MSSS), as performed by Li and Robertson (2015). The MSSS is based on the MSE, an accuracy measure similar to the RMSE (see Section 4.1), with the main difference being that the square root needed for the RMSE is not computed for the MSE. In the examples shown in Figs. 13–15 of Li and Robertson (2015), the reference set of forecasts used to compute the MSSS consisted of climatological forecasts given by the climatological average rainfall for a given weekly average. The maximum MSSS equals unity and is obtained for perfect forecasts with null MSE. Negative values indicate that the forecasts are less accurate than the reference climatological forecast.

5.2 Probabilistic S2S Forecast Verification Practices

A common procedure in S2S probabilistic forecast verification is to construct ROC plots and reliability diagrams, as described in Section 4.3, or to compute the RPSS and construct reliability diagrams as in Vigaud et al. (2017a,b). Fig. 12 of Vitart and Molteni (2010) and Fig. 3 of Hudson et al. (2011) show additional ROC plots and reliability diagrams of S2S examples for a collection of forecasts aggregated over a number of grid points within a predefined area/region. The area under the ROC plot provides an indication of the ability of the forecasting system in successfully discriminating occurrence from nonoccurrence of the event of interest (i.e., how forecast probabilities vary when stratified on the observations).

The reliability diagram provides a graphical interpretation of probabilistic forecast quality in terms of reliability (how well forecast probabilities match the observed frequency of the event of interest) and resolution (how the observed frequency varies when the data are stratified by the forecast probabilities).

By computing the ROC area at each grid point and mapping the collection of obtained values one can have a spatial idea of forecast discrimination ability, particularly for regions exhibiting ROC area above 0.5 [the reference value for unskillful forecasts with equal (50%) probability of distinguishing/discriminating events from nonevents]. Fig. 2 of Hudson et al. (2011) shows examples of ROC area maps for probabilistic forecasts of precipitation averaged over the first and second fortnight for events defined as precipitation in the lower and upper terciles.

5.3 Madden and Julian Oscillation (MJO) Forecast Verification

A specific type of sub-seasonal forecast is the forecast of the MJO (Madden and Julian, 1971, 1972, 1994; Zhang, 2005), which usually emerges as enhanced convection over the tropical Indian Ocean and propagates eastward along the equator. MJO forecasts and the associated verification are important to both model developers and forecasters in order to provide information about model behavior and performance in representing tropical precipitation.

MJO forecasts are distinct from traditional weather and climate forecasts because they are displayed in a two-dimensional (2D) phase space represented by two so-called Real-time Multivariate MJO indices (RMM1 and RMM2), as defined by Wheeler and Hendon (2004). Fig. 3A shows an example of an MJO forecast initialized on January 1, 1986, for the following 41 days. The initial points for the observations (blue line) and ensemble mean forecasts (red dashed line) are indicated with large brown dots. The small black dots are separated by 5 days. Counterclockwise progression indicates eastward MJO signal propagation. The MJO strength is measured by the distance of each point in the phase-space diagram to the origin. The central circle represents one standard deviation and is usually considered as the threshold for defining an active MJO signal. The observed RMM1 and RMM2 are the principal component time series of the first and second leading modes of the combined empirical orthogonal function (EOF) analysis of daily outgoing longwave radiation (OLR), with 850-hPa and 200-hPa zonal wind anomalies latitudinally averaged from 15°S to 15°N. Both RMM1 and RMM2 are normalized by the observational standard deviation, resulting in indices with zero mean and unit variance. See Rashid et al. (2011) and Gottschalck et al. (2010) for additional information on how the forecast RMM1 and RMM2 are computed.

Fig. 3B shows the schematic for a pair of points [$O(t)$ and $F(t,\tau)$] in the 2D MJO phase space represented by the RMM1 and RMM2 indices (horizontal and vertical axis, respectively). The point $O(t)$, highlighted with a blue dot, represents the location of the observed MJO signal at time t. The point $F(t,\tau)$, highlighted with a red dot, represents the forecast MJO signal at time t produced τ days earlier. For example, the points $O(t)$ and $F(t,\tau)$ in Fig. 3B illustrate the fourth black dots after the initial large brown dots shown in Fig. 3A representing a forecast for time t equal to January 20, 1986, produced the previous January 1, 1986 ($\tau = 20$ days lead forecast). The blue solid line connecting the origin of the phase-space plot to the point $O(t)$ graphically illustrates the observed MJO signal. The projections of this signal along the horizontal and

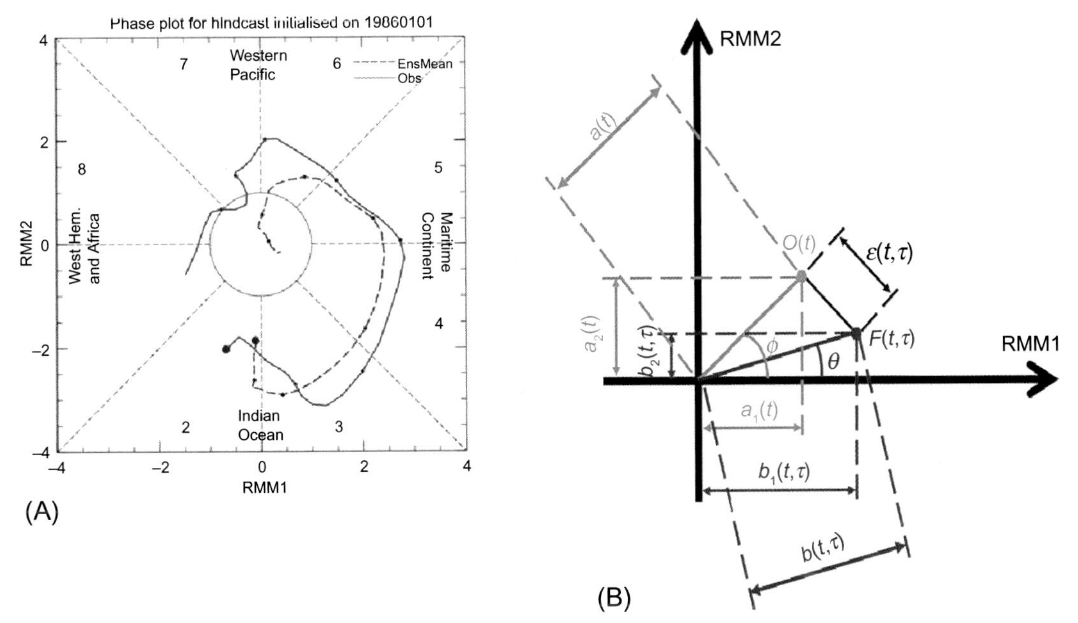

FIG. 3 (A) Phase-space plots of RMM1 and RMM2, computed from National Centers for Environmental Prediction/National Center for Atmospheric Research (NCEP/NCAR) reanalysis and satellite OLR *(blue)* and the ensemble-mean POAMA hindcast initialized on January 1, 1986 *(red)*, for the period January 1–February 10, 1986, as shown in Fig. 3 of Rashid et al. (2011). The *black dots* represent every 5 days. Each octant of the phase diagram is numbered (from 1 to 8) according to the phase definitions of Wheeler and Hendon (2004) Also labeled are the approximate locations of the enhanced convective signal of the MJO for that location of the phase space (e.g., the Indian Ocean for phases 2 and 3). The RMM1 and RMM2 values were smoothed with a 1–2–1 filter in time prior to plotting. (B) Schematic representation of an MJO forecast $F(t,\tau)$ in the RMM1 versus RMM2 phase space for a particular time t produced τ days in advance (i.e., lead time of τ days), with the corresponding observed MJO signal $O(t)$.

vertical axes are illustrated in Fig. 3B as $a_1(t)$ and $a_2(t)$ and represent the observed RMM1 and RMM2, respectively. The red solid line connecting the origin of the phase-space plot to the point $F(t,\tau)$ graphically illustrates the forecast MJO signal for time t produced τ days earlier. The projection of this signal along the horizontal and vertical axis is illustrated in Fig. 3B as $b_1(t,\tau)$ and $b_2(t,\tau)$ and represents the τ days lead RMM1 and RMM2 forecasts for time t, respectively.

As the RMM1 and RMM2 axis in Fig. 3B are orthogonal, the observed $a(t)$ and forecast $b(t,\tau)$ MJO amplitudes are expressed as

$$a(t) = \left[a_1(t)^2 + a_2(t)^2\right]^{1/2} \tag{1}$$

$$b(t,\tau) = \left[b_1(t,\tau)^2 + b_2(t,\tau)^2\right]^{1/2} \tag{2}$$

and the observed (φ) and forecast (Θ) MJO phases, represented by the angles between the blue and red lines and the horizontal RMM1 axis, respectively, are expressed as

$$\phi(t) = \tan^{-1}\left(\frac{a_2(t)}{a_1(t)}\right) \tag{3}$$

$$\theta(t,\tau) = \tan^{-1}\left(\frac{b_2(t,\tau)}{b_1(t,\tau)}\right) \tag{4}$$

Following Rashid et al. (2011), the amplitude $A(\tau)$ and phase $P(\tau)$ errors for a collection of N forecast and observed MJO pairs as a function of forecast lead time τ are defined as

$$A(\tau) = \frac{1}{N}\sum_{t=1}^{N}[b(t,\tau) - a(t)] \tag{5}$$

$$P(\tau) = \frac{1}{N}\sum_{t=1}^{N}\tan^{-1}\left(\frac{a_1 b_2 - a_2 b_1}{a_1 b_1 + a_2 b_2}\right) \tag{6}$$

The amplitude error verification metric $A(\tau)$ is similar to the linear bias (see Section 4.1). Both $A(\tau)$ and $P(\tau)$ measure accuracy in terms of the average error. $A(\tau)$ is negatively oriented (best forecasts have $A(\tau) = 0$). $P(\tau)$ expresses the mean angle difference $(\Theta - \varphi)$ of the forecast Θ and observed φ MJO phases over the N available pairs. $P(\tau)$ is positive if the forecast phase on average leads the observed phase. Note that to obtain Eq. (6), one needs to use cross- and dot-product properties in the process of finding the angle $(\Theta - \varphi)$ between the observed (blue line) and forecast (red line) MJO signals.

Lin et al. (2008) introduced the following metrics for evaluating the quality of the bivariate MJO forecasts displayed in the RMM1 versus RMM2 phase space: the bivariate correlation $r(\tau)$, the RMSE(τ), and the mean square skill score MSSS(τ), following Murphy (1988a).

$$r(\tau) = \frac{\sum_{t=1}^{N}[a_1(t)b_1(t,\tau) + a_2(t)b_2(t,\tau)]}{\left(\sum_{t=1}^{N}\left[a_1(t)^2 + a_2(t)^2\right]\right)^{1/2}\left(\sum_{t=1}^{N}\left[b_1(t,\tau)^2 + b_2(t,\tau)^2\right]\right)^{1/2}} \tag{7}$$

$$\mathrm{RMSE}(\tau) = \left(\frac{1}{N}\sum_{t=1}^{N}\varepsilon(t,\tau)^2\right)^{1/2} \tag{8}$$

where

$$\varepsilon(t,\tau)^2 = [a_1(t) - b_1(t,\tau)]^2 + [a_2(t) - b_2(t,\tau)]^2 \tag{9}$$

See Fig. 3B for a graphical representation of $\varepsilon(t,\tau)$.

Finally,

$$\mathrm{MSSS}(\tau) = 1 - \frac{\mathrm{MSE}(\tau)}{\mathrm{MSE}_C} \tag{10}$$

where

$$\mathrm{MSE}(\tau) = \frac{1}{N}\sum_{t=1}^{N}[a_1(t) - b_1(t,\tau)]^2 + [a_2(t) - b_2(t,\tau)]^2 \quad (11)$$

and

$$MSE_C = \frac{1}{N}\sum_{t=1}^{N}\left[a_1(t)^2 + a_2(t)^2\right] \quad (12)$$

is the mean squared error for the climatological (unskillful) forecast that always issues an absent MJO signal RMM1 = RMM2 = 0 for all t and τ [$b_1(t,\tau) = b_2(t,\tau) = 0$], and is equivalent to the observed (climatological) variance of the MJO.

The bivariate correlation $r(\tau)$ is an association measure examining the strength of agreement (or disagreement) between the observed (φ) and forecast (Θ) MJO phases, but it is insensitive to MJO amplitude errors (biases in the magnitude of the forecast MJO signal). The RMSE(τ) is a simultaneous accuracy measure of both the phase and amplitude of the MJO, similar to the RMSE earlier introduced in Section 4.1.

The upper limit for the bivariate correlation $r(\tau)$ is obtained for perfect forecasts indicating an exact match between the forecast and observed phases of the MJO (when $\Theta = \varphi$), and it equals unity. The lower limit for $r(\tau)$ is obtained for forecasts indicating an opposite match between the forecast and observed phases, and it is equal to -1 (when $\Theta = \varphi + 180°$). For perfect forecasts with $a_1(t) = b_1(t,\tau)$ and $a_2(t) = b_2(t,\tau)$, the bivariate RMSE(τ) equals zero. For the climatological forecasts [in the absence of an MJO signal, and thus $b_1(t,\tau) = b_2(t,\tau) = 0$], the bivariate RMSE($\tau$) equals $2^{1/2}$ because the variance of each of the two observed RMM indices [$a_1(t)$ and $a_2(t)$] is equal to 1. Forecasts are generally considered skillful if their RMSE(τ) values are $<2^{1/2}$ (the RMSE(τ) for climatological MJO forecasts). For forecasts with observed amplitude but completely random phase [a persistence forecast at very long lead time such that $a_1(t) = b_1(t,\tau)$ and $a_2(t) = -b_2(t,\tau)$], the RMSE(τ) asymptotes to 2.

MSSS(τ) provides a relative measure of skill for the MJO forecasts compared to the climatological forecast that indicates an absent MJO signal [$b_1(t,\tau) = b_2(t,\tau) = 0$]. Perfect forecasts with MSE(τ) = 0 have MSSS(τ) = 1. Forecasts with errors as large as the climatological variance [MSE(τ) = $\mathrm{MSE_C}$] have a null SS [MSSS(τ) = 0], and forecasts performing worse than the climatological forecast (i.e., MSE(τ) > $\mathrm{MSE_C}$) have a negative SS (MSSS(τ) < 0).

It is common practice (Lin et al., 2008; Lin and Brunet, 2011; Rashid et al., 2011) to present all MJO forecast verification metrics discussed here as a graph of each metric as a function of forecast lead time τ. For positively oriented metrics [e.g., $r(\tau)$], with larger values indicating better forecast performance, such graphs usually display a decreasing curve with large values of the metric for shorter forecast lead times and smaller values for longer forecast lead times. The opposite feature is generally noticed for negatively oriented metrics, with smaller values indicating better forecast performance [e.g., RMSE(τ)]. For these metrics, the graphs usually display an increasing curve, with small values for shorter forecast lead times and large values for longer forecast lead times.

Finally, it is worth noting that this section has addressed MJO forecast verification from a deterministic (ensemble mean) perspective. The reader is encouraged to look at Marshall et al. (2016a), which recently proposed a methodology for probabilistic MJO forecast verification.

6 SUMMARY, CHALLENGES, AND RECOMMENDATIONS IN S2S VERIFICATION

This chapter presented an overview of forecast verification methods relevant to S2S, including current practices. Deterministic and probabilistic verification metrics commonly used for weather and seasonal forecast verification are also used for sub-seasonal forecast verification. However, a number of challenges still need to be addressed, including the following:

- Advancing seamless verification practices to allow a smooth and comparative quality assessment across different timescales (Wheeler et al., 2017; Zhu et al., 2014)
- Dealing with different ensemble sizes in S2S retrospective forecasts, which are usually much reduced, and real-time forecasts when computing forecast probabilities and verification scores (Weigel et al., 2008)
- Advancing the treatment of observational uncertainty in S2S verification (Bellprat et al., 2017)
- Application of spatial verification methods in generally coarse resolution S2S models

Here are some recommendations for advancing S2S forecast verification research and practices:

- Identify the most relevant forecast quality attributes for the target audience and verification questions of interest, and choose the appropriate scores for a thorough assessment.
- Develop an S2S forecast verification framework for comparing real-time and retrospective forecast skill levels. In the light of the richness of the S2S project database (Vitart et al., 2017) in terms of available retrospective forecasts and near-real-time forecasts from several modeling centers, and the need for the production of verification information in support of future routine sub-seasonal forecast delivery, there is clearly the need for producing verification information to help forecasters and users in various sectors to acquire knowledge about the strengths and weaknesses of these forecasts in order to build confidence in S2S forecast products (Coelho et al., 2018).
- Use verification metrics meaningful to users (e.g., use user-relevant thresholds when verifying probabilistic forecasts).
- Move beyond traditional weekly/fortnightly verification to more user-oriented procedures (e.g., active and break rainfall phases, dry/wet spells, heat wave forecast verification). Various application sectors usually require detailed weather within climate information, which is not traditionally verified. The S2S project database (Vitart et al., 2017) provides an excellent opportunity to assess the forecast quality of these longstanding demands of various sectors.
- Use appropriate verification measures when dealing with extreme events (e.g., Ferro and Stephenson, 2011; Stephenson et al., 2008) such as heat waves, cold snaps, droughts, and extended rainy conditions.
- Use novel verification measures adequate for S2S forecasts (e.g., probabilistic measures such as the generalized discrimination score, Weigel et al., 2008; Weigel and Mason, 2011) and spatial methods that provide performance information for forecasts with coherent

structures (Gilleland et al., 2009) if the spatial resolution of the forecasts allows such detailed spatial verification.

- Explore the novel concept of fair scores in S2S forecast verification (Ferro, 2014; Fricker et al., 2013).
- Address sampling uncertainties when computing scores using, for example, bootstrap procedures (Doblas-Reyes et al., 2009) for generating confidence intervals for verification measures and producing statistically meaningful comparisons among forecasting systems. Due to the generally limited number of available S2S retrospective forecasts and near-real-time forecasts, it becomes important to have strategies for estimating the uncertainties around the computed verification scores. The bootstrap procedure, which allows the computation of a large number of verification scores by resampling the limited number of available forecasts, is an interesting alternative for this purpose.
- Further explore the framework for probabilistic 2D phase-space MJO forecast verification (Marshall et al., 2016a). Until very recently, MJO forecast verification has been performed using deterministic scores based on ensemble mean forecasts. Again, the S2S project database (Vitart et al., 2017), which contains a very rich amount of ensemble retrospective forecasts and near-real-time forecasts from various modeling centers, provides an excellent opportunity for advancing probabilistic MJO forecast verification practice.
- Advance conditional verification practices such as verification conditional on an element such as the MJO, and the El Niño–Southern Oscillation (ENSO) phases, as well as on particular weather regimes. As MJO and ENSO are recognized as important predictability sources on the S2S timescale, more studies aiming to diagnose the impact of these two phenomena on the prediction ability of current S2S models with a wide range of variables, such as precipitation and near-surface temperature, are required.

PART IV

S2S APPLICATIONS

CHAPTER

17

Sub-seasonal to Seasonal Prediction of Weather Extremes

Frédéric Vitart, Christopher Cunningham†, Michael DeFlorio‡, Emanuel Dutra§, Laura Ferranti*, Brian Golding¶, Debra Hudson‖, Charles Jones#, Christophe Lavaysse**, Joanne Robbins¶, Michael K. Tippett††*

*European Centre for Medium-Range Weather Forecasts (ECMWF), Reading, United Kingdom †National Centre for Monitoring and Early Warning of Natural Disasters (CEMADEN), São José dos Campos, Brazil ‡NASA Jet Propulsion Laboratory/California Institute of Technology, Pasadena, CA, United States §Instituto Dom Luiz, IDL, Faculty of Sciences, University of Lisbon, Lisbon, Portugal ¶Met Office, Exeter, United Kingdom ‖Bureau of Meteorology, Melbourne, Australia #University of California, Santa Barbara, (UCSB), Santa Barbara, CA, United States **University of Grenoble Alpes, CNRS, IRD, G-INP, IGE, Grenoble, France ††Columbia University, New York, NY, United States

OUTLINE

Sub-seasonal to Seasonal Prediction
https://doi.org/10.1016/B978-0-12-811714-9.00017-6

1 INTRODUCTION

There is increasing interest in extreme weather and climate events, both in terms of gaining a better understanding of the impacts of climate change and developing early warning systems for better preparedness. Extreme weather and climate events pose a serious threat to health and welfare. For instance, between 2011 and 2013, the United States experienced 32 weather events that each caused at least $1 billion in damage, with a total of more than $110 billion in damages in 2012 (NOAA, 2013). According to Munich Re (2011a,b), the world's largest reinsurance company, >90% of all disasters and 65% of associated economic damage in 2010 were weather and climate related (i.e., high winds, flooding, heavy snowfall, heat waves, droughts, wildfires), although there were far more deaths from geological disasters that year (almost entirely from the Haiti earthquake). In all, 874 weather- and climate-related disasters resulted in 68,000 deaths and $99 billion in damage worldwide in 2010. These extreme weather events can also damage critical infrastructure, such as roads, railways, and power and telecommunication grids. For instance, during flooding in the United Kingdom at the end of 2015, some 20,000 homes were left without power. Therefore, the prediction of weather extremes is one of the main duties of national services to allow appropriate mitigating action to be taken and contingency plans to be put into place by the authorities and the public.

In this chapter, the terminology *extreme weather* is not restricted to climatologically rare events (i.e., a return period larger than a few decades); it also covers weather and climate events that have a strong societal impact (e.g., tropical cyclones, tornadoes, heat or cold waves, and flooding).

Historically, extreme weather numerical prediction started in the 1960s, with the emergence of computers and operational numerical weather prediction (NWP). It was originally exclusively a short-range forecasting problem. The main goal was to predict a specific extreme weather event a few minutes to a few days in advance. In the 1990s, extreme weather became an increasingly popular research topic for climate change prediction. Here, the main goal is to determine how much climate change will affect the statistics (frequency, intensity, duration, etc.) of extreme weather (see, e.g., Knutson et al., 2015 for the impact of global warming on tropical cyclone activity). Therefore, the short to medium range and climate forecasting of extreme weather events address very different questions—the former being an initial value problem and the latter a boundary condition problem. S2S prediction lies between these points. Depending on the nature and predictability of a specific extreme weather event, S2S forecasting systems could be used to predict a specific extreme weather event or to predict changes to the statistics of the event in the coming few weeks or months compared to climatology.

Small-scale, short-lived extreme weather events cannot be predicted individually weeks in advance, but it may be possible to predict changes to their probability of occurrence over large areas and long periods due to their relationship to the large-scale weather circulation. It also may be possible to predict the genesis, duration, and decay of large-scale, long-lasting extreme climate events (defined here as extreme events lasting more than a few weeks), such as heat waves, a few weeks in advance.

Hence, this chapter has been divided into two main sections: the first will explore the predictability of large-scale, long-lasting extreme events (e.g., heat waves, cold spells, droughts) and the second will explore the S2S predictability of smaller-scale extreme events (e.g., tropical cyclones, tornadoes, windstorms, and flooding). The last section of this chapter will discuss current ways of displaying probabilities of extreme events in S2S predictions. This chapter will focus on the predictability and predictive skill of extreme weather events at the S2S timescale. The important question of the usefulness of S2S forecasts of extreme events will be discussed in the following chapters of this book, which are dedicated to applications.

2 PREDICTION OF LARGE-SCALE, LONG-LASTING EXTREME EVENTS

The first type of extreme events that are expected to exhibit some predictability at the S2S timescale are long-lasting, large-scale extreme weather hazards that occur on a scale of >1000 km, with a lifetime ranging from a week to a few months. These events can be particularly destructive and deadly. For instance, the 2003 heat wave over Europe killed an estimated 70,000 people across Europe (Robine et al., 2008), and the Russian heat wave of 2010 killed an estimated 55,000 people (Hoag, 2014), causing fires throughout the country, the worst drought in nearly 40 years, and the loss of millions hectares of crops. According to Buizza and Leutbecher (2015), extended-range forecast skill could be expected for extreme events with similar timescales and space scales. Of course, S2S models will probably not have the required level of skill to predict the daily variations of these extreme events, but they can be expected to provide guidance on their genesis, time evolution, intensity, and decay on at least a weekly basis. Even if some of these events have lifetimes exceeding a few months (which is more relevant for seasonal prediction), S2S forecasts could be useful to predict their weekly evolution.

2.1 Heat Waves/Cold Spells

Heat waves in summer and cold spells in winter have significant socioeconomic impacts, such as temperature-related mortality, financial loss due to crop failure, and damage to infrastructure, industries, and transport. In the context of global warming, a change in these extreme events in terms of severity, frequency, and duration is expected (Perkins, 2015). It is, therefore, very relevant to explore the ability of the current models to predict heat waves and cold spells up to 2–4 weeks in advance, giving sufficient time for preparedness and mitigation strategies to be carried out.

What are the physical drivers that lead and maintain the occurrence of extreme high or low temperatures for more than a week? Synoptic systems, such as persisting high-pressure systems (e.g., blocking), land surface interactions, and climate variability phenomena such as the Madden-Julian Oscillation (MJO) and El Niño–Southern Oscillation (ENSO), play a major role in the life cycle of heat-wave events. For instance, Teng et al. (2013) found that anomalous atmospheric planetary waves, which are not necessarily linked to tropical heating, tend to precede heat waves in the United States. Relationships among those physical drivers and extreme events differ from one event to the next, depending on the extreme temperatures of the events and on their geographical locations. This section will discuss two examples of heat wave and cold spell prediction.

2.1.1 Heat Wave/Cold Spell Prediction Over Europe

Since 2000, a few extreme temperature events have affected various parts of Europe, with devastating consequences. In particular, the deadly heat waves that struck Western Europe in August 2003 and Russia in 2010, and the very cold winter of 2009/2010 made the prediction of these high-impact events at the S2S timescale a matter of the highest priority. Because the variability of the European climate is mostly controlled by the latitudinal positions of the jet stream, intraseasonal-to-seasonal fluctuations are often described as the alternation between a limited number of flow configurations (e.g., weather regimes). For example, summer extremes in European temperatures are generally associated with the exceptional persistence of blocking anticyclones (Cassou et al., 2005), and cold spells in winter are generally dominated by the persistence of a negative phase of the North Atlantic Oscillation (NAO), as seen during the winter of 2009/2010 (Cattiaux et al., 2010). The structure of the NAO negative phase includes an anticyclonic anomaly over Greenland, a cyclonic anomaly over the Azores, a strong reduction of westerly flow across the Atlantic, and the reinforcement of northerly winds from the Arctic (Walker and Bliss, 1932). It follows that the accurate prediction of the life cycle of regimes such as the NAO is instrumental for early warnings of extreme temperatures over Europe. Severe temperatures are also influenced by the establishment of persistent high pressure (anticyclonic anomaly) over Scandinavia—the so-called European blocking.

During the European winter, negative NAO and European blocking are the circulation types that more likely lead to persisting cold conditions. Fig. 1 shows the skill of several sub-seasonal forecasting systems to predict negative NAO and European blocking in winter over the period 1999–2010. Depending on the model, the probabilistic skill score drops to zero (a score equal to or lower than zero indicates no better skill than climatology) between days 12 and 24 for negative NAO and between days 10 and 17 for European blocking. Therefore, Fig. 1 suggests that sub-seasonal forecasting systems have some ability to predict these weather regimes >10 days in advance, which could lead to some potential for predicting extreme cold conditions over Europe associated with the negative NAO and European blocking (BL) regime circulation in winter.

In summer, while blocking/persistent high-pressure systems are necessary synoptic ingredients for heat waves, coupling with the land surface also plays an important role. Soil moisture/temperature interactions increase summer temperature variability, resulting in extreme temperatures when soil moisture is low (Seneviratne et al., 2006; Lorenz et al., 2010). When soil moisture decreases in these regions, extreme temperatures are more likely. For extreme summertime temperatures to occur over Europe, the preceding winter and spring must be dry (Quesada et al., 2012). This causes antecedent dry soil moisture conditions, and, when combined with blocking highs, the positive feedback amplifies. It follows that a realistic representation of the land surface conditions in the sub-seasonal system can enhance the skill at predicting summer heat waves over Europe (Prodhomme et al., 2016).

2.1.2 Heat-Wave Prediction in Australia

Heat waves are a regular feature of Australian climate, with often severe impacts on a wide range of sectors within the society, such as agriculture and health. They are responsible for more deaths in Australia than any other natural hazard, including bush fires, floods, and storms (Price Waterhouse Coopers, 2011). The Australian Bureau of Meteorology (BoM) has been investigating the capability of its seasonal prediction system to forecast heat extremes on sub-seasonal timescales, including the so-called out-of-summer-season warm waves, which can have a major impact on agriculture (Hudson and Marshall, 2016; Hudson et al., 2015a,b; Marshall et al., 2014; White et al., 2014). This research includes attempts to

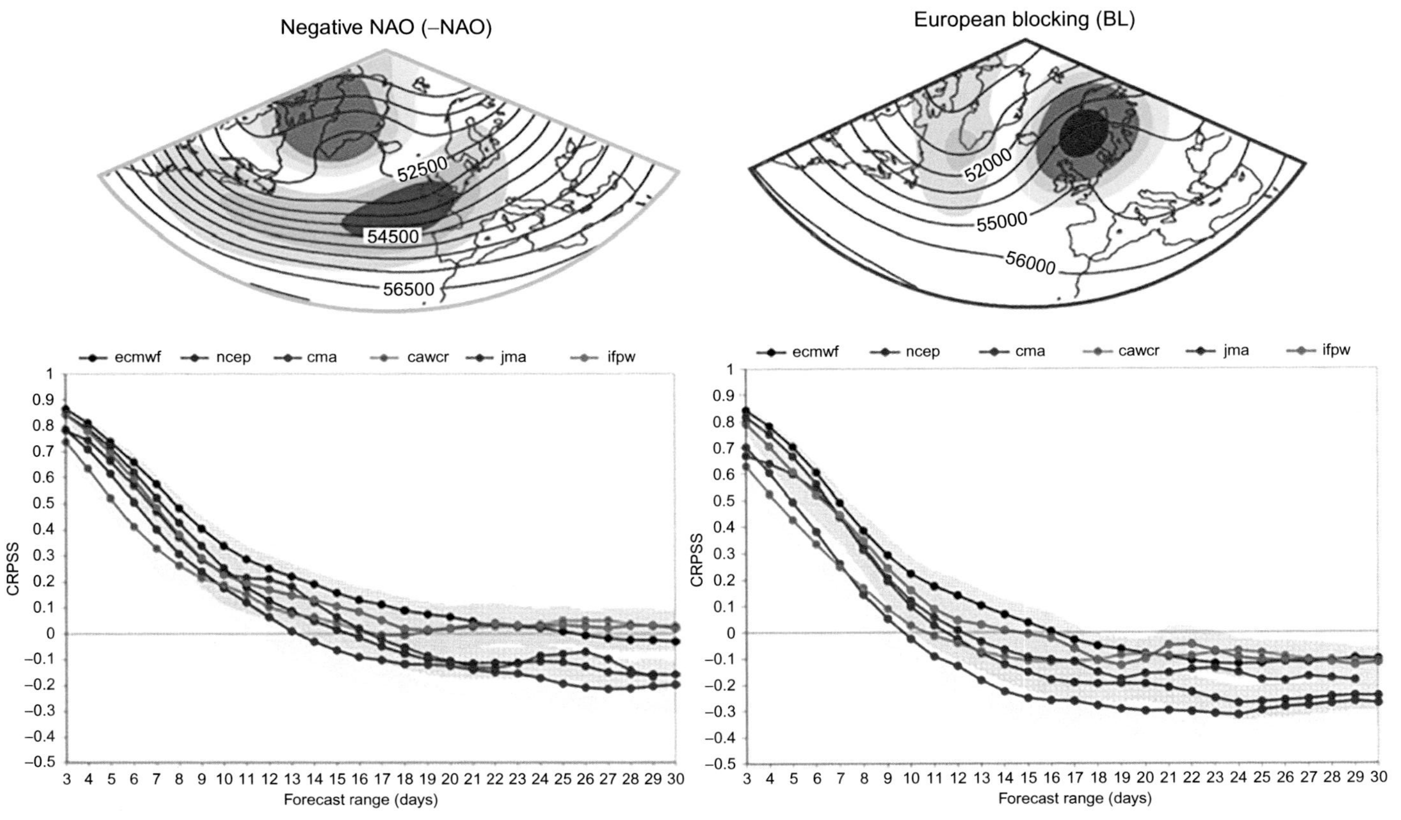

FIG. 1 The top panels show the spatial patterns associated with the negative NAO (left) and European blocking (BL, right). Bottom panels: Continuous Ranked Probability Skill Score (CRPSS) as a function of forecast lead time for the prediction of the negative NAO (bottom left) and European blocking (BL, bottom right). As the sub-seasonal forecast does not target the weather on an individual day, but rather weather conditions averaged over 4–7 days, a 5-day running mean has been applied to the forecasts and verifying areas. The areas around each curve represent the 95% level of confidence using a 10,000 bootstrap resampling procedure. The *black, red, blue, green, brown, and magenta curves* represent the scores of the European Centre for Medium-range weather forecasts (ECMWF), the National Centers for Environmental Prediction (NCEP), the Chinese Meteorological Administration (CMA), the Australian Bureau of Meteorology (BoM), the Japanese Meteorological Agency (JMA), and Météo France models, respectively.

understand the role of key drivers of climate variability in producing heat extremes over Australia in both observations and models.

There is a clear tendency for Australia's extreme heat events to cluster with phases of the ENSO, the Indian Ocean Dipole (IOD), the MJO, the Southern Annular Mode (SAM), atmospheric blocking, and persistent high pressure over the Tasman Sea (Marshall et al., 2014; White et al., 2014). For example, El Niño doubles the probability of having an extremely hot week (i.e., the weekly mean maximum temperature anomaly is in the upper 10% of events for that time of year based on a 30-year climatology) across much of southern Australia in austral spring (September, October, and November). The largest impact of the MJO on heat extremes over extratropical Australia occurs when the MJO is located over the Indian Ocean in austral spring. The chance of having a hot week over southeastern Australia more than triples in this situation compared to climatology (Marshall et al., 2014). The teleconnection is explained by a Rossby wave train that results in a strong, anticyclonic anomaly in the midlevel circulation and subsidence over the southeastern regions (Wheeler et al., 2009).

In general, the seasonal forecasting system captures the relationships between the drivers of heat waves and the regional impacts, but usually underestimates their strength. This type of knowledge contributes to the evaluation of the forecast system, aids in the interpretation of forecasts, and can assist in the understanding and communication of a particular heat event. A comprehensive verification of summer heat-wave forecasts using the 30-year reforecast set from the BoM seasonal forecast system showed that the system has generally good skill in detecting the occurrence of heat waves at all lead times over Australia (i.e., week 2, week 3, fortnight 1, fortnight 1.5, fortnight 2, and month 1 of the forecast) (Hudson and Marshall, 2016; Fig. 2). These and other results suggest that there is significant potential to augment traditional weather forecast warnings for heat waves in Australia to include guidance on subseasonal timescales.

2.2 Drought Prediction

Decision-makers and end users require adapted and robust forecast indicators that are capable of informing about the onset, possible duration, intensity, and demise of drought conditions, ranging from weeks to months. There are numerous drought definitions depending on temporal and spatial scale and the hydrological/environmental component considered (e.g., precipitation, soil moisture, and vegetation) (Wilhite and Glantz, 1985). In general, drought can be defined as "a prolonged period of dryness" driven by a precipitation deficit. This precipitation deficit (meteorological drought) then can lead to other hydrological deficits in soil moisture, river discharge, groundwater, and vegetation state. High temperatures also can increase the evaporative demand, which together with dry soils can bring about positive feedback, potentially leading to heat waves. Therefore, precipitation forecasts are fundamental in a drought-forecasting system. Due to the slow-moving nature of drought onset, reliable and accurate observations of near-real-time precipitation are also important. This is challenging over many regions of the world that lack high-density in situ networks (Dutra et al., 2014a).

The concept of flash-droughts has gained visibility in the last years and is mainly associated with a fast intensification of drought conditions, often during the summer, due to increased evaporative demand (Otkin et al., 2017). Several studies have shown the potential to use statistical approaches such as stochastic or neural networks (Kim and Valdés, 2003; Mishra et al., 2007) in drought forecasting with 1–2 months lead time. Statistical downscaling

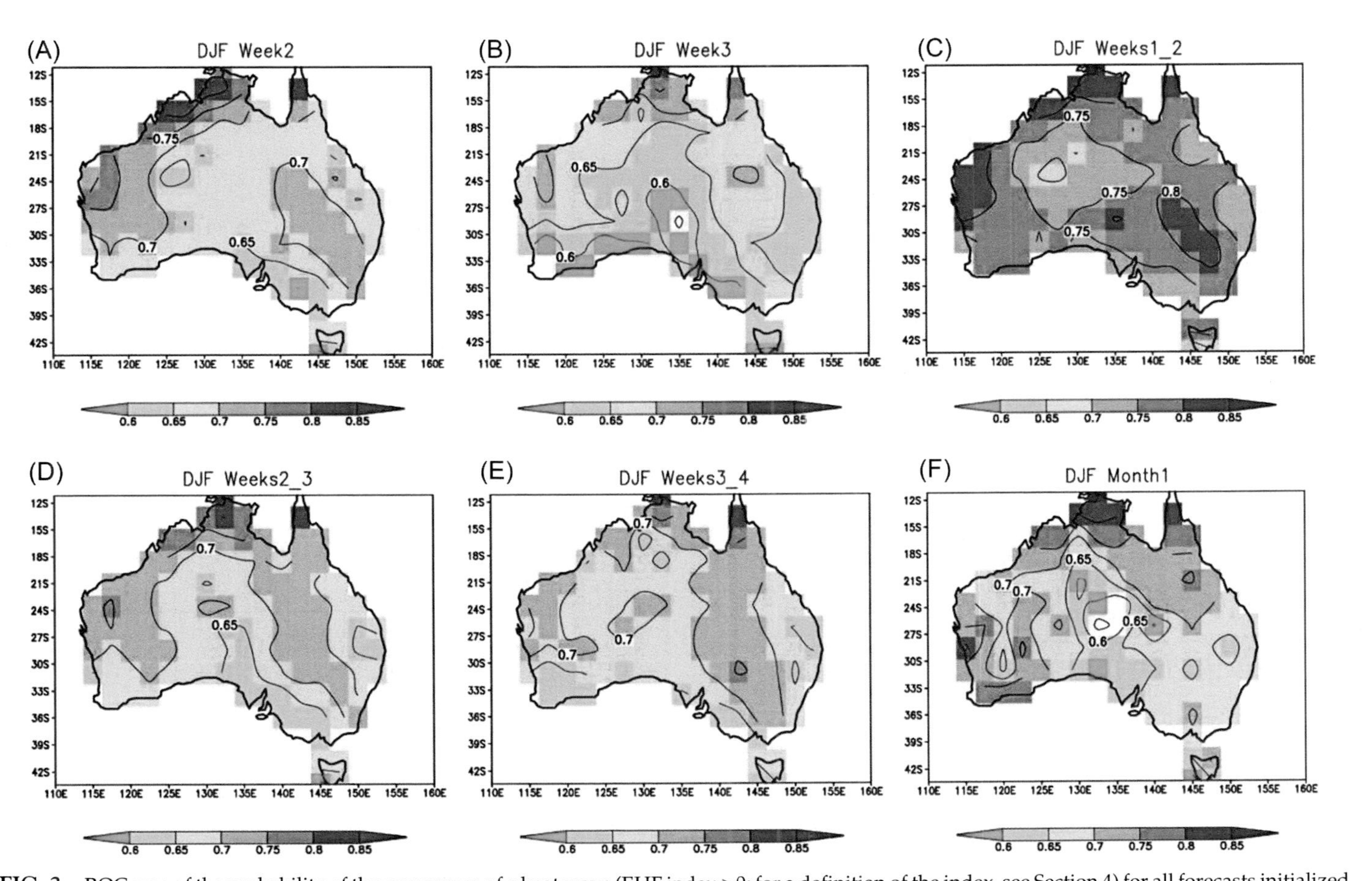

FIG. 2 ROC area of the probability of the occurrence of a heat wave (EHF index >0; for a definition of the index, see Section 4) for all forecasts initialized in the months of December, January, and February. The verification at various lead times is shown: (A) week 2 (days 8–14 of the forecast), (B) week 3 (days 15–21 of the forecast), (C) fortnight 1 (weeks 1 and 2), (D) fortnight 1.5 (weeks 2 and 3), (E) fortnight 2 (weeks 3 and 4), and (F) month 1 (i.e., the first calendar month; for instance, if the forecast starts on November 11, then December is verified). ROC areas that are significantly better than a climatological forecast at the 5% significance level are shown in color. *From Hudson, D., Marshall, A.G., 2016. Extending the Bureau's heatwave forecast to multi-week timescales. Bureau Research Report, No. 16. Bureau of Meteorology Australia. Available online at* http://www.bom.gov.au/research/research-reports.shtml.

methods using weather types also can be used (Lavaysse et al., 2017). Eshel et al. (2000), for example, used the North Atlantic sea level pressure precursors to forecast drought over the eastern Mediterranean. Forecasts of droughts also can be produced using the precipitation forecasts of NWP models. Deterministic forecasts are highly uncertain due to the chaotic nature of the atmosphere, which is particularly strong on a sub-seasonal timescale (Vitart, 2014). Moreover, these deterministic forecasts have a reduced skill for lead times above a few days (Richardson et al., 2013; Weisheimer and Palmer, 2014).

Coupled atmosphere-ocean ensemble prediction systems have become the accepted methodology to address the initial conditions and model uncertainty. These provide an estimated uncertainty of the forecasts, which are fundamental for end users. These probabilistic forecasts become particularly important for assessing the likelihood of high-impact and rare weather events such as tropical cyclones or droughts (Hamill et al., 2012; Dutra et al., 2013, 2014b). In the case of drought, an analysis including both the numerical forecasting skill and the possibilities for binary decisions to issue drought warnings has shown that 40% of droughts can be correctly forecasted 1 month in advance over Europe (Lavaysse et al., 2015). Due to the accumulated nature of the dryness on the drought onset, for short lead times (up to 1 month), the near-real-time precipitation also plays an important role, acting as "initial conditions" to the drought forecast. Once a drought is established, its demise will be driven by a return to normal or excess precipitation, which is difficult to forecast for long lead times. Using the ECMWF seasonal forecasts, Dutra et al. (2014b) found that for long lead times (>3 months), it was very difficult to improve on the use of climatological forecasts. However, on short lead times, drought onset forecasts are feasible and skillful in several regions of the world (Mo and Lyon, 2015).

The lead time and duration of drought forecasts need to reflect the needs of the users and decision-makers in different regions. For example, in regions where resilience and capacity (i.e., improved water management and irrigation techniques) is higher, variations in the frequency of long-term drought may be most relevant to the water sector, as it allows sustainable

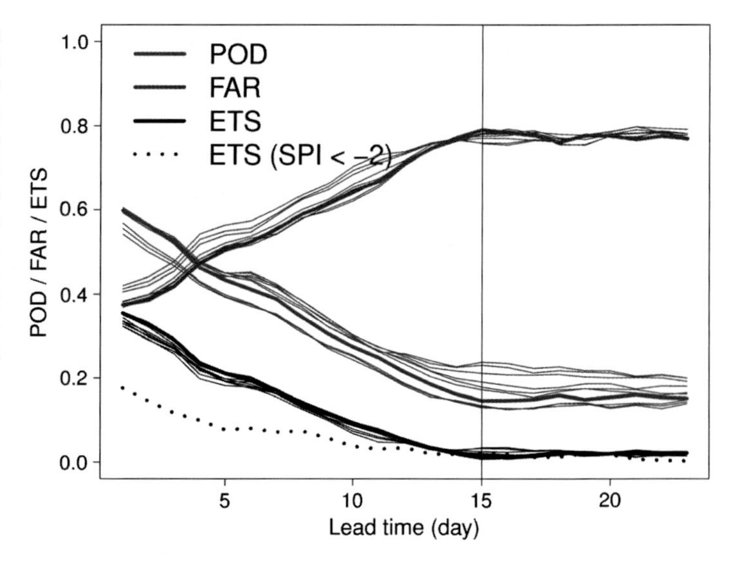

FIG. 3 Probability of detection (POD; *blue*), false alarm rate (FAR; *red*), and equitable threat score (ETS; *black*) scores of the dry spells (defined as standardized precipitation index (SPI) −10d lower than −1, calculated with moving windows of lead time) of the ECMWF forecasts compared to the EOBS data sets over Europe from 1995 to 2015. The different lines of each score represent the scores by using various methods to extract dichotomous forecasts from the ensemble (see Lavaysse et al., 2015).

preparedness activities to be identified in the long term. For those regions that depend on rain-fed agriculture, a short-term deficit of precipitation also constitutes higher risk. In that case, forecasts of intraseason dry spells (short-term droughts of about 10 days) are also important (Winsemius et al., 2014). An example of the predictability of these dry spells is provided in Fig. 3. The scores are calculated using 20 years of hindcasts of precipitation derived from the extended ensemble forecast of ECMWF (which is also part of the S2S database; see Vitart et al., 2017) and validated using *E*-OBS data sets (http://www.ecad.eu/download/ensembles/download.php). These results highlight the capacity of the model to provide substantially reliable forecasts for cumulated precipitation from $t + 15$ to $t + 25$ days lead time. Beyond this lead time, the benefit of using forecasts appears only with longer, cumulated periods of precipitation (e.g., SPI-1 month, to reduce the temporal uncertainties) or with coarser horizontal resolutions (to reduce the spatial uncertainties).

3 PREDICTION OF MESOSCALE EVENTS

Mesoscale hazardous weather events occur on scales of <1 km and up to hundreds of kilometers and have lifetimes of a few minutes to a few days. Their intensity is usually enhanced by diabatic energy sources (mainly latent heating due to condensation and freezing). These weather events generally form part of a larger-scale disturbance, such as an extratropical cyclone or a tropical wave. Weather-related hazards that can be considered to be caused by mesoscale events include:

- Floods that occur on a wide range of timescales and space scales, which are the most common cause of weather-related disasters worldwide (Doocy et al., 2013). Here, we are concerned with flash floods and surface water flooding resulting from convective storms, tropical cyclones, fronts, warm conveyor belts, and atmospheric rivers in extratropical cyclones. Short-duration floods lead to considerable numbers of deaths, devastation to property and infrastructure, and major economic losses. They also cause long-term health impacts in those affected.
- Extreme winter weather, including sudden drops in temperature, snow and ice, and thick fog, which may be associated with cold fronts in extratropical cyclones, polar lows, etc. The main impacts are on health and infrastructure, especially transportation.
- Extreme heat and humidity, including sudden rises in temperature associated with warm fronts in extratropical cyclones. Such events typically lead to peaks in death and illness, especially in unacclimatized communities at the beginning of the hot season and in persons with preexisting health conditions.
- Extreme fire weather associated with strong winds and low humidity within a more extensively hot, dry period. The impacts of extreme fire weather include larger, more rapidly developing, more erratic, and faster-moving fires, which are more likely to cause fatalities, as well as property and infrastructure damage, economic losses, and personal suffering.
- Extreme winds, associated with tropical cyclones, extratropical cyclones, downslope windstorms, and convective storms (including tornadoes). The impacts of such storms include direct property and infrastructure damage, loss of life and property damage

caused by storm surge–induced coastal flooding, and health and economic impacts arising from loss of infrastructure, services, and personal suffering.

The prediction of mesoscale hazardous weather at S2S timescales requires that (1) deviations from the climatological frequency distribution of the larger-scale wave disturbances are predictable and (2) there is a statistical relationship between extreme mesoscale weather events and some predictable aspect of the larger-scale waves in which they are embedded. The relationship between the extreme weather event and its subsequent impact also may affect the value of such connections. For instance, the preconditioning effect of repeated rainfall events in a wet season makes severe flooding from a subsequent individual extreme event more likely.

Research addressing the first point is advancing rapidly, showing that variability in the frequency of these larger-scale atmospheric waves is in many cases predictable on S2S timescales, at least in some parts of the world, and skill in predicting them is increasing (e.g., Vitart, 2014). There is less evidence addressing the second point, which is almost certainly dependent upon the weather event in question. The following sections present examples of events in which there is a statistical relationship with a predictable aspect of the larger-scale waves in which they are embedded.

3.1 Tropical Cyclones

Short- and medium-range forecasts of tropical storm tracks and intensities have been available for a few decades. These forecasts predict the trajectory of tropical cyclones that are already present in the initial conditions (e.g., Kurihara et al., 1998), and more recently, also from tropical storms that are predicted to develop in a few days (Yamaguchi et al., 2015). More recently, seasonal forecasts of tropical storms have been developed (e.g., Vitart and Stockdale, 2001) that predict whether a tropical cyclone season over a specific basin will be more or less active than normal. These seasonal forecasts are justified by the fact that tropical cyclone genesis is sensitive to its large-scale environment, especially the vertical wind shear between the upper and lower tropospheres, midlevel humidity, and low-level vorticity (Gray, 1979). As a consequence, the ENSO and local sea surface temperatures (SSTs) that affect these environmental parameters play a major role in modulating tropical cyclone genesis and activity over some ocean basins.

S2S, which lies between medium-range and seasonal time ranges, can include both aspects of tropical cyclone prediction. There are cases where extended-range prediction of a specific tropical storm might be possible if the genesis can be predicted a few days in advance (e.g., tropical cyclone Nargis over the North Indian Ocean in 2010; Belanger et al., 2012) combined with the fact that some tropical storms can last for >2 weeks. However, successful tropical storm track forecasting beyond 2 weeks remains very rare, despite significant improvement in short- and medium-range tropical storm track forecasting over the past decades. At the S2S timescale, most of the tropical storm predictability is provided by changes to large-scale circulation that are predictable a few weeks in advance. ENSO and IOD can be a source of predictability for tropical cyclones in some ocean basins, as they are for seasonal forecasting, and indeed they are used as predictors for statistical sub-seasonal forecast models (e.g., Leroy and Wheeler, 2008). However, the main source of predictability for tropical storm activity at the sub-seasonal timescale is the MJO (e.g., Nakazawa, 1986), particularly over the Southern Hemisphere, where the tropical cyclone season coincides with the strongest MJO activity (boreal winter and spring), mostly through its impact on low-level absolute vorticity (Camargo et al., 2009) and vertical wind shear (Jiang et al., 2012). The modulation of tropical cyclone numbers by the phase of the MJO has been quoted to be as high as 4:1 in some locations (e.g., Maloney and Hartmann, 2000). Other sources of predictability at the intraseasonal timescale include equatorial Rossby (ER) waves, mixed Rossby gravity (MRG) waves, easterly waves, extratropical waves, and equatorial Kelvin waves (Frank and Roundy, 2006).

State-of-the-art operational S2S models display some skill in predicting the main sources of tropical cyclone variability (MJO, ENSO, SST anomalies, etc.), and therefore they could be used in conjunction with a statistical model to predict tropical cyclone activity in the next weeks or months. The numerical model would predict the large-scale conditions, and the statistical model would translate these large-scale forecasts into tropical cyclone activity forecasts. An alternative is to use the numerical models to predict tropical cyclone activity directly if they satisfy the following two conditions:

- They are able to produce tropical cyclone–like vortices with a climatology and seasonal variability consistent with observations.
- They are able to simulate the observed modulation of tropical cyclone activity by the various sources of tropical cyclone predictability (MJO, ENSO, etc.).

Because numerical models can handle nonlinearities and events that have never been previously recorded, they have the potential to outperform statistical models. State-of-the-art numerical sub-seasonal forecasting systems are currently able to produce a reasonable climatology of tropical cyclone activity and simulate the interaction between MJO and tropical cyclones (e.g., Vitart, 2009). As a consequence, some operational centers are already issuing sub-seasonal forecasts of tropical cyclones (e.g., Vitart et al., 2012a). These probabilistic forecasts can be issued in several forms:

- Predicted number of tropical storms and hurricanes of accumulated cyclone energy (integral of the square of the maximum velocity over the storm's lifetime) over a tropical cyclone basin and over a period of time (week or month)
- Tropical cyclone strike probability map; that is, the probability of a tropical cyclone passing within a certain distance (e.g., 300 km) within a period of time (week or month)
- Tropical cyclone track cluster, which gives an indication of the most likely tropical cyclone tracks in the coming weeks or months (Elsberry et al., 2010)

3.2 Heavy Precipitation/Flooding

While operational forecasting of precipitation in the short-to-medium range has a long history in meteorology, research dedicated to understanding the predictability of heavy precipitation in the sub-seasonal range is much more recent. Because the MJO is a major source of sub-seasonal predictability, some studies have investigated specifically how the MJO influences the variability of precipitation and therefore affects its predictive skill (see review in Jones, 2017). Jones et al. (2004), for example, investigated the potential predictability of heavy precipitation (i.e., precipitation greater than the 90th percentile) associated with the MJO using perfect model experiments with an uncoupled global atmospheric model. Their study indicated increased predictability of heavy precipitation during an active MJO. Overall, the predictability experiments indicated that the mean number of correct forecasts of heavy precipitation during active MJO periods was nearly twice the correct number of events during quiescent days. Although the skill in forecasting precipitation has been a focus of several studies, the results do not yet provide a well-consolidated view.

Janowiak et al. (2010) showed that during a period of active MJO, global models have forecast skill of daily precipitation extending to a 9-day lead. Jones et al. (2011) found that probabilistic forecasts of heavy precipitation (i.e., greater than the 75th percentile) show improvements over climatology of 0%–40% at 1-day lead and 0%–5% at 7-day leads. Tian et al. (2017), using reforecasts from Climate Forecast System version 2 (CFSv2) to analyze the forecast skill of 10 indices of daily temperature and precipitation over the contiguous United States during the summer and winter seasons, found temperature forecasts to be skillful up to 3–4 weeks. They also concluded that forecast skill varies significantly with the precipitation metric used.

Sub-seasonal forecasts of mean precipitation, frequency, and duration of precipitation extremes were skillful up to 1 week but modest in week 2. Jones and Dudhia (2017) conducted perfect model predictability experiments during the 2004–05 boreal winter. That winter season was characterized by an MJO event, weak El Niño, strong NAO, and extremely wet conditions over the contiguous United States. They also showed that errors in initial conditions in the tropics have a significant impact on the potential predictability of precipitation over the United States. Errors on small scales relative to the large-scale characteristics of the MJO grow quickly, propagating to the extratropics and degrading forecast skill. The potential predictability of daily precipitation extends to 1–5 days over most of the United States, but to longer leads (7–12 days) over regions of the country with orographic forcing of precipitation, such as the Sierra Nevada in California.

Associated with heavy precipitation events, some studies have identified links between sub-seasonal climate variability, especially the MJO, and occurrences of floods, despite the challenges of attributing the direct causes of floods (Zhang, 2013). Nevertheless, statistically significant patterns of worldwide major flood events can be related to the eastward propagation of the MJO. An interesting example of the potential use of the S2S forecast database is discussed by Cunningham (2017). ECMWF's ensemble precipitation forecasts were used to drive hydrological model ensembles to forecast extreme runoff in the upper São Francisco river basin in Southeastern Brazil (SEB). The median of the runoff ensemble showed consistent skill only for runs with lead times of 15 days or shorter. Nonetheless, evaluating from a different point of view, outliers, indicating extreme runoff at the target period, showed recurrent skill through various initializations at lead times longer than 2 weeks. This might suggest that prognostic information about the extreme event was contained within the ensemble. The examples given here illustrate that probabilistic forecasts of heavy precipitation and floods in the sub-seasonal range have yet to be fully explored with obvious benefits in terms of economic and emergency preparedness.

Some extreme floods can be associated with atmospheric rivers. Atmospheric rivers (ARs) are responsible for the majority of horizontal water vapor in the middle latitudes and can have both beneficial (e.g., replenishing water reservoirs) and detrimental (e.g., floods and landslides) impacts on regional economics and public safety. ARs can intensify downstream precipitation and influence flooding, snowpack, and water availability (Zhu and Newell, 1998; Ralph et al., 2004; Ralph et al., 2006; Paltan et al., 2017), and the abundance or deficit of ARs strongly influences the availability or lack of fresh water in a given water year, respectively (Guan et al., 2010; Dettinger et al., 2011; Dettinger, 2013). In addition, heavy precipitation events, flooding conditions, and hazardous winds are strongly correlated to the occurrence of ARs in many regions around the globe (Ralph et al., 2006; Ralph and Dettinger, 2011; Ralph and Dettinger, 2012; Lavers and Villarini, 2015; Nayak et al., 2016; Ralph et al., 2016; Waliser and Guan, 2017).

In DeFlorio et al. (2018a), the extent to which ARs can be forecast reliably in advance, as well as the extent to which forecast skill depends on season, lead time, and climate pattern variations, are evaluated globally for the first time in a state-of-the-art ECMWF ensemble hindcast. Mean values of AR prediction skill extend to about 7–10 days, with seasonal variations highest over the Northern Hemispheric ocean basins, where AR prediction skill increases by 15%–20% at 7-day lead times during winter compared to summer. For some regions, AR skill increases significantly at longer lead times during particular phases of the ENSO, Arctic Oscillation (AO), and Pacific–North America (PNA) patterns, which are determined by calculating the relative operating characteristic (ROC) curves. DeFlorio et al. (2018a) provides the first global quantification of AR prediction skill and presents simple reference plots for various regions (Fig. 4) that may be developed over time into products of interest to water managers and regional forecasters that show the dependence of AR forecast skill on region, season, and allowable distance between an observed and forecasted AR.

DeFlorio et al. (2018b) builds upon the methodology and results from DeFlorio et al. (2018a) and extends the global assessment of AR prediction skill into the sub-seasonal timescale using hindcast data from the ECMWF S2S forecast system. An aggregate quantity is used to assess AR sub-seasonal prediction skill, defined as the number of AR days occurring over a 2-week period (AR2wk). The observed pattern of seasonal, mean AR2wk strongly resembles the general pattern of AR frequency, but with a range from 0 to 2.5 AR days per 2 weeks for the former, rather than 0%–15% for the latter. ECMWF AR2wk forecast skill outperforms a reference forecast based on monthly climatology of AR2wk at 1-week (7–21-day) lead over a number of subtropical to midlatitude regions (see Fig. 5, right) with slightly better skill evident in wintertime. AR2wk is modulated during certain phases of the ENSO, AO, and PNA teleconnection patterns, and MJO, and statistically significant differences in AR2wk forecast skill are shown during positive and negative phases of these modes, such as during +PNA relative to −PNA at 0-week and 1-week leads over the North Pacific/West United States.

3.3 Tornadoes/Thunderstorms

Annual average insured losses from thunderstorms (tornado, large hail, or damaging straight-line wind) in the United States were $11.23 billion (in 2016 USD) during the period 2003–15, comparable to the losses from hurricanes during the same period (Gunturi and Tippett, 2017). Specialized forecasts, forecasters and report collection criteria for tornadoes have existed in the United States since the 1950s (Galway, 1989).

U.S. tornado forecasting is a multistage process in which the longest lead tornado products being issued by the storm prediction center of the National Oceanographic and Atmospheric Administration (NOAA) in so-called convective outlooks that assess the potential for tornado occurrence in the subsequent 1, 2, 3, and 4–8 days. Local weather forecast offices issue tornado warnings when tornadoes have been observed or are expected within the hour. The average warning time is about 15 min. Tornadoes are less frequent in other parts of the world, resulting

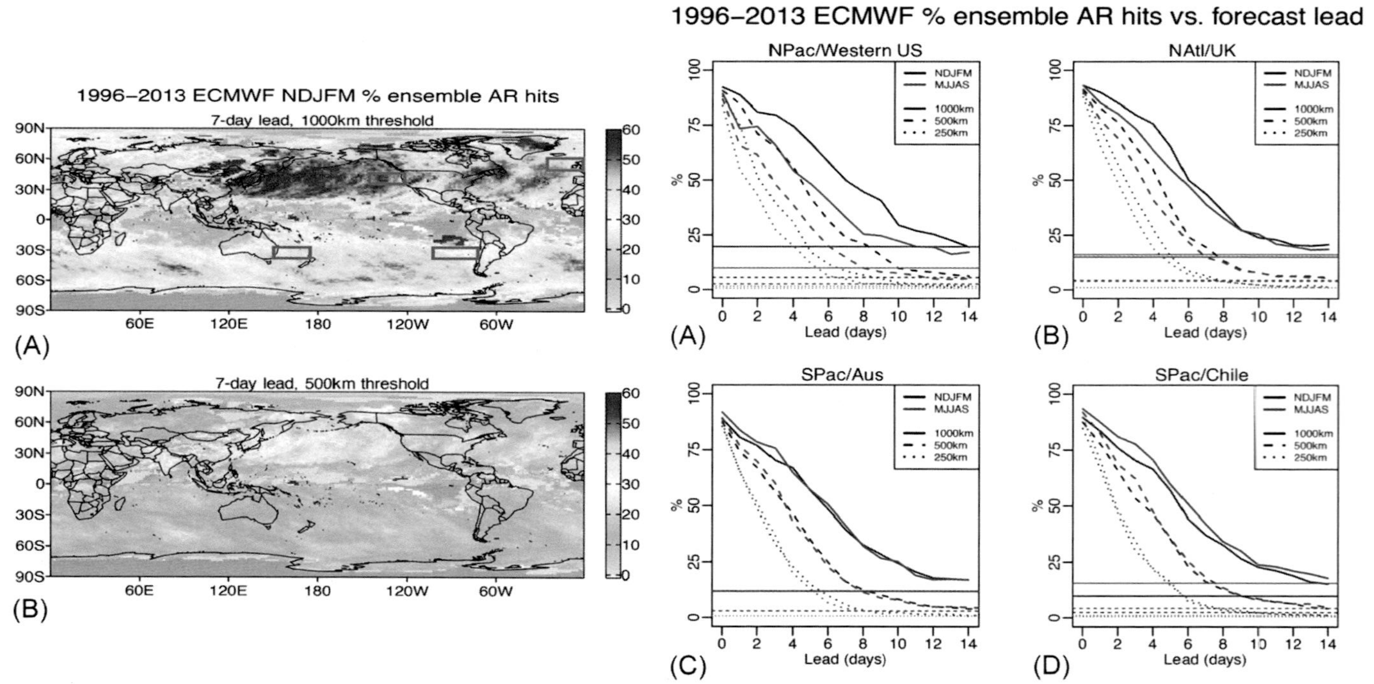

FIG. 4 (Left) 1996–2013 ECMWF November–December–January–February–March (NDJFM) average percentage of ensemble members that forecast AR hits at 7-day lead time using 1000-km (A) and 500 km (B) distance thresholds. The blue polygons in (A) denote the area average domains used in the right panel. *Gray shading* indicates regions where no ARs are detected. (Right) 1996–2013 ECMWF average percentage of ensemble member AR hits versus lead time over the North Pacific/Western United States. (A), North Atlantic/United Kingdom, (B), South Pacific–Australia, (C), and South Pacific–Chile (D) regions during NDJFM (*black*) and May–June–July–August–September (MJJAS, *blue*) and using distance thresholds of 1000 km (*solid*), 500 km (*dashed*) and 250 km (*dotted*). Horizontal solid lines represent ERA-I reference climatology of AR frequency for the given region, season, and distance threshold.

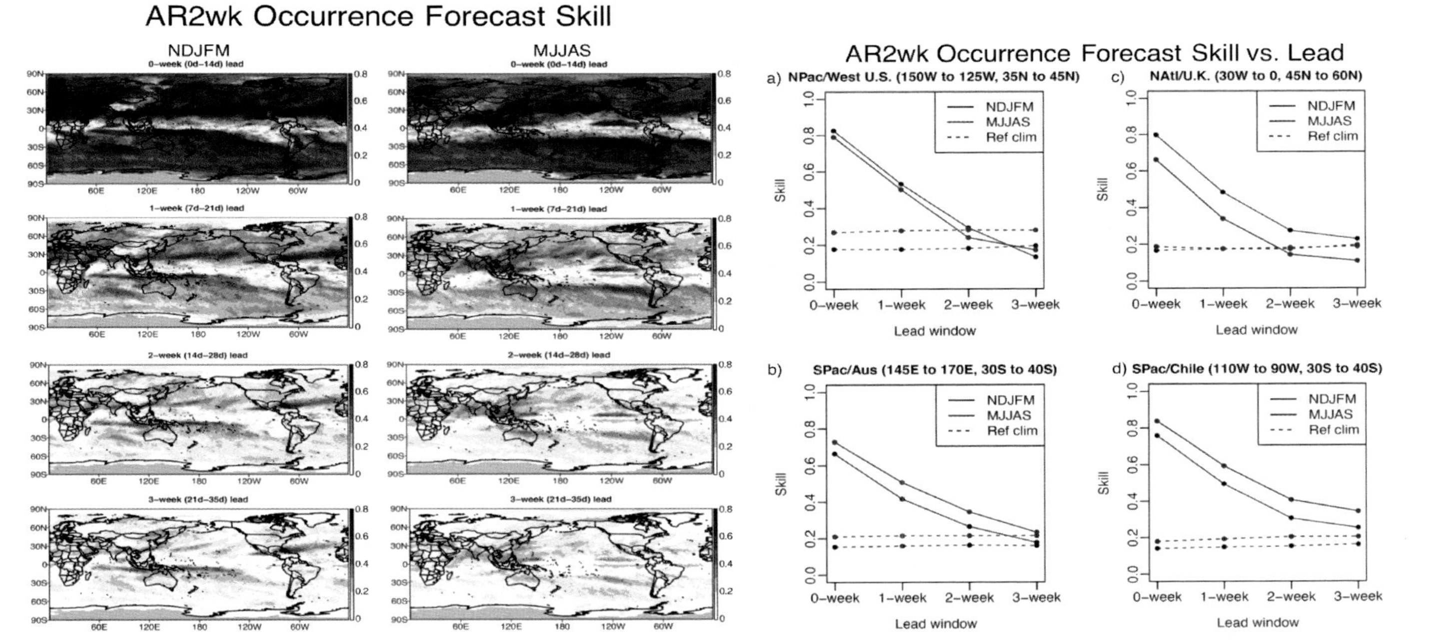

FIG. 5 (Left) AR2wk (number of AR days per 2 weeks) anomaly forecast skill during 2-week windows centered at 0 weeks (top row), 1 week (second row), 2 weeks (third row), and 3 weeks (bottom row) for the lead times in NDJFM (left column) and MJJAS (right column) during the period 1996–2014. Here, forecast skill is defined as the correlation coefficient between ERA-I and ECWMF AR2wk monthly anomaly values during NDJFM and MJJAS. (Right) AR2wk (number of AR days per 2 weeks) area-mean anomaly forecast skill as a function versus lead window over (A) the North Pacific/Western United States, (B) South Pacific/Australia, North Atlantic/United Kingdom (C), and South Pacific/Chile (D) regions in NDJFM (black) and MJJAS (blue), averaged during the period 1996–2014. The forecast skill can be compared against a reference forecast (dashed lines) defined as the correlation of monthly climatology of the sum of the observed AR occurrence anomaly time series monthly climatology with the observed AR occurrence monthly climatology.

in fewer tornado forecasts. A survey of European national meteorological services found that only 31% (8 respondents) had issued tornado warnings of any kind (Rauhala and Schultz, 2009). The longest lead time of most official and experimental severe thunderstorm forecasts in Europe is 24 h (Rauhala and Schultz, 2009; Brooks et al., 2011). So far, investigation of extended and long-range tornado activity forecasting has been limited to the United States.

A particular challenge facing tornado forecasters is that NWP models do not resolve tornadoes. The lack of direct NWP guidance for tornadoes is in contrast with the situation for other types of severe weather such as tropical cyclones or windstorms. Forecasters base their tornado predictions to a large extent on their understanding of the synoptic and mesoscale features associated with severe thunderstorms. An important advance in the 1990s was the development of severe thunderstorm parameters or indices (functions of winds and instability) that summarize the favorability of the atmosphere for the occurrence of tornadoes and severe (especially supercell) thunderstorms (Davies and Johns, 1993; Davies et al., 1993). Convective available potential energy (CAPE) and deep-layer vertical wind shear are typical ingredients in severe thunderstorm indices. The ingredient-based forecast methodology (Doswell III et al., 1996) is an important part of tornado forecasting, but tornadoes are still relatively rare even under favorable conditions. With increased computational power and resolution, convection-permitting regional models (typically 4 km or less) provide information that is closer to the scale of the thunderstorm, but they still do not resolve tornadoes. Storm-scale parameters such as maximum updraft and updraft helicity from ensembles of regional models provide forecast guidance out to about 36 h (Gallo et al., 2016).

Although the lead times of official tornado forecasts and outlooks are limited to about a week, a growing body of work suggests that there is a scientific basis for extended and long-range predictability of tornado activity. On the seasonal timescale, ENSO, in addition to modulating seasonal precipitation and near-surface temperature, also affects tornado activity, explaining a modest fraction of the variability of U.S. tornado activity in winter (January–March; Cook and Schaefer, 2008) and spring (March–May; Allen et al., 2015).

Overall, La Niña conditions are associated with enhanced U.S. tornado activity, but more detailed aspects of ENSO also may be relevant (Lee et al., 2012). The Gulf of Mexico SST is negatively correlated with the Tropical Pacific SST. Warm Gulf of Mexico SSTs in spring enhance low-level moisture transport and southerly flow and are associated with enhanced U.S. tornado and hail activity (Molina et al., 2016). Seasonal (May–July) averages of Gulf of Mexico SSTs can be predicted with some skill (Jung and Kirtman, 2016). On the sub-seasonal timescale, recent studies point to modulation of tornado activity by the MJO phase (Barrett and Gensini, 2013; Thompson and Roundy, 2013). Atmospheric angular momentum also shows the impact of tropical forcing on tornado activity (Gensini and Marinaro, 2016). However, for the most part, the demonstrated statistical relations between tornado activity and predictable climate signals are simultaneous and not predictive, although a few statistical methods for forecasting seasonal tornado activity have been proposed (Elsner and Widen, 2013; Allen et al., 2015).

Recently, the ingredient-based forecast methodology has been applied to longer lead forecasts of tornado activity. Severe weather indices are computed from the output of extended and long-range (up to a month) global circulation model (GCM) forecasts. The advantage of this approach over purely statistical forecasts is that the GCM integrates all the predictable signals that modulate tornado activity. This approach depends on the ability of the GCM to predict the severe weather indices and the degree to which the indices correspond with tornado activity. Carbin et al. (2016) computed daily averages of a standard severe weather index, the supercell composite parameter (SCP), which is a function of CAPE, storm relative helicity (SRH), and vertical wind shear (near-surface to 500 hPa) from four-member CFSv2 (Saha et al., 2014) forecasts during 2014. Despite the relatively coarse spatial resolution, the day-1 forecast number of grid points with SCP ≥ 1 correlates well ($r^2 = 0.58$) with the number of tornado and hail reports. SCP forecasts capture some tornado events a week or more in advance, with some indications of skill at longer lead times as well. Experimental products are available at http://www.spc.noaa.gov/exper/cfs_dashboard/.

Tippett et al. (2012) developed a monthly tornado environment index (TEI), which is a function of monthly averages of convective precipitation and SRH. TEI computed from reanalysis captures many aspects of the climatological and interannual variability of U.S. tornado reports (Tippett et al., 2012, 2014). Tippett et al. (2012) demonstrated skillful prediction

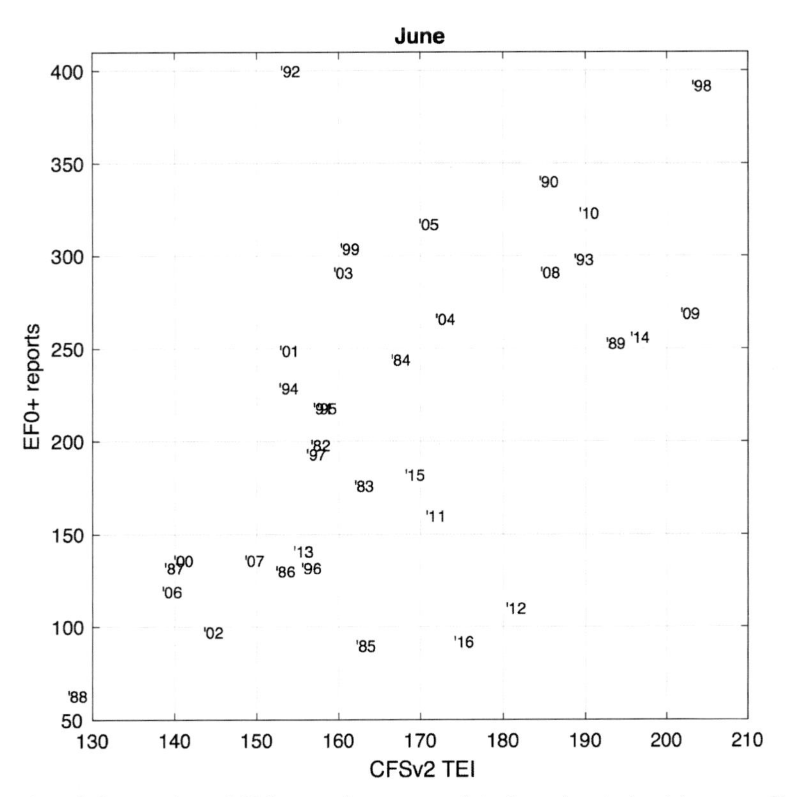

FIG. 6 Scatterplot of the number of U.S. tornadoes reported in June (vertical axis) versus CFSv2 reforecast (1982–2010) and operational forecast (2011–16) values of the TEI (horizontal axis). The two-digit markers indicate the year. The correlation is 0.58. *Updated and adapted from Tippett, M.K., Sobel, A.H., Camargo, S.J., 2012. Association of U.S. tornado occurrence with monthly environmental parameters. Geophys. Res. Lett., 39, L02 801.*

of monthly tornado activity using the TEI computed from CFSv2 reforecasts for the period 1982–2010. Fig. 6 shows that the number of June U.S. tornado reports (preliminary tornado report numbers for 2016) correlates well (0.58) with CFSv2 forecasts (zero-lead; initialized at the beginning of June) values of the TEI for the period 1982–2016. Forecast skill is lower in other months and varies regionally. Most of the monthly skill is due to skill in the first 2 weeks (Lepore et al., 2017).

3.4 Windstorms

Strong surface winds are associated with many forms of atmospheric disturbance, the most violent being tropical cyclones and tornadoes, which have been examined in previous sections. Leaving these aside, windstorms are otherwise mainly associated with traveling low-pressure systems in the middle latitudes, especially in the storm tracks over the major ocean basins and their continental margins. The kinetic energy of these weather systems comes from baroclinic conversion of the potential energy of the temperature gradient between cold polar and hot tropical air masses. Over land, friction reduces surface wind strength, but topographic interactions may produce locally enhanced winds, such as in the case of lee waves and downslope windstorms.

Prediction of midlatitude storms has been a core focus of NWP development since computers first developed in the late 1940s. The skill of deterministic prediction has gradually extended from 1 to 4–5 days since that time. Using the ECMWF model, Jung et al. (2004) assessed the (then) current predictability of three extreme windstorms of the 20th century as being up to 4 days. Using ensemble prediction, it is possible to extend the useful storm-specific predictability range further. Buizza and Leutbecher (2015) considered the predictability of NWP ensemble predictions, focusing on the Northern Hemisphere extratropics. Using the ECMWF T639 ensemble (with about 32-km grid spacing) for 107 cases in 2012–13, they found that instantaneous, full-resolution fields of geopotential height, temperature, and wind had some predictability up to 15–20 days, consistent with earlier estimates based on simplified models (Lorenz, 1969a). However, using temporal averaging, they achieved significantly longer periods of predictability, up to 1 month or more, indicating that synoptic-scale weather systems are embedded in more slowly varying atmospheric disturbances with much greater predictability.

In an analysis based on the National Center for Atmospheric Research (NCAR)–NCEP reanalysis for 1948–2012, Feliks et al. (2016) showed that variability in the storminess of the North Atlantic is related to the orientation and strength of the jet stream, and that these are both closely correlated with the NAO. In recent years, several researchers have demonstrated skill in the seasonal prediction of winter storminess. Renggli et al. (2011) used the multimodel DEMETER (Palmer et al., 2004) and ENSEMBLES (Weisheimer et al., 2009) ensemble reforecast data sets covering the period 1960–2001. They compared windstorms in the reforecasts with those in the ERA-40 data set at 2.5° resolution (and about 250-km grid spacing) relative to the local 98th percentile of the wind climatology, and tracked individual storms to assess storminess. At this resolution, they found significant skill, especially in the more recent period, 1980–2001.

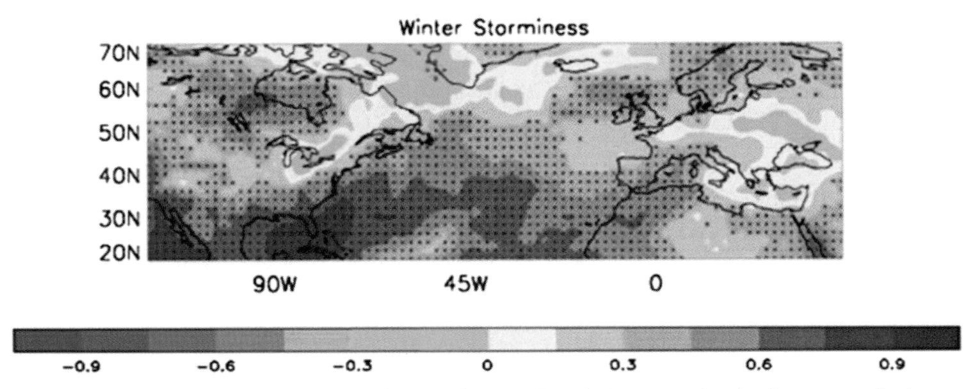

FIG. 7 Forecast skill of surface winter weather conditions. Correlation score for the frequency of winter storms (measured by 10th percentile daily sea level pressure minima). Observed storminess, from the ERA-Interim reanalysis (Dee et al., 2011), correlated with forecast storminess. Hatching indicates values above 90% statistical significance according to a student *t*-test and allowing for autocorrelation.

A more recent study by Scaife et al. (2014a) used data for 1993–2012 from the 0.83° × 0.55° resolution (about 60-km grid spacing) Glosea5 seasonal forecasting system to demonstrate skill at 1–4 months lead time in prediction of the NAO and of storminess, where they defined storminess as the frequency of daily minima of surface pressure falling below the 10th percentile of the climatological distribution. Fig. 7 (their Fig. 4A) shows the correlation between forecast and analyzed (ERA-Interim) storminess over the 20-year period. The stippled areas are significant at the 90% level. Scaife et al. (2014a) also examined the sources of predictability in their results and identified contributions from ENSO, North Atlantic SSTs, Arctic sea ice and the Quasi-Biennial Oscillation (QBO).

These results indicate that there is significant S2S predictability in the NAO, and that this is reflected in the predictability of storminess. The gains achieved with enhanced resolution also suggest that further progress should be expected. However, with a few notable exceptions, such as the insurance and wind power industries, there is no clear understanding at present of how to use such information. Further work is needed to develop the link between these temporally averaged quantities, the associated socioeconomic impacts, and the mitigating actions that might be taken.

The impact of windstorms on land depends on the local wind climatology because ecosystems respond naturally to the wind climate, while the built environment is constructed according to safety constraints, dependent on the observed climate. Thus, a modest windstorm in a low-wind climate may cause more damage and disruption than a more intense storm in a higher-wind speed climate. In Western Europe, the mean wind speed tends to increase to the north, so storms taking an unusually southerly track are likely to have a particularly large impact. Such unusual tracks may pose particular difficulties in being recognized in S2S predictions.

Storm surges are a particularly dangerous hazard associated with some windstorms. They depend on the relationship between the wind direction and the orientation of the ocean basin and coastline and have the most impact when they coincide with spring high tides (see, e.g., Horsburgh and Wilson, 2007). A S2S prediction of a period of enhanced storminess along a particular track may be sufficient to provide early warning of a major surge, although inevitably, the likelihood of the orientation and timing being right will be low.

4 DISPLAY AND VERIFICATION OF SUB-SEASONAL FORECASTS OF EXTREME EVENTS

Sub-seasonal prediction of mesoscale extreme events is generally communicated using a total number of events or the probability of having more or less extreme events than in climatology over a sufficiently large area and period of time (as outlined in Section 3.1 earlier in this chapter). The choice of the time window and the region where the statistics of mesoscale extreme events are performed presents a trade-off between predictability and usefulness to the user. Smaller regions and time windows are preferable from the perspective of users, but if a region/window is too small, there will be insufficient numbers of events for robust verification. In the case of tropical cyclones, the domain and time window need to be large enough to demonstrate that statistical or dynamical sub-seasonal forecasts are skillful, and small enough to provide more useful information than predicting the total number of storms over a monthly period, for a whole ocean basin.

For large-scale events such as heat waves, cold spells, or droughts, sub-seasonal forecasts can be displayed in several ways:

- The probability that a relevant parameter (e.g., the maximum 2-m temperature for heat waves) will be in the top quintile or decile in the forthcoming weeks, fortnights, or months. This is often the most straightforward and common way to issue extended-range forecasts of weather extremes. For short- and medium-range forecasts, where model reforecasts are often not available, the boundaries of the tercile and decile probabilities are generally defined from observations or reanalysis and are often replaced by fixed thresholds that have societal implications. However, numerical models drift very quickly toward their own climate, which can be very different from observations. Therefore, for extended-range forecasts, the preferred option is to define the quintile or decile boundaries from reforecasts. Reforecasts, also called hindcasts, are past forecasts made over a long period of time and are ideally performed using the same assimilation and modeling system as in the operational framework (Hamill et al., 2013; Hagedorn et al., 2012). They provide a climatology of the forecasting system and have been shown to be useful for the calibration of operational ensemble forecast systems (Hagedorn et al., 2008; Hamill et al., 2008).
- EFI. Within the ensemble forecasting framework, ECMWF has developed the EFI (Lalaurette, 2003) and the shift of tails (SOT; Zsótér, 2006) to better quantify exceptional meteorological situations with respect to climate models. These indices measure the difference between the cumulative distribution function of the model climatology and the actual ensemble forecast, with a stronger weight on the tail of the distribution. The complementary SOT provides information about how extreme an event might be. Positive values of the SOT indicate that at least 10% of the ensemble members are above the 99th percentile of the model climate. The higher the SOT value is, the further this top 10% of the ensemble forecast is beyond the model climate. The strength of the EFI and

SOT lies in the absence of a prerequisite climatology of observations, as well as the fact that they also can be used in areas where observations are unavailable (Prates and Buizza, 2011).

- Return periods, also referred to as recurrence intervals, used to estimate the likelihood and severity of extreme events (such as cyclones, hurricanes, or flooding) (e.g., van den Brink et al., 2005). They are based on the statistical analysis of data (such as historical climatic records, flood measurements) and used to provide a probability that an event of any given magnitude will occur in any given year. The value of the return period of specific type of extreme event changes from one location to another. Because the ensemble size of operational sub-seasonal forecasts is generally too small (typically fewer than 50 members) to produce a well-defined tail of the model ensemble distribution, methods based on the statistical theory of extreme values are used to extrapolate probabilities of extreme events (see Coles, 2001 on this topic).
- A histogram product showing the distribution of forecast daily values over the selected forecast period. A change in the tails of the histogram (compared to climatology) provides an indication of possible changes in extremes (e.g., see Hudson et al., 2015b).
- The number of days with extreme weather (expressed as the chance of having more than the usual number of days of extreme weather). For example, see Hudson et al., 2015b for forecasts of hot days.
- *Indices specifically designed for extreme weather.* For instance, the BoM has developed a warning service issuing short-range forecasts for heat waves, severe heat waves, and extreme heat waves in Australia. This has been extended, on a trial basis, to sub-seasonal timescales. This trial product provides probabilities of heat waves in the upcoming weeks or months (Hudson and Marshall, 2016). The product uses the excess heat factor (EHF), designed by Nairn and Fawcett (2013, 2015), which combines the effect of excess heat (maximum and subsequent minimum temperatures averaged over a 3-day period and compared against a climate reference value) and heat stress (calculated as the maximum and subsequent minimum temperatures averaged over 3 days and compared against a 30-day mean temperature value). The EHF methodology has been modified and extended by the Met Office in the United Kingdom to forecast severe heat waves globally. In this instance, EHF and excess cold factor (ECF) forecasts are provided through the Global Hazard Map, which aims to forecast high-impact weather out to 7 days ahead. The tool allows probabilistic forecast layers, for a range of meteorological variables (e.g., heat waves, cold spells, precipitation, wind gusts, tropical cyclone strike probability and track, and snowfall), to be overlain with vulnerability and exposure (e.g., population density, Fragile States Index) data sets in order to inform users such as operational forecasters) of the likely impacts. Current research has focused on identifying and testing a new, semiautomated approach for verifying high-impact weather, using traditional verification scores and community impact databases, for both heavy rainfall (Robbins and Titley, 2018) and heat waves and cold spells.

Verifying extreme weather is a challenging task due to the rarity of the events. Even when the verification is based on ensemble reforecasts, the typically small ensemble size (usually much smaller than that in the real-time system) and the relatively short period covered (generally <30 years) present strong limitations. Larger reforecast data sets covering longer periods with larger ensemble sizes than in current operational systems would be needed to produce robust verification of extreme events. Because the forecasts of extreme events are probabilistic, the verification has to be probabilistic as well. Methods used include reliability diagrams, ROC curves (e.g., Hudson and Marshall, 2016, for heat waves). More details on the verification methods can be found in Chapter 16.

5 CONCLUSIONS

Sub-seasonal prediction of extreme events is still in its infancy. This chapter has provided several examples of extreme weather events. In some cases, extreme weather prediction uses large-scale patterns (e.g., weather regimes) predicted by the model and a statistical relationship between the large-scale patterns, and the extreme event is then used to issue the sub-seasonal forecast of extreme events. In other cases, the extreme event forecast is produced directly from the numerical model's output. This is the case only when the model is able to simulate realistically the relationship between large-scale events and the statistics of the extreme event (e.g., the impact of MJO and ENSO on tropical cyclones). Overall, these examples suggest that current state-of-the-art sub-seasonal forecasting systems can display skill in predicting several types of extreme events. However, much more work is needed to explore the predictability of other types of extreme events. Future progress will be conditioned on the improvement of large-scale circulation in numerical weather forecasts, as well as the better representation of smaller-scale extreme events through higher model resolution and better model physics.

Some operational weather centers are starting to issue sub-seasonal forecasts of extreme events. For example, the BoM produces the probabilities of the occurrence of heat waves (discussed earlier in this chapter in Section 2.1.2) at the sub-seasonal timescale on a limited, trial basis. However, sub-seasonal forecasts of extreme events are still rarely used in applications. An important issue is that, unlike short- and medium-range weather forecasting, in the extended range, the probabilities of extreme events are generally very small and the ensemble spread very large (forecasts are often very close to climatology), making their use for decision-making particularly challenging.

CHAPTER

18

Pilot Experiences in Using Seamless Forecasts for Early Action: The "Ready-Set-Go!" Approach in the Red Cross

Juan Bazo, Roop Singh*, Mathieu Destrooper[†], Erin Coughlan de Perez[‡,§]*

*Red Cross Red Crescent Climate Centre, The Hague, Netherlands [†]German Red Cross, Berlin, Germany [‡]Red Cross Red Crescent Climate Centre, International Research Institute for Climate and Society, Columbia University, Palisades, NY, United States [§]Institute for Environmental Studies, VU University Amsterdam, Netherlands

OUTLINE

1 INTRODUCTION

Originally, the humanitarian sector was founded to provide disaster response, both for conflict and natural disaster situations. However, the best way to save lives and protect the dignity of all people often involves action in advance of the disaster itself. Increasingly,

Sub-seasonal to Seasonal Prediction
https://doi.org/10.1016/B978-0-12-811714-9.00018-8

humanitarian organizations around the world pay attention to climate information to support their work in disaster risk management. This includes investments in awareness, preparedness, understanding uncertainty, and responses to extreme events, forecasts, and warnings. The various timescales of prediction (e.g., seasonal, sub-seasonal, or short-term) can support disaster risk reduction (DRR) activities, emergency management and response, prepositioning relief items, and strengthening community resilience.

Weather and climate span a continuum of timescales, with different lead times relevant to a range of end-user requirements. In the context of disaster preparedness, the Red Cross Red Crescent Climate Centre and the International Research Institute (IRI) have proposed a concept called "Ready-Set-Go!" for using forecasts from the seasonal to the weather timescales based on a seamless "weather-to-climate" prediction approach (Goddard et al., 2014)—for more, see Fig. 4 in Chapter 22. This concept could be used in flood preparedness, for example, whereby large-scale climate indexes, catchment wetness, and observed rainfall events at the seasonal lead time can encourage humanitarians to update their contingency plans and early warning systems in the "Ready" phase. Then, sub-monthly sub-seasonal to seasonal (S2S) forecasts can enter the "Set" phase, where they are used to preposition materials, alert volunteers, and warn communities and decision-makers about an increased risk of flooding; and numerical weather prediction (NWP) weather forecasts and warnings would fuel the "Go!" phase, as they are used to activate volunteers, distribute instructions to communities, and evacuate areas if needed.

In this approach, extended flood warning (on S2S lead times) can be an intuitive component to encouraging early action, part of a seamless extreme event–prediction system for decision-makers. The use and value of weather forecast information for societal impacts has been demonstrated on various forecasting lead times (Morss et al., 2008a; Demuth et al., 2013). Further research on flood risk, risk perception, and forecast uncertainty on the S2S timescale can help decision-makers use weather information more effectively (White et al., 2013).

2 WHY SUB-SEASONAL?

A forecast creates a window of opportunity for the humanitarian actor, who can use this information to act even before extreme weather events happen. Humanitarian actors and the governments in low- and middle-income countries need this time for three main reasons:

(1) The more time the humanitarian actor or the community has, the better the preparedness activities can be. With an S2S forecast, such as for El Niño–Southern Oscillation (ENSO), houses can be strengthened, disaster response units trained, and relief items purchased. A midterm forecast of 7 days, on the other hand, only allows humanitarian actions to work on early warning, evacuation, and timely distributions.

(2) As the adage says, "Time is money"—actions can be financed and carried out even before an event occurs. Time also allows for protecting valuable assets in vulnerable areas.

(3) Unlike Europe, the United States, and Japan, for instance, many low- and middle-income countries do not have the warning systems or means of deployment to reach out quickly to a city or community to take preventive action. In places with limited personnel and infrastructure, last-minute activities will take days or even weeks to be carried out.

The more lead time, the longer the preparedness time can be. S2S forecasts are extremely useful in this sense, but their use has been generally limited to droughts, El Niño, and La Niña. On the other hand, for cold waves, heat waves, snowfall, heavy rain, hurricanes, and flooding, short-term forecasts are among the only products available that provide information about the risk of these extreme events. Therefore, National Met offices only offer a midterm or short-term forecast for those events. This only allows humanitarian actors and national disaster risk management authorities to inform, warn, and evacuate; it makes it difficult for them to distribute information or perform other preparatory acts early, well before the event.

Therefore, there is a constant tension between lead time and useful preparedness time, or between confidence and time to act. A sub-seasonal forecast can fill this gap, giving information about the probable location and intensity of an event. Based on this information, a government or humanitarian actor can shift budgets, purchase, preposition, strengthen houses and dams, dig out river benches, and warn the most remote areas of the country.

Even if all entities involved recognize that the confidence of sub-seasonal forecasting needs to be improved, it seems much less an issue for the humanitarian actor in the field. From a humanitarian perspective, a sub-seasonal forecast will never come alone; it will be evaluated and used together with observations (e.g., water level upstream) or a shorter-term forecast. The observations can raise confidence, and the forecast can trigger high-cost activities, but the sub-seasonal forecast brings it all together.

3 CASE STUDY: PERU EL NIÑO

El Niño is a complex interaction of the tropical Pacific and the atmosphere, resulting in cyclical episodes (every 4–7 years) of changes in ocean and weather patterns in many parts of the world. Often, these episodes have considerable impacts occurring over several months, such as altered marine habitats, rainfall, floods, droughts, and changes in storm patterns.

In the 1982–83 and 1997–98 El Niño cycles, northern Peru (Tumbes, Piura, and Lambayeque) suffered flooding from heavy rain, while the south of the country suffered severe drought. In this context, the Peruvian Red Cross (PRC), the German Red Cross (GRC), and the Red Cross Red Crescent Climate Centre designed a project that used scientific observations and forecasts to implement early action in the most vulnerable areas before the potential disaster. Forecast based finance (FbF) is still in its pilot phase, and the 2015–16 El Niño was one of the first applications of this mechanism.

FbF aims to improve the effectiveness and efficiency of humanitarian preparedness by acting on national and international hydrometeorological forecasts. The system is based on calculations of regional danger levels (thresholds) and predefined early actions. These actions are triggered when a forecast exceeds a danger level in a vulnerable intervention area (e.g., a specified amount of rain that makes rivers and communities flood). It also makes financing available for predefined actions to be taken automatically, without need for an emergency appeal. Hence, actions can be taken before the impact of a disaster to strengthen the resilience of both communities and institutions.

3.1 Outline of the Early Action Protocols (EAPs) and Products Used

With the FbF El Niño pilot, the Red Cross developed intervention protocols of early actions to be implemented when forecast thresholds are reached. The combination of probabilistic forecast thresholds and early actions is called an *Early Action Protocol (EAP)*.

The Red Cross first identified highly vulnerable, disaster-prone intervention areas (districts and communities) and their priority needs. In participative regional workshops with civil defense agencies, regional governments, and representatives of all sectors (housing, agriculture, education, health, etc.) those needs were translated into actions. The Red Cross and its partners opted to design a comprehensive intervention program that includes early warning; first aid; community-based health, drinking water, and hygiene promotions; and housing in 12 flood-prone communities with little means to anticipate and absorb these shocks. Every set of actions, identified beforehand, was set up to be triggered (and financed) based on a probability that heavy rain or flooding would occur in the selected intervention area. For more costly actions, a higher probability of the forecast was used in order to limit the risk of acting in vain. Every district, therefore, had a specific set of actions based on their needs.

3.2 Preconditions for FbF Action

To set up a successful early warning system, there is a need for trained staff and regional volunteers, good knowledge of the intervention area, and constructive participation by the communities. This setup should be the first step in any FbF intervention, even before there is a specific, probabilistic indication that a disaster will happen (in other words, before an EAP is triggered by forecasts).

This preparedness phase includes recruiting staff, training volunteers and staff, identifying community brigades, carrying out vulnerability capacity assessment at the community level, developing the community's risk mapping (including the identification of evacuation roads and safe construction grounds), and preparing awareness-raising materials for the communities in community-based health and water/sanitation/hygiene. Moreover, it includes the identification of cars and warehouses to rent, the preparation of purchasing procedures, and the organization of preparedness workshops detailing the logistics and distribution plans.

3.3 Forecast Thresholds

S2S forecasts are available from different international forecasting agencies, such as the European Centre for Medium-range Weather Forecasting (ECMWF) and the National Oceanographic and Atmospheric Administration (NOAA) in the United States. They can be selected depending on both what is suitable and what is available.

For Peru, these models show that when the water off the coast is warmer, it is very likely to rain a lot. This warm water is one of the features of an El Niño event, and such events have been linked to severe flooding and heavy rainfall in Peru, with serious impacts observed in recent extraordinary events in 1982–83 and 1997–98.

When the models show high coastal sea surface temperatures (SSTs), this will set in motion clouds and changes in the atmosphere over the following 2–3 months, which is why we can act 2–3 months in advance. Therefore, the Peruvian team developed three thresholds that correspond to the probability of this rainfall index and flood return period happening: low, medium, and high, based on observations and forecasts.

S2S Model Climate Forecast System version 2 (CFSv2; Saha et al., 2011), also forecasts rainfall closer to the event, and the team decided to use these forecasts 1 month in advance. Because this is a large-scale relationship, we do not know at 1 month in advance exactly where the impacts would be, but we can say that there is a low, medium, or high chance of extreme rainfall across the area. See Box 1 for more detailed explanation of the triggers.

By the time it comes to 1 week before a potential flood, we have weather models like Global Forecast System (GFS) and Global Flood Awareness System (GLoFAS), which can give specific risk information for regions, districts, and major rivers. Using these models, we can forecast rainfall on specific days, and we have low, medium, or high thresholds for how likely it is that the extreme rainfall will happen on a specific day in the coming 5 days.

The low, medium, and high levels for thresholds were defined thoroughly based on detailed analysis of the available forecast products and expert opinions.

BOX 1

THE THRESHOLDS THAT TRIGGER EARLY ACTIONS OF THE 2015–16 EAPS

In 2015, EAPs were developed according to the following triggers. Based on the lessons learned, a revision process began after the 2016 El Niño event, and new EAPs are being developed. Here, we reproduce the original EAPs to demonstrate how the seamless forecasting system was linked to action on the ground, and offer reflections and lessons learned from the process. In each section, the sub-seasonal threshold is built based on the CFSv2 (NOAA) forecasts of precipitation. The critical threshold has been set up as an anomaly of the values mentioned in Table 1.

Trigger 1: Low Probability Threshold

Two EAPs (EAP 1 on training and awareness raising and EAP 2 on early warning) will be triggered based on seasonal (up to 3 months lead time) or sub-seasonal forecasts (up to 1 month lead time) with a low probability (see Table 1).

The seasonal threshold will be reached if 3 out of 4 of the conditions in Table 1 are met at any point during the time window from November 1 to December 30. The threshold is built with a forecast that focuses on the anomaly of sea surface temperature next to Peru; the consensus-based El Niño forecasts of the Peruvian ENFEN, and the rainfall forecasts issued by the U.S. IRI and the European EUROSIP.

EAP 1 on training/awareness raising and EAP 2 on early warning can be triggered and fully implemented by either the seasonal or the sub-seasonal forecasts. The flood Global Flood Awareness System (GloFAS; Alfieri et al., 2013). The low probability threshold triggers the most basic and cost-effective actions.

Continued

TABLE 1 A matrix of the forecast triggers used in Peru, organized by lead time and probability

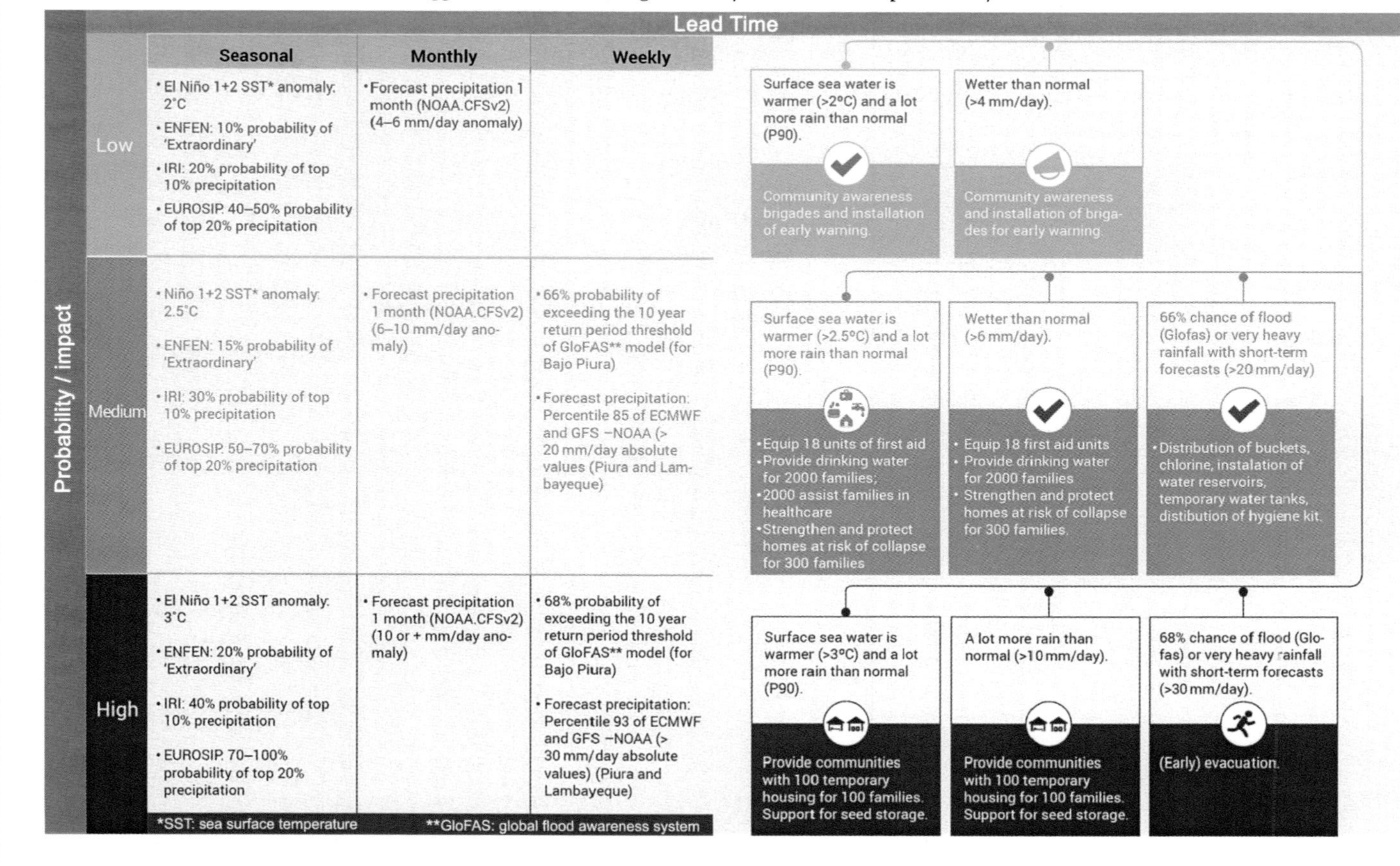

Probability / impact	Seasonal	Monthly	Weekly	Lead Time		
Low	• El Niño 1+2 SST* anomaly: 2°C • ENFEN: 10% probability of 'Extraordinary' • IRI: 20% probability of top 10% precipitation • EUROSIP: 40–50% probability of top 20% precipitation	• Forecast precipitation 1 month (NOAA.CFSv2) (4–6 mm/day anomaly)		Surface sea water is warmer (>2°C) and a lot more rain than normal (P90). Community awareness brigades and installation of early warning.	Wetter than normal (>4 mm/day). Community awareness and installation of brigades for early warning.	
Medium	• Niño 1+2 SST* anomaly: 2.5°C • ENFEN: 15% probability of 'Extraordinary' • IRI: 30% probability of top 10% precipitation • EUROSIP: 50–70% probability of top 20% precipitation	• Forecast precipitation 1 month (NOAA.CFSv2) (6–10 mm/day anomaly)	• 66% probability of exceeding the 10 year return period threshold of GloFAS** model (for Bajo Piura) • Forecast precipitation: Percentile 85 of ECMWF and GFS –NOAA (> 20 mm/day absolute values (Piura and Lambayeque)	Surface sea water is warmer (>2.5°C) and a lot more rain than normal (P90). • Equip 18 units of first aid • Provide drinking water for 2000 families; • 2000 assist families in healthcare • Strengthen and protect homes at risk of collapse for 300 families	Wetter than normal (>6 mm/day). • Equip 18 first aid units • Provide drinking water for 2000 families • Strengthen and protect homes at risk of collapse for 300 families.	66% chance of flood (Glofas) or very heavy rainfall with short-term forecasts (>20 mm/day) • Distribution of buckets, chlorine, instalation of water reservoirs, temporary water tanks, distibution of hygiene kit.
High	• El Niño 1+2 SST anomaly: 3°C • ENFEN: 20% probability of 'Extraordinary' • IRI: 40% probability of top 10% precipitation • EUROSIP: 70–100% probability of top 20% precipitation	• Forecast precipitation 1 month (NOAA.CFSv2) (10 or + mm/day anomaly)	• 68% probability of exceeding the 10 year return period threshold of GloFAS** model (for Bajo Piura) • Forecast precipitation: Percentile 93 of ECMWF and GFS –NOAA (> 30 mm/day absolute values) (Piura and Lambayeque)	Surface sea water is warmer (>3°C) and a lot more rain than normal (P90). Provide communities with 100 temporary housing for 100 families. Support for seed storage.	A lot more rain than normal (>10 mm/day). Provide communities with 100 temporary housing for 100 families. Support for seed storage.	68% chance of flood (Glofas) or very heavy rainfall with short-term forecasts (>30 mm/day). (Early) evacuation.

*SST: sea surface temperature **GloFAS: global flood awareness system

BOX 1 (*cont'd*)

EAP 1 on training and awareness raising includes, among other actions: community volunteer training (early warning early action committees), awareness raising in the communities (on community-based health and water/sanitation by community sessions and household visits), and organization of a clean-community campaign. In EAP 2, the community will be trained in early warning and evacuation based on community risk maps.

Trigger 2: Medium Probability Threshold

EAPs 3 to 6 will be triggered based on seasonal (up to 3 months lead time) or sub-seasonal (up to 1 month lead time) forecasts with a medium probability. The midrange forecast (1 week lead time) will be used to identify where to act, and it will trigger the distribution in the communities (see Table 1).

The seasonal threshold will be reached if 3 out of 4 of the conditions in Table 1 are met at any point during the time window from November 1 to December 30. The threshold is built with a forecast that focuses on the anomaly of surface temperature of sea, the consensus-based El Niño forecast of the Peruvian ENFEN (the permanent technical committee for El Niño in Peru), and the rainfall forecast issued by the U.S. IRI and the EUROSIP. The forecast sources are the same as for the low probability threshold, but threshold values are set to represent higher probability of occurrence.

For medium range, a 66% of probability of exceeding the 10-year return period threshold of GloFAS model is used (for Bajo Piura). For medium range, we used the ensemble model (GFS) NOAA for Piura and Lambayeque and use the 85th percentile = 20 mm/day.

EAP 3 on first aid, EAP 4 on safe drinking water, EAP 5 on health/sanitation/hygiene, and EAP 6 on strengthening and protecting existing houses can be triggered by both the seasonal and the sub-seasonal forecast and fully implemented. The medium-range forecast (at 1 week lead time) would identify the locations where the actions and distribution should start. If the medium-range forecast threshold is not reached, EAP 6 will be further implemented, but EAPs 3–5 will have no distributions or actions that reach out to the communities. The items for EAPs 3–5 will be prepositioned (50% in Lambayeque and 50% in Piura), and they will only be distributed based on a medium-range forecast for the districts. The protocol of early action will detail the distribution of relief items.

At a medium probability threshold, the FbF mechanism will provide clean drinking water and first aid units, and it will assist the district government and the Ministry of Health in key activities which are crucial for disaster-prone areas: fumigation against dengue, distribution of hygiene kits, sanitation of the community, and avoiding obstructions of canals. Moreover, 300 vulnerable houses will be strengthened in the community to withstand flooding and heavy rain, and a prototype for a temporary house in a safe area will be built.

Trigger 3: High Probability Forecast

EAP 7 will be triggered and fully implemented based on seasonal (up to

Continued

BOX 1 *(cont'd)*

3 months lead time) or sub-seasonal (up to 1 month lead time) forecasts with a high probability. The seasonal threshold will be reached if 3 out of 4 conditions are met at any point during the time window from November 1 to December 30. The threshold is built with a forecast that focuses on the anomaly of surface temperature of sea, the consensus-based El Niño forecast of the Peruvian ENFEN, and the rainfall forecasts issued by the U.S. IRI and the European EUROSIP. The forecast sources are the same as for the low probability, but threshold values are set to represent higher probability of occurrence.

For medium range, we use the ensemble (GFS) NOAA for Piura and Lambayeque and use 93th percentile = 30 mm/day. Based in our discussion with Peruvian Meteorological and Hydrological Service, our trigger early warning and evacuation will start. It also triggers the distribution of EAPs 3 to 5.

EAP 7 on building temporary houses can be triggered by both the seasonal and the sub-seasonal forecasts and will be fully implemented. The medium-range forecast GFS (1 week lead time) will give the starting point for early warning/evacuation or for activating the first aid units for the respective locations. The decision on where to build the temporary houses will be identified beforehand and specified in the protocol of early action.

4 REFLECTIONS ON THE USE OF S2S FORECASTS

S2S forecasts provide information at a valuable lead time, filling a critical gap in information which, combined with seasonal and short-term forecasts, can facilitate humanitarian action before a potential extreme event. However, because these S2S forecasts are currently in the experimental phase for several users in different countries, using them for operational purposes is difficult and can lead to unintended consequences. Here, we offer some reflections for the future use of such forecasts.

The forecast used experimentally in the Peru case study, based on the CFSv2 model, had the potential to generate useful signals for increasing rain in northern Peru 1 month in advance. However, there was no available verification of the forecast to indicate skill and reliability, making it very difficult to link appropriate actions to the forecast information.

In the first weeks of January 2016, the CFSv2 model varied from day to day; one day, forecasts issued were above the trigger, and the following day, forecasts issued were below it. This lack of consistency is likely related to the low skill and reliability of the forecast and made it very difficult for decision-makers to interpret, especially since it was not anticipated in advance. In the case study, we used a target period of a 1-month lead-time forecast and obtained the real-time forecast from NCEP (http://nomads.ncep.noaa.gov/pub/data/nccf/com/cfs/prod/cfs/), the forecast ensemble mean were presented in terms of millimeters per day, and all automatic scripts were working in the servers from the Peruvian Meteorological and Hydrological Service. In February 2016, the CFSv2 model predicted rainfall above the set danger level in the areas of Morropon, Picsi, and Morrope (>+6 mm/day), triggering action. However, an extreme rainfall event above the danger level was only observed in Morropon.

This experience underscores the importance of using verified forecasts for forecast-based actions. Verification of S2S forecasts for the variables of interest is critical to ensuring that humanitarian actors are aware of the skill and reliability of forecasts, and thus the risk of acting in vain. The risk of acting in vain is a large hurdle in the effective use of forecasts. Given the incentive structures pervasive in the humanitarian community, donors and policymakers clearly see the negative repercussions of acting in vain, while there are no clear negative consequences of the wait-and-see approach (Bailey, 2012). This has resulted in inaction and large-scale humanitarian emergencies such as the 2011 famine in Somalia, despite having forecast information in advance. Forecast verification allows technical experts to quantify the risk of acting in vain and of taking appropriate actions, calibrated to the risk tolerance of government and donor partners, ensuring their buy-in. As the humanitarian community moves beyond pilot projects and implements FbF mechanisms at scale, the importance of using forecasts that are verified and calibrated for reliability will become paramount.

As the science of forecasting on the S2S timescale develops, the scale of forecasts and the variables forecasted need to match the needs of users. In the Peru case study, the 1° grid of the CFSv2 model made it difficult to use at the district or community level. In general, the humanitarian community would benefit from forecasts of extreme precipitation (instead of mean precipitation), and at a scale between that of available seasonal and weather forecasts to provide increasingly local information as lead time progresses. Leaders in the use of forecasts in the humanitarian sector indicate a greater focus and need for forecasts that correspond to the risk of impacts instead of absolute temperature or precipitation variables. In East Africa, for example, delayed, erratic, and failed rains typically begin to have a significant impact on people following a second failed rainy season in a row. Thus, forecasts that are presented in a way that takes into account the observations from previous seasons, plus the forecast for the upcoming 2 months or weeks, would be most useful for humanitarian actors in this region. This tailoring of forecast information to the determinants of risk is a key opportunity for S2S forecasts to better meet the needs of users, and in turn to be useful for managing risk (Box 2).

5 CONCLUSIONS

Ultimately, S2S forecasts represent a critical section of seamless forecasting, which can enable a variety of disaster-preparedness actions that would not be possible otherwise. From the experimental use of such forecasts in the humanitarian sector, we recommend that forecast skill evaluations be conducted and published for extreme events, so that decision-makers can better understand the ability of the forecast to predict the events of interest to them. In terms of thresholds, extreme rainfall events and accumulated precipitation (5 days, for instance) are relevant variables that are not always captured by forecasting institutes. For example, when El Niño hits in Peru, the extreme rainfall is not only lasting 1 or 2 days. In addition, actors who are intending to use such forecasts should clearly define actions to be taken at specific forecasts, including examples of what a triggering forecast looks like, in order to ensure rapid action in real time.

BOX 2

THE PILOT TRIGGER

For the sub-seasonal forecast, the pilot trigger was written as follows:

1 month lead time (medium
probability)
Sub-seasonal

- Forecast precipitation 1 month (NOAA.CFSv2)
- Medium = percentile 80-mean
  80% of climatology
  (6–10 mm/day anomaly)

Sub-seasonal EAP: Here is an example of a triggering forecast:

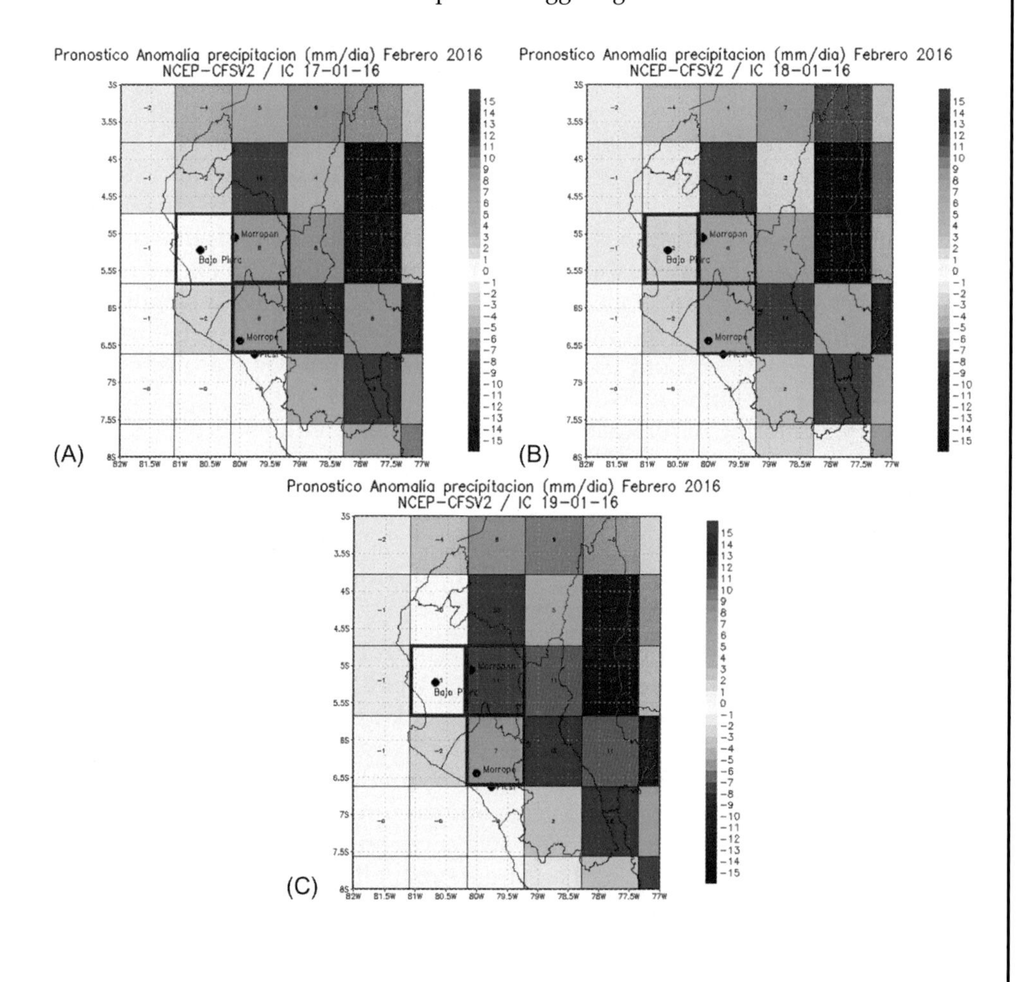

Continued

BOX 2 (*cont'd*)

The figure here shows the forecasts of three different CFS runs (January 16, 17, and 18, 2016) for a monthly average of February in terms anomaly (mm/day). The green-blue areas show positive anomalies and the orange-red areas show negative anomalies. In the grid box, we put the mean value of the grid; the red squares show the communities that were selected.

Name of forecast: CFSv2

How often forecasts were issued: Every day

Predicted variable: Millimeters of rainfall per day, averaged over a calendar month

Lead time: Days before the upcoming calendar month; ranges from 30-day to 1-day lead times

Danger level: 6 mm/day (medium green color)

Trigger-level probability: Deterministic forecast, and therefore no probability; just use the danger level

In what areas: One or more of the red-circled grid boxes

How many forecasts need to exceed the trigger: 3 consecutive days of forecasts above the trigger for a specific grid box

In February 2016, the model predicted rainfall above the trigger listed here in the areas of Morropon, Picsi, and Morrope (>+6 mm/day). However, these were not reached in observations except in Morropon (see below).

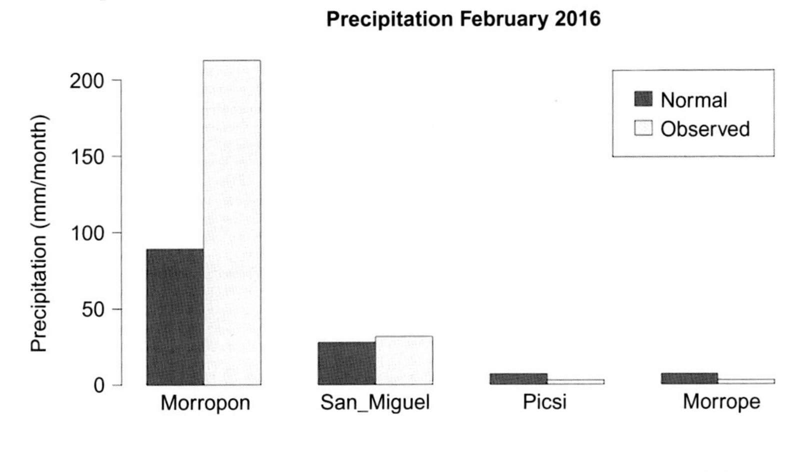

Seasonal and short-term forecasts are currently widely disseminated to the humanitarian sector, and efforts are ongoing to use the information that is available. Statistical seasonal forecasts and consensus forecasts are developed operationally by many meteorological services (Mason and Tippett, 2017). To increase the use of these forecasts by disaster managers, forecasting institutes should publish a skill evaluation for extreme events, so that humanitarians can evaluate whether to do low-cost or high-cost actions. Skill evaluations also should include measures of sharpness, so that disaster managers can anticipate the frequency of specific probabilities being reached. In the Peru case study, this information was unknown and caused uncertainty among the project teams about how frequently the trigger would happen. One obstacle to forecast verification is the lack of historical model output, and we recommend archiving historical forecasts or running reforecasts for this purpose.

As seen in the case study, global models like IRI or ECMWF have the ability to predict seasonal total rainfall in the northern part of Peru. To activate humanitarian actions, this needs to be associated with actions at the spatial and temporal scales of the forecast. At the resolution of the global models, humanitarians could trigger budgets, purchases, or prepositioning. Then they could use another kind of forecast, like a short-term forecast, to distribute relief items, which can take 1–2 weeks to reach vulnerable communities. In this sense, the lead times afforded by a S2S forecast for weeks 2–3 would support the ability to take action—that is, if skillful forecasts could be provided.

CHAPTER

19

Communication and Dissemination of Forecasts and Engaging User Communities

Joanne Robbins, Christopher Cunningham†, Rutger Dankers*, Matthew DeGennaro‡, Giovanni Dolif†, Robyn Duell§, Victor Marchezini†, Brian Mills¶, Juan Pablo Sarmiento‡, Amber Silver||, Rachel Trajber‡, Andrew Watkins§*

*Met Office, Exeter, United Kingdom †National Centre for Monitoring and Early Warning of Natural Disasters (CEMADEN), São José dos Campos, Brazil ‡Florida International University, Miami, United States §Australian Bureau of Meteorology, Melbourne, VIC, Australia ¶Meteorological Research Division, Environment and Climate Change Canada, Waterloo, ON, Canada ||University of Albany, New York, United States

OUTLINE

Sub-seasonal to Seasonal Prediction
https://doi.org/10.1016/B978-0-12-811714-9.00019-X

1 INTRODUCTION

Material presented in previous chapters and recent syntheses (National Research Council, 2010; National Academies of Sciences, Engineering and Medicine, 2016; Robertson et al., 2015), show that our ability to predict weather and climate conditions at sub-seasonal to seasonal (S2S) scales (i.e., between 2 weeks to several months into the future) has improved dramatically over recent decades. Aided in part by advances in computing, observation, and telecommunication technologies, research has unlocked many sources of predictability, from the large-scale climate drivers, such as the El Niño–Southern Oscillation (ENSO), the Indian Ocean Dipole (IOD), and the Madden-Julian Oscillation (MJO), to soil moisture, carbon dioxide, and other external forcings. Prospects for further forecast improvement remain high.

Since the early 1990s, several national meteorological and hydrometeorological services (NMHSs), academic-based research organizations, and private-sector enterprises have been actively translating this knowledge into experimental, demonstration, and operational information products and services, with the intention of supporting weather- and climate-sensitive decisions. Assuming these decisions result in actions, behaviors, and tangible outcomes (e.g., lives saved or economic efficiencies gained), they collectively generate value to the individual user and society as a whole.

Despite notable enhancements in science and the proliferation of seasonal prediction applications, recent reviews have observed that S2S knowledge remains largely underutilized in decision-making, leaving much societal value unrealized as yet (National Academies of Sciences, Engineering, and Medicine, 2016; see also Chapters 21 and 22 of this volume). Indeed, this untapped potential value is used to justify and advocate for expanded development and investment in S2S science and services. However, lessons from health, technological, and environmental risk-related fields suggest that this knowledge-value gap is not unique to S2S weather and climate prediction and is not the result of a deficit in available scientific understanding (White et al., 2001; Cash et al., 2006; Coffey et al., 2015). Therefore, any investment in developing or improving weather and climate forecasts should be coupled with investment to improve the services, communication, and decision-making processes through the entire information value chain (Fig. 1; Fischhoff, 1995) and across research-practitioner and provider-user boundaries.

Fortunately, many in the S2S community acknowledge this requirement and have made significant strides toward addressing issues around the communication, delivery, and usability of forecast information. This chapter highlights and examines the progress of this work. A brief review of the S2S application literature and readily available public products and services is provided, as well as more detailed descriptions of individual applications, sectors, and decision problems. A synthesis of these descriptions is provided to characterize the treatment of communication challenges and discern broad sectoral and geographic gaps and trends.

2 SECTOR-SPECIFIC METHODS AND PRACTICES IN S2S FORECAST COMMUNICATION, DISSEMINATION, AND ENGAGEMENT

2.1 Availability to the Public

A review of public-facing NMHS websites for content related to forecast information at the S2S timescale was completed by the authors in March 2017, and it provided a snapshot of

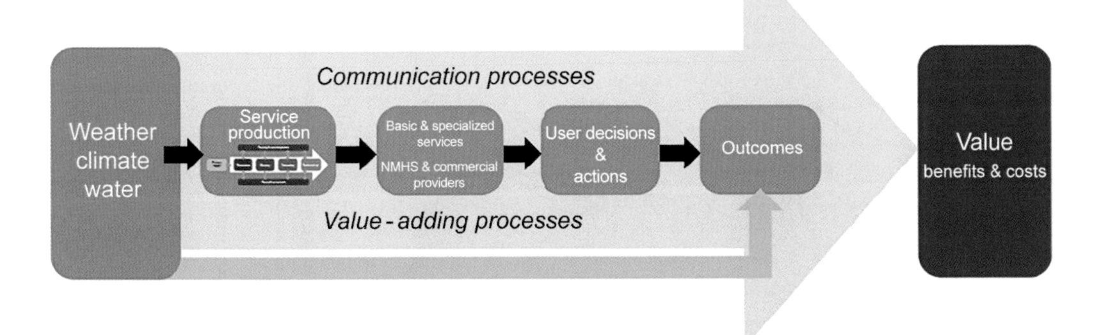

FIG. 1 A simplified schematic of a meteorological/hydrological service information value chain (WMO, 2015), which aims to model the value of information through weather and climate service production and delivery systems. Value, which is typically described in economic terms, can be added or lost at each step in this nonlinear process.

current information availability. A total of 34 distinct sites were reviewed, with particular attention to the following attributes: accessibility, content (e.g., variables, spatiotemporal coverage, and resolution), format, language/terminology, quality/verification, and availability of supporting guidance. Many NMHSs produce forecasts in the S2S time range, but there is only a patchwork distribution of S2S forecasts available to the public, with most focusing on the shorter range (up to 7–10 days). Where information is available, it is presented under a variety of headings, such as "medium-range," "extended-range," "long-range," "monthly," and "seasonal" forecasts or outlooks; and these headings represent different time ranges, such as 6–15 days, 7–45 days, 10–30 days, 7–60 days, monthly (i.e., expected conditions over the next month), and seasonal (i.e., expected conditions over the next 3 months). Additionally, there is inconsistency among NMHSs in how they classify the forecast windows, with some making S2S forecast products available through weather pages and others through climate pages. In only a few cases were monthly or seasonal forecasts directly obtained from the NMHS home page. It was noted that none of the websites searched made forecasts available under the term "sub-seasonal forecast."

Where seasonal information is available to the public, it is typically provided at coarse resolution as either text-based descriptions or tercile (above, below, or near normal) plots and maps (Fig. 2). Climate Outlooks, Climate Advisories, and Monthly Outlooks provide a textural narrative of the expected weather conditions over the coming month, contextualized with descriptions of the large-scale drivers of the climate and the preceding conditions (e.g., observed rainfall accumulations). They are typically descriptions that relate to the country as a whole, with limited local detail making it difficult for users making local-level decisions. The lack of local detail is often a reflection of limited skill at this scale, resulting in a mismatch between user expectations and the available science.

Tercile plots and maps provide gridded deterministic and/or probabilistic information on the likelihood of above, below, or near-normal conditions and almost universally focus on two predominant variables (precipitation and temperature). This approach assumes a background knowledge about past climate conditions, which the user may not have if the

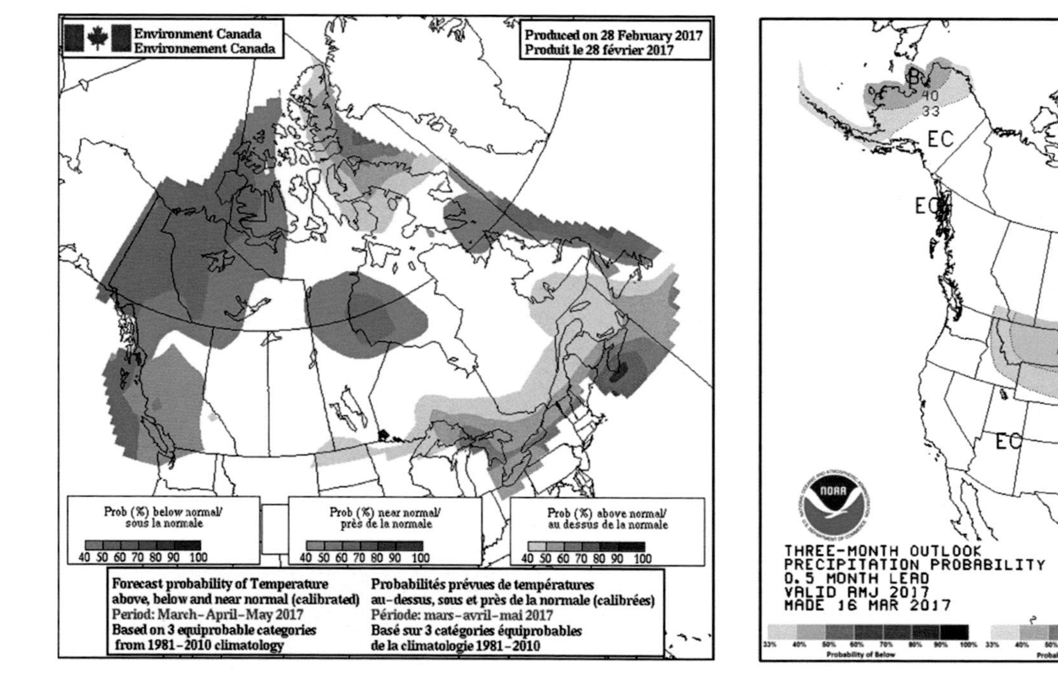

FIG. 2 Examples of seasonal forecast plots. The left plot is from Environment Canada and shows the probabilities of temperature above, below, and near normal (period: March–April–May 2017; produced February 28, 2017). The right plot is from the National Oceanographic and Atmospheric Administration (NOAA) and shows the probabilities of precipitation above, below, and near normal (period: April–May–June; produced March 16, 2017). *Source: (Left plot) https://weather.gc.ca/saisons/prob_e.html. (Right plot) http://www.cpc.noaa.gov/.*

information is not provided alongside the outlook or forecast (White et al., 2017). It also fails to provide any details on high-impact quantities, such as the timing, frequency, and intensity of extreme rainfall or temperature, or the resulting cascade of hazards and impacts that such events can trigger (e.g., flooding, heat waves, food insecurity, disease, or malnutrition) and can therefore be perceived as lacking relevance to user requirements and decisions (Marshall et al., 2011b; Hartmann et al., 2002). Many users want to understand how well the forecast performs in different situations so that they can decide which actions should be implemented based on the information provided. However, forecast reliability and skill, which are sometimes communicated through supplementary plots such as relative operating characteristic (ROC) maps, ROC plots, and reliability plots, have limited comprehension among the user community, and in many cases, they are arbitrary when it comes to informing users about whether a particular forecast is suitable for their specific use or decision-making requirements. For more information on this topic, see http://www.cawcr.gov.au/projects/verification/#ROC.

S2S forecasts are often used in commercial products, with very few available to the public, resulting in poor visibility for other potential users, which prevents them from assessing and evaluating the benefits of S2S forecasts relative to their own requirements. Where subseasonal information is available to the public, it is normally provided in textural format as outlooks. In a limited number of cases, map-based products are available using a similar structure as the seasonal forecast. The Japan Meteorological Agency (JMA), for example, provides maps showing the probability of below-, above-, and near-normal conditions for a number of variables, including temperature, precipitation, sunshine, and snowfall. Information is provided for 1-month (including weekly subdivisions) and 3-month forecasts (with maps available for individual months or the whole 3-month period). In this example, forecast information is provided at the prefecture level. The snapshot of NMHS S2S forecast availability indicates that in the majority of cases, the availability, location, and style of S2S forecast information on public websites are determined by the positioning of S2S research within the organizational structure of the NMHS rather than by the requirements of the end user. Given competing demands on time and resources within NMHSs, S2S information clearly has taken the proverbial backseat up to now.

2.1.1 Improving S2S Public Service Through Community Engagement: Example From the Australian Bureau of Meteorology

Unlike other NMHSs to date, the Australian Bureau of Meteorology (BoM) has dedicated considerable time and effort to enhance its S2S public service, engaging extensively with users to improve their public offerings. The BoM has undertaken two major consultations with users in recent years, one in 2010–11 and another in 2015–16. The objective of these consultations was to determine (1) user needs for S2S forecasts, (2) how this information feeds into decision-making, and (3) user satisfaction and comprehension about its current S2S public services. The BoM used a range of methods to collect the information, including in-depth interviews, focus groups, and online surveys. The research found that user satisfaction with the current service was high; however, comprehension of probabilistic Climate Outlooks and model skill maps could be improved. Satisfaction was related to the comprehension level, where those who answered comprehension questions correctly were three times more likely to be satisfied with the BoM's service than those who answered the questions incorrectly.

The most common reasons cited for user dissatisfaction with the BoM's service were inaccuracy, both perceived and actual, and limited resolution of the outlooks. Tercile outlooks were found to be challenging for users to interpret, with 49% of a recent survey (completed by 117 people) feeling "uncomfortable" with the word *tercile*. Despite this, 65% of users correctly interpreted the test map. This may indicate that using terms that are unfamiliar to those working outside of science may be part of the barrier to users engaging with outlooks described in this way. It also may indicate that providing a map or image is more useful than just a textural description.

One of the most fundamental outcomes of the research was that users are diverse regarding their preferred means of receiving S2S information. While some preferred maps, others wanted values at their location, data on grids, graphs, text summaries, or audio/video briefings. In 2014, the BoM redesigned its website in an effort to refresh the way that Climate Outlooks were presented and improve user satisfaction and comprehension of its service (Fig. 3; www.bom.gov.au/climate/ahead). The new website included features such as

- An at-a-glance summary of the rainfall and temperature outlooks
- The ability for users to tailor the forecasts to their location and critical thresholds (e.g., pop-up boxes for locations, pan and zoom maps, user-selected probability, and rainfall thresholds)
- Better provision of model accuracy information (e.g., easier-to-understand language, better color schemes, supporting educational material)
- A short monthly video with a climatologist explaining the outlook and putting information in the context of recent conditions and current climate drivers
- Supporting explanatory information about the service and educational material, including videos and infographics about the major climate drivers

The BoM partnered with user-centered design specialists to create the new website based on user feedback. Over 40 design concepts were considered, and focus-group sessions and interviews were held through the design process to ensure that the final designs were consistent with user preferences.

The new Web portal was a leap forward toward improving availability and communication of S2S forecasts, but it did not address the requirements for more accurate and higher-resolution outlooks. In 2013, the BoM made a modest improvement in the accuracy of its S2S forecasts by transitioning from a statistical seasonal forecast model to a dynamical seasonal prediction model, the Predictive Climate Ocean Atmosphere Model for Australia (POAMA). This provided a more skillful overall forecast for Australia than the previous statistical system (Charles et al., 2015) and provided the potential for higher-resolution modeling and the ability to forecast more variables than just temperature and rainfall across more timescales. The transition to the dynamical model in 2014 meant that, for the first time, the BoM offered the public forecasts for rainfall and temperature for separate months, in addition to the seasonal forecast, in a user-oriented way. In 2018, the BoM is transitioning to a new dynamical model (ACCESS-S), which will operate at a higher resolution (60 km rather than the current 250 km) and is expected to deliver outlooks more accurately. It also will allow the BoM to issue outlooks more frequently (weekly, instead of the current monthly frequency) and provide outlooks for the fortnights ahead (e.g. forecast for the two week periods of week 2 and 3

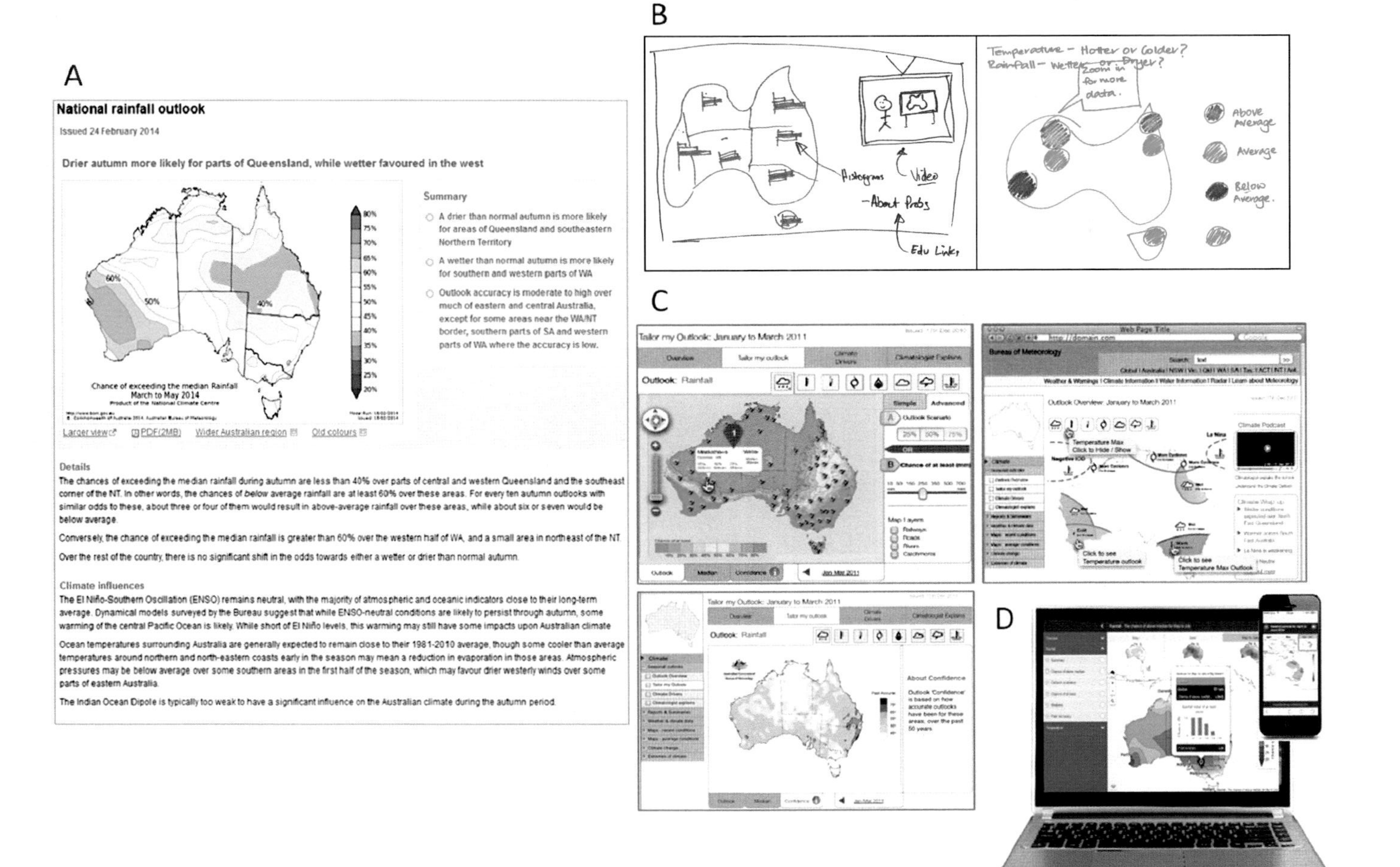

FIG. 3 (A) Example of how rainfall and temperature seasonal outlooks were presented by the BoM in 2014 (prior to web interface redesign). (B) Two of the 40 design concepts that were investigated as a way to deliver Climate Outlooks to decision-makers during the 2010 consultation period. (C) Final design suggestions for the Web portal for S2S forecasts developed by a user-centered design company in consultation with BoM experts and end users. (D) Image of the rebuilt web portal (2014) that displays the BoM's Climate Outlooks.

and week 3 and 4) in order to fill the gap between the current 7-day weather forecast and monthly and seasonal outlooks.

To ensure the effective communication and dissemination of S2S information, the BoM uses a holistic approach that is broader than just delivery through a website. A comprehensive communication strategy sets out ways to engage with users when outlooks are issued. The strategy revolves around consistency of message and recognition that different communication channels are required to reach different user groups. Some users prefer to get their information from a website, others from mobile apps, television, newspaper, podcasts, social media, radio, or even phone text/notifications. Where possible, the BoM matches audiences with information that meets their specific needs. Monthly dissemination of the service includes:

- Updates to the website.
- A Climate and Water Outlook video provided monthly on the BoM website and YouTube, and also broadcast on a national television program (*Landline*), where it reaches an estimated audience of 500,000. The program gets a free segment, while the BoM outlook gets "free" dissemination to a target audience (agriculture).
- A National Climate and Water Briefing, whose primary audience is public-sector senior officials who make medium- to long-term policy and planning decisions.
- Webinars; the primary audience is state-based extension officers who work with decision-makers from the agricultural sector and want more detailed explanations of likely conditions.
- *E*-mail direct marketing, including a subscription service that provides users an e-alert whenever new Climate Outlooks are issued; there are more than 5000 subscribers to this list, with industry stakeholders and media making up a large component of the subscribers.
- Media activities; Outlooks are posted in the BoM's Media Newsroom, and key spokespeople are available for more in-depth media inquiries.
- Social media campaigns (on sites such as Facebook, Twitter, and LinkedIn), in which social media posts are carefully crafted to reach a large and diverse audience and include links to the BoM's website.
- Distribution of key messages and talking points, for BoM staff and forecasters who may be performing media or having discussions with stakeholders. This document is not for the public but gives staff a consistent and clear message to communicate to media and stakeholders.

2.2 Current S2S Research and Applications for Weather- and Climate-Sensitive Sectors

To assess the current body of research on S2S applications and communication, an extensive review of the peer-reviewed literature was undertaken in 2011 and updated in 2017 (Silver and Mills, 2018). This review accessed articles through several prominent databases, including Google Scholar, JSTOR, and EBSCO, utilizing a number of search terms related to S2S forecasts/predictions (e.g., *sub-seasonal forecast, monthly forecast, seasonal forecast,* and

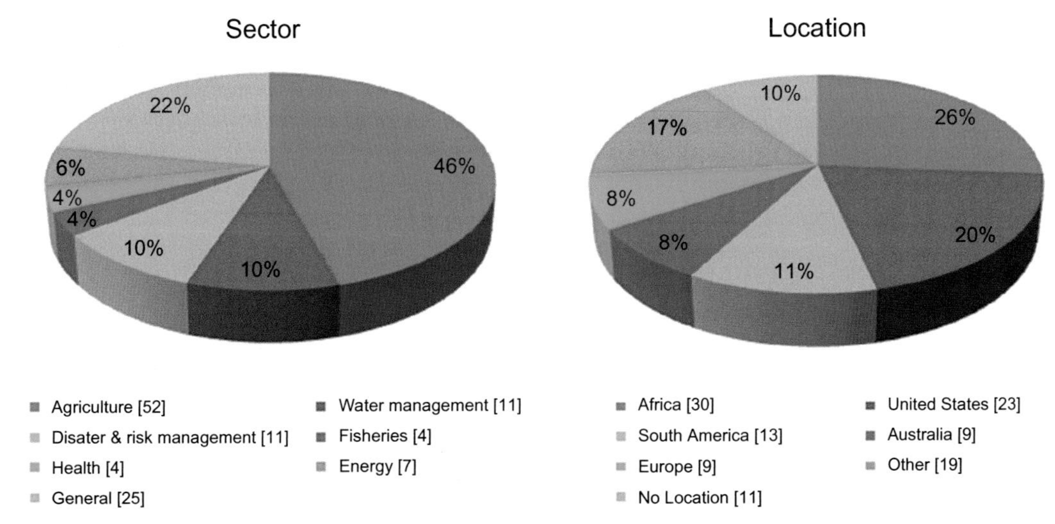

FIG. 4 The distribution of peer-reviewed S2S literature, collected within the updated bibliography (Silver and Mills, 2018) over the period 1976–2016, by (A) sector and (B) geographical research focus.

medium-range forecast). Reference lists within the identified articles were screened to identify additional publications. The original search gathered 84 articles related to S2S research and applications, which were entered into an annotated bibliography that described the study area, methods, and casual findings of each article. The 2017 update added another 30 articles to the list, bringing the total number of articles reviewed to 114.

Although this review is by no means exhaustive, it does provide insight into the distribution of S2S research between 1976 and 2016 (Fig. 4). Unsurprisingly, agriculture is the most prominent sector represented in the review, with 52 articles exploring the value, applications, and limitations of S2S information. Other sectors, including water management (11 articles), hazard and risk management (11 articles), fisheries (4 articles), and energy (7 articles), are featured less prominently. The bibliography also includes a substantial number of articles that assessed S2S forecasting systems in general, without focusing on any particular sector. In terms of geographic location. Africa and the United States were featured prominently, as well as South America and Australia. This is not entirely surprising, given the historic prevalence of severe droughts and floods that have affected many parts of these countries and the role that large-scale circulation modes have in influencing these events in many of these regions. It also should be noted that the articles show a predominant focus on the seasonal forecast timescale, with a smaller proportion focusing on the monthly time frame and only a small fraction looking specifically at the 2–4-week period.

The review also sought to assess the relative research emphasis of the articles on the list, focusing on whether articles considered communication, dissemination, and engagement approaches to address how users obtained, interpreted, and utilized S2S forecast information for decision-making. The treatment of communication aspects was very small or even nonexistent in the vast majority of articles, most of which focused instead on the verification of S2S forecasts at the operational level or on the anticipated economic benefits of improved forecasting to end users. Approximately 24% of the articles contained substantive contributions to understanding the communication and application of S2S forecasts to end users' decision-making. These publications examined how individuals obtained S2S forecasts, how they understood and interpreted the information they received, and how this information was used in decision-making. A substantial proportion of this literature emphasized the need for participatory, collaborative communication between providers and end users to encourage trust, comprehension, and utilization.

2.2.1 *Agricultural Sector*

The majority of agricultural-based research at the S2S timescale focuses on (1) enhancing the forecast skill of variables that are particularly relevant to the agriculture sector (e.g., rainfall amount, rainfall onset, and within-season dry/wet spell occurrence); and (2) identifying the social and economic benefits of forecasts to the agricultural community (Adams et al., 2003; Baethegen et al., 2009). The article list highlighted a smaller body of work that addressed the translation of forecast information into decisions and approaches for engaging end users to encourage sustainable uptake. Several researchers have observed that despite improvements in the forecast skill, reliability, and applicability of S2S forecasts, S2S information is generally underused by the agricultural community (Carberry et al., 2002; PytlikZillig et al., 2010; McCown et al., 2012; Prokopy et al., 2013). The reasons for this include but are not restricted to (1) uncontextualized information dissemination; (2) information dissemination without understanding, experience, or practice; (3) lack of ownership and contribution by farming communities to the knowledge and tools developed; (4) lack of integration with the day-to-day activities of farmers and their resource requirements (Ingram et al., 2002); (5) diversity of user needs; (6) skepticism by the user community of the value of models to their specific problems; (7) inadequate or inappropriate information for their requirements; and (8) lack of trust in the information source (Hu et al., 2006).

The gap between useful information, provided by improvements in and availability of S2S forecasts, and useable information, which can be integrated within agricultural practices, decisions, and requirements (White et al., 2017; Artikov et al., 2006) is being increasingly addressed through research and enterprise. Decision support tools and portals such as AgroClimate (Breuer et al., 2008, 2009; Dinon et al., 2012), FARMSCAPE (Carberry et al., 2002; McCown et al., 2009), Africa RiskView (UN-SPIDER, 2012), Uruguay's National Agricultural Information System (IDSS; Hansen and Coffey, 2011); Citrus Copper Application Scheduler (Dewdney et al., 2012); and the Strawberry Advisory System (Pavan et al., 2011) aim to reduce complexity, improve information availability, enhance efficient decision-making, and provide more actionable information based on weather and climate forecasts. This is accomplished by translating a range of specialized data sets (e.g., weather/climate, agricultural, economic) and displaying the information in graphics, tables, dashboards, maps, and text relevant to specific user questions (e.g., "What is the risk of late onset rainfall?" "When is the optimum time for fertilizer/pesticide application?" "What if I did a certain action—how would this affect my productivity?"). The tools often allow and/or require users to add their own information specific to their crops, management practices, and location.

Research illustrates that the uptake of these types of tools are improved substantially through coproduction activities with end users. Participatory action research has been used in a number of initiatives to engage with the user community (Sivakumar et al., 2014; Breuer et al., 2009; Cash et al., 2006) and identify high-priority actions (Tadesse et al., 2015) to ensure the usability of forecast information. One example of this type of work is the Droughts Wiki Project, being conducted by the National Early Warning and Monitoring Centre for Natural Disasters (Cemaden) in Brazil, in partnership with the Austrian Institute for Applied Systems Analysis (IIASA). The project has developed a database platform for monitoring agricultural drought impacts, supported by a mobile application called AgriSupport. This application is used to collect agricultural data (e.g., location of cultivated area, planting dates, and crop type) from users and provide weather-based, agricultural activity–specific warnings, as well as information in support of cultivation and management decisions for smallholder farmers.

A strategy to encourage uptake and use of AgriSupport is being guided by existing educational outreach activities, such as the Cemaden Education Network of schools and communities for disaster risk prevention, which focuses on citizen science and crowdsourcing. Workshops and seminars have been run with high school students and teachers (Marchezini and Trajber, 2016), smallholder producers, community leaders, representatives from government organizations, and weather and climate experts. The aim of these workshops is to draw together experiences, map the distribution of knowledge, and extend the understanding of weather and climate information, as well as encourage participants to engage with model data and learn techniques for using this data in decision-making. These efforts are currently being further expanded so that younger generations living in rural and urban communities can support and encourage the uptake of the app through combined workshops and in-community endorsement.

To ensure that the perceived benefits of S2S forecast information (Adams et al., 2003; Jones et al., 2000b) are realized, Artikov et al. (2006), Hansen (2002), and Hu et al. (2006) described the importance of the human dimension in decision-making (i.e., personal attitudes; social norms; perceived behavioral controls, such as technical and economical ability to use forecasts; and perceived and actual obstacles to use, such as the ability to choose alternative options based on forecast information). To fully realize this dimension requires sustained and inclusive engagement with the user community through surveys, workshops, interviews, and collaborative dialogue, as well as research using economic techniques (e.g., derived demand modeling) and social psychology models of human behavior (e.g., theory of planned behavior).

The BoM is currently expanding on its earlier consultation approaches to further improve its S2S service (beyond the 2014 public web portal, as described in Section 2.1) through its Improved Seasonal Forecasting Services project. A round of face-to-face meetings and interactive workshops (2015–16) has been conducted to understand changes in user requirements as science capabilities have improved (e.g., availability of higher-resolution forecasts and more accurate and frequent outlooks, as well as new fortnightly outlooks; see Fig. 5). The majority of participants came from the agricultural sector, although government, health, energy, construction, mining, and emergency service sectors were also represented to a lesser degree. The high representation of agricultural users prompted later workshops to include on-farm visits in which BoM staff traveled to working farms and discussed weather, climate, and water issues with each owner (Fig. 5) while walking or driving around the property. Such detailed engagement proved invaluable and added significantly to the results obtained through the office-based workshops, as key farming tasks could be discussed.

(A)

(B)

FIG. 5 (A) User workshop held by the BoM in Darwin, Australia in 2016 and (B) an on-site farm visit as part of a user workshop in Longreach, Queensland, in 2016. *Photos taken by Jenny Metcalfe, Econnect Communications.*

The agricultural user community showed a clear preference for forecasts of the chance of extremes and the chance of a user-selected threshold at the sub-seasonal timescale. While the chance of a user-selected threshold forecast has more relevance at the sub-seasonal timescale, forecasts of the chance of extremes were a clear user requirement on the seasonal scale as well. The engagement also made it possible to identify decisions that varied with lead time. For example, at the sub-seasonal time frame, agricultural users are-making decisions such as when to fertilize, when to plant or harvest, and when to move stock or water in preparation for certain weather events or conditions. Meanwhile, with the seasonal time frame, different types of decisions were being made, such as how much to plant, what crop variety to plant, and when to restock and destock. Although these decisions are different, all of them can have significant social and economic implications for the user, and therefore communication of forecast uncertainty and skill is essential so that users can make appropriate decisions based on their risk appetite. However, this can be successful only if the probabilities being provided in S2S forecasts are reliable. If probabilities are inaccurate (e.g., a 60% forecast probability is actually only 30% when tested in the real world), then users may have made a better decision had they had no forecast at all.

Overall, the information collected from these detailed engagements identified clear priorities for service development (e.g., chance of extremes and chance of user-selected threshold forecasts). It also identified user preferences for information dissemination. Access to information on mobile devices is important, as is information specific to their location. Web pages or apps that integrate more information than just S2S forecasts are desired, as most decisions depend on more than just weather/climate information. It was also clear that the way that uncertainty and model skill is communicated needs improvement. These preferences are now being factored into future development plans.

2.2.2 *Energy and Water Management Sectors*

The energy and water management sectors are strongly linked, particularly where reservoir systems are operated to address multiple functions, such as hydropower production, flood control, and irrigation. Both sectors tend to have preexisting decision-making processes and knowledge and interaction with weather and climate forecast information (White et al., 2017). It has been suggested that it may be easier to develop successful relationships with these sectors (Brunet et al., 2010), although Rayner et al. (2005) highlighted that organizational conservatism and complexity, reliance of traditional large-built infrastructure, political disincentives to innovation, and regulatory constraints have hindered the uptake and use of forecast information by water managers in the United States. Building relationships across both sectors requires engagement with a wide range of stakeholders, from water managers and individual utility company operators, to national distributors, energy traders, government departments, and policymakers. In addition, private-sector interaction tends to be higher across both of these sectors, with a number of commercial companies offering business solutions (e.g., Steadysun, IrSOLav, RENES, EuroWind, Prewind, and AleaSoft) to facilitate improved decision-making. The opportunities and challenges of increased private-sector interaction, including service delivery models and future trends, were described by Gunasekera et al. (2014), and the study highlighted the similarities and differences of communication and engagement strategies that are necessary to ensure information uptake within this community. It was also noted that with increasing awareness and understanding of S2S forecast information, the relative role of NMHSs is likely to decline as engagement, collaboration, and service delivery models increase and improve.

With the expansion of renewable energy markets there is a growing body of research assessing the ability (Torralba et al., 2017; Ely et al., 2013), value (Hamlet et al., 2002), accuracy, and skill (Clark et al., 2017; Lynch et al., 2014; Brayshaw et al., 2011; Buontempo et al., 2010) of S2S forecasts. There is generally a focus on the seasonal scale, with a smaller but growing literature around the sub-seasonal time frame. Despite this growth, there remains limited published material on industry use (De Felice et al., 2015, is a rare example) and the strategies used to engage and communicate with the energy sector. However, work by Bruno Soares and Dessai (2015, 2016) has illustrated how expert elicitation workshops and semistructured interviews can be used to engage with this community. Between 2013 and 2014, they used these techniques to identify the users of seasonal forecasts, the provider-user information flows, and the barriers and solutions to seasonal forecast use. The results of this engagement indicated that the energy sector featured prominently as early adopters of these forecasts. The majority of these early adopters were large companies working at the international or national level, with a level of capacity and expertise that allow for ingesting weather and climate information (Bruno Soares and Dessai, 2015). The in-house expertise of some of these early adopters means that postprocessing and/or tailoring of seasonal forecast information can be performed without significant additional support, in terms of making the forecast information communicable, from NMHSs. Such a process utilizes the expertise of all the organizations in the information dissemination chain, allowing improved efficiency and increased profitability for the end user. However, interviews identified that the main barriers to using seasonal forecasts were the perceived lack of reliability, linked to perceptions of high levels of uncertainty and lack of accuracy (Bruno Soares and Dessai, 2016), and the lack of relevance to their particular way of working. The main enablers were existing relationships and collaborations, which improved the access to and understanding of seasonal forecasts, and resource and expertise in the user organization.

There is a greater distribution of research and applications related to streamflow, rainfall, and ENSO forecasting and water management at the S2S timescale. However, the articles collected in the updated bibliography (Silver and Mills, 2018) remain predominantly focused on assessing skill (Shukla et al., 2012), value (Broad et al., 2007) and uncertainty in the forecasts (Kwon et al., 2012; Sharma and Gosain, 2010) and their ability to capture drought/flood extremes (climate shocks). There are almost no articles representing the sub-seasonal time frame, with almost all reviewing the value, skill, and uncertainty of seasonal forecasts. There is a general assumption that S2S forecasts will provide value to the water sector by informing short-term planning (i.e., water allocation and restrictions) and enabling the setup of flexible and adaptive contingency measures (i.e., infrastructure and management systems) that can continue to be efficient during extreme dry or wet years (White et al., 2017; Sharma and Gosain, 2010).

There is also a growing recognition of the water-energy-food nexus and that forecast advancements in one sector can both benefit and deter activities in others (Conway et al., 2015). However, the effectiveness and usability of such forecasts across this nexus is influenced by science (i.e., actual and perceived skill, timeliness, resolution, and relevance), policy, and politics, particularly where management actions in one country or region result in consequences in another. Lemos et al. (2002) highlighted that forecast information for water and drought policymaking risks distortion, misinterpretation, and political manipulation and that rapid uptake within policymaking, without clear guidance on the uncertainties and utility within the user context, can have broad-scale negative impacts on uptake within the user community. For example, policymakers exaggerating the potential usefulness of a tool can result in culture dissonance between science, local knowledge (rain prophets), and belief systems, and this can quickly erode the value of the information and, in some cases, cause end users to view the forecasts in an adversarial light (Lemos et al., 2002). The driving message behind such research is that S2S forecast information can provide benefit to the water sector and the water-energy-food nexus as a whole, but that this benefit can be realized only if it is included within broader development aims to reduce socioeconomic exposure and vulnerability.

2.2.3 Natural Hazards and Disaster Risk Reduction (DRR)

There is a growing body of evidence that identifies S2S forecasts as valuable in supporting enhanced preparedness activities and emergency response (White et al., 2017; Tadesse et al., 2016; Tall et al., 2012). This value is best described by the Red Cross Red Crescent Climate Centre's "Ready-Set-Go!" approach (Coughlan de Perez and Mason, 2014), whereby seasonal forecasts are used to trigger the "Ready" stage, sub-seasonal forecasts the "Set" phase, and short-range weather forecasts the "Go" phase. Each phase requires specific response actions to be completed based on the increasing or decreasing likelihood of the event occurring. The methodology has the potential to be used across a range of sectors, but initial examples focus on implementation for disaster risk reduction (DRR). The aim is to empower DRR managers to adapt and respond more efficiently in order to improve response and reduce the impacts (severity, duration, and geographical extent) of severe events upon communities.

Although still in its infancy, forecast-based action and financing follows this conceptual approach by translating forecasts from a range of timescales into defensible decision-making for humanitarian action (Coughlan De Perez et al., 2015). Forecast probability thresholds are matched with appropriate actions at different lead times so that in the long term, the benefits of taking preventive action outweigh the cost of acting in vain when an event does not materialize. A key element in this process is the disbursement of funds and the development of standard operating procedures (SOPs) that mandate action once the forecast threshold is

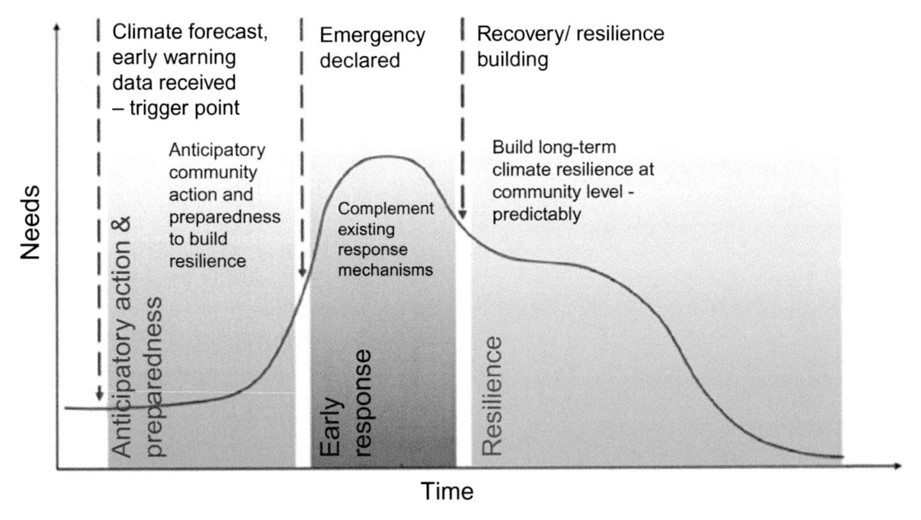

FIG. 6 Diagrammatic illustration of the Food Security Climate Resilience (FoodSECuRE) facility developed by the WFP to support community-centered action to build and reinforce resilience. Similar to the concept of FbF, FoodSECuRE links climate and hazard forecasting with flexible multiuser financing, providing governments with the means to unlock funding quickly to scale up food and nutrition responses, as well as DRR activities before an event occurs. *From Food Security Climate Reslience Facility (FoodSECuRE) URL: http://www.wfp.org/climate-change/initiatives/foodsecure, Copyright: World Food Programme.*

reached (Fig. 6). This can be achieved only through a codesigned and implemented strategy, with a commitment from all contributors (e.g., meteorologists, climate scientists, humanitarians, and donors).

The Red Cross Red Crescent Climate Centre, together with the German Federal Foreign Office and the World Food Programme (WFP), unveiled the first operational use of the forecast-based finance (FbF) approach in December 2015. In Uganda, FbF allowed the distribution of water-purification tablets, jerry cans, storage sacks, and soap to households likely to be affected by flooding that had been forecast by the European Commission's Global Flood Awareness System (GloFAS) and the Uganda National Meteorological Authority. Once the forecast exceeded prespecified thresholds, the Uganda Red Cross Society was able to brief district authorities on the planned distribution of aid and worked with doctors to show people how to use water purification tablets and identify early signs of malaria. Ensuring forecast thresholds match appropriate responses and that funding and operating procedures are sufficiently supported is achieved through consistent (twice yearly) and collaborative dialogue platforms.

The Ready-Set-Go concept is also being increasingly used by NMHSs to manage the flow of information when issuing early warnings. In this context, information is disseminated to user and responder groups based on the forecast confidence and lead time. The Met Office in the United Kingdom has adopted this strategy when communicating about potential severe coastal flood events. This has been made possible by the development of Coastal Decider (Fig. 7), a new medium- to long-range operational forecasting tool to highlight periods with an increased likelihood of coastal flooding. The tool is used by the Flood Forecasting Centre

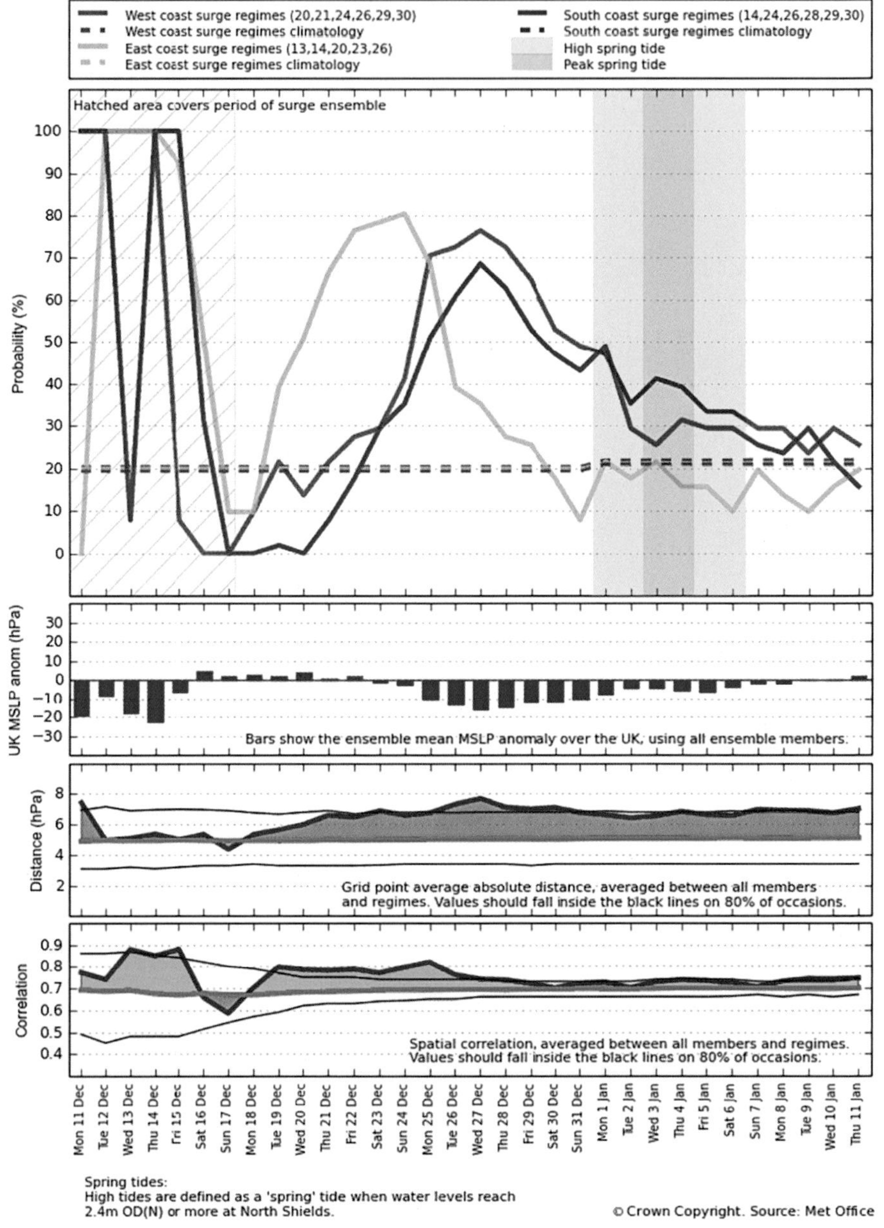

FIG. 7 A Coastal Decider monthly regional summary forecast for England and Wales from the 00UTC run on December 11, 2017, using the monthly system of the European Centre for Medium-range Weather Forecasts (ECMWF). The regional forecasts show the probability of high-risk weather patterns that are relevant for surge events along different coastal stretches (top). Climatologies of these high-risk surge-related weather patterns (*dashed lines* in top plot) and the UK mean sea level pressure anomaly forecast is provided for context. Distance and correlation plots (bottom) are used to assess how good the match is between ensemble members and their assigned weather pattern. For full descriptions of these plots, see Neal et al. (2018).

(FFC) to provide a heads-up about a heightened probability of high-impact coastal flooding associated with the occurrence of specific synoptic weather patterns (Neal et al., 2018). Forecasts are available from 1 to 7 weeks, extending the forecast horizon beyond the 7-days available from the Met Office storm surge and wave ensemble models. Ensemble members from a range of ensemble prediction systems are objectively assigned to the closest matching weather pattern, using a set of 30 objectively derived patterns for the United Kingdom and surrounding area of Europe (Neal et al., 2016). These are then grouped to provide probabilities of a broader cluster of weather patterns that, based on a historical analysis, are associated with high-impact flooding along a specific stretch of the UK coastline.

A range of forecast outputs have been tested and coproduced with FFC hydrometeorologists to ensure the efficient and effective assessment of the evolving hazard. Coastal Decider feeds into a new briefing product designed to communicate the risk of significant coastal flooding to the Environment Agency at longer lead times. This is particularly useful when proactive, large-scale, and multiregional responses, such as evacuations, need to be considered. Initial indications of a significant event are used internally within the FFC to monitor the situation, but as confidence in the forecast increases, the information will be passed to Monitoring and Forecasting Duty Officers within the Environment Agency through regular (daily) teleconference calls and briefing notes. Once the forecast event falls within the 1–5-day lead time, higher-resolution models are used to produce more detailed risk assessments. These in turn support the issuing of warnings to the responder community and the public.

2.2.4 Health Sector

The updated bibliography (Silver and Mills, 2018) highlighted that the use of S2S forecasts within the health sector was still emerging, including in areas where S2S forecasts could be expected to have some useable skill, such as extreme heat and cold (Hudson and Marshall, 2016; Teng et al., 2013). The majority of publications focused on quantifying the forecast skill of various malaria-relevant climate variables. Thomson et al. (2006b), Jones et al. (2007), Jones and Morse (2012), and MacLeod et al. (2015) discuss and demonstrate the success of seasonal forecasts to predict the climate anomalies associated with malaria epidemics and malaria transmission. Such research is primarily carried out in Africa, where climate drives both mosquito-vector dynamics and parasite development rates (Thomson et al., 2006b).

The link between vector-borne disease, key environmental factors, and social, economic, and political aspects has been demonstrated more recently by the Zika episode that emerged in South America coincident with the lead-up to the summer Olympic Games in Brazil and which subsequently spread through Central America and the continental United States (US) by early 2016. The World Health Organization (WHO) declared a Public Health Emergency of International Concern related to Zika on February 1, 2016. Despite environmental health programs being framed within seasonal windows, mosquito control practices are driven by traditional protocols, which are strongly influenced by short-term weather forecasts. This is often because the effectiveness of environmental health interventions can be strongly controlled by short-term atmospheric conditions.

Since the WHO declaration, U.S. health authorities have undertaken epidemiological surveillance, case control, and follow-up actions, as well as an aggressive environmental health campaign that includes, among other measures, the elimination of water reservoirs where the mosquito can reproduce and extensive truck spray treatments to eliminate mosquito larvae and adult mosquitoes. This has been complemented by aerial spraying to help control the mosquito population. Spraying protocols in particular rely on short-term weather forecasts, as seen in Florida, where there was a need to suspend aerial pesticide spraying because of unexpected strong winds and heavy precipitation. Similarly, for some applications like liquid larviciding, a certain level of humidity is desired to ensure proper deposition; and for spatial adulticiding, low wind speed increases effectiveness.

Recent studies on spatial and seasonal variability of *Aedes aegypti* (the mosquito species responsible for transmission of Zika) in the United States and the periods of higher risk of Zika introduction offer insights that could influence public health decision-making related to vector-borne diseases. For example, Monaghan et al. (2016), with support from the National Institutes of Health (NIH), National Aeronautics and Space Administration (NASA), and the National Science Foundation (NSF), used meteorologically driven models for the period 2006–15 to simulate the potential seasonal abundance of adult *Ae. aegypti* for 50 cities across the United States. These models include variables associated with human exposure and susceptibility, such as travel and socioeconomic factors. They employed temperature, precipitation, and humidity conditions to drive two process-based life-cycle models and simulated the daily potential abundance of *Ae. aegypti* for the most recent 10 years. The simulations revealed that meteorological conditions for *Ae. aegypti* are suitable across all 50 cities during the peak summer months (July–September), although the mosquito has not been observed in all cities. Meteorological conditions are largely unsuitable for the species during the winter months (December–March), except in southern portions of Florida and Texas that can sustain low-to-moderate potential mosquito abundance compared to summer. Model predictions confirmed diminishing populations of the *Ae. aegypti* mosquito and a halt in Zika transmission as cold weather approached by mid-December 2016 (Cohen, 2016). To ensure that public health decisions (e.g., planning/resourcing spraying protocols, disseminating purification tablets and hygiene kits to communities at risk of flooding, and activating awareness campaigns) take advantage of the benefits of the new forecast products available on the S2S time range, a determined effort and close collaboration among public and environmental health officers, entomologists, and climate-weather scientists is required to build a solid evidence base for the appropriateness of their use and subsequently adjust their health policies, programs, and protocols for intervention.

3 GUIDING PRINCIPLES FOR IMPROVED COMMUNICATION PRACTICES

The preceding examples highlight a substantial body of research illustrating the potential benefits of the improving S2S forecast capability and availability. However, user communities can realize these benefits only if the information is useable. Usability is a function of many competing factors, including the skill, accuracy, timeliness, and resolution of the forecast, as well as many social, economic, and environmental factors. The importance of any of these factors varies in time and space and depends on the user community and their requirements.

One of the many challenges in communicating S2S forecasts is that they are inherently uncertain and often provide only limited information (if anything) on the timing, location, scale, or frequency of individual weather events that may be of interest to the user. Communicating this uncertainty and these limitations is key to developing user confidence in the forecast. In experimental settings, probabilistic weather forecasts that convey uncertainty information have been shown to lead to better decisions when the probabilities provided are reliable (Joslyn and LeClerc, 2012; Ramos et al., 2013; Roulston et al., 2006), and yet their uptake in operational forecasting and decision-making has not been as widespread as expected (Demeritt et al., 2013). For example, many publicly available weather forecasts still do not include uncertainty information (Joslyn and LeClerc, 2012; Morss et al., 2008a), in spite of over a decade of research on ensemble weather forecasting techniques and their applications.

There are many challenges associated with the communication of uncertainty and skill, including (1) operational forecasters' skepticism about the ability of forecast recipients to understand or use probabilistic forecasts appropriately (Demeritt et al., 2010); (2) some users have low comprehension of skill maps and a poor understanding of uncertainty (e.g., low expertise in risk management, low comprehension of probabilities, and little or no understanding of model skill and reliability metrics); (3) some users do not take the time to look at information about forecast skill, as they expect all issued forecasts to have skill; and (4) accessibility and style of provided information can override how skillful a forecast is (e.g., users that need information quickly will use a website or app that is easy to use or that provides all the information they need in one location, rather than analyzing which provider has the forecasts that are most skillful). It is essential, therefore, that forecast skill is described from the users' perspective with user-oriented metrics, in order to ensure an understanding of the model's performance relative to their specific needs.

Despite these challenges, emerging experience in the communication of uncertainty from the development of climate services (Otto et al., 2016) highlights the need to be transparent about all known sources of uncertainty, including knowledge gaps and issues relating to the methodology and processing of the forecast. Additional educational outreach and supplementary material explaining the forecast and its applicability for different user groups is increasingly used to ensure this transparency and extend knowledge across research-practitioner and provider-user boundaries. However, this should be accomplished through two-way communication to ensure that the users' expectations are realistic with respect to the current science, their needs are met to the best ability of the forecast, and the information provided is relevant, understandable, and available through channels of preference. In a similar way, forecast accuracy and reliability need to be communicated to ensure effective use of the information. If forecasts lack reliability (e.g., if a 60% chance is not really a 60% chance), they could be more damaging than useful, and users need to be told this in plain language.

The main aim of forecast information, tools, and early warning systems is to build awareness and improve decision-making. The S2S time frame is different from weather and climate time frames, in that it can support both sustainable, long-term preparedness actions (e.g., educational awareness campaigns) and shorter-term response activities (e.g., distribution of medication and hygiene kits) through prepreparedness and early action. The audience for short-term forecasts and S2S forecasts also differs. Seasonal and sub-seasonal forecasts are mainly used by industry, while the public rarely make decisions on S2S probability forecasts but are big consumers of weather forecasts. This has an impact on both the channels through which information is provided, at the different timescales, and the content of the information. Key industry users, for example, may be able to understand and digest complex forecast information and its associated uncertainty and prefer raw forecast information for ingestion to their own systems or bespoke products. The general public, meanwhile, need forecasts to be easier to digest and accessible through a generic Web interface or media and mobile apps.

S2S forecasts typically inform decisions that have a significant economic, safety, or environmental value associated with them. Of course, the same can be true for weather forecasts—especially in high-impact weather situations—but at shorter timescales; and especially during benign conditions, many forecast-informed decisions by the public may be (or can be perceived to be) less consequential (e.g., whether to hang the washing on the line or not). This does not necessarily mean that the stakes are lower, however, as poor forecast performance or failure to communicate all aspects of a forecast during benign conditions may erode users' trust in forecasts during extreme weather, and in fact, that also may influence their perception of the skill and usefulness of S2S forecasts.

Communication and dissemination of S2S forecast information and engagement, through participatory action, with user communities provide an opportunity for users to transition from passive recipients of information to actively engaged contributors. Indeed, successful communication, dissemination, and engagement should allow vulnerable communities to become more empowered and more resilient. Increased understanding and participation in the development of forecast products, through educational outreach, coproduction, open dialogue, and collaboration, are essential so that user communities are encouraged to identify how improving science can benefit them. Furthermore, it allows users to look, build, and lobby for alternative options and choices prior to potentially impactful weather.

It also should be acknowledged that NMHSs are experts in model development and forecasting, while the local communities and sectoral users are experts in their local context, their day-to-day decision-making, and their abilities and options should weather affect them. This should not be taken for granted, as acknowledgment and inclusion of user skills and knowledge can improve the relevance of forecast information (Chengula and Nyambo, 2016). Coproduction approaches also offer NMHSs the ability to educate users about the skill and limitations of the forecasts and the potential long-term benefits of embedding S2S forecast information into their operations.

4 SUMMARY AND RECOMMENDATIONS FOR FUTURE RESEARCH

This chapter has revealed the low level of maturity of S2S communications in current published work and operational systems, with communication, dissemination, and engagement strategies rarely being considered in the design, development, and rollout of these services. Based on this review, the following recommendations emerge, many of which are applicable beyond the S2S time frame:

- Engagement with users is essential to produce actionable forecast information from the S2S time range. Engagement must be iterative to account for advances in model performance and forecast product development, as well as the evolving requirements of users, which shift with changes in financial markets, policies, and politics and technological advances within their sector. Current strategies for user engagement include workshops, surveys, seminars, and one-to-one interviews, as well as consultation exercises and collaborative discussion sessions, which bring together scientists working to develop the forecast products and those who will ultimately use the information.

- Educational outreach is a critical component to improve user comprehension and sustained uptake of S2S services. Educational outreach at a variety of levels (e.g., individual, youth, sector, and government) also opens up new dissemination pathways.
- Collaborative and coproduced tools to aid decision-making are essential and must extend across research disciplines and research-practitioner and provider-user boundaries. The focus for these tools needs to be led by the questions that users need to answer in order to, for example, protect themselves or become economically efficient.
- Successful communication, dissemination, and engagement should improve understanding and empower user communities. As user engagement and codevelopment of applications and services increase, the relative role of NMHSs will necessarily decline, while new capacity and resilience will emerge through an extended reach of knowledge translators and brokers that can talk across disciplinary, researcher-practitioner, and expert-public boundaries.
- Given the uncertainties associated with the S2S timescale, it is important to consider information management and the structure that this should take when applying S2S forecasts in early warning and DRR. Consideration of the appropriate recipient at each step in the information management chain is also critical; it helps improve efficiency and trust and ultimately provides tangible outcomes for society.
- There remains very little access to privileged applications known or thought to be in place in the private sector (e.g., insurance, energy, or military applications). Greater effort should be made by NMHSs, university researchers, and non-government groups to engage and partner with private sector organizations to identify areas of knowledge that could benefit public health and safety that may be shared without compromising commercial services.
- For the development and integration of communication aspects both in research and operationally, there is a need for explicit and systematic documentation, evaluation, and publication of all aspects of S2S information communication and use.
- Furthermore, there needs to be a sustained effort to design, pilot, and operationalize new S2S services and measure their effectiveness and usability over years, not months or the life of a traditional research demonstration project.

CHAPTER

20

Seamless Prediction of Monsoon Onset and Active/Break Phases

A.K. Sahai, Rajib Chattopadhyay*, Susmitha Joseph*, Phani M. Krishna*, D.R. Pattanaik†, S. Abhilash‡*

*Indian Institute of Tropical Meteorology, Pune, India †India Meteorological Department, New Delhi, India ‡Department of Atmospheric Science, Cochin University of Science and Technology, Kochi, India

OUTLINE

https://doi.org/10.1016/B978-0-12-811714-9.00020-6

1 INTRODUCTION

The Indian summer monsoon (ISM) is spectrally characterized by the dominance of a 1–7-day synoptic mode (monsoon lows and depressions), a 10–20-day mode (westward-propagating supersynoptic mode), and a 20–80-day mode (northward-propagating monsoon intraseasonal mode). The rainfall distribution over the Indian subcontinent is presumed to receive important contributions from all these modes. These intraseasonal or sub-seasonal fluctuations of ISM rainfall are an integral part of the regional hydrological cycle and have been a reliable source of consumable water for centuries on the Indian subcontinent. Efficient usage of this water resource is essential for dam and reservoir management, agricultural crop management, health services, disaster management, and other services for the benefit of the community during the monsoon season. Meteorological forecast of the sub-seasonal fluctuation of monsoon rainfall on different spatial and temporal scales is an essential component of this process.

Although capabilities to forecast rainfall and temperature on different scales depend on many factors, a streamlined multiscale and seamless approach spanning different processes and incorporating several stakeholders on a single platform is slowly evolving based on the availability of improved forecasting resources and improved understanding of the monsoonal dynamics. The aim of this chapter is to review and evaluate this forecasting strategy based on our understanding of monsoonal variability on the sub-seasonal scale in the extended range (forecast extending up to 3–4 weeks of lead time, but more than medium range, which is, about 10 days). Sub-seasonal variability during the monsoon season (June–September) displays a broad spectral range from synoptic scale (<7 days) to intraseasonal scale (about 90 days), and forecasts in different scales require different modeling strategies. This chapter will focus on the extended range.

The life cycle of the ISM has a prominent seasonal cycle with a marked onset and subsequent northward propagation of rainfall during June–September (i.e., JJAS). This northward propagation is an integral part of the sub-seasonal variability of ISM rainfall (ISMR). The enhanced (subdued) precipitation associated with the northward propagation of intraseasonal rainbands manifests as the active (break) spells of ISMR, and the first spell of this northward propagation marks the onset of ISM (Goswami, 2005). Although the date of onset of ISM is not correlated with the subsequent progression of the monsoon or the seasonal mean, its timing has a considerable impact on the agricultural productivity and in turn on the Indian economy (Gadgil and Rupa Kumar, 2006).

Conversely, the duration and frequency of active/break cycles largely determine the sign of the seasonal mean monsoon anomaly (i.e., above normal, normal, or below normal). Therefore, the predictions of monsoon onset and active/break cycles well in advance are of great importance. Although the prediction of monsoon withdrawal during September is also important for agriculture, it is not discussed here in the absence of a clear and well-defined index of withdrawal.

To predict the monsoon rainfall variability over the Indian region, it is customary to define the Monsoon Zone of India (MZI), designated by the Indian Meteorological Department (IMD) (Rajeevan et al., 2010) as comprising most of central India and adjoining regions (Fig. 1A), which experience the monsoon intraseasonal fluctuations. A spatially averaged daily index of MZI rainfall is used to optimally identify the active/break spells of rainfall.

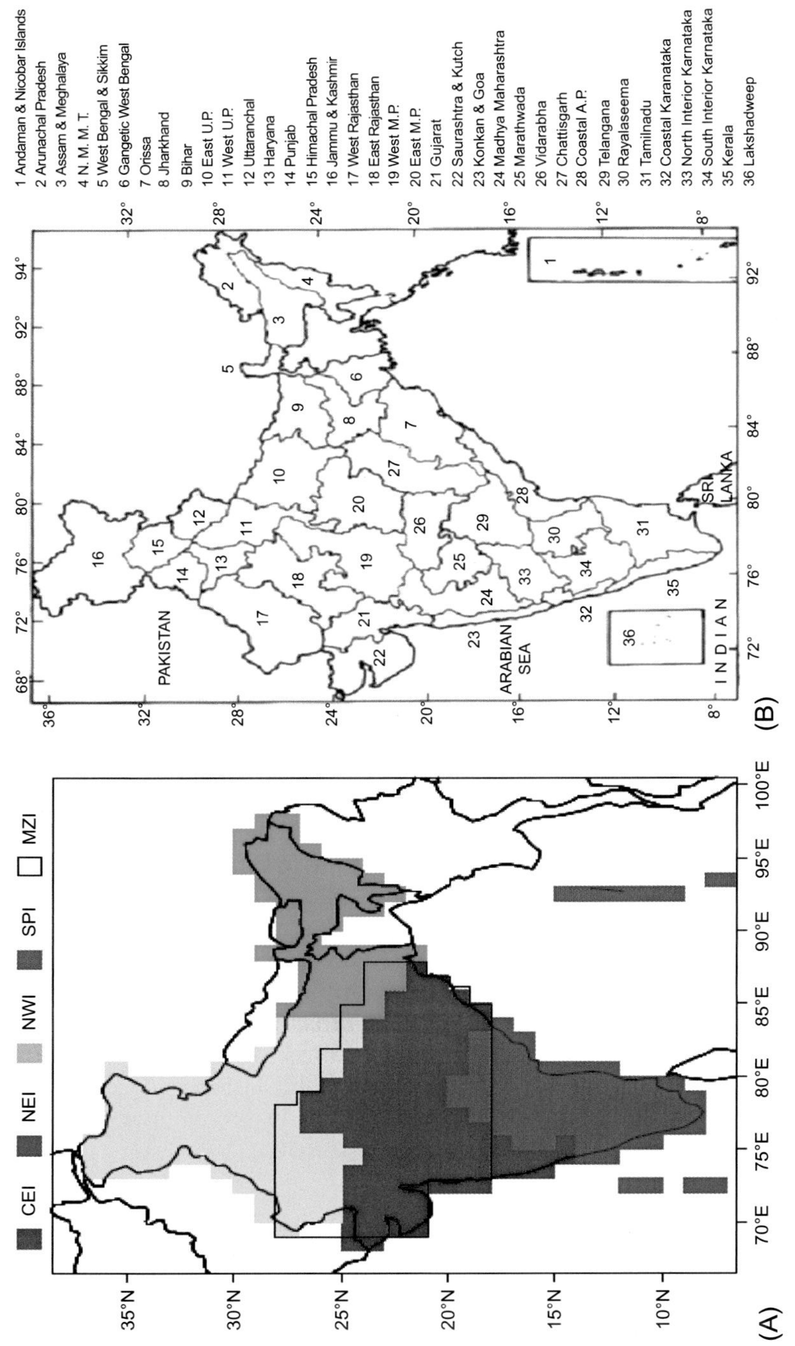

FIG. 1 (A) Homogeneous regions and (B) meteorological subdivisions, as defined by IMD.

In addition to the temporal variability, ISM is characterized by spatial inhomogeneity. Even though the composite peak active or break pattern has a spatial structure in terms of rainfall distribution, it differs from event to event (Chattopadhyay et al., 2008), and several individual events do not conform to this structure. During normal monsoon years, some parts of India receive higher rainfall and other parts receive lower rainfall; the spatially averaged All-India time series in such cases smooths out the regional variability. Dynamical sub-seasonal forecasts should be tuned to capture the regional rainfall patterns across several zones, and the multiweek lead time of these forecasts would help increase the preparedness toward large and small rainfall events. Prediction of rainfall at the highest possible spatial resolution and at multiple lead times from the short range to the seasonal scale during the JJAS monsoon season is currently operationally tested by IMD.

The importance of sub-seasonal, extended-range forecasts was primarily appreciated by the hydroclimatic and agrometeorological stakeholders when the experimental forecasts were released earlier in pseudo-operational mode. Skillful forecasts of onset date and the active/break spells at multiple spatial scales could provide essential information to the agrarian community and water managers about the pattern of decision support systems to be implemented in case of likely extremes forecasted over a small region. An example with regard to onset date may be helpful. The onset is awaited eagerly by the farming community, whose arrival and intensity over Kerala gives a confidence boost regarding the sowing pattern, crop selection, watering pattern, pesticide spraying pattern, etc. The lead time of 2–3 weeks provides sufficient lead time for alternative planning and solutions in the agricultural and health sectors (e.g., selection of alternate crops).

It is a well-known fact that the skill of the forecast is higher if the region considered is larger, while the demand for forecasts has risen tremendously for shorter spatial scales at all lead times. While the Indian landmass is divided into five homogeneous regions (Fig. 1A) to facilitate the large-scale forecasts at the seasonal and extended range, there is now a demand for forecasts from the smallest administrative division in India, which include urban municipalities and rural blocks (aka *tehsils, Mandals,* or *talukas*). Also, district-level forecasts (equivalent to a few blocks) are sought from time to time for various reasons. IMD, on the other hand, issues forecasts for meteorological subdivisions (almost comparable to a few districts; Fig. 1B) to help the state and local agencies that manage floods, droughts, and other weather emergencies. Thus, current-generation forecasts are required to be tailor-made and be tuned to serve operational requirements on multiple spatiotemporal scales in a seamless manner for administrative units that may be smaller than the meteorological subdivisions. In this chapter, the skill of an indigenous extended-range forecast system will be discussed starting from large scale and then slowly zooming to subdivision scale to assess the practical feasibility seamless monsoon prediction with the currently available resources.

2 EXTENDED-RANGE FORECAST OF MONSOON SUB-SEASONAL VARIABILITY

The sub-seasonal predictability of summer monsoon rainfall over the Indian region, as well as over several tropical domains in southeast Asia, is quite well established, and operational prediction has been attempted by many studies owing to the existence of low-frequency sub-seasonal variability beyond synoptic scales. Prediction of the sub-seasonal rainfall spells on various spatial scales in a seamless manner has several operational interests, especially for flood and disaster forecasting and agricultural planning over the Indian subcontinent. Thus, in the past decade operational prediction in the sub-seasonal scale became a necessity, in addition to theoretical knowledge. Statistical prediction models of monsoon intraseasonal oscillations (MISOs) and tropical intraseasonal oscillations generally have been well established over the past 20 years (Waliser et al., 1999; Jones et al., 2000a; Xavier and Goswami, 2007; Chattopadhyay et al., 2008; Jiang et al., 2008). Its importance was realized by the application of the forecasts in hydrometeorological and agrometeorological domains. One of the earlier statistical, extended-range forecasts specifically deals with hydrological forecasting (Webster and Hoyos, 2004), based on the implicit assumption of the periodicity of MISO in the 40-day range. Later, other linear and nonlinear models were introduced for real-time predictions (Xavier and Goswami, 2007; Chattopadhyay et al., 2008). These statistical prediction models have led to increased optimism about the dynamical prediction on the sub-seasonal scale in recent years.

Earlier works using dynamical models (Waliser et al., 2003; Liess et al., 2005) have shown that the limit of potential prediction skill in the sub-seasonal range could be extended beyond 15 days in advance using dynamical models. In recent years, several studies validated this assessment of the skill of sub-seasonal forecasts. Liu et al. (2014a) indicated that the NCEP Coupled Forecast System version 2 (NCEP CFSv2) showed reasonable skill at sub-seasonal prediction of summer monsoon rainfall over several tropical Asian ocean domains.

Sub-seasonal predictability also has been studied in the context of prediction of the Boreal Summer Intraseasonal Oscillation (BSISO) modes. Lee et al. (2015) examined the predictability and prediction skill of BSISO over the Asian monsoon region in the Intraseasonal Variability Hindcast Experiment (ISVHE), and found the multimodel mean BSISO predictability and prediction skill with strong initial BSISO amplitude are about 45 and 22 days, respectively. Recently, Neena et al. (2017) evaluated the BSISO prediction skill based on 20 years of hindcasts using 27 general circulation models (GCMs). Many of these GCMs showed BSISO propagation, but significant biases were present in the precipitation simulations. The northward and eastward propagation were captured with better fidelity in most of these hindcasts. Both studies indicate that there is considerable room for improvement in BSISO prediction.

As part of India's National Monsoon Mission (NMM) program toward improving Indian monsoon region prediction capabilities, a multimodel ensemble (MME) prediction system has been developed at the Indian Institute of Tropical Meteorology (IITM) and is now operational at IMD. The dynamical model is derived from an operational climate model developed at the National Centers for Environmental Prediction (NCEP) in the United States for the extended range (2–3 weeks). For generating the MME, we used a version of CFSv2 (Saha et al., 2006, 2014), run at two horizontal resolutions: T126 (110 km) and T382 (38 km). The MME also consisted of the stand-alone atmospheric component of CFSv2, the Global Forecast System (GFS) model, also run at T126 and T382 horizontal resolutions. GFS is coupled to an ocean model, a sea ice model, and a land surface model. For its ocean component, the CFSv2 uses the GFDL Modular Ocean Model, version 4p0d (MOM4; Griffies et al., 2004). The stand-alone GFS, with slightly different physics options than CFSv2, is forced with daily, bias-corrected, forecasted sea surface temperatures (SSTs) from CFSv2. Bias correction involves subtracting the mean bias as a function of a calendar day and lead time, with

Optimum Interpolation Sea Surface Temperature (OISST) observations (Reynolds et al., 2007) as the reference. We denote this two-tier, stand-alone GFS forecast as *GFSbc*, with *bc* indicating bias-corrected boundary conditions (Abhilash et al., 2015).

Choosing a proper method to generate initial conditions is a prerequisite for MME prediction. Although there are several approaches to generate ensembles of different initial conditions, we use a method similar to the complex-and-same-model environment group, as classified in Buizza (2008). An ensemble of perturbed atmospheric initial conditions is developed, as well as an actual initial condition. The actual (i.e., unperturbed) initial condition is prepared from NCEP's coupled data assimilation system (CDAS) with T574L64 (about 23-km horizontal resolution, with 64 vertical levels) resolution atmospheric assimilation and MOM4-based oceanic assimilation, a real-time extension of the CFSR (Saha et al., 2010). These are generated in real time and made available for operational implementation by the National Centre for Medium-range Weather Forecasting (NCMRWF), New Delhi, and the Indian National Centre for Ocean Information Services (INCOIS), Hyderabad, through an ongoing collaboration. Each ensemble member's forecast was generated by perturbing these actual initial climatic conditions. We perturb the wind, temperature, and moisture fields, and the amplitude of perturbation for all variables are consistent with the magnitude of the variance of each variable at a given vertical level. More details about the ensemble generation techniques can be found in Abhilash et al. (2014). More model and experimental details and skill assessments of GFSbc, CFST126, and CFST382 can be found in Abhilash et al. (2013, 2014) and Sahai et al. (2015).

The MME, the initialization, and the modeling strategy are created in this way to increase model diversity within available computing resources. This MME uses different variants of the CFSv2 and its atmosphere component, GFS, with different resolutions, parameters, and coupling configurations (to address coupled SST biases), which is motivated by the various physical mechanisms thought to influence monsoon forecast errors in the extended range (i.e., 10–20-day lead times) using the same dynamical core. Based on performance tests, and aiming to maximize the operational skill for our available computer resource, a 16-member forecast ensemble was chosen for operational purposes. Using CFS, we ran 4 members of CFST126 (about 100 km) and 4 members of CFST382 (about 38 km). Similarly, for the GFS, we ran 4 members each from GFSbc at two resolutions, forced with bias-corrected, forecasted SST from CFS. IMD adopted this system for operational, extended-range prediction (ERP) over the Indian region from the 2017 monsoon season (see Fig. 2 for a flowchart of the model and prediction strategy). The 4-week lead-time forecast is operationally made once a week (specifically, every Wednesday) at present.

Various customized forecast products are being disseminated from this ERP, which includes extended-range prediction of active-break spells of ISM, monsoon onset, progression, withdrawal, heat and cold waves, monitoring of MISOs and Madden-Julian Oscillations (MJOs), cyclogenesis, and many other events. Now, this ERP system is capable of generating extended outlooks for various sector-specific applications, such as agriculture, hydrology, energy, insurance, reinsurance, urban planning, and health. Fig. 3 shows the schematic of the end-to-end forecast and dissemination system that was developed and implemented for operational ERP systems. Because these sectors rely on forecasts on various spatial and temporal scales, a seamless sub-seasonal forecast scheme spanning multiple spatial and temporal scales is required for operational reasons.

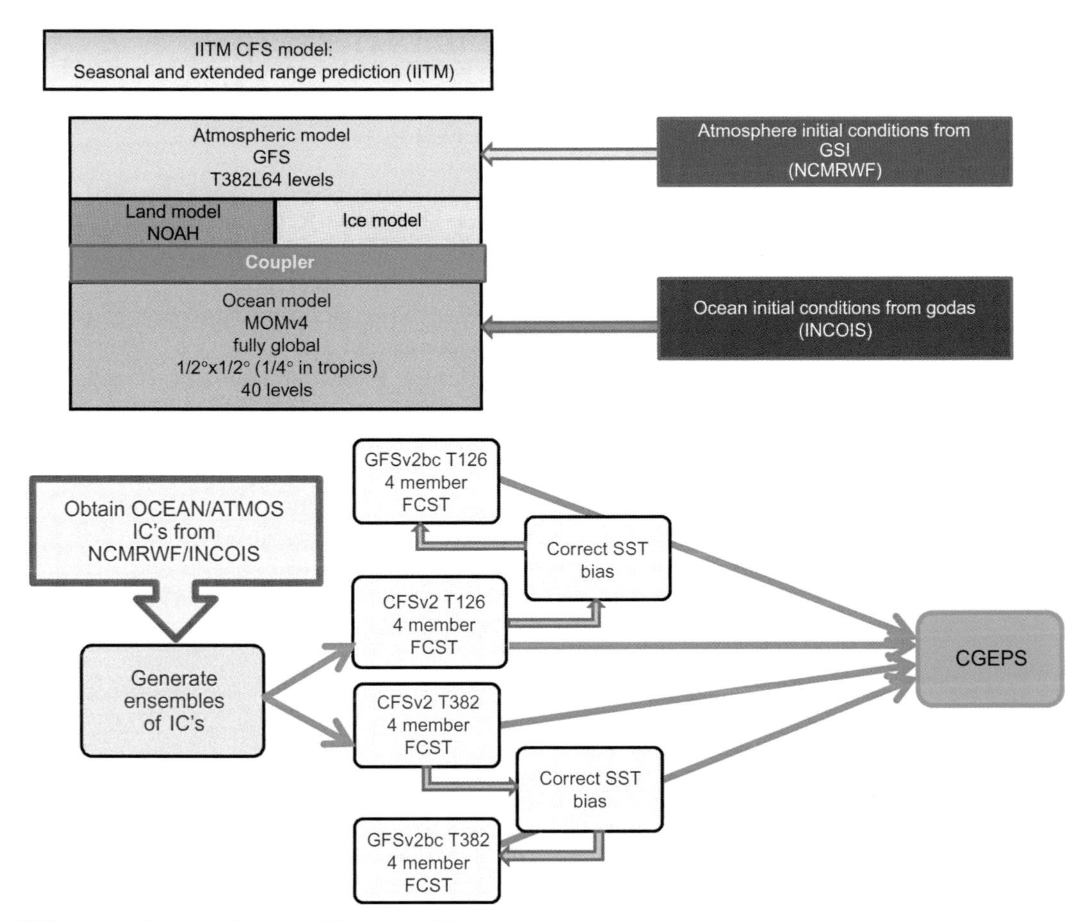

FIG. 2 A schematic diagram of the recent IMD forecast system.

The forecast scheme has shown promise and fits into a seamless forecast framework spanning from the larger Indian homogeneous regions down to the smaller block levels. The modeling framework and its applications has been reported on in detail in several studies and shows improvement over other models (Abhilash et al., 2014; Pattanaik and Kumar, 2014; Sahai et al., 2015; Joseph et al., 2015). It compares well with the state-of-the-art ECMWF medium-range forecasting schemes when larger homogeneous regions are considered (Chattopadhyay et al., 2017)). The skill of forecasting at different spatial scales during the 2003–14 hindcast period will be described in the remaining part of this chapter. Operational predictions of large-scale conditions for recent years also will be shown based on these hindcasts. A large to small-scale verification approach will be taken, in which forecasts of large-scale features will be described first, and then forecast skills for smaller and smaller spatial scales will be presented.

3 MONSOON ONSET AND IDENTIFICATION OF ACTIVE/BREAK SPELLS

3.1 Criteria for Monsoon Onset Over Kerala (MOK)

The onset of ISM happens over the southern peninsular state of Kerala, and hence Monsoon Onset over Kerala (MOK) has tremendous operational importance. Deterministic forecasts have been generated in real time for the last few years. The criteria for MOK have been defined from model hindcasts (Joseph et al., 2015) and forecasted MOK for the period 2001–14, as shown in Table 1. The main characteristics that differentiate the monsoon rains from the premonsoon thundershowers (i.e., the seasonal reversal of the low-level winds, the persistence of both rainfall and low-level wind after the MOK date) have been considered while formulating the criteria to avoid the occurrence of "bogus onsets" that are unrelated to the large-scale monsoon system. The onset dates (MOK) based on the rainfall over Kerala (ROK) and the strength of the zonal wind over the Arabian Sea (uARAB) are plotted in the left panel of Fig. 3 based on forecasts run from May 16 initial conditions for two selected years (2003 and 2005). Also, composites of wind and rainfall patterns based on the MME forecast ensemble means are shown in the right panel of Fig. 4 and are compared with observations (OBS) during MOK for the 14-year period. The plot indicates that the MME has good fidelity in matching the observed spatial pattern during the onset phase.

TABLE 1 Comparison of IMD Declared and Model Forecasted MOK for 2001–14

Year	Actual MOK	Forecasted MOK	Standard Deviation Among Ensemble Members	Difference Between Actual and Forecasted MOK
2001	May 23	May 25	2	2
2002	May 29	May 21	5	8
2003	Jun 8	May 30	5	9
2004	May 18	May 18	1	0
2005	Jun 5	Jun 5	3	0
2006	May 26	May 25	2	1
2007	May 28	Jun 2	8	5
2008	May 31	Jun 2	7	1
2009	May 23	May 24	2	1
2010	May 31	May 30	5	1
2011	May 29	Jun 1	2	3
2012	Jun 5	Jun 4	4	1
2013	Jun 1	May 29	2	3
2014	Jun 6	Jun 5	6	1

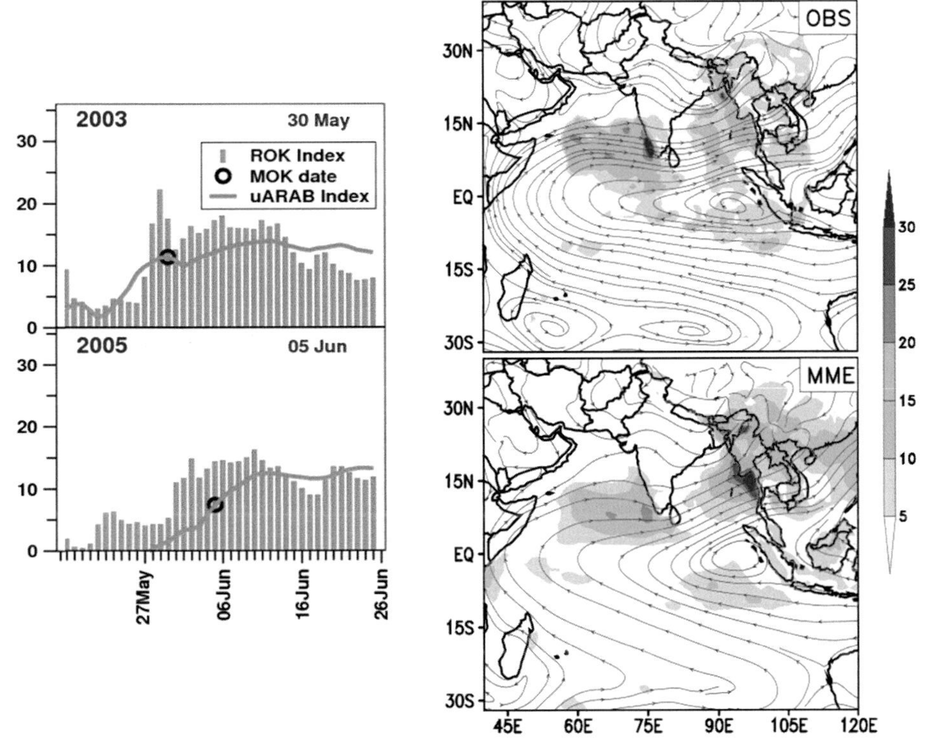

FIG. 3 An example of extended-range prediction of MOK for selected years is shown in the left panel. The composite of rainfall (mm/day) and 850 hPa wind (m/s) during MOK day of 2001–14 in observations and MME is shown in the right panel.

3.2 Active/Break Spells Associated With MISOs

MISOs, with 20–80-day periodicity, are large-scale systems that require particular attention, as they produce rainfall over substantial parts of central India and play a dominant role in the maintenance of monsoon troughs. For predicting and tracking MISOs, we use the MISO index, defined by Suhas et al. (2012) based on an extended empirical orthogonal function (EEOF) analysis of daily rainfall, following the method proposed by Wheeler and Hendon (2004) for MJO monitoring and verification. The scatterplot of the first two principal components (PCs), MISO1(t) and MISO2(t) (where t is the running time), of EEOFs gives the qualitative geographical location of convective clouds in the north-south direction (i.e., from the equator to the foothills of Himalaya).

Based on the phase location, we define the monsoon active and break spells associated with MISO. The active spells are identified based on the location of a point (*MISO1, MISO2*) in the scatterplot. If the normalized MISO amplitude (with respect to the standard deviation), i.e., $\sqrt{(\mathrm{MISO1}^2(t) + \mathrm{MISO2}^2(t))}$ is >1, and the phase point [i.e., *(MISO1(t), MISO2(t)*] lies in a location of the phase space where the convection is located over central India, which broadly represents the MZI the calendar date corresponding to the phase point is taken as an active

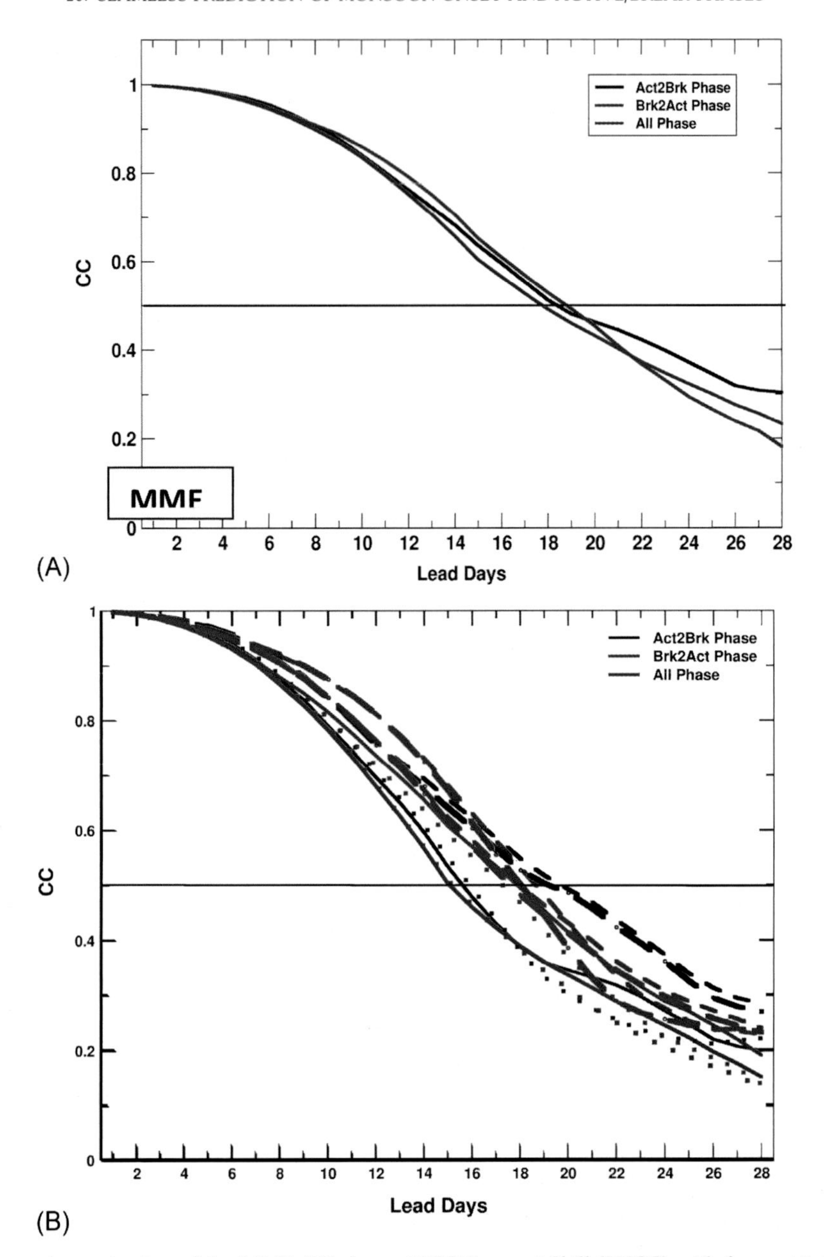

FIG. 4 Phasewise evaluation of the MME: (A) shows MISO Forecast Skill (BVCC) with forecast initialized from days clustered in Active phases (Act2brk), Break phases (Brk2act) and ALL phases for the MME. (B) shows the same skill for ensemble means of individual components. These results are based on 2003–14 hindcast runs for CFST126 *(solid)*, CFST382 *(dot)*, GFST126 *(dash)*, and GFST382 *(dot-dash)* models.

day. Similarly, based on the seesaw pattern of the active and break phases, when the convection is over the Indian Ocean, it is defined as a *break phase* over central India. It is essential for extended-range models to display skill in predicting these spells on this timescale. Additionally, deterministic as well as probabilistic forecasts of heavy rainfall events at longer lead times are desired by users. The active and break phases, of which the forecasts will be shown in Section 4.1, are identified based on this criteria. Although the estimates may have some spatiotemporal errors for heavy rainfall events, it still can be used as guidance for the impending events.

4 DEMONSTRATION OF SEAMLESS SUB-SEASONAL PREDICTION

4.1 Phase-Dependent Skill of Large-Scale MISO Indices

The skill of MISO forecasting starting from different MISO phases of the initial atmospheric states are presented in terms of bivariate correlation coefficients (BVCCs) in Fig. 4. BVCC(τ) is defined as follows (following Rashid et al., 2011; see Chapter 16 on forecast verification):

$$\mathrm{BVCC}(\tau) = \frac{\sum_{t=1}^{N}[a_1(t)b_1(t,\tau) + a_2(t)b_2(t,\tau)]}{\sqrt{\sum_{t=1}^{N}[a_1{}^2(t) + a_2{}^2(t)]}\sqrt{\sum_{t=1}^{N}\left[b_1{}^2(t,\tau) + b_2{}^2(t,\tau)\right]}}.$$

Here, $a1(t)$ and $a2(t)$ refer to the observed MISO1 and MISO2 time series at time t, and $b1(t,\tau)$ and $b2(t,\tau)$ are the respective forecasts for time t for a lead time of τ days; N is the number of forecasts. The initial atmospheric states are classified based on the observed MISO phases—namely, active to break (Act2Brk), break to active (Brk2Act), and initial conditions (ICs) starting from all phases (All). The MISO forecast skill in terms of BVCCs for the MME and other participating single-model ensembles (SMEs) shows that the limit of useful prediction, which is measured by a threshold correlation coefficient (CC) value of 0.5, for the MME is about 18 days for All IC phases (blue curves in Fig. 4). The forecast skill of MME is not significantly different for Act2Brk and Brk2Act ICs (black and red curves). The CC values of both CFST126 and CFST382 drop below 0.5 by 16 days for All ICs and the limit of useful prediction extended slightly to 18 days for Brk2Act ICs. However, all three ICs show an extended prediction skill of 18–19 days for GFSbc at both T126 and T382 resolutions. The skill is comparable to the BSISO skills of ECMWF and BoM forecast models, as shown in Fig. 8 of Jie et al. (2017).

The All-phase BVCCs of MME and the participating models [i.e., the atmospheric (GFS) and coupled (CFS) for the selected years 2013, 2014, 2015, and 2016] are given in Fig. 5. The correlation skill of MISO indices forecast at different lead times varies from year to year, indicating the substantial year-to-year variability and extreme-year variability. The MME, as well as the individual models, show negligible variation during 2015, whereas the maximum difference is seen during 2016. In any case, MME skill represents the best possible scenario when all years are considered (see Chapter 15 for a more general discussion of the multimodel approach).

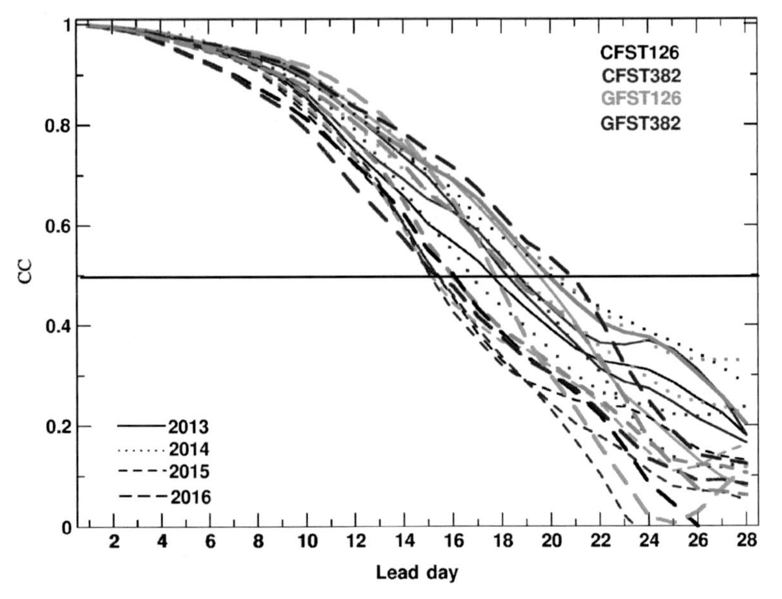

FIG. 5 BVCC of MISO indices for CFST126 *(black)*, CFST382 *(red)*, GFST126 *(light green)*, GFST382 *(blue)*, and MME *(dark green)* for 2013 *(line)*, 2014 *(dot)*, 2015 *(short dash)*, and 2016 (long dash) of operational data. Climatologies are based on 2003–14 hindcast data.

4.2 MISO Forecast Ensemble Spread Versus RMSE

The ensemble spread and root-mean-squared error (RMSE) at a given forecast lead time measures the fidelity of the MME forecast: for a well-calibrated forecast, these should be equal. The BVCC, spread, and bivariate RMSE of MME and individual models, as a function of lead time, are shown in Fig. 6. The bivariate RMSE is defined using the same nomenclature as BVCC:

$$\mathrm{RMSE}(\tau)=\sqrt{\frac{1}{N}\sum_{t=1}^{N}\left[(a_1(t)-b_1(t,\tau))^2+(a_2(t)-b_2(t,\tau))^2\right]}.$$

It may be seen that the MME BVCC skill remains above the significance level for up to 18 days, achieving the objective of extended-range prediction (Fig. 6A). From Fig. 6B, it is clear that the RMSE (solid line) has improved and the spread (interensemble standard deviation, shown as a dotted line) has increased in MME compared to the individual models. This indicates that the MME is superior to the individual models and the skill is improved by the combination of the variants of the same model. This improvement in the skill of forecasting is attributed to the MME, which is based on a single-model dynamical core, with multiple configurations of physical parameterizations, model resolutions, and boundary- and initial-value bias corrections. The multimodel combination improves the spread among the ensemble members, as seen in Fig. 7, leading to improvement in skill. Thus, the MME formulation provides better (i.e., skillful) prediction of intraseasonal oscillations during the monsoon season.

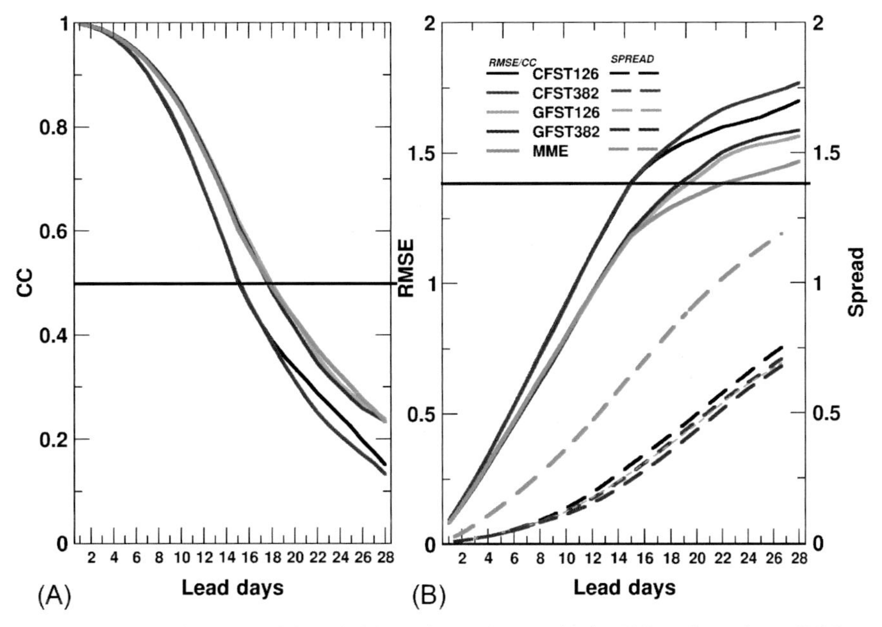

FIG. 6 Bivariate (A) correlation BVCC and (B) RMSE and spread of MISO indices from CFST126, CFST382, GFST126, GFST382, and MME. These results are based on 2003–14 hindcast data.

4.3 Forecast Skill of Five Homogeneous Regions

Although the large-scale forecast is useful for monitoring MISOs, it cannot provide information on the regional rainfall distribution pattern over Indian regions. Also, because day-to-day rainfall forecasts cannot currently be extended beyond a few days to a week, the forecast and the observations are averaged on a weekly basis, and the skill is described over weekly means. Based on the rainfall distribution and regional climatology, the Indian landmass is divided into five homogeneous regions (see Fig. 1A). The hindcast correlation skill for weekly data for individual component models and MME over various homogenous regions is shown in Fig. 7. It may be seen that while the correlation skill is similar for MME and individual-component models in the first week (i.e., forecasts averaged from days 1–7), the MME forecast skill is comparable to (or sometimes better than) the skill of GFST382 at longer lead times. Hence, the MME forecast improves the skill in the extended range. The forecasts have little skill with the South Peninsular India (SPI), and Northeast India (NEI) areas at longer lead times. However, for the central Indian region in particular, agriculture and water management activities could potentially benefit from these forecasts.

4.4 The Forecast Skill of Active and Break Spells for Meteorological Subdivisions

As there is increasing demand from the user community for forecasts on smaller spatial scales, it is essential to assess the skill of the MME over various meteorological subdivisions of India, so as to understand the confidence level in utilizing the forecasts over these regions.

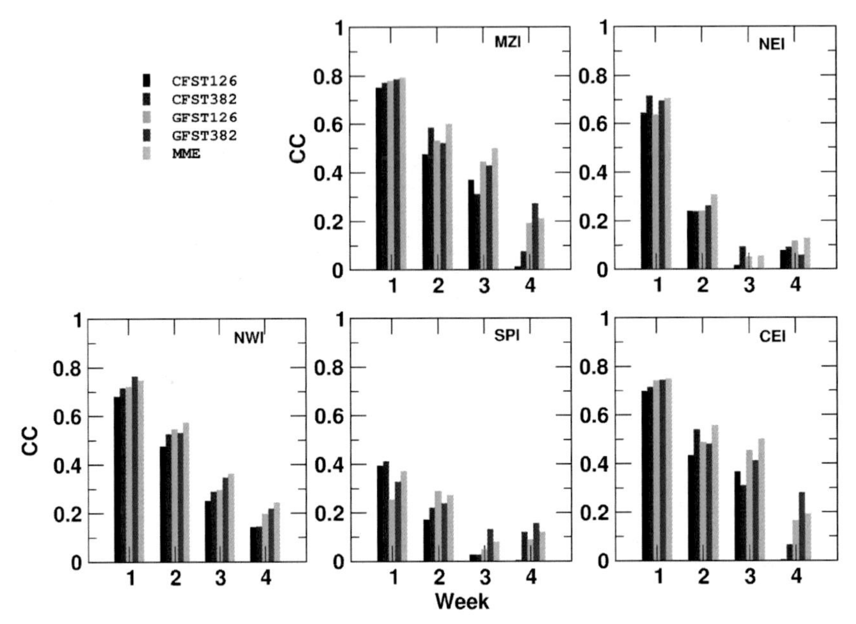

FIG. 7 The deterministic correlation skill (CC) of weekly lead prediction of area-averaged rainfall with observations for the hindcast period (2003–14) over the homogeneous zones of India.

Fig. 8 shows the anomaly correlation coefficient (ACC) of rainfall for the hindcast period for all pentads during JJAS. Correlations significant at the 99.9% significance level are shaded on the maps. It is clear that the MME has remarkable skill over almost the whole country except for the Tamil Nadu region (which is not much affected by monsoon rains in JJAS) in week 1 lead. Although the skill reduces with increased lead time, most parts of the country have considerable skill until week 3 lead. This provides a good sign that this MME can be used for operational purposes at smaller spatial scales.

4.5 Feasibility of MME Prediction to Further Smaller Spatial Scales

Although operational demand is high for very-high-resolution prediction, it is essential to investigate the operational predictability before giving a forecast to very high-resolution spatial scale at the block or *taluka* level. The Fractions Skill Score (FSS; Mittermaier and Roberts, 2010; Mittermaier et al., 2013; Sahai et al., 2015) determines the spatial scale of predictability for a given forecast lead time. The FSSs for the current-generation forecast system for different lead times for the above-normal (i.e., active) and below-normal (i.e., break) tercile categories are shown in Fig. 9. The abscissa indicates the grid-box size ($1° \times 1°$, $2° \times 2°$, $3° \times 3°$, etc., indexed as 1, 2, 3, …). The reference (target) skill is taken as 0.67, below which it is insignificant. The target is defined as $FSS_{target} = 0.5 + \frac{f_0}{2}$, where f_0 is the base value [=0.33 for tercile (active, normal, and break) categories (Sahai et al., 2015)]. For week 1 (days 1–7 lead time) forecasts, 2×2 or 3×3 degrees (where the FSS curve crosses the target) is the most skillful location. As the lead time increases, the spatial scale when the forecast is skillful increases, which is expected. The MISOs can be forecasted at longer lead times but cannot be forecasted

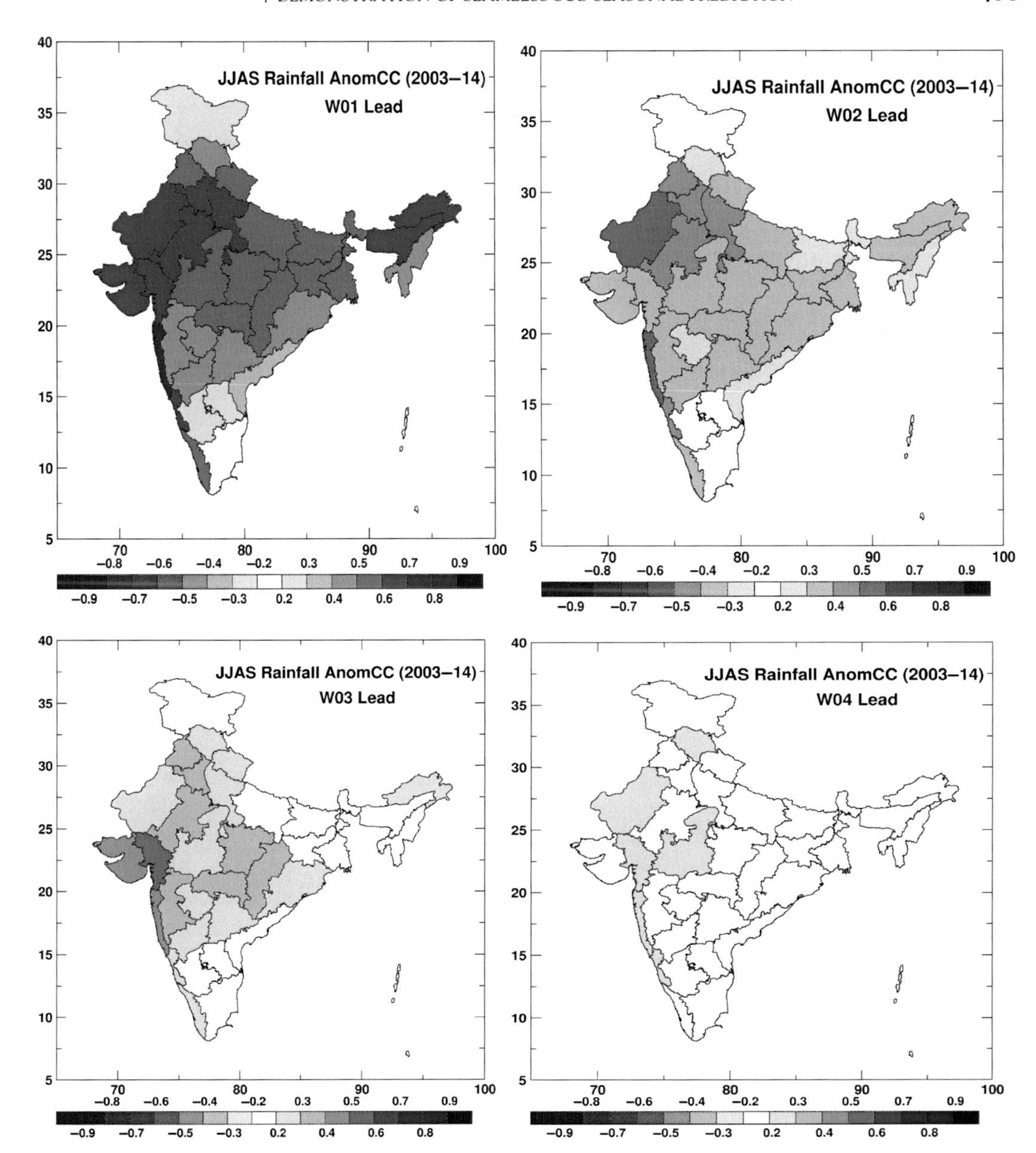

FIG. 8 The deterministic skill of weekly lead prediction of area-averaged JJAS rainfall with observations for the hindcast period over the meteorological subdivisions of India.

at smaller spatial scales at longer lead times. A comparison of active and break categories (left and right panels) shows that the spatial scale of the skillful active forecast is much smaller than the spatial scale of the skillful break forecast. Thus, to forecast at the district ($2° \times 2°$ grid size) or the block level ($1° \times 1°$) week 1 forecast is still the best, as compared to longer lead times.

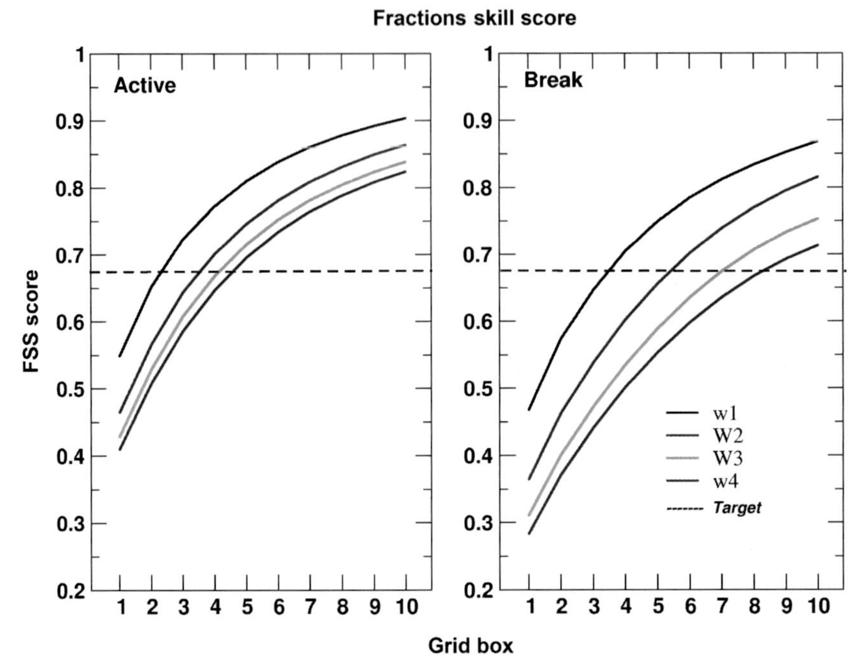

FIG. 9 FSS for the JJAS Active and Break Spells during the 2003–14 hindcast period.

4.6 Application of MME to the Forecast of Extreme Events: An Example

Any unexpected, unusual, and unseasonal weather event, such as heavy rainfall, heat or cold waves, or cyclones, could be defined as an extreme weather event. If such events are modulated by large-scale forcing, their probabilities ought to be predictable over the extended range. The application of the MME in predicting the heavy rainfall event that occurred in June 2013 over the north Indian state of Uttarakhand is shown in Fig. 10. The solid black line

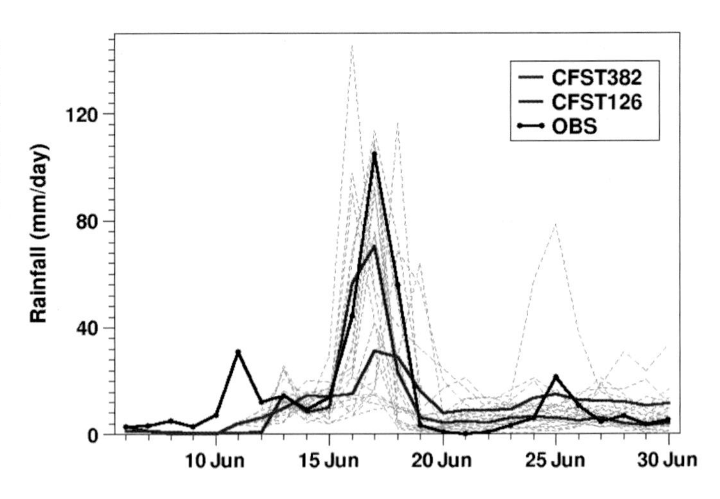

FIG. 10 Example of extreme weather forecasts for the Uttarakhand heavy rainfall event of June 2013. The *dotted line* shows the evolution of rainfall from individual ensemble members of both CFST382 and CFST126, and the *solid red (blue) line* shows the same for the ensemble mean of CFST382 (CFST126).

with dots represents the observed rainfall amount over Uttarakhand region (78°–80°E; 29°–31°N), while the solid red (blue) line indicates the T382 (T126) ensemble mean forecast intialized from June 5, 2013, IC, and the dashed brown lines indicate individual ensemble members from T382 (11 members) and T126 (11 members), run from the same date's IC. It is clear that the T382 MME predicted the event at least 10–15 days in advance, while T126 MME grossly underestimated the rainfall magnitude, suggesting the need for adequately high model spatial resolution.

5 CONCLUSIONS

5.1 Prospects and Problems

This chapter reviewed the extended-range prediction skill of an MME-based ensemble prediction system, developed at IITM and recently implemented operationally at IMD. A few examples in the discussion showed that a spatially seamless sub-seasonal prediction made with the state-of-the-art IMD prediction system can be operationally deployed, although its skill requires a good deal of improvement at the block or *taluka* level, which roughly corresponds to a spatial scale less than 1° x 1° grid box.

It was found that the MME has good fidelity in predicting the rainfall during the monsoon season on the larger spatial scale. The active and break phases based on large-scale indices (i.e., MISO1 and MISO2) showed significant correlation skill 15–20 days in advance. The MISO amplitude has a correlation skill above the significance level for up to 15 days of lead time. The active and break phases showed similar variations in prediction skill in the current-generation models. The forecasts over homogenous/subdivision regions indicate that most regions in the plains of central India can have predictability up to a 3-week lead time, while the South Peninsular and Northeast Indian regions have a lead time of 1–2 weeks with the current-generation forecast system.

The low skill over some regions could be related to the fact that the distribution of rainfall during the northward-propagating pulse of the MISO does not follow a simple periodic pattern with fixed spatial structure; instead, it is an evolving convective system with multiscale structure. Experiments with convective parameterizations and microphysics scheme, for example, showed that the skill could vary widely under different schemes, as the representation of convective systems in the monsoon region is extemely complex. It is recommended, therefore, that the MME model be improved through the inclusion of multiphysics options (e.g., using multiple convective parameterizations at multiple spatial resolutions). Also, the role of the diurnal cycle in the coupled SST runs is not clear, as CFSv2 has a significant bias when it comes to representing a proper diurnal cycle of convection.

5.2 Future Directions for Spatially Seamless Sub-seasonal Prediction

While the experiments with model physics and model dynamics could be continued for research purposes, the demand of skillful operational predictions over smaller spatial scales is increasing. It is recommended, therefore, that the model output should be postprocessed further to improve regional forecasts, especially for the extreme rainfall events that affect smaller regions. Such extreme events are often associated with the clustering of synoptic-scale systems with large-scale MISOs. To obtain highly localized forecasts, the GCMs must have a very high spatial resolution, which requires substantial computational resources. In such cases, downscaling of the low-resolution model based on statistical and dynamical models could be a way to provide useful forecasts. The focus of IMD and Indian research agencies is now to improve the downscaling methods to find out how the regional and microlevel forecasts could be improved. It is unlikely that a single method of downscaling would work. Hence, a suite of statistical and dynamical downscaling techniques is essential to fulfill the demands of stakeholders. A probabilistic downscaled forecast with a decision support system has to evolve in order to improve regional monsoon forecasting for the Indian subcontinent.

Acknowledgments

IITM is fully supported by the Ministry of Earth Sciences of the Government of India, New Delhi. We thank NCEP for analysis data sets and technical support on the CFS model. We also thank IMD for providing TRMM and rain-gauge-merged daily rainfall data for the study period, and NCMRWF and INCOIS for providing real-time operational and past ICs for retrospective model forecasts.

CHAPTER

21

Lessons Learned in 25 Years of Informing Sectoral Decisions With Probabilistic Climate Forecasts

Rafael Terra, Walter E. Baethgen*[†]

[*]Department of Fluid Mechanics and Environmental Engineering, School of Engineering, Universidad de la República, Montevideo, Uruguay [†]International Research Institute of Climate and Society (IRI), Columbia University, Palisades, NY, United States

OUTLINE

Sub-seasonal to Seasonal Prediction
https://doi.org/10.1016/B978-0-12-811714-9.00021-8

1 INTRODUCTION

Human endeavors always have been exposed to the perils of climate variability and societies have developed a variety of mechanisms to cope with such events since ancient times. Since the early 1980s, the science of predicting climate fluctuations at the seasonal timescale has made major improvements, and nowadays, most prediction centers routinely include seasonal climate forecasts (SCFs).

Given the wide array of activities of large socioeconomic importance that are greatly affected by climate variability, there was a general expectation that these climate forecasts would be smoothly integrated into the decision-making process and lessen the burden of climate risk. However, the assimilation of probabilistic climate forecasts into actual decisions proved to be more challenging than expected, and it has not always been successful (Lemos, 2003).

From the climate science perspective, the attention naturally has been directed toward the characteristics of the climate forecasts per se, including their lead time, skill, used predictors, and physical causality. The format for presenting the forecasts, which are inherently probabilistic, also has been a focus of interest. Seasonal forecasts thus have been expressed using probability of terciles (above, below, or close to normal), considering the full expected distributions, etc. All these issues are crucial when attempting to introduce climate forecasts into actual decisions. In this chapter, however, we focus on the process of embedding climate forecasts into decision processes, identify and describe some difficulties usually encountered, give a couple of examples to better visualize them, and propose options to best approach the task. The chapter is based on the authors' 25 years of experience and lessons learned in the process of embedding SCFs in actual planning and decision-making in various socioeconomic sectors.

Research conducted on establishing climate change adaptation initiatives (Lesnikowski et al., 2011) and on building climate services (Vaughan et al., 2017) identified a common set of activities that usually progress in a somewhat orderly way to create the necessary groundwork that may include (1) establishing adequate institutional arrangements, (2) identifying champions within those institutions that exercise leadership in the use of climate information, and (3) promoting constructive collaborations between the climate and sectoral communities that result in a productive critical mass of actors directly involved in both, the generation and use of climate information.

In the last few years, the international climate science community started investing significant efforts in exploring the predictability of climate at the sub-seasonal to seasonal (S2S) temporal scale, that is, the gap between medium-range weather forecasts (up to 2 weeks) and seasonal outlooks (3–6 months). Concurrent efforts are ongoing to investigate the utility and relevance of S2S forecasts for informing decisions and planning in operational mode (Vitart et al., 2012b). We propose that the process of integrating S2S climate forecasts into actual decision-making, planning, and the elaboration of policy will present similarities to the process of embedding SCFs that has been going on during the last few decades.

This chapter discusses lessons learned in the endeavor of incorporating SCFs in sectoral decisions and plans, with the expectation that some of those lessons will inform similar applications of S2S forecasts. Section 2 provides an overview of the aspects of the decision process within institutions that need to be fully understood before any attempt is made to embed climate forecasts, a step that is addressed in Section 3. Section 4 presents two examples of long-term processes in which climate information was slowly intertwined in sectorial decision-making. Lessons learned and conclusions are drawn in Section 5.

2 LEARNING AND UNDERSTANDING THE STATUS QUO

Before exploring the process of integrating climate forecasts into decisions, we should realize that there is always a preexisting way of making decisions that has been shaped by experience. It is critical first to learn and understand the status quo, its own logic, and how it evolved historically. It is only on this basis that there can be any hope that changes will be implemented. We should be aware that decision-making under uncertainty almost always involves some financial, personal, and/or political risks, and stakeholders are unlikely to adopt measures that do not account for the existing conditions and incentives that have shaped the current state of affairs. These are some of the issues that must be addressed before even attempting to use climate predictions.

2.1 Characterization of Uncertainties and Associated Exposure

When interacting with decision-makers for embedding climate forecasts into their routine work, a first needed step consists of explicitly showing the full distribution of the most relevant climate variables to which the system is exposed. Decision-makers' awareness of climate uncertainty is frequently implicit and qualitative. Moreover, they are often not used to dealing with graphical or quantitative representations of relevant climate variability. Assisting stakeholders to become familiar with climate distributions lays the groundwork for adequately using climate forecasts, which creates a shifted version of the expected climate.

A second and more complex step consists of determining how climate inherent uncertainties propagate through the system of interest and affect outcomes downstream. Some of these propagated impacts can be modeled and programmed into an algorithm that can readily show the effect of the predicted climate shifts on certain relevant variables of interest to stakeholders. However, it is crucial to always keep in mind that decisions involve other quantitative and qualitative information that cannot be programmed into an algorithm, but can nevertheless heavily influence decisions.

Another key element to consider is the lead time of decisions compared to that of predictions. The former depends on the dynamics and inherent timescales of the system of interest; in some cases, the timing of decisions may be rigid, while in others, there may be room for flexibility. The lead time of predictions is determined by the underlying physical processes from which predictability arises. Skills may increase with improvements in the understanding, monitoring, and modeling, but essentially, the upper bound at each lead time is controlled by the laws of physics.

2.2 Explicitness of Options and Associated Upside and Downside

Closely related to the understanding of how climate uncertainty affects the system of interest is assessing the management options available, their upsides and downsides, and their sensitivity to climate variability. Even when the impact of climate shifts on a system is large, the lack of management options to respond to those effects can hamper the usefulness of climate predictions. Thus, if nothing can be done about the expected effects, then there is nothing to decide; consequently, there is no room for using climate predictions. This may sound obvious, but it is sometimes overlooked: The goal of climate forecasts is to provide information that may assist in selecting among alternatives; therefore, when these options are not clearly defined or nonexistent, the exercise is futile.

In certain cases, very few options may be available to minimize physical and financial losses. Still, there might be possible interventions for damage control or for designing a public communications strategy in relation to the impending situation. In those cases, the incorporation of forecast information has to be framed in the context of selecting among such interventions.

2.3 Identification of Stakeholders and Associated Incentives

The decision-making process can vary widely in its degree of formalization. At one extreme, there are institutions with sufficient experience to allow them to establish a complex and strict protocol to manage climate risks. At the other extreme, there are single individuals making decisions based on intuition and personal experience. Most cases lie between these extremes, with some protocols in place, but with a sizeable, subjective final synthesis by a small number of people that are ultimately responsible for the decision. It is imperative to understand the decision-making process and the incentives that drive each of the individuals involved.

2.4 Summary

The reader may recognize that what has been described so far closely resembles the elements required in a risk management system: institutional dimension, procedures, risk assessment, and risk management. Baethgen (2010) proposes an approach to improve management of climate-related risks based on four pillars:

(a) Identify vulnerabilities and/or potential opportunities due to climate variability for a given system (agriculture, water, energy, public health, etc.), in close collaboration with stakeholders.

(b) Characterize, quantify, and, if possible, reduce climate uncertainties to facilitate decision-making in various socioeconomic sectors. The associated information may include historical analyses (understanding the past), monitoring systems (observing the present), and climate forecasts (predicting the future) at different timescales.

(c) Identifying technologies and practices that optimize results in normal or favorable years, reduce vulnerabilities to climate variability and change, or both. Examples in agriculture include crop diversification, improved tillage systems for increasing water soil storage, improved plant water use efficiency, and drought-resistant cultivars.

(d) Identifying policies and institutional arrangements that reduce exposure to climate hazards and enable one to take advantage of favorable climatic conditions. Exposure reduction can be achieved, for example, with improved early warning and response systems and by transferring portions of the existing risks with various forms of insurance.

A fifth pillar recently was added to this approach (Goddard and Buizer, 2016), which involves monitoring and evaluation of outcomes. Whenever possible, the interventions mentioned in the first four pillars should be implemented within a monitoring and evaluation framework that includes assessing baseline conditions, as well as changes in conditions that followed the interventions. Evaluation helps one to identify impacts that are attributable to the interventions and to adjust processes as needed.

In many situations, none of these components is explicit, nor are they expressed in these terms. Still, if an ongoing business is affected by climate variability, some sort of climate risk management, no matter how implicit or simple, is taking place. In large institutions, the risk management system may be spread out across many departments, not always aligned. In very small organizations, where decisions are often made by a single individual, the whole risk management strategy may lie in that person's own judgment. A key message in this regard is that before any attempt is made to incorporate climate prediction to decision-making in any institution (small or large), the preexisting climate risk management strategy has to be fully understood.

Experience shows that the integration of climate forecasts to decision-making requires that all the components of the risk management strategy be transparent. Moreover, if any of the elements is too weak, it is best to strengthen it before attempting to apply climate forecasts.

3 EMBEDDING A PROBABILISTIC CLIMATE FORECAST INTO DECISIONS

As we have seen in the previous section, climate uncertainty is already incorporated into decisions with varying degree of explicitness. Thus, stakeholders who have survived and succeeded in an environment exposed to climate variability must possess the relevant know-how to adapt to the expected climate. It could even be said that an unshifted climate forecast (i.e., the climatological distribution) is already embedded into the decision system. However, this is usually not the way that stakeholders conceptualize or understand their own actions. Therefore, when stakeholders are presented with a climate forecast, which is merely a slight shift of the expected climatic distribution already under consideration, they may have a hard time perceiving it in familiar terms.

A proper way to frame the assimilation of climate forecasts into a decision-making process is, indeed, as a subtle modification of such a process in order to account for the slight shift in the expected climate that the forecast often represents. However subtle, such adjustments are not easy to assess, much less implement. In order to describe the difficulties that generally emerge, we will go through the three steps presented in the previous section, namely, characterization of uncertainties and their predicted shifts, identification of management options, and associated changes in risk and involvement of stakeholders in order to align incentives.

3.1 Shifting Climate Uncertainty

In our experience, two misunderstandings usually arise in early stages of the process and should be elucidated. One is that SCFs somehow eliminate climate uncertainty, an ever-present desire in stakeholders conscious of their exposure to climate variability. The second misconception, the opposite of the previous one, occurs in cases where stakeholders are insufficiently aware of the uncertainties that they normally face and wrongly associate their existence with the introduction of the climate forecasts. The interaction with decision-makers should clarify that forecasts modify uncertainties—and can even reduce them in some cases—but can neither eliminate nor create them.

Emphasis must be made on framing climate predictions as a variation on business-as-usual practices, rather than something entirely new. This entails that the decision system already explicitly considers climate uncertainties and its propagated effect on variables closer to the decision-makers; otherwise, there is no place for probabilistic forecasting. Whenever possible, the quantifiable component of climate risk needs to be computed explicitly. This quantitative estimation should be performed and tested before introducing any forecast-related shift. Shortcomings that might emerge from the modeling of the propagation of climate uncertainty constitute a problem that is independent from the incorporation of forecasts and should be addressed accordingly. Once adopted, such an algorithm is the natural place to connect smoothly with the forecasts. Eventually, an algorithm that transparently includes climate forecasts may become the new norm.

Similar thinking holds for the more qualitative and subjective components of such an uncertainty. Every possible effort must be made to render it explicit and to familiarize everyone involved at all stages of the decision process with the existence of the uncertainty, its properties, and its impact on the activities of interest. Only then can climate forecasts be introduced as additional information to be considered in the decision-making process.

3.2 Assessing Changes in Risk and Options

Options should be in place as an inherent component of the risk management system prior to the incorporation of climate information. Moreover, alternatives should be presented as shifts in a continuous set of possibilities: proportion of option B rather than A, changing the date when a certain action is taken, relevant physical quantities within a range, etc. Whenever possible, all-or-nothing decisions should be reframed. They may reveal the absence of a risk management approach to business, which constitutes a prerequisite to introducing climate forecasts. A shift of the expected climate, usually slight, cannot reasonably trigger the total reversal of a standard practice, unless the impact of the inherent variability was unaccounted for in the original strategy. In a way, all-or-nothing decisions load the climate forecast with the burden of the entire uncertainty, rather than the expected shift or reduction of climate uncertainties that they claim.

Stakeholders in the real world routinely make decisions while facing uncertainties that originate from a variety of sources. Climate is never the only source, but it often can be quite significant. The application of a climate prediction will be meaningful only if the associated shift, which is necessarily much less than the entire distribution, is expected to produce a significant impact on the activity of interest.

Flexibility in the decision-making processes also can be achieved by spreading optionality in time. The International Research Institute for Climate and Society (IRI) at Columbia University, together with the Red Cross Red Crescent Climate Centre, developed a "Ready-Set-Go" framework to illustrate the various decisions that are made across timescales, prompted by climate information. At the earliest stage, SCFs may trigger stakeholders to get ready by updating preparedness plans, and also in general by establishing typical no-regret actions that can help in unfavorable situations and have minimal negative consequences if acting in vain. The second phase, "get set," focuses on midrange forecasts with relatively high resolution and skill, which can prompt specific actions for enhancing preparedness. The third phase, "go," is activated by short-range weather forecasts at high resolution with the highest accuracy, and stakeholders typically take action to reduce risk in the near term (e.g., evacuate a region due to the high chance of flooding). Such an approach to risk management can use diagnostic and predictive climate information over a wide range of timescales: from monitoring to weather and S2S and seasonal prediction. Our experience shows that the integration of decisions and relevant climate information at different timescales facilitates the inclusion of climate information. The classic predictability desert (Vitart et al., 2012b) between the extended weather forecast and seasonal climate prediction has operated as a barrier to the incorporation of the latter into decision-making. This constitutes both a challenge and an opportunity for S2S prediction, which could guide the adjustments in preparedness as the season pans out, and in this way eliminate the gap between seasonal prediction and action.

3.3 Involvement With Stakeholders

On a technical level, stakeholder involvement in the course of embedding climate information in a decision-making process is required to provide the needed expertise about the specifics of the risk management system: uncertainties, options, associated upsides and downsides, etc. Moreover, it is also critical that stakeholders gain familiarity and trust with the information and tools being developed. The type of decisions that climate forecasts attempt to inform always involve risk: personal, financial, and/or political. It is human nature that people do not easily change the way that they behave when they have personal, vested interests. Experience shows that the trust required to incorporate the climate information into actual decisions is developed among the stakeholders and the scientists and other technical personnel who participate in the coproduction. Without this trust-building, implementation and adoption are rare.

An additional challenge arises when the institution has a principal-agent problem or an agency dilemma (Laffont and Martimort, 2002). The incentives of administrators may not always align with the bottom line of the principal or the interest of the institution per se. Introducing changes to business as usual may risk the job if the evaluation is unfavorable. Climate seasonal forecasts, which are inherently probabilistic, are difficult to evaluate, especially for short periods of time, with few realizations. This difficulty is inherited and magnified further when forecasts are embedded into a decision system exposed to uncertainties of other origins.

When the institution is large, there might even be a number of agents with diverse incentives, related to the differential impact of climate risk on the various sectors of the activity.

In these circumstances, it is key to gain full institutional commitment with the process, involving a large number of people, before attempting to implement changes.

4 EXAMPLES

In the following subsections, we present two examples of successful incorporation of climate information (including seasonal forecasts) to make decisions and elaborate on policy. One example is in the electricity sector, and the other is in the agricultural sector; both are from Uruguay, where precipitation variability in some seasons is greatly conditioned by El Niño–Southern Oscillation (ENSO; Pisciottano et al., 1994).

4.1 The Management of the Interconnected Electric System

A severe drought (Fig. 1) affected the southern South America region during the 1988–89 period, associated with an extremely cold phase of ENSO (Rivera and Penalba, 2014). At the time, the Uruguayan electric system comprised 1199 MW of hydroelectric capacity, able to generate approximately 140% of the annual energy demand (approximately 5000 GWh) in a normal year, and 255 MW of thermal power capacity for backup (DNE, 2017). If the backup capacity were overwhelmed, the only remaining options would be energy imports from neighboring countries—when available—and blackouts. Given that Uruguay has no fossil fuel deposits, all options other than the hydroelectric plants entailed large costs for the national economy. Due to the extended drought, the thermal generation required in 1988

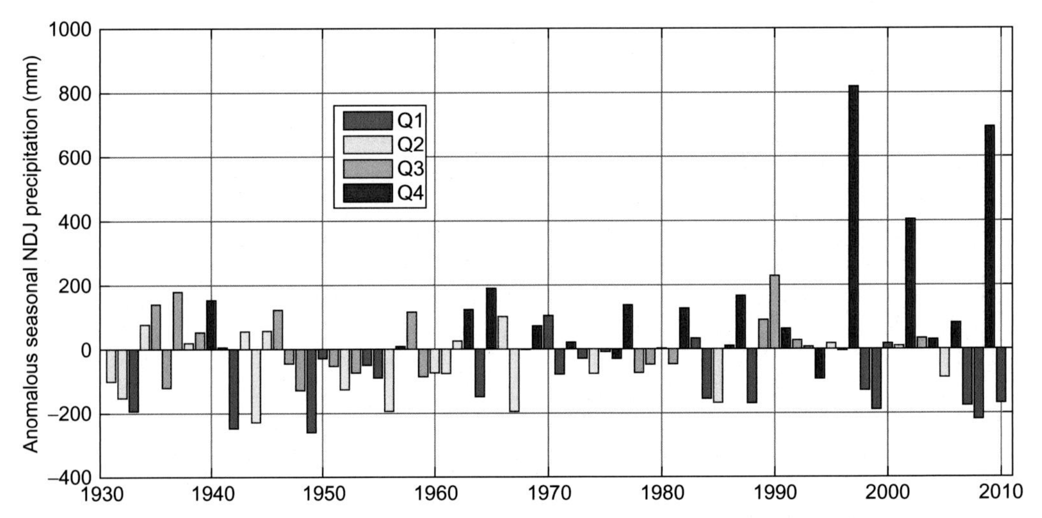

FIG. 1 Anomalies of seasonal (November-December-January (NDJ)) precipitation averaged in two stations (Salto and Artigas) in northwestern Uruguay and conditioned by ENSO. Quartiles are defined based on the NDJ N3.4 index for the period of interest; the N3.4 index was taken from https://www.esrl.noaa.gov/psd/gcos_wgsp/Timeseries/Data/nino34.long.anom.data.

and 1989 amounted to 27.5% and 39.4% of the total annual energy, respectively (INE, 2017), with severe consequences for the national economy (INE, 2017). (See Figs. 2 and 3.)

This event prompted UTE, the government utility that managed the entire electric system (generation, transmission, and distribution), to work with the University of Uruguay and

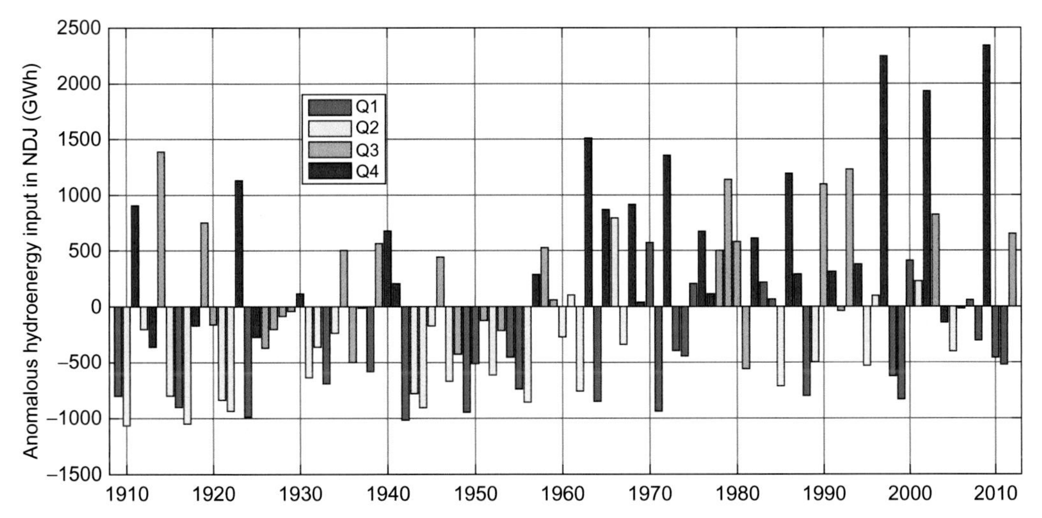

FIG. 2 Cumulative NDJ inflow to the hydroelectric reservoirs weighted by energy factors, topped by the capacity to turbine, and conditioned by ENSO. It is expressed in energy units, but it is not a time series of generated energy. Rather, it actually predates hydroelectric plants. ENSO quartiles are defined as in Fig. 1, but it may not coincide exactly due to the difference in periods.

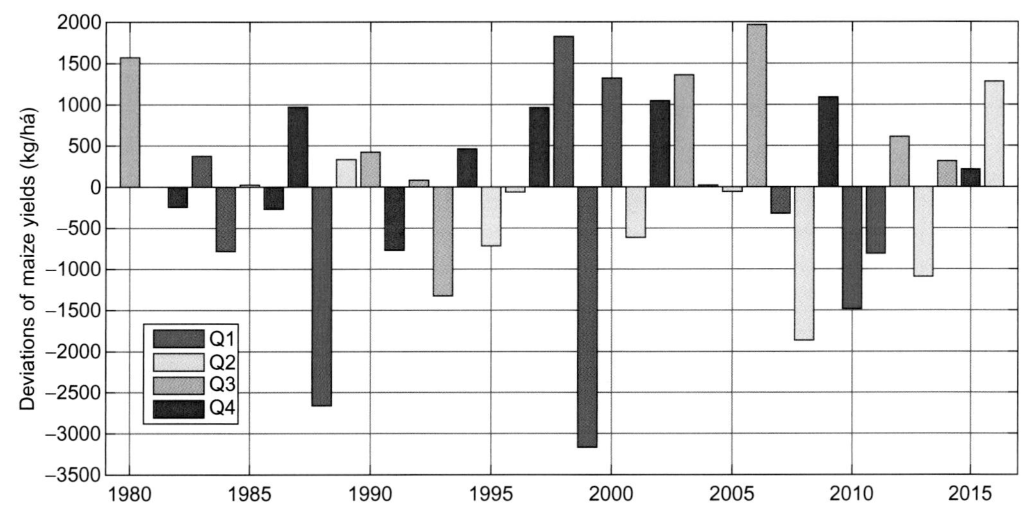

FIG. 3 Uruguay national maize yields conditioned by ENSO quartiles. The yields are expressed as deviations from the expected values of maize yields adjusted to the technological level of the 2016–17 growing season. ENSO quartiles are defined as in Fig. 1, but they may not coincide exactly due to the difference in periods.

fund the development of a climate group that would focus on regional climate variability (in particular ENSO phenomena) and its impact on the energy sector. A series of studies were undertaken in the following years that characterized the variability of precipitation and inflows to the dams and assessed predictability at different timescales, mainly associated with ENSO.

The advances in climate knowledge achieved at the time helped to understand the climate uncertainty faced by the energy sector and to link the shifts in the expected climate distribution with ENSO phases. In turn, this new knowledge contributed to establishing the first step of the process described in Sections 2 and 3 (Pisciottano et al., 1994). Presumably, these shifts in turn would translate into shifts in management decisions when predictions of the evolution of ENSO were robust. Indeed, the information associated with ENSO and its impact on the local climate eventually became mainstream. However, it was never clearly established whether any decision in UTE was affected by seasonal climate prediction.

In retrospect, several aspects of the process described in Sections 2 and 3 remained absent through the 1990s as climate scientists at the University of Uruguay and decision-makers at UTE continued to interact. The studies were limited to climate itself; the propagation of uncertainties through the system was never addressed, at least not in a coproduction mode. The available alternatives and their upsides and downsides were never explicit. The algorithms used to assess risk and optimize decisions were not subject to analysis.

UTE is a large institution, with approximately 6700 employees and multiple departments. Throughout the years, the focal points of the collaboration with climate scientists of the university changed several times. There were interactions with several departments of UTE, including the Hydrological Unit (which provides hydrological forecast information for internal use), the Dispatch Unit, and the Planning Unit. The decisions, risks, and timescales involved in each unit are vastly different, and they can even have conflicting interests: import of oil, dispatch of hydroelectric as opposed to thermal generators to save the water in the reservoir, scheduling of blackouts, negotiations with neighboring countries, definition of investments in generation capacity, and other issues. It became evident, however, that the main concern for most stakeholders is the ability to predict severe droughts, with less interest in the optimization of operation during nonextreme conditions. Because the predictability of extreme cases is low and the margin of action limited, there has been little use for SCF.

A series of changes took place around 2005 that collectively enabled the successful embedding of SCFs into decision-making. After a gap of a few years, the joint work between the engineers at UTE and the researchers at the University of Uruguay resumed. The trust-building process continued with a tailored course offered by the university at UTE's request on climate predictability and the participation of UTE staff in collaborative research projects with university researchers. Slowly, the common ground of knowledge on both climate science and the management of the electric system grew.

In parallel, several changes took place at the institutional level concerning the installed capacity and evolution of the electric system. A new regulatory framework opened the system to private generators (with transmission and distribution remaining centralized) and created a new nongovernmental public agency, Administración del Mercado Eléctrico (ADME), to manage the electric system and operate the dispatch. Because dispatch now involves several generators (a dominant public one and several private ones), transparency is required to define transaction costs and prices in the spot market.

In addition, most of the new capacity added to the system is wind generation, which dramatically changed the characteristics and operation of the electric system. As of the end of 2016, the installed capacity consisted of 1538 MW hydroelectric, 650 MW thermal, 1146 MW wind (from 0 MW as recently as 2008), and 329 MW from other sources (DNE, 2017). The uncertainties faced by the system have greatly diversified, and hydroelectric power helps filter short-term variability of wind generation, which is used at the base of the system to save water and thus alleviate the impact of longer-term climate variability on its availability. The greater complexity of the system dramatically changed the rationales behind dispatch and short-, medium-, and long-term planning (Chaer et al., 2012). An entire new algorithm for optimization and simulation of the integrated electric system, SimSEE (Chaer, 2008), was developed by ADME and the Electrical Engineering department at the University of Uruguay. It consists of a dynamic stochastic optimization algorithm that minimizes direct costs (i.e., those incurred in meeting current demand), plus future expected costs. The latter, of course, depend on the uncertainties faced by the system, which are modeled through stochastic processes. The characterization of climate exposure is quantitatively represented in SimSEE through a stochastic model of inflows to the reservoirs that is designed to reproduce climatological variability and short-term autocorrelations, but lacked any other dynamics. The optimization done within SimSEE explicitly considers all available options and computes the optimal operational policy. SimSEE is an open, modular code in continuous development (http://simsee.adme.uy/simsee_principal/simsee.php). Starting in 2007, a course on SimSEE has been offered annually at the university with participants from UTE, ADME, private generators, government agencies, and university researchers. It is now the official model on which ADME bases the computation of the spot markets and costs of other transactions.

This platform enabled coproduction work, involving staff from ADME, UTE, and the university, which finally resulted in the embedding of ENSO information into the stochastic process that models inflows within SimSEE (Maciel et al., 2015) in order to shift uncertainties associated with ENSO signals. The solution involved minor changes in the model which, in turn, is a small component of SimSEE. This is consistent with the stated principle that incorporating climate forecast into decisions should not imply a major overhaul of decision systems, but rather small adjustments. In November 2016, 25 years after the 1988–89 drought and for the first time, the official biannual seasonal planning performed by ADME included climate prediction associated with ENSO (http://www.adme.com.uy/dbdocs/Docs_secciones/nid_230/PES_Nov2015_Abr2016_v8.pdf). The incorporation of ENSO signals into SimSEE has normalized the use of climate prediction for seasonal planning because it is wired into the code of the standard model used, and the community of users has become familiar with this notion.

Moreover, with adequate adaptations, SimSEE is the model platform used for decisions at a wide range of timescales, from management of the spot market to long-term analysis of the evolution of the generation capacity, with intermediate seasonal and nested weekly planning. The successful incorporation of climate prediction information into SimSEE constitutes an asset that could benefit seamless climate prediction, from weather to seasonal. Indeed, work is underway to embed diagnostic and prognostic climate information in SimSEE: short-term wind energy and flow forecast; climatic covariability of wind, solar, and flow; and the relation to energy demand.

ENSO information, which is already incorporated, is most useful for managing the largest reservoir of the system (Gabriel Terra dam), with approximately 5 months of storage capacity. The second-largest reservoir, Salto Grande dam, has a few weeks of storage capacity, and therefore, its management could potentially benefit from skillful S2S predictions. Given that it represents 25% of the country's generation capacity, the impact on the system's operations could be quite significant.

4.2 The Ministry of Agriculture and Fisheries and Three Recent Droughts

Uruguay's economy is largely dependent on the agricultural sector: about 70% of the total exports are agricultural commodities or agro-industrial products. During the last decade, the country's food exports (i.e., beef, soybean, rice, dairy products, and cellulose) increased by 400% (Duran, 2014), stimulated by good prices for agricultural commodities.

Most annual crops except rice are rainfed, and consequently, climate variability drastically affects both the agricultural productivity and the volume of export goods. Given the importance of agricultural production on the agro-industrial activity and on the transportation/logistics sector, reductions in production have multiplying effects on the country's economy. For example, the impact of the drought of 2008–09 resulted in direct losses of about $300 million in the beef production sector. The associated losses in the national economy were about $1 billion—that is, a multiplying factor of more than 3 due to impacts on employment, logistics, and other areas (Paolino et al., 2010).

The drought of 1988–89, commented upon in the previous section, found that the Uruguayan agricultural sector had no institutions or special policies/programs in place to respond to droughts. At that time, droughts were viewed as very-low-frequency phenomena that did not justify the development of special structures or programs. Consequently, governments had historically reacted to droughts with traditional "crisis management" responses such as special aid programs to affected regions.

In mid-1989, the government of Uruguay hired a consultancy to assist in the development of strategies to confront droughts. The consultancy resulted in a long list of recommendations for improving risk management in future droughts, emphasizing the need to create adequate institutions and institutional arrangements (e.g., drought commissions, emergency systems, etc.).

As stated in the previous section, research on teleconnections and the impacts of ENSO on rainfall in Uruguay was incipient in 1988. Ropelewski and Halpert (1987) had just published the first article showing the correlations between ENSO anomalies and rainfall patterns in southeastern South America. Climate scientists from the University of Uruguay were starting the first research studies on ENSO impacts, which were published in the early 1990s (e.g., Pisciottano et al., 1994).

In summary, the 1988–89 drought found Uruguay without the needed groundwork: no institutional structures, no capabilities to assess or monitor climate, and incipient research on the ENSO impacts on rainfall. As a result, 16% of the total cattle population was lost (Bartaburu et al., 2009), and the direct losses in the livestock sector attributed to the drought were approximately $640 million (adjusted to 2016 values). However, the losses were much larger because the reduction in the population of breeding animals was felt for several years after 1988–89. Important losses also were reported in the forestry sector due to frequent fires and in the summer crops, where yields at the national level were reduced by more than 40%.

Almost a decade later (1999–2000), another drought hit the agricultural sector of Uruguay. Several changes had occurred in the country since the previous drought. First, the government had created institutions to deal with emergencies, including droughts: a National Emergency System, appointed by the Office of the Presidency; and the National Commission for Drought, created under the Ministry of Agriculture with researchers (agriculture and climate), some governmental offices, and several private-sector organizations.

Further, in December 1997, during a strong El Niño event, Uruguay hosted the first Regional Climate Outlook Forum (RCOF) for Southeast South America (SESA). Meteorologists from Argentina, Brazil, Paraguay, and Uruguay met with colleagues from the International Research Institute for Climate and Society (IRI) at Columbia University, National Oceanic and Atmospheric Agency (NOAA), and other international institutions and produced the first consensual SCF for the region. Since that time, the SESA region has uninterruptedly organized RCOFs at least once per year.

Finally, the National Institute for Agricultural Research (INIA) had established an interdisciplinary team (GRAS) that started collaborative research with international research institutes (including the IRI). The work involved research in applications of SCFs in the agricultural sector and developing an information and decision support system for the agricultural sector. The collaboration also included the creation of a Technical Working Group (TWG) for improving the generation, dissemination, and application of climate information. The TWG was constituted by researchers (agriculture and climate), and representatives of the major farmer organizations, agribusinesses, and government offices. The TWG met periodically to discuss the seasonal climate outlooks produced by the IRI and the current situation of the vegetation. During the TWG meetings, climate scientists presented the most recent climate outlooks, and agricultural scientists presented advances on tools for applying climate information to improve decisions. Stakeholders from the public and private sectors discussed the possible uses and shortcomings of the information they received. The work of this interdisciplinary group is a good example of coproduction of useful information, and it greatly contributed to develop the needed trust between researchers (climate and agricultural sciences) and agricultural stakeholders, as discussed earlier in this chapter.

All the information produced in the collaborative project was published and continuously updated on INIA's Web page, which is visited by farmers, agribusiness representatives, agronomists, and government officers. In addition, researchers gave several live presentations and teleconferences, in collaboration with the Ministry of Agriculture and Fisheries, which reached all the major regions of the country. Unfortunately, no surveys or research studies were conducted at that time, and therefore it is impossible to document how the climate information was used in the sector. However, informal interactions with farmer advisors, agronomists, and farmers revealed that the climate information (from monitoring and forecasts) was widely used to assist in decisions.

However, the actions of the Ministry of Agriculture and Fisheries during the 1999–2000 drought are easier to document. During the early months of 2000, the impacts of the drought were already quite evident. The Ministry of Agriculture decided to establish an emergency plan and defined priorities for distributing aid to the various regions of the country. Before the 1999–2000 drought, the definition of such priorities had mainly been based on reports prepared by the ministry's staff working in the field. However, the field staff was usually not able to cover the entire country, and consequently, several regions were ignored in the distribution of aid.

In the 1999–2000 drought, the ministry based its prioritization of aid on objective information: namely, the vegetation status [i.e., the normalized difference vegetation index (NDVI)] and the seasonal forecasts produced by the IRI. In the words of Juan E. Notaro, the Minister of Agriculture and Fisheries at the time,

> ... The results of [the research team's] work during the drought were useful for making both operational and political decisions. From the operational standpoint, [the] work allowed us to concentrate our efforts in the regions highlighted as being the ones with the worst and longest water deficit.
>
> ... From the strictly political standpoint, [the] work provided us with objective information to defend our prioritization of regions, in a moment in which every governor, politician, and farmer in the country was asking for aid. In the same line, the work also allowed to mitigate pressures since we provided the press and the general public with transparent, technically sound, and precise information.
>
> ... And most importantly, the most feared threat of significant cattle deaths due to the drought never occurred thanks to the celerity of the actions taken, which would have been impossible without the information you provided to us.

This was the first time that climate information was embedded in actual agricultural policy in Uruguay. During the following years, the IRI, INIA, and the university continued producing climate information and products that were gradually considered in decisions, plans, and policies.

In September 2010, the climate seasonal forecast for October–November–December elaborated by the IRI showed high probabilities for lower-than-normal rainfall, and the ENSO outlook showed 98% probability of a La Niña year. By October 2010, satellite data started showing negative anomalies in the vegetation, and the INIA-IRI soil water balances showed low plant available water. Based on the information produced by the INIA-IRI team, the Minister of Agriculture declared a national emergency in December 2010 for all police districts in most of the northern half of Uruguay. The declaration triggered a set of interventions, including the exoneration of taxes and the establishment of a line of subsidized credit to help acquire animal feed and to implement "water solutions" for small farmers (wells, reservoirs, irrigation, etc.).

In February 2011, when the rainfall was still below normal, the Minister of Agriculture was convened to the Parliament to provide a report of the current drought situation and to explain the concrete actions that his ministry had taken to reduce the impacts on agricultural production. The minister explained that in August and September 2010, his ministry had made public IRI's seasonal forecasts showing the high chances of low rainfall in the local spring (October, November, and December). The main objective of sharing those forecasts was to induce the agricultural sector to take preventative action. He ended his presentation with a slide presenting the current situation of the vegetation and soil water content (INIA-IRI's work), as well as the IRI's latest seasonal rainfall forecast for February–March–April 2011, which still showed some chances of below-normal rain.

In 2012, the Ministry of Agriculture implemented a project funded with a loan from the World Bank entitled "Development and Adaptation to Climate Change–DACC," which included establishing a National Agricultural Information System (NAIS) to assist the private sector in their planning and decisions and the government in elaborating public policy. The NAIS was established to take advantage of the work that had been developed in the previous years by INIA, IRI, and collaborators, including the tools that had been used successfully in the 1999–2000 drought. In addition, the NAIS integrated numerous databases of the Ministry of Agriculture in ways that encouraged their interoperability for assisting decisions and plans (Baethgen et al., 2016). Finally, the IRI worked with the Uruguayan Meteorological Institute to improve their capabilities to establish SCFs.

5 FINAL REMARKS

The successful incorporation of SCFs into decision-making has proved to be a slow process that entails much more than the elaboration of tailored climate information. A required first step is to fully understand the preexisting risk management strategy, regardless whether it is explicit or implicit. In the latter case, every effort should be made to gain transparency about how risk is considered in decisions. Incorporation of climate forecasts should assist in fine-tuning risk management, but it will not replace or conceal preexisting weaknesses in how the decision process deals with risk. Even situations that end in severe climate events such as droughts may require small shifts in decisions as such an event is developing. Unawareness of how climate uncertainty propagates through the system and interacts with other sources of uncertainty, lack of optionality that can be gradually adjusted in response to an expected climate shift, and agency problems should be addressed prior to the incorporation of climate information to decisions.

Therefore, full disclosure is required in sensitive aspects of the decision-making process that will occur only in a relationship of trust. Moreover, any proposed change to business as usual will inherently affect exposure to risk. Thus, it is not likely to be adopted unless stakeholders fully own the adjustments to be implemented. It follows that a committed engagement of stakeholders in a coproduction process is the most desirable way to address the process of incorporating climate forecasts into decision-making. Developing this kind of trust can be a lengthy process. For example, in the Uruguay cases described in this chapter, such relationships of trust were achieved after more than a decade of collaboration between the counterparts involved in INIA, IRI, the university, and stakeholders in the agriculture and energy sectors.

Once trust has been established, a foundation is in place, and there is a successful experience of embedding SCFs in a decision-making process, it is much simpler to take the next step (e.g., incorporating S2S products). Moreover, climate information at different timescales can readily complement each other in a synergistic way following the Ready-Steady-Go framework described previously. Skillful S2S predictions could thus enhance the usefulness and applicability of SCFs, bridging the gap with weather forecasts or providing useful information in places and seasons where the skill of seasonal forecasts is low. Furthermore, subseasonal predictions can be helpful for supporting decisions that require shorter lead times, more detailed information than the expected aggregated rainfall or mean temperature provided by seasonal forecasts, or both.

CHAPTER

22

Predicting Climate Impacts on Health at Sub-seasonal to Seasonal Timescales

Adrian M. Tompkins, Rachel Lowe†,‡, Hannah Nissan§, Nadège Martiny¶, Pascal Roucou¶, Madeleine C. Thomson§, Tetsuo Nakazawa|*

*Abdus Salam International Center for Theoretical Physics (ICTP), Trieste, Italy †Department of Infectious Disease Epidemiology, Center for the Mathematical Modelling of Infectious Diseases, London School of Hygiene and Tropical Medicine, London, United Kingdom ‡Climate and Health Programme, Barcelona Institute for Global Health (ISGLOBAL), Barcelona, Spain §International Research Institute for Climate and Society (IRI), Columbia University, Palisades, NY, United States ¶Center de Recherches de Climatologie, Biogéosciences, UMR 6282 CNRS, Université Bourgogne Franche-Comté, Besançon, France |Typhoon Research Department, Meteorological Research Institute, Japan Meteorological Agency, Tokyo, Japan

OUTLINE

https://doi.org/10.1016/B978-0-12-811714-9.00022-X

1 INTRODUCTION

Variations in climate can have wide-ranging impacts on human health. These include direct impacts, involving immediate danger to human life resulting from weather extremes such as high winds, floods, storm surges, and weather-related accidents. Indirect impacts include nutritional deficiencies due to crop failures resulting from climate-induced pest outbreaks or drought in regions relying on rain-fed agriculture (World Health Organization, 2012). A wide range of diseases are also affected by climate due to the sensitivity of disease pathogens, vectors, or hosts to variations in climate.

Health impacts may be derived as a result of variations in rainfall, temperature, and humidity (and related factors) over multiple timescales, from weather variations (days to weeks) through seasonal variability (weeks to months) to decadal variability and climate change (years to centuries). Here, the focus is on the prediction of weather and its impacts over sub-seasonal to seasonal (S2S) time frames of weeks to approximately 2 months. This timescale is situated between short-term weather forecasts, which predict the evolution from accurate atmospheric initial conditions, and seasonal climate forecasts (SCFs). At each prediction timescale, modes of atmospheric and oceanic variability can provide predictability, such as the El Niño-Southern Oscillation (ENSO) at a seasonal timescale or the Madden-Julian Oscillation at sub-seasonal scales.

In this chapter, we explore the potential value of S2S forecasting for health decision makers and seek to identify opportunities for adding value to current uses of weather and SCFs in the development of health early warning systems (HEWSs). Operational requirements for effective S2S-based forecasts for health are introduced using four case studies contributed by the coauthors, where prior work has shown the potential for weather and SCFs to inform decision making. Our discussion compares the results of heat stress early warning with that of infectious diseases, including viral (dengue) and parasitic (malaria) vector-borne diseases, as well as an airborne bacterial infection (meningococcal meningitis). Despite the early stage of applying S2S predictions to support public health decision making, we believe that valuable lessons can be learned and applied to a wide range of climate-sensitive diseases.

While variations in climate affect health in both developed and developing countries, developing countries are more vulnerable. These countries lack protection against exposure to extreme temperatures in the home and workplace, and vector-borne disease outbreaks are more prevalent. Therefore, the case studies presented here predominantly focus on the developing regions of the globe.

1.1 Climate Impacts on Health

Variations in weather and climate affect health outcomes, and climate information is already implicitly or explicitly accounted for in health decisions, such as incorporating climate into the derivation of spatial risk maps for malaria (Hay and Snow, 2006; Omumbo et al., 2013) or accounting for the seasonal cycle of disease in intervention planning (Jancloes et al., 2014). Accurate predictions of climate thus can lead to valuable information that can help with preparation for a specific health consequence, provided that the link between climate and health is well understood. Often, the relationship between climate and health outcomes is determined empirically, using a univariate or bivariate analysis with rainfall, temperature, or both as key determinants of many health outcomes, particularly those associated with vector-borne diseases. For example, documented malaria outbreaks in highland areas of Africa have long been known to be associated with ENSO events and have been attributed to both enhanced rainfall and warmer temperatures (Kilian et al., 1999; Lindblade et al., 1999). A simple empirical relationship between rainfall and malaria was used in an early, groundbreaking prototype of a seasonal forecast system (Thomson et al., 2006a). More recent work with statistical models has evolved to use general additive/linear models (Lowe et al., 2011, 2013a,b, 2016a; Colón-González et al., 2016).

An advantage of using statistical models is the ability to simultaneously consider the complex interaction of climate hazards, socioeconomic disparities, and human vulnerability with predictive disease risk in space and time (Lowe and Rodó, 2016). This approach attempts to predict health case data using a range of socioeconomic and environmental indicators, some of which can act as proxies for climatic variables, such as altitude (a proxy for temperature), location, or month of the year (a proxy for seasonality). These fields are static and do not vary from year to year. The model is then shown linear and nonlinear functions of temperature and rainfall to determine if the fit to the health data is improved. This approach thus focuses on the relationship between interannual variations in climate and health outcomes, which could be predicted at multiple timescales, including by S2S climate-forecasting systems. The drawback of statistical methods is the need for long datasets of high quality, as well as the fact that such a model is restricted to areas of similar social and eco-epidemiological conditions to those for which the model was developed. Often, models are shown to be skillful in the locale in which they are derived and are not necessarily transferable to new locations. This is particularly true for models driven by rainfall, which can have nonlinear and complex relationships with health outcomes depending on the context.

Table 1 gives examples of how rainfall has been found to affect some common health outcomes, and in each case, opposing impacts can be found depending on the environmental setting. For example, malaria is generally assumed to be associated with the rainy season,

TABLE 1 Example Health Outcomes Due to Variability in Rainfall, Emphasizing How the Relationship Can Be Complex and Depend on the Environmental Setting

Health Outcome	Link With Precipitation	Reference
Cholera	Rain increases risk in dry areas and decreases it in estaurine regions	Pascual et al. (2002)
Rift Valley Fever	Rain increases risk, which suggests that extreme rain separated by a dry period may trigger outbreaks	Caminade et al. (2011)
Malaria	Rainfall provides breeding sites for vectors; however, drought can lead to breeding sites as rivers dry	Bomblies et al. (2009), Haque et al. (2010), Kusumawathie et al. (2006)
Dengue	Rainfall can provide breeding sites, but low rainfall may increase urban storage of water	Brown et al. (2014), El-Badry and Al-Ali (2010), Padmanabha et al. (2010)

as rainfall provides the water for temporary breeding sites for many of the key mosquito vectors. A study of malaria in a Sahelian village shows cases tracking rainfall with a delay of approximately 2 months (Bomblies et al., 2009). In contrast, in the vicinity of rivers, transmission can flare up during periods of drought. As river flow slows or halts, still ponds are created as a result (Kusumawathie et al., 2006; Haque et al., 2010). Extended drought can also affect population vulnerability through malnutrition, loss of herd immunity, or even through the loss of vector predators, causing more intense outbreaks once rains and transmission resume (Gagnon et al., 2002).

The relationship between rainfall and cholera also depends on the hydrological setting. Rainfall sometimes increases risk in dry regions, while the dilution effect of rainfall in estaurine locations may result in lower cholera incidence (Pascual et al., 2002). In East Africa, Rift Valley Fever (RVF) epidemics often follow rainfall anomalies during the short rainy season (October-December) associated with El Niño. In West Africa, the relationship between rainfall and RVF appears to be more complex, with a dry period between rainy events being identified as an outbreak trigger (Caminade et al., 2011). This would suggest a highly nonlinear dependence on sub-seasonal rainfall variability, which would be very challenging for S2S forecasting systems to predict accurately beyond the range of a few days. Finally, rainfall also can provide open-air breeding sites for dengue vectors, but dry periods increase water storage in urban areas with unreliable supplies, which can actually increase vector density if these facilities are poorly managed (Brown et al., 2014; El-Badry and Al-Ali, 2010; Padmanabha et al., 2010). Indeed, water storage and harvesting may have implications for other health outcomes, such as schistosomiasis and malaria (Boelee et al., 2013).

All these examples show why a statistical approach to modeling climate-disease interactions may be restricted to the area in which it is derived. If an improved understanding of the disease biology can be attained, empirical statistical approaches can be supplemented by mathematical models of diseases, which may be more generic and transferable in theory. However, such models also have their drawbacks, as the uncertainty of many biological processes can be large, partly related to the representativeness of controlled laboratory conditions to real-world situations and the fact that single laboratory experiments give little indication of parameter uncertainty. Moreover, location adaption of vector species also may hinder the ability of the model to be transferred from one location to another. Mathematical dynamical disease models thus often suffer from considerable structural and parameter uncertainty (Tompkins and Thomson, 2018).

1.2 Toward S2S Predictions in Health

Apart from disease model (statistical or dynamical) uncertainties, the efficacy of any climate-driven early warning system for health depends on the underlying skill of the driving climate-forecasting system. Approaching the climate-health nexus from the meteorological perspective, the focus of S2S system skill for health application is on near-surface meteorology, although of course, the skill in predicting these parameters depends on the whole meteorological system. Accurate representation of the near-surface temperature in S2S systems is the result not only of a good representation of the land surface and soil moisture, but also of the turbulence, convection, cloud microphysics, and large-scale dynamical processes.

The key near-surface parameters are rainfall, temperature, relative humidity, winds, and radiative fluxes. Reliable weather and climate predictions of these parameters in S2S forecast systems from 1 to 60 days ahead (see Chapter 1) could be used to drive statistical and/or dynamical modeling systems for health outcomes. The potential advance warning available will vary substantially according to the specific health impact. This may be contemporaneous with the meteorological anomaly (e.g., flooding or heat waves) or involve a delay of weeks to months (e.g., vector-borne disease risk). In the case of malaria, the early warning (lead time) of 1–2 months available from climate monitoring could be potentially extended by a further 2 months with the use of skillful S2S prediction systems. Indeed, SCFs have already shown predictive capacity in certain regions and seasons (Doblas-Reyes et al., 2013) and research has demonstrated the opportunity to incorporate forecast climate information into early warning systems for climate-sensitive diseases such as dengue and malaria (Connor and Mantilla, 2008; Thomson et al., 2006a; Lowe et al., 2014; Ballester et al., 2016). The advantage of S2S systems is that they are more frequently initialized, generally have a higher spatial resolution than the equivalent seasonal forecasting system from the same producing center, and in some cases, offers more frequent model upgrades (Vitart et al., 2008). A direct comparison between the European Center for Medium-range Weather Forecasts (ECMWF) S2S and seasonal forecast systems for Africa showed a significant improvement in skill, especially for 2-m temperature (Tompkins and Di Giuseppe, 2015).

In summary, an HEWS requires the identification of critical meteorological thresholds associated with adverse (or beneficial) health impacts, the ability to predict these meteorological conditions accurately in advance and translate them reliably to a particular potential health outcome. Forecasts should be issued to decision makers, indicating whether key thresholds are likely to be reached while effectively communicating the uncertainty associated with this prediction. Further, a set of standard operating procedures should be established, to implement risk reduction measures based on the forecast information provided. The system needs to be robustly evaluated over a set of past events to ensure its overall benefit. All of these steps represent significant research and operational challenges. Despite the obvious potential of S2S forecasts and the fact that policymakers are often well aware of the relationship between climate variations and health outcomes, it is perhaps unsurprising that climate information is still rarely exploited to help prevent and control such health risks. The four case studies described here attempt to determine the possible added value of incorporating climate information at S2S timescales in health early warning systems. The chapter highlights some of the barriers and bottlenecks that exist, which need to be overcome to effectively operationalize such systems. Possible steps are suggested to address these and ensure that the potential of present state-of-the-art climate observational and prediction systems are fully realized to complement public health decision-making and ultimately improve well-being.

2 CASE STUDIES

2.1 Malaria (Tompkins and Thomson)

Malaria has long been linked to the environment, with the association of pestilence with drying marshes identified in Roman times (O'Sullivan et al., 2008). After Ross discovered that mosquitoes were the vectors for the disease, the direct link with rainfall that provides vector breeding sites was firmly established. By the World War I, environmental engineering techniques were employed to control the disease until the 1960s, when the focus was switched to the use of the insecticide dichlorodiphenyltrichloroethane (DDT) application (Konradsen et al., 2004; Tompkins et al., 2016b). While engineering solutions such as swamp drainage schemes can be classified as long-term abatement interventions, annual interventions such as the application of DDT are often carried out according to the seasonality of the disease.

One of the earliest studies that attempted to use rainfall anomalies to predict malaria transmission anomalies for the season ahead for guidance was conducted in the early 1920s (Gill, 1923), with the method employed for several decades afterward with claimed success in linking May rainfall in the Punjab province to late autumn transmission (Swaroop, 1949). The separation of India and Pakistan was cited as the cause of the cessation of this particular forecasting system in the province (Swaroop, 1949). However, after 1955, the introduction of insecticides and newly developed medicines aimed at malaria eradication resulted in decreased interest in forecasting and early warning systems (Rogers et al., 2002).

After the collapse of intervention efforts in the 1970s and the subsequent rebound of the disease, interest in early warning systems was rekindled in the 1980s and 1990s (Connor et al., 1998). Effective early warning of an outbreak in epidemic-prone areas or anomalous transmission in malaria-endemic regions could be used to manage decisions concerning indoor residual spraying (IRS) interventions, bed net distribution, ensuring that appropriate medical supplies in health clinics are available, and conducting public information campaigns, with the appropriate action or combination of actions depending on the disease transmission setting in question. Although it was known that temperature and humidity could impact the transmission of the disease (Mayne, 1930), the majority of studies into early warning were univariate, focusing on rainfall variability as the predictand for transmission. The widespread outbreaks in the African highlands resulting from the extreme El Niño event of 1998 greatly increased interest in the potential of malaria early warning systems, and while the focus was still on rainfall, some studies also noted that the warmer tropical temperatures associated with El Niño also could act to push transmission to higher altitudes and encourage highland outbreaks (Kilian et al., 1999; Lindblade et al., 1999; Hay et al., 2001). The delay between the rains and the peak in transmission of 1–2 months meant that many early efforts for early warning were based on rainfall monitoring, using in situ and/or remotely sensed data (Grover-Kopec et al., 2005), which provided additional advance warning compared to simply attempting to monitor cases. Case monitoring, especially prior to the implementation of the first- and second-generation digitized health management and information systems, was rarely close to real time due to the delay involved in collating paper records centrally.

Considering the use of dynamical monthly and seasonal forecasts for climate, skillful forecasts could potentially extend the available lead time for planning purposes (Cox and Abeku, 2007). While this was discussed in the context of rainfall predictions, to maximize the potential benefit of dynamical climate predictions, a multivariate approach would be beneficial, accounting for all climate variables that affect the disease. Of course, rainfall provides breeding sites for malaria vectors, although the relationship is far from straightforward because extreme events on sub-seasonal timescales can flush breeding sites of early-stage larvae and even reduce vector density in the short term (Paaijmans et al., 2007). This might explain why statistical analysis of the malaria-rainfall relationship often reveals a highly nonlinear parametric relationship, with transmission peaking at a rainfall rate of approximately 3–5 mm/day over a month (Thomson et al., 2006a; Lowe et al., 2013a; Colón-González et al., 2016). Temperatures affect both the parasite and vector development rates, while higher temperatures affect vector mortality in both the larval and adult phases (Craig et al., 1999). Humidity also affects vector survival (Mayne, 1930; Thomson, 1938; Lyons et al., 2014) although the impact is less well understood. Uncertainties in these relationships are large, due to lack of experimentation, and more important, complex topography and heterogeneous land cover can imply considerable variations in climate spatially, and thus satellite or station measurements may not accurately reflect variations in microclimates in which adult mosquitoes and their larvae reside.

Recent reviews of early warning systems for malaria show that the majority of them are based on climate monitoring, and a small number use SCFs (Mabaso and Ndlovu, 2012; Zinszer et al., 2012). The connection to malaria in these systems consists of statistical models (Lowe et al., 2013a), mathematical compartmental Susceptible, Exposed, Infected, Recovered (SEIR) models that incorporate climate in idealized and sometimes ad hoc ways (Laneri et al., 2010), or full-process models that attempt to model the key elements of the full-vector and parasite life cycles, relying mostly on results from experiments using controlled laboratory settings to set the life-cycle parameters (Hoshen and Morse, 2004; Bomblies et al., 2009; Tompkins and Ermert, 2013).

One of the first attempts to demonstrate a prototype forecasting system using dynamic climate predictions used multimodel precipitation forecasts from the DEMETER seasonal forecasting project (Palmer et al., 2004) to drive a simple, univariate statistical model that fitted the national total laboratory-confirmed malaria cases in Botswana to monthly rainfall data (Thomson et al., 2006a). Cross-validation showed the system to be skillful in this region, which exhibits significant teleconnections with ENSO. The Botswana data were also recently used to test a system using newer climate forecasts to drive a dynamical malaria model, which had some success at prediction in high-transmission years (MacLeod et al., 2015). Other studies demonstrated the potential for malaria early warning systems driven by the new ENSEMBLES seasonal forecast system (Weisheimer et al., 2009; Jones and Morse, 2010, 2012). However, these demonstrations consist of a so-called tier 2 validation, where the malaria forecasts are validated against the malaria model driven by climate reanalysis (Dee et al., 2011) rather than actual case data.

Most dynamical malaria forecast demonstrations have used seasonal dynamical systems, but what is the potential of S2S systems that are the subject of this book? As stated previously, these systems have a number of configuration advantages that should imply enhanced skill relative to their seasonal forecast cousins at their 1- to 2-month lead times. As these systems are relatively new, there has been limited evaluation of their potential for malaria forecasting to date. One study (Tompkins and Di Giuseppe, 2015) has employed the S2S system from the ECMWF model (Vitart et al., 2008) in tandem with its seasonal forecasting system (Molteni et al., 2011). The study showed an increase in both precipitation and temperature skill for the monthly system relative to the system 4 seasonal forecast over Africa. It used the S2S system for the first 32 days of the forecast, which was then supplemented by the seasonal system. A novel aspect of the study was that, rather than spinning up the forecast system from an artificial state, the forecasts were initialized from a simple malaria analysis, created by driving the model with reanalysis data. In this way, wetter/warmer-than-average conditions in the weeks and months preceding the forecast would lead to enhanced vector densities and higher parasite ratios in the initial conditions. The study showed that the S2S system skill extended the malaria advance warning from 1 to 2 months (available from simply using climate monitoring) out to 2–3 months; and in limited areas the seasonal system extended the lead time further still (Fig. 1). Analyzing the S2S skill separately demonstrated that most of the prediction skill with malaria was actually derived from the temperature forecasts, which tended to be more skillful than with rainfall. One limitation of this study, however, was that the evaluation was again of the tier 2 type; that is, the forecasts were evaluated against a reanalysis-driven model. A more recent study instead evaluated the forecast system against actual confirmed case data in Uganda and demonstrated significant skill out to 4 months (Tompkins et al., 2016a).

In summary, while some limited studies have demonstrated the potential use of one S2S system for the prediction of malaria, there has been limited uptake of S2S systems so far. Further use of S2S systems is expected as the database grows. Open access to the real-time forecasts with no operational delay may also increase the evaluation and development of S2S-based malaria early warning systems.

2.2 Dengue (Lowe)

Dengue is a mosquito-transmitted viral infection, widespread in tropical and subtropical regions (Guzman and Harris, 2015). Dengue epidemics in Brazil often occur without warning and can overwhelm the public health services (Lowe et al., 2016b). SCFs combined with early data from a dengue surveillance system provide the opportunity for public health services to anticipate dengue outbreaks several months in advance. This could improve the allocation of intervention measures, such as targeted vector control activities and medical provisions, to those areas most at risk.

In a recent study, Lowe et al. (2014) developed a prototype dengue early warning system for Brazil. Real-time SCFs and disease surveillance data were integrated into a spatiotemporal model framework to produce probabilistic dengue forecasts. The model was used to predict the risk of dengue 3 months ahead of the 2014 FIFA World Cup in Brazil, a mass gathering of more than 3 million spectators.

Brazil is divided into more than 550 microregions. The authors assessed the potential for dengue epidemics during the tournament by providing probabilistic forecasts of dengue risk for each microregion, with risk-level warnings issued for the 12 cities where the matches were played. The dengue early warning system was formulated using a Bayesian spatiotemporal model framework (Lowe et al., 2011, 2013b), allowing specific public health issues to be addressed in terms of probability. It was driven by real-time SCFs for the period March-April-May and the dengue cases reported to the Brazilian Ministry of Health in February 2014. This information was combined to produce a dengue forecast at the start of March 2014.

Predicted probability distributions of dengue incidence rates (DIRs) were summarized and translated into risk warnings, which were determined using dengue risk thresholds of 100 and 300 cases per 100,000 inhabitants, defined by the National Dengue Control Programme of the Brazilian Ministry of Health. The probability of dengue incidence falling into predefined categories of low, medium, and high risk was mapped using a visualization

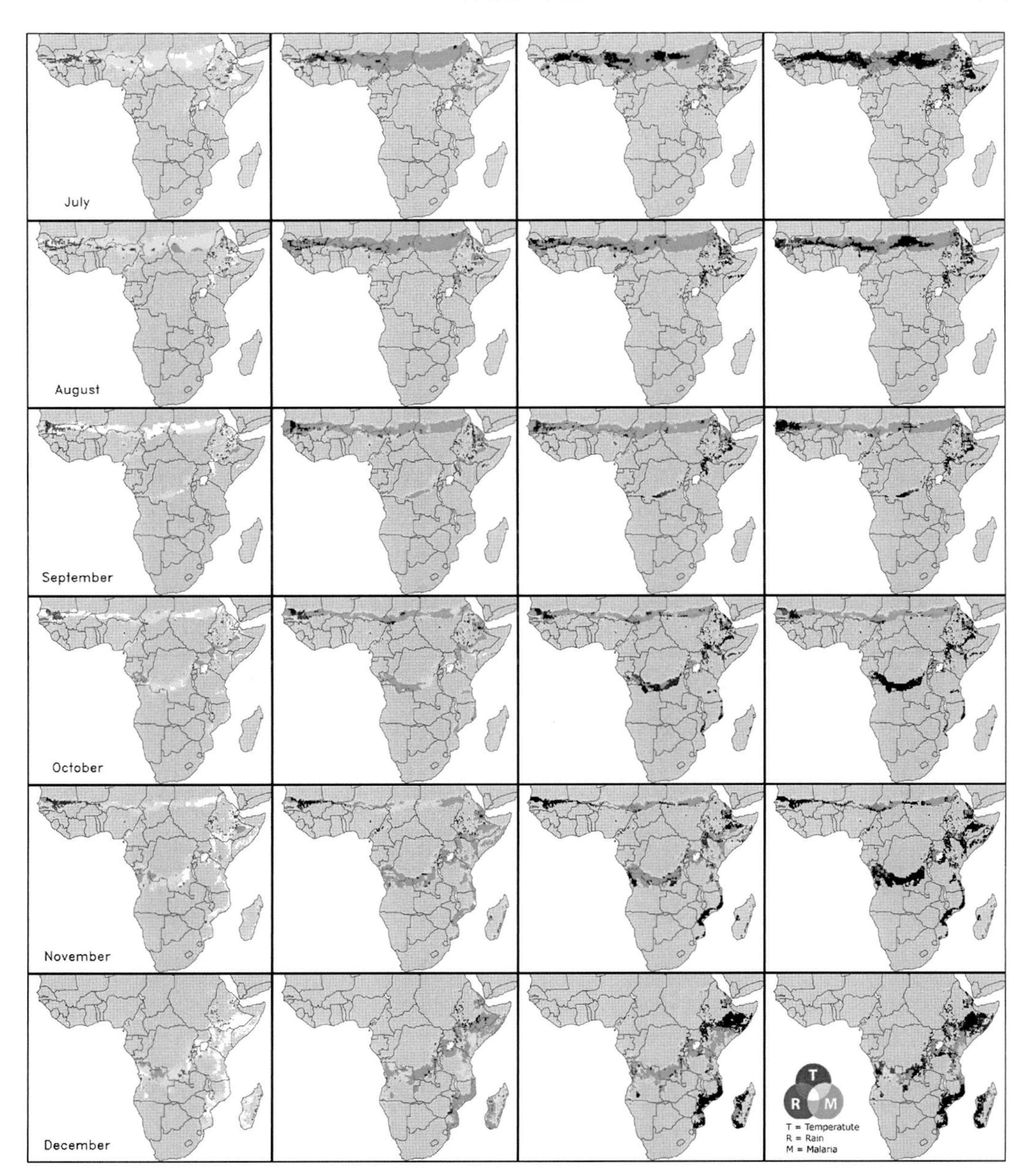

FIG. 1 Points where forecast skill is significant for temperature, rainfall, and an analog of malaria cases at lead times from 1 to 4 months over Africa, using reanalysis-derived data for evaluation (see Tompkins and Di Giuseppe (2015) for details of the calculation methodology). Points are filtered to show only those areas where malaria transmission is highly variable for the month in question (i.e., where malaria is epidemic, or where the onset of the transmission is highly variable). The graph shows that in the first month, often all three variables are skillfully predicted, or just temperature and malaria are. From month 2 onward, only malaria is successfully predicted, due to the lag between rains and malaria. This highlights the importance of correctly initializing the malaria modeling system. This is a challenge because real-time information on vector densities, puddle availability, and parasite prevalence is not available. It also shows that the successful climate forecasts in months 1 and 2 extend the malaria predictability range. *Reproduced with permission from the American Meteorological Society.*

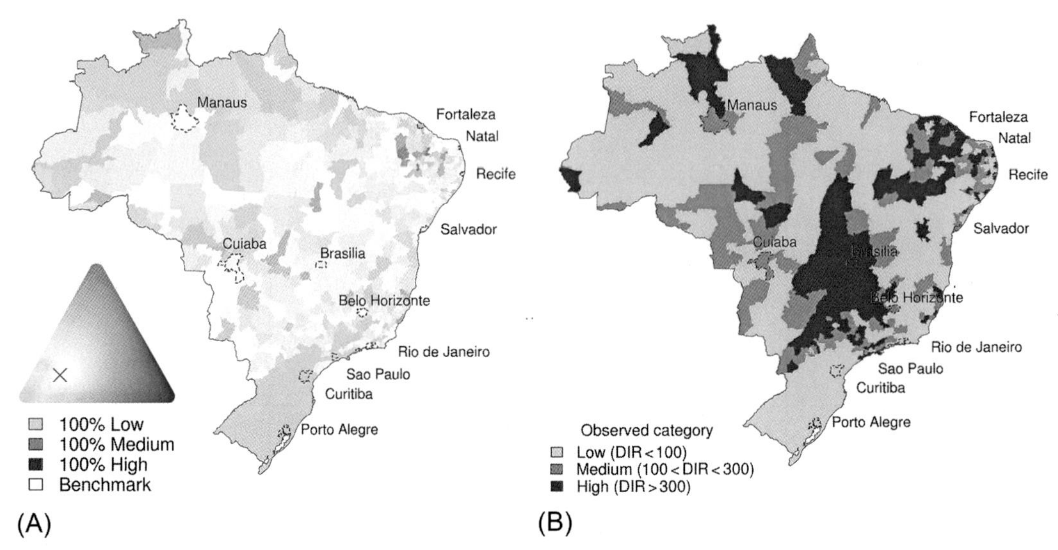

FIG. 2 Probabilistic dengue forecast and observed dengue incidence rate categories for Brazil, June 2014. (A) Probabilistic dengue forecast for Brazil, June 2014. The continuous color palette (ternary phase diagram) conveys the probabilities assigned to low-, medium-, and high-risk dengue categories. Category boundaries defined as 100 cases per 100,000 inhabitants and 300 cases per 100,000 inhabitants. The greater the color saturation, the more certain is the forecast of a particular outcome. Strong *red* shows a high probability of high dengue risk. Strong *blue* indicates a high probability of low dengue risk. *Colors close to white* indicate a forecast similar to the benchmark (long-term average distribution of dengue incidence in Brazil), marked by a *cross*. (B) Observed dengue incidence rate categories for June 2014. *Lowe, R., et al., 2016. Evaluating probabilistic dengue risk forecasts from a prototype early warning system for Brazil. elife 5, e11285.*

technique in which color saturation expressed forecast certainty (Jupp et al., 2012). A verification map, expressing the past performance of the model, was provided along with the forecast. This indicated how "trustworthy" the model was in different areas of Brazil.

Following the tournament, the authors evaluated the prototype against the actual reported cases of dengue during the event. Fig. 2 shows a ternary probabilistic forecast map and the corresponding observed DIR categories (low, medium, and high). The model correctly predicted a high probability of low dengue risk in South Brazil and large areas of the Amazon (with certainty depicted by color saturation). High-risk microregions were also correctly detected in parts of Northeast Brazil. Observed DIRs were greater than expected in Brasília, although the likelihood of observing high levels of dengue for the surrounding region was relatively greater than observing lower levels. For some microregions in the state of São Paulo, the model was uncertain about the most likely category (depicted by pale colors). Some of these areas experienced high DIRs in June 2014.

The forecast model correctly predicted the dengue risk level in 7 of the 12 cities that hosted the World Cup games. The forecast model was compared to a null model based on seasonal averages of previously observed dengue incidence. When considering the ability of the two models to predict high dengue risk across Brazil, the forecast model produced more hits and fewer missed events than the null model, with a hit rate of 57% for the forecast model as opposed to 33% for the null model. Therefore, the forecast model, based on SCFs and early surveillance data, outperformed a simple seasonal profile, upon which decisions are typically based. The implementation of SCFs and early reports of dengue cases into an early warning system is now a priority for public health authorities (Lowe et al., 2016b). This action is likely to help them to prepare for and minimize epidemics of dengue and other diseases transmitted by the same mosquito vector, including Zika and chikungunya.

The prototype dengue early warning system for Brazil developed from a Leverhulme network project: EURO-Brazilian Initiative for Improving South American Seasonal Forecasts (EUROBRISA), which explored how SCFs could be better exploited to improve climate resilience in South America (Coelho et al., 2006). EUROBRISA is an operational forecasting system designed to produce 3-month average forecasts, issued 1 month ahead. To accommodate the design of the seasonal forecasting system in the dengue model framework, precipitation and temperature variables were averaged over the 3 months prior to the month in which dengue predictions were valid. However, climate data at the monthly, weekly, or daily scale, such as diurnal temperature range or number of days in a month with rainfall exceeding set thresholds, may better resolve the climate sensitivity of the mosquito life cycle (Lambrechts et al., 2011; Chen et al., 2012). Indeed, one of the challenges of moving toward the use of S2S systems is the need to use weekly rather than monthly averaged climate data as a driver.

In Ecuador, lag times ranging from several weeks to 2 months between dengue and rainfall and minimum temperature were found to be important for dengue prediction (Stewart-Ibarra and Lowe, 2013; Stewart-Ibarra et al., 2013). A recent study used SCFs to predict the evolution of the 2016 dengue season in the coastal city of Machala in Ecuador, following one of the strongest El Niño events on record (Lowe et al., 2017). The forecasts were obtained from the Climate Forecast System (CFSv2) model, developed by the National Center for Environmental Research (NCAR), using the grid point most representative of the climate in the coastal area of Machala. The forecasts were arranged as a 24-member ensemble initiated January 1, 2016, producing monthly averages of precipitation and daily minimum temperatures for the 10 months following the forecast start date.

Given a delay of 1 month between anomalies in climate variables and dengue incidence, dengue forecasts were produced up to 11 months ahead. These climate-driven forecasts successfully anticipated the peak in dengue incidence to occur 3 months earlier than expected, based on seasonal averages (Lowe et al., 2017). The temporal resolution of the dengue forecasts could be improved even further by incorporating sub-SCF information in the dengue model framework, which is an area of active investigation.

2.3 Meningitis (Martiny, Roucou, and Nakazawa)

Bacterial meningitis (meningococcus, or *Neisseria meningitidis*) is a highly contagious, person-to-person, infectious disease of the meninges, the thin layers that cover the brain and spinal cord. Some regions of the world are particularly exposed to epidemic risks, like the "African meningitis Belt" (Lapeyssonnie, 1963) a semiarid region that extends from Senegal to Ethiopia. This region suffers the highest incidence in the world, with both annual local epidemics and multiannual large waves affecting the whole meningitis belt (Broutin et al., 2007).

The Meningitis Environmental Risk Information Technologies (MERIT) initiative (Thomson et al., 2013) was established in 2007 as a multidisciplinary community of partners who conduct research to advance the use of climate-related information in strengthening public health strategies for the control of epidemic meningitis in Africa. Coinciding with this initiative, a new conjugate vaccine called MenAfriVac (Frasch et al., 2012), against the dominant serogroup A (NmA), was tested and rolled out across the region. The vaccine rapidly suppressed case numbers, but it has not eliminated the need for a meningitis early warning system or a better understanding of environmental drivers of disease. Serogroups other than A pose a significant epidemic threat and currently cannot be controlled by preventative vaccination (Agier et al., 2017).

The transmission and dynamics of meningitis can be influenced at different scales by host immunity, coinfections, household lifestyle, population density and dynamics, and socioeconomic conditions (Agier et al., 2017). Additionally, a link with climate was found in the 1960s, and the disease is now recognized as one of the most climate-sensitive in Africa. The World Health Organization (2012) stated in 2012 "While the temporal association between climate and meningitis is evident, what triggers or ends an epidemic is as yet unknown". One assumption made by epidemiologists is that extremely low air humidity combined with high dust loadings that sometimes persist over many weeks may contribute to host susceptibility, including physical damage to the mucosa to the point where the colonizing meningococci are more likely to invade the nasopharyngeal epithelium (Mueller and Gessner, 2010). In any case, despite remaining uncertainty in the infection mechanism itself, the recognized climate risk factors of hot, dry, and dusty conditions are potentially predictable at S2S timescales, given accurate enough forecasting systems.

The geographic boundaries of the Belt are limited by the isohyet 300 mm to the north (approximately 15 degrees N) and 1100 mm to the south (7 degrees N) (Lapeyssonnie, 1963), which corresponds to the climate definition of the Sahel. It is well documented in previous studies that meningitis outbreaks occur in the dry season, from January to May, dominated by strong, warm, dry, and dust-laden northeasterly Harmattan winds, which blow from the Sahara Desert to the sub-Sahel regions, and that the premonsoon rainfall halts the outbreak (Sultan et al., 2005; Thomson et al., 2006b; Dukić et al., 2012; Broman, 2013; Pérez García-Pando et al., 2014; Pérez et al., 2014; Cuevas et al., 2015; Pandya et al., 2015; Diokhane et al., 2016). A recent review (Agier et al., 2017) reported the statistical methods (regression models, disease mapping, hypothetical explanatory models, mathematical modeling) employed to explain the meningitis incidences at different spatial (country/region/district/individual) and time (year/season/month/week) scales. The most commonly used climate factors are wind, humidity, temperature, and dust.

Climatic variability explains 25% of the year-to-year variability of meningitis according to previous analysis (Yaka et al., 2008). The meridional wind component in October, November, and/or December seems particularly important: An enhancement of the Harmattan flow at the beginning of the dry season may affect the number of cases in meningitis and provoke epidemics several months later. The climate conditions in the October-to-December season sometimes may permit early cases in meningitis that could, in turn, influence the final size of the epidemics through contacts between people and the increased size of the reservoir. This statistical model was refined at the health sanitary district scale in Niger by adding as new predictors dust at the beginning of the dry season, early cases of meningitis, and population density (Pérez et al., 2014). The next key challenge is to explain the meningitis incidence at specific points of the epidemics (most notably the onset) in order to get closer to the operational needs.

Burkina Faso is one of the countries most affected by meningitis in the Belt (Agier et al., 2008). Based on the attack rate (number of cases in meningitis/population × 100,000), 7 years are identified as epidemic in the country for the period 1979–2014: 1984, 1996, 1997, 2001, 2002, 2006, and 2007. Monthly climate composites (epidemic years minus all years) highlight specific synoptic conditions in epidemic years. The November composite highlights a significant reinforcement of the surface wind in the northeasterly direction (Fig. 3A). This affects relative humidity, which experiences significant dry anomalies in the area (Fig. 3B).

The work of Agier et al. (2008) then used a standarized WHO weekly meningitis incidence dataset at the sanitary district scale for the 1997–2007 period to determine the onset week of the meningitis epidemic years. The average onset date falls between the fifth and sixth weeks of the year (±1.6 weeks). Weekly climate composites (epidemic years minus all years) were constructed to highlight specific conditions that may occur prior to the epidemic onset. A significant Harmattan wind reinforcement was found in the weeks before the averaged onset date.

Sultan et al. (2005) developed an HEWS meningitis index in Mali, demonstrating that the week of the epidemic onset is highly correlated to the week of the winter maximum (the sixth week of the year, ±2 weeks). Their strategy for an HEWS involves both a longer lead-time seasonal prediction made in November preceding the transmission season, supplemented by weekly updated predictions based on wind, temperature, and humidity predictions from the end of January onward. It is in this second phase of shorter lead-time weekly updates where the newly available S2S systems would naturally fit. Indeed, more recent analysis of the disease in Niger demonstrated that both zonal wind and dustiness could be used to predict cases on a local and district scale (Pérez García-Pando et al., 2014), emphasize the role that S2S systems could play in forecast meningitis incidence at a finer weekly timescale, initially using the winds produced by the systems directly. Because most S2S systems do not presently provide aerosol information in their output, the S2S system winds also could be used to drive an offline dust model as an input to the early warning system.

A differential equation for meningitis incidence was applied to the multivariate log-linear regression analysis using four meteorological variables (northeasterly surface wind, relative humidity, rainfall, and temperature, and one of four dust products over Burkina Faso), to determine their respective contribution to the number of cases of meningitis (Nakazawa and Matsueda, 2017). They showed that northeasterly wind makes a major contribution to the rate of change of the number of cases. Although the highest correlation coefficient between the estimated and observed tendency was for the regression models using all four meteorological variables in addition to the dust surface mass concentration data, even a single-parameter model using just wind was significantly skillful, simplifying the development of HEWSs. Other recent studies have compared dust to other climate variables as drivers of meningitis in the January-March season (JFM)—(i.e., concomitantly with the epidemics. At the country scale (Niger and Mali), the variability in the weekly number of cases in meningitis seems to be triggered by that of dust, with a 1- to 2-week time-lag between dust and meningitis (Martiny and Chiapello, 2013). Humidity observations also have been used in early warning systems

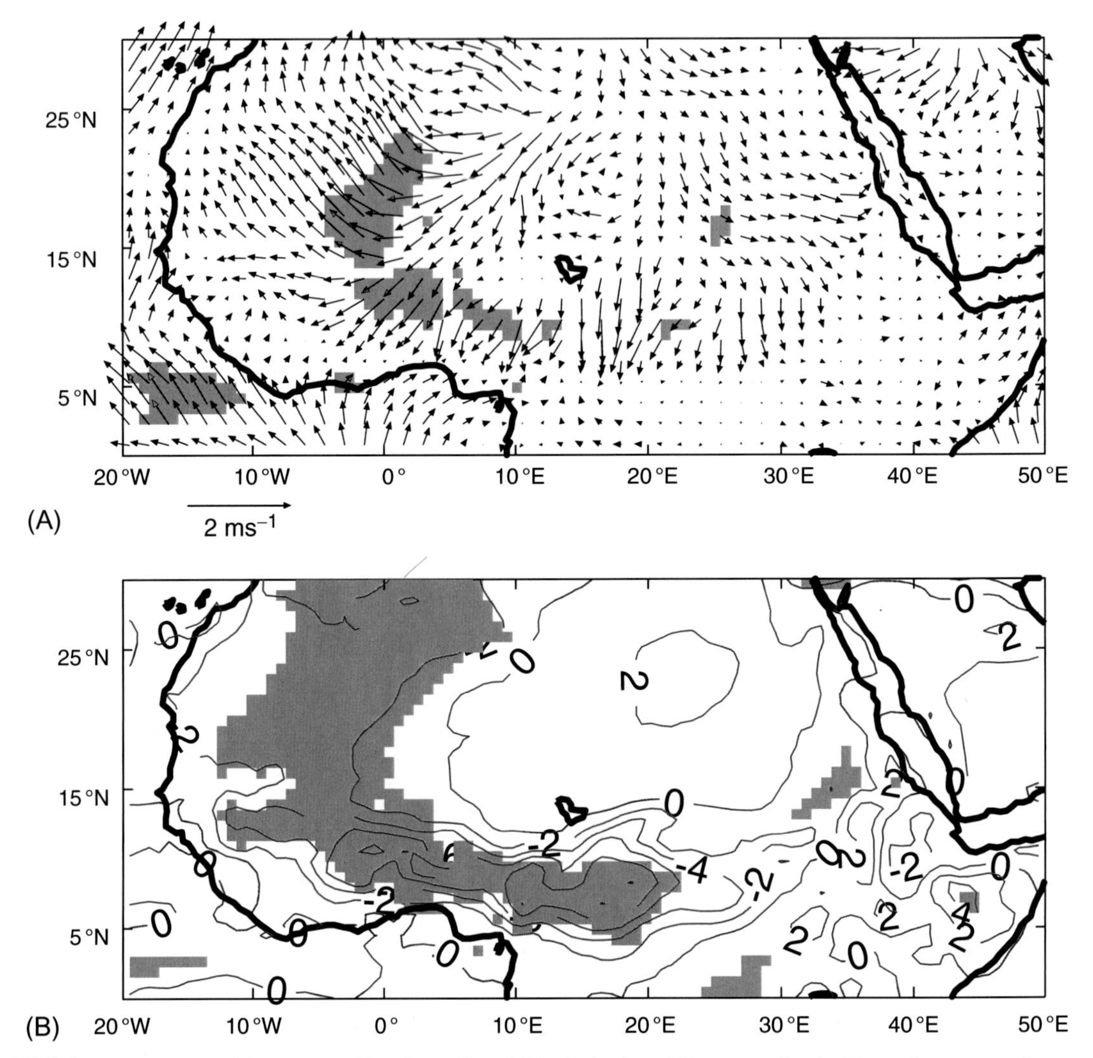

FIG. 3 ERA Interim. (A) 10-m wind (see legend) and (B) relative humidity anomalies for November averaged over meningitis epidemic years 1984, 1996, 1997, 2001, 2002, 2006, and 2007, relative to the 1979–2016 mean. *Shaded areas* mark statistical significance ($P = .1$).

with a 2-week lead time (Pandya et al., 2015). This time-lag, confirmed at the district spatial scale in Niger (Agier et al., 2008) and using different aerosol datasets (Deroubaix et al., 2013) is consistent with the incubation time period of the Nm bacteria of around 14 days (Stephens et al., 2007). In India, Sinclair et al. (2010) highlighted the relationship between the end of the meningitis epidemics and the increase in relative humidity, and it is possible that S2S-driven forecast systems may have applications outside the African continent, subject to rigorous local epidemiological study and evaluation.

If observations of meteorological conditions and aerosol loadings can predict case numbers on a 2-week lead time, then the skillful prediction of these parameters by S2S systems may extend the lead time by a further 4–8 weeks, depending on the limitation of the S2S skill. Thus, relative to diseases such as malaria that usually have longer lags with climate, the impact of S2S on the length of the advance early warning could be significant for meningitis.

2.4 Heat Waves (Nissan and Lowe)

Extreme heat is the leading cause of weather-related deaths in the United States and Europe (Changnon et al., 1996; Klinenberg, 2015; Lowe et al., 2015), and was responsible for 4 of the 10 deadliest natural disasters worldwide in 2015 (UNISDR and USAID and Centre for Research on the Epidemiology of Disasters, 2015). However, heat-related illness and deaths are largely preventable. Heat action plans (HAPs) are operational in many locations worldwide and have been shown to save lives by improving preparedness (Ebi et al., 2004; McGregor et al., 2014; Natural Resources Defense Council et al., 2017; Bittner et al., 2013).

Reliable weather and climate forecasts can facilitate better short- to medium-term resource management when incorporated into a broader strategy to implement measures over a range of timescales, from long-term disaster risk reduction and seasonal preparedness to early warning systems and emergency response and recovery. A heat HEWS, when paired with appropriate actions to reduce risk, can be an effective component of such a strategy. As stated in the introduction, a heat HEWS involves the identification of critical meteorological thresholds associated with adverse health impacts for the location in question, the issuance of a forecast indicating whether these thresholds are likely to be reached, and a set of standard operating procedures to implement risk-reduction measures.

An effective heat HEWS is tailored to the local context, with thresholds determined according to the vulnerability of the local population and forecasts relating to these thresholds issued at lead times appropriate to the local capacity to take action. Vulnerable sections of the population can be identified through research on heat-health relationships (where sufficient data are available) and through surveys and community consultations. Appropriate actions to take in response to an alert vary depending on the target vulnerable populations, capacity of local government and stakeholders to respond, available resources, and skill of the forecast.

The city of Ahmedabad, in India, is the site of South Asia's first HAP (Knowlton et al., 2014a). The plan was developed in response to a devastating heat wave in 2010, during which nearly 1350 additional people died, compared with the same period in previous years (Azhar et al., 2014). Similar HAPs are now being rolled out across 13 other cities in India (Council, 2016). Analyses of daily maximum temperature data, all-cause mortality, hospital admissions, and ambulance calls, plus focus groups and sampling surveys, identified people living in slums, outdoor workers, the elderly, and neonates as the most vulnerable population groups. Critical maximum temperature thresholds for different levels of alert, from "no alert" to "extreme heat health alert" were determined by consensus after analyses of temperature and mortality data during the 2010 heat wave (Knowlton et al., 2014a; Azhar et al., 2014). Once an alert is issued, the plan outlines a series of actions to be taken, which include alerting key stakeholders such as community workers and hospitals, opening cooling centers and circulating warnings via text messages and radio broadcasts. In addition to an early warning system to respond to forecasts of an imminent heat wave, the HAP involves seasonal activities including a community outreach and awareness raising program, training of key personnel, and procurement of ice packs and other supplies.

The Ahmedabad HAP now uses 7-day deterministic forecasts for maximum temperatures issued daily by the Indian Meteorological Department (IMD). Warnings are issued at the discretion of a dedicated nodal officer from 1 to 7 days in advance, in consultation with forecasters. More established heat HEWSs utilize probabilistic forecasts. In the United Kingdom (UK), heat alerts are triggered when there is a 60% probability of critical daytime and nighttime temperature thresholds being reached on at least 2 consecutive days (England et al., 2015). False alarms can be costly and damaging to the credibility of the key individuals and institutions responsible for implementing adaptation measures (Coughlan De Perez et al., 2015). Using a probabilistic forecast can maximize the time available to prepare, while minimizing the risk of a false alarm to a level considered acceptable by decision-makers. A compromise is struck between having confidence in the forecast and obtaining adequate lead time to take action. UK forecasts usually reach the minimum 60% confidence level approximately 2–3 days before a heat-wave hit.

Many risk-reduction actions require more advanced warning than the few days provided by weather forecasts. Fig. 4 illustrates some of the measures that could be taken to prevent the health impacts of extreme heat if the lead time of forecasts could be extended. For this reason, extending the lead time of forecasts beyond the weather timescale is an important method of improving disaster preparedness for a range of hazards (Knowlton et al., 2014a; IFRC, 2008; England et al., 2015; Letson et al., 2007; Vitart et al., 2012). At longer lead times, forecast reliability becomes critical for early warning systems. Deterministic forecasts give no indication of the degree of confidence in the prediction, which is a particular concern with sub-seasonal timescales, where skill is lower. A reliable probabilistic forecast is one where the probability assigned to a certain outcome corresponds to the observed probability of that outcome. For example, when heat-wave conditions are forecast with 70% probability, a heat wave should occur exactly 70% of the time (refer to chapter 16 for further detail). Reliable forecasts are essential because they allow decision makers to pair appropriate actions with forecast skill. Low-regret interventions, such as recapping emergency response procedures or closely monitoring weather forecasts, can be paired with low-probability forecasts. Actions that incur a higher cost, which might include rescheduling outdoor sporting events or setting up emergency drinking water stations, can be contingent on a higher probability trigger.

In some ways, heat waves present a good test case for the use of sub-seasonal forecasts in early warning systems. Adaptation measures for hydrometeorological hazards like flash floods and cyclones can involve high-cost interventions such as evacuations and flood defense reinforcements. By contrast, there are many low-cost, no-regret adaptation options appropriate to coping with extreme heat, from refreshing training for medical staff and community outreach officers to procuring emergency drinking water (Indian Institute of Public Health Ganghinagar et al., 2017).

At longer lead times it becomes difficult to specify the precise location of hazards, so forecasts must be issued for a larger region than is possible for high resolution weather forecasts for a few days ahead (Vitart et al., 2014). A forecast covering a larger area presents a challenge for early warnings of hydro-meteorological hazards like flooding or heavy rainfall, which are often highly localized. By comparison, hot and cold conditions do not occur in

FIG. 4 Preparedness and risk-reduction strategies for extreme heat that could be undertaken in a city based on short-term, S2S forecasting systems. Specific measures would vary according to local needs, but these examples were taken from heat action plans in India, the UK and elsewhere, and from a climate services workshop in South Asia (Knowlton et al., 2014b; McGregor et al., 2015; PHE, 2015). Seasonal preparedness measures also can be initiated without climate forecasts, based on an understanding of the seasonality of heat-wave risk in a given location. Items in gray indicate actions that may become possible as sub-seasonal forecasts improve in skill and become operationally available.

adjacent neighborhoods of a city. For heat waves, vulnerability and exposure are more important factors in determining risk than the hazard, which will be felt across a fairly broad spatial area. Therefore, assistance can be targeted toward the most vulnerable (e.g., the elderly or young children) and exposed (e.g., construction workers), reducing pressure on long-range forecasts to provide information at a fine spatial scale.

Furthermore, predictability of heat waves has been demonstrated on sub-seasonal timescales for some regions. Strong coupling between atmospheric temperatures and land surface conditions has been noted in several studies, with low soil moisture playing a role in the major European and Russian heat waves of 2003 and 2010 (Miralles et al., 2014; Hirschi et al., 2011). The persistent memory of the land surface suggests that forecast skill could be extended beyond weather timescales where soil moisture plays a role. In Bangladesh, low soil moisture conditions are detectable for several weeks on average before a heat wave (Nissan et al., in press).

A recent study developed a climate-driven mortality model to simulate the probability of exceeding emergency mortality thresholds for heat-wave and cold-spell scenarios in 54 European regions across 16 countries (Lowe et al., 2015). The model showed good skill when driven by observed apparent temperature data from reanalyses (in other words, to simulate heat-wave impacts in near-real time). Next, the extent to which S2S climate forecasts could be incorporated into HAPs to support timely public health decision-making ahead of imminent heat-wave events in Europe was assessed (Lowe et al., 2016). Forecasts of apparent temperatures at different lead times, ranging from 1 day up to 3 months, were used to generate probabilistic mortality forecasts for the 2003 heat-wave event in Europe and compared with the mortality forecasts driven by observed apparent temperatures. The model showed that after 1 week, skill decreased rapidly at forecast lead times longer than 1 week. Nonetheless, in some regions in Spain and the United Kingdom, excess mortality was detected up to 3 months ahead. In general, the skill of the mortality forecast was not limited by the mortality model, but rather by the predictability of the climate variables at S2S timescales over Europe.

In conclusion, these studies suggest that predictability for extreme heat exists on subseasonal timescales in some regions, indicating that incorporating S2S forecasts into existing heat-health action plans could improve preparedness. Skillful and reliable forecasts are an important component of an HAP, but they can be effective only when paired with a coordinated awareness campaign and engagement with a range of institutions and individuals across different sectors of society.

3 OPERATIONALIZATION: CHALLENGES AND OPPORTUNITIES

Despite the obvious and growing potential to apply S2S forecasts of surface climate parameters to predicting health outcomes and plan mitigating actions, there has been a dearth of operational implementations to date. Many demonstration projects that successfully show the potential of such systems have not developed beyond the research phase or led to benefits for health and more efficient allocation of health resources. Some of the potential reasons for these operational bottlenecks are discussed next.

3.1 Data Access and Usage

One of the key impediments to operationalizing health prediction systems is a simple lack of access to high-quality observations for climate and health. Pay-for-access information hampers operational development in two ways. Not only does it prevent the operationalization of health forecasting systems directly, it inhibits the research needed to develop these systems in the first place. Research is held back by the difficulty in accessing in situ data for national networks of station data, which is often difficult to obtain for a health ministry from the national agency of the country. Similarly, while some countries lead the way in making aggregated health datasets openly available and easily accessible through the health ministry websites, this remains the exception rather than the rule, and access to health data for research remains elusive. Often health data are available only through specific research projects and are subject to strict third-party conditions despite the possibility of applying rigorous anonymization techniques to ensure that ethical privacy concerns are met.

Several developments are underway that are in the process of improving this situation. Concerning climate observations, most reanalysis datasets produced by operational centers, which combine a variety of observations, have been free to use for research, such as those by National Centers for Environmental Prediction (NCEP) (Kistler et al., 2001) and ERA-Interim (Dee et al., 2011). While invaluable, these products are available on coarse grids on the order of 100 km; the high-resolution operational analyses (8 km at ECMWF as of 2017) remain closed. However, improved reanalysis products will soon become available as part of the climate services of the Copernicus program funded by the European Union (EU) (e.g., ERA-5). Another current development is the Enhancing National Climate Services (ENACTS) project, an initiative of the international Research Institute for climate and society (IRI) (Dinku et al., 2014; Thomson et al., 2014). This project aims to merge satellite data with all available national station data in order to produce a high-quality, gridded rainfall dataset that can be used by national services. The ENACTS products are already available for Rwanda, Kenya, Madagascar, Tanzania, Ethiopia, Zambia, Mali, Ghana, and soon Uganda as well.

Access to health data also should improve as more countries are introducing second-generation, web-based, digital information systems (Karuri et al., 2014). In addition to improving the reliability of data, these systems allow health data to be quickly and easily aggregated for wider use, which should facilitate their use for research, ultimately benefiting health organizations.

Concerning the forecasts themselves, the commercial value of weather forecast information has resulted in operational centers outside the United States implementing a closed-access data policy the short-range forecasts, which has been extended to S2S products as these were progressively added to the operational catalog over the past two decades. Where forecast information is made available, it often takes the form of graphical maps, which cannot be used to drive downstream statistical or dynamical models. Again, several developments are underway to address this. Sub-seasonal and seasonal systems must run large sets of forecasts for past periods to allow them to be calibrated for biases and errors. These datasets represent an invaluable resource for research, and several efforts have been made to release them from multiple centers in a single location. The World Climate Research Programme's (WCRP) Working Group on Sub-seasonal to Interdecadal Prediction (WGSIP) has the flagship Climate-system Historical Forecast Project (CHFP), which makes seasonal hindcasts publicly available from a large number of operational centers (Tompkins et al., 2017). Since 2015, the S2S database has been established to provide access to sub-seasonal prediction systems (Vitart et al., 2016). Since late 2017, the EU program known as the Copernicus Climate Change Service (C3S) has made digital data from three operational seasonal forecast systems available in real time, in addition to the associated datasets of hindcasts. We are thus entering an era of unprecedented climate forecast data access for both research and operational activities.

3.2 Operationalization of Climate Information

In some respects, the greatest impediment to the uptake of climate information in health planning may not be the barriers to developing the systems, but the adaptation of operational strategies required to make effective use of them. While the seasonality of many diseases is well understood and the annual cycle of disease interventions already takes account of this known history (Tompkins et al., 2016b), the use of predictions of monthly or seasonal anomalies of the occurrence of a specific health impact represents a new operational paradigm. Endemic diseases with a well-defined seasonal cycle are tackled with a regular set of interventions; or where a disease may be subject to irregular epidemic outbreaks, the focus is on surveillance and reaction. Using forecasts to plan intervention or stockpile medicines for the months ahead requires unfamiliar operational procedures to be adopted.

To integrate climate information into planning, effective communication about the spatial scales and uncertainties of the forecasts may be required. In our experience of running workshops interacting with both academics and health officials, a frequently raised objection to the use of climate information concerns the spatial scale of the information. The argument is that health outcomes are usually highly heterogeneous over small spatial scales, and thus the coarse spatial scale of the forecast system driven by O (10–100 km) scale climate information may provide limited useful guidance at the local level. This argument confounds mean risk with risk anomalies. The climate-driven anomaly, which will likely occur on a scale of tens or even hundreds of kilometers, is superimposed upon a heterogeneous vulnerability surface that reflects changes in the vicinity of water, income levels, availability of health care and so on. Mapping efforts such as Kienberger and Hagenlocher (2014) aim to help project the climate-driven hazard to smaller spatial scales. Moreover, at the local and district scales, health district officers are familiar with the disease incidence landscape within their jurisdiction. Naturally, factors that affect both vulnerability and transmission intensity also change from year to year, but these are usually inherently unpredictable, providing one of many sources of error in model predictions. One of the tasks of an S2S forecast system, therefore, is to attempt to determine the locations and months where climate may determine a high proportion of the interannual variability in transmission (Tompkins and Di Giuseppe, 2015).

A second aspect relates to the confidence that can be placed in forecasts. Information at S2S timescales is uncertain and is assessed using ensemble forecasting techniques, as introduced in this book, which could also be applied to models of impacts (Ruiz et al., 2014; Caminade et al., 2014). Accounting for uncertainty in decision-making is highly challenging, and accurately assessing and effectively communicating forecast uncertainty are crucial. One aid to this process is the concept of cost-loss economic analysis (Murphy, 1977). This assesses ensemble or deterministic skill in terms of the cost benefit if a mitigating action is correctly taken on the basis of the forecast, therefore avoiding a loss (e.g., preventing an outbreak by rolling out interventions, or mitigating the impacts of heat waves). This is offset against the losses incurred when the forecast was incorrect, leading to the wasted cost of an intervention, or worse still, an outbreak that was not prepared for. While such an analysis can offer guidance, it is often difficult to precisely define a threshold for a mitigating action and to associate costs by translating health outcomes into equivalent disability-adjusted life years (DALYs).

In health, not only economic losses (e.g., lost productivity) arise, but also serious and possibly fatal health implications. If funds are adequate, a health officer may always err on the side of caution and take precautionary action, rather than risk appearing unprepared for a heat wave or disease outbreak. Moreover, if action is taken, it is difficult to certify whether a health crisis consequentially has been avoided, or it would not have happened in the first place (Lowe et al., 2016b). While forecasts of extreme temperatures can be validated subsequently, it is not possible to say whether interventions have prevented an otherwise predicted outbreak of a disease such as dengue or malaria, unless neighboring regions with similar climate anomalies and no interventions are subjected to an anomalous disease outbreak. Our present understanding of the efficacy of disease interventions still rests heavily on idealized modeling, which cannot easily account for the situation on the ground.

This disconnect between the information potentially provided by climate-driven early warning systems and the reality of a health officer's decision-making process is perhaps one reason why there is slow uptake of climate information in early warning systems in most cases. Heat-wave early warning systems are perhaps the exception, as they can be instantly and easily validated and have high proven skill in the shorter lead times of a few days, which still can provide useful warning. More effort is required to project climate-based forecast information onto health decision processes. Does the probability of an upper-tercile event or outcome (a commonly used criterion to designate "above normal" conditions in meteorological ensemble forecasts) relate to a particular decision process in health? In general, this has not been demonstrated. This disconnect can partially explain low demand from the health sector for early warning products. This relates not only to the coarse spatial scale of the forecast data, but also to the limitations in terms of the lead time provided at an adequate skill.

As an operational example, the time taken to acquire stocks and subsequently distribute vaccinations in the field during the rainy season for the 2006/2007 RVF outbreak that occurred in Kenya implied that the intervention came too late to affect the outbreak significantly (Jost et al., 2010). The difficulty in acquiring sufficient vaccines (with a limited shelf life) indicates that accurate S2S-based prediction at a 1 month lead time would not provide sufficient lead time for intervention. In this case, seasonal forecasts with longer lead times of several months would be required to be effective with presently available vaccines and distribution infrastructure.

3.3 Interaction Through Workshops

Underlying some of these bottlenecks to the uptake of climate information is a communication breakdown between the multifaceted players in this interdisciplinary issue. Health officials may lack understanding concerning the potential of climate data to improve public health outcomes, and how to access/interpret climate data or climate-based forecast products. Additionally, small-scale research demonstration projects may be dismissed as unfeasible to implement over larger scales (Awoonor-Williams et al., 2004). Likewise, climate and weather experts may often have a poor understanding of health-planning needs and the decision-making processes within a national or regional health hierarchy. Health planning is frequently top-down in nature, with interventions and mitigation actions planned at the national or regional level based on the advice of international bodies such as the WHO, or funding agencies and donors. This leaves limited scope to benchmark and possibly introduce new methodologies and technologies on smaller national or subnational scales without explicit WHO backing.

Within this context, the WHO is increasingly recognizing the importance of integrating climate into planning and has established a joint office with the World Meteorological Organization (WMO) to sustain these efforts. The WMO has also made health a key priority in its Global Framework for Climate Services (GFCS), a program that aims to coordinate and communicate research and observational developments to stakeholders at national, regional and global scales. S2S and seasonal forecasting may gain a foothold by building on timescales that are already familiar. Examples include short-range forecasts, prediction of weather extremes, or indeed long-term climate projections that are increasingly incorporated into the National Adaptation Programme of Action (NAPA) plans in least-developed countries (Füssel, 2007; Kalame et al., 2011). Overall, it is difficult to derive decision-relevant information from SCFs due to their lack of skill in many regions. However, shorter-range forecasts may add value due to their enhanced forecast quality, as demonstrated with heat-related mortality forecasts (Lowe et al., 2016). The focus on sub-seasonal predictions with better forecast skill may help to increase the uptake and use of climate forecasts in Europe. Focusing on top-down, national-level decision support, however, ignores many of the potential gains that can be made using information at health-district levels, where health decisions can have a direct and immediate impact on local populations. While health district officers have busy schedules and rigid targets, local public officials in developed countries are increasing becoming climate aware (Bedsworth, 2009) and there is no reason why this trend should not occur in tandem in developing countries.

Forecasts always need to be considered in context. If a hypothetical forecast is issued for "above-normal" conditions, then the user must have a good understanding of what "normal" is and how the observed climate influences the health issues that they care about. Forecasts for vector-borne diseases can be made with 2 months or more lead time using observed rainfall and temperature, due to the time that it takes for vector populations and pathogen cycles to build an epidemic cycle once climate conditions are conducive. In these predictions, uncertainty results from the quality of the observations rather than the inherent stochasticity of predictions. Until recently, the importance of health-sector researcher and practitioner access to historical and monitoring climate data has been ignored by the climate community, with restrictive data policies operating in many countries. We argue that familiarity with and use of observed data at appropriate temporal and spatial scales (such as provided by ENACTS) provide the foundation for the uptake of less certain, but forward-looking forecast information.

Progress in this matter can occur through increased communication, which is most readily facilitated through workshops. The development of new curricula for health practitioners that introduce climate information as a possible resource is also needed. It is imperative to stress the necessary two-way communication that must take place, so that experts in climate can understand the ways in which their forecasts might be potentially used, while at the same time attempting to communicate the usage and uncertainties that climate products have. Once again, involvement of, and sponsorship from, pan-national organizations such as WHO are invaluable through the support and coordination of the GFCS. Scaling up such events to increase their value and penetration may require further involvement of e-training resources in tandem with traditional in situ workshop settings (Barteit et al., 2015). Documentation of successful events provides helpful blueprints for duplication (Tompkins et al., 2012; Lowe et al., 2016c).

4 OUTLOOK

This chapter has attempted to summarize the opportunities that now exist to increase the uptake of S2S forecast information on the 2-week to 2-month timescale in health applications.

Four case studies have highlighted potential areas in which S2S could be employed or where pilot or preoperational systems are already in place. Often, the final step from a pilot to an operational system is the most challenging, and some of the key impediments for this uptake have been discussed. Despite this, S2S and related initiatives will lead to ever-increasing access to forecast and hindcasts in (near) real time. This will help demonstrate the potential and raise awareness of these systems and aid the rethinking of health policy to take advantage of this untapped source of predictability of health outcomes. The successful implementation of climate services for health requires political will and close collaboration among a range of partners, including climate scientists and practitioners, data managers, medical professionals, hospitals, public health agencies, and governments.

Acknowledgments

Adrian M. Tompkins would like to thank David Taylor, Mark Booth, and Jeffrey Mariner for discussions that took place during the course of the EU-funded HEALTHY FUTURES project (2011–14), which helped shape some of the arguments presented on the topic of forecast operationalization.

CHAPTER

23

Epilogue

Andrew W. Robertson[*], *Frédéric Vitart*[†]

[*]International Research Institute for Climate and Society (IRI), Columbia University, Palisades, NY, United States [†]European Centre for Medium-Range Weather Forecasts (ECMWF), Reading, United Kingdom

The aim of this book has been to provide an introduction, background, and overview of the current status of research, operational prediction, and applications in sub-seasonal to seasonal forecasting (S2S). We conclude here with some thoughts on future perspectives and the need for additional studies, many of which are raised at the end of the individual chapters.

Many sources of S2S predictability are discussed in Parts 1 and II, highlighting the important progress that has been made in recent years to better understand and predict these various sources of S2S predictability. One of the reasons for the recent increased interest in S2S forecasts is the significant skill displayed by statistical and dynamical models in representing these sources of predictability, most especially the Madden-Julian Oscillation (MJO; skill up to 5 weeks) and sudden stratospheric warmings. However, there is still a great deal of room for improvement in their predictions. For instance, the difficulty of dynamical models to propagate the MJO across the Maritime Continent and the strong underestimation of MJO teleconnections in the northern extratropics are important obstacles to the dynamical models fully exploiting these windows of opportunity. This could be viewed as good news because it suggests that the skill of current operational S2S forecasting systems can get a lot better. Other sources of S2S predictability beyond the ones described in this book may exist. For instance, atmospheric composition (aerosols, ozone, etc.), which is often represented with climatological values in S2S dynamical models, interacts with the meteorology and may affect S2S forecasts.

Although each chapter in Part II describes a single source of predictability, there is increasing evidence that these sources of predictability may not always be independent. For instance, Chapter 11, on sub-seasonal predictability and the stratosphere, highlighted the possible interaction between the Quasi-Biennal Oscillation (QBO) and the MJO, while some recent publications suggest a link between the MJO and sudden stratospheric warmings of the polar

Sub-seasonal to Seasonal Prediction
https://doi.org/10.1016/B978-0-12-811714-9.00023-1

vortex. These links are not yet fully understood, and their physical mechanisms are still hypothetical. Nevertheless, research on the interactions among various sources of predictability is likely to increase in momentum in the coming years. These interactions can also extend across timescales: for instance, because MJO and El Niño–Southern Oscillation (ENSO) affect each other, MJO variability may affect seasonal timescales. Longer timescales also can modulate S2S sources of predictability and their impacts, such as the impact of ENSO and the Pacific Decadal Oscillation (PDO).

Because the development of S2S prediction is much more recent than for medium-range and seasonal predictions, the methods used to produce S2S forecasts are largely borrowed from medium-range weather forecasting and seasonal prediction. Although in principle, there is no reason why seamless forecasts, from medium-range to seasonal, could not be produced from the same forecasting system, in practice medium-range and seasonal forecast systems have generally been developed separately, with the specificity of each timescale in mind. Some operational centers use their seasonal forecasts to issue sub-seasonal forecasts, while other centers extend their medium-range forecasts to produce S2S forecasts. It is not clear which strategy is better, and neither approach is likely to be optimal for S2S forecasts. For instance, ensemble generation is often tuned to produce reliable seasonal or medium-range forecasts. Sub-seasonal forecasts of the MJO are often strongly underdispersive. Very little research has been done to date on several modeling issues, such as the benefit of lag versus burst ensemble generation and the impact of horizontal and vertical resolution of the atmosphere and ocean in S2S forecasting systems. At the S2S time range, predictability comes from the initial conditions, but also from the atmospheric boundary conditions. It is important, therefore, to have a well-initialized and complex Earth-system model that includes interactive ocean and the cryosphere. Currently, most operational systems are initializing these Earth-system components separately. Coupled data assimilation of the atmosphere-ocean-land-cryosphere system is likely to bring more consistent S2S forecast initialization in the near future. Forecast calibration and multimodel combination are currently using techniques based mostly on previous experience with seasonal forecasting. S2S verification is also in its infancy, as discussed in Chapter 17, on S2S prediction of extreme weather.

Part III of this book has documented several areas where S2S prediction could benefit society through new weather/climate services, early warning/early action of extreme events, and climate-smart decision-making: disaster early warning, hydrology, agriculture, health applications, energy, and more. Research on the potential use of S2S forecasts for applications has started only recently, and S2S-based early warning systems, fully integrated with decision support, have yet to emerge. Many challenges remain, including the need to develop better skill estimates and user-friendly forecast products such as probabilities of exceedance that maximize appropriate actions and minimize the danger of acting in vain.

Much more research is required to develop user-relevant quantities (e.g., to inform alternative actions as shifts in a continuous set of possibilities, rather than all-or-nothing decisions; see Chapter 19). Building on well-established experience in seasonal forecast applications, both Chapters 19 and 21 stress that successful incorporation of S2S forecasts into actual decisions will require strong and sustained interaction between S2S forecast developers and stakeholders in a coproduction process within which trust can be established. In the health sector, Chapter 22 recommends creating partnership platforms through the Global

Framework for Climate Services (GFCS) and related mechanisms in order to enable the academic and operational communities in climate and health to work together on real-time health early warning systems, especially in developing countries where climate-driven health outcomes can be severe. Therefore, one of the main objectives of the World Weather Research Programme/World Climate Research Programme (WWRP/WCRP) S2S prediction project has been to establish demonstration projects involving potential users to demonstrate the usefulness of S2S forecasts.

References

Abbe, C., 1901. The physical basis of long-range weather forecasts. Mon. Weather Rev. 29, 551–561.

Abhilash, S., Sahai, A.K., Pattnaik, S., De, S., 2013. Predictability during active break phases of Indian summer monsoon in an ensemble prediction system using climate forecast system. J. Atmos. Sol. Terr. Phys. 100–101, 13–23. https://doi.org/10.1016/j.jastp.2013.03.017.

Abhilash, S., Sahai, A.K., Borah, N., et al., 2014. Prediction and monitoring of monsoon intraseasonal oscillations over Indian monsoon region in an ensemble prediction system using CFSv2. Clim. Dyn. 42, 2801–2815. https://doi.org/10.1007/s00382-013-2045-9.

Abhilash, S., Sahai, A.K., Borah, N., et al., 2015. Improved spread-error relationship and probabilistic prediction from CFS based grand ensemble prediction system. J. Appl. Meteorol. Climatol., 1569–1578.

Adames, A.F., Kim, D., 2016. The MJO as a dispersive, convectively coupled moisture wave: theory and observations. J. Atmos. Sci. 73, 913–941.

Adames, A.F., Wallace, J.M., 2014. Three dimensional structure and evolution of the MJO and its relation to the mean flow. J. Atmos. Sci. 71, 2007–2026.

Adams, R.M., Houston, L.L., McCarl, B.A., Tiscareno, M.L., Matus, J.G., Weiher, R.F., 2003. The benefits to Mexican agriculture of an El Niño-southern oscillation (ENSO) early warning system. Agric. For. Meteorol. 115, 183–194.

Agier, L., et al., 2013. Timely detection of bacterial meningitis epidemics at district level: a study in three countries of the African meningitis belt. Trans. R. Soc. Trop. Med. Hyg. 107, 30–36.

Agier, L., et al., 2017. Towards understanding the epidemiology of Neisseria meningitidis in the African meningitis belt: a multi-disciplinary overview. Int. J. Infect. Dis. 54, 103–112.

Agustí-Panareda, A., et al., 2014. Forecasting global atmospheric CO_2. Atmos. Chem. Phys. 14, 11959–11983.

Ahn, M.-S., Kang, I.-S., 2018. A practical approach to scale-adaptive deep convection in a GCM by controlling the cumulus base mass flux. npj Clim. Atmos. Sci. https://doi.org/10.1038/s41612-018-0021-0.

Ahn, M.-S., Kim, D., Sperber, K.R., Kank, I.-S., Maloney, E., Waliser, D., Hendon, H., 2017. MJO simulation in CMIP5 climate models: MJO skill metrics and process oriented diagnostics. Clim. Dyn. 49, 4023–4045.

Albergel, C., de Rosnay, P., Balsamo, G., Isaksen, L., Muñoz-Sabater, J., 2012. Soil moisture analyses at ECMWF: evaluation using global ground-based in situ observations. J. Hydrometeorol. 13, 1442–1460.

Albers, J.R., Birner, T., 2014. Vortex preconditioning due to planetary and gravity waves prior to sudden stratospheric warmings. J. Atmos. Sci. 71, 4028–4054. https://doi.org/10.1175/JAS-D-14-0026.1.

Aldrian, E., 2008. Dominant factors of Jakarta's three largest floods. J. Hidrosfir Indones 3, 105–112.

Alessio, S.M., 2016. Digital Signal Processing and Spectral Analysis for Scientists: Concepts and Applications. Springer International Publishing, Switzerland. http://www.springer.com/us/book/9783319254661.

Alfieri, L., Burek, P., Dutra, E., Krzeminski, B., Muraro, D., Thielen, J., Pappenberger, F., 2013. GloFAS—global ensemble streamflow forecasting and flood early warning. Hydrol. Earth Syst. Sci. 17, 1161–1175. https://doi.org/10.5194/hess-17-1161-2013.

Allen, J.T., Tippett, M.K., Sobel, A.H., 2015. Influence of the El Nino/Southern Oscillation on tornado and hail frequency in the United States. Nat. Geosci. 8, 278–283. https://doi.org/10.1038/ngeo2385.

Ambaum, M.H.P., Hoskins, B.J., 2002. The NAO troposphere–stratosphere connection. J. Clim. 15, 1969–1978. https://doi.org/10.1175/1520-0442(2002)015<1969:TNTSC>2.0.CO;2.

AMS, 2000. In: Glickman, T.S. (Ed.), Glossary of Meteorology. American Meteorological Society. Available at: http://glossary.ametsoc.org/wiki/.

Andersen, J.A., Kuang, Z., 2012. Moist static energy budget of MJO-like disturbances in the atmosphere of a zonally symmetric aquaplanet. J. Clim. 25, 2782–2804. https://doi.org/10.1175/JCLI-D-11-00168.1.

Anderson, J.L., 1997. The impact of dynamical constraints on the selection of initial conditions for ensemble predictions: low-order perfect model results. Mon. Weather Rev. 125 (11), 2969–2983.

Anderson, J.L., 2001. An ensemble adjustment Kalman filter for data assimilation. Mon. Weather Rev. 129, 2884–2903.

Anderson, J.L., Van den Dool, H.M., 1994. Skill and return of skill in dynamic extended-range forecasts. Mon. Weather Rev. 122, 507–516.

Anderson, D., Stockdale, T., Balmaseda, A., Ferranti, L., Vitart, F., Molteni, F., Doblas-Reyes, F., Mogensen, K., Vidard, A., 2007. Development of the ECMWF seasonal forecast System 3. In: ECMWF Research Department Technical Memorandum n. 503, p. 58. Available from ECMWF, Shinfield Park, Reading RG2-9AX, UK.

Anderson, G., Carson, J., Clements, J., Fleming, G., Frei, T., Kootval, H., Kull, D., Lazo, J., Letson, D., Mills, B., Perrels, A., Vaughan, C., Zillman, J., 2015. Valuing Weather and Climate: Economic Assessment of Meteorological and Hydrological Services. World Meteorological Organization, The World Bank, and Climate Services Partnership, Geneva. WMO No. 1153.

Andreae, M.O., et al., 2002. Biogeochemical cycling of carbon, water, energy, trace gases, and aerosols in Amazonia: the LBA-EUSTACH experiments. J. Geophys. Res. 107. LBA33-1–LBA33-25.

Andrews, D.G., Holton, J.R., Leovy, C.B., 1987. Middle Atmosphere Dynamics. Academic Press, Cambridge, MA. 489 pp.

Annamalai, H., Slingo, J.M., 2001. Active/break cycles: diagnosis of the intraseasonal variability of the Asian summer monsoon. Clim. Dyn. 18, 85–102.

Anstey, J.A., Shepherd, T.G., 2014. High-latitude influence of the quasi-biennial oscillation. Q. J. R. Meteorol. Soc. 140, 1–21. https://doi.org/10.1002/qj.2132.

Anstey, J.A., Scinocca, J.F., Keller, M., 2016. Simulating the QBO in an atmospheric general circulation model: sensitivity to resolved and parameterized forcing. J. Atmos. Sci. 73, 1649–1665. https://doi.org/10.1175/JAS-D-15-0099.1.

Arakawa, A., Jung, J.-H., Wu, C.-M., 2011. Toward unification of the multiscale modeling of the atmosphere. Atmos. Chem. Phys. 11 (8), 3731–3742.

Arguez, A., Vose, R.S., 2011. The definition of the standard WMO climate normal: the key to deriving alternative climate normals. Bull. Am. Meteorol. Soc. 92, 699–704.

Arribas, A., Glover, M., Maidens, A., Peterson, K., Gordon, M., MacLachlan, C., Graham, R., Fereday, D., Camp, J., Scaife, A.A., Xavier, P., McLean, P., Colman, A., Cusack, S., 2011. The GloSea4 ensemble prediction system for seasonal forecasting. Mon. Weather Rev. 139, 1891–1910.

Artikov, I., Hoffman, S.J., Lynne, G.D., PytlikZillig, L.M., Hu, Q., Tomkins, A.J., Hubbard, K.G., Hayes, M.J., Waltman, W., 2006. Understanding the influence of climate forecasts on farmer decisions as planned behavior. J. Appl. Meteorol. Climatol. 45, 1202–1214.

Asaadi, A., Brunet, G., Yau, P., 2016a. On the dynamics of the formation of the Kelvin cat's eye in tropical cyclogenesis: Part I: Climatological investigation. J. Atmos. Sci. 73, 2317–2338.

Asaadi, A., Brunet, G., Yau, P., 2016b. On the dynamics of the formation of the Kelvin cat's eye in tropical cyclogenesis: Part II: Numerical simulation. J. Atmos. Sci. 73, 2339–2359.

Asaadi, A., Brunet, G., Yau, P., 2017. The importance of critical layer in differentiating developing from non-developing easterly waves. J. Atmos. Sci. 74, 409–417.

Atger, F., 2001. Verification of intense precipitation forecasts from single models and ensemble prediction systems. Nonlinear Process. Geophys. 8, 401–417.

Awoonor-Williams, J.K., et al., 2004. Bridging the gap between evidence-based innovation and national health-sector reform in Ghana. Stud. Fam. Plan. 35, 161–177.

Ayarzagüena, B., Orsolini, Y.J., Langematz, U., Abalichin, J., Kubin, A., 2015. The relevance of the location of blocking highs for stratospheric variability in a changing climate. J. Clim. 28, 531–549. https://doi.org/10.1175/JCLI-D-14-00210.1.

Azhar, G.S., et al., 2014. Heat-related mortality in India: excess all-cause mortality associated with the 2010 Ahmedabad heat wave. PLoS ONE 9.

Baddeley, A.J., 1992a. Errors in binary images and an Lp version of the Hausdorff metric. NieuwArch. Wiskunde 10, 157–183.

Baddeley, A.J., 1992b. An error metric for binary images. In: Forstner, W., Ruwiedel, S. (Eds.), Robust Computer Vision: Quality of Vision Algorithms. Wichmann, pp. 59–78.

Baethegen, W.E., Carriquiry, M., Ropelewski, C., 2009. Tilting the odds in maize yields: how climate information can help manage risks. Bull. Am. Meteorol. Soc. 90 (2), 179–183.

Baethgen, W.E., 2010. Climate risk management for adaptation to climate variability and change. Crop Sci. 50 (2), 70–76.

Baethgen, W.E., Berterretche, M., Gimenez, A., 2016. Informing decisions and policy: the national agricultural information system of Uruguay. Agrometeoros 24, 97–112.

Bai, Z., Demmel, J., Dongarr, J., Ruh, A., Van Der Vorst, H. (Eds.), 2000. Generalized Hermitian Eigenvalue Problems. Templates for the Solution of Algebraic Eigenvalue Problems: A Practical Guide. SIAM, Philadelphia, ISBN: 978-0-89871-471-5.

Bailey, R., 2012. Famine Early Warning and Early Action: The Cost of Delay. Chatham House, London.

Baldwin, M.P., Dunkerton, T.J., 2001. Stratospheric harbingers of anomalous weather regimes. Science 294, 581–584. https://doi.org/10.1126/science.1063315.

Baldwin, M.P., Gray, L.J., Dunkerton, T.J., Hamilton, K., Haynes, P.H., Randel, W.J., Holton, J.R., Alexander, M.J., Hirota, I., Horinouchi, T., Jones, D.B.A., Kinnersley, J.S., Marquardt, C., Sato, K., Takahashhi, M., 2001. Quasi-biennial oscillation. Rev. Geophys. 39, 179–229.

Baldwin, M.P., Stephenson, D.B., Thompson, D.W.J., Dunkerton, T.J., Charlton, A.J., O'Neill, A., 2003. Stratospheric memory and skill of extended-range weather forecasts. Science (80-) 301, 636–640. https://doi.org/10.1126/science.1087143.

Ballester, J., Lowe, R., Diggle, P.J., Rodó, X., 2016. Seasonal forecasting and health impact models: challenges and opportunities. Ann. N. Y. Acad. Sci. 1382, 8–20.

Balsamo, G., Beljaars, A., Scipal, K., Viterbo, P., van den Hurk, B., Hirschi, M., Betts, A.K., 2009. A revised hydrology for the ECMWF model: verification from field site to terrestrial water storage and impact in the Integrated Forecast System. J. Hydrometeorol. 10, 623–643.

Balsamo, G., Dutra, E., Stepanenko, V.M., Viterbo, P., Miranda, P., Mironov, D., 2010. Deriving an effective lake depth from satellite lake surface temperature data: a feasibility study with MODIS data. Boreal Environ. Res. 15, 178–190.

Balsamo, G., Pappenberger, F., Dutra, E., Viterbo, P., van den Hurk, B.J.J.M., 2011. A revised land hydrology in the ECMWF model: a step towards daily water flux prediction in a fully-closed water cycle. Hydrol. Proc. 25, 1046–1054.

Balsamo, G., Salgado, R., Dutra, E., Boussetta, S., Stockdale, T., Potes, M., 2012. On the contribution of lakes in predicting near-surface temperature in a global weather forecasting model. Tellus A 64, 15829.

Balsamo, G., Agustì-Panareda, A., Albergel, C., Beljaars, A., Boussetta, S., Dutra, E., Komori, T., Lang, S., Muñoz-Sabater, J., Pappenberger, F., de Rosnay, P., Sandu, I., Wedi, N., Weisheimer, A., Wetterhall, F., Zsoter, E., 2014. Representing the Earth surfaces in the Integrated Forecasting System: recent advances and future challenges. In: ECMWF Research Department Technical Memorandum n. 729, p. 50 Available from ECMWF, Shinfield Park, Reading RG2-9AX, UK.

Balsamo, G., et al., 2015. ERA-Interim/Land: a global land surface reanalysis data set. Hydrol. Earth Syst. Sci. 19, 389–407.

Bao, M., Hartmann, D.L., 2014. The response to MJO-like forcing in a nonlinear shallow-water model. Geophys. Res. Lett. 41, 1322–1328. https://doi.org/10.1002/2013GL057683.

Barbu, A.L., Calvet, J.C., Mahfouf, J.F., Lafont, S., 2014. Integrating ASCAT surface soil moisture and GEOV1 leaf area index into the SURFEX modelling platform: a land data assimilation application over France. Hydrol. Earth Syst. Sci. 18, 173–192.

Barlow, M., Salstein, D., 2006. Summertime influence of the Madden-Julian Oscillation on daily rainfall over Mexico and Central America. Geophys. Res. Lett. 33L21708.

Barnes, E.A., Screen, J.A., 2015. The impact of Arctic warming on the midlatitude jetstream: Can it? Has it? Will it? WIREs Clim. Change 6, 277–286.

Barnett, D.G., 1980. A long-range ice forecasting method for the north coast of Alaska. Sea Ice Process. Models, 402–409.

Barnston, A.G., Livezey, R.E., 1987. Classification, seasonality and persistence of low-frequency atmospheric circulation patterns. Mon. Weather Rev. 115, 1083–1126. https://doi.org/10.1175/1520-0493(1987)115<1083:CSAPOL>2.0.CO;2.

Barnston, A.G., Tippett, M.K., 2017. Do statistical pattern corrections improve seasonal climate predictions in the North American multimodel ensemble models? J. Clim. 30 (20), 8335–8355.

Barnston, A.G., Leetmaa, A., Kousky, V.E., Livezey, R.E., O'Lenic, E.A., Van den Dool, H., Wagner, A.J., Unger, D.A., 1999. NCEP forecasts for the El Nino of 1997–98 and its U.S. impacts. Bull. Am. Meteorol. Soc. 80, 1829–1852.

Barnston, A.G., Li, S., DeWitt, D., Goddard, L., Gong, X., 2010. Verification of the first 11 years of IRI's seasonal climate forecasts. J. Clim. Appl. Meteorol. 49, 493–520.

Barnston, A.G., Tippett, M.K., L'Heureux, M.L., et al., 2012. Skill of real-time seasonal ENSO model predictions during 2002-2011: is our capability increasing? Bull. Am. Meteorol. Soc. 93, 631–651. https://doi.org/10.1175/BAMS-D-11-00111.1. http://journals.ametsoc.org/doi/abs/10.1175/BAMS-D-11-00111.1.

Barrett, B.S., Gensini, V.A., 2013. Variability of central United States April–May tornado day likelihood by phase of the Madden-Julian Oscillation. Geophys. Res. Lett. 40, 2790–2795. https://doi.org/10.1002/grl.50522.

Barsugli, J.J., Battisti, D.S., 1998. The basic effects of atmosphere–ocean thermal coupling on midlatitude variability. J. Atmos. Sci. 55, 477–493. https://doi.org/10.1175/1520-0469(1998)055,0477:TBEOAO.2.0.CO;2.

Bartaburu, D., Duarte, E., Montes, E., Morales Grosskopf, H., Pereira, M., 2009. Las sequías: un evento que afecta la trayectoria de las empresas y su gente. In: Grosskopf, M., Cameroni, D. (Eds.), Familias y campo. Rescatando estrategias de adaptación. IPA, Montevideo, pp. 155–168. www.planagropecuario.org.uy/publicaciones/libros/Familias_y_campo/Capitulo_4_155.pdf.

Barteit, S., et al., 2015. Self-directed e-learning at a tertiary hospital in Malawi—a qualitative evaluation and lessons learnt. GMS Z. Med. Ausbild. 32.

Bates, J.R., 1981. A dynamical mechanism through which variations in solar ultraviolet radiation can influence tropospheric climate. Sol. Phys. 74, 399–415. https://doi.org/10.1007/BF00154526.

Batté, L., Déqué, M., 2012. A stochastic method for improving seasonal predictions. Geophys. Res. Lett. 39(9).

Bauer, P., Thorpe, A., Brunet, G., 2015. The quiet revolution of numerical weather prediction. Nature 525, 47–55.

Baxter, S., Weaver, S., Gottschalck, J., Xue, Y., 2014. Pentad evolution of wintertime impacts of the Madden–Julian oscillation over the contiguous United States. J. Clim. 27, 7356–7367. https://doi.org/10.1175/JCLI-D-14-00105.1.

Bechtold, P., Köhler, M., Jung, T., Doblas-Reyes, F., Leutbecher, M., Rodwell, M.J., Vitart, F., Balsamo, G., 2008a. Advances in simulating atmospheric variability with the ECMWF model: from synoptic to decadal timescales. Q. J. R. Meteorol. Soc. 137, 553–597.

Bechtold, P., Koehler, M., Jung, T., Doblas-Reyes, P., Leutbecher, M., Rodwell, M., Vitart, F., 2008b. Advances in simulating atmospheric variability with the ECMWF model: from synoptic to decadal time-scales. Q. J. R. Meteorol. Soc. 134, 1337–1351.

Becker, E.J., Berbery, E.H., Higgins, R.W., 2011. Modulation of cold-season U.S. daily precipitation by the Madden–Julian oscillation. J. Clim. 24, 5157–5166. https://doi.org/10.1175/2011JCLI4018.1.

Bedsworth, L., 2009. Preparing for climate change: a perspective from local public health officers in California. Environ. Health Perspect. 117, 617.

Belanger, J.I., Webster, P.J., Curry, J.A., Jelinek, M.T., 2012. Extended prediction of North Indian Ocean tropical cyclones. Weather Forecast. 27, 757–769.

Beljaars, A.C., Viterbo, P., Miller, M.J., Betts, A.K., 1996. The anomalous rainfall over the United States during July 1993: sensitivity to land surface parameterization and soil moisture anomalies. Mon. Weather Rev. 124, 362–383.

Bell, C.J., Gray, L.J., Charlton-Perez, A.J., Joshi, M.M., Scaife, A.A., 2009. Stratospheric communication of El Niño teleconnections to European winter. J. Clim. 22, 4083–4096.

Bellprat, O., Massonnet, F., Siegert, S., Prodhomme, C., Macias-Gomez, M., Guemas, V., Doblas-Reyes, F.J., 2017. Exploring observational uncertainty in verification of climate model predictions. Remote Sens. Environ. Under review.

Bender, M.A., Ginis, I., 2000. Real-case simulations of hurricane-ocean interaction using a high-resolution coupled model: effects on hurricane intensity. Mon. Weather Rev. 128, 917–946.

Benedict, J.J., Randall, D.A., 2009. Structure of the Madden–Julian Oscillation in the superparameterized CAM. J. Clim. 66, 3277–3296.

Benedict, J.J., Maloney, E.D., Sobel, A.H., Frierson, D.M.W., 2014. Gross moist stability and MJO simulation skill in three full-physics GCMs. J. Atmos. Sci. 71, 3327–3349.

Bengtsson, L., 1991. Advances and prospects in numerical weather prediction. Q. J. R. Meteorol. Soc. 117, 855–902.

Benzi, R., Malguzzi, P., Speranza, A., Sutera, A., 1986. The statistical properties of general atmospheric circulation: observational evidence and a minimal theory of bimodality. Q. J. R. Meteorol. Soc. 112, 661–674. https://doi.org/10.1256/smsqj.47305.

Berbery, E.H., Nogués-Paegle, J., 1993. Intraseasonal interactions between the tropics and extratropics in the Southern Hemisphere. J. Atmos. Sci. 50, 1950–1965.

Bergthorsson, P., Doos, B., 1955. Numerical weather map analysis. Tellus 7, 329–340.

Berhane, F., Zaitchik, B., 2014. Modulation of daily precipitation over East Africa by the Madden–Julian Oscillation. J. Clim. 27, 6016–6034.

Berner, J., Shutts, G., Leutbecher, M., Palmer, T.N., 2008. A spectral stochastic kinetic energy backscatter scheme and its impact on flow-dependent predictability in the ECMWF ensemble prediction system. J. Atmos. Sci. 66, 603–626.

Bernie, D.J., Guilyardi, E., Madec, G., Slingo, J.M., Woolnough, S.J., Cole, J., 2008. Impact of resolving the diurnal cycle in an atmosphere-ocean GCM. Part 2: A diurnally coupled CGCM. Clim. Dyn. 31, 909–925.

Best, M.J., Pryor, M., Clark, D.B., Rooney, G.G., Essery, R.L.H., Ménard, C.B., Edwards, J.M., Hendry, M.A., Porson, A., Gedney, N., Mercado, L.M., Sitch, S., Blyth, E., Boucher, O., Cox, P.M., Grimmond, C.S.B., Harding, R.J., 2011. The Joint UK Land Environment Simulator (JULES), model description. Part 1: Energy and water fluxes. Geosci. Model Dev. 4, 677–699.

Best, M., Lock, A., Santanello, J., Svensson, G., Holtslag, B., 2013. A new community experiment to understand land-atmosphere coupling processes. GEWEX News 23 (2), 3–5.

Best, M.J., et al., 2015. The plumbing of land surface models: benchmarking model performance. J. Hydrometeorol. 16, 1425–1442.

Betts, A.K., 1994. Relation between equilibrium evaporation and the saturation pressure budget. Bound.-Layer Meteorol. 71, 235–245.

Betts, A.K., Ball, J.H., Beljaars, A.C.M., Miller, M.J., Viterbo, P.A., 1996. The land surface-atmosphere interaction: a review based on observational and global modeling perspectives. J. Geophys. Res. 101, 7209–7225.

Bhattacharya, K., Ghil, M., Vulis, I.L., 1982. Internal variability of an energy-balance model with delayed albedo effects. J. Atmos. Sci. 39, 1747–1773.

Bhumralkar, C.M., 1975. Numerical experiments on the computation of ground surface temperature in an atmospheric general circulation model. J. Appl. Meteorol. 14, 1246–1258.

Biello, J.A., Majda, A.J., 2005. A new multiscale model for the Madden-Julian Oscillation. J. Atmos. Sci. 62, 1694–1721.

Bishop, C.H., Etherton, B.J., Majumdar, S.J., 2001. Adaptive sampling with the ensemble transform Kalman filter. Part I: theoretical aspects. Mon. Weather Rev. 129, 420–436.

Bittner, M.-I., Matthies, E.F., Dalbokova, D., Menne, B., 2013. Are European countries prepared for the next big heat-wave? Eur. J. Pub. Health 24, 615–619.

Bitz, C.M., Roe, G.H., 2004. A mechanism for the high rate of sea ice thinning in the Arctic Ocean. J. Clim. 17 (18), 3623–3632.

Bitz, C.M., Battisti, D.S., Moritz, R.E., Beesley, J.A., 1996. Low-frequency variability in the Arctic atmosphere, sea ice, and upper-ocean climate system. J. Clim. 9 (2), 394–408.

Bitz, C.M., Holland, M.M., Hunke, E.C., Moritz, R.E., 2005. Maintenance of the sea ice edge. J. Clim. 18 (15), 2903–2921.

Bjerknes, V., 1904. Das Problem der Wettervorhersage betrachtet vom Standpunkt der Mechanik und Physik. Meteorol. Z. 21, 1–7.

Blackburn, M., Methven, J., Roberts, N., 2008. Large-scale context for the UK floods in summer 2007. Weather 63, 280–288.

Blackmon, M.L., 1976. A climatological spectral study of the 500 mb geopotential height of the Northern Hemisphere. J. Atmos. Sci. 33, 1607–1623.

Blackmon, M.L., Lee, Y.-H., Wallace, J.M., 1984. Horizontal structure of 500 mb height fluctuations with long, intermediate and short time scales. J. Atmos. Sci. 41, 961–979.

Blackport, R., Kushner, P.J., 2017. Isolating the atmospheric circulation response to Arctic Sea ice loss in the coupled climate system. J. Clim. 30, 2163–2185. https://doi.org/10.1175/JCLI-D-16-0257.1.

Bladé, I., Hartmann, D.L., 1995. The linear and nonlinear extratropical response to tropical intraseasonal heating. J. Atmos. Sci. 52, 4448–4471. https://doi.org/10.1175/1520-0469(1995)052<4448:TLANER>2.0.CO;2.

Blanchard-Wrigglesworth, E., Bitz, C.M., 2014. Characteristics of Arctic sea ice thickness variability in GCMs. J. Clim. 27 (21), 8244–8258.

Blanchard-Wrigglesworth, E., Armour, K.C., Bitz, C.M., DeWeaver, E., 2011a. Persistence and inherent predictability of Arctic sea ice in a GCM ensemble and observations. J. Clim. 24 (1), 231–250.

Blanchard-Wrigglesworth, E., Bitz, C.M., Holland, M.M., 2011b. Influence of initial conditions and climate forcing on predicting Arctic sea ice. Geophys. Res. Lett. 38(18).

Blanchard-Wrigglesworth, E., Cullather, R.I., Wang, W., Zhang, J., Bitz, C.M., 2015. Model forecast skill and sensitivity to initial conditions in the seasonal Sea Ice Outlook. Geophys. Res. Lett. 42 (19), 8042–8048.

Blondin, C., 1991. Parameterization of land-surface processes in numerical weather prediction. In: Schmugge, T.J., Andre, J.C. (Eds.), Land Surface Evaporation—Measurement and Parameterization. Springer-Verlag, pp. 31–54.

Blyth, E.M., et al., 2006. JULES: a new community land surface model. In: Global Change Newsletter. vol. 66. IGBP, Stockholm, Sweden, pp. 9–11.

Boelee, E., et al., 2013. Options for water storage and rainwater harvesting to improve health and resilience against climate change in Africa. Reg. Environ. Chang. 13, 509–519.

Boer, G.J., 2000. A study of atmosphere-ocean predictability on long time scales. Clim. Dyn. 16 (6), 469–477.

Boer, G.J., 2003. Predictability as a function of scale. Atmosphere-Ocean 41 (3), 203–215. https://doi.org/10.3137/ao.410302.

Boer, G.J., Hamilton, K., 2008. QBO influence on extratropical predictive skill. Clim. Dyn. 31, 987–1000. https://doi.org/10.1007/s00382-008-0379-5.

Boffetta, G., Guliani, P., Paladin, G., Vulpiani, A., 1998. An extension of the Lyapunov analysis for the predictability problem. J. Atmos. Sci. 55, 3409–3416.

Bomblies, A., Duchemin, J.B., Eltahir, E.A.B., 2009. A mechanistic approach for accurate simulation of village scale malaria transmission. Malar. J. 8, 223. https://doi.org/10.1186/1475-2875-8-223.

Bonan, G., 2008. Ecological Climatology—Concepts and Applications. Cambridge University Press, second ed. 550 pp.

Bonavita, M., Hólm, E., Isaksen, L., Fisher, M., 2016. The evolution of the ECMWF hybrid data assimilation system. Q. J. R. Meteorol. Soc. 142, 287–303.

Bonavita, M., Trémolet, Y., Holm, E., Lang, S.T.K., Chrust, M., Janiskova, M., Lopez, P., Laloyaux, P., De Rosnay, P., Fisher, M., Hamrud, M., English, S., 2017. A Strategy for Data Assimilation. ECMWF Research Department Tech. Memorandum n. 800, .p. 44. Available from ECMWF, Shinfield Park, Reading, RG2 9AX UK, or from the ECMWF Web Site, https://www.ecmwf.int/en/elibrary/technical-memoranda.

Bond, N.A., Vecchi, G.A., 2003. The influence of the Madden–Julian oscillation on precipitation in Oregon and Washington. Weather Forecast. 18, 600–613. https://doi.org/10.1175/1520-0434(2003)018<0600:TIOTMO>2.0.CO;2.

Boone, A., Calvet, J.C., Noilhan, J., 1999. Inclusion of a third layer in a land surface scheme using the force restore. J. J. Appl. Meteorol. 38, 1611–1630.

Booth, J.F., Thompson, L., Patoux, J.K., Kelly, K.A., Dickinson, S., 2010. The signature of midlatitude tropospheric storm tracks in the surface winds. J. Climate 23, 1160–1174.

Bouillon, S., Morales Maqueda, M.A., Legat, V., Fichefet, T., 2009. An elastic-viscous-plastic sea-ice model formulated on Arakawa B and C grids. Ocean Model. 27, 174–184.

Bourke, R.H., Garrett, R.P., 1987. Sea ice thickness distribution in the Arctic Ocean. Cold Reg. Sci. Technol. 13 (3), 259–280.

Bourke, W., Hart, T., Steinle, P., Seaman, R., Embery, G., Naughton, M., Rikus, L., 1995. Evolution of the Bureau of Meteorology's Global Assimilation and Prediction system. Part 2: resolution enhancements and case studies. Aust. Meteorol. Mag. 44, 19–40.

Bourke, W., Buizza, R., Naughton, M., 2004. Performance of the ECMWF and the BoM ensemble systems in the Southern Hemisphere. Mon. Weather Rev. 132, 2338–2357.

Boussetta, S., et al., 2013a. Natural land carbon dioxide exchanges in the ECMWF Integrated Forecasting System: implementation and offline validation. J. Geophys. Res. 118, 5923–5946.

Boussetta, S., Balsamo, G., Baljaars, A., Kral, T., Jarlan, L., 2013b. Impact of a satellite-derived leaf area index monthly climatology in a global numerical weather prediction model. Int. J. Remote Sens. 34, 3520–3542.

Bouttier, F., Mahfouf, J., Noilhan, J., 1993a. Sequential assimilation of soil-moisture from atmospheric low-level parameters. 1. Sensitivity and calibration studies. J. Appl. Meteorol. 32, 1335–1351.

Bouttier, F., Mahfouf, J., Noilhan, J., 1993b. Sequential assimilation of soil-moisture from atmospheric low-level parameters. 2. Implementation in a mesoscale model. J. Appl. Meteorol. 32, 1352–1364.

Boville, B.A., 1984. The influence of the polar night jet on the tropospheric circulation in a GCM. J. Atmos. Sci. 41, 1132–1142. https://doi.org/10.1175/1520-0469(1984)041<1132:TIOTPN>2.0.CO;2.

Bowler, N.E., Arribas, A., Mylne, K.R., Robertson, K.B., 2007. Numerical weather prediction: the MOGREPS short-range ensemble prediction system. Part I: system description. In: UK Met. Office NWP Technical Report No. 497, p. 18.

Bowler, N.E., Arribas, A., Mylne, K.R., Robertson, K.B., Shutts, G.J., 2008. The MOGREPS short-range ensemble prediction system. Q. J. R. Meteorol. Soc. 134, 703–722.

Branković, Č., Palmer, T.N., 1997. Atmospheric seasonal predictability and estimates of ensemble size. Mon. Weather Rev. 125, 859–874.

Brankovic, C., Molteni, F., Palmer, T.N., Cubasch, U., 1988. In: Extended range ensemble forecasting at ECMWF.Proc. ECMWF Workshop on Predictability in the Medium and Extended Range, Reading, United Kingdom, ECMWF, pp. 45–87.

Branstator, G., 1985. Analysis of general circulation model sea-surface temperature anomaly simulations using a linear model. Part I: Forced solutions. J. Atmos. Sci. 42, 2225–2241.

Branstator, G., 1987. A striking example of the atmosphere's leading traveling pattern. J. Atmos. Sci. 44, 2310–2323. https://doi.org/10.1175/1520-0469(1987)044<2310:aseota>2.0.co;2.

Brassington, G.B., Martin, M.J., Tolman, H.L., Akella, S., Balmeseda, M., Chambers, C.R.S., Chassignet, E., Cummings, J.A., Drillet, Y., Jansen, P.A.E.M., Laloyaux, P., Lea, D., Mehra, A., Mirouze, I., Ritchie, H., Samson, G., Sandery, P.A., Smith, G.C., Suarez, M., Todling, R., 2015. Progress and challenges in short- to medium-range coupled prediction. J. Oper. Oceanogr. 8 (Suppl. 2), 2015.

Brayshaw, D.J., Troccoli, A., Fordham, R., Methven, J., 2011. The impact of large scale atmospheric circulation patterns on wind power generation and its potential predictability: a case study over the UK. Renew. Energy 36, 2087–2096.

Breiman, L., 2001. Random forests. Mach. Learn. 45, 5–32.

Bretherton, F.P., 1966. Critical layer instability in baroclinic flows. Q. J. R. Meteorol. Soc. 92, 325–334.

Breuer, N.E., Cabrera, V.E., Ingram, K.T., Broad, K., Hildebrand, P.E., 2008. AgClimate: a case study in participatory decision support system development. Clim. Chang. 87 (3-4), 385–403.

Breuer, N.E., Fraisse, C.W., Hildebrand, P.E., 2009. Molding the pipeline into a loop: the participatory process for developing AgroClimate, a decision support system for climate risk reduction in agriculture. J. Serv. Climatol. 3 (1), 1–12.

Brier, G.W., 1950. Verification of forecasts expressed in terms of probability. Mon. Weather Rev. 78, 1–3.

Briggs, R.J., Daugherty, J.D., Levy, R.H., *1970*. Role of Landau damping in crossed-field electron beams and inviscid shear flow. Phys. Fluids *13*, *421–433*.

Bright, D.R., Mullen, S.L., 2002. Short-range ensemble forecasts of precipitation during the southwest monsoon. Weather Forecast. 17, 1080–1100.

Broad, K., Pfaff, A., Taddei, R., Sankarasubramanian, A., Lall, U., de Assis de Souza Filho, F., 2007. Climate stream flow prediction and water management in northeast Brazil: societal trends and forecast value. Clim. Chang. 84, 217–239.

Broman, D.P., 2013. Spatio-Temporal Variability and Predictability of Relative Humidity and Meningococcal Meningitis Incidence in the West African Monsoon Region. Ph.D. thesis, Architectural Engineering.

Brooks, H.E., et al., 2011. Evaluation of European Storm Forecast Experiment (ESTOFEX) forecasts. Atmos. Res. 100, 538–546. https://doi.org/10.1016/j.atmosres.2010.09.004.

Broutin, H., et al., 2007. Comparative study of meningitis dynamics across nine African countries: a global perspective. Int. J. Health Geogr. 6, 29.

Brown, A.R., et al., 2002. Large-eddy simulation of the diurnal cycle of shallow cumulus convection over land. Q. J. R. Meteorol. Soc. 128, 1075–1093.

Brown, L., Medlock, J., Murray, V., 2014. Impact of drought on vector-borne diseases—how does one manage the risk? Public Health 128, 29–37.

Brunet, G., 1994. Empirical normal mode analysis of atmospheric data. J. Atmos. Sci. 51, 932–952.

Brunet, G., Methven, J., 2018. Identifying wave processes associated with predictability across time scales: an empirical normal mode approach. In: Robertson, A.W., Vitart, F. (Eds.), The Gap Between Weather and Climate Forecasting: Sub-Seasonal to Seasonal Prediction. Elsevier. (Chapter 4). 40 pp.

Brunet, G., Vautard, R., 1996. Empirical normal modes versus empirical orthogonal functions for statistical prediction. J. Atmos. Sci. 53, 3468–3489.

Brunet, G., Shapiro, M., Hoskins, B., Moncrieff, M., Dole, R.M., Kiladis, G.N., et al., 2010. Collaboration of the weather and climate communities to advance subseasonal-to-seasonal prediction. Bull. Am. Meteorol. Soc. 91, 1397–1406. https://doi.org/10.1175/2010BAMS3013.1.

Bruno Soares, M., Dessai, S., 2015. Exploring the use of seasonal climate forecasts in Europe through expert elicitation. Clim. Risk Manag. 10, 8–16.

Bruno Soares, M., Dessai, S., 2016. Barriers and enablers to the use of seasonal climate forecasts amongst organisations in Europe. Clim. Chang. 137 (1), 89–103.

Bryan, G.H., Morrison, H., 2012. Sensitivity of a simulated squall line to horizontal resolution and parameterization of microphysics. Mon. Weather Rev. 140 (1), 202–225.

Bryan, G.H., Wyngaard, J.C., Fritsch, J.M., 2003. Resolution requirements for the simulation of deep moist convection. Mon. Weather Rev. 131 (10), 2394–2416.

Bryan, F.O., Tomas, R., Dennis, J.M., Chelton, D.B., Loeb, N.G., McClean, J.L., 2010. Frontal scale air-sea interaction in high-resolution coupled climate models. J. Clim. 23, 6277–6291. https://doi.org/10.1175/2010JCLI3665.1.

Buizza, R., 2001. Accuracy and economic value of categorical and probabilistic forecasts of discrete events. Mon. Weather Rev. 129, 2329–2345.

Buizza, R., 2008. The value of probabilistic prediction. Atmos. Sci. Lett. 9, 36–42. https://doi.org/10.1002/asl.170.

Buizza, R., 2010. The value of a variable resolution approach to numerical weather prediction. Mon. Weather Rev. 138, 1026–1042.

Buizza, R., 2014. The TIGGE medium-range, global ensembles. In: ECMWF Research Department Technical Memorandum n. 739. ECMWF, Shinfield Park, Reading, p. 53. http://www.ecmwf.int/sites/default/files/elibrary/2014/7529-tigge-global-medium-range-ensembles.pdf.

Buizza, R., Leutbecher, M., 2015. The forecast skill horizon. Q. J. R. Meteorol. Soc. 141, 3366–3382. https://doi.org/10.1002/qj.2619.

Buizza, R., Palmer, T.N., 1995. The singular-vector structure of the atmospheric general circulation. J. Atmos. Sci. 52 (9), 1434–1456.

Buizza, R., Tribbia, J., Molteni, F., Palmer, T.N., 1993. Computation of optimal unstable structures for a numerical weather prediction model. Tellus 45A, 388–407.

Buizza, R., Miller, M., Palmer, T.N., 1999. Stochastic representation of model uncertainties in the ECMWF Ensemble Prediction System. Q. J. R. Meteorol. Soc. 125, 2887–2908.

Buizza, R., Houtekamer, P.L., Toth, Z., Pellerin, G., Wei, M., Zhu, Y., 2005. A comparison of the ECMWF, MSC, and NCEP global ensemble prediction systems. Mon. Weather Rev. 133, 1076–1097.

Buizza, R., Bidlot, J.-R., Wedi, N., Fuentes, M., Hamrud, M., Holt, G., Vitart, F., 2007a. The new ECMWF VAREPS (variable resolution ensemble prediction system). Q. J. R. Meteorol. Soc. 133, 681–695.

Buizza, R., Cardinali, C., Kelly, G., Thepaut, J.-N., 2007b. The value of observations—Part II: the value of observations located in singular vectors-based target areas. Q. J. R. Meteorol. Soc. 133, 1817–1832.

Buizza, R., Leutbecher, M., Isaksen, L., 2008. Potential use of an ensemble of analyses in the ECMWF Ensemble Prediction System. Q. J. R. Meteorol. Soc. 134, 2051–2066. https://doi.org/10.1002/qj.346.

Buizza, R., Leutbecher, M., Thorpe, A., 2015. Leaving with the butterfly effect: a seamless view of predictability. In: ECMWF Newsletter n. 145. ECMWF, Shinfield Park, Reading, pp. 18–23.

Bunzel, F., Notz, D., Baehr, J., Müller, W.A., Fröhlich, K., 2016. Seasonal climate forecasts significantly affected by observational uncertainty of Arctic sea ice concentration. Geophys. Res. Lett. 43 (2), 852–859.

Buontempo, C., Brookshaw, A., Arribas, A., Mylne, K., 2010. Multi-Scale Projections of Weather and Climate at the UK Met Office. Management of Weather and Climate Risk in the Energy Industry. Springer, Netherlands, pp. 39–50.

Bushuk, M., Msadek, R., Winton, M., Vecchi, G.A., Gudgel, R., Rosati, A., Yang, X., 2017. Summer enhancement of Arctic sea ice volume anomalies in the September-ice zone. J. Clim. 30 (7), 2341–2362.

Bushuk, M., Giannakis, D., 2017. The seasonality and interannual variability of Arctic Sea ice reemergence. J. Clim.

Businger, J., Wyngaard, J., Izumi, Y., Bradley, E., 1971. Flux-profile relationships in the atmospheric surface layer. J. Atmos. Sci. 28, 181–189.

Butler, A.H., Seidel, D.J., Hardiman, S.C., Butchart, N., Birner, T., Match, A., 2015. Defining sudden stratospheric warmings. Bull. Am. Meteorol. Soc. 96, 1913–1928. https://doi.org/10.1175/BAMS-D-13-00173.1.

Butler, A.H., et al., 2016. The climate-system historical forecast project: do stratosphere-resolving models make better seasonal climate predictions in boreal winter? Q. J. R. Meteorol. Soc. 142. https://doi.org/10.1002/qj.2743.

Butler, A.H., Sjoberg, J.P., Seidel, D.J., Rosenlof, K.H., 2017. A sudden stratospheric warming compendium. Earth Syst. Sci. Data 9, 63–76. https://doi.org/10.5194/essd-9-63-2017.

Cagnazzo, C., Manzini, E., 2009. Impact of the stratosphere on the winter tropospheric teleconnections between ENSO and the North Atlantic and European Region. J. Clim. 22, 1223–1238. https://doi.org/10.1175/2008JCLI2549.1.

Cai, M., Yu, Y., Deng, Y., van den Dool, H.M., Ren, R., Saha, S., Wu, X., Huang, J., 2016. Feeling the pulse of the stratosphere: an emerging opportunity for predicting continental-scale cold-air outbreaks 1 month in advance. Bull. Am. Meteorol. Soc. 97, 1475–1489. https://doi.org/10.1175/BAMS-D-14-00287.1.

Calvet, J.-C., Rivalland, V., Picon-Cochard, C., Guehl, J.-M., 2004. Modelling forest transpiration and CO_2 fluxes-response to soil moisture stress. Agric. For. Meteorol. 124, 143–156.

Calvo, N., Marsh, D.R., 2011. The combined effects of ENSO and the 11 year solar cycle on the Northern Hemisphere polar stratosphere. J. Geophys. Res. Atmos. 116. https://doi.org/10.1029/2010JD015226.

Calvo, N., Giorgetta, M.A., Garcia-Herrera, R., Manzini, E., 2009. Nonlinearity of the combined warm ENSO and QBO effects on the Northern Hemisphere polar vortex in MAECHAM5 simulations. J. Geophys. Res. 114, D13109. https://doi.org/10.1029/2008JD011445.

Calvo, N., Polvani, L.M., Solomon, S., 2015. On the surface impact of Arctic stratospheric ozone extremes. Environ. Res. Lett. 10, 94003. https://doi.org/10.1088/1748-9326/10/9/094003.

Camargo, S.J., Wheeler, M.C., Sobel, A.H., 2009. Diagnosis of the MJO modulation of tropical cyclogenesis using an empirical index. J. Atmos. Sci. 66, 3061–3074.

Camberlin, P., Moron, V., Okoola, R., Philippon, N., Gitau, W., 2009. Components of rainy seasons' variability in Equatorial East Africa: onset, cessation, rainfall frequency and intensity. Theor. Appl. Climatol. 98, 237–249.

Caminade, C., et al., 2011. Mapping Rift Valley fever and malaria risk over West Africa using climatic indicators. Atmos. Sci. Lett. 12, 96–103.

Caminade, C., et al., 2014. Impact of climate change on global malaria distribution. Proc. Natl. Acad. Sci. 111, 3286–3291.

Camp, C.D., Tung, K.K., 2007. Surface warming by the solar cycle as revealed by the composite mean difference projection. Geophys. Res. Lett. 34, L14703. https://doi.org/10.1029/2007GL030207.

Candille, G., 2009. The multi-ensemble approach: the NAEFS example. Mon. Weather Rev. 137, 1655–1665.

Candille, G., Côté, C., Houtekamer, P.L., Pellerin, G., 2007. Verification of an ensemble prediction system against observations. Mon. Weather Rev. 135, 1140–1147.

Carberry, P.S., Hochman, Z., McCown, R.L., Dalgliesh, N.P., Foale, M.A., Poulton, P.L., Hargreaves, J.N.G., Hargreaves, D.M.G., Cawthray, S., Hollcoat, N., Robertson, M.J., 2002. The FARMSCAPE approach to decision support: farmers', advisers', researchers' monitoring, simulation, communication and performance evaluation. Agric. Syst. 74 (1), 141–177.

Carbin, G.W., Tippett, M.K., Lillo, S.P., Brooks, H.E., 2016. Visualizing long-range severe thunderstorm environment guidance from CFSv2. Bull. Am. Meteorol. Soc. 97, 1021–1031. https://doi.org/10.1175/BAMS-D-14-00136.1.

Carmago, S.J., 2009. Diagnosis of the MJO modulation of tropical cyclogenesis using an empirical index. J. Atmos. Sci. 66, 3061–3074.

Carton, J.A., Giese, B.S., 2008. A reanalysis of ocean climate using Simple Ocean Data Assimilation (SODA). Mon. Weather Rev. 136, 2999–3017.

Carvalho, L.M.V., Jones, C., Ambrizzi, T., 2005. Opposite phases of the Antarctic Oscillation and relationships with intraseasonal to interannual activity in the Tropics during austral summer. J. Clim. 18, 702–718.

Casati, B., 2010. New developments of the intensity-scale technique within the Spatial Verification Methods Intercomparison Project. Weather Forecast. 25, 113–143.

Cash, D.W., Borck, J.C., Patt, A.G., 2006. Countering the loading-dock approach to linking science and decision making: comparative analysis of El Niño/Southern Oscillation (ENSO) forecasting systems. Sci. Technol. Hum. Values 31 (4), 465–494.

Cassou, C., 2008. Interannual interaction between the Madden-Julian Oscillation and the North Atlantic Oscillation. Nature 455 (7212), 523–527. https://doi.org/10.1038/nature07286.

Cassou, C., Terray, L., Phillips, A.S., 2005. Tropical Atlantic influence on European heat waves. J. Clim. 18 (15), 2805–2811.

Castanheira, J.M., Barriopedro, D., 2010. Dynamical connection between tropospheric blockings and stratospheric polar vortex. Geophys. Res. Lett. 37. https://doi.org/10.1029/2010GL043819.

Cattiaux, J., Vautard, R., Cassou, C., Yiou, P., Masson-Delmotte, V., Codron, F., 2010. Winter 2010 in Europe: a cold extreme in a warming climate. Geophys. Res. Lett. 37. https://doi.org/10.1029/2010GL044613.

Chaer, R., 2008. Simulación del Sistema de Energía Eléctrica. Tesis de maestría en Ingeniería Eléctrica. Universidad de la República, Montevideo, Facultad de Ingeniería.https://iie.fing.edu.uy/publicaciones/2008/Cha08/Cha08.pdf.

Chaer, R., Cornalino, E., Coppes, E., 2012. In: Modeling and simulation of the power energy system of Uruguay in 2015 with high penetration of wind energy.XII SEPOPE, Rio de Janeiro, Brazil, 20–23 May, SP082, 10. http://iie.fing.edu.uy/publicaciones/2012/CCC12/CCC12.pdf.

Chan, C.J., Plumb, R.A., 2009. The response to stratospheric forcing and its dependence on the state of the troposphere. J. Atmos. Sci. 66, 2107–2115. https://doi.org/10.1175/2009JAS2937.1.

Chang, C.-H., Johnson, N.C., 2015. The continuum wintertime Southern Hemisphere Atmospheric teleconnection patterns. J. Clim. 28, 9507–9529. https://doi.org/10.1175/JCLI-D-14-00739.s1.

Chang, C.P., Lau, K.-M., 1980. Northeasterly cold surges and near-equatorial disturbances over the winter MONEX area during December 1974. Part II: Planetary-scale aspects. Mon. Weather Rev. 108, 298–312.

Chang, E.K.M., Lee, S.Y., Swanson, K.L., 2002. Storm track dynamics. J. Clim. 15, 2163–2183. https://doi.org/10.1175/1520-0442(2002)015,02163:STD.2.0.CO;2.

Chang, C.-P., Harr, P.A., J., C. H., 2005. Synoptic disturbances over the equatorial South China Sea and western Maritime Continent during boreal winter. Mon. Weather Rev. 133, 489–503.

Chang, C.P., Ghil, M., Latif, M., Wallace, J.M. (Eds.), 2015. Climate Change: Multidecadal and Beyond. World Scientific, Singapore.

Changnon, S.A., et al., 1996. Impacts and responses to the 1995 heat wave: a call to action. Bull. Am. Meteorol. Soc. 77, 1497–1506.

Charles, A.N., Duell, R.E., Robyn, E., Wang, X.D., Watkins, A.B., Andrew, B., 2015. Seasonal forecasting for Australia using a dynamical model: improvements in forecast skill over the operational statistical model. Aust. Meteorol. Oceanogr. J. 65 (3-4), 356–375.

Charlton, A.J., Polvani, L.M., 2007. A new look at stratospheric sudden warmings. Part I: Climatology and modeling benchmarks. J. Clim. 20, 449–469. https://doi.org/10.1175/JCLI3996.1.

Charlton, A.J., O'Neill, A., Stephenson, D.B., Lahoz, W.A., Baldwin, M.P., 2003. Can knowledge of the state of the stratosphere be used to improve statistical forecasts of the troposphere? Q. J. R. Meteorol. Soc. 129, 3205–3224. https://doi.org/10.1256/qj.02.232.

Charlton, A.J., Oneill, A., Lahoz, W.A., Massacand, A.C., 2004. Sensitivity of tropospheric forecasts to stratospheric initial conditions. Q. J. R. Meteorol. Soc. 130, 1771–1792. https://doi.org/10.1256/qj.03.167.

Charlton, A.J., O'Neill, A., Berrisford, P., Lahoz, W.A., 2005. Can the dynamical impact of the stratosphere on the troposphere be described by large-scale adjustment to the stratospheric PV distribution? Q. J. R. Meteorol. Soc. 131, 525–543. https://doi.org/10.1256/qj.03.222.

Charlton-Perez, A.J., et al., 2013. On the lack of stratospheric dynamical variability in low-top versions of the CMIP5 models. J. Geophys. Res. Atmos. 118, 2494–2505. https://doi.org/10.1002/jgrd.50125.

Charney, J.G., DeVore, J.G., 1979. Multiple flow equilibria in the atmosphere and blocking. J. Atmos. Sci. 36, 1205–1216.

Charney, J.G., Drazin, P.G., 1961. Propagation of planetary-scale disturbances from the lower into the upper atmosphere. J. Geophys. Res. 66, 83–109. https://doi.org/10.1029/JZ066i001p00083.

Charney, J., Straus, D., 1980. Form-drag instability, multiple equilibria and propagating planetary waves in baroclinic, orographically forced, planetary wave systems. J. Atmos. Sci. 37, 1157–1176. https://doi.org/10.1175/1520-0469(1980)037<1157:FDIMEA>2.0.CO;2.

Charney, J., Fjørtoft, R., von Neumann, J., 1950. Numerical integration of the barotropic vorticity equation. Tellus 2 (4), 237–254.

Charney, J.G., Shukla, J., Mo, K.C., 1981. Comparison of a barotropic blocking theory with observation. J. Atmos. Sci. 38, 762–779. https://doi.org/10.1175/1520-0469(1981)038<0762:coabbt>2. 0.co;2.

Charron, M., Brunet, G., 1999. Gravity wave diagnosis using empirical normal modes. J. Atmos. Sci. 56, 2706–2727.

Charron, M., Pellerin, G., Spacek, L., Houtekamer, P.L., Gagnon, N., Mitchell, H.L., Michelin, L., 2010. Toward random sampling of model error in the Canadian ensemble prediction system. Mon. Weather Rev. 138, 1877–1901.

Chattopadhyay, R., Sahai, A.K., Goswami, B.N., 2008. Objective identification of nonlinear convectively coupled phases of monsoon intraseasonal oscillation: implications for prediction. J. Atmos. Sci. 65, 1549–1569. https://doi.org/10.1175/2007JAS2474.1.

Chattopadhyay, R., Phani, M., Susmitha, J., et al., 2017. A Comparison of Extended-Range Prediction of Monsoon in the IITM-CFSv2 with ECMWF S2S Forecast System. Indian Institute of Tropical Meteorology, Pune (accepted).

Chekroun, M.D., Kondrashov, D., 2017. Data-adaptive harmonic spectra and multilayer Stuart-Landau models. Chaos. 27, 093110. https://doi.org/10.1063/1.4989400.

Chekroun, M., Simonnet, E., Ghil, M., 2011a. Stochastic climate dynamics: random attractors and time-dependent invariant measures. Physica D 240, 1685–1700. https://doi.org/10.1016/j.physd.2011.06.005.

Chekroun, M.D., Kondrashov, D., Ghil, M., 2011b. Predicting stochastic systems by noise sampling, and application to the El Niño-Southern Oscillation. Proc. Natl. Acad. Sci. 108, 11766–11771. https://doi.org/10.1073/pnas.1015753108.

Chekroun, M., Neelin, J., Kondrashov, D., McWilliams, J., Ghil, M., 2014. Rough parameter dependence in climate models: the role of Ruelle-Pollicott resonances. Proc. Natl. Acad. Sci. 111, 1684–1690. https://doi.org/10.1073/pnas.1321816111.

Chelton, D.B., Xie, S.P., 2010. Coupled ocean-atmosphere interactions at oceanic mesoscales. Oceanography 23, 52–69.

Chelton, D.B., Schlax, M.G., Freilich, M.H., Milliff, R.F., 2004. Satellite measurements reveal persistent small-scale features in ocean winds. Science 303, 978–983.

Chelton, D.B., Schlax, M.G., Samelson, R.M., 2011. Global observations of nonlinear mesoscale eddies. Prog. Oceanogr. 91, 167–216.

Chen, T.-C., Alpert, J.C., 1990. Systematic errors in the annual and intraseasonal variations of the planetary-scale divergent circulation in NMC medium-range forecasts. Mon. Weather Rev. 118, 2607–2623.

Chen, F., Dudhia, J., 2001. Coupling an advanced land surface—hydrology model with the Penn State—NCAR MM5 modeling system. Part I: Model implementation and sensitivity. Mon. Weather Rev. 129, 569–585.

Chen, D., Yuan, X., 2004. A Markov model for seasonal forecast of Antarctic sea ice. J. Clim. 17 (16), 3156–3168.

Chen, F., Mitchell, K., Schaake, J., Xue, Y., Pan, H.-L., Koren, V., Duan, Q.Y., Ek, M., Betts, A., 1996a. Modeling of land-surface evaporation by four schemes and comparison with FIFE observations. J. Geophys. Res. 101, 7251–7268.

Chen, S.S., Houze, R.A., Mapes, B.E., 1996b. Multiscale variability of deep convection in relation to large-scale circulation in TOGA-COARE. J. Atmos. Sci. 53, 1380–1409.

Chen, F., Janjic, Z., Mitchell, K., 1997. Impact of atmospheric surface-layer parameterizations in the new land-surface scheme of the NCEP mesoscale Eta model. Bound.-Layer Meteorol. 85, 391–421.

Chen, Y., Brunet, G., Yau, P., 2003. Spiral bands in a simulated hurricane. Part II: Wave activity diagnostics. J. Atmos. Sci. 60, 1239–1256.

Chen, M.-J., et al., 2012. Effects of extreme precipitation to the distribution of infectious diseases in Taiwan, 1994–2008. PLoS One 7. e34651.

Chen, M., Wang, W., Kumar, A., 2013. Lagged ensembles, forecast configuration, and seasonal predictions. Mon. Weather Rev. 141, 3477–3497.

Chen, C., Cane, M.A., Henderson, N., Lee, D.E., Chapman, D., Kondrashov, D., Chekroun, M.D., 2016. Diversity, nonlinearity, seasonality, and memory effect in ENSO simulation and prediction using empirical model reduction. J. Clim. 29, 1809–1830. https://doi.org/10.1175/JCLI-D-15-0372.1.

Cheng, X., Wallace, J.M., 1993. Cluster analysis of the Northern Hemisphere wintertime 500-hPa height field: spatial patterns. J. Atmos. Sci. 50, 2674–2696. https://doi.org/10.1175/1520-0469(1993) 050<2674:caotnh>2.0.co;2.

Chengula, F., Nyambo, B., 2016. The significance of indigenous weather forecast knowledge and practices under weather variability and climate change: a case study of smallholder farmers on the slopes of Mount Kilimanjaro. Int. J. Agric. Educ. Exten. 2 (2), 031–043.

Chevallier, M., Salas y Mélia, D., Voldoire, A., Déqué, M., & Garric, G., 2013. Seasonal forecasts of the pan-Arctic sea ice extent using a GCM-based seasonal prediction system. J. Clim. 26 (16), 6092–6104.

Chevallier, M., Salas-Mélia, D., 2012. The role of sea ice thickness distribution in the Arctic sea ice potential predictability: a diagnostic approach with a coupled GCM. J. Clim. 25 (8), 3025–3038.

Chevallier, M., Smith, G.C., Dupont, F., Lemieux, J.F., Forget, G., Fujii, Y., et al., 2017. Intercomparison of the Arctic sea ice cover in global ocean–sea ice reanalyses from the ORA-IP project. Clim. Dyn., 1–30.

Chorin, A.J., Hald, O.H., 2006. Stochastic tools in mathematics and science. In: Number 147 in Surveys and Tutorials in the Applied Mathematical Sciences. Springer, New York.

Chorin, A.J., Hald, O.H., Kupferman, R., 2002. Optimal prediction with memory. Physica D 166, 239–257. https://doi.org/10.1016/s0167-2789(02)00446-3.

Christensen, H.M., Moroz, I.M., Palmer, T.N., 2014. Simulating weather regimes: impact of stochastic and perturbed parameter schemes in a simple atmospheric model. Clim. Dyn. 44, 2195–2214. https://doi.org/10.1007/s00382-014-2239-9.

Christiansen, B., 2005. Downward propagation and statistical forecast of the near-surface weather. J. Geophys. Res. Atmos. 110. https://doi.org/10.1029/2004JD005431.

Chu, P.C., 1999. Two kinds of predictability in the Lorenz system. J. Atmos. Sci. 56, 1427–1432.

Clapp, R.B., Hornberger, G.M., 1978. Empirical equations for some soil hydraulic properties. Water Resour. Res. 14, 601–604.

Clark, M., Gangopadhyay, S., Hay, L., Rajagopalan, B., Wilby, R., 2004. The schaake shuffle: a method for reconstructing space–time variability in forecasted precipitation and temperature fields. J. Hydrometeorol. 5 (1), 243–262.

Clark, R.T., Bett, P.E., Thornton, H.E., Scaife, A.A., 2017. Skilful seasonal predictions for the European energy industry. Environ. Res. Lett. 12 (2), 024002. https://doi.org/10.1088/1748-9326/aa57ab.

Coelho, C., Stephenson, D., Balmaseda, M., Doblas-Reyes, F., van Oldenborgh, G., 2006. Toward an integrated seasonal forecasting system for South America. J. Clim. 19, 3704–3721.

Coelho, C.A.S., Firpo, M.A.F., de Andrade, F.M., 2018. A verification framework for South American sub-seasonal precipitation predictions. Meteorol. Z. https://doi.org/10.1127/metz/2018/0898.

Coffey, K., Haile, M., Halperin, M., Wamukoya, G., Hansen, J., Kinyangi, J., Fantaye, K.T., 2015. Expanding the Contribution of Early Warning to Climate-Resilient Agricultural Development in Africa. CCAFS Working Paper no. 115. Copenhagen, Denmark. CGIAR Research Program on Climate Change, Agriculture and Food Security. Available online at:www.ccafs.cgiar.org.

Cohen, J., 2016. Yes, Zika will soon spread in the United States. But it won't be a disaster. Science. https://doi.org/10.1126/science.aaf9988.

Cohen, J., Entekhabi, D., 1999. Eurasian snow cover variability and northern hemisphere climate variability. Geophys. Res. Lett. 26, 345–348.

Cohen, J., Jones, J., 2011. Tropospheric precursors and stratospheric warmings. J. Clim. 24, 6562–6572. https://doi.org/10.1175/2011JCLI4160.1.

Cohen, J., Barlow, M., Kushner, P.J., Saito, K., 2007. Stratosphere-troposphere coupling and links with Eurasian land surface variability. J. Clim. 20, 5335–5343. https://doi.org/10.1175/2007JCLI1725.1.

Cohen, J., Foster, J., Barlow, M., Saito, K., Jones, J., 2010. Winter 2009–2010: a case study of an extreme Arctic Oscillation event. Geophys. Res. Lett. 37, L17707.

Cohen, J., Screen, J.A., Furtado, J.C., Barlow, M., Whittleston, D., Coumou, D., et al., 2014. Recent Arctic amplification and extreme mid-latitude weather. Nat. Geosci. 7 (9), 627–637.

Cohen-Tannoudji, C., Diu, B., Laloe, F., 1973. Mécanique Quantique (tome 1 et 2). Hermann, p. 889.

Coles, S.G., 2001. An Introduction to Statistical Modeling of Extreme Values. Springer Verlag, New York.

Collimore, C.C., et al., 2003. On the relationship between the QBO and tropical deep convection. J. Clim. 16, 2552–2568. https://doi.org/10.1175/1520-0442(2003)016<2552:OTRBTQ>2.0.CO;2.

Colón-González, F.J., Tompkins, A.M., Biondi, R., Bizimana, J.P., Namanya, D.B., 2016. Assessing the effects of air temperature and rainfall on malaria incidence: an epidemiological study across Rwanda and Uganda. Geospat. Health 11 (1s), 18–37. https://doi.org/10.4081/gh.2016.379.

Colucci, S.J., Kelleher, M.E., 2015. Diagnostic comparison of tropospheric blocking events with and without sudden stratospheric warming. J. Atmos. Sci. 72, 2227–2240. https://doi.org/10.1175/JAS-D-14-0160.1.

Comiso, J.C., 1995. SSM/I Ice Concentrations Using the Bootstrap Algorithm. NASA Report 1380.

COMNAP, 2015. COMNAP Sea Ice Challenges Workshop, Hobart, Tasmania, Australia-13 May 2015 Workshop Report. Council of Managers of National Antarctic Programs (COMNAP).https://www.comnap.aq/Publications/Comnap%20Publications/COMNAP_Sea_Ice_Challenges_BKLT_Web_Final_Dec2015.pdf.

Connor, S.J., Mantilla, G.C., 2008. Integration of seasonal forecasts into early warning systems for climate-sensitive diseases such as malaria and dengue. In: Thomson M.C., Garcia-Herrera R., Beniston M. (Eds.), Seasonal Forecasts, Climatic Change and Human Health. Advances in Global Change Research, vol. 30, 2008, Springer, Dordrecht, 71–84.

Connor, S.J., Mantilla, G.C., 2008. Integration of seasonal forecasts into early warning systems for climate-sensitive diseases such as malaria and dengue. In: Seasonal Forecasts, Climatic Change and Human Health. Springer, pp. 71–84.

Connor, S.J., Thomson, M.C., Flasse, S.P., Perryman, A.H., 1998. Environmental information systems in malaria risk mapping and epidemic forecasting. Disasters 22, 39–56.

Conway, D., Archer van Garderen, E., Derying, D., Dorling, S., Krueger, T., Landman, W., Lankford, B., Lebek, K., Osborn, T., Ringler, C., Thurlow, J., Zhu, T., Dalin, C., 2015. Climate and southern Afirca's water-energy-food nexus. Nat. Clim. Chang. 5, 837–846.

Cook, A.R., Schaefer, J.T., 2008. The relation of El Nino–Southern Oscillation (ENSO) to winter tornado outbreaks. Mon. Weather Rev. 136, 3121–3137. https://doi.org/10.1175/2007MWR2171.1.

Cooper, F.C., Haynes, P., 2011. Climate sensitivity via a nonparametric fluctuation-dissipation theorem. J. Atmos. Sci. 68, 937–953.

Corti, S., Molteni, F., Palmer, T.N., 1999. Signature of recent climate change in frequencies of natural atmospheric circulation regimes. Nature 398, 799–802. https://doi.org/10.1038/19745.

Cotton, W.R., Lin, M.-S., McAnelly, R.L., Tremback, C., 1989. A composite model of mesoscale convective complexes. Mon. Weather Rev. 117, 765–783.

Coughlan de Perez, E., Mason, S.J., 2014. Climate information for humanitarian agencies: some basic principles. Earth Perspect. 1, 11. https://doi.org/10.1186/2194-6434-1-11.

Coughlan De Perez, E., et al., 2015. Forecast-based financing: an approach for catalyzing humanitarian action based on extreme weather and climate forecasts. Nat. Hazards Earth Syst. Sci. 15, 895–904.

Coumou, D., Petoukhov, V., Rahmstorf, S., 2014. Quasi-resonant circulation regimes and hemispheric synchronization of extreme weather in boreal summer. PNAS 111, 12331–12336.

Courtier, P., Thépaut, J.N., Hollingsworth, A., 1994. A strategy for operational implementation of 4D-VAR, using an incremental approach. Q. J. R. Meteorol. Soc. 120, 1367–1388.

Coutinho, M.M., 1999. Ensemble Prediction Using Principal-Component-Based Perturbations. Thesis in Meteorology, National Institute for Space Research (INPE), p. 136 (in Portuguese).

Coutinho, M.M., Hoskins, B.J., Buizza, R., 2004. The influence of physical processes on extratropical singular vectors. J. Atmos. Sci. 61, 195–209.

Cox, J., Abeku, T.A., 2007. Early warning systems for malaria in Africa: from blueprint to practice. Trends Parasitol. 23, 243–246.

Cox, P.M., Betts, R.A., Bunton, C.B., Essery, R.L.H., Rowntree, P.R., Smith, J., 1999. The impact of new land surface physics on the GCM simulation of climate and climate sensitivity. Clim. Dyn. 15, 183–203.

Coy, L., Eckermann, S., Hoppel, K., 2009. Planetary wave breaking and tropospheric forcing as seen in the stratospheric sudden warming of 2006. J. Atmos. Sci. 66, 495–507. https://doi.org/10.1175/2008JAS2784.1.

Craig, M.H., Snow, R.W., le Sueur, D., 1999. A climate-based distribution model of malaria transmission in sub-Saharan Africa. Parasitol. Today 15, 105–111.

Crommelin, D.T., 2003. Regime transitions and heteroclinic connections in a barotropic atmosphere. J. Atmos. Sci. 60, 229–246. https://doi.org/10.1175/1520-0469(2003)060<0229:rtahci>2.0.co;2.

Crommelin, D.T., 2004. Observed nondiffusive dynamics in large-scale atmospheric flow. J. Atmos. Sci. 61, 2384–2396. https://doi.org/10.1175/1520-0469(2004)061<2384:ondila>2.0.co;2.

Crooks, S.A., Gray, L.J., 2005. Characterization of the 11-year solar signal using a multiple regression analysis of the ERA-40 dataset. J. Clim. 18, 996–1015. https://doi.org/10.1175/JCLI-3308.1.

Cuevas, E., et al., 2015. The MACC-II 2007–2008 reanalysis: atmospheric dust evaluation and characterization over northern Africa and the Middle East. Atmos. Chem. Phys. 15, 3991–4024.

Cunningham, C.A., 2017. Subseasonal Prediction of Extreme Runoff Over the South America Monsoon Region—A Case Study Over the Upper São Francisco Watershed. Unpublished manuscript.

Curran, M.A.J., van Ommen, T.D., Morgan, V.I., Phillips, K.L., Palmer, A.S., 2003. Ice core evidence for Antarctic Sea ice decline since the 1950s. Science 302 (5648), 1203–1206. https://doi.org/10.1126/science.1087888.

D'Andrea, F., 2002. Extratropical low-frequency variability as a low-dimensional problem. II: Stationarity and stability of large-scale equilibria. Q. J. R. Meteorol. Soc. 128, 1059–1073. https://doi.org/10.1256/003590002320373201.

D'Andrea, F., Vautard, R., 2001. Extratropical low-frequency variability as a low-dimensional problem. I: A simplified model. Q. J. R. Meteorol. Soc. 127, 1357–1374. https://doi.org/10.1256/ smsqj.57412.

Dahlin, K.M., Fisher, R.A., Lawrence, P.J., 2015. Environmental drivers of drought deciduous phenology in the Community Land Model. Biogeosciences 12, 5061–5074.

Dai, A., Lin, X., Hsu, K.-L., 2009. The frequency, intensity, and diurnal cycle of precipitation in surface and satellite observations over low-to mid-latitudes. Clim. Dyn. 29, 727–744.

Daley, R., 1991. Atmospheric Data Analysis. Cambridge University Press.

Danabasoglu, G., Yeager, S.G., Bailey, D., Behrens, E., Bentsen, M., Bi, D., et al., 2014. North Atlantic simulations in coordinated ocean-ice reference experiments phase II (CORE-II). Part I: mean states. Ocean Model. 73, 76–107.

D'Andrea, F., Gentine, P., Betts, A.K., Lintner, B.R., 2014. Triggering deep convection with a probabilistic plume model. J. Atmos. Sci. 71, 3881–3901.

Davies, J.M., Johns, R.H., 1993. The Tornado: its structure, dynamics, pre- diction, and hazards. Some wind and instability parameters associated with strong and violent tornadoes. 1. Wind shear and helicity. Geophys. Monogr., Am. Geophys. Union vol. 79. 573–582.

Davies, T., Warrilow, D., 1986. Soil model and surface temperatures. In: Boer, G. (Ed.), Research Activities in Atmospheric and Oceanic Modelling, Report No. 9. In: vol. 141. WMO/TD, pp. 4.50–4.53.

Davies, J.M., Johns, R.H., Leftwich, P.W., 1993. The Tornado: its structure, dynamics, prediction, and hazards. Some wind and instability parameters associated with strong and violent tornadoes. 2. Variations in the combinations of wind and instability parameters. Geophys. Monogr., Am. Geophys. Union 79, 573–582.

Davis, R.E., 1976. Predictability of sea surface temperature and sea level pressure anomalies over the North Pacific Ocean. J. Phys. Oceanogr. 6, 249–266.

Davis, M., 2000. Late Victorian Holocausts: El Niño Famines and the Making of the Third World. Verso. 464 pp., ISBN 1-85984-739-40.

Davis, C.A., Bosart, L.F., *2004*. Forecasting the tropical transition of cyclones. Bull. Am. Meteorol. Soc., *1657–1662.*

Davis, C., Brown, B., Bullock, R., 2006a. Object-based verification of precipitation forecasts. Part I: Methods and application to mesoscale rain areas. Mon. Weather Rev. 134, 1772–1784.

Davis, C.A., Brown, B.G., Bullock, R.G., 2006b. Object-based verification of precipitation forecasts, Part II: Application to convective rain systems. Mon. Weather Rev. 134, 1785–1795.

Dawson, A., Palmer, T.N., 2014. Simulating weather regimes: impact of model resolution and stochastic parameterization. Clim. Dyn. 44, 2177–2193. https://doi.org/10.1007/s00382-014-2238-x.

Day, J.J., Tietsche, S., Hawkins, E., 2014. Pan-Arctic and regional sea ice predictability: initialization month dependence. J. Clim. 27 (12), 4371–4390.

Day, J.J., Tietsche, S., Collins, M., Goessling, H.F., Guemas, V., Guillory, A., Hurlin, W.J., Ishii, M., Keeley, S.P., Matei, D., Msadek, R., Sigmond, M., Tatebe, H., Hawkins, E., 2016. The Arctic predictability and prediction on seasonal-to-interannual timescales (APPOSITE) data set version 1. Geosci. Model Dev. 9 (6), 2255.

De Felice, M., Alessandri, A., Catalano, F., 2015. Seasonal climate forecasts for medium-term electricity demand forecasting. Appl. Energy 137, 435–444.

De Lannoy, G.J.M., Reichle, R.H., 2016. Assimilation of SMOS brightness temperatures or soil moisture retrievals into a land surface model. Hydrol. Earth Syst. Sci. 20, 4895–4911.

de Leeuw, J., Methven, J., Blackburn, M., 2016. Variability and trends in England and Wales precipitation. Int. J. Climatol. 36, 2823–2836.

de Rosnay, P., Balsamo, G., Albergel, C., Muñoz-Sabater, J., Isaksen, L., 2014. Initialisation of land surface variables for numerical weather prediction. Surv. Geophys. 35, 607–621.

Deardorff, J.W., 1979. Prediction of convective mixed-layer entrainment for realistic capping inversion structure. J. Atmos. Sci. 36, 424–436.

Deardorff, J.W., Willis, G., 1980. Laboratory studies of the entrainment zone of a convectively mixed layer. J. Fluid Mech. 100, 41–64.

Decharme, B., Douville, H., 2006. Introduction of a sub-grid hydrology in the ISBA land surface model. Clim. Dyn. 26, 65–78.

Dee, D., 2009. Representation of climate signals in reanalysis.Presentation at the Fifth International Symposium on Data Assimilation, Melbourne, Australia, 5–9 October 2009. https://www.dropbox.com/s/ifge2r5wimiyc3h/Dee_2009_Melbourne.pdf?dl=0.

Dee, D.P., et al., 2011. The ERA-Interim reanalysis: configuration and performance of the data assimilation system. Q. J. R. Meteorol. Soc. 137, 553–597.

DeFlorio, M.J., Waliser, D.E., Guan, B., Lavers, D.A., Ralph, F.M., Vitart, F., 2018a. Global prediction skill of atmospheric rivers. J. Hydrometerol. submitted.

DeFlorio, M.J., Waliser, D.E., Guan, B., Ralph, F.M., Vitart, F., 2018b. Global evaluation of atmospheric river subseasonal prediction skill. Clim. Dyn. in revision.

Delle Monache, L., Nipen, T., Liu, Y., Roux, G., Stull, R., 2011. Kalman filter and analog schemes to postprocess numerical weather predictions. Mon. Weather Rev. 139 (11), 3554–3570.

Deloncle, A., Berk, R., D'Andrea, F., Ghil, M., 2007. Weather regime prediction using statistical learning. J. Atmos. Sci. 64, 1619–1635. https://doi.org/10.1175/jas3918.1.

DelSole, T., 2007. A Bayesian framework for multimodel regression. J. Clim. 20 (12), 2810–2826.

Demeritt, D., Nobert, S., Cloke, H., Pappenberger, F., 2010. Challenges in communicating and using ensembles in operational flood forecasting. Meteorol. Appl. 17, 209–222.

Demeritt, D., Nobert, S., Cloke, H., Pappenberger, F., 2013. The European Flood Alert System and the communication, perception, and use of ensemble predictions for operational flood risk management. Hydrol. Process. 27 (1), 147–157.

DeMott, C.A., Randall, D.A., Khairoutdinov, M., 2007. Convective precipitation variability as a tool for general circulation model analysis. J. Clim. 20, 91–112.

DeMott, C.A., Klingaman, N.P., Woolnough, S.J., 2015. Atmosphere-ocean coupled processes in the Madden–Julian Oscillation. Rev. Geophys. 53, 1099–1154.
DeMott, C.A., Benedict, J.J., Klingaman, N.P., Woolnough, S.J., Randall, D.A., 2016. Diagnosing ocean feedbacks to the MJO: SST modulated surface fluxes and the moist static energy budget. J. Geophys. Res. Atmos. 121, 8350–8373.
Demuth, J.L., Morss, R.E., Lazo, J.K., Hilderbrand, D.C., 2013. Improving effectiveness of weather risk communication on the NWS Point-and-Click web page. Weather Forecast. 28, 711–726.
Denis, B., Côté, J., Laprise, R., 2002. Spectral decomposition of two-dimensional atmospheric fields on limited-area domains using the discrete cosine transform (DCT). Mon. Weather Rev. 130, 1812–1829.
Denis, B., Laprise, R., Caya, D., 2003. Sensitivity of a regional climate model to the resolution of the lateral boundary conditions. Clim. Dyn. 20, 107–126.
Déqué, M., 1997. Ensemble size for numerical seasonal forecasts. Tellus A 49, 74–86.
Déqué, M., Royer, J.F., 1992. The skill of extended-range extratropical winter dynamical forecasts. J. Clim. 5, 1346–1356.
Deremble, B., Simonnet, E., Ghil, M., 2012. Multiple equilibria and oscillatory modes in a midlatitude ocean-forced atmospheric model. Nonlinear Process. Geophys. 19, 479–499. https://doi.org/10.5194/ npg-19-479-2012.
Derome, J., Lin, H., Brunet, G., 2005. Seasonal forecasting with a simple general circulation model: predictive skill in the AO and PNA. J. Clim. 18, 597–609.
Deroubaix, A., Martiny, N., Chiapello, I., Marticoréna, B., 2013. Suitability of OMI aerosol index to reflect mineral dust surface conditions: preliminary application for studying the link with meningitis epidemics in the Sahel. Remote Sens. Environ. 133, 116–127.
Deser, C., Timlin, M.S., 1997. Atmosphere–ocean interaction on weekly timescales in the North Atlantic and Pacific. J. Clim. 10, 393–408.
Deser, C., Walsh, J.E., Timlin, M.S., 2000. Arctic sea ice variability in the context of recent atmospheric circulation trends. J. Clim. 13 (3), 617–633.
Deser, C., Tomas, R., Alexander, M., Lawrence, D., 2010. The seasonal atmospheric response to projected Arctic sea ice loss in the late 21st century. J. Clim. 23, 333–351. https://doi.org/10.1175/2009JCLI3053.1.
Deser, C., Simpson, I.R., McKinnon, K.A., Phillips, A.S., 2017. The northern hemisphere extratropical atmospheric circulation response to ENSO: how well do we know it and how do we evaluate models accordingly? J. Clim. 30, 5069–5082.
Dettinger, M.D., 2013. Atmospheric rivers as drought busters on the U.S. West Coast. J. Hydrometeorol. 14, 1721–1732. https://doi.org/10.1175/JHM-D-13-02.1.
Dettinger, M.D., Ralph, F.M., Das, T., Neiman, P.J., Cayan, D.R., 2011. Atmospheric rivers, floods and the water resources of California. Water 3, 445–478. https://doi.org/10.3390/w3020445.
Deutsche Meteorologische Gesellshaft e.V, 2000. 50th Anniversary of Numerical Weather Prediction Commemorative Symposium, Postdam, 9-10 March 2000. Book of Lectures, published by Deutsche Meteorologische Gesellshaft, ISBN: 3-928903-22-5.
Dewdney, M.M., Fraisse, C.W., Zortea, T., Burrow, J., 2012. A Web-Based Tool for Timing Copper Applications in Florida Citrus. PP289. Plant Pathology Department, Florida Cooperative Extension Service, Institute of Food and Agricultural Sciences, University of Florida, Gainesville. Available at: http://agroclimate.org/tools/ Citrus-Copper-Application-Scheduler/PP28900.pdf. [(Accessed 24 March 2017)].
Diamond, J., 2005. Collapse. Penguin Books, New York. 616 pp.
Dickey, J.O., Ghil, M., Marcus, S.L., 1991. Extratropical aspects of the 40–50 day oscillation in length-of-day and atmospheric angular momentum. J. Geophys. Res. Atmos. 96, 22643–22658. https://doi.org/10.1029/91jd02339.
Dickinson, M., Molinari, J., 2002. Mixed Rossby—gravity waves and western Pacific tropical cyclogenesis. Part I: Synoptic evolution. J. Atmos. Sci. 59, 2183–2196.
Dijkstra, H.A., 2005. Nonlinear Physical Oceanography: A Dynamical Systems Approach to the Large Scale Ocean Circulation and El Niño, second ed. Springer Science & Business Media, Berlin/Heidelberg, Germany.
Dijkstra, H.A., 2013. Nonlinear Climate Dynamics. Cambridge University Press, Cambridge.
Dijkstra, H.A., Ghil, M., 2005. Low-frequency variability of the large-scale ocean circulation: a dynamical systems approach. Rev. Geophys. 43. RG3002 https://doi.org/10.1029/2002RG000122.
Dinku, T., et al., 2014. Bridging critical gaps in climate services and applications in Africa. Earth Perspect. 1, 15.
Dinon, H., Breuer, N., Boyles, R., Wilkerson, G., 2012. North Carolina Extension Agent Awareness of and Interest in Climate Information for Agriculture. Southeast Climate Consortium Tech. Rep. 12–003, 44 pp. Available at: http://www.seclimate.org/pdfpubs/SECCsurveyReportFinal.pdf. [(Accessed 24 March 2017)].

Diokhane, A.M., Jenkins, G.S., Manga, N., Drame, M.S., Mbodji, B., 2016. Linkages between observed, modeled Saharan dust loading and meningitis in Senegal during 2012 and 2013. Int. J. Biometeorol. 60, 557–575.

Dirkson, A., Merryfield, W.J., Monahan, A., 2017. Impacts of sea ice thickness initialization on seasonal Arctic Sea ice predictions. J. Clim. 30 (3), 1001–1017.

Dirmeyer, P.A., 1999. Assessing GCM sensitivity to soil wetness using GSWP data. J. Meteorol. Soc. Jpn. 77, 367–385.

Dirmeyer, P.A., 2006. The hydrologic feedback pathway for land-climate coupling. J. Hydrometeorol. 7, 857–867.

Dirmeyer, P.A., 2013. Characteristics of the water cycle and land-atmosphere interactions from a comprehensive reforecast and reanalysis data set: CFSv2. Clim. Dyn. 41, 1083–1097.

Dirmeyer, P.A., Halder, S., 2017. Application of the land-atmosphere coupling paradigm to the operational Coupled Forecast System (CFSv2). J. Hydrometeorol. 18, 85–108.

Dirmeyer, P.A., Dolman, A.J., Sato, N., 1999. The pilot phase of the Global Soil Wetness Project. Bull. Am. Meteorol. Soc. 80, 851–878.

Dirmeyer, P.A., Gao, X., Zhao, M., Guo, Z., Oki, T., Hanasaki, N., 2006a. GSWP-2: multimodel analysis and implications for our perception of the land surface. Bull. Am. Meteorol. Soc. 87, 1381–1397.

Dirmeyer, P.A., Koster, R.D., Guo, Z., 2006b. Do global models properly represent the feedback between land and atmosphere? J. Hydrometeorol. 7, 1177–1198.

Dirmeyer, P.A., Schlosser, C.A., Brubaker, K.L., 2009. Precipitation, recycling and land memory: an integrated analysis. J. Hydrometeorol. 10, 278–288.

Dirmeyer, P.A., Gochis, D.J., Hogue, T.S., Barros, A., Duffy, C.J., Friedrich, K., Hughes, M., Krajewski, W., Molotch, N.P., 2014. In: Advancing hydrometeorological-hydroclimatic-ecohydrological process understanding and predictions.White Paper: Hydrologic-Atmospheric Community Workshop, Golden, Colorado. 12 pp. Available at: http://inside.mines.edu/~thogue/nsf-hydro-atmo-workshop/NSFHydroAtmosWorkshopWhitePaper 120314FINAL.pdf.

Dirmeyer, P.A., Peters-Lidard, C., Balsamo, G., 2015. Land-atmosphere interactions and the water cycle. In: Brunet, G., Jones, S., Ruti, P.M. (Eds.), Seamless Prediction of the Earth System: From Minutes to Months. World Meteorological Organization (WMO-No. 1156), Geneva (Chapter 8).

Dirmeyer, P.A., et al., 2016. Confronting weather and climate models with observational data from soil moisture networks over the United States. J. Hydrometeorol. 17, 1049–1067.

DNE (National Energy Agency of Uruguay), 2017. Statistical Time Series of Electric Energy. http://www.dne.gub.uy/web/energia/-/series-estadisticas-de-energia-electrica.

Doblas-Reyes, F.J., Hagedorn, R., Palmer, T.N., 2005. The rationale behind the success of multi-model ensembles in seasonal forecasting-II. Calibration and combination. Tellus A 57 (3), 234–252.

Doblas-Reyes, F.J., Weisheimer, A., Déqué, M., Keenlyside, N., McVean, M., Murphy, J.M., Rogel, P., Smithd, D., Palmer, T.N., 2009. Addressing model uncertainty in seasonal and annual dynamical ensemble forecasts. Q. J. R. Meteorol. Soc. 135, 1538–1559.

Doblas-Reyes, F.J., García-Serrano, J., Lienert, F., Biescas, A.P., Rodrigues, L.R., 2013. Seasonal climate predictability and forecasting: status and prospects. Wiley Interdiscip. Rev. Clim. Chang. 4, 245–268.

Dodson, J.B., Randall, D.A., Suzuki, K., 2013. Comparison of observed and simulated tropical cumuliform clouds by CloudSat and NICAM. J. Geophys. Res. Atmos. 118, 1852–1867.

Dole, R.M., Gordon, N.D., 1983. Persistent anomalies of the extratropical Northern Hemisphere wintertime circulation: geographical distribution and regional persistence characteristics. Mon. Weather Rev. 111, 1567–1586. https://doi.org/10.1175/1520-0493(1983)111<1567:paoten>2.0.co;2.

Dole, R., Hoerling, M., Perlwitz, J., Eischeid, J., Pegion, P., Zhang, T., Quan, X.-W., Xu, T., Murray, D., 2011. Was there a basis for anticipating the 2010 Russian heat wave? Geophys. Res. Lett. 38, L06702. https://doi.org/10.1029/2010GL046582.

Domeisen, D.I.V., Butler, A.H., Fröhlich, K., Bittner, M., Müller, W.A., Baehr, J., 2015. Seasonal predictability over Europe arising from El Niño and stratospheric variability in the MPI-ESM seasonal prediction system. J. Clim. 28, 256–271. https://doi.org/10.1175/JCLI-D-14-00207.1.

Donald, A., Meinke, H., Power, B., Maia, A.d.H.N., Wheeler, M.C., White, N., Stone, R.C., Ribbe, J., 2006. Near-global impact of the Madden-Julian Oscillation on rainfall. Geophys. Res. Lett. 33, L09704.

Doocy, S., Daniels, A., Murray, S., Kirsch, T.D., 2013. The human impact of floods: a historical review of events 1980-2009 and systematic literature review. In: PLOS Currents Disasters, first ed. Apr 16.

Doswell III, C.A., Brooks, H.E., Maddox, R.A., 1996. Flash flood forecasting: an ingredients-based methodology. Weather Forecast. 11, 560–581. https://doi.org/10.1175/1520-0434(1996)0110560:FFFAIB2.0.CO;2.

Douville, H., 2009. Stratospheric polar vortex influence on Northern Hemisphere winter climate variability. Geophys. Res. Lett. 36, L18703. https://doi.org/10.1029/2009GL039334.

Douville, H., Chauvin, F., 2000. Relevance of soil moisture for seasonal climate prediction: a preliminary study. Clim. Dyn. 16, 719–736.

Douville, H., Royer, J.-F., Mahfouf, J.-F., 1995. A new snow parameterization for the Meteo-France climate model. Part II: Validation in a 3-D GCM experiment. Clim. Dyn. 12, 37–52.

Douville, H., Chauvin, F., Broqua, H., 2001. Influence of soil moisture on the Asian and African monsoons. Part I: mean monsoon and daily precipitation. J. Clim. 14, 2381–2403.

Douville, H., Conil, S., Tyteca, S., Voldoire, A., 2007. Soil moisture memory and West African monsoon predictability: artefact or reality? Clim. Dyn. 28, 723–742.

Downes, S.M., Farneti, R., Uotila, P., Griffies, S.M., Marsland, S.J., Bailey, D., et al., 2015. An assessment of Southern Ocean water masses and sea ice during 1988–2007 in a suite of interannual CORE-II simulations. Ocean Model. 94, 67–94.

Draper, C., Reichle, R., 2015. The impact of near-surface soil moisture assimilation at subseasonal, seasonal, and interannual timescales. Hydrol. Earth Syst. Sci. 19, 4831–4844.

Dreybrodt, W., 1988. Processes in Karst Systems: Physics, Chemistry, and Geology. Springer, Berlin 288 pp.

Drobot, S.D., 2007. Using remote sensing data to develop seasonal outlooks for Arctic regional sea ice minimum extent. Remote Sens. Environ. 111 (2), 136–147.

Drobot, S.D., Maslanik, J.A., 2002. A practical method for long-range forecasting of ice severity in the Beaufort Sea. Geophys. Res. Lett. 29(8).

Drobot, S.D., Maslanik, J.A., Fowler, C., 2006. A long-range forecast of Arctic summer sea-ice minimum extent. Geophys. Res. Lett. 33(10).

Drótos, G., Bódai, T., Tél, T., 2015. Probabilistic concepts in a changing climate: a snapshot attractor picture. J. Clim. 28, 3275–3288. https://doi.org/10.1175/jcli-d-14-00459.1.

Dubuisson, M.-P., Jain, A.K., 1994. A modified Hausdorff distance for object matching.Proc. International Conference on Pattern Recognition, Jerusalem, Israel, pp. 566–568.

Duchon, C., 1979. Lanczos filtering in one and two dimensions. J. Appl. Meteorol. 18, 1016–1022.

Duffy, C., et al., 2006. Towards and Integrated Observing Platform for the Terrestrial Water Cycle: From Bedrock to Boundary Layer. Available online, http://www.usgcrp.gov/usgcrp/Library/watercycle/ssg-whitepaper-dec2006.pdf.

Dukhovskoy, D.S., Ubnoske, J., Blanchard-Wrigglesworth, E., Hiester, H.R., Proshutinsky, A., 2015. Skill metrics for evaluation and comparison of sea ice models. J. Geophys. Res. Oceans 120, 5910–5931. https://doi.org/10.1002/2015JC010989.

Dukić, V., et al., 2012. The role of weather in meningitis outbreaks in Navrongo, Ghana: a generalized additive modeling approach. J. Agric. Biol. Environ. Stat., 1–19.

Dunn-Sigouin, E., Shaw, T.A., 2015. Comparing and contrasting extreme stratospheric events, including their coupling to the tropospheric circulation. J. Geophys. Res. Atmos. 120, 1374–1390. https://doi.org/10.1002/2014JD022116.

Dunstone, N., Smith, D., Scaife, A., Hermanson, L., Eade, R., Robinson, N., Andrews, M., Knight, J., 2016. Skilful predictions of the winter North Atlantic Oscillation one year ahead. Nat. Geosci. 9, 809–814. https://doi.org/10.1038/ngeo2824.

Duran, V., 2014. Situación y perspectivas de las cadenas agroindustriales. In: OPYPA Yearbook. MGAPwww.mgap.gub.uy/sites/default/files/anuario_opypa_2014.pdf.

Dutra, E., Balsamo, G., Viterbo, P., Miranda, P.M.A., Beljaars, A., Schär, C., Elder, K., 2010a. An improved snow scheme for the ECMWF land surface model: description and offline validation. J. Hydrometeorol. 11, 899–916.

Dutra, E., Stepaneko, V.M., Balsamo, G., Viterbo, P., Miranda, P., Mironov, D., Schär, C., 2010b. An offline study of the impact of lakes in the performance of the ECMWF surface scheme. Boreal Environ. Res. 15, 100–112.

Dutra, E., Di Giuseppe, F., Wetterhall, F., Pappenberger, F., 2013. Seasonal forecasts of droughts in African basins using the Standardized Precipitation Index. Hydrol. Earth Syst. Sci. 17, 2359–2373. https://doi.org/10.5194/hess-17-2359-2013.

Dutra, E., Wetterhall, F., Di Giuseppe, F., Naumann, G., Barbosa, P., Vogt, J., Pozzi, W., Pappenberger, F., 2014a. Global meteorological drought—Part 1: Probabilistic monitoring. Hydrol. Earth Syst. Sci. 18, 2657–2667. https://doi.org/10.5194/hess-18-2657-2014.

Dutra, E., Pozzi, W., Wetterhall, F., Di Giuseppe, F., Magnusson, L., Naumann, G., Barbosa, P., Vogt, J., Pappenberger, F., 2014b. Global meteorological drought—Part 2: Seasonal forecasts. Hydrol. Earth Syst. Sci. 18, 2669–2678. https://doi.org/10.5194/hess-18-2669-2014.

Ebert, E.E., 2008. Fuzzy verification of high resolution gridded forecasts: a review and proposed framework. Meteorol. Appl. 15, 51–64.

Ebert, E.E., McBride, J.L., 2000. Verification of precipitation in weather systems: determination of systematic errors. J. Hydrol. 239, 179–202.

Ebi, K.L., Teisberg, T.J., Kalkstein, L.S., Robinson, L., Weiher, R.F., 2004. Heat watch/warning systems save lives: estimated costs and benefits for Philadelphia 1995–98. Bull. Am. Meteorol. Soc. 85, 1067–1073.

Ebisuzaki, W., Kalnay, E., 1991. Ensemble experiments with a new lagged average forecasting scheme. WMO, Research activities in atmospheric and oceanic modeling, pp. 6.31–6.32. Report #15. (Available from WMO, C.P. No 2300, CH1211, Geneva, Switzerland).

ECMWF, 2010. SAT/DA Training Course. Available at: https://www.google.com/url?sa=i&rct=j&q=&esrc=s&source=images&cd=&ved=2ahUKEwikpMiH3cTZAhUJHqwKHaqJCKwQjRx6BAgAEAY&url=http%3A%2F%2Fslideplayer.com%2Fslide%2F752902%2F&psig=AOvVaw0Q8FnypURbKUMOL-w31gxR&ust=1519773849498096.

ECMWF, 2016. ECMWF Strategy 2016-2025. Available from ECMWF, Shinfield Park, Reading RG2-9AX, UK. See also: https://www.ecmwf.int/sites/default/files/ECMWF_Strategy_2016-2025.pdf.

Edinburgh, T., Day, J.J., 2016. Estimating the extent of Antarctic summer sea ice during the Heroic Age of Antarctic Exploration. Cryosphere 10 (6), 2721–2730.

Egger, J., 1978. Dynamics of blocking highs. J. Atmos. Sci. 35, 1788–1801. https://doi.org/10.1175/1520-0469(1978)035<1788:dobh>2.0.co;2.

Ek, M.B., Holstlag, A.A.M., 2004. Influence of soil moisture on boundary layer cloud development. J. Hydrometeorol. 5, 86–99.

Ek, M.B., Mitchell, K.E., Lin, Y., Rogers, E., Grunmann, P., Koren, V., Gayno, G., Tarpley, J.D., 2003. Implementation of the Noah land surface model advances in the National Centers for Environmental Prediction operational mesoscale Eta model. J. Geophys. Res. 108.

El-Badry, A., Al-Ali, K., 2010. Prevalence and seasonal distribution of dengue mosquito, *Aedes aegypti* (Diptera: Culicidae) in Al-Madinah Al-Munawwarah, Saudi Arabia. J. Entomol. 7, 80–88.

Elsberry, R.L., Jordan, M.S., Vitart, F., 2010. Predictability of tropical cyclone events on intraseasonal timescales with the ECMWF monthly forecast model. Asia-Pacific J. Atmos. Sci. 46, 135. https://doi.org/10.1007/s13143-010-0013-4.

Elsner, J.B., Widen, H.M., 2013. Predicting spring tornado activity in the Central Great Plains by 1 March. Mon. Weather Rev. 142, 259–267. https://doi.org/10.1175/ MWR-D-13-00014.1.

Ely, C.R., Brayshaw, D.J., Methven, J., Cox, J., Pearce, O., 2013. Implications of the North Atlantic Oscillation for a UK-Norway renewable power system. Energy Policy. https://doi.org/10.1016/j.enpol.2013.06.037.

Emanuel, K.A., 1987. An air-sea interaction model of intra-seasonal oscillation in the tropics. J. Atmos. Sci. 44, 2324–2340.

Emanuel, K., et al., 2013. Influence of tropical tropopause layer cooling on Atlantic hurricane activity. J. Clim. 26, 2288–2301. https://doi.org/10.1175/JCLI-D-12-00242.1.

Entekhabi, D., Nakamura, H., Njoku, E., 1994. Solving the inverse problems for soil-moisture and temperature profiles by sequential assimilation of multifrequency remotely-sensed observations. IEEE Trans. Geosci. Remote Sens. 32, 438–448.

Entekhabi, D., et al., 2010. The Soil Moisture Active Passive (SMAP) mission. Proc. IEEE 98, 704–716.

Epstein, E.S., 1969a. A scoring system for probability forecasts of ranked categories. J. Appl. Meteorol. 8, 985–987.

Epstein, E.S., 1969b. Stochastic dynamic prediction. Tellus A 21, 739–759.

Errico, R.M., 1997. What is an adjoint model? Bull. Am. Meteorol. Soc. 78, 2577–2591.

Errico, R., Baumhefner, D., 1987. Predictability experiments using a high-resolution limited-area model. Mon. Weather Rev. 115, 488–504.

Errico, R.M., Prive, N.C., 2014. An estimate of some analysis-error statistics using the Global Modeling and Assimilation Office observing-system simulation framework. Q. J. R. Meteorol. Soc. 140 (680), 1005–1012. https://doi.org/10.1002/qj.2180.

Errico, R.M., Rasch, P.J., 1988. A comparison of various normal-mode initialization schemes and the inclusion of diabatic processes. Tellus 40A, 1–25.

Eshel, G., Cane, M.A., Farrell, B.F., 2000. Forecasting eastern Mediterranean droughts. Mon. Weather Rev. 128, 3618–3630.

Esler, J.G., Scott, R.K., 2005. Excitation of transient Rossby waves on the stratospheric polar vortex and the barotropic sudden warming. J. Atmos. Sci. 62, 3661–3682. https://doi.org/10.1175/JAS3557.1.

European Centre for Medium-Range Weather Forecasts, 2014. ECMWF ERA—Interim: Reduced N256 Gaussian Gridded Pressure Level Analysis Time Parameter Data (ggap). NCAS British Atmospheric Data Centre. 12th November 2017. http://catalogue.ceda.ac.uk/uuid/cae68b35c821ce036f28eda09e7d3a7c.

Evans, S., Marchand, R., Ackerman, T., 2014. Variability of the Australian monsoon and precipitation trends at Darwin. J. Clim. 27, 8487–8500.

Evensen, G., 1994. Sequential data assimilation with a non-linear quasi-geostrophic model using Monte Carlo methods to forecast error statistics. J. Geophys. Res. 99 (C5), 10143–10162.

Evensen, G., 2003. The Ensemble Kalman Filter: theoretical formulation and practical implementation. Ocean Dyn. 53, 343–367.

Eyring, V., Cionni, I., Arblaster, J., Sedlacek, J., Perlwitz, J., Young, P., Bekki, S., Bergmann, D., Cameron-Smith, P., Collins, W.J., Faluvegi, G., Gottschaldt, K.-D., Horowitz, L.W., Kinnison, D., Lamarque, J.-F., Marsh, D.R., Saint-Martin, D., Shindell, D.T., Sudo, K., Szopa, S., Watanabe, S., 2013. Long-term changes in tropospheric and stratospheric ozone and associated climate impacts in CMIP5 simulations. J. Geophys. Res. Atmos. 118 (10), 5029–5060.

Famiglietti, J.S., Devereaux, J.A., Laymon, C.A., Tsegaye, T., Houser, P.R., Jackson, T.J., Graham, S.T., Rodell, M., van Oevelen, P.J., 1999. Ground-based investigation of soil moisture variability within remote sensing footprints during the Southern Great Plains 97 (SGP97) hydrology experiment. Water Resour. Res. 35, 1839–1851.

Farneti, R., Downes, S.M., Griffies, S.M., Marsland, S.J., Behrens, E., Bentsen, M., et al., 2015. An assessment of Antarctic Circumpolar Current and Southern Ocean meridional overturning circulation during 1958–2007 in a suite of interannual CORE-II simulations. Ocean Model. 93, 84–120.

Faucher, M., Roy, F., Ritchie, H., Desjardins, S., Fogarty, C., Smith, G., Pellerin, P., 2010. Coupled atmosphere-ocean-ice forecast system for the gulf of St-Lawrence, Canada. Q. Newsl. 23.

Fauchereau, N., Pohl, B., Lorrey, A., 2016. Extratropical impacts of the Madden-Julian Oscillation over the New Zealand from a weather regime perspective. J. Clim. 2. https://doi.org/10.1175/JCLI-D-15-0152.1.

Feldmann, K., Scheuerer, M., Thorarinsdottir, T.L., 2015. Spatial postprocessing of ensemble forecasts for temperature using nonhomogeneous Gaussian regression. Mon. Weather Rev. 143 (3), 955–971.

Feldstein, S.B., 2003. The dynamics of NAO teleconnection pattern growth and decay. Q. J. R. Meteorol. Soc. 129, 901–924.

Feliks, Y., Ghil, M., Robertson, A.W., 2010. Oscillatory climate modes in the Eastern Mediterranean and their synchronization with the North Atlantic Oscillation. J. Clim. 23, 4060–4079. https://doi.org/10.1175/2010jcli3181.1.

Feliks, Y., Robertson, A.W., Ghil, M., 2016. Interannual variability in North Atlantic weather: data analysis and a quasigeostrophic model. J. Atmos. Sci. 73 (8), 3227–3248.

Feng, J., Liu, P., Chen, W., Wang, X., 2015. Contrasting Madden-Julian Oscillation activity during various stages of EP and CP El Niños. Atmos. Sci. Lett. 16, 32–37.

Ferranti, L., Palmer, T.N., Molteni, F., Klinker, E., 1989. Tropical-extratropical interaction associated with the 30–60 day oscillation and its impact on medium and extended range prediction. J. Atmos. Sci. 47, 2177–2199.

Ferreira, D., Marshall, J., Bitz, C.M., Solomon, S., Plumb, A., 2015. Antarctic ocean and sea ice response to ozone depletion: a two-time-scale problem. J. Clim. 28 (3), 1206–1226. https://doi.org/10.1175/JCLI-D-14-00313.1.

Ferro, C.A.T., 2007. Comparing probabilistic forecasting systems with the Brier score. Weather Forecast. 22, 1076–1088.

Ferro, C.A.T., 2014. Fair scores for ensemble forecasts. Q. J. R. Meteorol. Soc. 140, 1917–1923. https://doi.org/10.1002/qj.2270.

Ferro, C.A.T., Stephenson, D.B., 2011. Extremal dependence indices: improved verification measures for deterministic forecasts of rare binary events. Weather Forecast. 26, 699–713. https://doi.org/10.1175/WAF-D-10-05030.1.

Ferro, C.A.T., Richardson, D.S., Weigel, A.P., 2008. On the effect of ensemble size on the discrete and continuous ranked probability scores. Meteorol. Appl. 15, 19–24. https://doi.org/10.1002/met.45.

Fetterer, F., Knowles, K., Meier, W., Savoie, M., 2002. Sea Ice Index. Natl Snow and Ice Data Center, Boulder, CO. Available at http://nsidc.org/data/g02135.htm. [(Accessed 9 February 2009)].

Fichefet, T., Morales Maqueda, M.A., 1997. Sensitivity of a global sea ice model to the treatment of ice thermodynamics and dynamics. J. Geophys. Res. 102 (12), 612–646.

Findell, K.L., Eltahir, E.A.B., 2003. Atmospheric controls on soil moisture-boundary layer interactions. Part I: Framework development. J. Hydrometeorol. 4, 552–569.

Findell, K., Gentine, P., Lintner, B.R., Kerr, C., 2011. Probability of afternoon precipitation in eastern United States and Mexico enhanced by high evaporation. Nat. Geosci. 4, 434–439.

Finger, F.G., Teweles, S., 1964. The mid-winter 1963 stratospheric warming and circulation change. J. Appl. Meteorol. 3, 1–15. https://doi.org/10.1175/1520-0450(1964)003<0001:TMWSWA>2.0.CO;2.

Fischhoff, B., 1995. Risk perception and communication unplugged: twenty years of process. Risk Anal. 15 (2), 137–145.

Fitzjarrald, D.R., Acevedo, O.C., Moore, K.E., 2001. Climatic consequences of leaf presence in the eastern United States. J. Clim. 14, 598–614.

Flatau, M., Kim, Y.-J., 2013. Interaction between the MJO and polar circulations. J. Clim. 26, 3562–3574. https://doi.org/10.1175/JCLI-D-11-00508.1.

Flato, G.M., 1995. Spatial and temporal variability of Arctic ice thickness. Ann. Glaciol. 21 (1), 323–329.

Flato, G., Marotzke, J., Abiodun, B., Braconnot, P., Chou, S.C., Collins, W., Cox, P., Driouech, F., Emori, S., Eyring, V., Forest, C., Gleckler, P., Guilyardi, E., Jakob, C., Kattsov, V., Reason, C., Rummukainen, M., 2013. Evaluation of climate models. In: Stocker, T.F., Qin, D., Plattner, G.-K., Tignor, M., Allen, S.K., Boschung, J., Nauels, A., Xia, Y., Bex, V., Midgley, P.M. (Eds.), Climate Change 2013: The Physical Science Basis. Contribution of Working Group I to the Fifth Assessment Report of the Intergovernmental Panel on Climate Change. Cambridge University Press, Cambridge, United Kingdom and New York, NY, USA.

Fleming, R.J., 1971. On stochastic dynamic prediction: I. The energetics of uncertainty and the question of closure. Mon. Weather Rev. 99, 851–872.

Fletcher, C.G., Cassou, C., 2015. The dynamical influence of separate teleconnections from the Pacific and Indian Oceans on the northern annular mode. J. Clim. 28, 7985–8002. https://doi.org/10.1175/JCLI-D-14-00839.1.

Fletcher, C.G., Kushner, P.J., 2011. The role of linear interference in the annular mode response to tropical SST forcing. J. Clim. 24, 778–794. https://doi.org/10.1175/2010JCLI3735.1.

Food and Agriculture Organization, 1988. FAO-UNESCO Soil Map of the World. World Soil Resources Report 60. FAO, Rome.

Ford, T.W., Quiring, S.M., 2014. In situ soil moisture coupling with extreme temperatures: a study based on the Oklahoma Mesonet. Geophys. Res. Lett. 41, 4727–4734.

Frame, T.H.A., Methven, J., Roberts, N.M., Titley, H., 2015. Predictability of frontal waves and cyclones. Weather Forecast. 30, 1291–1302.

Frank, W.M., Roundy, P.E., 2006. The relationship between tropical waves and tropical cyclogenesis. Mon. Weather Rev. 134, 2397–2417.

Frankenberg, C., et al., 2011. New global observations of the terrestrial carbon cycle from GOSAT: patterns of plant fluorescence with gross primary productivity. Geophys. Res. Lett. 38, L17706.

Frankenberg, C., O'Dell, C., Guanter, L., McDuffie, J., 2012. Remote sensing of near-infrared chlorophyll fluorescence from space in scattering atmospheres: implications for its retrieval and interferences with atmospheric CO_2 retrievals. Atmos. Meas. Tech. 5, 2081–2094.

Frankenberg, C., et al., 2014. Prospects for chlorophyll fluorescence remote sensing from the Orbiting Carbon Observatory-2. Remote Sens. Environ. 147, 1–12.

Frankignoul, C., Hasselmann, K., 1977. Stochastic climate models, Part II Application to sea-surface temperature anomalies and thermocline variability. Tellus 29 (4), 289–305.

Frankignoul, C., Sennéchael, N., Kwon, Y.-O., Alexander, M.A., 2011. Influence of the meridional shifts of the Kuroshio and the Oyashio Extensions on the atmospheric circulation. J. Clim. 24, 762–777. https://doi.org/10.1175/2010JCLI3731.1.

Frasch, C., Preziosi, M.-P., LaForce, F.M., 2012. Development of a group A meningococcal conjugate vaccine, MenAfriVacTM. Hum. Vaccin. Immunother. 8, 715–724.

Frederiksen, J.S., 1982. A unified three-dimensional instability theory of the onset of blocking and cyclogenesis. J. Atmos. Sci. 39, 969–982.

Frederiksen, J.S., 1983. A unified three-dimensional instability theory of the onset of blocking and cyclogenesis. II. Teleconnection patterns. J. Atmos. Sci. 40, 2593–2609.

Frederiksen, J.S., 2002. Genesis of intraseasonal oscillations and equatorial waves. J. Atmos. Sci. 59, 2761–2781.
Frederiksen, J.S., 2007. Instability theory and predictability of atmospheric disturbances. In: Frontiers in Turbulence and Coherent Structures, pp. 29–58 (Chapter 2).
Frederiksen, J.S., Bell, R.C., 1987. Teleconnection patterns and the roles of baroclinic, barotropic and topographic instability. J. Atmos. Sci. 44, 2200–2218.
Frederiksen, C.S., Frederiksen, J.S., 1992. Northern Hemisphere storm tracks and teleconnection patterns in primitive equation and quasi-geostrophic models. J. Atmos. Sci. 49, 1443–1458.
Frederiksen, J.S., Frederiksen, C.S., 1993. Monsoon disturbances, intraseasonal oscillations, teleconnection patterns, blocking and storm tracks of the global atmosphere during January 1979: linear theory. J. Atmos. Sci. 50, 1349–1372.
Frederiksen, J.S., Frederiksen, C.S., 1997. Mechanism of the formation of intraseasonal oscillations and Australian monsoon disturbances: the roles of convection, barotropic and baroclinic instability. Contrib. Atmos. Phys. 70, 39–56.
Frederiksen, J.S., Frederiksen, C.S., 2011. Twentieth century winter changes in Southern Hemisphere synoptic weather modes. Adv. Meteorol. 353829 16 pp. https://doi.org/10.1155/2011/353829.
Frederiksen, J.S., Lin, H., 2013. Tropical-extratropical interactions of intraseasonal oscillations. J. Atmos. Sci. 70, 3180–3197.
Frederiksen, J.S., Webster, P.J., 1988. Alternative theories of atmospheric teleconnections and low-frequency fluctuations. Rev. Geophys. 26, 459–494.
Frenger, I., Gruber, N., Knutti, R., Munnich, M., 2013. Imprint of Southern Ocean eddies on winds, clouds and rainfall. Nat. Geosci. 6, 608–612. https://doi.org/10.1038/ngeo1863.
Fricker, T.E., Ferro, C.A.T., Stephenson, D.B., 2013. Three recommendations for evaluating climate predictions. Meteorol. Appl. 20, 246–255. https://doi.org/10.1002/met.1409.
Fu, X., Wang, B., Waliser, D.E., Tao, L., 2007. Impact of atmosphere-ocean coupling on the predictability of monsoon intraseasonal oscillations. J. Atmos. Sci. 64, 157–174.
Fu, X., Lee, J., Hsu, P.-C., Taniguchi, H., Wang, B., Wng, W., 2013. Multi-model MJO forecasting during DYNAMO/CINDY period. Clim. Dyn. 41, 1067–1081.
Fukutomi, Y., Yasunari, T., 2009. Cross-equatorial influences of submonthly scale southerly surges over the eastern Indian Ocean during Southern Hemisphere winter. J. Geophys. Res. 114D20119. https://doi.org/10.1029/2008JD011441.
Fukutomi, Y., Yasunari, T., 2014. Extratropical forcing of tropical wave disturbances along the Indian Ocean ITCZ. J. Geophys. Res. 119, 1154–1171. https://doi.org/10.1002/2013JD020696.
Füssel, H.-M., 2007. Adaptation planning for climate change: concepts, assessment approaches, and key lessons. Sustain. Sci. 2, 265–275.
Gadgil, S., 2003. The Indian monsoon and its variability. Annu. Rev. Earth Planet. Sci. 31, 429–467.
Gadgil, S., Rupa Kumar, K., 2006. The Asian monsoon—agriculture and economy. In: The Asian Monsoon. Springer, Berlin, Heidelberg, pp. 651–683.
Gagné, M.-È., Gillett, N.P., Fyfe, J.C., 2015. Observed and simulated changes in Antarctic sea ice extent over the past 50 years. Geophys. Res. Lett. 42, 90–95. https://doi.org/10.1002/2014GL062231.
Gagnon, A.S., Smoyer-Tomic, K.E., Bush, A.B., 2002. The El Niño southern oscillation and malaria epidemics in South America. Int. J. Biometeorol. 46, 81–89.
Gagnon, N., et al., 2007. In: An update on the CMC ensemble medium-range forecast system.Proc. ECMWF 11th Workshop on Meteorological Operational Systems, Shinfield Park, Reading, United Kingdom, ECMWF, pp. 55–59.
Gagnon, N., et al., 2014a. Improvements to the Global Ensemble Prediction System (GEPS) from version 3.1.0 to version 4.0.0. In: Canadian Meteorological Centre Technical Note. Available on request from Environment Canada, Centre Météorologique Canadien, Division du Dévelopement, 2121 route Transcanadienne, 4e étage, Dorval, Québec, H9P1J3 or via the following web site:http://collaboration.cmc.ec.gc.ca/cmc/CMOI/product_guide/docs/changes_e.html#20141118_geps_4.0.0.
Gagnon, N., et al., 2014b. Improvements to the Global Ensemble Prediction System (GEPS) Reforecast System From Version 3.1.0 to Version 4.0.0. Canadian Meteorological Centre Technical Note. Available on request from Environment Canada, Centre Météorologique Canadien, Division du Développement, 2121 route Transcanadienne, 4e étage, Dorval, Québec, H9P1J3 or via the following web site:http://collaboration.cmc.ec.gc.ca/cmc/cmoi/product_guide/docs/lib/Tech_Note_GEPS400_reforecast_v1.1_E.pdf.

Gagnon, N., et al., 2015. Improvements to the Global Ensemble Prediction System (GEPS) From Version 4.0.1 to Version 4.1.1. Canadian Meteorological Centre Technical Note. Available on request from Environment Canada, Centre Météorologique Canadien, Division du Développement, 2121 route Transcanadienne, 4e étage, Dorval, Québec, H9P1J3 or via the following web site:http://collaboration.cmc.ec.gc.ca/cmc/cmoi/product_guide/docs/changes_e.html#20151215_geps_4.1.1.

Gallo, B.T., Clark, A.J., Dembek, S.R., 2016. Forecasting tornadoes using convection-permitting ensembles. Weather Forecast. 31, 273–295. https://doi.org/10.1175/ WAF-D-15-0134.1.

Galway, J.G., 1989. The evolution of severe thunderstorm criteria within the Weather Service. Weather Forecast. 4, 585–592. https://doi.org/10.1175/1520-0434(1989) 0040585:TEOSTC2.0.CO;2.

Gandin, L.S., 1963. Objective analysis of meteorological fields. Gidrometerologicheskoe Izdatelstvo, Leningrad English translation by Israeli Program for Scientific Translations, Jerusalem, 1965.

Garfinkel, C.I., Hartmann, D.L., 2008. Different ENSO teleconnections and their effects on the stratospheric polar vortex. J. Geophys. Res. 113, D18114. https://doi.org/10.1029/2008JD009920.

Garfinkel, C.I., Hartmann, D.L., 2011a. The influence of the Quasi-Biennial Oscillation on the troposphere in winter in a hierarchy of models. Part I: Simplified Dry GCMs. J. Atmos. Sci. 68, 1273–1289. https://doi.org/10.1175/2011JAS3665.1.

Garfinkel, C.I., Hartmann, D.L., 2011b. The influence of the Quasi-Biennial Oscillation on the troposphere in winter in a hierarchy of models. Part II: Perpetual winter WACCM runs. J. Atmos. Sci. 68, 2026–2041. https://doi.org/10.1175/2011JAS3702.1.

Garfinkel, C.I., Schwartz, C., 2017. MJO-related tropical convection anomalies lead to more accurate stratospheric vortex variability in subseasonal forecast models. Geophys. Res. Lett. 44, 10,054–10,062. https://doi.org/10.1002/2017GL074470.

Garfinkel, C.I., Hartmann, D.L., Sassi, F., 2010. Tropospheric precursors of anomalous northern hemisphere stratospheric polar vortices. J. Clim. 23, 3282–3299. https://doi.org/10.1175/2010JCLI3010.1.

Garfinkel, C.I., Butler, A.H., Waugh, D.W., Hurwitz, M.M., Polvani, L.M., 2012a. Why might stratospheric sudden warmings occur with similar frequency in El Niño and La Niña winters? J. Geophys. Res. Atmos. 117. https://doi.org/10.1029/2012JD017777.

Garfinkel, C.I., Feldstein, S.B., Waugh, D.W., Yoo, C., Lee, S., 2012b. Observed connection between stratospheric sudden warmings and the Madden-Julian Oscillation. Geophys. Res. Lett. 39. https://doi.org/10.1029/2012GL053144.

Garfinkel, C.I., Shaw, T.A., Hartmann, D.L., Waugh, D.W., 2012c. Does the Holton–Tan mechanism explain how the quasi-biennial oscillation modulates the arctic polar vortex? J. Atmos. Sci. 69, 1713–1733. https://doi.org/10.1175/JAS-D-11-0209.1.

Garfinkel, C.I., Waugh, D.W., Oman, L.D., Wang, L., Hurwitz, M.M., 2013. Temperature trends in the tropical upper troposphere and lower stratosphere: connections with sea surface temperatures and implications for water vapor and ozone. J. Geophys. Res. Atmos. 118, 9658–9672. https://doi.org/10.1002/jgrd.50772.

Garfinkel, C.I., Son, S.-W., Song, K., Aquila, V., Oman, L.D., 2017. Stratospheric variability contributed to and sustained the recent hiatus in Eurasian winter warming. Geophys. Res. Lett. 44, 374–382. https://doi.org/10.1002/2016GL072035.

Gedney, N., Cox, P.M., 2003. The sensitivity of global climate model simulations to the representation of soil moisture heterogeneity. J. Hydrometeorol. 4, 1265–1275.

Geer, A., 2016. Significance of changes in medium-range forecast score. Tellus A 68, 30229. https://doi.org/10.3402/tellusa.v68.30229.

Geller, M.A., et al., 2016. Modeling the QBO-improvements resulting from higher-model vertical resolution. J. Adv. Model. Earth Syst. 8, 1092–1105. https://doi.org/10.1002/2016MS000699.

Gensini, V.A., Marinaro, A., 2016. Tornado frequency in the United States related to global relative angular momentum. Mon. Weather Rev. 144, 801–810. https://doi.org/10.1175/MWR-D-15-0289.1.

Gentine, P., Entekhabi, D., Chehbouni, A., Boulet, G., Duchemin, B., 2007. Analysis of evaporative fraction diurnal behaviour. Agr. For. Meteorol. 143, 13–29.

Gentine, P., Entekhabi, D., Polcher, J., 2011a. The diurnal behavior of evaporative fraction in the soil-vegetation-atmospheric boundary layer continuum. J. Hydrometeorol. 12, 1530–1546.

Gentine, P., Polcher, J., Entekhabi, D., 2011b. Harmonic propagation of variability in surface energy balance within a coupled soil-vegetation-atmosphere system. Water Resour. Res. 47, 1–21.

Gentine, P., Holtslag, A.A.M., D'Andrea, F., Ek, M., 2013a. Surface and atmospheric controls on the onset of moist convection over land. J. Hydrometeorol. 14, 1443–1462.

Gentine, P., Betts, A.K., Lintner, B.R., 2013b. A probabilistic bulk model of coupled mixed layer and convection. Part I: Clear-sky case. J. Atmos. Sci. 70, 1543–1556.

Gentine, P., Betts, A.K., Lintner, B.R., 2013c. A probabilistic bulk model of coupled mixed layer and convection. Part II: Shallow convection case. J. Atmos. Sci. 70, 1557–1576.

Gentine, P., Chhang, A., Rigden, A., Salvucci, G., 2016. Evaporation estimates using weather station data and boundary layer theory. Geophys. Res. Lett. 43, 11,661–11,670.

Gerber, E.P., Polvani, L.M., 2009. Stratosphere-troposphere coupling in a relatively simple AGCM: the importance of stratospheric variability. J. Clim. 22, 1920–1933. https://doi.org/10.1175/2008JCLI2548.1.

Gerber, E.P., et al., 2010. Stratosphere-troposphere coupling and annular mode variability in chemistry-climate models. J. Geophys. Res. 115, D00M06. https://doi.org/10.1029/2009JD013770.

Germann, U., Zawadzki, I., 2004. Scale-dependence of the predictability of precipitation from continental radar images. Part II: Probability forecasts. J. Appl. Meteorol. 43, 74–89.

Gettleman, A., Forster, P.M., 2002. A climatology of the tropical tropopause layer. J. Meteorol. Soc. Jpn. 80, 911–924. https://doi.org/10.2151/jmsj.80.911.

Ghil, M., 1987. Dynamics, statistics and predictability of planetary flow regimes. In: Nicolis, C., Nicolis, G. (Eds.), Irreversible Phenomena and Dynamical Systems Analysis in the Geosciences. D. Reidel, Dordrecht/Boston/Lancaster, pp. 241–283.

Ghil, M., 2001. Hilbert problems for the geosciences in the 21st century. Nonlinear Process. Geophys. 8, 211–222. https://doi.org/10.5194/npg-8-211-2001.

Ghil, M., 2014. Climate variability: nonlinear and random aspects. In: North, G.R., Pyle, J., Zhang, F. (Eds.), Encyclopedia of Atmospheric Sciences, second ed. vol. 2. Elsevier, pp. 38–46.

Ghil, M., 2017. The wind-driven ocean circulation: applying dynamical systems theory to a climate problem. Discrete Cont. Dyn. Syst. Ser. A 37, 189–228. https://doi.org/10.3934/dcds.2017008.

Ghil, M., Childress, S., 1987. Topics in Geophysical Fluid Dynamics: Atmospheric Dynamics, Dynamo Theory, and Climate Dynamics. Springer, New York.

Ghil, M., Mo, K., 1991. Intraseasonal oscillations in the global atmosphere. Part I: Northern hemisphere and tropics. J. Atmos. Sci. 48, 752–779. https://doi.org/10.1175/1520-0469(1991)048<0752: ioitga>2.0.co;2.

Ghil, M., Robertson, A.W., 2000. Solving problems with GCMs: general circulation models and their role in the climate modeling hierarchy. In: Randall, D. (Ed.), General Circulation Model Development: Past, Present and Future. Academic Press, San Diego, pp. 285–325.

Ghil, M., Robertson, A.W., 2002. "Waves" vs. "particles" in the atmosphere's phase space: a pathway to long-range forecasting? Proc. Natl. Acad. Sci. 99 (Suppl. 1), 2493–2500.

Ghil, M., Ide, K., Bennet, A., Courtier, P., Kimoto, M., Nagata, M., Saiki, M., Sato, N. (Eds.), 1997. Data Assimilation in Meteorology and Oceanography. Meteor. Soc. Japan, Tokyo, Japan.

Ghil, M., Allen, M.R., Dettinger, M.D., Ide, K., Kondrashov, D., Mann, M.E., Robertson, A.W., Saunders, A., Tian, Y., Varadi, F., Yiou, P., 2002. Advanced spectral methods for climatic time series. Rev. Geophys. 40, 1003. https://doi.org/10.1029/2000GR000010.1029.

Ghil, M., Kondrashov, D., Lott, F., Robertson, A.W., 2003. In: Intraseasonal oscillations in the mid-latitudes: observations, theory and GCM results.Proceeding of the ECMWF/CLIVAR Workshop on Simulation and Prediction of Intra-Seasonal Variability With Emphasis on the MJO, 3–6 Nov. 2003, ECMWF, Reading, UK. 35–53.

Ghil, M., Chekroun, M., Simonnet, E., 2008. Climate dynamics and fluid mechanics: natural variability and related uncertainties. Physica D 237, 2111–2126. https://doi.org/10.1016/j.physd.2008.03.036.

Ghil, M., Read, P., Smith, L., 2010. Geophysical flows as dynamical systems: the influence of Hide's experiments. Astron. Geophys. 51 (4), 4.28–4.35. https://doi.org/10.1111/j.1468-4004.2010.51428.x.

Ghil, M., Chekroun, M.D., Stepan, G., 2015. A collection on 'climate dynamics: multiple scales and memory effects'. In: Ghil, M., Chekroun, M.D., Stepan, G. (Eds.), Proceedings of the Royal Society A. The Royal Society.In: vol. 471. p. 20150097.

Gill, C.A., 1923. The prediction of malaria epidemics. Indian J. Med. Res. 10, 1136–1143.

Gill, A.E., 1980. Some simple solutions for heat induced tropical circulations. Q. J. R. Meteorol. Soc. 106, 447–462.

Gill, A.E., 1982. Atmosphere-Ocean Dynamics. Academic Press, Orlando. 662 pp.

Gilleland, E., 2011. Spatial forecast verification: Baddeley's delta metric applied to the ICP test cases. Weather Forecast. 26, 409–415.

Gilleland, E., Ahijevych, D., Brown, B.G., Casati, B., Ebert, E., 2009. Intercomparison of spatial forecast verification methods. Weather Forecast. 24, 1416–1430.

Gilleland, E., Lindstrom, J., Lindgren, F., 2010. Analyzing the image warp forecast verification method on precipitation fields from the ICP. Weather Forecast. 25, 1249–1262.

Giorgetta, M.A., Bengtsson, L., Arpe, K., 1999. An investigation of QBO signals in the east Asian and Indian monsoon in GCM experiments. Clim. Dyn. 15, 435–450. https://doi.org/10.1007/s003820050292.

Glahn, H.R., Lowry, D.A., 1972. The use of model output statistics (MOS) in objective weather forecasting. J. Appl. Meteorol. 11 (8), 1203–1211.

Glahn, B., Peroutka, M., Wiedenfeld, J., Wagner, J., Zylstra, G., Schuknecht, B., Jackson, B., 2009. MOS uncertainty estimates in an ensemble framework. Mon. Weather Rev. 137 (1), 246–268.

Gloersen, P., 1995. Modulation of hemispheric sea-ice cover by ENSO events. Nature 373 (6514), 503–506.

Gneiting, T., Raftery, A.E., Westveld III, A.H., Goldman, T., 2005. Calibrated probabilistic forecasting using ensemble model output statistics and minimum crps estimation. Mon. Weather Rev. 133 (5), 1098–1118.

Gneiting, T., Balabdaoui, F., Raftery, A.E., 2007. Probabilistic forecasts, calibration and sharpness. J. R. Stat. Soc. Ser. B Stat Methodol. 69 (2), 243–268.

Gochis, D.J., et al., 2015. In: Operational, hyper-resolution hydrologic modeling over the contiguous U.S. using the multi-scale, multi-physics WRF-Hydro Modeling and Data Assimilation System.American Geophysical Union, Fall Meeting, H52A-02.

Goddard, L., Buizer, J., 2016. Integrating Climate Information and Decision Processes for Regional Climate Resilience. http://irapclimate.org/documents.

Goddard, L., Mason, S., Zebiak, S., Ropelewski, C., Basher, R., Cane, M., 2001. Current approaches to seasonal-to-interannual climate predictions. Int. J. Climatol. 21, 1111–1152.

Goddard, L., Bethgen, W.E., Bhojwani, H., Robertson, A.W., 2014. The International Research Institute for Climate & Society: why, what and how. Earth Perspect. 1–10.

Goessling, H.F., Tietsche, S., Day, J.J., Hawkins, E., Jung, T., 2016. Predictability of the Arctic sea ice edge. Geophys. Res. Lett. 43 (4), 1642–1650.

Gong, X., Barnston, A., Ward, M., 2003. The effect of spatial aggregation on the skill of seasonal precipitation forecasts. J. Clim. 16, 3059–3071.

Goo, T.-Y., Moon, S.-O., Cho, J.-Y., Cheong, H.-B., Lee, W.-J., 2003. Preliminary results of medium-range ensemble prediction at KMA: implementation and performance evaluation as of 2001. Korean J. Atmos. Sci. 6, 27–36.

Goosse, H., Zunz, V., 2014. Decadal trends in the Antarctic sea ice extent ultimately controlled by ice–ocean feedback. Cryosphere 8 (2), 453–470.

Goosse, H., Arzel, O., Bitz, C.M., de Montety, A., Vancoppenolle, M., 2009. Increased variability of the Arctic summer ice extent in a warmer climate. Geophys. Res. Lett. 36(23).

Goswami, B.N., 2005. South Asian monsoon. In: Intraseasonal Variability in the Atmosphere-Ocean Climate System. Springer, Berlin, Heidelberg, pp. 19–61.

Goswami, B.N., Shukla, J., 1991. Predictability and variability of a coupled ocean-atmosphere model. J. Mar. Syst. 1, 217–228.

Goswami, B.N., Ajayamohan, R.S., Xavier, P.K., Sengupta, D., 2003. Clustering of low pressure systems during the Indian summer monsoon by intraseasonal oscillations. Geophys. Res. Lett. 30, 1431. https://doi.org/10.1029/2002GL016734.

Goswami, B.B., Mani, N.J., Mukhopadhyay, P., Waliser, D.E., Benedict, J.J., Maloney, E.D., Khairoutdinov, M., Goswami, B.N., 2011. Monsoon intraseasonal oscillations as simulated by the superparameterized Community Atmosphere Model. J. Geophys. Res. 116, D22104. https://doi.org/10.1029/2011JD015948.

Gottschalck, J., Wheeler, M., Weickmann, K., Vitart, F., Savage, N., Lin, H., Hendon, H.H., Waliser, D.E., Sperber, K., Nakagawa, M., Prestrelo, C., Flatau, M., Higgins, W., 2010. A framework for assessing operational MJO forecasts: a CLIVAR MJO working group project. Bull. Am. Meteorol. Soc. 91, 1247–1258.

Gottschalck, J., Roundy, P.E., Shreck III, C.J., Vintzileos, A., Zhang, C., 2013. Largescale atmosphere and oceanic conditions during the 2011–2012 DYNAMO field campaign. Mon. Weather Rev. 141, 4173–4196.

Grabowski, W.W., Wu, X., Moncrieff, M.W., Hall, W.D., 1998. Cloud-resolving modeling of cloud systems during Phase III of GATE. Part II: Effects of resolution and the third spatial dimension. J. Atmos. Sci. 55 (21), 3264–3282.

Graham, R.J., Yun, W.-T., Kim, J., Kumar, A., Jones, D., Bettio, L., Gagnon, N., Kolli, R.K., Smith, D., 2011. Long-range forecasting and the Global Framework for Climate Services. Clim. Res. 47, 47–55.

Gray, W.M., 1979. Hurricanes: their formation, structure and likely role in the tropical circulation. In: Shaw, D.B. (Ed.), Meteorology Over Tropical Oceans. Roy. Meteor. Soc., James Glaisher House, Grenville Place, Bracknell, Berkshire, pp. 155–218.

Gray, W.M., 1984. Atlantic seasonal hurricane frequency: Part I: El Niño and 30-mb quasi-bienniel oscillation influences. Mon. Weather Rev. 112, 1649–1668.

Gray, M.E.B., Shutts, G.J., 2002. A stochastic scheme for representing convectively generated vorticity sources in general circulation models. In: APR Turbulence and Diffusion Note No. 285. Met Office, UK.

Gray, W.M., Sheaffer, J.D., Knaff, J.A., 1992. Hypothesized mechanism for stratospheric QBO influence on ENSO variability. Geophys. Res. Lett. 19, 107–110. https://doi.org/10.1029/91GL02950.

Green, J.K., Konings, A.G., Alemohammad, S.H., Berry, J., Entekhabi, D., Kolassa, J., Lee, J.-E., Gentine, P., 2017. Hotspots of terrestrial biosphere-atmosphere feedbacks. Nat. Geosci. NGS-2016-08-01557A.

Griffies, S., Harrison, M., Pacanowski, R., Rosati, A., 2004. A Technical Guide TO MOM4. GFDL Ocean Group, NOAA GFDL.

Groth, A., Ghil, M., 2011. Multivariate singular spectrum analysis and the road to phase synchronization. Phys. Rev. E. 84, 036206. https://doi.org/10.1103/PhysRevE.84.036206.

Groth, A., Dumas, P., Ghil, M., Hallegatte, S., 2015. Impacts of natural disasters on a dynamic economy. In: Chavez, M., Ghil, M., Urrutia-Fucugauchi, J. (Eds.), Extreme Events: Observations, Modeling and Economics. In: Geophysical Monographs, vol. 214. American Geophysical Union & Wiley, pp. 343–359 (Chapter 19).

Groth, A., Feliks, Y., Kondrashov, D., Ghil, M., 2017. Interannual variability in the North Atlantic ocean's temperature field and its association with the wind stress forcing. J. Clim. 30, 2655–2678. https://doi.org/10.1175/jcli-d-16-0370.1.

Grover-Kopec, E., et al., 2005. An online operational rainfall-monitoring resource for epidemic malaria early warning systems in Africa. Malar. J. 4, 6. https://doi.org/10.1186/1475-2875-4-6.

Gruber, A., 1974. Wavenumber-frequency spectra of satellite measured brightness in the tropics. J. Atmos. Sci. 31, 1675–1680.

Guan, B., Molotch, N.P., Waliser, D.E., Fetzer, E.J., Neiman, P.J., 2010. Extreme snowfall events linked to atmospheric rivers and surface air temperature via satellite measurements. Geophys. Res. Lett. 37, L20401. https://doi.org/10.1029/2010GL044696.

Gudkovich, Z.M., 1961. Relation of the ice drift in the Arctic Basin to ice conditions in the Soviet Arctic seas. Tr. Okeanogr. Kom. Akad. Nauk SSSR 11, 14–21.

Guémas, V., Blanchard-Wrigglesworth, E., Chevallier, M., Day, J., Déqué, M., Doblas-Reyes, F., Fuckar, N., Germe, A., Hawkins, E., Keeley, S., Koenigk, T., Salas y Mélia, D., Tietsche, S., 2016. A review on Arctic sea ice predictability and prediction on seasonal-to-decadal timescales. Q. J. R. Meteorol. Soc. 142, 546–561. https://doi.org/10.1002/qj.2401.

Guillod, B.P., Orlowsky, B., Miralles, D.G., Teuling, A.J., Seneviratne, S.I., 2015. Reconciling spatial and temporal soil moisture effects on afternoon rainfall. Nat. Commun. 6, 6443.

Gunasekera, D., Troccoli, A., Boulahya, M.S., 2014. Energy and meteorology: partnership for the future. In: Weather Matters for Energy. Springer, New York, pp. 497–511.

Gunturi, P., Tippett, M.K., 2017. Managing Severe Thunderstorm Risk: Impact of ENSO on U.S. Tornado and Hail Frequencies. Tech. rep WillisRe. http://www.willisre.com/Media_Room/Press_Releases_(Browse_All)/2017/WillisRe_Impact_of_ENSO_on_US_Tornado_and_Hail_frequencies_Final.pdf

Guo, Z., et al., 2006. GLACE: the global land-atmosphere coupling experiment. 2. Analysis. J. Hydrometeorol. 7, 611–625.

Guzman, M.G., Harris, E., 2015. Dengue. Lancet 385, 453–465.

Hagedorn, R., Hamill, T.M., Whitaker, J.S., 2008. Probabilistic forecast calibration using ECMWF and GFS ensemble reforecasts. Part I: 2-meter temperature. Mon. Weather Rev. 136, 2608–2619.

Hagedorn, R., Buizza, R., Hamill, M.T., Leutbecher, M., Palmer, T.N., 2012. Comparing TIGGE multi-model forecasts with re-forecast calibrated ECMWF ensemble forecasts. Q. J. R. Meteorol. Soc. 138, 1814–1827.

Hagos, S., Zhang, C., Tao, W.-K., Lang, S., Takayabu, Y.N., Shige, S., Katsumata, M., Olson, B., L'Ecuyer, T., 2009. J. Clim. 23, 542–558.

Haidvogel, D.B., Arango, H.G., Budgell, W.P., Cornuelle, B.D., Curchitser, E., Di Lorenzo, E., Fennel, K., Geyer, W.R., Hermann, A.J., Lanerolle, L., Levin, J., McWilliams, J.C., Miller, A.J., Moore, A.M., Powell, T.M., Shchepetkin, A.F., Sherwood, C.R., Signell, R.P., Warner, J.C., Wilkin, J., 2008. Regional ocean forecasting in terrain-following coordinates: model formulation and skill assessment. J. Comput. Phys. 227, 3595–3624.

Haigh, J.D., 1996. The impact of solar variability on climate. Science (80-) 272, 981–984. https://doi.org/10.1126/science.272.5264.981.

Haigh, J.D., Blackburn, M., Day, R., 2005. The response of tropospheric circulation to perturbations in lower-stratospheric temperature. J. Clim. 18, 3672–3685. https://doi.org/10.1175/JCLI3472.1.

Hale, J.K., Verduyn Lunel, S.M., 1993. Introduction to Functional Differential Equations. Applied Mathematical Sciences, vol. 99. Springer-Verlag, New York.

Hall, A., Visbeck, M., 2002. Synchronous variability in the Southern Hemisphere atmosphere, sea ice, and ocean resulting from the annular mode. J. Clim. 15 (21), 3043–3057. https://doi.org/10.1175/1520-0442(2004)017<2249:COSVIT>2.0.CO;2.

Hall, N.M., Thibault, S., Marchesiello, P., 2017. Impact of the observed extratropics on climatological simulation of the MJO in a tropical channel model. Clim. Dyn. 48, 2541–2555. https://doi.org/10.1007/s00382-016-3221-5.

Ham, Y.-G., Jong-SeongKug, I.-S.K., Jin, F.-F., Timmermann, A., 2010. Impact of diurnal atmosphere-ocean coupling on tropical climate simulations using a coupled GCM. Clim. Dyn. 34, 905–917.

Hamill, T.M., Juras, J., 2006. Measuring forecast skill: is it real skill or is it the varying climatology? Q. J. R. Meteorol. Soc. 132, 2905–2923.

Hamill, T.M., Kildadis, G.N., 2014. Skill of the MJO and Northern Hemisphere blocking in GEFS medium-range forecasts. Mon. Weather Rev. 142, 868–885.

Hamill, T.M., Snyder, C., Morss, R.E., 2000. A comparison of probabilistic forecasts from bred, singular-vector, and perturbed observation ensembles. Mon. Weather Rev. 128, 1835–1851.

Hamill, T.M., Whitaker, J.S., Mullen, S.L., 2006. Reforecasts: an important dataset for improving weather predictions. Bull. Am. Meteorol. Soc. 87, 33–46.

Hamill, T.M., Hagedorn, R., Whitaker, J.S., 2008. Probabilistic forecast calibration using ECMWF and GFS ensemble reforecasts. Part II: precipitation. Mon. Weather Rev. 136, 2620–2632.

Hamill, T.M., Brennan, M.J., Brown, B., DeMaria, M., Rappaport, E.N., Toth, Z., 2012. NOAA's future ensemble-based hurricane forecast products, 2012. Bull. Am. Meteorol. Soc. 93, 209–220.

Hamill, T.M., Bates, G.T., Whitaker, J.S., Murray, D.R., Fiorino, M., Galarneau Jr., T.J., Zhu, Y., Lapenta, W., 2013. NOAA's second-generation global medium-range ensemble reforecast data set. Bull. Am. Meteorol. Soc. 94, 1553–1565.

Hamlet, A.F., Huppert, D., Lettenmaier, D.P., 2002. Economic value of long-lead streamflow forecasts for Columbia river hydropower. J. Water Resour. Plan. Manag. 128, 91–101.

Hannachi, A., Straus, D.M., Franzke, C.L.E., Corti, S., Woollings, T., 2017. Low-frequency nonlinearity and regime behavior in the Northern Hemisphere extratropical atmosphere. Rev. Geophys. 55, 199–234. https://doi.org/10.1002/2015rg000509.

Hansen, J.W., 2002. Realizing the potential benefits of climate prediction to agriculture: issues, approaches, challenges. Agric. Syst. 74, 309–330.

Hansen, J., Coffey, K., 2011. Agro-Climate Tools for a New Climate-Smart Agriculture. CCAFS Rep. 4 pp.

Hansen, A.R., Sutera, A., 1995. The probability density distribution of the planetary-scale atmospheric wave amplitude revisited. J. Atmos. Sci. 52, 2463–2472. https://doi.org/10.1175/1520-0469(1995) 052<2463:tpddot>2.0.co;2.

Hansen, M.C., DeFries, R.S., Townshend, J.R.G., Carroll, M., Dimiceli, C., Sohlberg, R.A., 2003. Global percent tree cover at a spatial resolution of 500 meters: first results of the MODIS vegetation continuous fields algorithm. Earth Interact. https://doi.org/10.1175/1087-3562(2003)007<0001:GPTCAA>2.0.CO;2.

Hansen, F., Greatbatch, R.J., Gollan, G., Jung, T., Weisheimer, A., 2017. Remote control of North Atlantic Oscillation predictability via the stratosphere. Q. J. R. Meteorol. Soc. 143, 706–719. https://doi.org/10.1002/qj.2958.

Haque, U., et al., 2010. The role of climate variability in the spread of malaria in Bangladeshi highlands. PLoS ONE 5, e14341. https://doi.org/10.1371/journal.pone.0014341.

Harnik, N., 2009. Observed stratospheric downward reflection and its relation to upward pulses of wave activity. J. Geophys. Res. 114, D08120. https://doi.org/10.1029/2008JD010493.

Harnik, N., Lindzen, R.S., 2001. The effect of reflecting surfaces on the vertical structure and variability of stratospheric planetary waves. J. Atmos. Sci. 58, 2872–2894. https://doi.org/10.1175/1520-0469(2001)058<2872:TEORSO>2.0.CO;2.

Harper, K., Uccellini, L.W., Kalnay, E., Carey, K., Morone, L., 2007. 50th anniversary of operational numerical weather prediction. Bull. Am. Meteorol. Soc. 88, 639–650.

Harr, P.A., Elsberry, R.L., 2000. Extratropical transition of tropical cyclones over the western North Pacific. Part I: Evolution of structural characteristics during the transition process. Mon. Weather Rev. 128, 2613–2633.

Hart, R.E., Evans, J.L., 2001. A climatology of extratropical transition of Atlantic tropical cyclones. J. Clim. 14, 546–564.

Hartley, D.E., Villarin, J.T., Black, R.X., Davis, C.A., 1998. A new perspective on the dynamical link between the stratosphere and troposphere. Nature 391, 471–474. https://doi.org/10.1038/35112.

Hartmann, D.L., Lo, F., 1998. Wave-driven zonal flow vacillation in the southern hemisphere. J. Atmos. Sci. 55, 1303–1315. https://doi.org/10.1175/1520-0469(1998)055<1303:WDZFVI>2.0.CO;2.

Hartmann, D.L., Moy, L.A., Fu, Q., 2001. Tropical convection and the energy balance at the top of the atmosphere. J. Clim. 14, 4495–4511. https://doi.org/10.1175/1520-0442(2001)014<4495:TCATEB>2.0.CO;2.

Hartmann, H.C., Pagano, T.C., Sorooshian, S., Bales, R., 2002. Confidence builders: evaluating seasonal climate forecasts from user perspectives. Bull. Am. Meteorol. Soc. 83 (5), 683–698.

Hashino, T., Satoh, M., Hagihara, Y., Kubota, T., Matsui, T., Nasuno, T., Okamoto, H., 2013. Evaluating cloud microphysics from NICAM against CloudSat and CALIPSO. J. Geophys. Res. Atmos. 118, 7273–7292.

Hasselman, K.H., 1988. PIPs and POPs: the reduction of complex dynamical systems using principal interaction and oscillation patterns. J. Geophys. Res. 93, 11015–11021.

Hasselmann, K., 1976. Stochastic climate models. I: Theory. Tellus 28, 473–485.

Hastie, T., Tibshirani, R., Friedman, J., 2009. The Elements of Statistical Learning, second ed. Springer Series in Statistics, New York.

Haughton, N., et al., 2016. The plumbing of land surface models: is poor performance a result of methodology or data quality? J. Hydrometeorol. 17, 1705–1723.

Haurwitz, B., 1940. The motion of atmospheric disturbances on the spherical earth. J. Mar. Res. 3, 254–267.

Hay, S.I., Snow, R.W., 2006. The malaria atlas project: developing global maps of malaria risk. PLoS Med. 3, e473. https://doi.org/10.1371/journal.pmed.0030473.

Hay, S.I., Rogers, D.J., Shanks, G.D., Myers, M.F., Snow, R.W., 2001. Malaria early warning in Kenya. Trends Parasitol. 17, 95–99.

Haylock, M., McBride, J., 2001. Spatial coherence and predictability of Indonesian wet season rainfall. J. Clim. 14, 3882–3887.

Haynes, P.H., 1988. Forced, dissipative generalizations of finite-amplitude wave-activity conservation relation for zonal and nonzonal flows. J. Atmos. Sci. 45, 2352–2362.

Haynes, P.H., 2005. Stratospheric dynamics. Annu. Rev. Fluid Mech. 37, 263–293.

Haynes, P.H., McIntyre, M.E., Shepherd, T.G., Marks, C.J., Shine, K.P., 1991. On the "downward control" of extratropical diabatic circulations by eddy-induced mean zonal forces. J. Atmos. Sci. 48, 651–678. https://doi.org/10.1175/1520-0469(1991)048<0651:OTCOED>2.0.CO;2.

Hazeleger, W., Wang, X., Severijns, C., Stefanescu, S., Yang, S., Wang, X., Wyser, K., Dutra, E., Baldasano, J.M., Bintanja, R., Bougeault, P., Caballero, R., AML, E., Christensen, J.H., van den Hurk, B., Jimenez, P., Jones, C., Kallberg, P., Koenigk, T., McGrath, R., Miranda, P., van Noije, T., Palmer, T., Parodi, J.A., Schmith, T., Selten, F., Storelvmo, T., Sterl, A., Tapamo, H., Vancoppenolle, M., Viterbo, P., Willen, U., 2010. EC-Earth: a seamless Earth system prediction approach in action. Bull. Am. Meteorol. Soc. 91, 1357–1363.

He, J., Lin, H., Wu, Z., 2011. Another look at influences of the Madden-Julian Oscillation on the wintertime East Asian weather. J. Geophys. Res. 116, D03109. https://doi.org/10.1029/2010JD014787.

Heatwave Plan for England, 2015. Public Health England. PHE publications gateway number: 2015049, Available from, https://www.gov.uk/government/uploads/system/uploads/attachmentdata/file/429384/Heatwave Main Plan 2015.pdf.

Heifetz, E., Bishop, C.H., Hoskins, B.J., Methven, J., 2004. The counter-propagating Rossby wave perspective on baroclinic instability. Part I: Mathematical basis. Q. J. R. Meteorol. Soc. 130, 211–231.

Held, I., *1985*. Pseudomomentum and the orthogonality of modes in shear flows. J. Atmos. Sci. *42*, *527–565*.

Held, I.M., 2005. The gap between simulation and understanding in climate modeling. Bull. Am. Meteorol. Soc. 86, 1609–1614. https://doi.org/10.1175/bams-86-11-1609.

Held, I.M., Lyons, S.W., Nigam, S., 1989. Transients and the Extratropical Response to El Nino. J. Atmos. Sci. 46, 163–174.

Hemri, S., Lisniak, D., Klein, B., 2015. Multivariate postprocessing techniques for probabilistic hydrological forecasting. Water Resour. Res. 51 (9), 7436–7451.

Henderson, G.R., Barrett, B.S., Lafleur, D.M., 2014. Arctic sea ice and the Madden–Julian Oscillation (MJO). Clim. Dyn. 43 (7-8), 2185–2196.

Henderson, S.A., Maloney, E.D., Son, S.-W., 2017. Madden-Julian Oscillation teleconnections: the impact of the basic state and MJO representation in general circulation models. J. Clim. **30**, 4567–4587.

Hendon, H.H., Salby, M.L., 1994. The life cycle of the Madden-Julian Oscillation. J. Atmos. Sci. 51, 2225–2237.

Hendon, H.H., Zhang, C., Glick, J.D., 1999. Interannual variations of the Madden-Julian Oscillation during austral summer. J. Clim. 12, 2538–2550.

Hendon, H.H., Liebmann, B., Newman, M., Glick, J., 2000. Medium-range forecast errors associated with active episodes of the Madden-Julian oscillation. Mon. Weather Rev. 128, 69–86.

Heygster, G., Alexandrov, V., Dybkjær, G., von Hoyningen-Huene, W., Girard-Ardhuin, F., Katsev, I.L., et al., 2012. Remote sensing of sea ice: advances during the DAMOCLES project. Cryosphere 6 (6), 1411.

Higgins, R.W., Mo, K.C., 1997. Persistent North Pacific circulation anomalies and the tropical intraseasonal oscillation. J. Clim. 10, 223–244.

Higgins, R.W., Schemm, J.-K.E., Shi, W., Leetmaa, A., 2000. Extreme precipitation events in the western United States related to tropical forcing. J. Clim. 13, 793–820. https://doi.org/10.1175/1520-0442(2000)013h0793:EPEITWi2.0.CO;2.

Hirons, L.C., Inness, P., Vitart, F., Bechtold, P., 2013. Understanding advances in the simulation of intraseasonal variability in the ECWMF model. Part I: The representation of the MJO. Q. J. R. Meteorol. Soc. 139, 1417–1426.

Hirschi, M., et al., 2011. Observational evidence for soilmoisture impact on hot extremes in southeastern Europe. Nat. Geosci. 4, 17–21. https://doi.org/10.1038/ngeo1032.

Hitchcock, P., Haynes, P.H., 2016. Stratospheric control of planetary waves. Geophys. Res. Lett. 43, 11,884–11,892. https://doi.org/10.1002/2016GL071372.

Hitchcock, P., Simpson, I.R., 2014. The downward influence of stratospheric sudden warmings. J. Atmos. Sci. 71, 3856–3876. https://doi.org/10.1175/JAS-D-14-0012.1.

Hitchcock, P., Simpson, I.R., 2016. Quantifying eddy feedbacks and forcings in the tropospheric response to stratospheric sudden warmings. J. Atmos. Sci. 73, 3641–3657. https://doi.org/10.1175/JAS-D-16-0056.1.

Hitchcock, P., Shepherd, T.G., Manney, G.L., 2013. Statistical characterization of Arctic Polar-Night Jet oscillation events. J. Clim. 26, 2096–2116. https://doi.org/10.1175/JCLI-D-12-00202.1.

Hoag, 2014. Russian summer tops 'universal' heatwave index. Nature. https://doi.org/10.1038/nature.2014.16250.

Hoffman, R.N., Kalnay, E., 1983. Lagged average forecasting, an alternative to Monte Carlo forecasting. Tellus A 35A, 100–118.

Hoke, J.E., Phillips, N.A., DiMego, G.J., Tuccillo, J.J., Sela, J.G., 1989. The regional analysis and forecast system of the national meteorological center. Weather Forecast. 4, 323–334.

Holland, P., 2014. The seasonality of Antarctic sea ice trends. Geophys. Res. Lett. 41, 4230–4237. https://doi.org/10.1002/2014GL060172.

Holland, M.M., Stroeve, J., 2011. Changing seasonal sea ice predictor relationships in a changing Arctic climate. Geophys. Res. Lett. 38(18).

Holland, M.M., Bitz, C.M., Hunke, E.C., 2005. Mechanisms forcing an Antarctic dipole in simulated sea ice and surface ocean conditions. J. Clim. 18 (12), 2052–2066.

Holland, M.M., Bailey, D.A., Vavrus, S., 2011. Inherent sea ice predictability in the rapidly changing Arctic environment of the Community Climate System Model, version 3. Clim. Dyn. 36 (7), 1239–1253.

Holland, M.M., Blanchard-Wrigglesworth, E., Kay, J., Vavrus, S., 2013. Initial-value predictability of Antarctic sea ice in the Community Climate System Model 3. Geophys. Res. Lett. 40 (10), 2121–2124.

Holland, M.M., Landrum, L., Kostov, Y., Marshall, J., 2016. Sensitivity of Antarctic sea ice to the Southern Annular Mode in coupled climate models. Clim. Dyn. 1–19. https://doi.org/10.1007/s00382-016-3424-9.

Hollingsworth, A., 1980. In: An experiment in Monte Carlo forecasting.Proceedings of the ECMWF Workshop on Stochastic Dynamic Forecasting, pp. 65–86 Available from ECMWF, Shinfield Park, Reading RG2 9AX, UK.

Holloway, C.E., Woolnough, S.J., Lister, G.M.S., 2013. The effects of explicit versus parameterized convection on the MJO in a large-domain high-resolution tropical case study. Part I: Characterization of large-scale organization and propagation. J. Atmos. Sci. 70, 1342–1369.

Holloway, C.E., Woolnough, S.J., Lister, G.M.S., 2015. The effects of explicit versus parameterized convection on the MJO in a large-domain high resolution tropical case study. Part II: Processes leading to differences in MJO development. J. Atmos. Sci. 72, 2719–2743.

Holton, J.R., Lindzen, R.S., 1972. An updated theory for the quasi-biennial cycle of the tropical stratosphere. J. Atmos. Sci. 29, 1076–1080. https://doi.org/10.1175/1520-0469(1972)029<1076:AUTFTQ>2.0.CO;2.

Holton, J.R., Mass, C., 1976. Stratospheric vacillation cycles. J. Atmos. Sci. 33, 2218–2225. https://doi.org/10.1175/1520-0469(1976)033<2218:SVC>2.0.CO;2.

Holton, J.R., Tan, H.-C., 1980. The influence of the equatorial quasi-biennial oscillation on the global circulation at 50 mb. J. Atmos. Sci. 37, 2200–2208. https://doi.org/10.1175/1520-0469(1980)037<2200:TIOTEQ>2.0.CO;2.

Holton, J.R., Haynes, P.H., McIntyre, M.E., Douglass, A.R., Rood, R.B., Pfister, L., 1995. Stratosphere-troposphere exchange. Rev. Geophys. 33, 403–439. https://doi.org/10.1029/95rg02097.

Hong, Y., Liu, G., Li, J.-L.F., 2016. Assessing the radiative effects of global ice clouds based on CloudSat and CALIPSO measurements. J. Clim. 29, 7651–7674. https://doi.org/10.1175/JCLI-D-15-0799.1.

Hong, C.-C., Hsu, H.-H., Tseng, W.-L., Lee, M.-Y., Chow, C.-H., Jiang, L.-C., 2017. Extratropical forcing triggered the 2015 Madden-Julian Oscillation- El Nino event. Sci. Rep. 7, 46692. https://doi.org/10.1038/srep46692.

Horel, J.D., 1985. Persistence of the 500 mb height field during Northern Hemisphere winter. Mon. Weather Rev. 113, 2030–2042. https://doi.org/10.1175/1520-0493(1985)113<2030:potmhf>2.0.co;2.

Horel, J.D., Wallace, J.M., 1981. Planetary-scale atmospheric phenomena associated with the Southern Oscillation. Mon. Weather Rev. 109, 813–829. https://doi.org/10.1175/1520-0493(1981)109<0813:PSAPAW>2.0.CO;2.

Horsburgh, K.J., Wilson, C., 2007. Tide–surge interaction and its role in the distribution of surge residuals in the North Sea. J. Geophys. Res. Oceans 112, C08003. https://doi.org/10.1029/2006JC004033.

Hoshen, M.B., Morse, A.P., 2004. A weather-driven model of malaria transmission. Malar. J. 3, 32. https://doi.org/10.1029/2012GL054040.

Hoskins, B., 2012. The potential for skill across the range of the seamless weather-climate prediction problem: a stimulus for our science. Q. J. R. Meteorol. Soc. 139, 573–584. https://doi.org/10.1002/qj.1991.

Hoskins, B.J., 2013. Review article: the potential for skill across the range of the seamless weather-climate prediction problem: a stimulus for our science. Q. J. R. Meteorol. Soc. 139, 573–584.

Hoskins, B.J., Ambrizzi, T., 1661-1671. Rossby wave propagation on a realistic longitudinally varying flow. J. Atmos. Sci. 50.

Hoskins, B.J., James, I.N., 2014. Fluid Dynamics of the Mid-Latitude Atmosphere. Wiley, p. 300.

Hoskins, B.J., Karoly, D.J., 1981. The steady linear response of a spherical atmosphere to thermal orographic forcing. J. Atmos. Sci. 38, 1179–1196.

Hoskins, B.J., Yang, G.-Y., 2000. The equatorial response to higher-latitude forcing. J. Atmos. Sci. 57, 1197–1213.

Hoskins, B.J., James, I.N., White, G.H., 1983. The shape, propagation and mean-flow interactions of large-scale weather systems. J. Atmos. Sci. 40, 1595–1612.

Hoskins, B.J., McIntyre, M.E., Robertson, A.W., 1985. On the use and significance of isentropic potential vorticity maps. Q. J. R. Meteorol. Soc. 111, 877–946.

Hou, D., Toth, Z., Zhu, Y., Yang, W., 2008. Impact of a stochastic perturbation scheme on NCEP global ensemble forecast system.Proceedings of the 19th AMS Conference on Probability and Statistics, 21-24 January 2008, New Orleans, Louisiana.

Houtekamer, P.L., Derome, J., 1995. Methods for ensemble prediction. Mon. Weather Rev. 123, 2181–2196.

Houtekamer, P.L., Lefaivre, L., 1997. Using ensemble forecasts for model validation. Mon. Weather Rev. 125, 2416–2426.

Houtekamer, P.L., Mitchell, H.L., 2005. Ensemble Kalman filtering. Q. J. R. Meteorol. Soc. 131, 3269–3289.

Houtekamer, P.L., Derome, J., Ritchie, H., Mitchell, H.L., 1996. A system simulation approach to ensemble prediction. Mon. Weather Rev. 124, 1225–1242.

Houtekamer, P.L., Mitchell, H.L., Deng, X., 2009. Model error representation in an operational ensemble Kalman filter. Mon. Weather Rev. 137, 2126–2143.

Houtekamer, P.L., Deng, X., Mitchell, H.L., Baek, S.-J., Gagnon, N., 2014. Higher resolution in an operational ensemble Kalman filter. Mon. Weather Rev. 142, 1143–1162.

Hsu, H.-H., 1996. Global view of the intraseasonal oscillation during northern winter. J. Clim. 9, 2386–2406.

Hsu, P.-C., Li, T., 2012. Role of the boundary layer moisture asymmetry in causing the eastward propagation of the Madden–Julian Oscillation. J. Clim. 25, 4914–4931.

Hu, Q., Pytlik Zillig, L.M., Lynne, G.D., Tomkins, A.J., Waltman, W.J., Hayes, M.J., Hubbard, K.G., Artikov, I., Hoffman, S.J., Wilhite, D.A., 2006. Understanding farmer's forecast use from their beliefs, values, social norms, and perceived obstacles. J. Appl. Meteorol. Climatol. 45, 1190–1201.

Huang, J., van den Dool, H.M., Georgakakos, K.P., 1996. Analysis of model-calculated soil moisture over the United States (1931–1993) and applications to long-range temperature forecasts. J. Clim. 9, 1350–1362.

Hudson, D., Marshall, A.G., 2016. Extending the Bureau's Heatwave Forecast to Multi-Week Timescales. Bureau Research Report, No. 16. Bureau of Meteorology, Australia. Available online at:http://www.bom.gov.au/research/research-reports.shtml.

Hudson, D., Alves, O., Hendon, H.H., Marshall, A.G., 2011. Bridging the gap between weather and seasonal forecasting: intraseasonal forecasting for Australia. Q. J. R. Meteorol. Soc. 137, 673–689. https://doi.org/10.1002/qj.769. http://onlinelibrary.wiley.com/doi/10.1002/qj.769/abstract.

Hudson, D., Marshall, A.G., Yin, Y., Alves, O., Hendon, H.H., 2013. Improving intraseasonal prediction with a new ensemble generation strategy. Mon. Weather Rev. 141, 4429–4449. https://doi.org/10.1175/MWR-D-13-00059.1.

Hudson, D., Marshall, A.G., Alves, O., Young, G., Jones, D., Watkins, A., 2015a. Forewarned is forearmed: extended range forecast guidance of recent extreme heat events in Australia. Weather Forecast. https://doi.org/10.1175/WAF-D-15-0079.1.

Hudson, D., Marshall, A., Alves, O., Shi, L., Young, G., 2015b. Forecasting Upcoming Extreme Heat on Multi-Week To Seasonal Timescales: POAMA Experimental Forecast Products. Bureau Research Report, No. 1Bureau of Meteorology, Australia. Available online at http://www.bom.gov.au/research/research-reports.shtml.

Huffman, G.J., Adler, R.F., Morrissey, M., Bolvin, D.T., Curtis, R., Joyce, R., McGavock, B., Susskind, J., 2001. Global precipitation at one-degree daily resolution from multi-satellite observations. J. Hydrometeorol. 2, 36–50.

Huffman, G.J., Adler, R.J., Boivin, D.T., 2009. Improving the global precipitation record: GPCP version 2.1. Geophys. Res. Lett. 36, L17808.

Hung, C.-W., Hsu, H.-H., 2008. The first transition of the Asian summer monsoon, intraseasonal oscillation, and Taiwan Meiyu. J. Clim. 21, 1552–1568.

Hung, M.-P., Lin, J.-L., Wang, W., Kim, D., Shinoda, T., Weaver, S.J., 2013. MJO and convectively coupled equatorial waves simulated by CMIP5 climate models. J. Clim. 26, 6185–6214.

Hunke, E.C., Lipscomb, W.H., 2010. CICE: The Sea Ice Model Documentation and Software User's Manual, Version 4.1. Technical report LA-CC-06-012Los Alamos National Laboratory, Los Alamos, NM.

Hunke, E.C., Lipscomb, W.H., Turner, A.K., 2010. Sea ice models for climate study: retrospective and new directions. J. Glaciol. 56 (200), 1162–1172.

Hurrell, J.W., Meehl, G.A., Bader, D., Delworth, T., Kirtman, B., Wielicki, B., 2009. A unified modeling approach to climate system prediction. Bull. Am. Meteorol. Soc. 90, 1819–1832.

Huth, R., Beck, C., Philipp, A., Demuzere, M., Ustrnul, Z., Cahynova, M., Kysely, J., Tveito, O.E., 2008. Classifications of atmospheric circulation patterns. Ann. N. Y. Acad. Sci. 1146, 105–152. https://doi.org/10.1196/annals.1446.019.

IFRC (International Federation of Red Cross and Red Crescent Societies), 2008. Early warning > Early action. https://doi.org/10.5771/9783845252698-123.

Indian Institute of Public Health Ganghinagar, Mount Sinai, CDKN, Emory University & NRDC, 2018. Expert committee recommendations for a heat action plan based on the Ahmedabad experience. Tech. Rep.https://wwwhttps://www.nrdc.org/sites/default/files/ahmedabad-expert-recommendations.pdf

Indian Institute of Public Health Ganghinagar, Natural Resources Defense Council, Rollins School of Public Health of Emory University & Icahn School of Medicine at Mount Sinai, 2017. Evaluation of Ahmedabad's Heat Action Plan: Assessing India's First Climate Adaptation and Early Warning System for Extreme Heat. Tech. Rep.https://www.nrdc.org/resources/rising-temperatures-deadly-threat-preparing-communities-india-extreme-heat-events. [(Accessed 30 March 2017)].

INE (National Statistics Institute, Uruguay), 2017. Energy, Gas and Water Historical Timeseries. http://www.ine.gub.uy/web/guest/energia-gas-y-agua.

Ineson, S., Scaife, A.A., 2009. The role of the stratosphere in the European climate response to El Niño. Nat. Geosci. 2, 32–36. https://doi.org/10.1038/ngeo381.

Ingram, K.T., Roncoli, M.C., Kirshen, P.H., 2002. Opportunities and constraints for farmers of West Africa to use seasonal precipitation forecasts within Burkina Faso as a case study. Agric. Syst. 74, 331–349.

Inness, P.M., Slingo, J.M., Guilyardi, E., Cole, J., 2003. Simulation of the Madden- Julian Oscillation in a coupled general circulation model. Part II: The role of the basic state. J. Clim. 16, 365–382.

Inoue, T., Satoh, M., Hagihara, Y., Miura, H., Schmetz, J., 2010. Comparison of high-level clouds represented in a global cloud system–resolving model with CALIPSO/CloudSat and geostationary satellite observations. J. Geophys. Res. 115, D00H22.

Inoue, M., Takahashi, M., Naoe, H., 2011. Relationship between the stratospheric quasi-biennial oscillation and tropospheric circulation in northern autumn. J. Geophys. Res. Atmos. 116. https://doi.org/10.1029/2011JD016040.

Iorio, J., Duffy, P., Govindasamy, B., Thompson, S., Khairoutdinov, M., Randall, D., 2004. Effects of model resolution and subgrid-scale physics on the simulation of precipitation in the continental United States. Clim. Dyn. 23 (3-4), 243–258.

Isaksen, L., Fisher, M., Berner, J., 2007. In: Use of analysis ensembles in estimating flow-dependent background error variance. Proceedings of the ECMWF Workshop on Flow-Dependent Aspects of Data Assimilation. ECMWF, pp. 65–86. Available online at:http://www.ecmwf.int/publications/.

Isaksen, L., Bonavita, M., Buizza, R., Fisher, M., Haseler, J., Leutbecher, M., Raynaud, L., 2010. Ensemble of data assimilations at ECMWF. In: ECMWF Research Department Technical Memorandum n. 636. Available from ECMWF, Shinfield Park, Reading RG2-9AX. See also, http://old.ecmwf.int/publications/.

Itoh, H., Kimoto, M., 1996. Multiple attractors and chaotic itinerancy in a quasigeostrophic model with realistic topography: implications for weather regimes and low-frequency variability. J. Atmos. Sci. 53, 2217–2231. https://doi.org/10.1175/1520-0469(1996)053<2217:maacii>2.0.co;2.

Itoh, H., Kimoto, M., 1997. Chaotic itinerancy with preferred transition routes appearing in an atmospheric model. Physica D 109, 274–292. https://doi.org/10.1016/s0167-2789(97)00064-x.

Ivy, D.J., Solomon, S., Calvo, N., Thompson, D.W.J., 2017. Observed connections of Arctic stratospheric ozone extremes to Northern Hemisphere surface climate. Environ. Res. Lett. 12, 24004. https://doi.org/10.1088/1748-9326/aa57a4.

Jackson, T.J., Hsu, A.Y., 2001. Soil moisture and TRMM microwave imager relationships in the Southern Great Plains 1999 (SGP99) experiment. IEEE Trans. Geosci. Remote Sens. 39, 1632–1642.

Jackson, R.B., Canadell, J., Ehleringer, J.R., Mooney, H.A., Sala, O.E., Schulze, E.D., 1996. A global analysis of root distributions for terrestrial biomes. Oecologia 108, 389–411.

Jacobs, C.M.J., et al., 2008. Evaluation of European Land Data Assimilation System (ELDAS) products using in situ observations. Tellus A 60, 1023–1037.

James, I.N., 1994. Introduction to Circulating Atmospheres. Cambridge University Press, p. 422.

Jancloes, M., et al., 2014. Climate services to improve public health. Int. J. Environ. Res. Public Health 11, 4555–4559.

Janowiak, J.E., Bauer, P., Wang, W., Arkin, P.A., Gottschalck, J., 2010. An evaluation of precipitation forecasts from operational models and reanalyses including precipitation variations associated with mjo activity. Mon. Weather Rev. 138, 4542–4560.

Janssen, P., Bidlot, J.-R., Abdalla, S., Hersbach, H., 2005. Progress in ocean wave forecasting at ECMWF. In: ECMWF Research Department Technical Memorandum n. 478. Available from ECMWF, Shinfield Park, Reading RG2-9AX. See also, http://old.ecmwf.int/publications/.

Janssen, P., Breivik, O., Mogensen, K., Vitart, F., Balmaseda, M., Bidlot, J., Keeley, S., Leutbecher, M., Magnusson, L., Molteni, F., 2013. Air-sea interaction and surface waves. ECMWF Tech. Memorandum. 712, 36 pp.http://www.ecmwf.int/sites/default/files/elibrary/2013/10238-air-sea-interaction-and-surface-waves.pdf

Jarlan, L., Mangiarotti, S., Mougin, E., Mazzega, P., Hiernaux, P., Le Dantec, V., 2008. Assimilation of SPOT/VEGETATION NDVI data into a Sahelian vegetation dynamics model. Remote Sens. Environ. 112, 1381–1394.

Jarvis, P.G., 1976. The interpretation of the variations in leaf water potential and stomatal conductance found in canopies in the field. Philos. Trans. R. Soc. Lond. B 273, 593–610.

Jeong, J.-H., Ho, C.-H., Kim, B.-M., 2005. Influence of the Madden-Julian Oscillation on wintertime surface air temperature and cold surges in east Asia. J. Geophys. Res. 110D11104. https://doi.org/10.1029/2004DJ005408.

Jeong, J.-H., Kim, B.-M., Ho, C.-H., Noh, Y.-H., 2008. Systematic variation in wintertime precipitation in East-Asia by MJO-induced extratropical vertical motion. J. Clim. 21, 788–801.

Jia, L., et al., 2017. Seasonal prediction skill of northern extratropical surface temperature driven by the stratosphere. J. Clim. https://doi.org/10.1175/JCLI-D-16-0475.1 JCLI-D-16-0475.1.

Jiang, X., 2017. Key processes for the eastward propagation of the Madden-Julian Oscillation based on multimodel simulations. J. Geophys. Res. Atmos. 122, 755–770.

Jiang, X., Waliser, D.E., Wheeler, M.C., et al., 2008. Assessing the Skill of an All-Season Statistical Forecast Model for the Madden–Julian Oscillation. Mon. Weather Rev. 136, 1940–1956. https://doi.org/10.1175/2007MWR2305.1.

Jiang, X., Waliser, D.E., Olson, W.S., Tao, W.-K., L'Ecuyer, T.S., Li, K.-F., Yung, Y.L., Shige, S., Lang, S., Takayabu, Y.N., 2011. Vertical diabatic heating structure of the MJO: intercomparison between recent reanalyses and TRMM estimates. Mon. Weather Rev. 139, 3208–3223.

Jiang, X., Zhao, M., Waliser, D.E., 2012. Modulation of tropical cyclones over the eastern Pacific by the Intraseasonal Variability Simulated in an AGCM. J. Clim. 25, 6524–6538.

Jiang, X., Waliser, D.E., Xavier, P.K., Petch, J., Klingaman, N.P., Woolnough, S.J., Guan, B., Bellon, G., Crueger, T., DeMott, C., Hannay, C., Lin, H., Hu, W., Kim, D., Lappen, C.-L., Lu, M.-M., Ma, H.-Y., Miyakawa, T., Ridout, J.A., Schubert, S.D., Scinocca, J., Seo, K.-H., Shindo, E., Song, X., Stan, C., Tseng, W.-L., Wang, W.,

Wu, T., Wyser, K., Zhang, G.J., Zhu, H., 2015. Vertical structure and physical processes of the Madden–Julian Oscillation: exploring key model physics in climate simulations. J. Geophys. Res. Atmos. 120, 4718–4748.

Jie, W., Vitart, F., Wu, T., Liu, X., 2017. Simulations of the Asian summer monsoon in the sub-seasonal to seasonal prediction project (S2S) database. Q. J. R. Meteorol. Soc. 143, 2282–2295. https://doi.org/10.1002/qj.3085.

Jin, F.F., Ghil, M., 1990. Intraseasonal oscillations in the extratropics: Hopf bifurcation and topographic instabilities. J. Atmos. Sci. 47, 3007–3022. https://doi.org/10.1175/1520-0469(1990)047<3007:ioiteh>2.0.co;2.

Jin, F., Hoskins, B.J., 1995. The direct response to tropical heating in a baroclinic atmosphere. J. Atmos. Sci. 52, 307–319.

Jin, F.F., Neelin, J.D., Ghil, M., 1994. El Niño on the devil's staircase: annual subharmonic steps to chaos. Science 264, 70–72. https://doi.org/10.1126/science.264.5155.70.

Johnson, C.M., Lemke, P., Barnett, T.P., 1985. Linear prediction of sea ice anomalies. J. Geophys. Res. Atmos. 90 (D3), 5665–5675.

Johnson, N.C., Collins, D.C., Feldstein, S.B., L'Heureux, M.L., Riddle, E.E., 2014. Skillful wintertime North American temperature forecasts out to 4 weeks based on the state of ENSO and the MJO. Weather Forecast. 29, 23–38. https://doi.org/10.1175/WAF-D-13-00102.1.

Jolliffe, I., Stephenson, D.B., 2012a. Jolliffe, I., Stephenson, D. (Eds.), Forecast Verification: A Practitioner's Guide. Wiley, p. 292.

Jolliffe, I.T., Stephenson, D.B., 2012b. Forecast Verification: A Practitioner's Guide in Atmospheric Science. John Wiley & Sons.

Jones, C., 2017. Predicting subseasonal precipitation variations based on the MJO. In: Wang, S. et al., (Ed.), In Climate Extremes: Patterns & Mechanisms. In: AGU Monograph, Wiley. ISBN: 1119067847.

Jones, C., Carvalho, L.M.V., 2012. Spatial-intensity variations in extreme precipitation in the contiguous United States and the Madden–Julian oscillation. J. Clim. 25, 4849–4913.

Jones, C., Dudhia, J., 2017. Potential predictability during a Madden-Julian Oscillation event. J. Clim. in review.

Jones, A.E., Morse, A.P., 2010. Application and validation of a seasonal ensemble prediction system using a dynamic malaria model. J. Clim. 23, 4202–4215.

Jones, A.E., Morse, A.P., 2012. Skill of ENSEMBLES seasonal re-forecasts for malaria prediction in West Africa. Geophys. Res. Lett. 39, L23707. https://doi.org/10.1029/2012GL054040.

Jones, C., Waliser, D.E., Schemm, J.-K.E., Lau, W.K.M., 2000a. Prediction skill of the Madden and Julian Oscillation in dynamical extended range forecasts. Clim. Dyn. 16, 273–289. https://doi.org/10.1007/s003820050327.

Jones, J.W., Hansen, J.W., Royce, F.S., Messina, C.D., 2000b. Potential benefits of climate forecasting to agriculture. Agric. Ecosyst. Environ. 82, 169–184.

Jones, C.L., Waliser, D.E., Lau, K.-M., Stern, W., 2004. Global occurences of extreme precipitation and the Madden-Julian Oscillation: observations and predictability. J. Clim. 17, 4575–4589.

Jones, A.E., Wort, U.U., Morse, A.P., Hastings, I.M., Gagnon, A.S., 2007. Climate prediction of El Niño malaria epidemics in north-west Tanzania. Malar. J. 6, 162.

Jones, C., Gottschalck, J., Carvalho, L.M.V., Higgins, W.R., 2011. Influence of the Madden-Julian Oscillation on forecasts of extreme precipitation in the contiguous United States. Mon. Weather Rev. 139, 332–350.

Joseph, S., Sahai, A.K., Abhilash, S., et al., 2015. Development and evaluation of an objective criterion for the real-time prediction of Indian summer monsoon onset in a coupled model framework. J. Clim. 28, 6234–6248. https://doi.org/10.1175/JCLI-D-14-00842.1.

Joslyn, S.L., LeClerc, J.E., 2012. Uncertainty forecasts improve weather-related decisions and attenuate the effects of forecast error. J. Exp. Psychol. Appl. 18, 126–140.

Jost, C.C., et al., 2010. Epidemiological assessment of the Rift Valley fever outbreak in Kenya and Tanzania in 2006 and 2007. Am. J. Trop. Med. Hyg. 83, 65–72.

Joyce, T.M., Kwon, Y.-O., Yu, L., 2009. On the relationship between synoptic wintertime atmospheric variability and path shifts in the Gulf Stream and the Kuroshio Extension. J. Clim. 22, 3177–3192. https://doi.org/10.1175/2008JCLI2690.1.

Julian, P.R., Labitzke, K.B., 1965. A Study of Atmospheric Energetics During the January–February 1963 Stratospheric Warming. J. Atmos. Sci. 22, 597–610. https://doi.org/10.1175/1520-0469(1965)022<0597:ASOAED>2.0.CO;2.

Jung, J.-H., Arakawa, A., 2004. The resolution dependence of model physics: illustrations from nonhydrostatic model experiments. J. Atmos. Sci. 61 (1), 88–102.

Jung, T., Barkmeijer, J., 2006. Sensitivity of the tropospheric circulation to changes in the strength of the stratospheric polar vortex. Mon. Weather Rev. https://doi.org/10.1175/MWR3178.1.

Jung, E., Kirtman, B.P., 2016. Can we predict seasonal changes in high impact weather in the United States? Environ. Res. Lett. 11, 074018.
Jung, T., Leutbecher, M., 2007. Performance of the ECMWF forecasting system in the Arctic during winter. Q. J. R. Meteorol. Soc. 133, 1327–1340. https://doi.org/10.1002/qj.99.
Jung, T., Leutbecher, M., 2008. Scale-dependent verification of ensemble forecasts. Q. J. R. Meteorol. Soc. 132, 2905–2923.
Jung, T., Klinker, E., Uppala, S., 2004. Reanalysis and reforecast of three major European storms of the twentieth century using the ECMWF forecasting system. Part I: analyses and deterministic forecasts. Meteorol. Appl. 11, 343–361. https://doi.org/10.1017/S1350482704001434.
Jung, T., Vitart, F., Ferranti, L., Morcrette, J.-J., 2011. Origin and predictability of the extreme negative NAO winter of 2009/10. Geophys. Res. Lett. 38. https://doi.org/10.1029/2011GL046786.
Jung, T., Kasper, M.A., Semmler, T., Serrar, S., 2014. Arctic influence on subseasonal midlatitude prediction. Geophys. Res. Lett. 41, 3676–3680. https://doi.org/10.1002/2014GL059961.
Jung, T., Doblas-Reyes, F., Goessling, H., Guemas, V., Bitz, C., Buontempo, C., Caballero, R., Jakobsen, E., Jungclaus, J., Karcher, M., Koenigk, T., Matei, D., Overland, J., Spengler, T., Yang, S., 2015. Polar-lower latitude linkages and their role in weather and climate prediction. Bull. Am. Meteorol. Soc. 96, 197–200. https://doi.org/10.1175/BAMS-D- 15-00121.1.
Jung, T., Gordon, N., Bauer, P., Bromwich, D., Chevallier, M., et al., 2016. Advancing polar prediction capabilities on daily to seasonal time scales. Bull. Am. Meteorol. Soc. 97, 1631–1647.
Jupp, T.E., Lowe, R., Coelho, C.A., Stephenson, D.B., 2012. On the visualization, verification and recalibration of ternary probabilistic forecasts. Phil. Trans. R. Soc. A 370, 1100–1120.
Juricke, S., Goessling, H.F., Jung, T., 2014. Potential sea ice predictability and the role of stochastic sea ice strength perturbations. Geophys. Res. Lett. 41 (23), 8396–8403.
Kai, J., Kim, H., 2014. Characteristics of initial perturbations in the ensemble prediction system of the Korea Meteorological Administration. Weather Forecast. 29, 563–581.
Kain, J.S., 2004. The Kain–Fritsch convective parameterization: an update. J. Appl. Meteorol. 43, 170–181. https://doi.org/10.1175/1520-0450(2004)043,0170:TKCPAU.2.0.CO;2.
Kalame, F.B., Kudejira, D., Nkem, J., 2011. Assessing the process and options for implementing National Adaptation Programmes of Action (NAPA): a case study from Burkina Faso. Mitig. Adapt. Strateg. Glob. Chang. 16, 535–553.
Kalnay, E., 2012. Atmospheric Modelling, Data Assimilation and Predictability, seventh ed. Cambridge University Press, p. 341.
Kalnay, E., Livezey, R., 1985. Weather predictability beyond a week: an introductory review. In: Ghil, M., Benzi, R., Parisi, G. (Eds.), Turbulence and Predictability in Geophysical Fluid Dynamics and Climate Dynamics. North-Holland, Amsterdam, pp. 311–346.
Kamsu-Tamo, P.-H., Janicot, S., Monkam, D., Lenouo, A., 2014. Convection activity over the Guinean coast and Central Africa during northern spring from synoptic to intra-seasonal timescales. Clim. Dyn. 43, 3377–3401.
Kanamitsu, M., et al., 2002. NCEP-DOE AMIP-II reanalysis (R-2). Bull. Am. Meteorol. Soc. 83, 1631–1643.
Kang, I.-S., Kim, H.-M., 2010. Assessment of MJO predictability for boreal winter with various statistical and dynamical models. J. Clim. 23, 2368–2378.
Kang, W., Tziperman, E., 2017. More frequent sudden stratospheric warming events due to enhanced MJO forcing expected in a warmer climate. J. Clim. 30, 8727–8743. https://doi.org/10.1175/JCLI-D-17-0044.1.
Kang, I.-S., Yang, Y.-M., Tao, W.-K., 2015. GCMs with implicit and explicit representation of cloud microphysics for simulation of extreme precipitation frequency. Clim. Dyn. 45, 325–335.
Kang, I.-S., Ahn, M.-S., Yang, Y.-M., 2016. A GCM with cloud microphysics and its MJO simulation. Geosci. Lett. 3, 16.
Karpechko, A.Y., Perlwitz, J., Manzini, E., 2014. A model study of tropospheric impacts of the Arctic ozone depletion 2011. J. Geophys. Res. Atmos. 119, 7999–8014. https://doi.org/10.1002/2013JD021350.
Karpechko, A.Y., Hitchcock, P., Peters, D.H.W., Schneidereit, A., 2017. Predictability of downward propogation of major sudden stratospheric warmings. Q. J. R. Meteorol. Soc. https://doi.org/10.1002/qj.3017.
Karuri, J., Waiganjo, P., Daniel, O., MANYA, A., 2014. DHIS2: the tool to improve health data demand and use in Kenya. J. Health Inf. Dev. Countries 8.
Kauker, F., Kaminski, T., Karcher, M., Giering, R., Gerdes, R., Voßbeck, M., 2009. Adjoint analysis of the 2007 all time Arctic sea-ice minimum. Geophys. Res. Lett. 36(3).

Keil, C., Craig, G.C., 2007. A displacement-based error measure applied in a regional ensemble forecasting system. Mon. Weather Rev. 135, 3248–3259.

Keil, C., Craig, G.C., 2009. A displacement and amplitude score employing an optical flow technique. Weather Forecast. 24, 1297–1308.

Kelly, G., Thepaut, J.-N., Buizza, R., Cardinali, C., 2007. The value of observations - Part I: data denial experiments for the Atlantic and the Pacific. Q. J. R. Meteorol. Soc. 133, 1803–1815.

Kelly, K.A., Small, R.J., Samelson, R.M., Qiu, B., Joyce, T.M., Kwon, Y.-O., Cronin, M.F., 2010. Western boundary currents and frontal air–sea interaction: Gulf Stream and Kuroshio Extension. J. Clim. 23, 5644–5667. https://doi.org/10.1175/2010JCLI3346.1.

Kerns, B.W., Chen, S.S., 2016. Large-scale precipitatino tracking and the MJO over the Maritime Continent and Indo-Pacific warm pool. J. Geophys. Res. Atmos. 121, 8755–8776.

Kerr, Y.H., et al., 2010. The SMOS mission: new tool for monitoring key elements of the global water cycle. Proc. IEEE 98, 666–687.

Kessler, W.S., 2001. EOF representation of the Madden-Julian Oscillation and its connection with ENSO. J. Clim. 14, 3055–3061.

Khairoutdinov, M.F., Kogan, Y.L., 1999. A large eddy simulation model with explicit microphysics: validation against aircraft observations of a stratocumulus-topped boundary layer. J. Atmos. Sci. 56 (13), 2115–2131.

Khairoutdinov, M.F., Randall, D.A., 2003. Cloud resolving modeling of the ARM summer 1997 IOP: model formulation, results, uncertainties, and sensitivities. J. Atmos. Sci. 60 (4), 607–625.

Khairoutdinov, M., Randall, D., 2006. High-resolution simulation of shallow-to-deep convection transition over land. J. Atmos. Sci. 63, 3421–3436.

Khairoutdinov, M., Randall, D., DeMott, C., 2005. Simulations of the atmospheric general circulation using a cloud-resolving model as a superparameterization of physical processes. R. Meteorol. Soc. Interface 62 (7), 2136–2154.

Kharin, V.V., Zwiers, F.W., Gagnon, N., 2001. Skill of seasonal hindcasts as a function of the ensemble size. Clim. Dyn. 17, 835–843.

Kidston, J., Gerber, E.P., 2010. Intermodel variability of the poleward shift of the austral jet stream in the CMIP3 integrations linked to biases in 20th century climatology. Geophys. Res. Lett. 37. https://doi.org/10.1029/2010GL042873.

Kienberger, S., Hagenlocher, M., 2014. Spatial-explicit modeling of social vulnerability to malaria in East Africa. Int. J. Health Geogr. 13, 29.

Kikuchi, K., Takayabu, Y.N., 2004. The development of organized convection associated with the MJO during TOGA COARE IOP: trimodal characteristics. Geophys. Res. Lett. 31L10101.

Kiladis, G.N., Weickmann, K.M., 1992. Extratropical forcing of tropical Pacific convection during northern winter. Mon. Weather Rev. 120, 1924–1938.

Kiladis, G.N., Straub, K.H., Haertel, P.N., 2005. Zonal and vertical structure of the Madden-Julian Oscillation. J. Atmos. Sci. 62, 2790–2809.

Kiladis, G.N., Wheeler, M.C., Haertel, P.T., Straub, K.H., Roundy, P.E., 2009. Convectively coupled equatorial waves. Rev. Geophys. 47, RG2003. https://doi.org/10.1029/2008RG000266.

Kiladis, G.N., Dias, J., Straub, K.H., Wheeler, M.C., Tulich, S.N., Kikuchi, K., Weickmann, K.M., Ventrice, M.J., 2014. A comparison of OLR and circulation based indices for tracking the MJO. Mon. Weather Rev. 142, 1697–1715.

Kilian, A., Langi, P., Talisuna, A., Kabagambe, G., 1999. Rainfall pattern, El Niño and malaria in Uganda. Trans. R. Soc. Trop. Med. Hyg. 93, 22–23.

Kim, H., 2017. The impact of the mean moisture bias on the key physics of MJO propagation in the ECMWF reforecast. J. Geophys. Res. Atmos. 122, 7772–7784.

Kim, D., Kang, I.-S., 2012. A bulk mass flux convection scheme for climate model: description and moisture sensitivity. Clim. Dyn. 38, 411–429.

Kim, T.-W., Valdés, J.B., 2003. Nonlinear model for drought forecasting based on a conjunction of wavelet transforms and neural networks. J. Hydrol. Eng. 8, 319–328.

Kim, D., et al., 2009. Application of MJO simulation diagnostics to climate models. J. Clim. 22, 6413–6436.

Kim, B.-M., Son, S.-W., Min, S.-K., Jeong, J.-H., Kim, S.-J., Zhang, X., Shim, T., Yoon, J.-H., 2014. Weakening of the stratospheric polar vortex by Arctic sea-ice loss. Nat. Commun. 5, 4646. https://doi.org/10.1038/ncomms5646.

Kim, D., Kug, J.S., Sobel, A.H., 2014a. Propagating versus nonprogagating Madden-Julian Oscillation events. J. Clim. 27, 111–125.

Kim, D., Xavier, P., Maloney, E., Wheeler, M., Waliser, D., Sperber, K., Hendon, H., Zhang, C., Neale, R., Yen-Tinh, H., Liu, H., 2014b. Process-oriented MJO simulation diagnostic: moisture sensitivity of simulated convection. J. Clim. 27, 5379–5395.

Kim, H.-M., Webster, P.J., Toma, V.E., Kim, D., 2014c. Predictability and prediction skill of the MJO in two operational forecasting systems. J. Clim. 27, 5364–5378.

Kim, H.-M., Kim, D., Vitart, F., Toma, V., Kug, J.-S., Webster, P.J., 2016. MJO propagation across the Maritime Continent in the ECMWF ensemble prediction system. J. Clim. 29, 3973–3988.

Kimoto, M., Ghil, M., 1993a. Multiple flow regimes in the northern hemisphere winter. Part I: Methodology and hemispheric regimes. J. Atmos. Sci. 50, 2625–2644. https://doi.org/10.1175/1520-0469(1993)050<2625:mfritn>2.0.co;2.

Kimoto, M., Ghil, M., 1993b. Multiple flow regimes in the northern hemisphere winter. Part II: Sectorial regimes and preferred transitions. J. Atmos. Sci. 50, 2645–2673. https://doi.org/10.1175/1520-0469(1993)050<2645:mfritn>2.0.co;2.

King, M.P., Hell, M., Keenlyside, N., 2016. Investigation of the atmospheric mechanisms related to the autumn sea ice and winter circulation link in the Northern Hemisphere. Clim. Dyn. 46, 1185–1195. https://doi.org/10.1007/s00382-015-2639-5.

Kinter, J.L., et al., 2013. Revolutionizing climate modeling with project Athena: a multi-institutional, international collaboration. Bull. Am. Meteorol. Soc. 94, 231–245.

Kirtman, B.P., Ming, D., Infanti III, J.M., Kinter, J.L., Paolino, D.A., Zhang, Q., van den Dool, H., Saha, S., Mendez, M.P., Becker, E., Peng, P., Tripp, P., Huang, J., Witt, D.G.D., Tippet, M.K., Barnston, A.G., Schubert, S.D., Rienecker, M., Suarez, M., Li, Z.E., Marshak, J., Lim, Y.-K., Tribbia, J., Pegion, K., Merryfield, W.J., Denis, B., Wood, E.F., 2014. The North American multimodel ensemble: Phase-1 Seasonal-to-interannual prediction; Phase-2 toward developing intraseasonal prediction. Bull. Am. Meteorol. Soc. 49, 585–601.

Kistler, R., et al., 2001. The NCEP-NCAR 50-year reanalysis: monthly means CD-ROM and documentation. Bull. Am. Meteorol. Soc. 82, 247–267.

Klasa, M., Derome, J., Sheng, J., 1992. On the interaction between the synoptic-scale eddies and the PNA teleconnection pattern. Beitr. Phys. Atmos. Contrib. Atmos. Phys. 65, 211–222.

Klinenberg, E., 2015. Heat Wave: A Social Autopsy of Disaster in Chicago. University of Chicago Press.

Klingaman, N.P., Woolnough, S.J., 2014a. The role of air-sea coupling in the simulation of the Madden-Julian Oscillation in the Hadley Centre model. Q. J. R. Meteorol. Soc. 140, 2272–2286.

Klingaman, N.P., Woolnough, S.J., 2014b. Using a case-study approach to improve the Madden–Julian Oscillation in the Hadley Centre model. Q. J. R. Meteorol. Soc. 140, 2491–2505.

Klotzbach, P.J., 2014. The Madden-Julian Oscillation's impact on worldwide tropical cyclone activity. J. Clim. 27, 2317–2330.

Knippertz, P., 2007. Tropical-extratropical interactions related to upper-level troughs at low latitudes. Dyn. Atmos. Ocean 43, 36–62.

Knowlton, K., et al., 2014. Development and implementation of South Asia's first heat-health action plan in Ahmedabad (Gujarat, India). Int. J. Environ. Res. Public Health 11, 3473–3492.

Knutson, T.R., Weickmann, K.M., 1987. 30-60 day atmospheric oscillations: composite life cycles of convection and circulation anomalies. Mon. Weather Rev. 115, 1407–1436.

Knutson, T.R., Sirutis, J., Zhao, M., Tuleya, R.E., Bender, M., Vecchi, G.A., Villarini, G., Chavas, D., 2015. Global projections of intense tropical cyclone activity for the late twenty-first century from dynamical downscaling of CMIP5/RCP4.5 scenarios. J. Clim. 28, 7203–7224.

Kobayashi, S., Ota, Y., Harada, Y., Ebita, A., Moriya, M., Onoda, H., Onogi, K., Kamahori, H., Kobayashi, C., Endo, H., Miyaoka, K., Takahashi, K., 2015. The JRA-55 reanalysis: general specifications and basic characteristics. J. Meteorol. Soc. Jpn. 93, 5–48.

Kodama, C., Yamada, Y., Noda, A.T., Kikuchi, K., Kajikawa, Y., Nasuno, T., Tomita, T., Yamaura, T., Takahashi, H.G., Hara, M., Kawatani, Y., Satoh, M., Sugi, M., 2015. A 20-year climatology of a NICAM AMIP-type simulation. J. Meteorol. Soc. Jpn. 93, 393–424.

Kodera, K., 1995. On the origin and nature of the interannual variability of the winter stratospheric circulation in the northern hemisphere. J. Geophys. Res. 100, 14077. https://doi.org/10.1029/95JD01172.

Koenigk, T., Mikolajewicz, U., 2009. Seasonal to interannual climate predictability in mid and high northern latitudes in a global coupled model. Clim. Dyn. 32 (6), 783–798.

Kolassa, J., Gentine, P., Prigent, C., Aires, F., 2016. Soil moisture retrieval from AMSR-E and ASCAT microwave observation synergy. Part 1: Satellite data analysis. Remote Sens. Environ. 173, 1–14.

Kondrashov, D., Berloff, P., 2015. Stochastic modeling of decadal variability in ocean gyres. Geophys. Res. Lett. 42, 1543–1553. https://doi.org/10.1002/2014GL062871.

Kondrashov, D., Ide, K., Ghil, M., 2004. Weather regimes and preferred transition paths in a three-level quasigeostrophic model. J. Atmos. Sci. 61, 568–587. https://doi.org/10.1175/1520-0469(2004) 061<0568: wraptp>2.0.co;2.

Kondrashov, D., Kravtsov, S., Robertson, A.W., Ghil, M., 2005. A hierarchy of data-based ENSO models. J. Clim. 18, 4425–4444. https://doi.org/10.1175/jcli3567.1.

Kondrashov, D., Kravtsov, S., Ghil, M., 2006. Empirical mode reduction in a model of extratropical low-frequency variability. J. Atmos. Sci. 63, 1859–1877. https://doi.org/10.1175/jas3719.1.

Kondrashov, D., Shen, J., Berk, R., D'Andrea, F., Ghil, M., 2007. Predicting weather regime transitions in Northern Hemisphere datasets. Clim. Dyn. 29, 535–551. https://doi.org/10.1007/s00382-007-0293-2.

Kondrashov, D., Kravtsov, S., Ghil, M., 2011. Signatures of nonlinear dynamics in an idealized atmospheric model. J. Atmos. Sci. 68, 3–12. https://doi.org/10.1175/2010jas3524.1.

Kondrashov, D., Chekroun, M.D., Robertson, A.W., Ghil, M., 2013. Low-order stochastic model and "past-noise forecasting" of the Madden-Julian oscillation. Geophys. Res. Lett. 40, 5305–5310. https://doi.org/10.1002/grl.50991.

Kondrashov, D., Chekroun, M.D., Ghil, M., 2015. Data-driven non-Markovian closure models. Physica D 297, 33–55. https://doi.org/10.1016/j.physd.2014.12.005.

Kondrashov, D., Chekroun, M.D., Ghil, M., 2018. Data-adaptive harmonic decomposition and prediction of Arctic sea ice extent. Dyn. Stat. Clim. Syst. 3 (1). https://doi.org/10.1093/climsys/dzy001.

Kondrashov, D., Chekroun, M.D., Yuan, X., Ghil, M., 2017. Data-adaptive harmonic decomposition and stochastic modeling of Arctic sea ice. In: Tsonis, A. (Ed.), Advances in Nonlinear Geosciences. Springer, Cham, Switzerland, pp. 179–205. https://doi.org/10.1007/978-3-319-58895-7_10.

Konings, A.G., Entekhabi, D., Moghaddam, M., Saatchi, S.S., 2013. The effect of variable soil moisture profiles on P-band backscatter. IEEE Trans. Geosci. Remote Sens. 52, 6315–6325.

Konings, A.G., Williams, A.P., Gentine, P., 2017. Sensitivity of grassland productivity to aridity controlled by stomatal and xylem regulation. Nat. Geosci. 7, 2193–2197.

Konradsen, F., van der Hoek, W., Amerasinghe, F.P., Mutero, C., Boelee, E., 2004. Engineering and malaria control: learning from the past 100 years. Acta Trop. 89, 99–108.

Koo, S., Robertson, A.W., Ghil, M., 2002. Multiple regimes and low-frequency oscillations in the Southern Hemisphere's zonal-mean flow. J. Geophys. Res. Atmos. 107, https://doi.org/10.1029/2001jd001353 ACL 14–1–13.

Kopp, T.J., Kiess, R.B., 1996. The air force global weather central snow analysis model. Preprints, 15th Conf. on Weather Analysis and Forecasting, Norfolk, VA. Amer. Meteor. Soc, pp. 220–222.

Koster, R.D., Suarez, M.J., 1995. Relative contributions of land and ocean processes to precipitation variability. J. Geophys. Res. 100, 13,775–13,790.

Koster, R.D., Suarez, M.J., 2001. Soil moisture memory in climate models. J. Hydrometeorol. 2, 558–570.

Koster, R.D., Dirmeyer, P.A., Hahmann, A.N., Ijpelaar, R., Tyahla, L., Cox, P., Suarez, M.J., 2002. Comparing the degree of land-atmosphere interaction in four atmospheric general circulation models. J. Hydrometeorol. 3, 363–375.

Koster, R.D., et al., 2004. Regions of strong coupling between soil moisture and precipitation. Science 305, 1138–1140.

Koster, R.D., et al., 2006. GLACE: the global land-atmosphere coupling experiment. 1. Overview and results. J. Hydrometeorol. 7, 590–610.

Koster, R.D., et al., 2010a. Contribution of land surface initialization to subseasonal forecast skill: first results from a multi-model experiment. Geophys. Res. Lett. 37L02402.

Koster, R.D., Mahanama, S.P.P., Yamada, T.J., Balsamo, G., Berg, A.A., Boisserie, M., Dirmeyer, P.A., Doblas-Reyes, F.J., Drewitt, G., Gordon, C.T., Guo, Z., Jeong, J.-H., Lee, W.-S., Li, Z., Luo, L., Malyshev, S., Merryfield, W.J., Seneviratne, S.I., Stanelle, T., van den Hurk, B.J.J.M., Vitart, F., Wood, E.F., 2010b. The second phase of the global land-atmosphere coupling experiment: soil moisture contributions to subseasonal forecast skill. J. Hydrometeorol. https://doi.org/10.1175/2011JHM1365.1.

Koster, R.D., Chang, Y., Schubert, S.D., 2014. A mechanism for land-atmosphere feedback involving planetary wave structures. J. Clim. 27, 9290–9301.

Kravtsov, S., Kondrashov, D., Ghil, M., 2005. Multi-level regression modeling of nonlinear processes: derivation and applications to climatic variability. J. Clim. 18, 4404–4424. https://doi.org/10.1175/JCLI3544.1.

Kravtsov, S., Kondrashov, D., Ghil, M., 2009. Empirical model reduction and the modeling hierarchy in climate dynamics and the geosciences. In: Palmer, T.N., Williams, P. (Eds.), Stochastic Physics and Climate Modeling. Cambridge University Press, Cambridge, UK, pp. 35–72.

Kren, A.C., Marsh, D.R., Smith, A.K., Pilewskie, P., 2014. Examining the stratospheric response to the solar cycle in a coupled WACCM simulation with an internally generated QBO. Atmos. Chem. Phys. 14, 4843–4856. https://doi.org/10.5194/acp-14-4843-2014.

Kren, A.C., Marsh, D.R., Smith, A.K., Pilewskie, P., 2016. Wintertime northern hemisphere response in the stratosphere to the pacific decadal oscillation using the whole atmosphere community climate model. J. Clim. 29, 1031–1049. https://doi.org/10.1175/JCLI-D-15-0176.1.

Kretschmer, M., Coumou, D., Donges, J.F., Runge, J., 2016. Using causal effect networks to analyze different arctic drivers of midlatitude winter circulation. J. Clim. 29, 4069–4081. https://doi.org/10.1175/JCLI-D-15-0654.1.

Krishnamurthy, V., Achuthavarier, D., 2012. Intraseasonal oscillations of the monsoon circulation over South Asia. Clim. Dyn. 38, 2335–2353. https://doi.org/10.1007/s00382-011-1153-7.

Krishnamurthy, V., Sharma, A.S., 2017. Predictability at intraseasonal time scale. Geophys. Res. Lett. 44, 8530–8537. https://doi.org/10.1002/2017GL074984.

Krishnamurthy, V., Shukla, J., 2000. Intraseasonal and interannual variability of rainfall over India. J. Clim. 13, 4366–4377.

Krishnamurthy, V., Shukla, J., 2008. Intraseasonal and seasonnally persisting patterns of Indian monsoon rainfall. J. Clim. 20, 3–20.

Kumar, A., 2009. Finite samples and uncertainty estimates for skill measures for seasonal prediction. Mon. Weather Rev. 137, 2622–2631.

Kumar, A., Chen, M., 2015. Inherent predictability, requirements on ensemble size, and complementarity. Mon. Weather Rev. 143, 3192–3203.

Kumar, A., Hoerling, M.P., 2000. Analysis of a conceptual model of seasonal climate variability and implications for seasonal prediction. Bull. Am. Meteorol. Soc. 81, 255–264.

Kumar, A., Barnston, A.G., Hoerling, M.P., 2001. Seasonal predictions, probabilistic verifications, and ensemble size. J. Clim. 14, 1671–1676.

Kumar, S.V., Peters-Lidard, C.D., Tian, Y., Reichle, R.H., Geiger, J., Alonge, C., Eylander, J., Houser, P., 2008. An integrated hydrologic modeling and data assimilation framework enabled by the Land Information System (LIS). IEEE Comput. 41, 52–59.

Kumar, A., et al., 2012. An analysis of the non-stationarity in the bias of sea surface temperature forecasts for the NCEP climate forecast system (CFS) version 2. Mon. Weather Rev. 140, 3003–3016.

Kumar, S.V., Dirmeyer, P.A., Peters-Lidard, C.D., Bindlish, R., 2017. Information theoretic evaluation of satellite soil moisture retrievals. Remote Sens. Environ. (submitted).

Kuo, H.L., 1974. Further studies of the parameterization of the effect of cumulus convection on large-scale flow. J. Atmos. Sci. 31, 1232–1240.

Kurihara, Y., Tuleya, R.E., Bender, M.A., 1998. The GFDL hurricane prediction system and its performance in the 1995 hurricane season. Mon. Weather Rev. 126 (5), 1306–1322.

Kuroda, Y., 2008. Role of the stratosphere on the predictability of medium-range weather forecast: a case study of winter 2003–2004. Geophys. Res. Lett. 35, L19701. https://doi.org/10.1029/2008GL034902.

Kuroda, Y., Kodera, K., 2001. Variability of the polar night jet in the northern and southern hemispheres. J. Geophys. Res. Atmos. 106, 20703–20713. https://doi.org/10.1029/2001JD900226.

Kushnir, Y., 1987. Retrograding wintertime low-frequency disturbances over the North Pacific ocean. J. Atmos. Sci. 44, 2727–2742. https://doi.org/10.1175/1520-0469(1987)044<2727:rwlfdo>2.0.co;2.

Kushnir, Y., Robinson, W.A., Bladé, I., Hall, N.M.J., Peng, S., Sutton, R., 2002. Atmospheric GCM response to extratropical SST anomalies: synthesis and evaluation. J. Clim. 15, 2233–2256. https://doi.org/10.1175/1520-0442(2002)015,2233:AGRTES.2.0.CO;2.

Kusumawathie, P., Wickremasinghe, A., Karunaweera, N., Wijeyaratne, M., Yapaban-dara, A., 2006. Anopheline breeding in river bed pools below major dams in Sri Lanka. Acta Trop. 99, 30–33.

Kwok, R., 2011. Satellite remote sensing of sea ice thickness and kinematics: a review. Ann. Glaciol. 56 (200), 1129–1140.

Kwok, R., Rothrock, D.A., 2009. Decline in Arctic sea ice thickness from submarine and ICESat records: 1958–2008. Geophys. Res. Lett. 36(15).

Kwon, Y.-O., Alexander, M.A., Bond, N.A., Frankignoul, C., Nakamura, H., Qiu, B., Thompson, L.A., 2010. Role of the Gulf Stream and Kuroshio–Oyashio systems in large-scale atmosphere–ocean interaction: a review. J. Clim. 23, 3249–3281. https://doi.org/10.1175/ 2010JCLI3343.1.

Kwon, H.-H., de Assis de Souza Filho, F., Block, P., Sun, L., Lall, U., Reis, D., 2012. Uncertainty assessment of hydrologic and climate forecast models in Northeastern Brazil. Hydrol. Process. 26 (25), 3875–3885.

L'Heureux, M.L., Higgins, R.W., 2008. Boreal winter links between the Madden–Julian oscillation and the Arctic Oscillation. J. Clim. 21, 3040–3050. https://doi.org/10.1175/2007JCLI1955.1.

Labitzke, K., 1977. Interannual variability of the winter stratosphere in the Northern Hemisphere. Mon. Weather Rev. 105, 762–770. https://doi.org/10.1175/1520-0493(1977)105<0762:IVOTWS>2.0.CO;2.

Labitzke, K., van Loon, H., 1992. Association between the 11-year solar cycle and the atmosphere. Part V: Summer. J. Clim. 5, 240–251. https://doi.org/10.1175/1520-0442(1992)005<0240:ABTYSC>2.0.CO;2.

Laffont, J.-J., Martimort, D., 2002. The Theory of Incentives: The Principal-Agent Model. Princeton University Press, Princeton, NJ.

Lalaurette, F., 2003. Early detection of abnormal weather conditions using a probabilistic extreme forecast index. Q. J. R. Meteorol. Soc. 129, 3037–3057.

Lambrechts, L., et al., 2011. Impact of daily temperature fluctuations on dengue virus transmission by *Aedes aegypti*. Proc. Natl. Acad. Sci. 108, 7460–7465.

Laneri, K., et al., 2010. Forcing versus feedback: epidemic malaria and monsoon rains in northwest India. PLoS Comput. Biol. https://doi.org/10.1371/journal.pcbi.1000898.

Lapeyssonnie, L., 1963. La meningite cerebro-spinale en Afrique.

Large, W.G., Yeager, S.G., 2009. The global climatology of an interannually varying air-sea flux data set. Clim. Dyn. 33, 341–364.

Latif, M., Anderson, D., Barnett, T., Cane, M., Kleeman, R., Leetmaa, A., et al., 1998. A review of the predictability and prediction of ENSO. J. Geophys. Res. Oceans 103 (C7), 14375–14393.

Lau, N.-C., 1988. Variability of the observed midlatitude storm tracks in relation to low-frequency changes in the circulation pattern. J. Atmos. Sci. 45, 2718–2743.

Lau, K.-M., Chen, P.H., 1986. Aspects of the 30–50 oscillation during summer as inferred from outgoing longwave radiation. Mon. Weather Rev. 114, 1354–1369.

Lau, K.M., Peng, L., 1987. Origin of low-frequency (intraseasonal) oscillations in the tropical atmosphere, Part I: Basic theory. J. Atmos. Sci. 44, 950–972.

Lau, K.M., Phillips, T.J., 1986. Coherent fluctuations of extratropical geopotential height and tropical convection in intraseasonal time scales. J. Atmos. Sci. 43, 1164–1181. https://doi.org/10.1175/1520-0469(1986)043<1164:CFOFGH>2.0.CO;2.

Lau, W.K.-M., Waliser, D.E., 2005. Intraseasonal Variability in the Atmosphere—Ocean Climate System. Springer.

Lau, W.K.-M., Waliser, D.E., 2012. Intraseasonal Variability in the Atmosphere—Ocean Climate System, second ed. Springer.

Lavaysse, C., Vogt, J., Pappenberger, F., 2015. Early warning of drought in Europe using the monthly ensemble system from ECMWF. Hydrol. Earth Syst. Sci. 19, 3273–3286. https://doi.org/10.5194/hess-19-3273-2015.

Lavaysse, C., Toreti, A., Vogt, J., 2017. On the use of atmospherical predictors to forecast meteorological droughts over Europe. JAMC. in revision.

Lavers, D.A., Villarini, G., 2015. The contribution of atmospheric rivers to precipitation in Europe and the United States. J. Hydrol. 522, 382–390. https://doi.org/10.1016/j.jhydrol.2014.12.010.

Le Barbé, L., Lebel, T., Tapsoba, D., 2002. Rainfall variability in West Africa during the years 1950-90. J. Clim. 15, 187–202.

Le Trent, H., Li, Z.-X., 1991. Sensitivity of an atmospheric general circulation model to prescribed SST changes: feedback effects associated with the simulation of cloud optical properties. Clim. Dyn. 5 (3), 175–187.

Lebel, T., Ali, A., 2009. Recent trends in the Central and Western Sahel rainfall regime (1990–2007). J. Hydrol. 375, 52–64.

Lee, H.-T., NOAA CDR Program, 2011. NOAA Climate Data Record (CDR) of Daily Outgoing Longwave Radiation (OLR), Version 1.2, 1996–2014. NOAA National Climatic Data Center. https://doi.org/10.7289/V5SJ1HH2 (Accessed: 12-11-2017).

Lee, M.-I., Kang, I.-S., Kim, J.-K., Mapes, B.E., 2001. Influence of cloud-radiation interaction on simulating tropical intraseasonal oscillation with an atmospheric general circulation model. J. Geophys. Res. 106 (14), 219–233.

Lee, S.-K., Atlas, R., Enfield, D., Wang, C., Liu, H., 2012. Is there an optimal ENSO pattern that enhances large-scale atmospheric processes conducive to tornado outbreaks in the United States? J. Clim. 26, 1626–1642. https://doi.org/10.1175/JCLI-D-12-00128.1.

Lee, J.-L., Wang, B., Wheeler, M.C., Fu, X., Waliser, D.E., Kang, I.S., 2013. Real-time multivariate indices for the boreal summer intraseasonal oscillation over the Asian summer monsoon region. Clim. Dyn. 40, 493–503.

Lee, S.-S., Wang, B., Waliser, D.E., et al., 2015. Predictability and prediction skill of the boreal summer intraseasonal oscillation in the Intraseasonal Variability Hindcast Experiment. Clim. Dyn. 45, 2123–2135. https://doi.org/10.1007/s00382-014-2461-5.

Lee, J.-Y., Fu, X., Wang, B., 2016. Predictability and prediction of the Madden-Julian oscillation: a review on progress and current status. In: Chang, C.-P., Kuo, H.-C., Lau, N.-C., Johnson, R.H., Wang, B., Wheeler, M.C. (Eds.), The Global Monsoon System: Research and Forecast, third ed. World Scientific Publishing Co, pp. 147–159.

Lefebvre, W., Goosse, H., Timmermann, R., Fichefet, T., 2004. Influence of the Southern Annular Mode on the sea ice-Ocean system. J. Geophys. Res. C: Oceans 109 (9), 1–12. https://doi.org/10.1029/2004JC002403.

Legras, B., Dritschel, D., 1993. Vortex stripping and the generation of high vorticity gradients in two-dimensional flows. Appl. Sci. Res. 51, 445–455.

Legras, B., Ghil, M., 1985. Persistent anomalies, blocking and variations in atmospheric predictability. J. Atmos. Sci. 42, 433–471. https://doi.org/10.1175/1520-0469(1985)042<0433:pabavi>2.0.co;2.

Lehtonen, I., Karpechko, A.Y., 2016. Observed and modeled tropospheric cold anomalies associated with sudden stratospheric warmings. J. Geophys. Res. Atmos. 121, 1591–1610. https://doi.org/10.1002/2015JD023860.

Leith, C.E., 1965. Theoretical skill of Monte Carlo forecasts. Mon. Weather Rev. 102, 409–418.

Leith, C.E., 1973. The standard error of time-average estimates of climatic means. J. Appl. Meteorol. 12, 1066–1069.

Leith, C.E., 1975. Climate response and fluctuation dissipation. J. Atmos. Sci. 32, 2022–2026. https://doi.org/10.1175/1520-0469(1975)032<2022:CRAFD>2.0.CO;2.

Leith, C., *1978*. Objective methods for weather prediction. Annu. Rev. Fluid Mech. *10*, *107–128*.

Lemieux, J.-F., Beaudoin, C., Dupont, F., Roy, F., Smith, G.C., Shlyaeva, A., Buehner, M., Caya, A., Chen, J., Carrieres, T., Pogson, L., DeRepentigny, P., Plante, A., Pestieau, P., Pellerin, P., Ritchie, H., Garric, G., Ferry, N., 2015. The Regional Ice Prediction System (RIPS): verification of forecast sea ice concentration. Q. J. R. Meteorol. Soc. 142 (695), 632–643.

Lemke, P., Trinkl, E.W., Hasselmann, K., 1980. Stochastic dynamic analysis of polar sea ice variability. J. Phys. Oceanogr. 10 (12), 2100–2120.

Lemos, M.C., 2003. A tale of two policies: the politics of seasonal climate forecast Use in Ceará, Brazil. Policy. Sci. 32 (2), 101–123.

Lemos, M.C., Finan, T.J., Fox, R.W., Nelson, D.R., Tucker, J., 2002. The use of seasonal climatic forecasting in policymaking: lessons from northeastern Brazil. Clim. Chang. 55, 479–507.

Lengaigne, M., Boulanger, J.P., Menkes, C., Delecluse, P., Slingo, J., 2004. Westerly Wind Events in the tropical Pacific and their influence on the coupled ocean-atmosphere system: a review. In: Earth's Climate: The Ocean-Atmosphere Interaction. Geophys. Monogr. Ser., vol. 147. AGU, Washington D C, pp. 49–69.

Lepore, C., Tippett, M.K., Allen, J.T., 2017. CFSv2 forecasts of severe weather parameters. Clim. Atmos. Sci. in preparation.

Leppäranta, M., 2011. The Drift of Sea Ice. Springer Science & Business Media.

Leroy, A., Wheeler, M.C., 2008. Statistical prediction of weekly tropical cyclone activity in the Southern Hemisphere. Mon. Weather Rev. 136, 3637–3654.

Lesnikowski, A.C., et al., 2011. Adapting to health impacts of climate change: a study of UNFCCC Annex I Parties. Environ. Res. Lett. 6(4).

Letson, D., Sutter, D.S., Lazo, J.K., 2007. Economic value of hurricane forecasts: an overview and research needs. Natural Hazards Rev. 8, 78–86.

Leutbecher, M., Palmer, T.N., 2008. Ensemble forecasting. J. Comp. Phys. 227, 3515–3539.

Leutbecher, M., et al., 2016. Stochastic representations of model uncertainties at ECMWF: state of the art and future vision. In: ECMWF Research Department Technical Memorandum n. 785, p. 52. Available from ECMWF, Shinfield Park, Reading RG2-9AX. See also, http://old.ecmwf.int/publications/.

Lewis, J.M., 2005. Roots of ensemble forecasting. Mon. Weather Rev. 133, 1865–1885.

Lhomme, J.-P., 1997. An examination of the Priestley-Taylor equation using a convective boundary layer model. Water Resour. Res. 33, 2571–2578.

Li, Y., Lau, N.-C., 2013. Influences of ENSO on stratospheric variability, and the descent of stratospheric perturbations into the lower troposphere. J. Clim. 26, 4725–4748. https://doi.org/10.1175/JCLI-D-12-00581.1.

Li, S., Robertson, A.W., 2015. Evaluation of submonthly precipitation forecast skill from global ensemble prediction systems. Mon. Weather Rev. 143, 2871–2889.

Li, W., Duan, Q., Miao, C., Ye, A., Gong, W., Di, Z., 2017. A review on statistical postprocessing methods for hydrometeorological ensemble forecasting. WIREs Water 4 (6), e1246. https://doi.org/10.1002/wat2.1246.

Liebmann, B., Hartmann, D.L., 1984. An observational study of tropical-midlatitude interaction on intraseasonal time scales during winter. J. Atmos. Sci. 41, 3333–3350.

Liebmann, B., Smith, C., 1996. Description of a complete (interpolated) outgoing longwave radiation dataset. Bull. Am. Meteorol. Soc. 77, 1275–1277.

Liebmann, B., Kiladis, G.N., Carvalho, L.M.V., Jones, C., Vera, C.S., Bladé, I., Allured, D., 2009. Origin of convectively coupled Kelvin waves over South America. J. Clim. 22, 300–315.

Liess, S., Geller, M.A., 2012. On the relationship between QBO and distribution of tropical deep convection. J. Geophys. Res. Atmos. 117. https://doi.org/10.1029/2011JD016317.

Liess, S., Waliser, D.E., Schubert, S.D., 2005. Predictability studies of the intraseasonal oscillation with the ECHAM5 GCM. J. Atmos. Sci. 62, 3320–3336. https://doi.org/10.1175/JAS3542.1.

Lim, S.Y., Marzin, C., Xavier, P., Chang, C.-P., Trimbal, B., 2017. Impacts of the boreal winter monsoon cold surges and the interaction with the MJO on Southeast Asia rainfall. J. Clim. 30, 4267–4281.

Lim, Y., Son, S.-W., Kim, D., 2018. MJO prediction skill of the sub-seasonal (S2S) models. J. Clim. 31, 4075–4094.

Limpasuvan, V., Hartmann, D.L., 2000. Wave-maintained annular modes of climate variability. J. Clim. 13, 4414–4429. https://doi.org/10.1175/1520-0442(2000)013<4414:WMAMOC>2.0.CO;2.

Limpasuvan, V., Thompson, D.W.J., Hartmann, D.L., 2004. The life cycle of the Northern Hemisphere sudden stratospheric warmings. J. Clim. 17, 2584–2596. https://doi.org/10.1175/1520-0442(2004)017<2584:TLCOTN>2.0.CO;2.

Limpasuvan, V., Hartmann, D.L., Thompson, D.W.J., Jeev, K., Yung, Y.L., 2005. Stratosphere-troposphere evolution during polar vortex intensification. J. Geophys. Res. 110D24101. https://doi.org/10.1029/2005JD006302.

Lin, S.-J., 2004. A "vertically Lagrangian" finite-volume dynamical core for global models. Mon. Weather Rev. 132 (10), 2293–2307.

Lin, H., Brunet, G., 2009. The influence of the Madden-Julian oscillation on Canadian wintertime surface air temperature. Mon. Weather Rev. 137 (7), 2250–2262.

Lin, H., Brunet, G., 2011. Impact of the North Atlantic Oscillation on the forecast skillof the Madden-Julian Oscillation. Geophys. Res. Lett. 38, L02802. https://doi.org/10.1029/2010GL046131.

Lin, H., Brunet, G., 2017. Extratropical response to the MJO: nonlinearity and sensitivity to initial state. J. Atmos. Sci. in press.

Lin, H., Derome, J., 1997. On the modification of the high and low-frequency eddies associated with PNA anomaly: an observational study. Tellus 49A, 87–99.

Lin, J.W.B., Neelin, J.D., 2000. Influence of a stochastic moist convective parameterization on tropical climate variability. Geophys. Res. Lett. 27, 3691–3694.

Lin, H., Wu, Z., 2011. Contribution of the autumn Tibetan Plateau snow cover to seasonal prediction of North American winter temperature. J. Clim. 24, 2801–2813.

Lin, Y.-L., Farley, R.D., Orville, H.D., 1983. Bulk parameterization of the snow field in a cloud model. J. Clim. Appl. Meteorol. 22 (6), 1065–1092.

Lin, J.L., Kiladis, G.N., Mapes, B.E., Weickmann, K.M., Sperber, K.R., Lin, W., Wheeler, M., Shubert, S.D., Del Genio, A., Donner, L.J., Emori, S., Gueremy, J.F., Hourdain, F., Rasch, P.J., Roeckner, E., Scinocca, J.F., 2006. Tropical intraseasonal variability in 14 IPCC AR4 climate models. Part I: Convective signals. J. Clim. 19, 2665–2690.

Lin, H., Brunet, G., Derome, J., 2007. Intraseasonal variability in a dry atmospheric model. J. Atmos. Sci. 64, 2441–2442.

Lin, H., Brunet, G., Derome, J., 2008. Forecast skill of the Madden–Julian Oscillation in two Canadian atmospheric models. Mon. Weather Rev. **136**, 4130–4149.

Lin, H., Brunet, G., Derome, J., 2009. An observed connection between the North Atlantic Oscillation and the Madden-Julian Oscillation. J. Clim. 22, 364–380.

Lin, H., Brunet, G., Fontecilla, J.S., 2010a. Impact of the Madden-Julian Oscillation on the intraseasonal forecast skill of the North Atlantic Oscillation. Geophys. Res. Lett. 37, L19803.

Lin, H., Brunet, G., Mo, R., 2010b. Impact of the Madden-Julian Oscillation on wintertime precipitation in Canada. Mon. Weather Rev. 138, 3822–3839.

Lin, H., Gagnon, N., Beauregard, S., Muncaster, R., Markovic, M., Denis, B., Charron, M., 2016. GEPS based monthly prediction at the Canadian Meteorological Centre. Mon. Weather Rev. 144, 4867–4883.

Lindblade, K.A., Walker, E.D., Onapa, A.W., Katungu, J., Wilson, M.L., 1999. Highland malaria in Uganda: prospective analysis of an epidemic associated with El Niño. Trans. R. Soc. Trop. Med. Hyg. 93, 480–487.

Lindley, D.V., Smith, A.F., 1972. Bayes estimates for the linear model. J. R. Stat. Soc. Ser. B Methodol., 1–41.

Lindsay, R., 2010. New unified sea ice thickness climate data record. EOS Trans. Am. Geophys. Union 91 (44), 405–406.

Lindsay, R.W., Zhang, J., Schweiger, A.J., Steele, M.A., 2008. Seasonal predictions of ice extent in the Arctic Ocean. J. Geophys. Res. Oceans 113(C2).

Lindsay, R., Wensnahan, M., Schweiger, A., Zhang, J., 2014. Evaluation of seven different atmospheric reanalysis products in the Arctic. J. Clim. 27 (7), 2588–2606.

Lindzen, R.S., 1986. Stationary planetary waves, blocking, and interannual variability. Adv. Geophys. 29, 251–273. https://doi.org/10.1016/s0065-2687(08)60042-4.

Lindzen, R.S., Holton, J.R., 1968. A theory of the Quasi-Biennial Oscillation. J. Atmos. Sci. 25, 1095–1107. https://doi.org/10.1175/1520-0469(1968)025<1095:ATOTQB>2.0.CO;2.

Lindzen, R., Nigam, S., 1987. On the role of sea surface temperature gradients in forcing low-level winds and convergence in the tropics. J. Atmos. Sci. 44, 2418–2436.

Lindzen, R.S., Farrell, B., Jacqmin, D., 1982. Vacillations due to wave interference: applications to the atmosphere and to annulus experiments. J. Atmos. Sci. 39, 14–23.

Liston, G.E., Sud, Y.C., Walker, G.K., 1993. Design of a global soil moisture initialization procedure for the simple biosphere model. NASA Tech. Memo. 104590. 138 pp.

Liu, Z., Alexander, M., 2007. Atmospheric bridge, ocean tunnel, and global climatic teleconnections. Rev. Geophys. 45, RG2007.

Liu, F., Wang, B., 2013. An air–sea coupled skeleton model for the Madden–Julian Oscillation. J. Atmos. Sci. 70, 3147–3156.

Liu, J., Curry, J.A., Hu, Y., 2004. Recent Arctic sea ice variability: connections to the Arctic Oscillation and the ENSO. Geophys. Res. Lett. 31(9).

Liu, Q., Lord, S., Surgi, N., Zhu, Y., Wobus, R., Toth, Z., Marchok, T., 2006. Hurricane relocation in global ensemble forecast system.Pre-prints, 27th Conf. on Hurricanes and Tropical Meteorology, Monterey, CA, Amer. Meteor. Soc. p. 5.13.

Liu, J., et al., 2009a. Validation of Moderate Resolution Imaging Spectroradiometer (MODIS) albedo retrieval algorithm: dependence of albedo on solar zenith angle. J. Geophys. Res. 114, D01106.

Liu, P., et al., 2009b. An MJO simulated by the NICAM at 14- and 7-km Resolutions. Mon. Weather Rev. 137, 3254–3268. https://doi.org/10.1175/2009MWR2965.1.

Liu, X., Yang, S., Li, Q., et al., 2014a. Subseasonal forecast skills and biases of global summer monsoons in the NCEP Climate Forecast System version 2. Clim. Dyn. 42, 1487–1508. https://doi.org/10.1007/s00382-013-1831-8.

Liu, C., Tian, B., Li, K.-F., Manney, G.L., Livesey, N.J., Yung, Y.L., Waliser, D.E., 2014b. Northern Hemisphere midwinter vortex-displacement and vortex-split stratospheric sudden warmings: influence of the Madden-Julian Oscillation and Quasi-Biennial Oscillation. J. Geophys. Res. Atmos. 119, 12,599–12,620. https://doi.org/10.1002/2014JD021876.

Liu, J., Song, M., Horton, R.M., Hu, Y., 2015. Revisiting the potential of melt pond fraction as a predictor for the seasonal Arctic sea ice extent minimum. Environ. Res. Lett. 10(5)054017.

Liu, X., Wu, T., Yang, S., et al., 2017. MJO prediction using the sub-seasonal to seasonal forecast model of Beijing Climate Center. Clim. Dyn. 48, 3283–3307.

Livina, V., Edwards, N., Goswami, S., Lenton, T., 2008. A wavelet-coefficient score for comparison of two-dimensional climatic-data fields. Q. J. R. Meteorol. Soc. 134 (633), 941–955.

Lloyd's report, 2012. In: House, C. (Ed.), Arctic Opening: Opportunity and Risk in the High North. Available at http://www.chathamhouse.org/publications/papers/view/182839.

Lorenz, E.N., 1956. Empirical Orthogonal Functions and Statistical Weather Prediction. Technical report, Statistical Forecast Project Report 1, Dep of Meteor, MIT: 49.

Lorenz, E.N., 1963. Deterministic nonperiodic flow. J. Atmos. Sci. 20 (130), 141.

Lorenz, E.N., 1969a. The predictability of a flow which possess many scales of motion. Tellus XXI (3), 289–307.

Lorenz, E.N., 1969b. Atmospheric predictability as revealed by naturally occurring analogues. J. Atmos. Sci. 26, 636–646.

Lorenz, E.N., 1969c. How much better can weather prediction become? Technol. Rev., 39–49. Accessible from the MIT Library, http://eaps4.mit.edu/research/Lorenz/publications.htm.

Lorenz, E.N., 1975. Climatic predictability. In: Bolin, B., et al., (Eds.), The Physical Basis of Climate and Climate Modelling. In: GARP Publication Series, vol. 16. World Meteorological Organization, Geneva, pp. 132–136.

Lorenz, E.N., 1982. Atmospheric predictability experiments with a large numerical model. Tellus 34, 505–513.

Lorenz, D.J., Hartmann, D.L., 2001. Eddy–Zonal flow feedback in the Southern Hemisphere. J. Atmos. Sci. 58, 3312–3327. https://doi.org/10.1175/1520-0469(2001)058<3312:EZFFIT>2.0.CO;2.

Lorenz, D.J., Hartmann, D.L., 2003. Eddy–Zonal flow feedback in the Northern Hemisphere winter. J. Clim. 16, 1212–1227. https://doi.org/10.1175/1520-0442(2003)16<1212:EFFITN>2.0.CO;2.

Lorenz, R., Jaeger, E.B., Seneviratne, S.I., 2010. Persistence of heat waves and its link to soil moisture memory. Geophys. Res. Lett. 37(9).

Lott, F., Robertson, A.W., Ghil, M., 2001. Mountain torques and atmospheric oscillations. Geophys. Res. Lett. 28, 1207–1210. https://doi.org/10.1029/2000gl011829.

Lott, F., Robertson, A.W., Ghil, M., 2004a. Mountain torques and Northern Hemisphere low-frequency variability. Part I: Hemispheric aspects. J. Atmos. Sci. 61, 1259–1271. https://doi.org/10.1175/1520-0469(2004)061<1259:mtanhl>2.0.co;2.

Lott, F., Robertson, A.W., Ghil, M., 2004b. Mountain torques and Northern Hemisphere low-frequency variability. Part II: Regional aspects. J. Atmos. Sci. 61, 1272–1283. https://doi.org/10.1175/1520-0469(2004)061<1272:mtanhl>2.0.co;2.

Loveland, T.R., Reed, B.C., Brown, J.F., Ohlen, D.O., Zhu, Z., Yang, L., Merchant, J.W., 2000. Development of a global land cover characteristics database and IGBP DISCover from 1 km AVHRR data. Int. J. Remote Sens. 21, 1303–1330.

Lowe, R., Rodó, X., 2016. Modelling Climate-Sensitive Disease Risk: A Decision Support Tool for Public Health Services. Springer International Publishing, pp. 115–130.

Lowe, R., et al., 2011. Spatio-temporal modelling of climate-sensitive disease risk: towards an early warning system for dengue in Brazil. Comput. Geosci. 37, 371–381.

Lowe, R., et al., 2013a. The development of an early warning system for climate-sensitive disease risk with a focus on dengue epidemics in Southeast Brazil. Stat. Med. 32, 864–883.

Lowe, R., Chirombo, J., Tompkins, A.M., 2013b. Relative importance of climatic, geographic and socio-economic determinants of malaria in Malawi. Malar. J. 12. https://doi.org/10.1186/1475-2875-12-416.

Lowe, R., et al., 2014. Dengue outlook for the World Cup in Brazil: an early warning model framework driven by real-time seasonal climate forecasts. Lancet Infect. Dis. 14, 619–626.

Lowe, R., et al., 2015. Evaluating the performance of a climate-driven mortality model during heat waves and cold spells in Europe. Int. J. Environ. Res. Public Health 12, 1279–1294.

Lowe, R., et al., 2016a. Evaluating probabilistic dengue risk forecasts from a prototype early warning system for Brazil. elife 5, e11285.

Lowe, R., et al., 2016b. Training a new generation of professionals to use climate information in public health decision-making. In: Shumake-Guillemot, J., Fernandez-Montoya, L. (Eds.), Climate Services for Health: Case Studies of Enhancing Decision Support for Climate Risk Management and Adaptation. WHO/WMO, Geneva, pp. 54–55.

Lowe, R., et al., 2016c. Evaluation of an early-warning system for heat wave-related mortality in Europe: implications for sub-seasonal to seasonal forecasting and climate services. Int. J. Environ. Res. Public Health 13, 206.

Lowe, R., Cazelles, B., Paul, R., Rodó, X., 2016d. Quantifying the added value of climate information in a spatio-temporal dengue model. Stoch. Env. Res. Risk A. 30, 2067–2078.

Lowe, R., et al., 2017. Climate services for health: predicting the evolution of the 2016 dengue season in Machala, Ecuador. Lancet Planet. Health.

Lu, L., Shuttleworth, W.J., 2002. Incorporating NDVI-derived LAI into the climate version of RAMS and its impact on regional climate. J. Hydrometeorol. 3, 347–362.

Lubis, S.W., Jacobi, C., 2015. The modulating influence of convectively coupled equatorial waves (CCEWs) on the variability of tropical precipitation. Int. J. Climatol. 35, 1465–1483.

Lubis, S.W., Matthes, K., Omrani, N.-E., Harnik, N., Wahl, S., 2016a. Influence of the Quasi-Biennial Oscillation and sea surface temperature variability on downward wave coupling in the Northern Hemisphere. J. Atmos. Sci. 73, 1943–1965. https://doi.org/10.1175/JAS-D-15-0072.1.

Lubis, S.W., Omrani, N.-E., Matthes, K., Wahl, S., 2016b. Impact of the Antarctic ozone hole on the vertical coupling of the stratosphere–mesosphere–lower thermosphere system. J. Atmos. Sci. 73, 2509–2528. https://doi.org/10.1175/JAS-D-15-0189.1.

Lukovich, J.V., Barber, D.G., 2007. On the spatiotemporal behavior of sea ice concentration anomalies in the Northern Hemisphere. J. Geophys. Res. Atmos. 112(D13).

Lüpkes, C., Gryanik, V.M., Witha, B., Gryschka, M., Raasch, S., Gollnik, T., 2008. Modelling convection over leads with LES and a non-eddy-resolving microscale model. J. Geophys. Res. 113, C09028. https://doi.org/10.1029/2007JC004099.

Lynch, P., 2002. Resonant motions of the three-dimensional elastic pendulum. Int. J. Nonlin. Mech. 37, 345–367.

Lynch, P., 2006. Weather prediction by numerical process. In: The Emergence of Numerical Weather Prediction. Cambridge University Press, ISBN: 978-0-521-85729-1, pp. 1–27.

Lynch, P., 2008. The origins of computer weather prediction and climate modeling. J. Comput. Phys. 227 (7), 3431–3444.

Lynch, K.J., Brayshaw, D.J., Charlton-Perez, A., 2014. Verification of European subseasonal wind speed forecasts. Mon. Weather Rev. 142 (8), 2978–2990.

Lyons, C.L., Coetzee, M., Terblanche, J.S., Chown, S.L., 2014. Desiccation tolerance as a function of age, sex, humidity and temperature in adults of the African malaria vectors Anopheles arabiensis Patton and Anopheles funestus Giles. J. Exp. Biol. https://doi.org/10.1242/jeb.104638.

Ma, X.H., Chang, P., Saravanan, R., Montuoro, R., Hsieh, J.S., Wu, D.X., Lin, X.P., Wu, L.X., Jing, Z., 2015. Distant influence of Kuroshio Eddies on North Pacific weather patterns? Sci. Rep. 5, 17785. https://doi.org/10.1038/srep17785.

Ma, X., Jing, Z., Chang, P., Liu, X., Montuoro, R., Small, R.J., Bryan, F.O., Greatbatch, R.J., Brandt, P., Wu, D., Lin, X., Wu, L., 2016. Western boundary currents regulated by interaction between ocean eddies and the atmosphere. Nature 535, 533–537. https://doi.org/10.1038/nature18640.

Ma, X.H., Chang, P., Saravanan, R., Montuoro, R., Nakamura, H., Wu, D.X., Lin, X.P., Wu, L.X., 2017. Importance of resolving Kuroshio Front and Eddy influence in simulating the North Pacific storm track. J. Clim. 30, 1861–1880.

Mabaso, M.L.H., Ndlovu, N.C., 2012. Critical review of research literature on climate-driven malaria epidemics in sub-Saharan Africa. Public Health 126, 909–919.

Maciel, F., Terra, R., Chaer, R., 2015. Economic impact of considering El Niño-southern oscillation on the representation of streamflow in an electric system simulator. Int. J. Climatol. 35, 4094–4102. https://doi.org/10.1002/joc.4269.

MacLachlan, C., Arribas, A., Peterson, K.A., Maidens, A., Fereday, D., Scaife, A.A., Gordon, M., Vellinga, M., Williams, A., Comer, R.E., Camp, J., Xavier, P., Madec, G., 2014. Global seasonal forecast system version 5 (GloSea5): a high-resolution seasonal forecast system. Q. J. R. Meteorol. Soc. 141, 1072–1084. https://doi.org/10.1002/qj.2396.

MacLeod, D.A., Jones, A., Di Giuseppe, F., Caminade, C., Morse, A.P., 2015. Demonstration of successful malaria forecasts for Botswana using an operational seasonal climate model. Environ. Res. Lett. 10, 044005.

Macleod, D., Torralba, V., Davis, M., Doblas-Reyes, F.J., 2017. Transforming climate model output to forecasts of wind power production: how much resolution is enough? Meteorol. Appl. https://doi.org/10.1002/met.1660.

Macron, C., Pohl, B., Richard, Y., Bessafi, M., 2014. How do Tropical Temperate Troughs form and develop over Southern Africa? J. Clim. 27, 1633–1647.

Madden, R., Julian, P., 1971. Detection of a 40–50-day oscillation in the zonal wind in the tropical Pacific. J. Atmos. Sci. 28, 702–708.

Madden, R., Julian, P., 1972. Description of global-scale circulation cells in the tropics with a 40–50-day period. J. Atmos. Sci. 29, 1109–1123.

Madden, R., Julian, P., 1994. Observations of the 40–50-day tropical oscillation: a review. Mon. Weather Rev. 112, 814–837.

Madec, G., 2008. NEMO Ocean Engine. Note du Pole de Modelisation. Institut Pierre-Simon Laplace (IPSL), Paris.

Magnusson, L., Leutbecher, M., Källén, E., 2008. Comparison between singular vectors and breeding vectors as initial perturbations for the ECMWF ensemble prediction system. Mon. Weather Rev. 136, 4092–4104.

Mahmood, R., et al., 2014. Land cover changes and their biogeophysical effects on climate. Int. J. Climatol. 34, 929–953.

Mahrt, L., Ek, M., 1984. The influence of atmospheric stability on potential evaporation. J. Clim. Appl. Meteorol. 23, 222–234.

Mahrt, L., Pan, H.-L., 1984. A two-layer model of soil hydrology. Bound.-Layer Meteorol. 29, 1–20.

Majda, A.J., Stechmann, S.N., 2009. The skeleton of tropical intraseasonal oscillations. Proc. Natl. Acad. Sci. 106, 8417–8422.

Majda, A.J., Stechmann, S.N., 2012. Multiscale theories for the MJO. In: Lau, W.K., Waliser, D.E. (Eds.), Intraseasonal Variability in the Atmosphere–Ocean Climate System, second ed. Springer Praxis, pp. 549–568.

Majda, A.J., Timofeyev, I., Vanden-Eijnden, E., 1999. Models for stochastic climate prediction. Proc. Natl. Acad. Sci. 96, 14687–14691. https://doi.org/10.1073/pnas.96.26.14687.

Majda, A.J., Franzke, C.L., Fischer, A., Crommelin, D.T., 2006. Distinct metastable atmospheric regimes despite nearly Gaussian statistics: a paradigm model. Proc. Natl. Acad. Sci. 103, 8309–8314. https://doi.org/10.1073/pnas.0602641103.

Malguzzi, P., Buzzi, A., Drofa, O., 2011. The meteorological global model GLOBO at the ISAC-CNR of Italy: assessment of 1.5 year of experimental use for medium-range weather forecasts. Weather Forecast. 26, 1045–1055.

Maloney, E.D., Hartmann, D.L., 2000. Modulation of eastern North Pacific hurricanes by the Madden–Julian oscillation. J. Clim. 13, 1451–1460.

Maloney, D.E., Zhang, C., 2016. Dr. Yanai's contributions to the discovery and science of the MJO. Meteorol. Monogr. 56, 4.1–4.18.

Manabe, S., 1969. Climate and the circulation. I. The atmospheric circulation and the hydrology of the earth's surface. Mon. Weather Rev. 97, 739–774.

Manney, G.L., et al., 2011. Unprecedented Arctic ozone loss in 2011. Nature 478, 469–475. https://doi.org/10.1038/nature10556.

Manzini, E., 2009. Atmospheric science: ENSO and the stratosphere. Nat. Geosci. 2, 749–750. https://doi.org/10.1038/ngeo677.

Mapes, B.E., 2000. Convective inhibition, subgrid-scale triggering energy, and stratiform instability in a toy tropical wave model. J. Atmos. Sci. 57, 1515–1535.

Marchezini, V., Trajber, R., 2016. Youth based learning in disaster risk reduction education: barriers and bridges to promote resilience. In: Companion, M., Chaiken, M.S. (Eds.), Responses to Disasters and Climate Change: Understanding Vulnerability and Fostering Resilience. CRC Press, Taylor and Francis Group, Boca Raton, FL, pp. 27–36.

Marcus, S.L., Ghil, M., Dickey, J.O., 1994. The extratropical 40-day oscillation in the UCLA general circulation model. Part I: Atmospheric angular momentum. J. Atmos. Sci. 51, 1431–1446. https://doi.org/10.1175/1520-0469(1994)051<1431:tedoit>2.0.co;2.

Marcus, S.L., Ghil, M., Dickey, J.O., 1996. The extratropical 40-day oscillation in the UCLA general circulation model. Part II: Spatial structure. J. Atmos. Sci. 53, 1993–2014. https://doi.org/10.1175/1520-0469(1996)053<1993:tedoit>2.0.co;2.

Marshall, J., Molteni, F., 1993. Toward a dynamical understanding of atmospheric weather regimes. J. Atmos. Sci. 50, 1993–2014.

Marshall, J.S., Palmer, W. McK., 1948. The distribution of raindrops with size. J. Meteor. 5, 165–166.

Marshall, A.G., Scaife, A.A., 2009. Impact of the QBO on surface winter climate. J. Geophys. Res. 114, D18110. https://doi.org/10.1029/2009JD011737.

Marshall, A.G., Scaife, A.A., 2010. Improved predictability of stratospheric sudden warming events in an atmospheric general circulation model with enhanced stratospheric resolution. J. Geophys. Res. 115, D16114. https://doi.org/10.1029/2009JD012643.

Marshall, A.G., Hudson, D., Wheeler, M.C., Hendon, H.H., Alves, O., 2011a. Assessing the simulation and prediction of rainfall associated with the MJO in the POAMA seasonal forecast system. Clim. Dyn. 37, 2129–2141.

Marshall, N.A., Gordon, I.J., Ash, A.J., 2011b. The reluctance of resource-users to adopt seasonal climate forecasts to enhance resilience to climate variability on the rangelands. Clim. Chang. 107 (3-4), 511–529.

Marshall, A.G., Hudson, D., Wheeler, M., Alves, O., Hendon, H.H., Pook, M.J., Risbey, J.S., 2014. Intra-seasonal drivers of extreme heat over Australia in observations and POAMA-2. Clim. Dyn. 43, 1915–1937.

Marshall, A.G., Hendon, H.H., Hudson, D., 2016a. Visualizing and verifying probabilistic forecasts of the Madden-Julian Oscillation. Geophys. Res. Lett. 43 (23), 12278–12286. https://doi.org/10.1002/2016GL071423.

Marshall, A.G., Hendon, H.H., Son, S.-W., Lim, Y., 2016b. Impact of the quasi-biennial oscillation on predictability of the Madden–Julian oscillation. Clim. Dyn. https://doi.org/10.1007/s00382-016-3392-0.

Marshall, A.G., Hendon, H.H., Son, S.-K., Lim, Y., 2017. Impact of the quasi-biennial oscillation on predictability of the Madden-Julian Oscillation. Clim. Dyn. 49, 1365–1377.

Marsigli, C., Boccanera, F., Montani, A., Paccagnella, T., 2005. The COSMO–LEPS ensemble system: validation of the methodology and verification. Nonlinear Process. Geophys. 12, 527–536.

Martineau, P., Son, S.-W., 2015. Onset of circulation anomalies during stratospheric vortex weakening events: the role of planetary-scale waves. J. Clim. 28, 7347–7370. https://doi.org/10.1175/JCLI-D-14-00478.1.

Martinez, Y., Brunet, G., Yau, P., 2010a. On the dynamics of two-dimensional hurricane-like vortex symmetrization. J. Atmos. Sci. 67, 3559–3580.

Martinez, Y., Brunet, G., Yau, P., 2010b. On the dynamics of two-dimensional hurricane-like concentric rings vortex formation. J. Atmos. Sci. 67, 3253–3268.

Martinez, Y., Brunet, G., Yau, P., 2011. On the dynamics of concentric eyewall genesis: space-time empirical normal modes diagnosis. J. Atmos. Sci. 68, 457–476.

Martiny, N., Chiapello, I., 2013. Assessments for the impact of mineral dust on the meningitis incidence in West Africa. Atmos. Environ. 70, 245–253.

Martius, O., Polvani, L.M., Davies, H.C., 2009. Blocking precursors to stratospheric sudden warming events. Geophys. Res. Lett. 36, L14806. https://doi.org/10.1029/2009GL038776.

Marzban, C., Sandgathe, S., 2010. Optical flow for verification. Weather Forecast. 25, 1479–1494.

Mason, I., 1982. A model for assessment of weather forecasts. Aust. Meteorol. Mag. 30, 291–303.

Mason, S.J., Graham, N.E., 2002. Areas beneath the relative operating characteristics (ROC) and relative operating levels (ROL) curves: statistical significance and interpretation. Q. J. R. Meteorol. Soc. 128, 2145–2166.

Mason, S.J., Tippett, M.K., 2017. Climate Predictability Tool Version 15.6.3. Columbia University Academic Commons. https://doi.org/10.7916/D8DJ6NDS.

Mason, S., Goddard, L., Graham, N., Yulaeva, E., Sun, L., Arkin, P., 1999. The IRI seasonal climate prediction system and the 1997/98 El Niño Event. Bull. Am. Meteorol. Soc. 80, 1853–1873. https://doi.org/10.1175/1520-0477(1999)080<1853:TISCPS>2.0.CO;2.

Massonnet, F., Fichefet, T., Goosse, H., 2015. Prospects for improved seasonal Arctic sea ice predictions from multivariate data assimilation. Ocean Model. 88, 16–25. https://doi.org/10.1016/j.ocemod.2014.12.013.

Mastrangelo, D., Malguzzi, P., Rendina, C., Drofa, O., Buzzi, A., 2012. First outcomes from the CNR-ISAC monthly forecasting system. Adv. Sci. Res. 8, 77–82.

Masunaga, H., Satoh, M., Miura, H., 2008. A joint satellite and global cloud-resolving model analysis of a Madden-Julian Oscillation event: model diagnosis. J. Geophys. Res. 113, D17210.

Masunaga, R., Nakamura, H., Miyasaka, T., Nishii, K., Qiu, B., 2016. Interannual modulations of oceanic imprints on the wintertime atmospheric boundary layer under the changing dynamical regimes of the Kuroshio Extension. J. Clim. 29, 3273–3296. https://doi.org/10.1175/JCLI-D-15-0545.1.

Matsueda, S., Takaya, Y., 2015. The global influence of the Madden-Julian Oscillation on extreme temperature events. J. Clim. 28, 4141–4151.

Matsuno, T., 1966. Quasi-geostrophic motions in the equatorial area. J. Meteorol. Soc. Jpn. 44, 25–42.

Matsuno, T., 1971. A dynamical model of the stratospheric sudden warming. J. Atmos. Sci. 28, 1479–1494. https://doi.org/10.1175/1520-0469(1971)028<1479:ADMOTS>2.0.CO;2.

Matthewman, N.J., Esler, J.G., 2011. Stratospheric sudden warmings as self-tuning resonances. Part I: Vortex splitting events. J. Atmos. Sci. 68, 2481–2504. https://doi.org/10.1175/JAS-D-11-07.1.

Matthews, A.J., 2000. Propagation mechanisms for the Madden–Julian Oscillation. Q. J. R. Meteorol. Soc. 126, 2637–2651.

Matthews, A.J., 2008. Primary and successive events in the Madden-Julian Oscillation. Q. J. R. Meteorol. Soc. 134, 439–453.

Matthews, A.J., Kiladis, G.N., 1999. The tropical-extratropical interaction between high-frequency Transients and the Madden-Julian Oscillation. Mon. Weather Rev. 127, 661–667.

Matthews, A.J., Hoskins, B.J., Masutani, M., 2004. The global response to tropical heating in the Madden–Julian oscillation during the northern winter. Q. J. R. Meteorol. Soc. 130, 1991–2011. https://doi.org/10.1256/qj.02.123.

Maury, P., Claud, C., Manzini, E., Hauchecorne, A., Keckhut, P., 2016. Characteristics of stratospheric warming events during Northern winter. J. Geophys. Res. Atmos. 121, 5368–5380. https://doi.org/10.1002/2015JD024226.

Maycock, A.C., Hitchcock, P., 2015. Do split and displacement sudden stratospheric warmings have different annular mode signatures? Geophys. Res. Lett. 42, 10,943–10,951. https://doi.org/10.1002/2015GL066754.

Maycock, A.C., Keeley, S.P.E., Charlton-Perez, A.J., Doblas-Reyes, F.J., 2011. Stratospheric circulation in seasonal forecasting models: implications for seasonal prediction. Clim. Dyn. 36, 309–321. https://doi.org/10.1007/s00382-009-0665-x.

Maykut, G.A., Untersteiner, N., 1971. Some results from a time-dependent thermodynamic model of sea ice. J. Geophys. Res. 76 (6), 1550–1575.

Mayne, B., 1930. A study of the influence of relative humidity on the life and infectibility of the mosquito. Indian J. Med. Res. 17, 1119–1137.

McColl, K.A., Alemohammad, S.H., Akbar, R., Konings, A.G., Yueh, S., Entekhabi, D., 2017. The global distribution and dynamics of surface soil moisture. Nat. Geosci. 10, 100–104.

McCown, R.L., Carberry, P.S., Hochman, Z., Dalgliesh, N.P., Foale, M.A., 2009. Reinventing model-based decision support with Australian dryland farmers. 1. Changing intervention concepts during 17 years of action research. Crop Pasture Sci. 60, 1017–1030.

McCown, R.L., Carberry, P.S., Dalgliesh, N.P., Foale, M.A., 2012. Farmers use intuition to reinvent analytic decision support for managing seasonal climatic variability. Agric. Syst. 106, 33–45.

McCusker, K.E., Fyfe, J.C., Sigmond, M., 2016. Twenty-five winters of unexpected Eurasian cooling unlikely due to Arctic sea-ice loss. Nat. Geosci. 9, 838–842. https://doi.org/10.1038/ngeo2820.

McGregor, G.R., Bessemoulin, P., Ebi, K., Menne, B., 2014. Heatwaves and Health: Guidance on Warning-System Development. Tech. Rep. WMO-No. 1142.

McGregor, G.R., Bessemoulin, P., Ebi, K.L., Menne, B., 2015. Heatwaves and Health: Guidance on Warning-System Development. World Meteorological Organization.

McIntyre, M.E., 1980. Towards a Lagrangian-mean description of stratospheric circulations and chemical transports. Philos. Trans. R. Soc., A Math. Phys. Sci. 296 (1418), 129–148.

McIntyre, M.E., Palmer, T.N., 1984. The 'surf zone' in the stratosphere. J. Atmos. Terr. Phys. 46, 825–849.

McIntyre, M.E., Shepherd, T., 1987. An exact local conservation theorem for finite-amplitude disturbances to non-parallel shear flows, with remarks on Hamiltonian structure and Arnold stability theorems. J. Fluid Mech. 181, 527–567.

McLandress, C., Shepherd, T.G., 2011. Separating the dynamical effects of climate change and ozone depletion. Part II: southern hemisphere troposphere. J. Clim. 24, 1850–1868.

McWilliams, J.C., 1980. An application of equivalent modons to atmospheric blocking. Dyn. Atmos. Oceans 5, 43–66.

Meehl, G., 1997. Influence of the land surface in the Asian summer monsoon: external conditions versus internal feedbacks. J. Clim. 7, 1033–1049.

Meier, W., Markus, T., 2015. Remote sensing of sea ice. In: Tedesco, M. (Ed.), Remote Sensing of the Cryosphere. John Wiley & Sons, Ltd.

Meinke, H., Nelson, R., Kokic, P., Stone, R., Selaraju, R., Baethgen, W., 2006. Actionable climate knowledge: from analysis to synthesis. Clim. Res. 33, 101–110.

Meng, J., Yang, R., Wei, H., Ek, M., Gayno, G., Xie, P., Mitchell, K., 2012. The land surface analysis in the NCEP climate forecast system reanalysis. J. Hydrometeorol. 13, 1621–1630.

Merryfield, W.J., Lee, W.S., Wang, W., Chen, M., Kumar, A., 2013. Multi-system seasonal predictions of Arctic sea ice. Geophys. Res. Lett. 40 (8), 1551–1556.

Methven, J., 2013. Wave activity for large amplitude disturbances described by the primitive equations on the sphere. J. Atmos. Sci. 70, 1616–1630.

Methven, J., Berrisford, P., *2015*. The slowly evolving background state of the atmosphere. Q. J. R. Meteorol. Soc. 141, 2237–2258.

Methven, J., Frame, T., Boljka, L., Cafaro, C., 2018. Identifying Dynamical Modes of Variability From Global Data Including Boundary Wave Activity. Personal communication.

Michelangeli, P.A., Vautard, R., Legras, B., 1995. Weather regimes: recurrence and quasi stationarity. J. Atmos. Sci. 52, 1237–1256. https://doi.org/10.1175/1520-0469(1995)052<1237:wrraqs> 2.0.co;2.

Miller, M.A., et al., 2007. SGP Cloud and Land Surface Interaction Campaign (CLASIC): Science and Implementation Plan. Office of Biological and Environmental Research Office of Science, U.S. Department of Energy DoE/SC-ARM-0703. 14 pp.

Milrad, S.M., Gyakum, J.R., Atallah, E.H., 2015. A meteorological analysis of the 2013 Alberta flood: antecedent large-scale flow patterns and synoptic-dynamic characteristics. Mon. Weather Rev. 143, 2817–2841.

Minobe, S., Kuwano-Yoshida, A., Komori, N., Xie, S.P., Small, R.J., 2008. Influence of the Gulf Stream on the troposphere. Nature 452, 206–209. https://doi.org/10.1038/nature06690.

Mintz, Y., Serafini, Y., 1981. Global fields of soil moisture and surface evapotranspiration. NASA Tech. Memo. 83907, 178–180.

Miralles, D., den Berg, M., Teuling, A., de Jeu, R., 2012. Soil moisture- temperature coupling: a multiscale observational analysis. Geophys. Res. Lett. 39, L21707.

Miralles, D.G., Teuling, A.J., van Heerwaarden, C.C., Vila-Guerau de Arellano, J., 2014. Mega-heatwave temperatures due to combined soil desiccation and atmospheric heat accumulation. Nat. Geosci. 7, 345–349. http://www.nature.com/doifinder/10.1038/ngeo2141.

Mironov, D., Heise, E., Kourzeneva, E., Ritter, B., Schneider, N., Terzhevik, A., 2010. Implementation of the lake parameterisation scheme FLake into the numerical weather prediction model COSMO. Boreal Environ. Res. 15, 218–230.

Mishra, A., Desai, V., Singh, V., 2007. Drought forecasting using a hybrid stochastic and neural network model. J. Hydrol. Eng. 12, 626–638.

Mitchell, H.L., Derome, J., 1983. Blocking-like solutions of the potential vorticity equation: their stability at equilibrium and growth at resonance. J. Atmos. Sci. 40, 2522–2536. https://doi.org/10.1175/1520-0469(1983)040<2522:blsotp>2.0.co;2.

Mitchell, K.E., et al., 2004. The multi-institution North American Land Data Assimilation System (NLDAS): utilizing multiple GCIP products and partners in a continental distributed hydrological modeling system. J. Geophys. Res. 109, D07S90.

Mitchell, D.M., Charlton-Perez, A.J., Gray, L.J., 2011. Characterizing the variability and extremes of the stratospheric polar vortices using 2D moment analysis. J. Atmos. Sci. 68, 1194–1213. https://doi.org/10.1175/2010JAS3555.1.

Mittermaier, M., Roberts, N., 2010. Intercomparison of spatial forecast verification methods: identifying skillful spatial scales using the fractions skill score. Weather Forecast. 25, 343–354. https://doi.org/10.1175/2009WAF2222260.1.

Mittermaier, M., Roberts, N., Thompson, S.A., 2013. A long-term assessment of precipitation forecast skill using the Fractions Skill Score. Meteorol. Appl. 20, 176–186. https://doi.org/10.1002/met.296.

Miura, H., Satoh, M., Tomita, T., Noda, A.T., Nasuno, T., Iga, S., 2007a. A short-duration global cloud-resolving simulation with a realistic land and sea distribution. Geophys. Res. Lett. 34, L02804.

Miura, H., Satoh, M., Nasuno, T., Noda, A.T., Oouchi, K., 2007b. A Madden-Julian oscillation event realistically simulated by a global cloud-resolving model. Science 318 (5857), 1763–1765.

Miura, H., Miyakawa, T., Nasuno, T., Satoh, M., 2012. In: Simulations of the MJO events during the field campaign of 2011-12 by a global cloud-resolving model NICAM.Abstract A13O-03 presented at 2012 Fall Meeting, AGU, San Francisco, California, 3-7 December.

Miyakawa, T., Satoh, M., Miura, H., Tomita, H., Yashiro, H., Noda, A.T., Yamada, Y., Kodama, C., Kimoto, M., Yoneyama, K., 2014. Madden-Julian Oscillation prediction skill of a new-generation global model. Nat. Commun. 5, 3769.

Miyakawa, T., Satoh, M., Miura, H., Tomita, H., Yashiro, H., Noda, A.T., Yamada, Y., Kodama, C., Kimoto, M., Yoneyama, K., 2015. Madden-Julian Oscillation prediction skill of a new-generation global model demonstrated using a supercomputer. Nat. Commun. 5, 3769.

Miyakoda, K., Gordon, T., Caverly, R., Stern, W., Sirutis, J., Bourke, W., 1983. Simulation of a blocking event in January 1977. Mon. Weather Rev. 111, 846–869.

Miyakoda, K., Sirutis, J., Ploshay, J., 1986. One month forecast experiments—without anomaly boundary forcings. Mon. Weather Rev. 114, 2363–2401.

Miyoshi, T., Sato, Y., 2007. Assimilating satellite radiances with a local ensemble transform Kalman filter (LETKF) applied to the JMA global model (GSM). SOLA 3, 37–40.

Mlawer, E.J., Taubman, S.J., Brown, P.D., Iacono, M.J., Clough, S.A., 1997. Radiative transfer for inhomogeneous atmospheres: RRTM, a validated correlated-k model for the longwave. J. Geophys. Res. 102, 16663–16682. https://doi.org/10.1029/ 97JD00237.

Mo, K.C., Ghil, M., 1987. Statistics and dynamics of persistent anomalies. J. Atmos. Sci. 44, 877–902. https://doi.org/10.1175/1520-0469(1987)044<0877:sadopa>2.0.co;2.

Mo, K., Ghil, M., 1988. Cluster analysis of multiple planetary flow regimes. J. Geophys. Res. 93, 10927–10952. https://doi.org/10.1029/jd093id09p10927.

Mo, K.C., Higgins, R.W., 1998. Tropical convection and precipitation regimes in the western United States. J. Clim. 11, 2404–2423.

Mo, K.C., Lyon, B., 2015. Global meteorological drought prediction using the North American Multi-Model Ensemble. J. Hydrometeorol. 16, 1409–1424. https://doi.org/10.1175/JHM-D-14-0192.1.

Mo, K.C., White, G.H., 1985. Teleconnections in the Southern Hemisphere. Mon. Weather Rev. 113, 22–37.

Mo, K.C., Paegle, J.N., Higgins, R.W., 1997. Atmospheric processes associated with summer floods and droughts in the central United States. J. Clim. 10, 3028–3046.

Mogensen, K., Alonso Balmaseda, M., Weaver, A., 2012a. The NEMOVAR ocean data assimilation system as implemented in the ECMWF ocean analysis for System 4. In: ECMWF Research Department Technical

Memorandum n. 668, p. 59. Available from ECMWF, Shinfield Park, Reading RG2-9AX. See also, http://old.ecmwf.int/publications/.

Mogensen, K., Keeley, S., Towers, P., 2012b. Coupling of the NEMO and IFS models in a single executable. In: ECMWF Research Department Technical Memorandum n. 673, p. 23. Available from ECMWF, Shinfield Park, Reading RG2-9AX. See also, http://old.ecmwf.int/publications/.

Molina, M.J., Timmer, R.P., Allen, J.T., 2016. Importance of the Gulf of Mexico as a climate driver for U.S. severe thunderstorm activity. Geophys. Res. Lett. 43, 12,295–12,304. https://doi.org/10.1002/2016GL071603.

Möller, A., Lenkoski, A., Thorarinsdottir, T.L., 2012. Multivariate probabilistic forecasting using ensemble bayesian model averaging and copulas. Q. J. R. Meteorol. Soc. 139 (673), 982–991.

Molteni, F., Corti, S., 1998. Long-term fluctuations in the statistical properties of low-frequency variability: dynamical origin and predictability. Q. J. R. Meteorol. Soc. 124, 495–526. https://doi.org/10.1002/qj.49712454607.

Molteni, F., Cubasch, U., Tibaldi, S., 1986. In: 30- and 60-day forecast experiments with the ECMWF spectral models. Proc. ECMWF Workshop on Predictability in the Medium and Extended Range, Reading, United Kingdom, ECMWF, pp. 51–107.

Molteni, F., Sutera, A., Tronci, N., 1988. The EOFs of the geopotential eddies at 500 mb in winter and their probability density distributions. J. Atmos. Sci. 45, 3063–3080.

Molteni, F., Tibaldi, S., Palmer, T.N., 1990. Regimes in the wintertime circulation over northern extratropics. I: Observational evidence. Q. J. R. Meteorol. Soc. 116, 31–67. https://doi.org/10.1256/smsqj.49102.

Molteni, F., Buizza, R., Palmer, T.N., Petroliagis, T., 1996. The ECMWF ensemble prediction system: methodology and validation. Q. J. R. Meteorol. Soc. 122, 73–119.

Molteni, F., Stockdale, T., Balmaseda, M., Balsamo, G., Buizza, R., Ferranti, L., Magnusson, L., Mogensen, K., Palmer, T., Vitart, F., 2011. The new ECMWF Seasonal Forecast System (System 4). ECMWF Research Department Technical Memorandum n. 656, ECMWF, Shinfield Park, Reading, p. 51.

Molteni, F., Stockdale, T.N., Vitart, F., 2015. Understanding and modelling extra-tropical teleconnections with the Indo-Pacific region during the northern winter. Clim. Dyn. 45, 3119–3140.

Monaghan, A.J., Morin, C.W., Steinhoff, D.F., Wilhelmi, O., Hayden, M., Quattrochi, D.A., Reiskind, M., Lloyd, A.L., Smith, K., Schmidt, C.A., Scalf, P.E., Ernst, K., 2016. On the seasonal occurrence and abundance of the Zika virus vector mosquito Aedes Aegypti in the contiguous United States. PLoS Curr. Outbreaks. https://doi.org/10.1371/currents.outbreaks.50dfc7f46798675fc63e7d7da563da76.

Moncrieff, M.W., Klinker, E., 1997. Organized convective systems in the tropical western Pacific as a process in general circulation models. Q. J. R. Meteorol. Soc. 123, 805–828.

Moorthi, S., Suarez, M.J., 1992. Relaxed Arakawa–Schubert: a parameterization of moist convection for general circulation models. Mon. Weather Rev. 120, 978–1002.

Mori, M., Watanabe, M., 2008. The growth and triggering mechanisms of the PNA: a MJO-PNA coherence. J. Meteorol. Soc. Jpn. 86, 213–236.

Moron, V., Robertson, A., 2014. Interannual variability of Indian summer monsoon rainfall onset date at local scale. Int. J. Climatol. 34, 1050–1061.

Moron, V., Robertson, A., Ward, M., 2006. Seasonal predictability and spatial coherence of rainfall characteristics in the tropical setting of Senegal. Mon. Weather Rev. 134, 3248–3262.

Moron, V., Robertson, A.W., Ward, M.N., 2007. Spatial Coherence of tropical rainfall at Regional Scale. J. Clim. 20, 5244–5263.

Moron, V., Lucero, A., Hilario, F., Lyon, B., Robertson, A., DeWitt, D., 2009a. Spatiotemporal variability and predictability of summer monsoon onset over the Philippines. Clim. Dyn. 33, 1159–1177.

Moron, V., Robertson, A., Boer, R., 2009b. Spatial coherence and seasonal predictability of monsoon onset over Indonesia. J. Clim. 22, 840–850.

Moron, V., Robertson, A., Qian, J., 2010. Local versus regional-scale characteristics of monsoon onset and post-onset rainfall over Indonesia. Clim. Dyn. 34, 281–299.

Moron, V., Robertson, A., Ghil, M., 2012. Impact of the modulated annuam cycle and intraseasonal oscillation on daily-to-interannual rainfall variability across monsoonal India. Clim. Dyn. 38, 2409–2435.

Moron, V., Camberlin, P., Robertson, A., 2013. Extracting sub-seasonal scenarios: an alternative method to analyze seasonal predictability of regional-scale tropical rainfall. J. Clim. 26, 2580–2600.

Moron, V., Boyard-Micheau, J., Camberlin, P., Hernandez, V., Leclerc, C., Mwongera, C., Philippon, N., Fossa-Riglos, F., Sultan, B., 2015a. Ethnographic context and spatial coherence of climate indicators for farming communities—a multi-regional comparative assessment. Clim. Risk Manag. 8, 28–46.

Moron, V., Robertson, A.W., Qian, J.-H., Ghil, M., 2015b. Weather types across the Maritime Continent: from the diurnal cycle to interannual variations. Front. Environ. Sci. 2, 65.

Moron, V., Robertson, A., Pai, D., 2017. On the spatial coherence of sub-seasonal to seasonal Indian rainfall anomalies. Clim. Dyn. 49, 3403–3423.

Morss, R.E., Demuth, J.L., Lazo, J.K., 2008a. Communicating uncertainty in weather forecasts: a survey of the U.S. public. Weather Forecast. 23, 974–991.

Morss, R., Lazo, J., Brooks, H., Brown, B., Ganderton, P., Mills, B., 2008b. Societal and economic research and application priorities for the North American THORPEX programme. Bull. Am. Meteorol. Soc. 89 (3), 335–346.

Msadek, R., Vecchi, G.A., Winton, M., Gudgel, R.G., 2014. Importance of initial conditions in seasonal predictions of Arctic sea ice extent. Geophys. Res. Lett. 41 (14), 5208–5215.

Mudelsee, M., 2014. Climate Time Series Analysis. vol. 42. Atmospheric and Oceanographic Sciences Library.

Mueller, J.E., Gessner, B.D., 2010. A hypothetical explanatory model for meningococcal meningitis in the African meningitis belt. Int. J. Infect. Dis. 14, e553–e559.

Mueller, B., Seneviratne, S.I., 2012. Hot days induced by precipitation deficits at the global scale. Proc. Natl. Acad. Sci. 109, 12398–12403.

Mueller, B., et al., 2013. Benchmark products for land evapotranspiration: LandFlux-EVAL multi-data set synthesis. Hydrol. Earth Syst. Sci. 17, 3707–3720.

Mukougawa, H., 1988. A dynamical model of "quasi-stationary" states in large-scale atmospheric motions. J. Atmos. Sci. 45, 2868–2888. https://doi.org/10.1175/1520-0469(1988)045<2868:admoss>2.0.co;2.

Mukougawa, H., Hirooka, T., Kuroda, Y., 2009. Influence of stratospheric circulation on the predictability of the tropospheric Northern Annular Mode. Geophys. Res. Lett. 36, L08814. https://doi.org/10.1029/2008GL037127.

Müller, W.A., Appenzeller, C., Doblas-Reyes, F.J., Liniger, M.A., 2005. A debiased ranked probability skill score to evaluate probabilistic ensemble forecasts with small ensemble sizes. J. Clim. 18, 1513–1523.

Munich Re, 2011a. Press Release. https://www.munichre.com/en/media-relations/publications/press-releases/2011/2011-01-03-press-release/index.html.

Munich Re, 2011b, February. Topics Geo Natural Catastrophes 2010: Analyses, Assessments, Positions. Retrieved May 19, 2011, from, http://bit.ly/i5zbut.

Muñoz, Á., Yang, X., Vecchi, G., Robertson, A., Cooke, W., 2017. A weather-type based cross-timescale diagnostic framework for coupled circulation models. J. Clim. 30, 8951–8972. https://doi.org/10.1175/jcli-d-17-0115.1.

Murakami, T., Nakazawa, T., He, J., 1984. On the 40-50 dat oscillation during the 1979 Northern Hemisphere summer, Part I: Phase propagation. J. Meteorol. Soc. Jpn. 62, 440–467.

Murphy, A.H., 1977. The value of climatological, categorical and probabilistic forecasts in the costloss ratio situation. Mon. Weather Rev. 105, 803–816.

Murphy, A.H., 1988a. Skill scores based on the mean square error and their relationships to the correlation coefficient. Mon. Weather Rev. 116, 2417–2424.

Murphy, J.M., 1988b. The impact of ensemble forecasts on predictability. Q. J. R. Meteorol. Soc. 114, 463–493.

Murphy, A.H., 1993. What is a good forecast? An essay on the nature of goodness in weather forecasting. Weather Forecast. 8, 281–293.

Murphy, A.H., 1996. The Finley Affair: a signal event in the history of forecast verification. Weather Forecast. 11, 4–20.

Murphy, A.H., Winkler, R.L., 1987. A general framework for forecast verification. Mon. Weather Rev. 115, 1330–1338.

NAEFS, 2018. The North American Ensemble Forecasting System, see NAEFS web page hosted by NCEP. http://www.emc.ncep.noaa.gov/gmb/ens/NAEFS.html. and web page hosted by MSC Canada:http://weather.gc.ca/ensemble/naefs/index_e.html.

Nairn, J., Fawcett, R., 2013. Defining Heatwaves: Heatwave Defined as a Heat-Impact Event Servicing All Community and Business Sectors in Australia. CAWCR Technical Report 60, 84 pp.

Nairn, J.R., Fawcett, R., 2015. The excess heat factor: a metric for heatwave intensity and its use in classifying heatwave severity. Int. J. Environ. Res. Public Health 12, 227–253.

Naito, Y., Taguchi, M., Yoden, S., 2003. A parameter sweep experiment on the effects of the equatorial QBO on stratospheric sudden warming events. J. Atmos. Sci. https://doi.org/10.1175/1520-0469(2003)060<1380:APSEOT>2.0.CO;2.

Nakajima, T., Tsukamoto, M., Tsushima, Y., Numaguti, A., 1995. Modelling of the radiative processes in an AGCM. Clim. Syst. Dyn. Model. 3, 104–123.

Nakamura, N., 1995. Modified Lagrangian-mean diagnostics of the stratospheric polar vortices. Part I: Formulation and analysis in GFDL, SKYHI and GCM. J. Atmos. Sci. 52, 2096–2108.

Nakamura, H., Sampe, T., Goto, A., Ohfuchi, W., Xie, S.-P., 2008. On the importance of midlatitude oceanic frontal zones for the mean state and dominant variability in the tropospheric circulation. Geophys. Res. Lett. 35, L15709. https://doi.org/10.1029/ 2008GL034010.

Nakazawa, T., 1986. Intraseasonal variations of OLR in the tropics during the FGGE year. J. Meteorol. Soc. Jpn. 64, 17–34.

Nakazawa, T., 1998. Tropical super clusters within intraseasonal variations over the western Pacific. J. Meteorol. Soc. Jpn. 66, 823–829.

Nakazawa, T., Matsueda, M., 2017. Relationship between meteorological variables/dust and the number of meningitis cases in Burkina Faso. Meteorol. Appl. 24, 423–431.

Namias, J., 1962. In: Influences of abnormal surface heat sources and sinks on atmospheric behavior.Proc. Int. Symp. on Numerical Weather Prediction, Tokyo, Meteor. Soc. Japan, pp. 615–627.

Namias, J., 1963. In: Surface-atmosphere interactions as fundamental causes of drought and other climatic fluctuations.Proc. Rome Symposium on Changes of Climate, UNESCO, Paris, pp. 345–359.

Namias, J., 1968. Long-range weather forecasting: history, current status and outlook. Bull. Am. Meteorol. Soc. 49, 438–470.

NAS, 2016. National Academies of Sciences, Engineering, and Medicine. Next Generation Earth System Prediction: Strategies for Subseasonal to Seasonal Forecasts. The National Academies Press, Washington, DC.https://doi.org/10.17226/21873.

National Academies of Sciences, Engineering and Medicine, 2016. Next Generation Earth System Prediction: Strategies for Subseasonal to Seasonal Forecasts. National Academies of Sciences, Engineering, and Medicine, The National Academies Press, Washington.https://doi.org/10.17226/21873.

National Academies of Sciences, Engineering, and Medicine, 2016. Next Generation Earth System Prediction: Strategies for Subseasonal to Seasonal forecasts., p. 350. Washington, DC.

National Academy of Sciences, 2016. Next Generation Earth System Prediction: Strategies for Subseasonal to Seasonal Forecasts. The National Academy Press, Washington, DC, 351 pp.

National Research Council, 2010. Assessment of Intraseasonal to Interannual Climate Prediction and Predictability. Committee on Assessment of Intraseasonal to Interannual Climate Prediction and Predictability, Board on Atmospheric Sciences and Climate, Division on Earth and Life Studies, National Research Council. National Academies Press, Washington.https://www.nap.edu/catalog/12878/assessment-of-intraseasonal-to-interannual-climate-prediction-and-predictability. [(Accessed October 2016)].

Natural Resources Defense Council, 2016. Expanding Heat Resilient Cities Across India Ahmedabad 2016 Heat Action Plan. NRDC International, India.https://www.nrdc.org/resources/rising-temperatures-deadly-threat-preparing-communities-india-extreme-heat-events. [(Accessed 30 March 2017)].

Naumann, G., Vargas, W.M., 2010. Joint diagnostic of the surface air temperature in southern South America and the Madden–Julian oscillation. Weather Forecast. 25, 1275–1280.

Nayak, M.A., Villarini, G., Bradley, A.A., 2016. Atmospheric rivers and rainfall during NASA's Iowa Flood Studies (IFloodS) campaign. J. Hydrometeorol. 17, 257–271. https://doi.org/10.1175/JHM-D-14-0185.1.

Neal, R., Fereday, D., Crocker, R., Comer, R.E., 2016. A flexible approach to defining weather patterns and their application in weather forecasting over Europe. Meteorol. Appl. 23 (3), 389–400.

Neal, R., Dankers, R., Saulter, A., Lane, A., Millard, J., Evans, G., Price, D., 2018. The use of probabilistic medium to long-range weather pattern forecasts for identifying periods with an increases likelihood of coastal flooding around the UK. Meteorol. Appl. 1–14. https://doi.org/10.1002/met.11719.

Nearing, G.S., Gupta, H.V., 2015. The quantity and quality of information in hydrologic models. Water Resour. Res. 51, 524–538.

Neelin, J.D., Yu, H.-Y., 1994. Modes of tropical variability under convective adjustment and the Madden-Julian Oscillation. Part I: Analytical theory. J. Atmos. Sci. 51, 1876–1894.

Neelin, J., Battisti, D., Hirst, A., Jin, F.F., Wakata, Y., Yamagata, T., Zebiak, S., 1998. ENSO theory. J. Geophys. Res. 104 (C7), 14261–14290. https://doi.org/10.1029/97jc03424.

Neelin, J.D., Held, I.M., Cook, K.H., 1987. Evaporation-wind feedback and low-frequency variability in the tropical atmosphere. J. Atmos. Sci. 44, 2341–2348.

Neena, J.M., Lee, J.-Y., Waliser, D., Wang, B., Jiang, X., 2014. Predictability of the Madden-Julian Oscillation in the intraseasonal variability hindcast experiment (ISVHE). J. Clim. 27, 4531–4543.

Neena, J.M., Waliser, D., Jiang, X., 2017. Model performance metrics and process diagnostics for boreal summer intraseasonal variability. Clim. Dyn. 48, 1661–1683. https://doi.org/10.1007/s00382-016-3166-8.

Nehrkorn, T., Hoffman, R.N., Grassotti, C., Louis, J.-F., 2003. Feature calibration and alignment to represent model forecast errors: empirical regularization. Q. J. R. Meteorol. Soc. 129, 195–218.

Newman, P.A., Nash, E.R., 2005. The unusual southern hemisphere stratosphere winter of 2002. J. Atmos. Sci. 62, 614–628. https://doi.org/10.1175/JAS-3323.1.

Newman, P.A., Coy, L., Pawson, S., Lait, L.R., 2016. The anomalous change in the QBO in 2015-16: The anomalous change in the 2015-16 QBO. Geophys. Res. Lett. 43.

Newman, M., Sardeshmukh, P.D., Winkler, C.R., Whitaker, J.S., 2003. A study of subseasonal predictability. Mon. Weather Rev. 131, 1715–1732.

Nie, J., Sobel, A.H., 2015. Responses of tropical deep convection to the QBO: cloud-resolving simulations. J. Atmos. Sci. 72, 3625–3638. https://doi.org/10.1175/JAS-D-15-0035.1.

Nie, Y., Zhang, Y., Chen, G., Yang, X.-Q., Burrows, D.A., 2014. Quantifying barotropic and baroclinic eddy feedbacks in the persistence of the Southern Annular Mode. Geophys. Res. Lett. 41, 8636–8644. https://doi.org/10.1002/2014GL062210.

Nishii, K., Nakamura, H., Miyasaka, T., 2009. Modulations in the planetary wave field induced by upward-propagating Rossby wave packets prior to stratospheric sudden warming events: a case-study. Q. J. R. Meteorol. Soc. 135, 39–52. https://doi.org/10.1002/qj.359.

Nishimoto, E., Yoden, S., 2017. Influence of the stratospheric Quasi-Biennial Oscillation on the Madden-Julian Oscillation during austral summer. J. Atmos. Sci. https://doi.org/10.1175/JAS-D-16-0205.1 JAS-D-16-0205.1.

Nissan, H., Burkart, K., Mason, S.J., Coughlan de Perez, E., van Aalst, M., 2017. Defining and predicting heat waves in Bangladesh. J. Appl. Meteorol. Climatol. https://doi.org/10.1175/JAMC-D-17-0035.1 in press.

Nitsche, G., Wallace, J.M., Kooperberg, C., 1994. Is there evidence of multiple equilibria in planetary wave amplitude statistics? J. Atmos. Sci. 51, 314–322. https://doi.org/10.1175/1520-0469(1994) 051<0314:iteome>2.0.co;2.

Niu, G.-Y., et al., 2011. The community Noah land surface model with multiparameterization options (Noah-MP): 1. Model description and evaluation with local-scale measurements. J. Geophys. Res. 116, D12109.

NOAA, 2013. Billion-Dollar Weather and Climate Disasters. National Oceanic and Atmospheric Administration, National Climatic Data Center.

Noda, A.T., et al., 2010. Importance of the subgrid-scale turbulent moist process: cloud distribution in global cloud-resolving simulations. Atmos. Res. 96, 208–217.

Noguchi, S., Mukougawa, H., Kuroda, Y., Mizuta, R., Yabu, S., Yoshimura, H., 2016. Predictability of the stratospheric polar vortex breakdown: an ensemble reforecast experiment for the splitting event in January 2009. J. Geophys. Res. Atmos. 121, 3388–3404. https://doi.org/10.1002/2015JD024581.

Noh, Y., Kim, H.J., 1999. Simulations of temperature and turbulence structure of the oceanic boundary layer with the improved near surface process. J. Geophys. Res. 104, 15621–15634.

Noilhan, J., Mahfouf, J.-F., 1996. The ISBA land surface parameterization scheme. Glob. Planet. Chang. 13, 145–159.

Noilhan, J., Planton, S., 1989. A simple parameterization of land surface processes for meteorological models. Mon. Weather Rev. 117, 536–549.

Nonaka, M., Sasai, Y., Sasaki, H., Taguchi, B., Nakamura, H., 2016. How potentially predictable are midlatitude ocean currents? Sci. Rep. 6, 20153. https://doi.org/10.1038/srep20153.

North, G.R., 1984. Empirical orthogonal functions and normal modes. J. Atmos. Sci. 41, 879–887.

Norton, W.a., 2003. Sensitivity of northern hemisphere surface climate to simulation of the stratospheric polar vortex. Geophys. Res. Lett. https://doi.org/10.1029/2003GL016958.

Norton, A.J., Rayner, P.J., Koffi, E.N., Scholze, M., 2017. Assimilating solar-induced chlorophyll fluorescence into the terrestrial biosphere model BETHY-SCOPE: model description and information content. Geosci. Model Dev. Discuss., 1–26.

Notz, D., Bitz, C., 2017. Sea ice in Earth system models. In: Thomas, D.N. (Ed.), Sea Ice. John Wiley & Sons.

Notz, D., Haumann, F.A., Haak, H., Jungclaus, J.H., Marotzke, J., 2013. Sea ice evolution in the Arctic as modeled by MPI-ESM. J. Adv. Model. Earth Syst. https://doi.org/10.1002/jame.20016.

O'Connor, R.E., Yarnal, B., Dow, K., Jocoy, C.L., Carbone, G.L., 2005. Feeling at risk matters: water managers and decision to use forecasts. Risk Anal. 25 (5), 1265–1275.

O'Neill, A., Oatley, C.L., Charlton-Perez, A.J., Mitchell, D.M., Jung, T., 2017. Vortex splitting on a planetary scale in the stratosphere by cyclogenesis on a subplanetary scale in the troposphere. Q. J. R. Meteorol. Soc. 143, 691–705. https://doi.org/10.1002/qj.2957.

O'Reilly, C.H., Czaja, A., 2015. The response of the Pacific storm track and atmospheric circulation to Kuroshio Extension variability. Q. J. R. Meteorol. Soc. 141, 52–66. https://doi.org/10.1002/ qj.2334.

O'Sullivan, L., Jardine, A., Cook, A., Weinstein, P., 2008. Deforestation, mosquitoes, and ancient Rome: lessons for today. Bioscience 58, 756–760.

Obled, C., Bontron, G., Gar¸con, R., 2002. Quantitative precipitation forecasts: a statistical adaptation of model outputs through an analogues sorting approach. Atmos. Res. 63 (3-4), 303–324.

Oglesby, R.J., Erickson III, D.J., 1989. Soil moisture and persistence of North American drought. J. Clim. 2, 1362–1380.

Oleson, K.W., Bonan, G.B., Schaaf, C., Gao, F., Jin, Y., Strahler, A., 2003. Assessment of global climate model land surface albedo using MODIS data. Geophys. Res. Lett. 30, 1443.

Omumbo, J.A., Noor, A.M., Fall, I.S., Snow, R.W., 2013. How well are malaria maps used to design and finance malaria control in Africa? PLoS ONE 8, e53198.

Oouchi, K., Noda, A.T., Satoh, M., Miura, H., Tomita, H., Nasuno, T., Iga, S., 2009. A simulated preconditioning of typhoon genesis controlled by a boreal summer Madden-Julian Oscillation event in a global cloud-system-resolving model. SOLA 5, 65–68.

Orlanski, I., 1975. A rationale subdivision of scales for atmospheric processes. Bull. Am. Meteorol. Soc. 56, 527–530.

Orsolini, Y.J., Senan, R., Balsamo, G., Doblas-Reyes, F.J., Vitart, F., Weisheimer, A., Carrasco, A., Benestad, R.E., 2013. Impact of snow initialization on sub-seasonal forecasts. Clim. Dyn. 41, 1969–1982. https://doi.org/10.1007/s00382-013-1782-0.

Orsolini, Y.J., Senan, R., Vitart, F., Balsamo, G., Weisheimer, A., Doblas-Reyes, F.J., 2016. Influence of the Eurasian snow on the negative North Atlantic Oscillation in subseasonal forecasts of the cold winter 2009/2010. Clim. Dyn. 47, 1325–1334. https://doi.org/10.1007/s00382-015-2903-8.

Orszag, S.A., 1969. Numerical methods for the simulation of turbulence. Phys. Fluids 12 (Suppl. II), 250–257.

Osprey, S.M., Gray, L.J., Hardiman, S.C., Butchart, N., Hinton, T.J., 2013. Stratospheric variability in twentieth-century CMIP5 simulations of the met office climate model: high top versus low top. J. Clim. 26, 1595–1606. https://doi.org/10.1175/JCLI-D-12-00147.1.

Osprey, S.M., Butchart, N., Knight, J.R., Scaife, A.A., Hamilton, K., Anstey, J.A., Schenzinger, V., Zhang, C., 2016. An unexpected disruption of the atmospheric quasi-biennial oscillation. Science 353, 1424–1427. https://doi.org/10.1126/science.aah4156.

Otkin, J.A., Svoboda, M., Hunt, E.D., Ford, T.W., Anderson, M.C., Hain, C., Basara, J.B., 2017. Flash droughts: a review and assessment of the challenges imposed by rapid onset droughts in the United States. Bull. Am. Meteor. Soc. 99, 911–919. https://doi.org/10.1175/BAMS-D-17-0149.1.

Otto, J., Brown, C., Buontempo, C., Doblas-Reues, F., Jacob, D., Juckes, M., Keup-Thiel, E., Kurnik, B., Schulz, J., Taylor, A., Verhoelst, T., Walton, P., 2016. Uncertainty: lessons learned for climate services. Bull. Am. Meteorol. Soc. 97 (12), ES265–ES269. https://doi.org/10.1175/bams-d-16-0173.1.

Overland, J., Francis, J., Hall, R., Hanna, E., Kim, S.J., Vihma, T., 2015. The melting Arctic and midlatitude weather patterns: are they connected? J. Clim. 28 (20), 7917–7932.

Owen, J.A., Palmer, T.N., 1987. The impact of El-Niño on an ensemble of extended-range forecasts. Mon. Weather Rev. 115, 2103–2117.

Paaijmans, K.P., Wandago, M.O., Githeko, A.K., Takken, W., 2007. Unexpected high losses of *Anopheles gambiae* larvae due to rainfall. PLoS ONE 2, e1146. https://doi.org/10.1371/journal.pone.0001146.

Padmanabha, H., Soto, E., Mosquera, M., Lord, C., Lounibos, L., 2010. Ecological links between water storage behaviors and *aedes aegypti* production: implications for dengue vector control in variable climates. EcoHealth 7, 78–90.

Pai, D., Sridhar, L., Rajeevan, M., Sreejith, O., Satbhai, N., Mukhopadhyay, B., 2014. Development of a new high spatial resolution (0.25 $\times$ 0.25) long period (1901–2010) daily gridded rainfall data set over India and its comparison with existing data sets over the region. Mausam 65, 1–18.

Palmer, T.N., 1999. A nonlinear dynamical perspective on climate prediction. J. Clim. 12, 575–591.

Palmer, T.N., 2012. Towards the probabilistic Earth-system simulator: a vision for the future of climate and weather prediction. Q. J. R. Meteorol. Soc. 138 (665), 841–861.

Palmer, T.N., Williams, P. (Eds.), 2009. Stochastic Physics and Climate Modeling. Cambridge University Press, Cambridge, UK.

Palmer, T.N., Brankovic, C., Molteni, F., Tibaldi, S., 1990. Extended range predictions with ECMWF models. I: Interannual variability in operational model integrations. Q. J. R. Meteorol. Soc. 116, 799–834.

Palmer, T.N., Molteni, F., Mureau, R., Buizza, R., Chapelet, P., Tibbia, J., 1993. In: Ensemble prediction. Proc. ECMWF Seminar Proc on Validation of Models over Europe. vol. 1. ECMWF, Shinfield Park, Reading, pp. 21–66.

Palmer, T.N., et al., 2004. Development of a European multimodel ensemble system for seasonal- to-interannual prediction (DEMETER). Bull. Am. Meteorol. Soc. 85, 853–872.

Palmer, T.N., Buizza, R., Doblas-Reyes, F., Jung, T., Leutbecher, M., Shutts, G.J., Steinheimer, M., Weisheimer, A., 2009. Stochastic parametrization and model uncertainty. In: ECMWF Research Department Technical Memorandum No. 598, p. 42. Available from ECMWF, Shinfield Park, Reading RG2-9AX, UK.

Paltan, H., Waliser, D., Lim, W.H., Guan, B., Yamazaki, D., Pant, R., Dadson, S., 2017. Global floods and water availability driven by atmospheric rivers. Geophys. Res. Lett. https://doi.org/10.1002/2017GL074882.

Pandya, R., et al., 2015. Using weather forecasts to help manage meningitis in the West African Sahel. Bull. Am. Meteorol. Soc. 96, 103–115.

Pang, B., Chen, Z., Wen, Z., Lu, R., 2016. Impacts of two types of El Niño on the MJO during boreal winter. Adv. Atmos. Sci. 33, 979–986.

Paolino, C., Methol, M., Quintans, D., 2010. Estimación del impacto de una eventual sequía en la ganadería nacional y bases para el diseño de políticas de seguros. In: OPYPA-Yearbook. MGAP. www.mgap.gub.uy/sites/default/files/anuario2010.zip.

Park, Y.Y., Buizza, R., Leutbecher, M., 2008. TIGGE: preliminary results on comparing and combining ensembles. Q. J. R. Meteorol. Soc. 134, 2029–2050.

Parkinson, C.L., Cavalieri, D.J., 2012. Antarctic sea ice variability and trends, 1979–2010. Cryosphere 6 (4), 871.

Parrish, D.F., Derber, J.C., 1992. The National Meteorological Center's spectral statistical interpolation analysis system. Monthly Weather Rev. 120 (8), 1747–1763. https://doi.org/10.1175/1520-0493(1992)120<1747:TNMCSS>2.0.CO;2.

Pascual, M., Bouma, M.J., Dobson, A.P., 2002. Cholera and climate: revisiting the quantitative evidence. Microbes Infect. 4, 237–245.

Patricola, C.M., Li, M.K., Xu, Z., Chang, P., Saravanan, R., Hsieh, J.S., 2012. An investigation of tropical Atlantic bias in a high-resolution coupled regional climate model. Clim. Dyn. 39, 2443–2463. https://doi.org/10.1007/s00382-012-1320-5.

Pattanaik, D.R., Kumar, A., 2014. Comparison of intra-seasonal forecast of Indian summer monsoon between two versions of NCEP coupled models. Theor. Appl. Climatol. 118, 331–345. https://doi.org/10.1007/s00704-013-1071-1.

Pauluis, O., Garner, S., 2006. Sensitivity of radiative-convective equilibrium simulations to horizontal resolution. J. Atmos. Sci. 63 (7), 1910–1923.

Pavan, W., Fraisse, C.W., Peres, N.A., 2011. Development of a web-based disease forecasting system for strawberries. Comput. Electron. Agric. 71 (1), 169–175.

Peatman, S.C., Matthews, A.J., Stevens, D.P., 2014. Propagation of the Madden-Julian Oscillation through the Maritime Continent and scale interaction with the diurnal cycle of precipitation. Q. J. R. Meteorol. Soc. 140, 814–825.

Pedlosky, J., 1987. Geophysical Fluid Dynamics, second ed. Springer, New York.

Peings, Y., Brun, E., Mauvais, V., Douville, H., 2013. How stationary is the relationship between Siberian snow and Arctic Oscillation over the 20th century? Geophys. Res. Lett. 40, 183–188. https://doi.org/10.1029/2012GL054083.

Pellerin, P., Benoit, R., Kouwen, N., Ritchie, H., Donaldson, N., Joe, P., Soulis, R., 2002. On the use of coupled atmospheric and hydrologic models at regional scale. In: High Performance Computing Systems and Applications. Springer US, pp. 317–322.

Peña, M., Toth, Z., 2014. Estimation of analysis and forecast error variances. Tellus A 66, 21767. https://doi.org/10.3402/tellusa.v66.21767.

Penland, C., 1989. Random forcing and forecasting using principal oscillation pattern analysis. Mon. Weather Rev. 117, 2165–2185.

Penland, C., 1996. A stochastic model of IndoPacific sea surface temperature anomalies. Physica D 98, 534–558. https://doi.org/10.1016/0167-2789(96)00124-8.

Penland, C., Ghil, M., 1993. Forecasting Northern Hemisphere 700-mb geopotential height anomalies using empirical normal modes. Mon. Weather Rev. 121, 2355–2372. https://doi.org/10.1175/1520-0493(1993) 121<2355:fnhmgh>2.0.co;2.

Penland, C., Sardeshmukh, P.D., 1995. The optimal growth of tropical sea surface temperature anomalies. J. Clim. 8, 1999–2024. https://doi.org/10.1175/1520-0442(1995)008<1999:togots>2.0.co;2.

Pérez García-Pando, C., et al., 2014. Soil dust aerosols and wind as predictors of seasonal meningitis incidence in Niger. Environ. Health Perspect. 122, 679–686.

Pérez, G.-P., et al., 2014. Meningitis and climate: from science to practice. Earth Perspect. 1, 14.

Perkins, S.E., 2015. A review on the scientific understanding of heat waves—their measurement, driving mechanisms, and changes at the global scale. Atmos. Res. 164–165 (2015), 242–267.

Perlwitz, J., Graf, H.-F., 1995. The statistical connection between tropospheric and stratospheric circulation of the Northern Hemisphere in winter. J. Clim. 8, 2281–2295. https://doi.org/10.1175/1520-0442(1995)008<2281: TSCBTA>2.0.CO;2.

Perlwitz, J., Harnik, N., 2003. Observational evidence of a stratospheric influence on the troposphere by planetary wave reflection. J. Clim. 16, 3011–3026. https://doi.org/10.1175/1520-0442(2003)016<3011:OEOASI>2.0.CO;2.

Persson, A., 2005. Early operational numerical weather prediction outside the USA: an historical introduction. Part 1: Internationalism and engineering NWP in Sweden, 1952–69. Meterol. Appl. 12, 135–159.

Persson, O., Vihma, T., 2017. The atmosphere over sea ice. In: Thomas, D.N. (Ed.), Sea ice. John Wiley & Sons.

Peters-Lidard, C.D., et al., 2007. High performance earth system modeling with NASA/GSFC's Land Information System. Innov. Syst. Softw. Eng. 3, 157.

Peterson, K.A., Arribas, A., Hewitt, H.T., Keen, A.B., Lea, D.J., McLaren, A.J., 2015. Assessing the forecast skill of Arctic sea ice extent in the GloSea4 seasonal prediction system. Clim. Dyn. 44 (1-2), 147–162.

Petoukhov, V., Semenov, V.A., 2010. A link between reduced Barents-Kara sea ice and cold winter extremes over northern continents. J. Geophys. Res. 115, D21111. https://doi.org/10.1029/2009JD013568.

Petoukhov, V., Rahmstorf, S., Petri, S., Schellnhuber, H.J., 2013. Quasi-resonant amplification of planetary waves and recent Northern Hemisphere weather extremes. PNAS 110, 5336–5341.

Petrich, C., Eicken, H., 2017. Growth, structure and properties of sea ice. In: Thomas, D.N. (Ed.), Sea Ice. In: vol. 2. John Wiley & Sons.

Pfister, L., Savenije, H.H.G., Fenicia, F., 2009. Leonardo da Vinci's water theory; on the origin and date of water. IAHS Spec. Publ. 9, 94 pp.

Phillips, N.A., 1956. The general circulation of the atmosphere: a numerical experiment. Q. J. R. Meteorol. Soc. 82, 123–154.

Piazza, M., Terray, L., Boé, J., Maisonnave, E., Sanchez-Gomez, E., 2016. Influence of small-scale North Atlantic sea surface temperature patterns on the marine boundary layer and free troposphere: a study using the atmospheric ARPEGE model. Clim. Dyn. 46, 1699–1717. https://doi.org/10.1007/ s00382-015-2669-z.

Pielke Jr., R., Carbone, R.E., 2002. Weather, impacts, forecasts, and policy: an integrated perspective. Bull. Am. Meteorol. Soc. 83 (3), 393–403.

Pisciottano, G., Díaz, A., Cazes, G., Mechoso, C.R., 1994. El Niño-Southern oscillation impact on rainfall in Uruguay. J. Clim. 7, 1286–1302.

Plaut, G., Vautard, R., 1994. Spells of low-frequency oscillations and weather regimes in the Northern Hemisphere. J. Atmos. Sci. 51, 210–236. https://doi.org/10.1175/1520-0469(1994)051<0210: solfoa>2.0.co;2.

Plumb, R.A., 1981. Instability of the distorted polar night vortex: a theory of stratospheric warmings. J. Atmos. Sci. 38, 2514–2531. https://doi.org/10.1175/1520-0469(1981)038<2514:IOTDPN>2.0.CO;2.

Plumb, R.A., Semeniuk, K., 2003. Downward migration of extratropical zonal wind anomalies. J. Geophys. Res. 108, 4223. https://doi.org/10.1029/2002JD002773.

Pohl, B., Camberlin, P., 2006. Influence of the Madden-Julian oscillation on East-African rainfall. Part I: Intraseasonal variability and regional dependency. Q. J. R. Meteorol. Soc. 132, 2521–2539.

Pohl, B., Camberlin, P., 2006a. Influence of the Madden-Julian Oscillation on East African rainfall, Part I: Intraseaonal variability and regional dependency. Q. J. R. Meteorol. Soc. 132, 2521–2539.

Pohl, B., Camberlin, P., 2006b. Influence of the Madden-Julian Oscillation on East African rainfall, Part II: March-May seasonal extremes and interannual variability. Q. J. R. Meteorol. Soc. 132, 2541–2558.

Pohl, B., Janicot, S., Fontaine, B., Marteau, R., 2009. Implication of the Madden-Julian Oscillation in the 40-50 day variability of the monsoon. J. Clim. 22, 3769–3785.

Polvani, L.M., Kushner, P.J., 2002. Tropospheric response to stratospheric perturbations in a relatively simple general circulation model. Geophys. Res. Lett. 29, 1114. https://doi.org/10.1029/2001GL014284.

Polvani, L.M., Smith, K.L., 2013. Can natural variability explain observed Antarctic sea ice trends? New modeling evidence from CMIP5. Geophys. Res. Lett. 40 (12), 3195–3199.

Polvani, L.M., Waugh, D.W., Correa, G.J.P., Son, S.-W., 2011. Stratospheric ozone depletion: the main driver of twentieth-century atmospheric circulation changes in the Southern Hemisphere. J. Clim. 24, 795–812. https://doi.org/10.1175/2010JCLI3772.1.

Polvani, L.M., Sun, L., Butler, A.H., Richter, J.H., Deser, C., 2017. Distinguishing stratospheric sudden warmings from ENSO as key drivers of wintertime climate variability over the North Atlantic and Eurasia. J. Clim. 30, 1959–1970.

Prates, F., Buizza, R., 2011. PRET, the Probability of RETurn: a new probabilistic product based on generalized extreme-value theory. Q. J. R. Meteorol. Soc. 137, 521–537. https://doi.org/10.1002/qj.759.

Price Waterhouse Coopers, 2011. Protecting Human Health and Safety During Severe and Extreme Heat Events: A National Framework. Commonwealth Government Report, Australia.

Privé, N., Errico, R.M., 2015. Spectral analysis of forecast error investigated with an observing system simulation experiment. Tellus 67. https://doi.org/10.3402/tellusa.v67.25977.

Prodhomme, C., Doblas-Reyes, F., Bellprat, O., Dutra, E., 2016. Impact of land-surface initialization on sub-seasonal to seasonal forecasts over Europe. Clim. Dyn. 47, 919–935. https://doi.org/10.1007/s00382-015-2879-4.

Prokopy, L.S., Haigh, T., Mase, A.S., Angel, J., Hart, C., Knutson, C., Lemos, M.C., Lo, Y., McGuire, J., Wright Morton, L., Perron, J., Todey, D., Widhalm, M., 2013. Agricultural advisors: a receptive audience for weather and climate information? Weather Clim. Soc. 5 (2), 162–167.

Proshutinsky, A., Aksenov, Y., Kinney, J.C., Gerdes, R., Golubeva, E., Holland, D., et al., 2011. Recent advances in Arctic ocean studies employing models from the Arctic Ocean Model Intercomparison Project.

Public Health England, NHS England, Local Government Association & UK Met Office, 2015. Heatwave Plan for England. Tech. Rep. PHE publications gateway number: 2015049.

Putrasahan, D.A., Miller, A.J., Seo, H., 2013. Isolating meso-scale coupled ocean–atmosphere interactions in the Kuroshio Extension region. Dyn. Atmos. Oceans 63, 60–78. https://doi.org/10.1016/ j.dynatmoce.2013.04.001.

PytlikZillig, L.M., Hu, Q., Hubbard, K.G., Lynne, G.D., Bruning, R.H., 2010. Improving farmers' perception and use of climate predictions in farming decisions: a transition model. J. Appl. Meteorol. Climatol. 49 (6), 1333–1340.

Qian, J.-H., Robertson, A.W., Moron, V., 2010. Interactions among ENSO, the monsoon, and diurnal cycle in rainfall variability over Java. J. Atmos. Sci. 67, 3509–3524.

Qiu, B., Chen, S., Schneider, N., Taguchi, B., 2014. A coupled decadal prediction of the dynamic state of the Kuroshio Extension system. J. Clim. 27, 1751–1764. https://doi.org/10.1175/JCLI-D-13-00318.1.

Quaife, T., Lewis, P., De Kauwe, M., Williams, M., Law, B.E., Disney, M., Bowyer, P., 2008. Assimilating canopy reflectance data into an ecosystem model with an Ensemble Kalman Filter. Remote Sens. Environ. 112, 1347–1364.

Quesada, B., Vautard, R., Yiou, P., Hirschi, M., Seneviratne, S.I., 2012. Asymmetric European summer heat predictability from wet and dry southern winters and springs. Nat. Clim. Chang. 2 (10), 736–741.

Quiroz, R.S., 1986. The association of stratospheric warmings with tropospheric blocking. J. Geophys. Res. 91, 5277. https://doi.org/10.1029/JD091iD04p05277.

R Core Team, 2017. R: A Language and Environment for Statistical Computing. R Foundation for Statistical Computing, Vienna, Austria.

Rajagopalan, B., Lall, U., Zebiak, S.E., 2002. Categorical climate forecasts through regularization and optimal combination of multiple gcm ensembles. Mon. Weather Rev. 130 (7), 1792–1811.

Rajeevan, M., Gadgil, S., Bhate, J., 2010. Active and break spells of the Indian summer monsoon. J. Earth Syst. Sci. 119, 229–247. https://doi.org/10.1007/s12040-010-0019-4.

Ralph, F.M., Dettinger, M.D., 2011. Storms, floods, and the science of atmospheric rivers. Eos. Trans. AGU 92, 265. https://doi.org/10.1029/2011EO320001.

Ralph, F.M., Dettinger, M.D., 2012. Historical and national perspectives on extreme west coast precipitation associated with atmospheric rivers during December 2012. Bull. Am. Meteorol. Soc. 93, 783–790. https://doi.org/ 10.1175/BAMS-D-11-00188.1.

Ralph, F.M., Neiman, P.J., Wick, G.A., 2004. Satellite and CALJET aircraft observations of atmospheric rivers over the eastern North-Pacific Ocean during the El Niño winter of 1997/98. Mon. Weather Rev. 132, 1721–1745. https://doi. org/10.1175/1520-0493(2004)132<1721:SACAOO>2.0.CO;2.

Ralph, F.M., Neiman, P.J., Wick, G.A., Gutman, S.I., Dettinger, M.D., Cayan, D.R., White, A.B., 2006. Flooding on California's Russian River: role of atmospheric rivers. Geophys. Res. Lett. 33, L13801. https://doi.org/10.1029/ 2006GL026689.

Ralph, F.M., et al., 2016. CalWater field studies designed to quantify the roles of atmospheric rivers and aerosols in modulating U.S. west coast precipitation in a changing climate. Bull. Am. Meteorol. Soc. 97, 1209–1228. https:// doi.org/10.1175/BAMS-D-14-00043.1.

Ramos, M.H., van Andel, S.J., Pappenberger, F., 2013. Do probabilistic forecasts lead to better decisions? Hydrol. Earth Syst. Sci. 17, 2219–2232.

Rampal, P., Weiss, J., Marsan, D., Lindsay, R., Stern, H., 2008. Scaling properties of sea ice deformation from buoy dispersion analysis. J. Geophys. Res. Oceans. 113(C3).

Randall, D.A., 2013. Beyond deadlock. Geophys. Res. Lett. 40, 5970–5976.

Randall, D., Khairoutdinov, M., Arakawa, A., Grabowski, W., 2003. Breaking the cloud parameterization deadlock. Bull. Am. Meteorol. Soc. 84, 1547–1564.

Rashid, H.A., Hendon, H.H., Wheeler, M.C., Alves, O., 2011. Prediction of the Madden–Julian oscillation with the POAMA dynamical prediction system. Clim. Dyn. 36, 649–661. https://doi.org/10.1007/s00382-010-0754-x.

Rasmusson, E., Carpenter, T., 1982. Variations in tropical sea surface temperature and surface wind fields associated with the Southern Oscillation El Nino. Mon. Weather Rev. 111, 517–528.

Rauhala, J., Schultz, D.M., 2009. Severe thunderstorm and tornado warnings in Europe. Atmos. Res. 93, 369–380. https://doi.org/10.1016/j.atmosres. 2008.09.026.

Ray, P., Zhang, C., 2010. A case study of the mechanisms of extratropical influence on the initiation of the Madden-Julian oscillation. J. Atmos. Sci. 67, 515–528. https://doi.org/10.1175/2009JAS3059.1.

Raymond, D.J., 2001. A new model of the Madden–Julian Oscillation. J. Atmos. Sci. 58, 2807–2819.

Raymond, D.J., Fuchs, Z., 2009. Moisture modes and the Madden-Julian Oscillation. J. Clim. 22, 3031–3046.

Rayner, S., Lach, D., Ingram, H., 2005. Weather forecasts are for wimps: why water resource managers do not use climate forecasts. Clim. Chang. 69, 197–227.

Redelsperger, J.L., Thorncroft, C., Diedhiou, A., Lebel, T., Parker, D.J., Polcher, J., 2006. African Monsoon Multidisciplinary Analysis (AMMA): an international research project and field campaign. Bull. Am. Meteorol. Soc. 87, 1739–1746.

Reichle, R., 2008. Data assimilation methods in the Earth sciences. Adv. Water Resour. 31, 1411–1418.

Reichle, R.H., Koster, R.D., Liu, P., Mahanama, S.P.P., Njoku, E.G., Owe, M., 2007. Comparison and assimilation of global soil moisture retrievals from the Advanced Microwave Scanning Radiometer for the Earth Observing System (AMSR-E) and the Scanning Multichannel Microwave Radiometer (SMMR). J. Geophys. Res. 112, D09108.

Reichle, R.H., Crow, W.T., Koster, R.D., Sharif, H.O., Mahanama, S.P.P., 2008. Contribution of soil moisture retrievals to land data assimilation products. Geophys. Res. Lett. 35, L01404.

Reid, G.C., Gage, K.S., 1985. Interannual variations in the height of the tropical tropopause. J. Geophys. Res. 90, 5629. https://doi.org/10.1029/JD090iD03p05629.

Reinhold, B., Pierrehumbert, R., 1982. Dynamics of weather regimes: quasi-stationary waves and blocking. Mon. Weather Rev. 110, 1105–1145. https://doi.org/10.1175/1520-0493(1982)110<1105:DOWRQS>2.0.CO;2.

Renggli, D., Leckebusch, G.C., Ulbrich, U., Gleixner, S.N., Faust, E., 2011. The skill of seasonal ensemble prediction systems to forecast wintertime windstorm frequency over the North Atlantic and Europe. Mon. Weather Rev. 139, 3052–3068.

Reynolds, R.W., Smith, T.M., Liu, C., et al., 2007. Daily high-resolution-blended analyses for sea surface temperature. J. Clim. 20, 5473–5496. https://doi.org/10.1175/2007JCLI1824.1.

Ricciardulli, L., Sardeshmukh, P., 2002. Local time- and space scales of organized tropical deep convection. J. Clim. 15, 2775–2790.

Richardson, L.F., 1922. Weather Prediction by Numerical Process. Cambridge University Press, Cambridge, MA. Reprinted in 2006 by Cambridge University Press with a new introduction by Peter Lynch.

Richardson, D.S., 2000. Skill and economic value of the ECMWF Ensemble Prediction System. Q. J. R. Meteorol. Soc. 126, 649–668.

Richardson, D.S., 2001. Measures of skill and value of ensemble prediction systems, their interrelationship and the effect of ensemble size. Q. J. R. Meteorol. Soc. 127, 2473–2489.

Richardson, D., Bidlot, J., Ferranti, L., Haiden, T., Hewson, T., Janousek, M., Prates, F., Vitart, F., 2013. Evaluation of ECMWF Forecasts, Including 2012–2013 Upgrades. Tech. rep ECMWF Technical Memo, Reading.

Richter, J.H., Matthes, K., Calvo, N., Gray, L.J., 2011. Influence of the quasi-biennial oscillation and El Niño–Southern Oscillation on the frequency of sudden stratospheric warmings. J. Geophys. Res. 116, D20111. https://doi.org/10.1029/2011JD015757.

Richter, J.H., Deser, C., Sun, L., 2015. Effects of stratospheric variability on El Niño teleconnections. Environ. Res. Lett. 10, 124021. https://doi.org/10.1088/1748-9326/10/12/124021.

Riddle, E.E., Butler, A.H., Furtado, J.C., Cohen, J.L., Kumar, A., 2013. CFSv2 ensemble prediction of the wintertime Arctic Oscillation. https://doi.org/10.1007/s00382-013-1850-5.

Rieck, M., Hohenegger, C., Gentine, P., 2015. The effect of moist convection on thermally induced mesoscale circulations. Q. J. R. Meteorol. Soc. 141, 2418–2428.

Riesz, N., Sz-Nagy, B., 1953. Functional Analysis. Frederick Ungar Publishing, New York.

Ring, M.J., Plumb, R.A., 2008. The response of a simplified GCM to axisymmetric forcings: applicability of the fluctuation–dissipation theorem. J. Atmos. Sci. 65, 3880–3898. https://doi.org/10.1175/2008JAS2773.1.

Ripa, P., 1983. General stability conditions for zonal flows in a one-layer model on the β-plane or the sphere. J. Fluid Mech. 126, 463–489.
Rivera, J.A., Penalba, O.C., 2014. Trends and spatial patterns of drought affected area in Southern South America. Climate 2, 264–278. https://doi.org/10.3390/cli2040264.
Rivest, C., Farrell, B.F., 1992. Upper-tropospheric synoptic-scale waves. Part II: Maintenance and excitation of quasi-modes. J. Atmos. Sci. 49, 2120–2138.
Roads, J.O., 1986. Forecasts of time averages with a numerical weather prediction model. J. Atmos. Sci. 43, 871–892.
Robbins, J.C., Titley, H.A., 2018. Evaluating high-impact weather forecasts from the Met Office Global Hazard Map using a global impact database. Meteorol. Appl. in press.
Roberts, N.M., Lean, H.W., 2008. Scale-selective verification of rainfall accumulations from high-resolution forecasts of convective events. Mon. Weather Rev. 136, 78–97.
Robertson, A.W., Metz, W., 1989. Three-dimensional linear instability of persistent anomalous large-scale flows. J. Atmos. Sci. 46, 2783–2801.
Robertson, A.W., Kumar, A., Peña, M., Vitart, F., 2015. Improving and promoting subseasonal to seasonal prediction. Bull. Am. Meteorol. Soc. 96 (3), ES49–53.
Robine, J.-M., Cheung, S.L.K., Le Roy, S., Van Oyen, H., Griffiths, C., Michel, J.-P., Herrmann, F.R., 2008. Death toll exceeded 70,000 in Europe during the summer of 2003. C. R. Biol. 331 (2), 171–178.
Robinson, W.A., 1988. Irreversible wave–mean flow interactions in a mechanistic model of the stratosphere. J. Atmos. Sci. 45, 3413–3430. https://doi.org/10.1175/1520-0469(1988)045<3413:IWFIIA>2.0.CO;2.
Robinson, W.A., 1991. The dynamics of the zonal index in a simple model of the atmosphere. Tellus A. https://doi.org/10.3402/tellusa.v43i5.11953.
Robinson, W.A., 1996. Does eddy feedback sustain variability in the zonal index? J. Atmos. Sci. 53, 3556–3569. https://doi.org/10.1175/1520-0469(1996)053<3556:DEFSVI>2.0.CO;2.
Rodell, M., et al., 2004. The global land data assimilation system. Bull. Am. Meteorol. Soc. 85, 381–394.
Rodney, M., Lin, H., Derome, J., 2013. Subseasonal prediction of wintertime North American surface air temperature during strong MJO events. Mon. Weather Rev. 141, 2897–2909. https://doi.org/10.1175/MWR-D-12-00221.1.
Rodrı´guez-Iturbe, I., Entekhabi, D., Bras, R.L., 1991a. Nonlinear dynamics of soil-moisture at climate scales. 1. Stochastic-analysis. Water Resour. Res. 27, 1899–1906.
Rodrı´guez-Iturbe, I., Entekhabi, D., Lee, J., Bras, R.L., 1991b. Nonlinear dynamics of soil-moisture at climate scales. 2. Chaotic analysis. Water Resour. Res. 27, 1907–1915.
Roebber, P.J., 2009. Visualizing multiple measures of forecast quality. Weather Forecast. 24, 601–608.
Roff, G., Thompson, D.W.J., Hendon, H., 2011. Does increasing model stratospheric resolution improve extended-range forecast skill? Geophys. Res. Lett. 38. https://doi.org/10.1029/2010GL046515.
Rogers, D.J., Randolph, S.E., Snow, R.W., Hay, S.I., 2002. Satellite imagery in the study and forecast of malaria. Nature 415, 710–715.
Roh, W., Satoh, M., Nasuno, T., 2017. Improvement of a cloud microphysics scheme for a global nonhydrostatic model using TRMM and a satellite simulator. J. Atmos. Sci. 74, 167–184.
Ropelewski, C., Halpert, M., 1987. Global and regional scale precipitation patterns associated with the El Nino Southern Oscillation. Mon. Weather Rev. 115, 1606–1626.
Ropelewski, C., Halpert, M., 1996. Quantifying Southern Oscillation-precipitation relationships. J. Clim. 9, 1043–1059.
Rossby, C.-G., 1939. Relation between variations in the intensity of the zonal circulation of the atmosphere and the displacements of the semi-permanent centers of action. J. Mar. Res. 2, 38–55. https://doi.org/10.1357/002224039806649023.
Rotunno, R., 1983. On the linear-theory of the land and sea breeze. J. Atmos. Sci. 40, 1999–2009.
Roulston, M.S., Bolton, G.E., Kleit, A.N., Sears-Collins, A.L., 2006. A laboratory study of the benefits of including uncertainty information in weather forecasts. Weather Forecast. 21, 116–122. https://doi.org/10.1175/WAF887.1.
Roundy, P.E., 2012. Tropical extratropical interactions. In: Lau, W.K.M., Waliser, D.E. (Eds.), Intraseasonal Variability in the Atmosphere-Ocean Climate System, second ed. Springer, pp. 497–512.
Roundy, P., 2014. Some aspects of western hemisphere circulation and the Madden-Julian Oscillation. J. Atmos. Sci. 71, 2027–2039.
Roundy, J.K., Wood, E.F., 2015. The attribution of land-atmosphere interactions on the seasonal predictability of drought. J. Hydrometeorol. 16, 793–810.

Rowell, D., 1998. Assessing potential seasonal predictability with an ensemble of multidecadal GCM Simulations. J. Clim. 11, 109–120.

Rowntree, P.R., Bolton, J.A., 1983. Effects of soil moisture anomalies over Europe in summer. In: Street-Perrott, A., et al., (Eds.), Variations in the Global Water Budget. D. Reidel, pp. 447–462.

Roy, F., Chevallier, M., Smith, G., Dupont, F., Garric, G., Lemieux, J.-F., Lu, Y., Davidson, F., 2015. Arctic sea ice and freshwater sensitivity to the treatment of the atmosphere-ice-ocean surface layer. J. Geophys. Res. Oceans 120, 4392–4417. https://doi.org/10.1002/2014JC010677.

Rui, H., Wang, B., 1990. Development characteristics and dynamic structure of tropical intraseasonal oscillations. J. Atmos. Sci. 47, 357–379.

Ruiz, D., et al., 2014. Testing a multi-malaria-model ensemble against 30 years of data in the Kenyan highlands. Malar. J. 13, 1.

Rutledge, S.A., Hobbs, P.V., 1983. The mesoscale and microscale structure and organization of clouds and precipitation in midlatitude cyclones VIII: a model for the "seeder-feeder" process in warm-frontal rain bands. J. Atmos. Sci. 40, 1185–1206.

Rutledge, S.A., Hobbs, P.V., 1984. The mesoscale and microscale structure and organization of clouds and precipitation in midlatitude cyclones XII: a diagnostic modeling study of precipitation development in narrow cold frontal rainbands. J. Atmos. Sci. 41, 2949–2972.

S2S, 2018. The WMO Sub-seasonal to Seasonal prediction research project: see the S2S web pages hosted by WMO. http://www.s2sprediction.net/. And the page hosted by ECMWF:http://www.ecmwf.int/en/research/projects/s2s.

Saha, S., Nadiga, S., Thiaw, C., et al., 2006. The NCEP climate forecast system. J. Clim. 19, 3483–3517. https://doi.org/10.1175/JCLI3812.1.

Saha, S., et al., 2010. The NCEP climate forecast system reanalysis. Bull. Am. Meteorol. Soc. 91, 1015–1057.

Saha, S., et al., 2011. NCEP Climate Forecast System Version 2 (CFSv2) 6-Hourly Products. updated dailyResearch Data Archive at the National Center for Atmospheric Research, Computational and Information Systems Laboratory.https://doi.org/10.5065/D61C1TXF. [(Accessed 19 August 2016)].

Saha, S., Moorthi, S., Wu, X., et al., 2014. The NCEP climate forecast system version 2. J. Clim. 27, 2185–2208. https://doi.org/10.1175/JCLI-D-12-00823.1.

Sahai, A.K., Chattopadhyay, R., Joseph, S., et al., 2015. Real-time performance of a multi-model ensemble-based extended range forecast system in predicting the 2014 monsoon season based on NCEP-CFSv2. Curr. Sci. 109, 1802–1813.

Salomon, J.G., Schaaf, C.B., Strahler, A.H., Gao, F., Jin, Y., 2006. Validation of the MODIS bidirectional reflectance distribution function and albedo retrievals using combined observations from the aqua and terra platforms. IEEE Trans. Geosci. Remote Sens. 44, 1555–1565.

Sansom, P.G., Stephenson, D.B., Ferro, C.A.T., Zappa, G., Shaffrey, L., 2013. Simple uncertainty frameworks for selecting weighting schemes and interpreting multimodel ensemble climate change experiments. J. Clim. 26 (12), 4017–4037.

Santanello, J.A., Peters-Lidard, C.D., Kumar, S.V., Alonge, C., Tao, W.-K., 2009. A modeling and observational framework for diagnosing local land-atmosphere coupling on diurnal time scales. J. Hydrometeorol. 10, 577–599.

Santanello, J.A., Peters-Lidard, C.D., Kumar, S.V., 2011a. Diagnosing the sensitivity of local land-atmosphere coupling via the soil moisture-boundary layer interaction. J. Hydrometeorol. 12, 766–786.

Santanello, J.A., et al., 2011b. Local land-atmosphere coupling (LoCo) research: status and results. GEWEX News 21 (4), 7–9.

Santanello, J.A., et al., 2015. The importance of routine planetary boundary layer measurements over land from space. In: White Paper in Response to the Earth Sciences Decadal Survey Request for Information (RFI) From the National Academy of Sciences Space Studies Board. 5 pp.

Saravanan, R., 1998. Atmospheric low-frequency variability and its relationship to midlatitude SST variability: studies using the NCAR climate system model. J. Clim. 11, 1386–1404.

Saravanan, R., McWilliams, J.C., 1998. Advective ocean–atmosphere interaction: an analytical stochastic model with implications for decadal variability. J. Clim. 11, 165–188.

Sardeshmukh, P.D., Hoskins, B.J., 1988. The generation of global rotational flow by steady idealized tropical divergence. J. Atmos. Sci. 45, 1228–1251. https://doi.org/10.1175/1520-0469(1988)045<1228:TGOGRF>2.0.CO;2.

Sardeshmukh, P.D., Penland, C., 2015. Understanding the distinctively skewed and heavy tailed character of atmospheric and oceanic probability distributions. Chaos 25, 036410. https://doi.org/10.1063/1.4914169.

Sato, N., Sellers, P.J., Randall, D.A., Schneider, E.K., Shukla, J., Kinter III, J.L., Hou, Y.-T., Albertazzi, E., 1989. Effects of implementing the Simple Biosphere model in a general circulation model. J. Atmos. Sci. 46, 2757–2782.

Satoh, M., Matsuno, T., Tomita, H., Miura, H., Nasuno, T., Iga, S., 2008. Nonhydrostatic icosahedral atmospheric model (NICAM) for global cloud resolving simulations. J. Comput. Phys. 227, 3486–3514.

Satoh, M., Iga, S., Tomita, H., Tsushima, Y., Noda, A.T., 2012. Response of upper clouds in global warming experiments obtained using a global nonhydrostatic model with explicit cloud processes. J. Clim. 25, 2178–2191. https://doi.org/10.1175/JCLI-D-11-00152.1.

Satoh, M., et al., 2014. The non-hydrostatic icosahedral atmospheric model: description and development. Prog Earth Planet Sci 1, 18.

Scaife, A.A., Knight, J.R., 2008. Ensemble simulations of the cold European winter of 2005-2006. Q. J. R. Meteorol. Soc. 134, 1647–1659. https://doi.org/10.1002/qj.312.

Scaife, A.A., et al., 2014a. Skilful long-range prediction of European and North American winters. Geophys. Res. Lett. 41, 2514–2519. https://doi.org/10.1002/2014GL059637.

Scaife, A.A., et al., 2014b. Predictability of the quasi-biennial oscillation and its northern winter teleconnection on seasonal to decadal timescales. Geophys. Res. Lett. 41, 1752–1758. https://doi.org/10.1002/2013GL059160.

Scaife, A.A., et al., 2016. Seasonal winter forecasts and the stratosphere. Atmos. Sci. Lett. 17, 51–56. https://doi.org/10.1002/asl.598.

Schecter, D.A., Montgomery, M.T., 2006. Conditions that inhibit the spontaneous radiation of spiral inertia-gravity waves from an intense mesoscale cyclone. J. Atmos. Sci. 63, 435–456.

Schecter, D.A., Dubin, D.H.E., Cass, A.C., Driscoll, C.F., Lansky, I.M., O'Neil, T.M., 2000. Inviscid damping of asymmetries on a two-dimensional vortex. Phys. Fluids 12, 2397–2412.

Schecter, D.A., Montgomery, M.T., Reasor, P.D., 2002. A theory for the vertical alignment of a quasigeostrophic vortex. J. Atmos. Sci. 59, 150–168.

Schefzik, R., 2017. Ensemble calibration with preserved correlations: unifying and comparing ensemble copula coupling and member-by-member postprocessing. Q. J. R. Meteorol. Soc. 143 (703), 999–1008.

Schefzik, R., Thorarinsdottir, T.L., Gneiting, T., 2013. Uncertainty quantification in complex simulation models using ensemble copula coupling. Stat. Sci. 28 (4), 616–640.

Schenzinger, V., Osprey, S., Gray, L., Butchart, N., 2016. Defining metrics of the Quasi-Biennial Oscillation in global climate models. Geosci. Model Dev. Discuss. https://doi.org/10.5194/gmd-2016-284.

Scher, S., Haarsma, R.J., de Vries, H., Drijfhout, S.S., van Delden, A.J., 2017. Resolution dependence of extreme precipitation and deep convection over the Gulf Stream. J. Adv. Model. Earth Syst. 9. https://doi.org/10.1002/2016MS000903.

Scheuerer, M., Hamill, T.M., Whitin, B., He, M., Henkel, A., 2017. A method for preferential selection of dates in the Schaake shuffle approach to constructing spatiotemporal forecast fields of temperature and precipitation. Water Resour. Res. 53 (4), 3029–3046.

Schiller, A., Godfrey, J.S., 2003. Indian ocean intraseasonal variability in an ocean general circulation model. J. Clim. 16, 21–39.

Schlax, M.G., Chelton, D.B., 1992. Frequency-domain diagnostics for linear smoothers. J. Am. Stat. Assoc. 87, 1070–1081. https://doi.org/10.1080/01621459.1992.10476262.

Schneider, S.H., Dickinson, R.E., 1974. Climate modeling. Rev. Geophys. Space Phys. 25, 447–493.

Schroeder, D., Feltham, D.L., Flocco, D., Tsamados, M., 2014. September Arctic sea ice minimum predicted by spring melt-pond fraction. Nat. Clim. Chang. 4 (5), 353–357.

Schubert, S.D., 1985. A statistical-dynamical study of empirically determined modes of atmospheric variability. J. Atmos. Sci. 42, 3–17.

Schubert, S., Dole, R., van den Dool, H., Suarez, M., Waliser, D., 2002. Prospects for Improved Forecasts of Weather and Short-Term Climate Variability on Subseasonal (2-Week to 2-Month) Time Scales. NASA Technical Report Series on Global Modeling and Data Assimilation, 23, NASA/TM-2002-104606. 171 pp.

Schwarz, G., 1978. Estimating the dimension of a model. Ann. Stat. 6 (2), 461–464.

Schwedler, B.R.J., Baldwin, M.E., 2011. Diagnosing the sensitivity of binary image measures to bias, location, and event frequency within a forecast verification framework. Weather Forecast. 26, 1032–1044.

Schweiger, A., Lindsay, R., Zhang, J., Steele, M., Stern, H., Kwok, R., 2011. Uncertainty in modeled Arctic sea ice volume. J. Geophys. Res. Oceans. 116(C8).

Scott, R.K., 2016. A new class of vacillations of the stratospheric polar vortex. Q. J. R. Meteorol. Soc. 142, 1948–1957. https://doi.org/10.1002/qj.2788.

Scott, R.K., Haynes, P.H., 2000. Internal vacillations in stratosphere-only models. J. Atmos. Sci. 57, 3233–3250. https://doi.org/10.1175/1520-0469(2000)057<3233:IVISOM>2.0.CO;2.

Scott, R.K., Polvani, L.M., 2004. Stratospheric control of upward wave flux near the tropopause. Geophys. Res. Lett. 31. https://doi.org/10.1029/2003GL017965.

Screen, J.A., 2017a. Simulated atmospheric response to regional and pan-arctic sea-ice loss. J. Clim. https://doi.org/10.1175/JCLI-D-16-0197.1 JCLI-D-16-0197.1.

Screen, J.A., 2017b. The missing Northern European cooling response to Arctic sea ice loss. Nat. Commun. 8, 14603.

Seiki, T., Kodama, C., Noda, A.T., Satoh, M., 2015. Improvement in global cloud-system-resolving simulations by using a double-moment bulk cloud microphysics scheme. J. Clim. 28, 2405–2419.

Sekiguchi, M., Nakajima, T., 2008. A k-distribution-based radiation code and its computational optimization for an atmospheric general circulation model. J. Quant. Spectrosc. Radiat. Transf. 109, 2779–2793.

Sellers, P.J., Dorman, J.L., 1987. Testing the Simple Biosphere model (SiB) using point micrometeorological and biophysical data. J. Clim. Appl. Meteorol. 26, 622–651.

Sellers, P.J., Mintz, Y., Sud, Y.C., Dalcher, A., 1986. A simple biosphere model (SiB) for use within general circulation models. J. Atmos. Sci. 43, 505–531.

Sellers, P.J., Hall, F.G., Asrar, G., Strebel, D.E., Murphy, R.E., 1992. An overview of the First International Satellite Land Surface Climatology Project (ISLSCP) Field Experiment (FIFE). J. Geophys. Res. 97, 18,345–18,372.

Sellers, P.J., et al., 1995. The Boreal Ecosystem-Atmosphere Study (BOREAS): an overview and early results from the 1994 field year. Bull. Am. Meteorol. Soc. 76, 1549–1577.

Sellers, P.J., et al., 1997. Modeling the exchanges of energy, water, and carbon between the continents and the atmosphere. Science 275, 502–509.

Semmler, T., Jung, T., Serrar, S., 2016a. Fast atmospheric response to a sudden thinning of Arctic winter sea ice from an ensemble of model simulations. Clim. Dyn. 46, 1015–1025. https://doi.org/10.1007/s00382-015-2629-7.

Semmler, T., Stulic, J., Jung, T., Tilinina, N., Campos, C., Gulev, S., Koracin, D., 2016b. Impact of reduced Arctic sea ice on the Northern Hemisphere atmosphere in an ensemble of coupled model simulations. J. Clim. 29, 5893–5913. https://doi.org/10.1175/JCLI-D-15-0586.1.

Semmler, T., Kasper, M.A., Jung, T., Serrar, S., 2016c. Remote impact of the Antarctic atmosphere on the southern mid-latitudes. Meteorol. Z. 25, 71–77. https://doi.org/10.1127/metz/2015/0685.

Semmler, T., Kasper, M.A., Jung, T., Serrar, S., 2017. Using NWP to assess the influence of the Arctic atmosphere on mid-latitude weather and climate. Adv. Atmos. Sci. https://doi.org/10.1007/s00376-017-6290-4 accepted.

Semtner Jr., A.J., 1976. A model for the thermodynamic growth of sea ice in numerical investigations of climate. J. Phys. Oceanogr. 6 (3), 379–389.

Seneviratne, S.I., Lüthi, D., Litschi, M., Schär, C., 2006. Land–atmosphere coupling and climate change in Europe. Nature 443 (7108), 205–209.

Seneviratne, S.I., et al., 2010. Investigating soil moisture-climate interactions in a changing climate: a review. Earth-Sci. Rev. 99, 125–161.

Seo, K.-H., Kumar, A., 2008. The onset and life span of the Madden-Julian Oscillation. Theor. Appl. Climatolol. 94, 13–24.

Seo, K.H., Son, S.W., 2012. The global atmospheric circulation response to tropical diabatic heating associated with the Madden–Julian oscillation during northern winter. J. Atmos. Sci. 69, 79–96. https://doi.org/10.1175/2011JAS3686.1.

Seo, K.-H., Wang, W., 2010. The Madden–Julian oscillation simulated in the NCEP Climate Forecast System model: the importance of stratiform heating. J. Clim. 23, 4770–4793.

Seo, K.-H., Wang, W., Gottschalck, J., Zhang, Q., Schemm, J.-K.E., Higgins, W.R., Kumar, A., 2009. Evaluation of MJO forecast skill from several statistical and dynamical forecast models. J. Clim. 22, 2372–2388.

Seo, J., Choi, W., Youn, D., Park, D.-S.R., Kim, J.Y., 2013. Relationship between the stratospheric quasi-biennial oscillation and the spring rainfall in the western North Pacific. Geophys. Res. Lett. 40, 5949–5953. https://doi.org/10.1002/2013GL058266.

Seviour, W.J.M., Hardiman, S.C., Gray, L.J., Butchart, N., MacLachlan, C., Scaife, A.A., 2014. Skillful seasonal prediction of the Southern Annular Mode and Antarctic Ozone. J. Clim. 27, 7462–7474. https://doi.org/10.1175/JCLI-D-14-00264.1.

Shapiro, M., et al., 2010. An earth-system prediction initiative for the 21st century. Bull. Am. Meteorol. Soc. 91, 1377–1388. https://doi.org/10.1175/2010BAMS2944.1.

Sharma, K.D., Gosain, A.K., 2010. Application of climate information and predictions in water sector: capabilities. Procedia Environ Sci 1, 120–129.

Shaw, T.A., Perlwitz, J., 2013. The life cycle of Northern Hemisphere downward wave coupling between the stratosphere and troposphere. J. Clim. 26, 1745–1763. https://doi.org/10.1175/JCLI-D-12-00251.1.

Shaw, T.A., Perlwitz, J., Harnik, N., 2010. Downward wave coupling between the stratosphere and troposphere: the importance of meridional wave guiding and comparison with zonal-mean coupling. J. Clim. 23, 6365–6381. https://doi.org/10.1175/2010JCLI3804.1.

Shaw, T.A., Perlwitz, J., Weiner, O., 2014. Troposphere-stratosphere coupling: links to North Atlantic weather and climate, including their representation in CMIP5 models. J. Geophys. Res. Atmos. 119, 5864–5880. https://doi.org/10.1002/2013JD021191.

Shchepetkin, A.F., McWilliams, J.C., 2005. The Regional Ocean Modeling System: a split-explicit, free-surface, topography-following coordinates ocean model. Ocean Model. 9, 347–404. https://doi.org/10.1016/j.ocemod.2004.08.002.

Shelly, A., Xavier, P., Copsey, D., Johns, T., Rodriguez, J.M., Milton, S., Klingaman, N.P., 2014. Coupled versus uncoupled hindcast simulations of the Madden–Julian oscillation in the year of tropical convection. Geophys. Res. Lett., 5670–5677.

Shepard, D., 1968. Two-dimentional interpolation function for irregularly spaced data. In: Proc. 1968 ACM Nat. Conf., pp. 517–524.

Shepherd, T.G., 2003. Ripa's theorem and its relatives. In: Velasco Fuentes, O.U., Sheinbaum, J., Ochoa, J. (Eds.), Nonlinear Processes in Geophysical Fluid Dynamics: A Tribute to the Scientific Work of Pedro Ripa. Kluwer Academic, Dordrecht, pp. 1–14.

Shukla, J., 1981. Dynamical predictability of monthly means. J. Atmos. Sci. 38 (12), 2547–2572.

Shukla, J., 1998. Predictability in the midst of chaos: a scientific basis for climate forecasting. Science 282, 728–731.

Shukla, J., Mintz, Y., 1982. Influence of land-surface evapotranspiration on the earth's climate. Science 215, 1498–1501.

Shukla, S., Voisin, N., Letternmaiser, D.P., 2012. Value of medium range weather forecasts in the improvement of seasonal hydrological prediction skill. Hydrol. Earth Syst. Sci. Discuss. 9, 1827–1857.

Shuttleworth, W.J., 2012. Terrestrial Hydrometeorology. John Wiley and Sons. 448 pp.

Shutts, G., 2005. A kinetic energy backscatter algorithm for use in ensemble prediction systems. Q. J. R. Meteorol. Soc. 131, 3079–3100.

Siegert, S., Sansom, P.G., Williams, R.M., 2016a. Parameter uncertainty in forecast recalibration. Q. J. R. Meteorol. Soc. 142 (696), 1213–1221.

Siegert, S., Stephenson, D.B., Sansom, P.G., Scaife, A.A., Eade, R., Arribas, A., 2016b. A Bayesian framework for verification and recalibration of ensemble forecasts: how uncertain is NAO predictability? J. Clim. 29 (3), 995–1012.

Sigmond, M., Scinocca, J.F., Kushner, P.J., 2008. Impact of the stratosphere on tropospheric climate change. Geophys. Res. Lett. 35. https://doi.org/10.1029/2008GL033573.

Sigmond, M., Scinocca, J.F., Kharin, V.V., Shepherd, T.G., 2013. Enhanced seasonal forecast skill following stratospheric sudden warmings. Nat. Geosci. https://doi.org/10.1038/ngeo1698.

Silver, A., Mills, B., 2018. The Value and Application of Sub-seasonal to Seasonal Weather Forecasts. Unpublished annotated bibliography, Department of Geography and Environmental Management, University of Waterloo, Waterloo, Canada. 173 pp.

Silverman, B.W., 1986. Density Estimation for Statistics and Data Analysis. CRC Press, New York, NY.

Simmons, A.J., Wallace, J.M., Branstator, G.W., 1983. Barotropic wave propagation and instability, and atmospheric teleconnection patterns. J. Atmos. Sci. 40, 1363–1392.

Simpson, I.R., Polvani, L.M., 2016. Revisiting the relationship between jet position, forced response, and annular mode variability in the southern midlatitudes. Geophys. Res. Lett. https://doi.org/10.1002/2016GL067989.

Simpson, I.R., Hitchcock, P., Shepherd, T.G., Scinocca, J.F., 2013. Southern annular mode dynamics in observations and models. Part I: The influence of climatological zonal wind biases in a comprehensive GCM. J. Clim. 26, 3953–3967. https://doi.org/10.1175/JCLI-D-12-00348.1.

Sinclair, D., Preziosi, M.-P., Jacob John, T., Greenwood, B., 2010. The epidemiology of meningococcal disease in india. Tropical Med. Int. Health 15, 1421–1435.

Sirovich, L., Everson, R., 1992. Management and analysis of large scientific datasets. J. Super. Appl. 6, 50–58.

Sivakumar, M., 1988. Predicting rainy season potential from the onset of rains in southern Sahelian and Sudanian climatic zones of West-Africa. Agric. For. Meteorol. 42, 295–305.

Sivakumar, M.V.K., Collins, C., Jay, A., Hansen, J., 2014. Regional Priorities for Strengthening Climate Services for Farmers in Africa and South Asia. CCAFS Working Paper no. 71, CGIAR Research Program on Climate Change, Agriculture and Food Security (CCAFS), Copenhagen, Denmark. Available online at:www.ccafs.cgiar.org.

Skamarock, W.C., et al., 2008. A Description of the Advanced Research WRF Version 3. NCAR Tech. Note NCAR/TN-4751STR, 113 pp. https://doi.org/10.5065/D68S4MVH.

Slingo, J.M., Sperber, K.R., Boyle, J.S., Ceron, J.-P., Dix, M., Dugas, B., Ebisuzaki, W., Fyfe, J., Gregory, D., Gueremy, J.-F., Hack, J., Harzallah, A., Inness, P., Kitoh, A., Lau, W.K.-M., McAvaney, B., Madden, A., R.and Matthews, Palmer, T. N., Parkas, C.-K., Randall, D., & Renno, N., 1996. Intraseasonal oscillations in 15 atmospheric general circulation models: results from an AMIP diagnostics subproject. Clim. Dyn. 12, 325–357.

Slingo, J.M., Rowell, D.P., Sperber, K.R., Nortley, F., 1999. On the predictability of the interannual behaviour of the Madden-Julian Oscillation and its relationship with El Niño. Q. J. R. Meteorol. Soc. 125, 583–609.

Slingo, J., Inness, P., Neale, R., Woolnough, S., Yang, G.-Y., 2003. Scale interactions on diurnal to seasonal time scales and their relevance to model systematic errors. Ann. Geophys. 46, 139–155.

Small, R.J., et al., 2008. Air–sea interaction over ocean fronts and eddies. Dyn. Atmos. Oceans 45, 274–319. https://doi.org/10.1016/j.dynatmoce.2008.01.001.

Small, R.J., Tomas, R.A., Bryan, F.O., 2014. Storm track response to ocean fronts in a global high-resolution climate model. Clim. Dyn. 43, 805–828. https://doi.org/10.1007/s00382-013-1980-9.

Smith, A.K., 1989. An investigation of resonant waves in a numerical model of an observed sudden stratospheric warming. J. Atmos. Sci. 46, 3038–3054. https://doi.org/10.1175/1520-0469(1989)046<3038:AIORWI>2.0.CO;2.

Smith, S., 2013. Digital Signal Processing: A Practical Guide for Engineers and Scientists. Elsevier, p. 672.

Smith, K.L., Kushner, P.J., 2012. Linear interference and the initiation of extratropical stratosphere-troposphere interactions. J. Geophys. Res. Atmos. 117. https://doi.org/10.1029/2012JD017587.

Smith, K.L., Polvani, L.M., 2014. The surface impacts of Arctic stratospheric ozone anomalies. Environ. Res. Lett. 9, 74015. https://doi.org/10.1088/1748-9326/9/7/074015.

Smith, K.L., Scott, R.K., 2016. The role of planetary waves in the tropospheric jet response to stratospheric cooling. Geophys. Res. Lett. 43, 2904–2911. https://doi.org/10.1002/2016GL067849.

Smith, L.C., Stephenson, S.R., 2013. New Trans-Arctic shipping routes navigable by midcentury. Proc. Natl. Acad. Sci. 110 (13), E1191–E1195.

Smith, D., Gasiewski, A., Jackson, D., Wick, G., 2005. Spatial scales of tropical precipitation inferred from TRMM microwave imager data. IEEE Trans. Geosci. Remote Sens. 43, 1542–1551.

Smith, K.L., Fletcher, C.G., Kushner, P.J., 2010. The role of linear interference in the annular mode response to extratropical surface forcing. J. Clim. 23, 6036–6050. https://doi.org/10.1175/2010JCLI3606.1.

Smith, G.C., Roy, F., Reszka, M., Colan, D.S., He, Z., Deacu, D., et al., 2014. Sea ice forecast verification in the Canadian Global Ice Ocean Prediction System. Q. J. R. Meteorol. Soc. 695 (142), 659–671.

Smyth, P., Ide, K., Ghil, M., 1999. Multiple regimes in Northern Hemisphere height fields via mixture model clustering. J. Atmos. Sci. 56, 3704–3723. https://doi.org/10.1175/1520-0469(1999)056<3704: mrinhh>2.0.co;2.

Snyder, A.D.S., Pu, H., Zhu, Y., 2010. Tracking and verification of east Atlantic tropical cyclone genesis in the NCEP global ensemble: case studies during the NASA African Monsoon multidisciplinary analyses. Weather Forecast. 25, 1397–1411.

Sobel, A., Wang, S., Kim, D., 2014. Moist static energy budget of the MJO during DYNAMO. J. Atmos. Sci. 71, 4276–4291.

Sobolowski, S., Gong, G., Ting, M., 2010. Modeled climate state and dynamic responses to anomalous North American snow cover. J. Clim. 23, 785–799.

Son, S.-W., et al., 2010. Impact of stratospheric ozone on Southern Hemisphere circulation change: a multimodel assessment. J. Geophys. Res. 115, D00M07. https://doi.org/10.1029/2010JD014271.

Son, S.-W., Purich, A., Hendon, H.H., Kim, B.-M., Polvani, L.M., 2013. Improved seasonal forecast using ozone hole variability? Geophys. Res. Lett. 40, 6231–6235. https://doi.org/10.1002/2013GL057731.

Son, S.-K., Lim, Y., Yoo, C., Hendon, H.H., Kim, J., 2017. Stratopsheric control of the Madden-Julian Oscillation. J. Clim. 30, 1909–1922.

Song, Y., Robinson, W.A., 2004. Dynamical mechanisms for stratospheric influences on the troposphere. J. Atmos. Sci. 61, 1711–1725. https://doi.org/10.1175/1520-0469(2004)061<1711:DMFSIO>2.0.CO;2.

Sontakke, N., Singh, N., Singh, H., 2008. Instrumental period rainfall series of the Indian region (AD 1813-2005): revised reconstruction, update and analysis. The Holocene 18, 1055–1066.

Souza, E.B., Ambrizzi, T., 2006. Modulation of the intraseasonal rainfall over tropical Brazil by the Madden-Julian Oscillation. Int. J. Cimatol. 26, 1759–1776.

Sperber, K.R., 2003. Propagation and the vertical structure of the Madden-Julian Oscillation. Mon. Weather Rev. 131, 3018–3037.

Stan, C., Straus, D.M., Frederiksen, J.S., Lin, E.D., Malooney, H., Schumacher, C., 2017. Review of tropical-extratropical teleconnections on intraseasonal time scales. Rev. Geophys. 55, 902–937.

Stephens, D.S., Greenwood, B., Brandtzaeg, P., 2007. Epidemic meningitis, meningococcaemia, and Neisseria meningitidis. Lancet 369, 2196–2210.

Stephenson, D., Kumar, K.R., Doblas-Reyes, F., Royer, J., Chauvin, F., Pezzulli, S., 1999. Extreme daily rainfall events and their impact on ensemble forecasts of the Indian monsoon. Mon. Weather Rev. 127, 1954–1966.

Stephenson, D.B., Hannachi, A., O'Neill, A., 2004. On the existence of multiple climate regimes. Q. J. R. Meteorol. Soc. 130, 583–605. https://doi.org/10.1256/qj.02.146.

Stephenson, D.B., Coelho, C.A.S., Doblas-Reyes, F.J., Balmaseda, M., 2005. Forecast assimilation: a unified framework for the combination of multi-model weather and climate predictions. Tellus A 57 (3), 253–264.

Stephenson, D.B., Casati, B., Ferro, C.A.T., Wilson, C.A., 2008. The extreme dependency score: a non-vanishing measure for forecasts of rare events. Meteorol. Appl. 15, 41–50.

Stewart-Ibarra, A.M., Lowe, R., 2013. Climate and non-climate drivers of dengue epidemics in southern coastal Ecuador. Am. J. Trop. Med. Hyg. 88, 971–981.

Stewart-Ibarra, A.M., et al., 2013. Dengue vector dynamics (*Aedes aegypti*) influenced by climate and social factors in Ecuador: implications for targeted control. PLoS One 8, e78263.

Stinis, P., 2006. A comparative study of two stochastic mode reduction methods. Physica D 213, 197–213. https://doi.org/10.1016/j.physd.2005.11.010.

Straub, K.H., 2013. MJO initiation in the real-time multivariate MJO index. J. Clim. 26, 1130–1151.

Straub, K.H., Kiladis, G.N., 2003. Extratropical forcing of convectively coupled Kelvin waves during austral winter. J. Atmos. Sci. 60, 526–543.

Straus, D.M., 1983. On the role of the seasonal cycle. J. Atmos. Sci. 40, 303–313.

Straus, D.M., Molteni, F., 2004. Circulation regimes and SST forcing: results from large GCM ensembles. J. Clim. 17, 1641–1656. https://doi.org/10.1175/1520-0442(2004)017<1641:crasfr>2.0.co;2.

Straus, D.M., Corti, S., Molteni, F., 2007. Circulation regimes: chaotic variability versus SST-forced predictability. J. Clim. 20, 2251–2272. https://doi.org/10.1175/jcli4070.1.

Straus, D.M., Molteni, F., Corti, S., 2017. Atmospheric regimes: the link between weather and the large-scale circulation. In: Franzke, C.L.E., OKane, T.J. (Eds.), Nonlinear and Stochastic Climate Dynamics. Cambridge University Press, Cambridge, UK, pp. 105–135. https://doi.org/10.1017/9781316339251.005.

Stroeve, J.C., Serreze, M.C., Holland, M.M., Kay, J.E., Malanik, J., Barrett, A.P., 2012. The Arctic's rapidly shrinking sea ice cover: a research synthesis. Clim. Chang. 110 (3), 1005–1027.

Stroeve, J., Hamilton, L.C., Bitz, C.M., Blanchard-Wrigglesworth, E., 2014. Predicting September sea ice: ensemble skill of the SEARCH sea ice outlook 2008–2013. Geophys. Res. Lett. 41 (7), 2411–2418.

Strong, C.M., Jin, F.F., Ghil, M., 1993. Intraseasonal variability in a barotropic model with seasonal forcing. J. Atmos. Sci. 50, 2965–2986. https://doi.org/10.1175/1520-0469(1993)050<2965: iviabm>2.0.co;2.

Strong, C., Jin, F.f., Ghil, M., 1995. Intraseasonal oscillations in a barotropic model with annual cycle, and their predictability. J. Atmos. Sci. 52, 2627–2642. https://doi.org/10.1175/1520-0469(1995) 052<2627:ioiabm>2.0.co;2.

Strounine, K., Kravtsov, S., Kondrashov, D., Ghil, M., 2010. Reduced models of atmospheric low-frequency variability: parameter estimation and comparative performance. Physica D 239, 145–166. https://doi.org/10.1016/ j.physd.2009.10.013.

Stull, R.B., 1988. An Introduction to Boundary Layer Meteorology. Springer. 666 pp.

Su, X., Yuan, H., Zhu, Y., Luo, Y., Wang, Y., 2014. Evaluation of TIGGE ensemble predictions of Northern Hemisphere summer precipitation during 2008-2012. J. Geophys. Res. Atmos. 119 (12), 7292–7310.

Suhas, E., Neena, J., Goswami, B., 2012. An Indian monsoon intraseasonal oscillations (MISO) index for real time monitoring and forecast verification. Clim. Dyn., 1–12. https://doi.org/10.1007/s00382-012-1462-5.

Sultan, B., Labadi, K., Guegan, J.-F., Janicot, S., 2005. Climate drives the meningitis epidemics onset in West Africa. PLoS Med. 2, e6.

Sun, L., Deser, C., Tomas, R.A., 2015. Mechanisms of stratospheric and tropospheric circulation response to projected arctic sea ice loss. J. Clim. 28, 7824–7845. https://doi.org/10.1175/JCLI-D-15-0169.1.

Sura, P., Newman, M., Penland, C., Sardeshmukh, P., 2005. Multiplicative noise and non-gaussianity: a paradigm for atmospheric regimes? J. Atmos. Sci. 62, 1391–1409. https://doi.org/10.1175/ jas3408.1.

Swaroop, S., 1949. Forecasting of epidemic malaria in the Punjab, India. Am. J. Trop. Med. Hyg. 1, 1–17.

Sweeney, C.P., Lynch, P., Nolan, P., 2011. Reducing errors of wind speed forecasts by an optimal combination of post-processing methods. Meteorol. Appl. 20 (1), 32–40.

Swenson, E.T., Straus, D.M., 2017. Rossby wave breaking and transient eddy forcing during Euro-Atlantic circulation regimes. J. Atmos. Sci. in press.

Szoter, E., 2006. Recent developments in extreme weather forecsting, 2006. ECMWF Newslett. 107, 8–17.

Szunyogh, I., Toth, Z., 2002. The effect of increased horizontal resolution on the NCEP global ensemble mean forecasts. Mon. Weather Rev. 130, 1125–1143.

Szunyogh, I., Kostelich, E.J., Gyarmati, G., Kalnay, E., Hunt, B.R., Ott, E., Satterfield, E., Yorke, J.A., 2008. A local ensemble transform Kalman filter data assimilation system for the NCEP global model. Tellus 60A, 113–130.

Tabatabaeenejad, A., Burgin, M., Duan, X., Moghaddam, M., 2014. P-Band radar retrieval of subsurface soil moisture profile as a second-order polynomial: first AirMOSS results. IEEE Trans. Geosci. Remote Sens. 53, 645–658.

Tadesse, T., Bathke, D., Wall, N., Petr, J., Haigh, T., 2015. Participatory research workshop on seasonal prediction of hydroclimatic extremes in the greater horn of Africa. Bull. Am. Meteorol. Soc. 96, ES139–142. https://doi.org/10.1175/BAMS-D-14-00280.1.

Tadesse, T., Haigh, T., Wall, N., Shiferaw, A., Zaitchik, B., Beyene, S., Berhan, G., Petr, J., 2016. Linking seasonal predictions to decision-making and disaster management in the greater horn of Africa. Bull. Am. Meteorol. Soc. 97 (4), ES89–ES92. https://doi.org/10.1175/BAMS-D-15-00269.1.

Taguchi, M., 2014. Predictability of major stratospheric sudden warmings of the vortex split type: case study of the 2002 Southern Event and the 2009 and 1989 Northern Events. J. Atmos. Sci. 71, 2886–2904. https://doi.org/10.1175/JAS-D-13-078.1.

Taguchi, M., 2015. Changes in frequency of major stratospheric sudden warmings with El Niño/Southern Oscillation and Quasi-Biennial Oscillation. J. Meteorol. Soc. Jpn. Ser. II 93, 99–115. https://doi.org/10.2151/jmsj.2015-007.

Takata, K., Emori, S., Watanabe, T., 2003. Development of minimal advanced treatments of surface interaction and runoff. Glob. Planet. Chang. 38, 209–222.

Takaya, K., Nakamura, H., 2001. A formulation of a phase-independent wave-activity flux for stationary and migratory quasigeostrophic eddies on a zonally varying basic flow. J. Atmos. Sci. 58, 608–627.

Takaya, Y., Hirahara, S., Yasuda, T., Matsueda, S., Toyoda, T., Fujii, Y., Sugimoto, H., Matsukawa, C., Ishikawa, I., Mori, H., Nagasawa, R., Kubo, Y., Adachi, N., Yamanaka, G., Kuragano, T., Shimpo, A., Maeda, S., Ose, T., 2017. Japan Meteorological Agency/Meteorological Research Institute-Coupled Prediction System version 2 (JMA/MRI-CPS2): atmosphere–land–ocean–sea ice coupled prediction system for operational seasonal forecasting. Clim. Dyn. https://doi.org/10.1007/s00382-017-3638-5.

Takayabu, Y.N., 1994. Large-scale disturbances associated with equatorial waves. Part I: spectral features of the cloud disturbances. J. Meteor. Soc. Japan 72, 433–448.

Tall, A., Mason, S.J., Van Aalst, M., Suarez, P., Ait-Chellouche, Y., Diallo, A.A., Braman, L., 2012. Using seasonal climate forecasts to guide disaster management: the red cross experience during the 2008 West Africa floods. Int. J. Geophys., 986016. https://doi.org/10.1155/2012/986016.

Tanaka, H.L., Tokinaga, H., 2002. Baroclinic instability in high latitudes induced by polar vortex: a connection to the Arctic Oscillation. J. Atmos. Sci. 59, 69–82. https://doi.org/10.1175/1520-0469(2002)059<0069:BIIHLI>2.0.CO;2.

Tantet, A., van der Burgt, F.R., Dijkstra, H.A., 2015. An early warning indicator for atmospheric blocking events using transfer operators. Chaos 25, 036406. https://doi.org/10.1063/1.4908174.

Tao, W.-K., Simpson, J., Baker, D., Braun, S., Chou, M.-D., Ferrier, B., Johnson, D., Khain, A., Lang, S., Lynn, B., 2003. Microphysics, radiation and surface processes in the Goddard Cumulus Ensemble (GCE) model. Meteorol. Atmos. Phys. 82 (1), 97–137.

Taub, L., 2003. Ancient Meteorology. Psychology Press. 271 pp.

Tawfik, A.B., Dirmeyer, P.A., 2014. A process-based framework for quantifying the atmospheric preconditioning of surface triggered convection. Geophys. Res. Lett. 41, 173–178.

Taylor, K.E., 2001. Summarizing multiple aspects of model performance in a single diagram. JGR 106, 7183–7192.

Taylor, K.E., Stouffer, R.J., Meehl, G.A., 2012. An overview of CMIP5 and the experiment design. Bull. Am. Meteorol. Soc. 93, 485–498.

The GLACE Team, Koster, R.D., Dirmeyer, P.A., Guo, Z., Bonan, G., Chan, E., Cox, P., Gordon, C.T., Kanae, S., Kowalczyk, E., Lawrence, D., Liu, P., Lu, C.-H., Malyshev, S., McAvaney, B., Mitchell, K., Mocko, D., Oki, T.,

Oleson, K., Pitman, A., Sud, Y.C., Taylor, C.M., Verseghy, D., Vasic, R., Xue, Y., Yamada, T., 2004. Regions of strong coupling between soil moisture and precipitation. Science 305, 1138–1140.

Teixeira, L., Reynolds, C.A., 2008. Stochastic nature of physical parameterizations in ensemble prediction: a stochastic convection approach. Mon. Weather Rev. 136, 483–496.

Teng, H., Branstator, G., Wang, H., Meehl, G.A., Washington, W.M., 2013. Probability of US heat waves affected by a subseasonal planetary wave pattern. Nat. Geosci. 6, 1056–1061.

Teufel, B., et al., 2016. Investigation of the 2013 Alberta flood from weather and climate perspectives. Clim. Dyn. 48, 2881–2899.

Teuling, A.J., et al., 2010. Contrasting response of European forest and grassland energy exchange to heatwaves. Nat. Geosci. 3, 722–727.

Theis, S.E., Hense, A., Damrath, U., 2005. Probabilistic precipitation forecasts from a deterministic model: a pragmatic approach. Meteorol. Appl. 12, 257–268.

The WAMDI Group, 1988. The WAM Model - a third generation ocean wave prediction model. J. Phys. Oceanogr. 18, 1775–1810.

Thompson, P.D., 1957. Uncertainty of initial state as a factor in the predictability of large scale atmospheric flow patterns. Tellus 9, 275–295.

Thompson, D.B., Roundy, P.E., 2013. The relationship between the Madden-Julian oscillation and U.S. violent tornado outbreaks in the spring. Mon. Weather Rev. 141, 2087–2095. https://doi.org/10.1175/MWR-D-12-00173.1.

Thompson, D.W.J., Solomon, S., 2002. Interpretation of recent Southern Hemisphere climate change. Science (80-). https://doi.org/10.1126/science.1069270.

Thompson, D.W.J., Wallace, J.M., 2000. Annular modes in the extratropical circulation. Part I: Month-to-month variability. J. Clim. 13, 1000–1016. https://doi.org/10.1175/1520-0442(2000)013<1000:AMITEC>2.0.CO;2.

Thompson, D.W., Baldwin, M.P., Wallace, J.M., 2002. Stratospheric connection to Northern Hemisphere wintertime weather: implications for prediction. J. Clim. 15, 1421–1427.

Thompson, D.W.J., Furtado, J.C., Shepherd, T.G., 2006a. On the tropospheric response to anomalous stratospheric wave drag and radiative heating. J. Atmos. Sci. 63, 2616–2629. https://doi.org/10.1175/JAS3771.1.

Thompson, M.C., Doblas-Reyes, F.J., Mason, S.J., Hagedorn, R., Connor, S.J., Phindela, T., Morse, A.P., Palmer, T.N., 2006b. Malaria early warnings based on seasonal climate forecasts from multi-model ensembles. Nature 439, 576–579.

Thomson, R.C., 1938. The reactions of mosquitoes to temperature and humidity. Bull. Entomol. Res. 29, 125–140.

Thomson, M.C., et al., 2006a. Malaria early warnings based on seasonal climate forecasts from multimodel ensembles. Nature 439, 576–579.

Thomson, M.C., et al., 2006b. Potential of environmental models to predict meningitis epidemics in Africa. Tropical Med. Int. Health 11, 781–788.

Thomson, M.C., et al., 2013. A climate and health partnership to inform the prevention and control of *meningoccocal meningitis* in sub-Saharan Africa: the MERIT initiative. In: Climate Science for Serving Society. Springer, pp. 459–484.

Thomson, M.C., et al., 2014. Climate and health in Africa. Earth Perspect. 1, 17.

Thorndike, A.S., 1992. Estimates of sea ice thickness distribution using observations and theory. J. Geophys. Res. 97 (C8), 12601–12605. https://doi.org/10.1029/92JC01199.

Thorndike, A.S., Rothrock, D.A., Maykut, G.A., Colony, R., 1975. The thickness distribution of sea ice. J. Geophys. Res. 80, 4501–4513.

THORPEX, 2018. THe Observing system Research and Predictability Experiment. See THORPEX web page hosted by WMO:http://www.wmo.ch/thorpex/.

Tian, D., Wood, E.F., Yuan, X., 2017. CFSv2-based sub-seasonal precipitation and temperature forecast skill over the contiguous United States. Hydrol. Earth Syst. Sci. 21, 1477–1490.

Tibaldi, S., Brankovic, C., Cubasch, U., Molteni, F., 1988. In: Impact of horizontal resolution on extended-range forecasts at ECMWF.Proc. ECMWF Workshop on Predictability in the Medium and Extended Range, Reading, United Kingdom, ECMWF, pp. 215–250.

Tibshirani, R., 1996. Regression shrinkage and selection via the lasso. J. R. Stat. Soc. Ser. B Methodol. 58 (1), 267–288.

Tiedtke, M., 1984. Sensitivity of The Time-Mean Large-Scale Flow to Cumulus Convection in the ECMWF Model., pp. 297–316.

Tietsche, S., Notz, D., Jungclaus, J.H., Marotzke, J., 2013. Predictability of large interannual Arctic sea ice anomalies. Clim. Dyn. 41 (9-10), 2511.

Tietsche, S., Day, J.J., Guemas, V., Hurlin, W.J., Keeley, S.P.E., Matei, D., Msadek, R., Collins, M., Hawkins, E., 2014. Seasonal to interannual Arctic sea ice predictability in current global climate models. Geophys. Res. Lett. 41 (3), 1035–1043.

TIGGE, 2005. The report of the 1st TIGGE Workshop, held at ECMWF in 2005. WMO/TD-No. 1273 WWRP-THORPEX No. 5, is available from the WMO web site at www.wmo.int/thorpex See also the ECMWF web site: and the, http://www.ecmwf.int/en/research/projects/tigge and the WMO web site:https://www.wmo.int/pages/prog/arep/wwrp/new/thorpex_gifs_tigge_index.html.

TIGGE_LAM, 2018. See the ECMWF web site, https://software.ecmwf.int/wiki/display/TIGL/Project.

Timmreck, C., Pohlmann, H., Illing, S., Kadow, C., 2016. The impact of stratospheric volcanic aerosol on decadal-scale climate predictions. Geophys. Res. Lett. 43, 834–842. https://doi.org/10.1002/2015GL067431.

Tippett, M.K., Barnston, A.G., DelSole, T., 2010. Comments on "Finite samples and uncertainty estimates for skill measures for seasonal prediction". Mon. Weather Rev. 138, 1487–1493.

Tippett, M.K., Sobel, A.H., Camargo, S.J., 2012. Association of U.S. tornado occurrence with monthly environmental parameters. Geophys. Res. Lett. 39, L02801. https://doi.org/10.1029/2011GL050368.

Tippett, M.K., Sobel, A.H., Camargo, S.J., Allen, J.T., 2014. An empirical relation between U.S. tornado activity and monthly environmental parameters. J. Clim. 27, 2983–2999. https://doi.org/10.1175/JCLI-D-13-00345.1.

Tokioka, T., 2000. Climate services at the Japan Meteorological Agency using a general circulation model: dynamical one-month prediction. In: Randall, D.A. (Ed.), General Circulation Model Development: Past, Present and Future. Academic Press, pp. 355–371.

Tokioka, T., Yamazaki, K., Kitoh, A., Ose, T., 1988. The equatorial 30–60 day oscillation and the Arakawa-Schubert penetrative cumulus parameterization. J. Meteorol. Soc. Jpn. 66, 883–901.

Tolstykh, M.A., Diansky, N.A., Gusev, A.V., Kiktev, D.B., 2014. Simulation of seasonal anomalies of atmospheric circulation using coupled atmosphere–ocean model. Izvestiya, Atmos. Ocean. Phys. 50 (2), 111–121.

Tomita, H., 2008. New microphysical schemes with five and six categories by diagnostic generation of cloud ice. J. Meteorol. Soc. Jpn. 86A, 121–142.

Tomita, H., Satoh, M., 2004. A new dynamical framework of nonhydrostatic global model using the icosahedral grid. Fluid Dyn. Res. 34, 357–400.

Tomita, H., Tsugawa, M., Satoh, M., Goto, K., 2001. Shallow water model on a modified icosahedral geodesic grid by using spring dynamics. J. Comput. Phys. 174, 579–613.

Tomita, H., Miura, H., Iga, S., Nasuno, T., Satoh, M., 2005. A global cloud-resolving simulation: preliminary results from an aqua planet experiment. Geophys. Res. Lett. 32, 1–4. https://doi.org/10.1029/2005GL022459.

Tompkins, A.M., Di Giuseppe, F., 2015. Potential predictability of malaria using ECMWF monthly and seasonal climate forecasts in Africa. J. Appl. Meteorol. Climatol. 54, 521–540.

Tompkins, A.M., Ermert, V., 2013. A regionalscale, high resolution dynamical malaria model that accounts for population density, climate and surface hydrology. Malar. J. 12 https://doi.org/10.1186/1475-2875-12-65.

Tompkins, A.M., Thomson, M.C., 2018. Uncertainty in malaria simulations due to initial condition, climate and malaria model parameter settings investigated using a constrained genetic algorithm. PLoS One. accepted, in revision.

Tompkins, A.M., et al., 2012. The Ewiem Nimdie summer school series in Ghana: capacity building in meteorological education and research, lessons learned, and future prospects. Bull. Am. Meteorol. Soc. 93, 595–601.

Tompkins, A.M., Di Giuseppe, F., Colon-Gonzalez, F.J., Namanya, D.B., 2016a. A planned operational malaria early warning system for Uganda provides useful district-scale predictions up to 4 months ahead. In: Shumake-Guillemot, J., Fernandez-Montoya, L. (Eds.), Climate Services for Health: Case Studies of Enhancing Decision Support for Climate Risk Management and Adaptation. WHO/WMO, Geneva, pp. 130–131.

Tompkins, A.M., Larsen, L., McCreesh, N., Taylor, D.M., 2016b. To what extent does climate explain variations in reported malaria cases in early 20th century Uganda? Geospat. Health 11 (1s), 38–48. https://doi.org/10.4081/gh.2016.407.

Tompkins, A., de Zarate, M.I.O., Saurral, R.I., Verra, C., Saulo, C., Merryfield, W.J., Sigmond, M., Lee, W.-S., Baehr, J., Braun, A., Butler, A., Deque, M., Doblas-Reyes, F.J., Gordon, M., Scaife, A.A., Imada, Y., Ose, T., Kirtman, B., Kumar, A., Muller, W.A., Pirani, A., Stockdale, T., Rixen, M., Yasuda, T., 2017. The Climate-System historical forecast project-providing open access to seasonal forecast ensembles from centers around the globe. Bull. Am. Meteorol. Soc. 49, 2293–2301.

Torralba, V., Doblas-Reyes, F.J., MacLeod, D., Christel, I., Davis, M., 2017. Seasonal climate prediction: a new source of information for the management of wind energy resources. J. Appl. Meteorol. Climatol. 56, 1231–1247. http://journals.ametsoc.org/doi/suppl/10.1175/JAMC-D-16-0204.1.

Toth, Z., 1995. Degrees of freedom in Northern Hemisphere circulation data. Tellus A 47, 457–472. https://doi.org/10.1034/j.1600-0870.1995.t01-3-00005.x.

Toth, Z., Kalnay, E., 1993. Ensemble forecasting at NMC: the generation of perturbations. Bull. Am. Meteorol. Soc. 74, 2317–2330.

Toth, Z., Kalnay, E., 1997. Ensemble forecasting at NCEP and the breeding method. Mon. Weather Rev. 125, 3297–3319.

Toth, Z., Pena-Mendez, M., Vintzileos, A., 2007. Bridging the gap between weather and climate forecasting: research priorities for intraseasonal prediction. Bull. Am. Meteorol. Soc. 88, 1427–1429.

Townsend, R.D., Johnson, D.R., 1985. A diagnostic study of the isentropic zonally averaged mass circulation during the first {GARP} global experiment. J. Atmos. Sci. 42, 1565–1579.

Tracton, M.S., Kalnay, E.S., 1993. Operational ensemble prediction at the National Meteorological Center: practical aspects. Weather Forecast. 8, 379–398.

Tracton, M.S., Mo, K., Chen, W., Kalnay, E., Kistler, R., White, G., 1989. Dynamical extended range forecast (DERF) at the National Meteorological Center. Mon. Weather Rev. 117, 1604–1635.

Trenary, L., DelSole, T., Tippett, M.K., Pegion, K., 2017. A new method for determining the optimal lagged ensemble. J. Adv. Model. Earth Syst. 9, 291–306.

Trenberth, K., Zhang, Y., Gehne, M., 2017. Intermittency in precipitation: duration, frequency, intensity and amounts using hourly data. J. Hydrometeorol. 18, 1393–1412.

Trevisan, A., Buzzi, A., 1980. Stationary response of barotropic weakly non-linear Rossby waves to quasi-resonant orographic forcing. J. Atmos. Sci. 37, 947–957. https://doi.org/10.1175/1520-0469(1980)037<0947:SROBWN>2.0.CO;2.

Tribbia, J., Baumhefner, D., 2003. Scale interactions and atmospheric predictability: an updated perspective. Mon. Weather Rev. 132, 703–713.

Tripathi, O.P., et al., 2015. The predictability of the extratropical stratosphere on monthly time-scales and its impact on the skill of tropospheric forecasts. Q. J. R. Meteorol. Soc. 141, 987–1003. https://doi.org/10.1002/qj.2432.

Tripathi, O.P., et al., 2016. Examining the predictability of the stratospheric sudden warming of January 2013 using multiple NWP systems. Mon. Weather Rev. 144, 1935–1960. https://doi.org/10.1175/MWR-D-15-0010.1.

Tsushima, Y., Iga, S., Tomita, H., Satoh, M., Noda, A.T., Webb, M.J., 2014. High cloud increase in a perturbed SST experiment with a global nonhydrostatic model including explicit convective processes. J. Adv. Model. Earth Syst. 6, 571–585.

Tung, K.K., Lindzen, R.S., 1979. A theory of stationary long waves. Part II: Resonant Rossby waves in the presence of realistic vertical shears. Mon. Weather Rev. 107, 735–750. https://doi.org/10.1175/1520-0493(1979)107<0735:ATOSLW>2.0.CO;2.

Tziperman, E., Stone, L., Cane, M., Jarosh, H., 1994. El Niño chaos: overlapping of resonances between the seasonal cycle and the Pacific ocean-atmosphere oscillator. Science 264, 72–74. https://doi.org/10.1126/science.264.5155.72.

UNISDR, 2015. USAID & Centre for Research on the Epidemiology of Disasters. 2015. Disasters in Numbers. Tech. Rep.

Uno, I., Cai, X.-M., Steyn, D.G., 1995. A simple extension of the Louis method for rough surface layer modelling. Boundary-Layer Meteorol. 76, 395–409.

UN-SPIDER, 2012. Africa RiskView Online: Climate and Disaster Risk Solutions. Available at http://www.un-spider.org/sites/default/files/AfricaRiskViewOnlineNewsletter.pdf. [(Accessed 23 March 2017)].

Vallis, G.K., 2006. Atmospheric and Oceanic Fluid Dynamics. Cambridge University Press, p. 745.

van den Brink, H.W., Können, G.P., Opsteegh, J.D., van Oldenborgh, G.J., Burgers, G., 2005. Estimating return periods of extreme events from ECMWF seasonal forecast ensembles. Int. J. Climatol. 25, 1345–1354. https://doi.org/10.1002/joc.1155.

Van den Dool, H.M., 1994. Searching for analogues, how long must one wait? Tellus A 46, 314–324.

Van den Dool, H.M., Barnston, A.G., 1995. In: Forecasts of global sea surface temperature out to a year using the constructed analogue method.*Proceedings of the 19th Annual Climate Diagnostics Workshop*, Nov. 14–18, 1994, College Park, Maryland, pp. 416–419.

van den Hurk, B.J.J.M., Viterbo, P., 2003. The Torne-Kalix PILPS 2 (e) experiment as a test bed for modifications to the ECMWF land surface scheme. Glob. Planet. Chang. 38, 165–173.

van den Hurk, B.J.J.M., Viterbo, P., Beljaars, A.C.M., Betts, A.K., 2000. Offline validation of the ERA40 surface scheme. ECMWF Tech. Memo 295. [Available from ECMWF, Shinfield Park, Reading, RG2 9AX, United Kingdom]. 42 pp.

van Heerwaarden, C.C., de Arellano, J.V.-G., Moene, A.F., Holtslag, A.A.M., 2009. Interactions between dry-air entrainment, surface evaporation and convective boundary-layer development. Q. J. R. Meteorol. Soc. 135, 1277–1291.

van Heerwaarden, C.C., Mellado, J.P., De Lozar, A., 2014. Scaling laws for the heterogeneously heated free convective boundary layer. J. Atmos. Sci. 71, 3975–4000.

Van Woert, M.L., Zou, C.Z., Meier, W.N., Hovey, P.D., Preller, R.H., Posey, P.G., 2004. Forecast verification of the Polar Ice Prediction System (PIPS) sea ice concentration fields. J. Atmos. Ocean. Technol. 21 (6), 944–957.

Vanneste, J., Vial, F., 1994. On the nonlinear interactions of geophysical waves in shear flows. Geophys. Astrophys. Fluid Dyn. **78**, 115–141.

Vaughan, C., Dessai, S., 2014. Climate services for society: origins, institutional arrangements, and design elements for an evaluation framework. WIREs Clim. Change 2014 (5), 587–603. https://doi.org/10.1002/wcc.290.

Vaughan, C., Dessai, S., Hewitt, C., Baethgen, W., Terra, R., Berterreche, M., 2017. Creating an enabling environment for investment in agricultural climate services: the case of Uruguay's National Agricultural Information System. Clim. Serv. 8, 62–71.

Vautard, R., 1990. Multiple weather regimes over the North Atlantic: analysis of precursors and successors. Mon. Weather Rev. 118, 2056–2081.

Vautard, R., Ghil, M., 1989. Singular spectrum analysis in nonlinear dynamics, with applications to paleo-climatic time series. Physica D 35, 395–424. https://doi.org/10.1016/0167-2789(89)90077-8.

Vautard, R., Legras, B., 1988. On the source of midlatitude low-frequency variability. Part II: Nonlinear equilibration of weather regimes. J. Atmos. Sci. 45, 2845–2867. https://doi.org/10.1175/ 1520-0469(1988)045<2845:otsoml>2.0.co;2.

Vautard, R., Mo, K.C., Ghil, M., 1990. Statistical significance test for transition matrices of atmospheric Markov chains. J. Atmos. Sci. 47, 1926–1931. https://doi.org/10.1175/1520-0469(1990) 047<1926:sstftm>2.0.co;2.

Vautard, R., et al., 2007. Summertime European heat and drought waves induced by wintertime Mediterranean rainfall deficit. Geophys. Res. Lett. 34, L07711.

Vecchi, G.A., Bond, N.A., 2004. The Madden-Julian Oscillation (MJO) and the northern high latitude wintertime surface air temperatures. Geophys. Res. Lett. 31L04104. https://doi.org/10.1029/2003GL018645.

Verseghy, D.L., 2000. The Canadian land surface scheme (CLASS): its history and future. Atmos. Ocean 38, 1–13.

Vigaud, N., Robertson, A.W., Tippett, M.K., 2017a. Multimodel ensembling of subseasonal precipitation forecasts over North America. Mon. Weather Rev. 145, 3913–3928. https://doi.org/10.1175/MWR-D-17-0092.1.

Vigaud, N., Robertson, A.W., Tippett, M.K., Acharya, N., 2017b. Subseasonal predictability of boreal summer monsoon rainfall from ensemble forecasts. Front. Environ. Sci. 5, 67. https://doi.org/10.3389/fenvs.2017.00067.

Vincent, L.A., Mekis, É., 2006. Changes in daily and extreme temperature and precipitation indices for Canada over the twentieth century. Atmosphere-Ocean 44 (2), 177–193. https://doi.org/10.3137/ao.440205.

Vitart, F., 2009. Impact of the Madden Julian Oscillation on tropical storms and risk of landfall in the ECMWF forecast system. Geophys. Res. Lett. 36, L15802. https://doi.org/10.1029/2009GL039089.

Vitart, F., 2014. Evolution of ECMWF sub-seasonal forecast skill scores. Q. J. R. Meteorol. Soc. 140, 1889–1899.

Vitart, F., 2017. Madden-Julian Oscillation prediction and teleconnection in the S2S database. Q. J. R. Meteorol. Soc. 143, 2210–2220.

Vitart, F., Jung, T., 2010. Impact of the Northern Hemisphere extratropics on the skill in predicting the Madden-Julian Oscillation. Geophys. Res. Lett. 37, L23805.

Vitart, F., Molteni, F., 2010. Simulation of the Madden-Julian Oscillation and its teleconnections in the ECMWF forecast system. Q. J. R. Meteorol. Soc. **136**, 842–855. https://doi.org/10.1002/qj.v136:649. http://onlinelibrary.wiley.com/doi/10.1002/qj.623/pdf.

Vitart, F., Stockdale, T.N., 2001. Seasonal forecasting of tropical storms using coupled GCM integrations. Mon. Weather Rev. 129 (10), 2521–2527.

Vitart, F., Woolnough, J., Balmaseda, M.A., Tompkins, A.M., 2007. Monthly forecast of the Madden–Julian Oscillation using a coupled GCM. Mon. Weather Rev. 135, 2700–2715.

Vitart, F., et al., 2008. The new VarEPS-monthly forecasting system: a first step towards seamless prediction. Q. J. R. Meteorol. Soc. 134, 1789–1799.

Vitart, F., Prates, F., Bonet, A., Sahin, C., 2012a. New tropical cyclone products on the web. ECMWF Newslett. 130, 17–23.

Vitart, F., Robertson, A.W., Anderson, D.L.T., 2012b. Subseasonal to seasonal prediction project, 2012: bridging the gap between weather and climate. WMO Bull. 61 (2), 23–28.

Vitart, F., Balsamo, G., Buizza, R., Ferranti, L., Keeley, S., Magnusson, L., Molteni, F., Weisheimer, A., 2014a. Sub-seasonal predictions. In: ECMWF Research Department Technical Memorandum No. 738, p. 45. Available from ECMWF, Shinfield Park, Reading, RG2-9AX, U.K. See also:http://www.ecmwf.int/sites/default/files/elibrary/2014/12943-sub-seasonal-predictions.pdf.

Vitart, F., Robertson, A.W., S2S Steering Group, 2014b. Sub-seasonal to seasonal prediction: linking weather and climate. In: Seamless Prediction of the Earth System: From Minutes to Months. World Meteorological Organisation, pp. 385–401. WMO-No.1156 (Chapter 20).

Vitart, F., Ardilouze, C., Bonet, A., et al., 2016. The subseasonal to seasonal (S2S) prediction project database. Bull. Am. Meteorol. Soc. 98, 163–173. https://doi.org/10.1175/BAMS-D-16-0017.1.

Vitart, F., Ardilouze, C., Bonet, A., Brookshaw, A., Chen, M., Codorean, C., Déqué, M., Ferranti, L., Fucile, E., Fuentes, M., Hendon, H., Hodgson, J., Kang, H.-S., Kumar, A., Lin, H., Liu, G., Liu, X., Malguzzi, P., Mallas, I., Manoussakis, M., Mastrangelo, D., MacLachlan, C., McLean, P., Minami, A., Mladek, R., Nakazawa, T., Najm, S., Nie, Y., Rixen, M., Robertson, A.W., Ruti, P., Sun, C., Takaya, Y., Tolstykh, M., Venuti, F., Waliser, D., Woolnough, S., Wu, T., Won, D.-J., Xiao, H., Zaripov, R., Zhang, L., 2017. The subseasonal to seasonal (S2S) prediction project data base. Bull. Am. Meteorol. Soc. 98, 163–173. http://journals.ametsoc.org/doi/pdf/10.1175/BAMS-D-16-0017.1.

Viterbo, P., Beljaars, A.C.M., 1995. An improved land surface parameterization scheme in the ECMWF model and its validation. J. Clim. 8, 2716–2748.

Von Neumann, J., 1955. Some remarks on the problem of forecasting climatic fluctuations. In: Pfeffer, R.L. (Ed.), Dynamics of Climate. Pergamon Press, Oxford, UK, pp. 9–11. https://doi.org/10.1016/b978-1-4831-9890-3.50009-8.

Vrac, M., Friederichs, P., 2015. Multivariate—intervariable, spatial, and temporal—bias correction. J. Clim. 28 (1), 218–237.

Waliser, D., 2005. Predictability and forecasting. In: Lau, W.K., Waliser, D.E. (Eds.), Intraseasonal Variability in the Atmosphere-Ocean Climate System. Springer Praxis, pp. 389–423.

Waliser, D.E., 2011. Predictability and forecasting. In: Lau, W.K.M., Waliser, D.E. (Eds.), Intraseasonal Variability of the Atmosphere-Ocean Climate System, second ed. Springer, Heidelberg, Germany, pp. 433–468.

Waliser, D.E., Guan, B., 2017. Extreme winds and precipitation during landfall of atmospheric rivers. Nat. Geosci. https://doi.org/10.1038/NGEO2894. in press.

Waliser, D.E., Jones, C., Schemm, J.-K.E., Graham, N.E., 1999. A statistical extended-range tropical forecast model based on the slow evolution of the Madden–Julian Oscillation. J. Clim. 12, 1918–1939. https://doi.org/10.1175/1520-0442(1999)012<1918:ASERTF>2.0.CO;2.

Waliser, D.E., Lau, K.M., Stern, W., Jones, C., 2003a. Potential predictability of the Madden–Julian Oscillation. Bull. Am. Meteorol. Soc. 84, 33–50. https://doi.org/10.1175/BAMS-84-1-33.

Waliser, D.E., Stern, W., Schubert, S., Lau, K.M., 2003b. Dynamic predictability of intraseasonal variability associated with the Asian summer monsoon. Q. J. R. Meteorol. Soc. 129, 2897–2925.

Waliser, D., Hendon, H., Kim, D., Maloney, E., Wheeler, M., Weickmann, K., Zhang, C., Donner, L., Gottschalck, J., Higgins, W., Kang, I.-S., Legler, D., Moncrieff, M., Schubert, S., Stern, W., Vitart, F., Wang, B., Wang, W., Woolnough, S., 2009. MJO simulation diagnostics. J. Clim. 22, 3006–3030.

Walker, G.T., Bliss, E.W., 1932. World weather V. Mem. Roy. Meteorol. Soc. 4, 53–84.

Wallace, J.M., 2000. North Atlantic Oscillation/annular mode: two paradigms-one phenomenon. Q. J. R. Meteorol. Soc. 126, 791–805. https://doi.org/10.1256/smsqj.56401.

Wallace, J.M., Gutzler, D.S., 1981. Teleconnections in the geopotential height field during the Northern Hemisphere winter. Mon. Weather Rev. 109 (4), 784–812. https://doi.org/10.1175/1520-0493.

Wallace, J., Mitchell, T., Deser, C., 1989. The influence of sea surface temperature on surface wind in the eastern equatorial pacific: seasonal and interannual variability. J. Clim. 2, 1492–1499.

Walsh, J.E., 1980. Empirical orthogonal functions and the statistical predictability of sea ice extent. In: Sea Ice Processes and Models. University of Washington Press, Seattle, pp. 373–384.

Walsh, J.E., Johnson, C.M., 1979. An analysis of Arctic sea ice fluctuations, 1953–77. J. Phys. Oceanogr. 9 (3), 580–591.

Walters, D.N., Best, M.J., Bushell, A.C., Copsey, D., Edwards, J.M., Falloon, P.D., Harris, C.M., Lock, A.P., Manners, J.C., Morcrette, C.J., Roberts, M.J., Stratton, R.A., Webster, S., Wilkinson, J.M., Willett, M.R., Boutle, I.A., Earnshaw, P.D., Hill, P.G., MacLachlan, C., Martin, G.M., Moufouma-Okia, W., Palmer, M.D., Petch, J.C., Rooney, G.G., Scaife, A.A., Williams, K.D., 2011. The Met Office unified model global atmosphere 3.0/3.1 and JULES global land 3.0/3.1 configurations. Geosci. Model Dev. 4, 919–941.

Wang, B., 1988. Dynamics of tropical low-frequency waves: an analysis of the moist Kelvin wave. J. Atmos. Sci. 2051–2065, 45.

Wang, B., 2012. Theories. In: Lau, W.K., Waliser, D.E. (Eds.), Intraseasonal Variability in the Atmosphere-Ocean Climate System, second ed. Springer Praxis, pp. 335–398.

Wang, X., Bishop, C.H., 2003. A comparison of breeding and ensemble transform Kalman filter ensemble forecast schemes. J. Atmos. Sci. 60, 1140–1158.

Wang, B., Chen, G., 2017. A general theoretical framework for understanding essential dynamics of Madden-Julian Oscillation. Clim. Dyn. 49, 2309–2328.

Wang, B., Rui, H., 1990a. Dynamics of the coupled moist Kelvin-Rossby wave on an equatorial beta-plane. J. Atmos. Sci. 47, 397–413.

Wang, B., Rui, H., 1990b. Synoptic climatology of transient tropical intraseasonal convection anomalies 1975-1985. Meteorog. Atmos. Phys. 44, 43–61.

Wang, W., Schlesinger, M.E., 1999. The dependence on convective parameterization of the tropical intraseasonal oscillation simulated by the UIUC 11-layer atmospheric GCM. J. Clim. 12, 1423–1457.

Wang, X., Bishop, C.H., Julier, S.J., 2004. Which is better, an ensemble of positive/negative pairs or a centered spherical simplex ensemble? Mon. Weather Rev. 132, 1590–1605.

Wang, B., Webster, P., Kikuchi, K., Yasunari, T., Qi, Y., 2006. Boreal summer quasi-monthly oscillation in the global tropics. Clim. Dyn. 27, 661–675.

Wang, X., Hamill, T.M., Whitaker, J.S., Bishop, C.H., 2007. A comparison of hybrid ensemble transform Kalman filter–optimum interpolation and ensemble square root filter analysis schemes. Mon. Weather Rev. 135, 1055–1076.

Wang, W., et al., 2014. MJO prediction in the NCEP climate forecast system version 2. Clim. Dyn. 42, 2509–2520.

Wang, Q., Ilicak, M., Gerdes, R., Drange, H., Aksenov, Y., Bailey, D.A., et al., 2016. An assessment of the Arctic Ocean in a suite of interannual CORE-II simulations. Part I: Sea ice and solid freshwater. Ocean Model. 99, 110–132.

Wang, L., Ting, M., Kushner, P.J., 2017. A robust empirical seasonal prediction of winter NAO and surface climate. Sci. Rep. 7, 279. https://doi.org/10.1038/s41598-017-00353-y.

Wang, B., Lee, S.-S., Waliser, D.E., Zhang, C., Sobel, A., Maloney, E., Li, T., Jiang, X., Ha, K.-J., 2018. Dynamics-oriented diagnostics for the Madden-Julian Oscillation. J. Clim. 31, 3117–3135.

Warrilow, D.A., Sangster, A.B., Slingo, A., 1986. Modelling of Land Surface Processes and Their Influence on European Climate. DCTN 38, Dynamical Climatology Branch, United Kingdom Meteorological Office, Bracknell, Berkshire.

Watt-Meyer, O., Kushner, P.J., 2015. Decomposition of atmospheric disturbances into standing and traveling components, with application to Northern Hemisphere planetary waves and stratosphere–troposphere coupling. J. Atmos. Sci. 72, 787–802. https://doi.org/10.1175/JAS-D-14-0214.1.

Waugh, D.W., Sisson, J.M., Karoly, D.J., 1998. Predictive skill of an NWP system in the southern lower stratosphere. Q. J. R. Meteorol. Soc. 124, 2181–2200. https://doi.org/10.1002/qj.49712455102.

Waugh, D.W., Sobel, A.H., Polvani, L.M., Waugh, D.W., Sobel, A.H., Polvani, L.M., 2017. What is the polar vortex and how does it influence weather? Bull. Am. Meteorol. Soc. 98, 37–44. https://doi.org/10.1175/BAMS-D-15-00212.1.

Weary, D.J., Doctor, D.H., 2014. Karst in the United States: a digital map compilation and database. US Department of the Interior, US Geological Survey. Report 2014–1156, 27pp.

Webster, P.J., Holton, J.R., 1982. Wave propagation through a zonally varying basic flow: the influences of mid-latitude forcing in the equatorial regions. J. Atmos. Sci. 39, 722–733.

Webster, P.J., Hoyos, C., 2004. Prediction of monsoon rainfall and river discharge on 15–30-day time scales. Bull. Am. Meteorol. Soc. 85, 1745–1765. https://doi.org/10.1175/BAMS-85-11-1745.

Webster, P.J., Lukas, R., 1992. TOGA-COARE: the coupled ocean-atmosphere response experiment. Bull. Am. Meteorol. Soc. 73, 1377–1416.

Weckwerth, T.M., et al., 2004. An overview of the International H2O Project (IHOP_2002) and some preliminary highlights. Bull. Am. Meteorol. Soc. 85, 253–277.

Wedi, N.P., Hamrud, M., Mozdzynski, G., 2013. A fast spherical harmonics transform for global NWP and climate models. Mon. Weather Rev. 141 (10), 3450–3461.

Wei, M., Toth, Z., Wobus, R., Zhu, Y., Bishop, C., Wang, X., 2006. Ensemble Transform Kalman Filter-based ensemble perturbations in an operational global prediction system at NCEP. Tellus A 58, 28–44.

Wei, M., Toth, Z., Wobus, R., Zhu, Y., 2008. Initial perturbations based on the ensemble transform (ET) technique in the NCEP global operational forecast system. Tellus A 60, 62–79.

Weickmann, K.M., Lussky, G.R., Kutzbach, J.E., 1985. Intraseasonal (30–60 day) fluctuations of outgoing longwave radiation and 250 mb streamfunction during northern winter. Mon. Weather Rev. 113, 941–961. https://doi.org/10.1175/1520-0493(1985)113<0941:idfool>2.0.co;2.

Weigel, A.P., Mason, S., 2011. The generalized discrimination score for ensemble forecasts. Mon. Weather Rev. 139, 3069–3074.

Weigel, A.P., Liniger, M.A., Appenzeller, C., 2007a. The discrete Brier and ranked probability skill scores. Mon. Weather Rev. 135, 118–124.

Weigel, A.P., Liniger, M.A., Appenzeller, C., 2007b. Generalization of the discrete brier and ranked probability skill scores for weighted multimodel ensemble forecasts. Mon. Weather Rev. 135, 2778–2785.

Weigel, A., Baggenstos, D., Liniger, M.A., Vitart, F., Appenzeller, C., 2008. Probabilistic verification of monthly temperature forecasts. Mon. Weather Rev. **136**, 5162–5182. https://doi.org/10.1175/2008MWR2551.1.

Weigel, A.P., Knutti, R., Liniger, M.A., Appenzeller, C., 2010. Risks of model weighting in multimodel climate projections. J. Clim. 23 (15), 4175–4191.

Weisheimer, A., Palmer, T., 2014. On the reliability of seasonal climate forecasts. J. R. Soc. Interface 11. https://doi.org/10.1098/rsif.2013.1162.

Weisheimer, A., et al., 2009. ENSEMBLES: a new multi-model ensemble for seasonal-to-annual predictions—skill and progress beyond DEMETER in forecasting tropical Pacific SST. Geophys. Res. Lett. 36, L21711. https://doi.org/10.1029/2009GL040896.

Weisman, M.L., Skamarock, W.C., Klemp, J.B., 1997. The resolution dependence of explicitly modeled convective systems. Mon. Weather Rev. 125 (4), 527–548.

Wheeler, M.C., Hendon, H.H., 2004. An all-season real-time multivariate MJO index: development of an index for monitoring and prediction. Mon. Weather Rev. 132, 1917–1932.

Wheeler, M., Kiladis, G., 1999. Convectively coupled equatorial waves: analysis of clouds and temperature in the wavenumber-frequency domain. J. Atmos. Sci. 56, 374–399.

Wheeler, M., Kiladis, G.N., Webster, P.J., 2000. Large-scale dynamical fields associated with convectively coupled equatorial waves. J. Atmos. Sci. 57, 613–640.

Wheeler, M.C., Hendon, H.H., Cleland, S., Meinke, H., Donald, A., 2009. Impacts of the Madden–Julian oscillation on Australian rainfall and circulation. J. Clim. 22, 1482–1498.

Wheeler, M.C., Zhu, H., Sobel, A.H., Hudson, D., Vitart, F., 2017. Seamless precipitation prediction skill comparison between two global models. Q. J. R. Meteorol. Soc. 143 (702), 374–383.

Whelan, J.A., Frederiksen, J.S., 2017. Dynamics of the perfect storms: La Niña and Australia's extreme rainfall and floods of 1974 and 2011. Clim. Dyn. 48, 3935–3948. https://doi.org/10.1007/s00382-016-3312-3.

Whitaker, J.S., Hamill, T.M., Wei, X., Song, Y., Toth, Z., 2008. Ensemble data assimilation with the NCEP global forecast system. Mon. Weather Rev. 136, 463–482.

White, G.F., Kates, R.W., Burton, I., 2001. Knowing better and losing even more: the use of knowledge in hazards management. Environ. Hazards 3 (3-4), 81–92.

White, C.J., Hudson, D., Alves, O., 2013. ENSO, the IOD and intraseasonal prediction of heat extremes across Australia using POAMA-2. Clim. Dyn. 43, 1791–1810. https://doi.org/10.1007/s00382-013-2007-2.

White, C.J., Hudson, D., Alves, O., 2014. ENSO, the IOD and the intraseasonal prediction of heat extremes across Australia using POAMA-2. Clim. Dyn. 43, 1791–1810.

White, C.J., Carlsen, H., Robertson, A.W., Klein, R.J.T., Lazo, J.K., Kumar, A., Vitart, F., Coughlan de Perez, E., Ray, A.J., Murray, V., Bharwani, S., MacLeod, D., James, R., Fleming, L., Morse, A.P., Eggen, B., Graham, R., Kjellström, E., Becker, E., Pegion, K.V., Holbrook, N.J., McEvoy, D., Depledge, M., Perkins-Kirkpatrick, S., Hodgson-Johnston, I., Buontempo, C., Lamb, R., Meinke, H., Arheimer, B., Zebiak, S.E., 2017. Potential applications of subseasonal-to-seasonal (S2S) predictions. Meteorol. Appl. https://doi.org/10.1002/met.1654.

WHO, 2012. Atlas of Health and Climate. WHO Press, World Health Organization, Geneva.

Wigneron, J.-P., Chanzy, A., Calvet, J.-C., Olioso, A., Kerr, Y., 2002. Modeling approaches to assimilating L-band passive microwave observations over land surfaces. J. Geophys. Res. 107. ACL 11-1–ACL 11-14.

Wigneron, J.P., et al., 2017. Modelling the passive microwave signature from land surfaces: a review of recent results and application to the L-band SMOS & SMAP soil moisture retrieval algorithms. Remote Sens. Environ. 192, 238–262.

Wilber, C.D., 1881. The Great Valleys and Prairies of Nebraska and the Northwest. Daily Republican Printing Company, Omaha, Nebraska, USA. 382 pp.

Wilcox, L.J., Charlton-Perez, A.J., 2013. Final warming of the Southern Hemisphere polar vortex in high- and low-top CMIP5 models. J. Geophys. Res. Atmos. 118, 2535–2546. https://doi.org/10.1002/jgrd.50254.

Wilhite, D.A., Glantz, M.H., 1985. Understanding: the drought phenomenon: the role of definitions. Water Int. 10, 111–120.

Wilks, D., 2005. Statistical Methods in the Atmospheric Sciences, second ed. vol. 100. Academic Press, ISBN: 9780080456225p. 648.

Wilks, D.S., 2011a. Statistical Methods in the Atmospheric Sciences, third ed. Elsevier. 676 pp.

Wilks, D., 2011b. Statistical Methods in the Atmospheric Sciences, third ed. Academic Press.

Williams, G., Maksym, T., Wilkinson, J., Kunz, C., Murphy, C., Kimball, P., Singh, H., 2015. Thick and deformed Antarctic sea ice mapped with autonomous underwater vehicles. Nat. Geosci. 8 (1), 61–67.

Willison, J., Robinson, W.A., Lackmann, G.M., 2013. The importance of resolving mesoscale latent heating in the North Atlantic storm track. J. Atmos. Sci. 70, 2234–2250. https://doi.org/10.1175/JAS-D-12-0226.1.

Wilson, L., 2000. Comments on "Probabilistic prediction of precipitation using the ECMWF ensemble prediction system" Weather Forecast. 15, 361–364.

Wilson, L., 2014. Forecast Verification for the African Severe Weather Forecasting Demonstration Projects, WMO Technical Document TD 1132. 38 pp. Available on the WMO website.

Wilson, L., Giles, A., 2013. A new index for the verification of accuracy and timeliness of weather warnings. Meteorol. Appl. 20, 206–216.

Winsemius, H.C., Dutra, E., Engelbrecht, F.A., Archer Van Garderen, E., Wetterhall, F., Pappenberger, F., Werner, M.G.F., 2014. The potential value of seasonal forecasts in a changing climate in southern Africa. Hydrol. Earth Syst. Sci. 18, 1525–1538. https://doi.org/10.5194/hess-18-1525-2014.

Wittman, M.A.H., Charlton, A.J., Polvani, L.M., 2007. The effect of lower stratospheric shear on Baroclinic instability. J. Atmos. Sci. 64, 479–496. https://doi.org/10.1175/JAS3828.1.

Wolock, D.M., 1997. STATSGO Soil Characteristics for the Conterminous United States. US Department of the Interior, US Geological Survey. Report 97–656, 200 pp.

Woo, S.-H., Sung, M.-K., Son, S.-W., Kug, J.-S., 2015. Connection between weak stratospheric vortex events and the Pacific Decadal Oscillation. Clim. Dyn. 45, 3481–3492. https://doi.org/10.1007/s00382-015-2551-z.

Woodgate, R.A., Weingartner, T., Lindsay, R., 2010. The 2007 Bering Strait oceanic heat flux and anomalous Arctic sea-ice retreat. Geophys. Res. Lett. 37(1).

Woolnough, S.J., Vitart, F., Balmaseda, M.A., 2007. The role of the ocean in the Madden-Julian Oscillation: implications for the MJO prediction. Q. J. R. Meteorol. Soc. 133, 117–128.

Worby, A.P., Geiger, C.A., Paget, M.J., Van Woert, M.L., Ackley, S.F., DeLiberty, T.L., 2008. Thickness distribution of Antarctic sea ice. J. Geophys. Res. Oceans. 113(C5).

World Meteorological Organization, 2010. Manual on the Global Data-Processing and Forecasting System, WMO No. 485, Geneva, Switzerland. Available online at http://www.wmo.int/pages/prog/www/DPFS/Manual/GDPFS-Manual.html.

World Meteorological Organization, 2013. Subseasonal to Seasonal Prediction Research Implementation Plan, p. 63. Available online athttp://s2sprediction.net/static/documents#publications.

World Meteorological Organization, 2015. Seamless Prediction of the Earth System: From Minutes to Months, WMO No. 1156, Geneva, Switzerland. Available online at https://public.wmo.int/en/resources/library/seamless-prediction-of-earth-system-from-minutes-months.

Wu, C.-H., Hsu, H.-H., 2009. Topographic influence on the MJO in the Maritime Continent. J. Clim. 22, 5433–5448.

Wulfmeyer, V., Turner, D., 2016. Land-Atmosphere Feedback Experiment (LAFE) Science Plan. US Dept. of Energy. DOE/SC-ARM-16-038. 44 pp.

Wyngaard, J., Moeng, C., 1992. Parameterizing turbulent-diffusion through the joint probability density. Bound.-Layer Meteorol. 60, 1–13.

Xavier, P.K., Goswami, B.N., 2007. An analog method for real-time forecasting of summer monsoon subseasonal variability. Mon. Weather Rev. 135, 4149–4160. https://doi.org/10.1175/2007MWR1854.1.

Xavier, P., Rahmat, R., Cheong, W.K., Wallace, E., 2014. Influence of Madden-Julian Oscillation on Southeast Asian rainfall extremes. Geophys. Res. Lett. 41, 4406–4412.

Xia, Y., Sheffield, J., Ek, M.B., Dong, J., Chaney, N., Wei, H., Meng, J., Wood, E.F., 2014. Evaluation of multimodel simulated soil moisture in NLDAS-2. J. Hydrol. 512, 107–125.

Xie, S.-P., 2004. Satellite observations of cool ocean–atmosphere interaction. Bull. Am. Meteorol. Soc. 85, 195–208. https://doi.org/10.1175/ BAMS-85-2-195.

Xie, P., Arkin, P., 1996. Analyses of global monthly precipitation using gauge observations, satellite estimates, and numerical model predictions. J. Clim. 9, 840–858.

Xie, F., et al., 2016. A connection from Arctic stratospheric ozone to El Niño-Southern oscillation. Environ. Res. Lett. 11, 124026. https://doi.org/10.1088/1748-9326/11/12/124026.

Xu, L., Dirmeyer, P., 2013. Snow-atmosphere coupling strength. Part II: Albedo effect versus hydrological effect. J. Hydrometeorol. 14, 404–418.

Yaka, P., et al., 2008. Relationships between climate and year-to-year variability in meningitis outbreaks: a case study in Burkina Faso and Niger. Int. J. Health Geogr. 7, 34.

Yamaguchi, M., Majumdar, S.J., 2010. Using TIGGE data to diagnose initial perturbations and their growth for tropical cyclone ensemble forecasts. Mon. Weather Rev. 138, 3634–3655.

Yamaguchi, M., Nakazawa, T., Aonashi, K., 2012. Tropical cyclone track forecasts using JMA model with ECMWF and JMA initial conditions. Geop. Res. Lett. 39, L09801.

Yamaguchi, M., Vitart, F., Lang, S.T.K., Magnusson, L., Elsberry, R.L., Elliott, G., Kyouda, M., Nakazawa, T., 2015. Global distribution on the skill of tropical cyclone activity forecasts from short- to medium-range time scales. Weather Forecast. 30, 1695–1709.

Yanai, M., Lu, M.-M., 1983. Equatorially trapped waves at the 200 mb level and their association with meridional convergence of wave energy flux. J. Atmos. Sci. 40, 2785–2803.

Yang, G.-Y., Slingo, J., 2001. The diurnal cycle in the tropics. Mon. Weather Rev. 129, 784–801.

Yang, G.-Y., Hoskins, B., Slingo, J., 2007. Convectively coupled equatorial waves. Part III: synthesis structure and their forcing and evolution. J. Atmos. Sci. 64, 3438–3451.

Yang, Q., Fu, Q., Hu, Y., 2010. Radiative impacts of clouds in the tropical tropopause layer. J. Geophys. Res. 115, D00H12. https://doi.org/10.1029/2009JD012393.

Yang, X.-Y., Yuan, X., Ting, M., 2016. Dynamical link between the Barents–Kara sea ice and the Arctic Oscillation. J. Clim. 29, 5103–5122. https://doi.org/10.1175/JCLI-D-15-0669.1.

Yao, W., Lin, H., Derome, J., 2011. Submonthly forecasting of winter surface air temperature in North America based on organized tropical convection. Atmosphere-Ocean 49, 51–60. https://doi.org/10.1080/07055900.2011.556882.

Yasunari, T., 1979. Cloudiness fluctuations associated with the Nothern Hemisphere summer monsoon. J. Meteorol. Soc. Jpn. 57, 227–242.

Yates, E., Anquetin, S., Ducrocq, V., Creutin, J.-D., Ricard, D., Chancibault, K., 2006. Point and areal validation of forecast precipitation fields. Meteorol. Appl. 13, 1–20.

Yin, Y., Alves, O., Oke, P.R., 2011. An ensemble ocean data assimilation system for seasonal prediction. Mon. Weather Rev. 139, 786–808.

Yoden, S., 1987. Bifurcation properties of a stratospheric vacillation model. J. Atmos. Sci. 44, 1723–1733. https://doi.org/10.1175/1520-0469(1987)044<1723:BPOASV>2.0.CO;2.

Yoo, C., Son, S.-W., 2016. Modulation of the boreal wintertime Madden-Julian oscillation by the stratospheric quasi-biennial oscillation. Geophys. Res. Lett. 43, 1392–1398. https://doi.org/10.1002/2016GL067762.

Yoo, C., Lee, S., Feldstein, S.B., 2012. Mechanisms of Arctic surface air temperature change in response to the Madden–Julian oscillation. J. Clim. 25, 5777–5790. https://doi.org/10.1175/JCLI-D-11-00566.1.

Yoo, C., Park, S., Kim, D., Yoon, J.-H., Kim, H.-M., 2015. Boreal winter MJO teleconnection in the community atmosphere model version 5 with the unified convection parameterization. J. Clim. 28, 8135–8150.

Yuan, X., 2004. ENSO-related impacts on Antarctic sea ice: a synthesis of phenomenon and mechanisms. Antarct. Sci. 16 (4), 415–425.

Yuan, X., Martinson, D.G., 2001. The Antarctic dipole and its predictability. Geophys. Res. Lett. 28 (18), 3609–3612.

Yuan, H., Toth, Z., Pena, M., 2018. Overview of weather and climate systems. In: Duan, Q., Yuan, H., Toth, Z. (Eds.), Handbook of Hydrometeorological Ensemble Forecasting. Springer. under review.

Zadra, A., 2000. Empirical Normal Mode Diagnosis of Reanalysis Data and Dynamical-Core Experiments. PhD Thesis, McGill University, p. 221.

Zadra, A., Brunet, G., Derome, J., 2002a. An empirical normal mode diagnostics algorithm applied to NCEP reanalyses. J. Atmos. Sci. 59, 2811–2829.

Zadra, A., Brunet, G., Derome, J., Dugas, B., 2002b. Empirical normal mode study of the GEM model's dynamical core. J. Atmos. Sci. 59, 2498–2510.

Zaitchik, B.F., 2017. Madden-Julian Oscillation impacts on tropical African precipitation. Atmos. Res. 184, 88–102.

Zangvil, A., 1975. Temporal and spatial behaviour of large-scale disturbances in tropical cloudiness deduced from satellite brightness data. Mon. Weather Rev. 103, 904–920.

Zangvil, A., Yanai, M., 1980. Upper tropospheric waves in the tropics. Part I: Dynamical analysis in the wavenumber-frequency domain. J. Atmos. Sci. 37, 283–298.

Zhang, C., 2005. Madden–Julian oscillation. Rev. Geophys. 43, 1–36.

Zhang, C.D., 2013. Madden-Julian oscillation bridging weather and climate. Bull. Am. Meteorol. Soc. 94, 1849–1870.

Zhang, Z., Krishnamurti, T.N., 1999. A perturbation method for hurricane ensemble predictions. Mon. Weather Rev. 127, 447–469.

Zhang, G.J., Mu, M., 2005. Simulation of the Madden–Julian Oscillation in the NCAR CCM3 using a revised Zhang–McFarlane convective parameterization scheme. J. Clim. 18, 4046–4064.

Zhang, C., Webster, P.J., 1989. Effects of zonal flows on equatorially trapped waves. J. Atmos. Sci. 46, 3632–3652.

Zhang, C., Webster, P.J., 1992. Laterally forced equatorial perturbations in a linear model. Part I: Stationary transient forcing. J. Atmos. Sci. 49, 585–607.

Zhang, J., Steele, M., Lindsay, R., Schweiger, A., Morison, J., 2008. Ensemble 1-year predictions of Arctic sea ice for the spring and summer of 2008. Geophys. Res. Lett. 35(8).

Zhang, L., Wang, B., Zeng, Q., 2009. Impact of the Madden-Julian oscillation on summer rainfall in southeast China. J. Clim. 22, 201–216.

Zhang, C., Gottschalck, J., Maloney, E.D., Moncrieff, M.W., Vitart, F., Waliser, D.E., Wang, B., Wheeler, M.C., 2013a. Cracking the MJO nut. Geophys. Res. Lett. 40, 1223–1230. https://doi.org/10.1002/grl.50244.

Zhang, Q., Shin, C.-S., van den Dool, H., Cai, M., 2013b. CFSv2 prediction skill of stratospheric temperature anomalies. Clim. Dyn. 41, 2231–2249. https://doi.org/10.1007/s00382-013-1907-5.

Zhang, J., Tian, W., Chipperfield, M.P., Xie, F., Huang, J., 2016. Persistent shift of the Arctic polar vortex towards the Eurasian continent in recent decades. Nat. Clim. Chang. 6, 1094–1099. https://doi.org/10.1038/nclimate3136.

Zhou, S., L'Heureux, M., Weaver, S., Kumar, A., 2012. A composite study of the MJO influence on the surface air temperature and precipitation over the continental United States. Clim. Dyn. 38, 1459–1471. https://doi.org/10.1007/s00382-011-1001-9.

Zhou, G., Latif, M., Greatbatch, R.J., Park, W., 2015. Atmospheric response to the North Pacific enabled by daily sea surface temperature variability. Geophys. Res. Lett. 42, 7732–7739. https://doi.org/10.1002/2015GL065356.

Zhou, X., Zhu, Y., Hou, D., Kleist, D., 2016. A comparison of perturbations from an ensemble transform and an ensemble Kalman filter for the NCEP global ensemble forecast system. Weather Forecast. 31, 2057–2074.

Zhu, Y., Newell, R.E., 1998. A proposed algorithm for moisture fluxes from atmospheric rivers. Mon. Weather Rev. 126, 725–735. https://doi.org/10.1175/15200493(1998)126<0725:APAFMF>2.0.CO;2.

Zhu, H., Hendon, H., Jakob, C., 2009. Convection in a parameterized and superparameterized model and its role in the representation of the MJO. J. Atmos. Sci. 66 (9), 2796–2811.

Zhu, H., Wheeler, M.C., Sobel, A.H., Hudson, D., 2014. Seamless precipitation prediction skill in the tropics and extratropics from a global model. Mon. Weather Rev. 142, 1556–1569.

Zinszer, K., et al., 2012. A scoping review of malaria forecasting: past work and future directions. BMJ Open 2, e001992.

Zsótér, E., 2006. Recent developments in extreme weather forecasting. ECMWF Newsletter 107, 8–17. http://old.ecmwf.int/publications/newsletters/pdf/107.pdf.

Zsoter, E., Buizza, R., Richardson, D., 2009. 'Jumpiness' of the ECMWF and UK Met Office EPS control and ensemble-mean forecasts. Mon. Weather Rev. 137, 3823–3836.

Zuo, H.M., Balmaseda, A., Mogensen, K., 2014. The ECMWF-MyOcean2 eddy-permitting ocean and sea-ice reanalysis ORAP5. Part 1: Implementation. In: ECMWF Research Department Technical Memorandum No. 736, p. 42. Available from ECMWF, Shinfield Park, Reading, RG2-9AX, U.K. See also:http://www.ecmwf.int/sites/default/files/elibrary/2014/12943-sub-seasonal-predictions.pdf.

Zuo, H., Balmaseda, M.A., Mogensen, K., 2015. The new eddy-permitting ORAP5 ocean reanalysis: description, evaluation and uncertainties in climate signals. Clim. Dyn.

Index

Note: Page numbers followed by *f* indicate figures, *t* indicate tables, and *b* indicate boxes.

F

N

Q

R

S

Printed in the United States
By Bookmasters